Quality Control of Cellular Protein in Neurodegenerative Disorders

Md. Sahab Uddin
Southeast University, Bangladesh

Ghulam Md. Ashraf
King Abdulaziz University, Saudi Arabia

IGI Global
DISSEMINATOR OF KNOWLEDGE

A volume in the Advances in Medical Diagnosis, Treatment, and Care (AMDTC) Book Series

Published in the United States of America by
IGI Global
Medical Information Science Reference (an imprint of IGI Global)
701 E. Chocolate Avenue
Hershey PA, USA 17033
Tel: 717-533-8845
Fax: 717-533-8661
E-mail: cust@igi-global.com
Web site: http://www.igi-global.com

Library of Congress Cataloging-in-Publication Data

Names: Sahab Uddin, Md., 1992- editor. | Ashraf, Ghulam Md, 1984- editor.
Title: Quality control of cellular protein in neurodegenerative disorders / Md. Sahab Uddin, Ghulam Md Ashraf.
Description: Hershey, PA : Medical Information Science Reference, [2020] | Includes bibliographical references and index. | Summary: "This book examines the diverse aspects of protein quality control systems and potential therapeutic approaches to halt the development and propagation of neurodegenerative disorders"--Provided by publisher.
Identifiers: LCCN 2019028368 (print) | LCCN 2019028369 (ebook) | ISBN 9781799813170 (hardcover) | ISBN 9781799813187 (ebook)
Subjects: MESH: Neurodegenerative Diseases--etiology | Proteostasis Deficiencies--etiology | Proteins--physiology | Protein Folding
Classification: LCC RC376.5 (print) | LCC RC376.5 (ebook) | NLM WL 358.5 | DDC 616.8/3--dc23
LC record available at https://lccn.loc.gov/2019028368
LC ebook record available at https://lccn.loc.gov/2019028369

This book is published in the IGI Global book series Advances in Medical Diagnosis, Treatment, and Care (AMDTC) (ISSN: 2475-6628; eISSN: 2475-6636)

British Cataloguing in Publication Data
A Cataloguing in Publication record for this book is available from the British Library.

All work contributed to this book is new, previously-unpublished material. The views expressed in this book are those of the authors, but not necessarily of the publisher.

For electronic access to this publication, please contact: eresources@igi-global.com.

Advances in Medical Diagnosis, Treatment, and Care (AMDTC) Book Series

ISSN:2475-6628
EISSN:2475-6636

Mission

Advancements in medicine have prolonged the life expectancy of individuals all over the world. Once life-threatening conditions have become significantly easier to treat and even cure in many cases. Continued research in the medical field will further improve the quality of life, longevity, and wellbeing of individuals.

The **Advances in Medical Diagnosis, Treatment, and Care (AMDTC)** book series seeks to highlight publications on innovative treatment methodologies, diagnosis tools and techniques, and best practices for patient care. Comprised of comprehensive resources aimed to assist professionals in the medical field apply the latest innovations in the identification and management of medical conditions as well as patient care and interaction, the books within the AMDTC series are relevant to the research and practical needs of medical practitioners, researchers, students, and hospital administrators.

Coverage

- Experimental Medicine
- Internal Medicine
- Emergency Medicine
- Disease Prevention
- Disease Management
- Patient Interaction
- Medical Procedures
- Patient-Centered Care
- Cancer Treatment
- Alternative Medicine

IGI Global is currently accepting manuscripts for publication within this series. To submit a proposal for a volume in this series, please contact our Acquisition Editors at Acquisitions@igi-global.com or visit: http://www.igi-global.com/publish/.

The Advances in Medical Diagnosis, Treatment, and Care (AMDTC) Book Series (ISSN 2475-6628) is published by IGI Global, 701 E. Chocolate Avenue, Hershey, PA 17033-1240, USA, www.igi-global.com. This series is composed of titles available for purchase individually; each title is edited to be contextually exclusive from any other title within the series. For pricing and ordering information please visit http://www.igi-global.com/book-series/advances-medical-diagnosis-treatment-care/129618. Postmaster: Send all address changes to above address.

Titles in this Series

For a list of additional titles in this series, please visit: https://www.igi-global.com/book-series/advances-medical-diagnosis-treatment-care/129618

Multidimensional Perspectives and Global Analysis of Universal Health Cverage
Yeter Demir Uslu (Istanbul Medipol University, Turkey) Hasan Dinçer (Istanbul Medipol University, Turkey) and Serhat Yüksel (İstanbul Medipol University, Turkey)
Medical Information Science Reference • © 2020 • 499pp • H/C (ISBN: 9781799823292) • US $265.00

Examining Biological Foundations of Human Behavior
Barre Vijaya Prasad (Dharwad Institute of Mental Health and Neurosciences, India)
Medical Information Science Reference • © 2020 • 300pp • H/C (ISBN: 9781799828600) • US $285.00

Handbook of Research on Medical Interpreting
Izabel E.T. de V. Souza (Osaka University, Japan) and Effrossyni (Effie) Fragkou (National and Kapodistrian University of Athens, Greece)
Medical Information Science Reference • © 2020 • 511pp • H/C (ISBN: 9781522593089) • US $345.00

Quality Assurance in the Era of Individualized Medicine
Anastasius S. Moumtzoglou (P. & A. Kyriakou Children's Hospital, Greece)
Medical Information Science Reference • © 2020 • 346pp • H/C (ISBN: 9781799823902) • US $285.00

Diagnosing and Managing Hashimoto's Disease Emerging Research and Opportunities
Mahmoud Sakr (Alexandria University, Egypt)
Medical Information Science Reference • © 2020 • 263pp • H/C (ISBN: 9781522596554) • US $245.00

Using Narrative Writing to Enhance Healing
Jennifer Lynne Bird (Oxbridge Academy, USA)
Medical Information Science Reference • © 2020 • 355pp • H/C (ISBN: 9781799819318) • US $275.00

Strategies to Overcome Superbug Invasions Emerging Research and Opportunities
Dimple Sethi Chopra (Punjabi University, India)
Medical Information Science Reference • © 2020 • 284pp • H/C (ISBN: 9781799803072) • US $225.00

Advanced Applications of Nanotechnology in Cancer Detection and Therapy
Alireza Heidari (Faculty of Chemistry, California South University, USA)
Medical Information Science Reference • © 2020 • 280pp • H/C (ISBN: 9781799801399) • US $275.00

IGI Global
DISSEMINATOR OF KNOWLEDGE

701 East Chocolate Avenue, Hershey, PA 17033, USA
Tel: 717-533-8845 x100 • Fax: 717-533-8661
E-Mail: cust@igi-global.com • www.igi-global.com

Md. Sahab Uddin would like to dedicate this book to his beloved parents, Sabikun Naher and Md. Helal Uddin.

Ghulam Md. Ashraf would like to dedicate this book to his family members.

Table of Contents

Detailed Table of Contents

Proteostasis is essential for regulating the integrity of the proteome. Disruption of proteostasis under some rigorous conditions leads to the aggregation and accumulation of misfolded toxic proteins, which plays a central role in the pathogenesis of protein conformational disorders. The protein quality control (PQC) system serves as a multi-level security system to shield cells from abnormal proteins. The intrinsic PQC systems maintaining proteostasis include the ubiquitin-proteasome system (UPS), chaperon-mediated autophagy (CMA), and autophagy-lysosome pathway (ALP) that serve to target misfolded proteins for unfolding, refolding, or degradation. Alterations of PQC systems in neurons have been implicated in the pathogenesis of various neurodegenerative disorders. This chapter provides an overview of PQC pathways to set a framework for discussion of the role of PQC in neurodegenerative disorders. Additionally, various pharmacological approaches targeting PQC are summarized.

Autophagy is a normal physiological process characterized by the degradation of complex cellular contents into a simpler one and reutilized them in biosynthetic pathways. Lysosomes are the cell organelle that participates in the process of autophagy. The brain is the most vulnerable organ in most lysosome disorders because neurons are inefficient in removing impaired organelles and waste materials. In the brain, autophagy suppresses the accumulation of ubiquitinated proteins that results in further damage to the neurons responsible for neurodegeneration. Autophagy mediates protective effects in age-related diseases. In the chapter, the authors describe the process of autophagy, the mechanism involved, and the implication of the autophagic pathways in the various neurodegenerative disorders.

Mitophagy is a selective autophagy process in which damaged or surplus mitochondria are removed to sustain normal homeostasis. The efficient removal of damaged or stressed mitochondria is crucial for cellular health. Recent literature emphasizes the role of PINK1-Parkin pathways in the pathogenesis process of various neurodegenerative disorders. Further, mitophagy has shown potential therapeutic activity in treating neurodegenerative diseases. Thus, mitophagy might be important in the field of pharmacotherapeutics. In the present chapter, the authors explain mitophagy, mitophagy signaling pathways, as well as mechanisms and roles of mitophagy in various neurodegenerative disorders.

Microglia are important in the regulation of the inflammatory response in regulating the release of proinflammatory mediators in the brain. Through their phagocytic actions, microglia are significant in the CNS when it comes to the body's response to physiological insults by promoting repair of impaired brain function. They do so by engulfing and degrading microbes as well as brain-derived debris and proteins such as myelin and axonal fragments, amyloid-beta, and apoptotic cells. This mitophagic activity of microglia is of importance in neurodegeneration. In most neurodegenerative disorders, mitophagy is impaired with resultant accumulation of dysfunctional mitochondria as well as processes such as lysosomal fusion and autophagosomes. In Parkinson's and Alzheimer's for example, impaired mitophagy accounts for the build-up of α-synuclein and amyloid respectively in affected individuals. The chapter discusses extensively the link between microglia mitophagy and neurodegeration and how dysfunctional mitophagy increases the likelihood of their occurrence.

Post-translational modifications (PTMs) increase proteome activity for controlling every feature of normal cell biology. PTMs such as phosphorylation, acetylation, glycosylation, fatty acylation, palmitoylation, myristoylation, ubiquitination, SUMOylation (small ubiquitin-like modifiers), methylation, deamidation, nitrosylation, etc. of proteins can regulate the properties of protein including intracellular distribution, functionality, stability, accumulation, as well as interactions. PTMs take place at any stage of the protein life cycle, regulating protein folding and activity in time and space, subcellular localization of the protein, and their activity. Hence, PTMs play a pivotal role in the regulation of numerous cellular processes. Abnormal PTMs of one or more culprit proteins might contribute to neurodegeneration, which is shown in some neurodegenerative disorders including Alzheimer's, Parkinson's, and prion disease. In this chapter, the authors focus on the most essential PTMs that are observed in neurodegenerative disorders and elucidating the pathogenesis wherein they are involved.

Proteostasis or protein homeostasis consists of a complex interrelated cellular system that controls several steps of protein quality and function from the initial step of synthesis as well as folding, and eventually degradation over enormous biochemical pathways. Proteostasis involves controlling protein folding, modification of the post-translational protein, and degradation of misfolded protein. However, the failure of proteostasis has resulted to produce a toxic protein that leads to disrupt aging and neurodegeneration. Additionally, endoplasmic reticulum degradation and autophagy dysfunction may outcome in cellular

Section 3
Proteins and Protein Complexes in Neurodegeneration

Beclin1 is the mammalian orthologue of yeast Atg6/vacuolar protein sorting-30 (VPS30). Beclin1 interacts with various biological macromolecules like ATG14, BIF-1, NRBF2, RUBICON, UVRAG, AMBRA1, HMGB1, PINK1, and PARKIN. Such interactions promote Beclin1-PI3KC3 complex formation. Autophagy is blocked in apoptosis owing to the breakdown of Beclin1 by caspase whereas autophagy induction inhibits effector caspase degradation, therefore, blocks apoptosis. Thus, the Beclin1 is an essential biomolecular species for cross-regulation between autophagy and apoptosis. Various studies carried out in neurodegenerative animal models associated with aggregated proteins have confirmed that multifunctional Beclin1 protein is necessary for neuronal integrity. The role of Beclin1 protein has been investigated and was reported in various human neurodegeneration disorders. This chapter aims to provide an insight into the role of Beclin1 in the development of neurodegenerative disorders.

The glial cells along with cells of hematopoietic origin and microvascular endothelia work together to maintain the normal development and/or functioning of the nervous system. Disruption in functional coordination among these cells interrupts the efficiency of the nervous system, leading to neurodegeneration. Various proteins in the nerve cells maintain the normal signaling mechanism with these cells and throughout the body. Structural/functional disorganization of these proteins causes neurodegenerative disorders. The molecular mechanisms involved in these phenomena are yet to be explored extensively from therapeutic perspectives. Through this chapter, the authors have elaborated on less known protein Bcl-2 associated athanogene 3 (BAG3) involved in neurodegeneration. They have explored BAG3 protein and its role in neurodegeneration, protein homeostasis, its mechanism of action, its uses as a drug target, and its uses as a possible diagnostic marker of neurodegeneration.

PTEN-induced kinase 1 (PINK1), a mitochondrial serine/threonine-protein kinase encoded by the PINK1 gene, is thought to protect cells from stress-induced mitochondrial dysfunction. The activity of PINK1 facilitates the binding of Parkin protein with depolarized mitochondria to induce autophagy. Mutations of PINK1causes a type of autosomal recessive early-onset Parkinson's disease. Cell depends on the surveillance systems or mechanisms like protein quality control to handle the alterations in the proteins that are induced because of these mutations. These mutant proteins are found to be pathogenic and are reported to be related to various neurodegenerative disorders. This chapter focuses on the role of PINK1/ Parkin in mitochondria quality control and its subsequent effect in neurodegeneration.

Oxidative stress is strongly linked to neurodegeneration and oxidative species can modify many amino acids and proteins in the brain. Cysteine amino acid is most susceptible to oxidative post-translational modifications (PTMs). Reversible or irreversible cysteine PTMs can cause dyshomeostasis, which further continued to cellular damage. Many cysteine dependent proteins and many non-proteins using cysteine as their structural components are affected by oxidative stress. Several cysteine dependent enzymes are acting as antioxidants. Cysteine is a major contributor to glutathione (GSH) and superoxide dismutase (SOD) synthesis. Cysteine precursor N-acetylcysteine (NAC) supplementation is proven as a potent free radical scavenger and increase brain antioxidants and subsequently potentiates the natural antioxidant cellular defense mechanism. Thus, in this chapter, the authors explore the linkage of cellular cysteine networks and neurodegenerative disorders.

Preface

Protein misfolding and aggregation are hallmarks of several neurodegenerative proteinopathies. Compelling proof suggests that pathogenic proteins trigger synaptic dysfunction and neuronal damage, which leads to neurodegenerative disorders. The mechanisms by which pathogenic proteins aggregate, misfold, and eventually cause neurotoxicity is an emerging area of research. Though multiple factors like aging, oxidative stress, mitochondrial dysfunction, proteotoxic insults, genetic inconsistency, etc. are responsible for the dysfunction of the neuronal protein quality control system, targeting protein quality control has become an auspicious approach to halt the propagation of neurodegeneration. *Quality Control of Cellular Protein in Neurodegenerative Disorders* provides diverse aspects exploring the role of the protein quality control in neurodegenerative disorders and potential therapeutic strategies to combat the development and propagation of neurodegeneration.

This book represents the copious set of specific research updates. All over the world numerous erudite, experienced and eminent academicians, researchers and scientists had participated to write the texts of this book to give a succinct and thorough understanding of the protein quality control system in neurodegeneration at a more advanced level with excellent presentation.

This book is suitable for professionals, academicians, students, researchers, scientists and industrialists around the world. Biomedical, health, and life science departments can use this book as a crucial textbook. Researchers and scientists from research institutes can use this book as efficient research info. Pharmacists, physicians, and other healthcare professionals can use this book as a complete reference book. Furthermore, for interested readers, this book is a storehouse of knowledge to comprehend the proteopathic neurodegenerative disorders complexity. The organizations of this book provided a profound knowledge and also maintain the reader's interest.

This book contains 14 chapters divided into four sections. The contents of the book cover the imperative cellular mechanisms that control protein homeostasis and protein quality control system in neurodegenerative disorders along with the utmost appropriate therapeutic approaches to alleviate neurodegeneration.

Section 1 discusses the organization of the cellular protein quality control system, their cytoprotective strategies under proteotoxic insults and auspicious therapeutic approaches to mitigate neurodegenerative disorders. This section consists of six chapters. A description of each chapter follows:

Chapter 1 is "Protein Quality Control in Neurodegeneration and Neuroprotection". This chapter clarifies the role of protein quality control in copious neurodegenerative disorders. Furthermore, highlights current putative therapeutic tactics to effectively remove misfolded and aggregated proteins from degenerating neurons.

Chapter 2 is "Autophagic Dysfunction in Neurodegeneration: Pathogenic Cellular and Molecular Mechanistic Approaches". This chapter explores the process of autophagy, the mechanism involved and the implication of the autophagic pathways in the various neurodegenerative disorders.

Chapter 3 is "Molecular Mechanisms Underlying the Role of Mitophagy in Neurodegeneration". This chapter explains the multifaced role of mitophagy and the impact of mitophagy signaling in neurodegenerative disorders.

Chapter 4 is "Microglial Mitophagy and Neurodegenerative Disorders". This chapter illustrates extensively the link between microglia mitophagy and its role in the misfolded protein associated with neurodegenerative disorders. The processes of how dysfunctional mitophagy increases the chance of neurodegeneration are also discussed.

Chapter 5 is "Post-Translational Modifications in Neurodegeneration: Tools of Protein Quality Control System". This chapter emphasizes on the most essential post-translational modifications that are observed in neurodegenerative disorders and elucidating the pathogenesis wherein they are involved.

Chapter 6 is "Proteostasis and Neurodegeneration: Perspectives in the Pathogenesis of Molecular and Cellular Mechanisms". This chapter describes essential factors that disturb proteostasis as well as the current knowledge regarding how cellular proteostasis interruption contributes to progress neurodegenerative disorders.

Section 2 focuses on the clear explanation of the proteotoxic stress in the pathogenesis of imperious neurodegenerative disorders with a crucial impact on disease management. This section consists of two chapters. A description of each chapter follows:

Chapter 7 is "ER Stress Signaling in Alzheimer's Disease: Molecular Mechanisms and Therapeutic Implications". This chapter mentions the role of several transmembrane kinase proteins responsible for the onset of Alzheimer's disease as well as show the interrelationship between ER stress and Alzheimer's pathogenesis. Moreover, it demonstrates therapeutic significance in inhibiting the pathological pathway that triggers ER stress.

Chapter 8 is "Molecular Interactions of α-Synuclein, Mitochondria and Cellular Degradation Pathways in Parkinson's Disease". This chapter provides the role of α-synuclein in the pathogenesis of Parkinson's disease. It also relates several molecular mechanisms, which are involved in regulating mitochondrial and lysosomal related pathways and are linked to α-synuclein. Lastly, it confers the importance of using α-synuclein as a diagnostic marker and therapeutic target for Parkinson's disease.

Section 3 offers the multidimensional interlinking of various proteins and protein complexes in neurodegenerative disorders as well as draw special attention in the current status, future opportunities, and challenges. This section consists of three chapters. A description of each chapter follows:

Chapter 9 is "Beclin 1 Complex and Neurodegenerative Disorders". This chapter highlights the molecular insights of the Beclin1 functioning in autophagy and apoptosis as well as portrays its role in the development of neurodegenerative disorders.

Chapter 10 is "Multifarious Role of BAG3 in Neurodegenerative Disorders". This chapter attempts to answer the molecular interplay of BAG3 in neurodegenerative disorders. Additionally, it presents the probable mechanism of BAG3 to control the protein homeostasis of neuronal cells through close interaction with various chaperones, anti-apoptotic factors, and other cellular components.

Chapter 11 is "PINK1/Parkin in Neurodegenerative Disorders: Crosstalk Between Mitochondrial Stress and Neurodegeneration". This chapter focuses on the role of PINK1/Parkin in mitochondria quality control and its subsequent effect in the pathogenesis of neurodegenerative disorders.

Section 4 explores the protein quality control network that acts to prevent protein misfolding, refold misfolded proteins or degrade them, thereby controlling neuronal proteostasis. This section consists of three chapters. A description of each chapter follows:

Chapter 12 is "Cellular Cysteine Network and Neurodegeneration". This chapter reveals the effect of oxidative stress on the antioxidant system of the brain specifically cysteine modifications due to redox dysregulation and its role in several neurodegenerative disorders.

Chapter 13 is "Cysteine in Alzheimer's Disease: Redox Regulation of Protein Functions". This chapter deals with the role of cysteine in handling reactive species during oxidative and nitrosative stress and post-translational modification of proteins causing protein misfolding or aggregation in Alzheimer's disease.

Chapter 14 is "Molecular Chaperones in Neurodegeneration: Mechanisms of Regulation in Cellular Proteostasis". This chapter represents the impact of protein misfolding and aggregation in neurodegeneration and then critically analyses some of the known cellular chaperones and their role in neurodegenerative disorders.

It is expected that readers shall find this book very informative and enormously useful. Since science is constantly changing; readers are strongly recommended to check the recent updates. The editors are ebulliently ready to accept any comment, suggestion, advice or critique.

Md. Sahab Uddin
Southeast University, Bangladesh

Ghulam Md. Ashraf
King Abdulaziz University, Saudi Arabia

Acknowledgment

Editing is a complex process, the editors would like to thank the people involved in this project for their quality time, expertise, and countless efforts. The editors are extremely indebted to the following:

The authors for the countless time and expertise that they have put into their texts.

Those authors who submitted works to contribute to this project but unfortunately rejected.

The reviewers for their constructive reviews in improving the quality and presentation of the contents.

The teachers, colleagues, friends, and supporters for their endless inspiration and assists.

Especially the IGI Global for giving us a great opportunity to edit this book and their aid in the preparation of this edition.

Last but not least, the editors would like to express the heartiest gratitude to almighty Allah and their parents.

Md. Sahab Uddin
Southeast University, Bangladesh

Ghulam Md. Ashraf
King Abdulaziz University, Saudi Arabia

Section 1
Cellular Mechanisms of Protein Quality Control in Neurodegeneration

Chapter 1
Protein Quality Control in Neurodegeneration and Neuroprotection

Yasmeena Akhter
https://orcid.org/0000-0002-3317-4280
Department of Pharmaceutical Sciences (Pharmacology Division), Faculty of Applied Sciences and Technology, University of Kashmir, Srinagar, India

Jahangir Nabi
https://orcid.org/0000-0002-5212-9063
Department of Pharmaceutical Sciences (Pharmacology Division), Faculty of Applied Sciences and Technology, University of Kashmir, Srinagar, India

Hinna Hamid
Department of Pharmaceutical Sciences (Pharmacology Division), Faculty of Applied Sciences and Technology, University of Kashmir, Srinagar, India

Nahida Tabassum
Department of Pharmaceutical Sciences (Pharmacology Division), Faculty of Applied Sciences and Technology, University of Kashmir, Srinagar, India

Faheem Hyder Pottoo
Department of Pharmaology, College of Clinical Pharmacy, Imam Abdulrahman Bin Faisal University, Saudi Arabia

Aashish Sharma
Centre for Research in Medical Devices (CURAM), National University of Ireland, Ireland & School of Medical and Allied Sciences, GD Goenka University, Gurgaon, India

ABSTRACT

Proteostasis is essential for regulating the integrity of the proteome. Disruption of proteostasis under some rigorous conditions leads to the aggregation and accumulation of misfolded toxic proteins, which plays a central role in the pathogenesis of protein conformational disorders. The protein quality control (PQC) system serves as a multi-level security system to shield cells from abnormal proteins. The intrinsic PQC systems maintaining proteostasis include the ubiquitin-proteasome system (UPS), chaperon-mediated autophagy (CMA), and autophagy-lysosome pathway (ALP) that serve to target misfolded proteins for unfolding, refolding, or degradation. Alterations of PQC systems in neurons have been implicated in the pathogenesis of various neurodegenerative disorders. This chapter provides an overview of PQC pathways to set a framework for discussion of the role of PQC in neurodegenerative disorders. Additionally, various pharmacological approaches targeting PQC are summarized.

DOI: 10.4018/978-1-7998-1317-0.ch001

INTRODUCTION

Proteins are regulatory factories that are responsible for maintaining the integrity and viability of cells. Systemic protein balance within cells offers proteome homeostasis (proteostasis)which is preserved by substituting newly synthesized proteins for old or damaged ones (Brandman et al., 2012). Protein synthesis is a highly organized process and is controlled specifically by a dynamic control system called the protein quality control (PQC) system (Gestwicki and Garza, 2012). Alterations of PQC systems in neurons under some pathological conditions (bacterial/viral infections, and tissue damage), environmental stress (elevated temperature, exposure to heavy metals or chemicals), metabolic stress (nutrient imbalance, reactive oxygen species [ROS] generation), aging and errors in transcription fidelity or protein synthesis translation amplify the accumulation of misfolded protein aggregates which is implicated in pathogenesis cascade of several neurodegenerative disorders, for example, Alzheimer's disease (AD), amyotrophic lateral sclerosis (ALS), Huntington's disease (HD), Parkinson's disease (PD), and Prion diseases (PrDs) (Sweeney et al., 2017).

The characteristic feature of neurodegenerative disorders includes a clinical disability that leads to the worst quality-of-life (QoL) (Batista and Pereira, 2016; Pottoo et al., 2020). These diseases represent one of the most significant medical and socio-economic challenges, affecting people of all ages (Nigar et al., 2016). The prevalence of these diseases has dramatically increased. Around 7-10 million people around the world are living with PD (Parkinson's Foundation, 2019), 50 million with AD (Alzheimer's Disease International, 2018), and an estimated 3-7 per 100,000 European population is living with HD (Huntington's Disease Foundation, 2019).

The challenge of neurodegenerative disorders needs to be addressed at several levels. Undoubtedly the critical unmet medical need is developing the drug therapies that are effective in slowing the progression and severity of these debilitating medical conditions (Pottoo et al., 2014). A better knowledge of molecular pathology and genetic basis of these debilitating conditions may provide a foundation for the emergence of new therapeutic approaches, targeting one or more pathogenesis sites (Pottoo et al., 2019). Therefore, the main aim of this chapter is to understand the recent insights of systematic cellular cytoprotective strategies under proteotoxic insults as well as current putative approaches to treatment to effectively remove misfolded and aggregated proteins from degenerating neuronal bodies in various neurodegenerative disorders.

PATHWAYS FOR REMOVAL OF ABERRANT PROTEINS

The classic characteristic of neurodegenerative disorders is the accumulation of soluble functional protein aggregates in the central nervous system (CNS). These protein-related disorders can be found both intracellularly and extracellularly in the endoplasmic reticulum (ER), nucleus and cytosol. During the 1980s, it was assumed that only the lysosomal pathway is involved in the disposal of aggregated proteins (Yang and Klionsky, 2010). There was not much definition of the role of other arms of PQC, as the proteasome and cytoplasmic proteases were deemed only to target cytoplasmic substrates. Also, there was no reason to think that these elements would contribute to PQC-based events, particularly given the central position of the lysosome as a component of the secretory pathway (Figure 1).

Figure 1. Degradation of misfolded proteins by protein quality control system

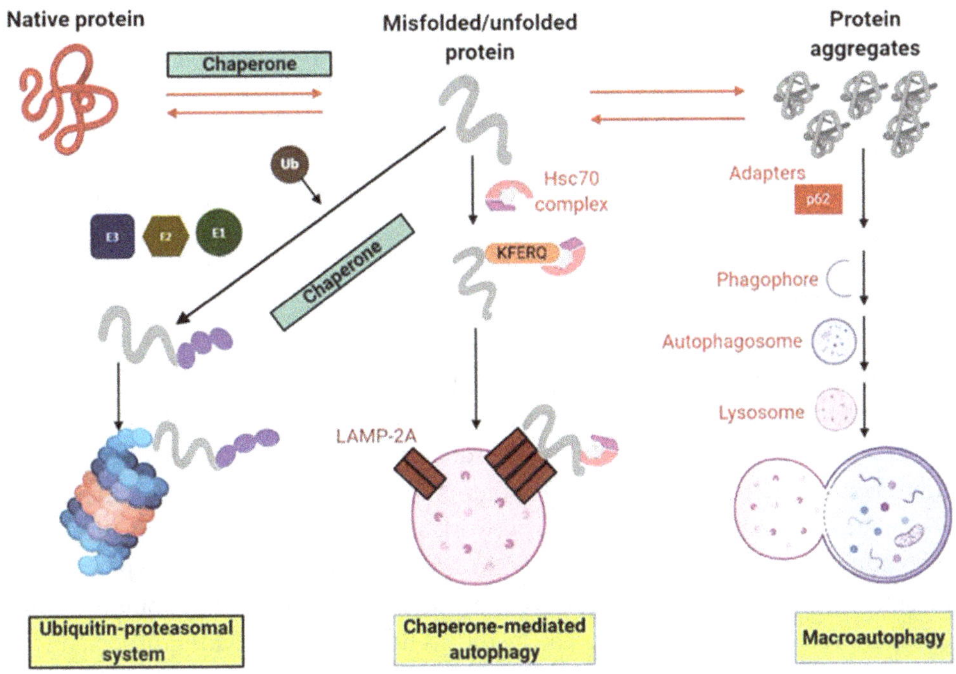

Molecular Chaperones

Molecular chaperones or heat-shock proteins (Hsps) act as a primary line of protection against misfolded aggregates. The expression of these proteins is heat/stress-sensitive, for example, elevated temperature and oxidative damage (Richter-Landberg and Goldbaum, 2003). The physiological function of molecular chaperones is to fold freshly produced polypeptides, help in their translocation across membranes, and to refold the denatured ones. Within eukaryotic cells, the functions of molecular chaperones are initiated by HSPs, which are stimulated by heat-shock factor (HCF) and function to shield the proteome from stress (Haslbeck, Franzmann, Weinfurtner and Buchner, 2005) and by chaperones linked to protein synthesis (CLIPS), that are physically and functionally related translation framework and help to fold freshly synthesized proteins.

Molecular chaperones are characterized into various classes depending on their molecular weights viz, Hsp 110, Hsp 100, Hsp 90 Hsp 70, Hsp 40, and small heat-shock proteins (sHsps). These classes are not homologous and seem to have distinct features. For example, the Hsp100-type chaperones contain AAA^{+} þ ATPase domain and are primarily associated with protein disaggregation and the primary folding of some selected substrates (Liberek, Lewandowska and Zietkiewicz, 2008). Hsp90 has been shown to bind folded substrates, for example, certain kinases and nuclear hormone receptors (NRs), to protect them from degradation and to serve in the presence of ligands in their remodeling (Pratt, Morishima, Murphy and Harrell, 2006). Hsp70, Hsp90, and Hsp110 degrade misfolded versions of the von Hippel-Lindau (VHL) and other proteins (McClellan, Scott and Frydman, 2005). Hsp70 is a group of 70-kDa chaperones that function to link molecular chaperones with UPS structures and autophagy in the main folding (Hageman and Kampinga, 2009; Höhfeld, Cyr and Patterson, 2001).

The various operations of chaperones arise from their immediate physical contact with substrates. Some chaperones can remove and refold proteins from aggregates, for example, sHsps have been seen to be strongly linked to aggregates (Haslbeck, Franzmann, Weinfurtner and Buchner, 2005). Upon detecting the accumulation of misfolded protein aggregates, the cytosol generates the heat-shock response (HSR) by titrating chaperones from heat-shock transcription factor 1 (HSF1), which is then translocated into the nucleus to induce the transcription of heat-shock-sensitive genes. Similarly, the unfolded protein response (UPR) is initiated in the ER after Hsp70 binding protein binds the misfolded aberrant proteins (Richter-Landberg and Goldbaum, 2003).

The Ubiquitin-Proteasome System

The ubiquitin-proteasome system (UPS) is the primary proteolytic house cleaning quality control network for the degradation of misfolded proteins. It plays the main role in regulating cellular functions (Hershko, 2005) and transcription (Auld and Silver, 2006). This quality control system acts by tagging and targeting short-lived transcription factors and abnormal proteins for degradation (Varshavsky, 2005). The UPS framework degradation includes an enzymatical network made up of ubiquitin-activating enzymes (E1s), ubiquitin-conjugating enzymes (E2s), and ubiquitin ligases (E3s) (Scheffner, Nuber and Huibregtse, 1995). The substrate for degradation is first targeted by the proteasome (multi-subunit complex) and then recognized, ubiquitinated, and turned over. The lysine 48 linked tetra-polyubiquitin chains are the main signals to direct the substrates to the proteasome for destruction (Huzil, Pannu, Ptak, Garen, and Ellison, 2007). The fundamental stage during this process is the initiation of E1 enzyme of ubiquitin (Ub), which establishes a strong-vitality thiol-ester link between both the final Ub glycine and the E1 cysteine. This thiol-linked Ub is subsequently aimed at the E2 enzyme. A special E3 ligase encourages the transport of the stimulated Ub to a lysine moiety on the desired protein. The arrangement of ubiquitylation is several leveled. About21 E1s, 38 E2s, and at least 1000 E3s are encoded by the human genome (Ye and Rape, 2009). The ubiquitination is promoted by the interaction of E3swithE2s and substrate. The E3 ligases are subtyped into four classes such as (i) homologous to the E6-AP carboxyl terminus (HECT) proteins, (ii) U-box proteins, (iii) zinc-binding RING proteins, and, PHD fingers (Nakayama and Nakayama, 2006).

The polyubiquitinated substrates are subjected to degradation through the proteasome. The most frequently known 26S proteasome consists of a 20S core and a 19S lid. There are two distal α rings and two focal β rings on the 20S core. The complex's proteolytic functions are arranged in the β-enriched rings, while the 19S lid serves as a controlling matrix containing S5a protein that binds polyubiquitin substrates. The 20S subunit includes a channel with trypsin and chymotrypsin-like hydrolyzing functions that collectively degrade substrate proteins that enter the 20S core. The 19S particles communicate with the two ends of the 20S particle by explicitly perceiving ubiquitin-modified substrates and other proteins labeled for destruction as gatekeepers. Also, these 19S particles have unfolding activity dependent on adenosine triphosphate (ATP). This activity is important because proteins need to be partially denatured to access the 20S particle's moderately narrow 50A° channel. Larger aggregates, such as those frequently found in neurodegenerative disorders, are resistant to the proteosome to a restricted extent due to their relative protection from unfolding (Elsasser and Finley, 2005; Waelter et al., 2001).

Figure 2. Integrated stress response triggered in response stress conditions

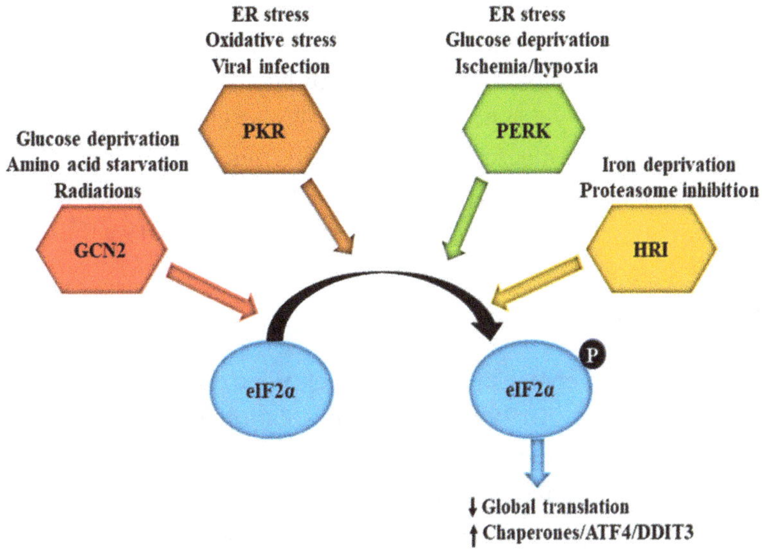

Unfolded Protein Response

The endoplasmic reticulum (ER) is an important organelle that mediates translation, protein folding and transfer of newly synthesized proteins across the cellular membrane (Hetz and Mollereau, 2014). In the ER lumen, vigilant homeostatic regulation of the region keeps the protein folding up. The accumulation of misfolded protein aggregates in the ER lumen triggers a defense mechanism called the integrated stress response (ISR) which eventually threatens the cell's stability and normal functions. To tackle this problem, the cell triggers a signaling reaction in the form of unfolded protein response (UPR) that functions to restore proteostasis (Hetz and Mollereau, 2014). The primary defense mechanism involved is the integrated stress response (ISR) which is activated when the translation initiation factor-2α (eIF2α) undergoes phosphorylation on the α-subunit by four types of protein kinases such as the PKR-like ER kinase (PERK), general control non-derepressible 2 kinase (GCN2), RNA-dependent protein kinase R (PKR) and heme-regulated eIF2α (HRI) kinase, thereby diminishing the global protein synthesis, while specifically upregulating the translation of cellular chaperones and other defensive ER proteins like the activating translation initiation factor4 (ATF4) (McConkey, 2017; Wang et al., 2018). Moreover, eIF2α phosphorylation may facilitate the expression of another transcription factor, DNA-inducible transcript 3 (DDIT3) which induces apoptosis (Jauhiainen et al., 2012). Hence, ISR acts to restore the ER stress and if it does not work, restores to killing the compromised cell as shown in Figure 2.

Autophagy Lysosomal Pathway

The autophagy lysosomal pathway (ALP) is an important regulatory framework for the removal of toxic proteins, the shortfalls of which leads to generation and accumulation of toxic protein species (Martini-Stoica, Xu, Ballabio and Zheng, 2016). Autophagy is a self-eating mechanism which carries the cytoplasmic materials to the lysosome for their degradation (Finkbeiner, 2019). This pathway essentially disposes of aggregates that are created through macromolecules, cytosolic parts, entire organelles or through lysosomes due to age-related, non-enzymatic, and/or stochastic post-translation changes (Finkbeiner, 2019; Nedić, Rattan, Grune and Trougakos, 2013). Autophagy is of critical importance for long-extension neuronal cells where aberrant proteins could not be degraded through cell division (Ariosa and Klionsky, 2016). Autophagy can be divided into three forms according to the method of cargo transport to the lysosome, including macroautophagy, CMA, and microautophagy. Each of these forms may be triggered by similar stimuli, such as infection, nutrient deprivation, stress conditions, or DNA impairment, despite the fact mechanisms may differ (Dong and Cui, 2019).

Macroautophagy

Macroautophagy acts as the main center of ALS in which the materials from cytoplasm are delivered to the vacuole (Galluzzi and Green, 2019). Normally, an isolation layer is wrapped across the cytoplasmic contents or other organelles of the cell, like aggresomes. The boundaries of the membrane fuse leading to the formation of a double-membrane framework called the phagophore which then develops into an autophagosome. The autophagosome framework is packed with autophagic proteins, including the autophagy-related protein 5 (Atg 5) and autophagy-related protein 6 (Atg 6). Upon delivery to the lysosome, the stored cargo is disposed of into fatty acids, nucleotides, and amino acids, which may be delivered back to the ER for recycling (Feng, He, Yao and Klionsky, 2014).

Microautophagy

Microautophagy is a route towards vacuolar degradation and begins with the lysosomal membrane invagination. Finally, invaginations carrying the cytoplasmic material, squeeze off into the vacuolar lumen, eventually resulting in the formation of vesicles. However, the extent of substrate sensitivity remains elusive (Bingol, 2018; Shpilka and Elazar, 2011).

Chaperone-Mediated Autophagy

Chaperone-mediated autophagy (CMA) is essentially specific for the removal of aberrant proteins bearing the KFERQ-like pentapeptide motif that is identified by heat-shock cognate 71 kDa protein (Hsp70) with the help of Hsp90 and Hsp40(Ciechanover and Kwon, 2017). The Hsc70 associated material is delivered to the lysosomal lumen via the interplay between Hsc70 and the lysosome-associated membrane protein 2A (LAMP2A) and is eventually degraded by lysosomal hydrolases. The CMA is crucial for the removal of cytotoxic proteins in neurodegenerative disorders and its failure is implicated in the neuronal pathological process (Ciechanover and Kwon, 2017; Cuervo and Wong, 2014).

NEUROTOXIC PROTEINS ASSOCIATED WITH NEURODEGENERATIVE DISORDERS

Different human neurodegenerative disorders such as AD, ALS, PD, and PrDs share a similar pathological mechanism of toxic protein aggregation (Lindholm, Wootz and Korhonen, 2006) as given in Figure 3.

Figure 3. Aberrant proteins involved in various neurodegenerative disorders.

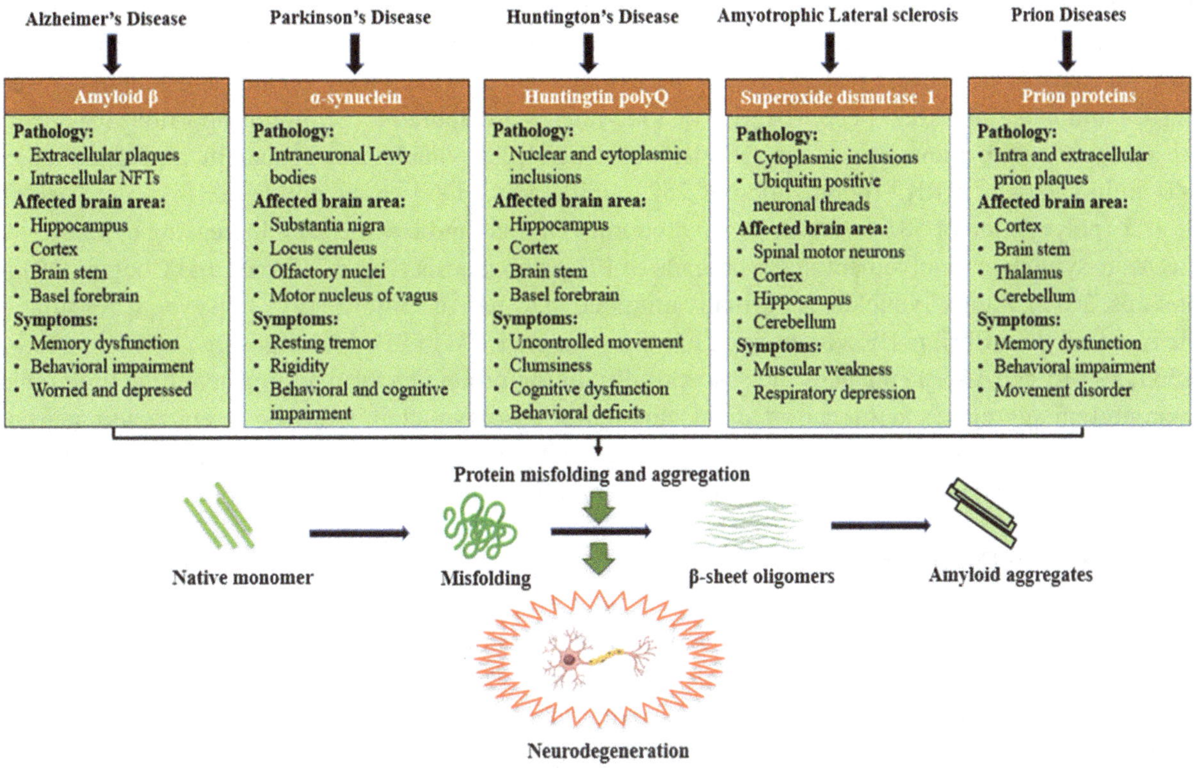

Alzheimer's Disease

Alzheimer's disease (AD) is an irreversible neurologic disorder associated with progressive degeneration of neuronal tissue, especially in the areas of the hippocampus, cortex and basal forebrain. The disease typically displays late-onset neuropsychiatric symptoms, including dementia, behavioral impairment and a loss of activities of daily living (ADLs), and ultimately, to losing higher brain function and motor control (Ashraf et al., 2014). The neuropathological cascade of AD involves the accumulation of toxic proteins, such as amyloid-beta (Aβ) within the brain, intracellular neurofibrillary tangles (NFTs) build-up of tau protein, and extracellular aggregation of Aβ within senile plaques (Hoozemans et al., 2005). Extracellular aggregation of Aβ seems to be the first lesion forming long before the clinical signs begin to appear; whereas NFT activation, synaptic dysfunction and neuroinflammation prompt self-supporting cycles that accelerate the progression of the disease (Serrano-Pozo, Frosch, Masliah, and Hyman, 2011).

Parkinson's Disease

Parkinson's disease (PD) is the second leading neurological disease that mostly affects people over the age of 60. The cardinal feature of the disease include the depletion of dopamine neurons in pars compacta region of the substantial nigra, however, other brain areas like olfactory nuclei, locus ceruleus and posterior motor nucleus of vagus may also be affected (Jankovic, 2008; Zarow, Lyness, Mortimer and Chui, 2003). PD often starts with resting tremor in one hand. Other symptoms include bradykinesia, rigidity, and irregular movements. Postural disability, disruption of the autonomic functions, cognitive impairment, as well as clinical depression can also be seen as the disease progresses (Welzel and Walsh, 2011). Although the loss of dopamine neurons appears to be the central point in PD neuropathogenesis, the role of cytoplasmic inclusions, known as Lewy bodies (LBs) also remains substantial (Power, Barnes and Chegini, 2017). The majority of PD patients are sporadic, while a proportion of patients are associated with common gene mutations, including alpha-synuclein (α-synuclein, a polypeptide of 140-amino acids), LRRK2 (polypeptide of 2527 amino acids, DJ-1 (polypeptide of 189-amino acids), PINK1 (polypeptide of 581-amino acids) and parkin (Thomas and Beal, 2007). Increasing evidence indicates α-synuclein involvement in the cascade of PD pathogenesis (Power, Barnes and Chegini, 2017; Stefanis, 2012). Alpha-synuclein is normally unfolded in its native conformation, however, it constitutes the main amyloid fibril portion (contained in intraneuronal LBs) in PD. Mutations in α-synuclein (e.g., A53T, E46K, andA30P) enhances the susceptibility of the protein to form aggregates. Like with other toxic protein aggregates, aggregation and accumulation of α-synuclein are toxic to cultivated neurons (Welzel and Walsh, 2011). Therefore, conditions increasing the generation or lowering the degradation of α-synuclein, contribute to its accumulation, and eventually in disease growth (Welzel and Walsh, 2011).

Huntington's Disease

Huntington's disease (HD) is an inherent and incurable neurological disease associated with impaired muscular activity and cognitive deficiencies (Finkbeiner, 2011). The key pathogenic feature of HD involves the aggregation and accumulation of mutant huntingtin (mHtt) protein. The wild form of the protein (Htt) comprises a string of polyglutamine (polyQ) tract that is encoded by CAG repeat of the *Htt* gene (Arrasate and Finkbeiner, 2012), the length of which ranges from 16-20 repeats (Warby et al., 2011). However, in affected patients, there is an abnormal CAG repeat expansion (>35 repeats), which results in abnormal expansion of the polyQ tract of mHtt proteins, resulting in their aggregation, and eventually to neuronal degeneration (Williams and Paulson, 2008). The polyQ inclusions may result from a defensive mechanism to sequestrate small mHtt oligomeric forms, that are highly toxic to neuronal cells (Williams and Paulson, 2008).

Amyotrophic Lateral Sclerosis

Amyotrophic lateral sclerosis (ALS) is a lethal neurological disorder associated with the gradual depletion of nerve cells that control voluntary muscular functions. ALS frequently manifests by muscular weakness and respiratory failure (Zarei et al., 2015). Death usually occurs within the first five years after diagnosis (Hobson and McDermott, 2016). ALS is divided into two types, viz the sporadic (SALS, 90% of cases), while, 5% of ALS forms are familial type (FALS) (Byrne et al., 2011). The central neuropathological feature of ALS is the formation of intracellular protein aggregates in affected motor neurons. The primary inclusions seen in ALS are the ubiquitinated hyaline-like inclusions (HLI) or skein-like inclusions (SLI) which are restricted not only to the spinal motor neurons but also to other brains areas like cortex (frontal and temporal), cerebellum and hippocampus (Blokhuis, Groen, Koppers, van den Berg and Pasterkamp, 2013).

Superoxide dismutase 1 (SOD1) was the first-ever protein aggregate detected in FALS patients bearing a mutation in *SOD1* gene (Rosen et al., 1993). Later, in a group of ALS cases, mutations in the *VAPB* gene were also identified to induce ALS (Nishimura et al., 2004). Due to rapid advancement in genetic techniques, several additional proteins, including FUS (Fused in Sarcoma), UBQLN2 (Ubiquilin 2), and OPTN (Optineurin) have been identified in ALS pathogenesis (Blokhuis, Groen, Koppers, van den Berg and Pasterkamp, 2013).

Prion Diseases

Prion diseases (PrDs) are a class of neurological disorder triggered by prions (misfolded proteins capable of transmitting their misfolded form to ordinary versions of the same protein). These diseases can be sporadic or obtained through the genetic or environmental transmission. The pathological hallmark of PrDs includes mutations in prion gene, resulting in infolding and aggregation of prion proteins (PrPs), that manifest as amyloid plaques both intracellularly and extracellularly (Ghaemmaghami, May, Renslo and Prusiner, 2010; Prusiner, 2001).

PROTEIN QUALITY CONTROL IN NEURODEGENERATIVE DISORDERS

Generally, when entering the ER lumen, the nascent polypeptide chains normally undergo unfolding, involving the use of glycosylation, several chaperones, and oxidoreductases. Failure of ER machinery leads to aberrant protein aggregation and eventually to an adapted secretary UPR which initiates ER-associated-protein degradation (ERAD) that causes the removal of toxic aggregates from ER lumen (Meusser, Hirsch, Jarosch and Sommer, 2005). The failure of ER functioning results in ER stress that encourages the aggregation and accumulation of misfolded proteins involved in neuronal pathologies (Figure 4). The central characteristic of several neurodegenerative disorders is the accumulation of protein aggregates that are usually ubiquitinated as has been evidenced by the finding that protein aggregation and subsequent neuronal degeneration in the woozy mutant mouse is triggered by *SIL1* mutation (Zhao, Longo-Guess, Harris, Lee and Ackerman, 2005). Hence, it important to study the role of PQC machinery in dealing with pathogenic cascades of neurodegenerative diseases.

Protein Quality Control in Alzheimer's Disease

The neuropathological cascade of AD involves Aβ aggregation, which induces stress reaction in the form of UPR. ER expression of Aβ leads to its delivery from the ER through ERAD indicating that ERs PQC frameworks recognize it as aberrant (Schmitz, Schneider, Kummer and Herzog, 2004). The UPR signaling pathway primarily manages the removal of misfolded proteins but changes to the apoptotic pathway when the stress in the ER is prolonged. The activation of the UPRs defensive and apoptotic pathways, for example, PERK/eIF2α/CHOP, Grp78/Bip, and caspase-4 show the protective impact in Aβ-mediated neuronal degeneration and/or death. PERK knock-down aggravates the neurotoxic effect of Aβ by decreasing Grp8/BiP and eIF2α neuronal activation (Lee et al., 2010). The UPR pathway associated activators of eIF2α up-regulate the expression of Grp78/BiP to maintain ER homeostasis and this approach may have a neuroprotective potential against Aβ neurotoxicity (Lee et al., 2010). Another possible mechanism is the effect of Aβ on calcium (Ca^{++}) homeostasis by blocking Ca^{++}entry into the plasma membrane or through interfering with calcium processing in the ER (Pierrot et al., 2004). The proteasome and Ca^{++} homeostasis disturbance may, in turn, trigger the UPR activation. Apart from promoting normal folding, preventing aggregate formation, and promoting degradation of misfolded proteins,

Figure 4. Failure of protein quality control system in neurodegenerative disorders.

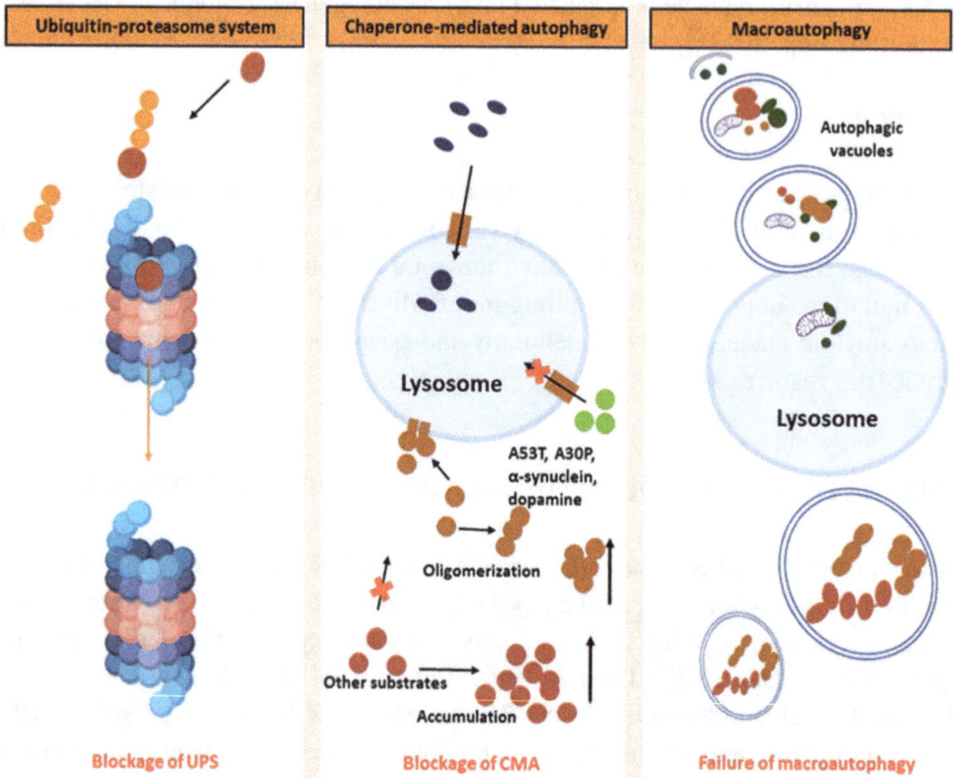

the Grp78/BiP also attaches Ca^{++} and serves as a regulator of ER stress signals (Lee, 2001). Evidence suggests that tau removal by UPS is induced by overexpression of Hsp70 (which binds misfolded proteins) as shown by the finding that inhibition of Hsp70 in transgenic mice decreases soluble tau levels and ameliorates cognition, suggesting that this chaperone plays a significant role in both controlling tau stability and determining proteotoxicity (O'Leary et al., 2010). Hsp27 also binds tau and regulates its stability (Abisambra et al., 2010). The co-chaperones BAG-1, BAG-2, FKBP52, and PP5 have additionally been proposed in helping chaperones to identify hyperphosphorylated tau (Chambraud et al., 2010). PP5 is an intriguing chaperone partner, as it is a protein phosphatase that has been shown to directly act on hyperphosphorylated tau in an Hsp90-coordinated fashion. In addition to chaperone-mediated clearance by the proteasome, tau also appears to be evacuated by macroautophagy and CMA, suggesting that various cellular pathways can act on this substrate (Wang et al., 2010). Together, perceptions propose a model wherein the cell levels of tau and its isoforms are firmly controlled and regulated by a network of PQC systems.

Protein Quality Control in Parkinson's Disease

The point mutations (e.g., A30P, A53T, and E46 K) associated with mutant α-synuclein activity makes α-synuclein susceptible to misfolding and accumulation (Singleton et al., 2003). In PD pathogenesis, non-fibrillary and monomeric mutant forms of α-synuclein can be more cytotoxic than LBs and fibrillary aggregates, which may be attributed to cytoprotective processes (Seidel et al., 2010). The wild form of α-synuclein is ubiquitinated and disposed of under proteasomal degradation which is promoted by its phosphorylation at Ser129 (Luk et al., 2012). Several UPS-regulatory proteins are involved in the cytoplasmic turnover of soluble α-synuclein, including CHIP (carboxyl-terminus of Hsc70 interacting protein) (Lopes da Fonseca, Villar-Piqué and Outeiro, 2015), MDM2 (mouse double minute 2 homolog) (Nair, McNaught, Gonzalez-Maeso, Sealfon and Olanow, 2006), SIAH (seven in absentia homolog) (Abeywardana, Lin, Rott, Engelender and Pratt, 2013), and the E3 Ub-ligase HRD1(Atkin and Paulson, 2014). Also, a phenotype of α-synuclein linked with endosome-lysosome pathway membranes is targeted by the E3 Ub-protein ligase NEDD4 (Tofaris, Kim, Hourez, Jung, Kim and Goldberg, 2011).

Accumulating evidence suggests that CMA can degrade α-synuclein via a specific CMA detection motif (KFERQ-like motif) (Ho et al., 2019; Mak, McCormack, Manning-Bog, Cuervo and Di Monte, 2010). The A30P and A53 T PD mutants, however, have a high affinity towards CMA adapter LAMP-2A and are not delivered adequately to the lysosomal lumen which in turn leads to CMA overload (Vogiatzi, Xilouri, Vekrellis and Stefanis, 2008) (Figure 4). Eventually, this can induce compensating response by macroautophagy et al., 2011). Although both CMA and UPS can target the monomeric or soluble oligomeric forms of α-synuclein, its aggregates are delivered to the lysosome by macroautophagy (Paxinou et al., 2001). In LBs, Hsp70 and crystalline are also involved in α-synuclein degradation (Bandopadhyay and de Belleroche, 2010). In addition, sHsp27 appears to be an efficient anti-α-synuclein agent (Willingham, Outeiro, DeVit, Lindquist and Muchowski, 2003). Macroautophagy activation using pharmacological agent rapamycin promoted both the mutant and wild-type α-synuclein degradation (Webb, Ravikumar, Atkins, Skepper and Rubinsztein, 2003). Further, removal of α-synuclein through macroautophagy has been found in transgenic mice overexpressing an autophagic regulator Beclin.

Protein Quality Control in Huntington's Disease

Numerous genetic and biochemical studies suggest the polyQ-related pathology in HD which can be blocked by enhancing the ability of PQC mechanisms, for example, Hsp70-Hsp40 complex was found to be associated selectively with tiny aggregates of huntingtin polyQ, indicating roles of these chaperones in critical oligomeric aggregation (Lotz et al., 2010). In yeast, worm, and fly models, polyQ-associated phenotypes are suppressed by concurrent molecular chaperone expression, including Hsp70 and Hsp40 (Bilen and Bonini, 2007; Cummings et al., 2001).

Evidence suggests that mHtt serves as a blocker of proteolytic pathways, for example, mHtt inclusions in the brains of HD mice and patients are abundant in UPS components (because mHtt species are firstly marked by Ub) (Juenemann et al., 2018). The aggregation and accumulation of mHtt inclusions do not result explicitly from proteasomal blockage, but rather from widespread PQC failure (Hipp et al., 2012). The wild-type Htt can be degraded by CMA (Jeong et al., 2009), during which Hsc70 is identified by two KFERQ-like motifs, for example by NEIKV at 248- 252 residues and by KDRVN at 99-103 residues (Qi et al., 2012). Similarly, mHtt can also be identified by Hsc70 for CMA associated destruction, but the polyQ expansion of mHtt delays the transfer of mHTT to lysosome because of mHtt displays a greater affinity for Hsc70 and LAMP-2A (Qi et al., 2012). This ultimately places a high load over CMA, leading to the accumulation of cytoplasmic and intraneuronal inclusions (Jeong et al., 2009). It has been shown that the macroautophagy components, for example, microtubule-associated protein1 light chain 3 (LC3), are usually overexpressed in neuronal cells of PD patients and several HD animal models (Martinez-Vicente et al., 2010). The apparent upregulation of macroautophagy is associated with the excessive formation of autophagic load-free vacuoles (Mehrpour, Botti and Codogno, 2012). It has been reported that when the autophagic flux is decreased, LC3-II, p62, and mTOR are accumulated in the striatum of transgenic HD mice (Lee et al., 2012). As a result, HD disease progression is amplified by Htt associated inhibition of macroautophagy.

Protein Quality Control in Amyotrophic Lateral Sclerosis

Mutations in ALS occur in genes encoding the core PQC components. This class of mutant ALS proteins involves dynactin and dynein (both are implicated in the backward transfer of autophagosomes from axons to the nucleus), p62, and the Ub containing proteins OPTN and UBQLN2 (Fecto and Siddique, 2012; Fecto et al., 2011). The second category of mutations occurs in genes, including *TDP-43, SOD1, and FUS/TLS*, resulting in the formation of abnormally folded proteins, that aggregate and subsequently form insoluble inclusions (Da Cruz and Cleveland, 2011; Sheng, Chattopadhyay, Whitelegge and Valentine, 2012). Around 20% of FALS cases are associated with point mutations in the *SOD1* (Beleza-Meireles and Al-Chalabi, 2009). Mutant SOD1 is highly dominant as it tends to form protease-resistant misfolded aggregates which are extremely toxic to affected motor neurons (Thangavelu, Tripathi, Arya, Mishra and Subramaniam, 2011). Initially, the misfolded mSOD1 and mTDP-43 are directed for UPS degradation, such as by Ub ligases, and degradation by molecular chaperones (Andersen and Al-Chalabi, 2011). However, these targeted mutants may escape the degradation process during their delivery to the proteasome, while some may be directed to autophagy. Altogether, the mutant forms resistant to the components of autophagy and UPS lead to aggregation and accumulation of intranuclear inclusions, as have been seen in human ALS patients and mutant SOD1 transgenic mice (Bendotti et al., 2004). The misregulation of autophagy in ALS is also substantial which can be explained by the evidence that the inclusions seen in ALS patients can disrupt PQC machinery by sequestering several components of UPS such as Ub ligases (e.g., Dorfin) and molecular chaperones (Hsp/Hsp70) and dynein motor protein that are involved in aggresomal cargo delivery(Rothenberg et al., 2010). Furthermore, decreased proteasomal activity may induce the aggregation and accumulation of aberrant ALS proteins (Crippa et al., 2010). Hence, ALS pathogenic cascade involves a vicious cycle between PQC pathways and misfolded aggregated proteins that stimulate the accumulation of excessive insoluble inclusions in the affected motor neurons, eventually leading to their degeneration and/or death.

Protein Quality Control in Prion Diseases

Prion diseases are a group of invariably fatal prion-associated neurodegenerative disorders, affecting both humans and animals. These diseases result in spongiform vacuolation and extensive neuronal degeneration (Geschwind, 2015). PrDs in humans include Creutzfeldt-Jakob disease, variant Creutzfeldt-Jakob disease, variably protease-sensitive prionopathy, Gerstmann-Sträussler-Scheinker syndrome fatal insomnia, and kuru (Geschwind, 2015). In animals, these protein misfolding disorders include Bovine spongiform encephalopathy (in goat and sheep), Feline spongiform encephalopathy in cats, transmissible mink encephalopathy, chronic wasting disease (in deer, elk, and moose) and scrapie in (mouflons, sheep, and goat) (Imran and Mahmood, 2011). The common pathological hallmark of PrDs involves the transformation of normal prion protein (PrP^C, C indicates a cellular form of protein) having a primary α-enriched structure into an abnormal misfolded form, called the scrapie prion protein (PrP^{Sc}) which is enriched with β-pleated sheet structure (Geschwind, 2015). The mature PrP^{Sc} generated in ER undergoes folding during its transportation through the Golgi-secretory system, during which a soluble fraction of the misfolded PrP^{Sc} is destroyed by several PQC systems, such as Ub-dependent ERAD. Since PrP^{Sc} primarily contains a β-pleated sheet structure, it results in the formation aggregates which are resistant to almost all the PQC machinery (Prusiner, 1982; Prusiner, 1998).

Further, PrP^{Sc} can come into contact with PrP^{C} and facilitate its transformation into PrP^{Sc}, resulting in two PrP^{Sc}, which in turn transform two more PrP^{C} into PrP^{Sc}, resulting in four PrP^{Sc}, and so forth, eventually resulting in the exponential accumulation of aggregated PrP^{Sc} (Geschwind, 2015). This transmissible nature of PrP^{Sc} was shown in a study where the induction of a small quantity of PrP^{Sc} into the mice resulted in typical features of the disease (Wang, Wang, Yuan and Ma, 2010). The disruption of the autophagic pathway PrDs was demonstrated by the generation of large autophagic vacuoles in scrapie-affected experimental hamsters (Boellaard, Kao, Schlote and Diringer, 1991). As the neuronal sage, these vacuoles usually grow in number and size, ultimately inhabiting the whole space of the affected neurons (Sikorska, Liberski and Brown, 2007).

THERAPEUTIC PERSPECTIVES

The therapeutic strategies in neurodegenerative disorders are often directed at targeting the aggregated proteins, with minimal success and associated side effects. Significant benefits of drug therapy could be obtained through the cooperative work of PQC systems (e.g., either by reducing harmful or increasing defensive responses) because this approach allows the use of the cell's quality control pathways which have evolved to explicitly address this issue.

Hsp levels can be induced to safeguard cells from aggregate toxicity. For instance, in a cell culture model of HD, low-dose geldanamycin treatment-induced HSR that inhibited huntingtin aggregation in a cell culture model of HD (Sittler et al., 2001). This can be further illustrated by the finding that arimoclomol (a co-inducer of Hsps) treatment of ALS mice prevented nerve cell depletion and subsequently enhanced the motor functions and lifespan (Kieran, Kalmar, Dick, Riddoch-Contreras, Burnstock and Greensmith, 2004). In another study, the upregulation of Hsp70 stimulated the proteasomal destruction of tau (Petrucelli et al., 2004).

Another approach of using PQC to tackle neuronal degeneration is by targeting the UPR. In a study, the pharmacological agent salubrinal (eIF2α phosphatase inhibitor) protected against ER stress (Hong, Wang, Sun, Xue, Li and Hou, 2016). Many autophagic inducers have been researched to expedite the elimination of toxic aggregates. For example, rapamycin (mTOR inhibitor) has been studied in various transgenic mouse models of neurodegeneration, for example, PD mice expressing mutant α-synuclein (Webb, Ravikumar, Atkins, Skepper and Rubinsztein, 2003), HD mice expressing mHtt (Sarkar and Rubinsztein, 2008), AD mice expressing Aβ (Spilman et al., 2010), and prion mice expressing PrP^{Sc} (Heiseke, Aguib, Riemer, Baier and Schatzl, 2009). All of these studies showed amelioration of aggregate induced neuropathology following induction with rapamycin. Also, rilmenidine improved motor functions and facilitated the removal of mHtt segment in the transgenic HD mouse model (Rose et al., 2010). Furthermore, mood stabilizer lithium, an inhibitor of phosphoinositol and inositol monophosphatase, induced the degradation of several misfolded protein aggregates, such as mutant α-synuclein of PD, mHtt of HD (Sarkar et al., 2005), and SOD1 of ALS (Feng, Leng, Ma, Zhang, Ren and Chuang, 2008). Trehalose, a disaccharide enhanced the clearance of α-synuclein, A53T and A30P mutants in cultured PD cell model (Sarkar, Davies, Huang, Tunnacliffe and Rubinsztein, 2007). The overall results of these studies suggest that autophagic inducers have neuroprotective potential in selected neurodegenerative disorders. However, further research is needed to validate these findings on a broader scale.

RECENT DEVELOPMENTS AND FUTURE RESEARCH DIRECTIONS

In this chapter, we reviewed recent findings underlying cellular PQC mechanisms and emphasized their central role in neuronal protein misfolding disorders. The selective elimination of misfolded and aggregated proteins and re-establishment of the proteostasis remains unclear and that some key issues remain unsolved. For example, how does a cell precisely identify a faulty protein within a crowded cellular context, or does the difference between normal proteins or damaged proteins occur under some basic principles? Based on the current information, it can be speculated that cells retain a PQC system that has the potential to combat the issue of protein aggregation. The goal is to obtain further information about the molecules of essential collaborators involved directly in this process and which molecular mechanisms control and assess its exquisite specificity and overall functionality? Indeed, the most critical question is whether these findings will offer potential clinical benefits in human protein conformation disorders.

New approaches against cellular toxicity by misfolded and aggregated proteins are essential, especially on pathways that cause late-onset neurological disorders. Another critical problem that needs to be addressed is to understand the mechanism of selectively recognizing misfolded proteins and how they can be promoted for destruction without disturbing normal cellular proteostasis. Until now, the induction of UPS has only been exploited *in vitro* primarily through the enhancement of proteolytic activities, while unfolding, Ub associated targeting, and de-ubiquitination of the substrate is the limiting measures in substrate disposition by this framework. Also, autophagy has only been induced via the upregulation of the lysosomal receptor. Therefore, a deeper understanding of the basic pathways to activate the cellular activities of autophagy and UPS to suppress the multi-factorial proteotoxic burden has to be achieved. Thereby, the greatest unmet need in this area is the development of drug candidates that can modulate and enhance the defense potential of cellular PQC systems, including UPS (especially that of E3 Ub ligases), CMA and autophagy. Nonetheless, the first consideration must be given to their safety and efficacy over a broader range of neuronal processing.

CONCLUSION

Accumulating evidence suggests the central role of cellular PQC systems in regulating the neuronal proteostasis, and close links have been identified between impaired PQC frameworks and the development of neurodegeneration. The common mechanism underlying the neuropathogenic cascade is the failure of PQC systems in removing the aberrant misfolded protein aggregates. Post-translational modifications of several substrates, including amyloid β, tau, Htt, α-synuclein, SOD1, and PrP^{Sc} generate pathogenic misfolded protein aggregates into β-sheet content-rich oligomers which subsequently develop into cytoplasmic inclusions or extracellular plaques, eventually impairing UPS and CMA. This results in reduced PQC capability and speeds up aggregate accumulation. The vicious cycle between the PQC pathways and defective proteins is highly toxic to nerve cells because their capability to address the issue of aberrant protein aggregation is naturally decreased with increasing age. Hence, neurodegenerative disorders are correlated with global impairment of all PQC systems.

REFERENCES

Abeywardana, T., Lin, Y. H., Rott, R., Engelender, S., & Pratt, M. R. (2013). Site-specific differences in proteasome-dependent degradation of monoubiquitinated α-synuclein. *Chemistry & Biology*, *20*(10), 1207–1213. doi:10.1016/j.chembiol.2013.09.009 PMID:24210006

Abisambra, J. F., Blair, L. J., Hill, S. E., Jones, J. R., Kraft, C., Rogers, J., ... Dickey, C. A. (2010). Phosphorylation dynamics regulate Hsp27-mediated rescue of neuronal plasticity deficits in tau transgenic mice. *The Journal of Neuroscience*, *30*(46), 15374–15382. doi:10.1523/JNEUROSCI.3155-10.2010 PMID:21084594

Andersen, P. M., & Al-Chalabi, A. (2011). Clinical genetics of amyotrophic lateral sclerosis: What do we really know? *Nature Reviews. Neurology*, *7*(11), 603–615. doi:10.1038/nrneurol.2011.150 PMID:21989245

Ariosa, A. R., & Klionsky, D. J. (2016). Autophagy core machinery: Overcoming spatial barriers in neurons. *Journal of Molecular Medicine (Berlin, Germany)*, *94*(11), 1217–1227. doi:10.100700109-016-1461-9 PMID:27544281

Arrasate, M., & Finkbeiner, S. (2012). Protein aggregates in Huntington's disease. *Experimental Neurology*, *238*(1), 1–11. doi:10.1016/j.expneurol.2011.12.013 PMID:22200539

Ashraf, G. M., Greig, N. H., Khan, T. A., Hassan, I., Tabrez, S., Shakil, S., ... Kamal, M. (2014). Protein misfolding and aggregation in Alzheimer's disease and type 2 diabetes mellitus. *CNS & Neurological Disorders - Drug Targets*, *13*(7), 1280–1293. doi:10.2174/1871527313666140917095514 PMID:25230234

Atkin, G., & Paulson, H. (2014). Ubiquitin pathways in neurodegenerative disease. *Frontiers in Molecular Neuroscience*, 7. PMID:25071440

Auld, K. L., & Silver, P. A. (2006). Transcriptional Regulation by the Proteasome as a Mechanism for Cellular Protein. *Cell Cycle*, *5*(14), 1503–1505. doi:10.4161/cc.5.14.2979 PMID:16861887

Bandopadhyay, R., & de Belleroche, J. (2010). Pathogenesis of Parkinson's disease: Emerging role of molecular chaperones. *Trends in Molecular Medicine*, *16*(1), 27–36. doi:10.1016/j.molmed.2009.11.004 PMID:20036196

Batista, P., & Pereira, A. (2016). Quality of Life in Patients with Neurodegenerative Diseases. *Journal of Neurology and Neuroscience*, *7*(1), 7. doi:10.21767/2171-6625.100074

Beleza-Meireles, A., & Al-Chalabi, A. (2009). Genetic studies of amyotrophic lateral sclerosis: Controversies and perspectives. *Amyotrophic Lateral Sclerosis*, *10*(1), 1–14. doi:10.1080/17482960802585469 PMID:19110986

Bendotti, C., Atzori, C., Piva, R., Tortarolo, M., Strong, M. J., DeBiasi, S., & Migheli, A. (2004). Activated p38MAPK is a novel component of the intracellular inclusions found in human amyotrophic lateral sclerosis and mutant SOD1 transgenic mice. *Journal of Neuropathology and Experimental Neurology*, *63*(2), 113–119. doi:10.1093/jnen/63.2.113 PMID:14989597

Bilen, J., & Bonini, N. M. (2007). Genome-wide screen for modifiers of ataxin-3 neurodegeneration in Drosophila. *PLOS Genetics*, *3*(10), e177. doi:10.1371/journal.pgen.0030177 PMID:17953484

Bingol, B. (2018). Autophagy and lysosomal pathways in nervous system disorders. *Molecular and Cellular Neurosciences*, *91*, 167–208. doi:10.1016/j.mcn.2018.04.009 PMID:29729319

Blokhuis, A. M., Groen, E. J. N., Koppers, M., van den Berg, L. H., & Pasterkamp, R. J. (2013). Protein aggregation in amyotrophic lateral sclerosis. *Acta Neuropathologica*, *125*(6), 777–794. doi:10.100700401-013-1125-6 PMID:23673820

Boellaard, J., Kao, M., Schlote, W., & Diringer, H. (1991). Neuronal autophagy in experimental scrapie. *Acta Neuropathologica*, *82*(3), 225–228. doi:10.1007/BF00294449 PMID:1927279

Brandman, O., Stewart-Ornstein, J., Wong, D., Larson, A., Williams, C. C., Li, G.-W., ... Weissman, J. S. (2012). A ribosome-bound quality control complex triggers degradation of nascent peptides and signals translation stress. *Cell*, *151*(5), 1042–1054. doi:10.1016/j.cell.2012.10.044 PMID:23178123

Byrne, S., Walsh, C., Lynch, C., Bede, P., Elamin, M., Kenna, K., ... Hardiman, O. (2011). Rate of familial amyotrophic lateral sclerosis: A systematic review and meta-analysis. *Journal of Neurology, Neurosurgery, and Psychiatry*, *82*(6), 623–627. doi:10.1136/jnnp.2010.224501 PMID:21047878

Chambraud, B., Sardin, E., Giustiniani, J., Dounane, O., Schumacher, M., Goedert, M., & Baulieu, E.-E. (2010). A role for FKBP52 in Tau protein function. *Proceedings of the National Academy of Sciences of the United States of America*, *107*(6), 2658–2663. doi:10.1073/pnas.0914957107 PMID:20133804

Ciechanover, A., & Kwon, Y. T. (2017). Protein Quality Control by Molecular Chaperones in Neurodegeneration. *Frontiers in Neuroscience*, *11*, 185. doi:10.3389/fnins.2017.00185 PMID:28428740

Crippa, V., Sau, D., Rusmini, P., Boncoraglio, A., Onesto, E., Bolzoni, E., ... Poletti, A. (2010). The small heat shock protein B8 (HspB8) promotes autophagic removal of misfolded proteins involved in amyotrophic lateral sclerosis (ALS). *Human Molecular Genetics*, *19*(17), 3440–3456. doi:10.1093/hmg/ddq257 PMID:20570967

Cuervo, A. M., & Wong, E. (2014). Chaperone-mediated autophagy: Roles in disease and aging. *Cell Research*, *24*(1), 92–104. doi:10.1038/cr.2013.153 PMID:24281265

Cummings, C. J., Sun, Y., Opal, P., Antalffy, B., Mestril, R., Orr, H. T., & ... . (2001). Over-expression of inducible HSP70 chaperone suppresses neuropathology and improves motor function in SCA1 mice. *Human Molecular Genetics*, *10*(14), 1511–1518. doi:10.1093/hmg/10.14.1511 PMID:11448943

Da Cruz, S., & Cleveland, D. W. (2011). Understanding the role of TDP-43 and FUS/TLS in ALS and beyond. *Current Opinion in Neurobiology*, *21*(6), 904–919. doi:10.1016/j.conb.2011.05.029 PMID:21813273

Dong, Z., & Cui, H. (2019). The autophagy-lysosomal pathways and their emerging roles in modulating proteostasis in tumors. *Cells*, *8*(1), 4. doi:10.3390/cells8010004 PMID:30577555

Elsasser, S., & Finley, D. (2005). Delivery of ubiquitinated substrates to protein-unfolding machines. *Nature Cell Biology*, *7*(8), 742–749. doi:10.1038/ncb0805-742 PMID:16056265

Fecto, F., & Siddique, T. (2012). UBQLN2/P62 cellular recycling pathways in amyotrophic lateral sclerosis and frontotemporal dementia. *Muscle & Nerve*, *45*(2), 157–162. doi:10.1002/mus.23278 PMID:22246868

Fecto, F., Yan, J., Vemula, S. P., Liu, E., Yang, Y., Chen, W., & ... . (2011). SQSTM1 mutations in familial and sporadic amyotrophic lateral sclerosis. *Archives of Neurology*, *68*, 1440–1446. doi:10.1001/archneurol.2011.250 PMID:22084127

Feng, H. L., Leng, Y., Ma, C. H., Zhang, J., Ren, M., & Chuang, D. M. (2008). Combined lithium and valproate treatment delays disease onset, reduces neurological deficits and prolongs survival in an amyotrophic lateral sclerosis mouse model. *Neuroscience*, *155*(3), 567–572. doi:10.1016/j.neuroscience.2008.06.040 PMID:18640245

Feng, Y., He, D., Yao, Z., & Klionsky, D. J. (2014). The machinery of macroautophagy. *Cell Research*, *24*(1), 24–41. doi:10.1038/cr.2013.168 PMID:24366339

Finkbeiner, S. (2019). The Autophagy Lysosomal Pathway and Neurodegeneration. *Cold Spring Harbor Perspectives in Biology*, a033993. doi:10.1101/cshperspect.a033993 PMID:30936119

Galluzzi, L., & Green, D. R. (2019). Autophagy-Independent Functions of the Autophagy Machinery. *Cell*, *177*(7), 1682–1699. doi:10.1016/j.cell.2019.05.026 PMID:31199916

Geschwind, M. D. (2015). Prion Diseases. *Continuum (Minneapolis, Minn.)*, *21*, 1612–1638. doi:10.1212/CON.0000000000000251 PMID:26633779

Gestwicki, J.E., & Garza, D. (2012). Protein quality control in neurodegenerative disease. In *Progress in molecular biology and translational science*. Elsevier.

Ghaemmaghami, S., May, B. C., Renslo, A. R., & Prusiner, S. B. (2010). Discovery of 2-aminothiazoles as potent antiprion compounds. *Journal of Virology*, *84*(7), 3408–3412. doi:10.1128/JVI.02145-09 PMID:20032192

Hageman, J., & Kampinga, H. H. (2009). Computational analysis of the human HSPH/HSPA/DNAJ family and cloning of a human HSPH/HSPA/DNAJ expression library. *Cell Stress & Chaperones*, *14*(1), 1–21. doi:10.100712192-008-0060-2 PMID:18686016

Haslbeck, M., Franzmann, T., Weinfurtner, D., & Buchner, J. (2005). Some like it hot: The structure and function of small heat-shock proteins. *Nature Structural & Molecular Biology*, *12*(10), 842–846. doi:10.1038/nsmb993 PMID:16205709

Heiseke, A., Aguib, Y., Riemer, C., Baier, M., & Schatzl, H. M. (2009). Lithium induces clearance of protease resistant prion protein in prion-infected cells by induction of autophagy. *Journal of Neurochemistry*, *109*(1), 25–34. doi:10.1111/j.1471-4159.2009.05906.x PMID:19183256

Hershko, A. (2005). The ubiquitin system for protein degradation and some of its roles in the control of the cell-division cycle (Nobel lecture). *Angewandte Chemie International Edition*, *44*(37), 5932–5943. doi:10.1002/anie.200501724 PMID:16142823

Hetz, C., & Mollereau, B. (2014). Disturbance of endoplasmic reticulum proteostasis in neurodegenerative diseases. *Nature Reviews. Neuroscience*, *15*(4), 233–249. doi:10.1038/nrn3689 PMID:24619348

Hipp, M. S., Patel, C. N., Bersuker, K., Riley, B. E., Kaiser, S. E., Shaler, T. A., ... Kopito, R. R. (2012). Indirect inhibition of 26S proteasome activity in a cellular model of Huntington's disease. *The Journal of Cell Biology*, *196*(5), 573–587. doi:10.1083/jcb.201110093 PMID:22371559

Ho, P. W., Leung, C. T., Liu, H., Pang, S. Y., Lam, C. S., Xian, J., & … . (2019). Age-dependent accumulation of oligomeric SNCA/alpha-synuclein from impaired degradation in mutant LRRK2 knockin mouse model of Parkinson disease: Role for therapeutic activation of chaperone-mediated autophagy (CMA). *Autophagy*, 1–24. PMID:30983487

Hobson, E. V., & McDermott, C. J. (2016). Supportive and symptomatic management of amyotrophic lateral sclerosis. *Nature Reviews. Neurology*, *12*(9), 526–538. doi:10.1038/nrneurol.2016.111 PMID:27514291

Höhfeld, J., Cyr, D. M., & Patterson, C. (2001). From the cradle to the grave: Molecular chaperones that may choose between folding and degradation. *EMBO Reports*, *2*(10), 885–890. doi:10.1093/embo-reports/kve206 PMID:11600451

Hong, Y., Wang, X., Sun, S., Xue, G., Li, J., & Hou, Y. (2016). Progesterone exerts neuroprotective effects against Abeta-induced neuroinflammation by attenuating ER stress in astrocytes. *International Immunopharmacology*, *33*, 83–89. doi:10.1016/j.intimp.2016.02.002 PMID:26878478

Hoozemans, J. J., Veerhuis, R., Van Haastert, E. S., Rozemuller, J. M., Baas, F., Eikelenboom, P., & Scheper, W. (2005). The unfolded protein response is activated in Alzheimer's disease. *Acta Neuropathologica*, *110*(2), 165–172. doi:10.100700401-005-1038-0 PMID:15973543

Huntington's Disease Foundation. (2019). *Life with Huntington's disease.*

Huzil, J. T., Pannu, R., Ptak, C., Garen, G., & Ellison, M. J. (2007). Direct catalysis of lysine 48-linked polyubiquitin chains by the ubiquitin-activating enzyme. *The Journal of Biological Chemistry*, *282*(52), 37454–37460. doi:10.1074/jbc.M705242200 PMID:17951259

Imran, M., & Mahmood, S. (2011). An overview of animal prion diseases. *Virology Journal*, *8*(1), 493. doi:10.1186/1743-422X-8-493 PMID:22044871

Jankovic, J. (2008). Parkinson's disease: Clinical features and diagnosis. *Journal of Neurology, Neurosurgery, and Psychiatry*, *79*(4), 368–376. doi:10.1136/jnnp.2007.131045 PMID:18344392

Jauhiainen, A., Thomsen, C., Strombom, L., Grundevik, P., Andersson, C., Danielsson, A., … Åman, P. (2012). Distinct cytoplasmic and nuclear functions of the stress induced protein DDIT3/CHOP/GADD153. *PLoS One*, *7*(4), e33208. doi:10.1371/journal.pone.0033208 PMID:22496745

Jeong, H., Then, F., Melia, T. J. Jr, Mazzulli, J. R., Cui, L., Savas, J. N., … Krainc, D. (2009). Acetylation targets mutant huntingtin to autophagosomes for degradation. *Cell*, *137*(1), 60–72. doi:10.1016/j.cell.2009.03.018 PMID:19345187

Juenemann, K., Jansen, A. H. P., van Riel, L., Merkx, R., Mulder, M. P. C., An, H., … Reits, E. A. (2018). Dynamic recruitment of ubiquitin to mutant huntingtin inclusion bodies. *Scientific Reports*, *8*(1), 1405. doi:10.103841598-018-19538-0 PMID:29362455

Kieran, D., Kalmar, B., Dick, J. R., Riddoch-Contreras, J., Burnstock, G., & Greensmith, L. (2004). Treatment with arimoclomol, a coinducer of heat shock proteins, delays disease progression in ALS mice. *Nature Medicine*, *10*(4), 402–405. doi:10.1038/nm1021 PMID:15034571

Lee, A. S. (2001). The glucose-regulated proteins: Stress induction and clinical applications. *Trends in Biochemical Sciences*, *26*(8), 504–510. doi:10.1016/S0968-0004(01)01908-9 PMID:11504627

Lee, H., Noh, J. Y., Oh, Y., Kim, Y., Chang, J. W., Chung, C. W., ... Jung, Y.-K. (2012). IRE1 plays an essential role in ER stress-mediated aggregation of mutant huntingtin via the inhibition of autophagy flux. *Human Molecular Genetics*, *21*(1), 101–114. doi:10.1093/hmg/ddr445 PMID:21954231

Lee, K.-S., Lee, H. J., Kim, D. H., Noh, Y. H., Yu, K., Jung, H.-Y., & ... . (2010). Activation of PERK signaling attenuates Aβ-mediated ER stress. *PLoS One*, *5*(5), e10489. doi:10.1371/journal.pone.0010489 PMID:20463975

Liberek, K., Lewandowska, A., & Zietkiewicz, S. (2008). Chaperones in control of protein disaggregation. *The EMBO Journal*, *27*(2), 328–335. doi:10.1038j.emboj.7601970 PMID:18216875

Lindholm, D., Wootz, H., & Korhonen, L. (2006). ER stress and neurodegenerative diseases. *Cell Death and Differentiation*, *13*(3), 385–392. doi:10.1038j.cdd.4401778 PMID:16397584

Lopes da Fonseca, T., Villar-Piqué, A., & Outeiro, T. F. (2015). The Interplay between Alpha-Synuclein Clearance and Spreading. *Biomolecules*, *5*(2), 435–471. doi:10.3390/biom5020435 PMID:25874605

Lotz, G. P., Legleiter, J., Aron, R., Mitchell, E. J., Huang, S.-Y., Ng, C., ... Muchowski, P. J. (2010). Hsp70 and Hsp40 functionally interact with soluble mutant huntingtin oligomers in a classic ATP-dependent reaction cycle. *The Journal of Biological Chemistry*, *285*(49), 38183–38193. doi:10.1074/jbc.M110.160218 PMID:20864533

Luk, K. C., Kehm, V., Carroll, J., Zhang, B., O'Brien, P., Trojanowski, J. Q., & Lee, V. M.-Y. (2012). Pathological alpha-synuclein transmission initiates Parkinson-like neurodegeneration in nontransgenic mice. *Science*, *338*(6109), 949–953. doi:10.1126cience.1227157 PMID:23161999

Mak, S. K., McCormack, A. L., Manning-Bog, A. B., Cuervo, A. M., & Di Monte, D. A. (2010). Lysosomal degradation of alpha-synuclein in vivo. *The Journal of Biological Chemistry*, 285. PMID:20200163

Martinez-Vicente, M., Talloczy, Z., Wong, E., Tang, G., Koga, H., Kaushik, S., ... Cuervo, A. M. (2010). Cargo recognition failure is responsible for inefficient autophagy in Huntington's disease. *Nature Neuroscience*, *13*(5), 567–576. doi:10.1038/nn.2528 PMID:20383138

Martini-Stoica, H., Xu, Y., Ballabio, A., & Zheng, H. (2016). The autophagy–lysosomal pathway in neurodegeneration: A TFEB perspective. *Trends in Neurosciences*, *39*(4), 221–234. doi:10.1016/j.tins.2016.02.002 PMID:26968346

McClellan, A. J., Scott, M. D., & Frydman, J. (2005). Folding and quality control of the VHL tumor suppressor proceed through distinct chaperone pathways. *Cell*, *121*(5), 739–748. doi:10.1016/j.cell.2005.03.024 PMID:15935760

McConkey, D. J. (2017). The integrated stress response and proteotoxicity in cancer therapy. *Biochemical and Biophysical Research Communications*, *482*(3), 450–453. doi:10.1016/j.bbrc.2016.11.047 PMID:28212730

Mehrpour, M., Botti, J., & Codogno, P. (2012). Mechanisms and regulation of autophagy in mammalian cells. *Atlas of Genetics and Cytogenetics in Oncology and Haematology*, (2). doi:10.4267/2042/46951

Meusser, B., Hirsch, C., Jarosch, E., & Sommer, T. (2005). ERAD: The long road to destruction. *Nature Cell Biology*, *7*(8), 766–772. doi:10.1038/ncb0805-766 PMID:16056268

Nair, V. D., McNaught, K. S., Gonzalez-Maeso, J., Sealfon, S. C., & Olanow, C. W. (2006). p53 mediates nontranscriptional cell death in dopaminergic cells in response to proteasome inhibition. *The Journal of Biological Chemistry, 281*(51), 39550–39560. doi:10.1074/jbc.M603950200 PMID:17060322

Nakayama, K. I., & Nakayama, K. (2006). Ubiquitin ligases: Cell-cycle control and cancer. *Nature Reviews. Cancer, 6*(5), 369–381. doi:10.1038/nrc1881 PMID:16633365

Nedić, O., Rattan, S., Grune, T., & Trougakos, I. (2013). Molecular effects of advanced glycation end products on cell signalling pathways, ageing and pathophysiology. *Free Radical Research, 47*(sup1), 28–38. doi:10.3109/10715762.2013.806798 PMID:23692178

Nigar, S., Pottoo, F. H., Tabassum, N., Verma, S. K., & Javed, M. N. (2016). Molecular Insights into the Role of Inflammation and Oxidative Stress in Epilepsy. *Journal of Advances in Medical and Pharmaceutical Sciences, 10*(1), 1–9. doi:10.9734/JAMPS/2016/24441

Nishimura, A. L., Mitne-Neto, M., Silva, H. C., Richieri-Costa, A., Middleton, S., Cascio, D., ... Zatz, M. (2004). A mutation in the vesicle-trafficking protein VAPB causes late-onset spinal muscular atrophy and amyotrophic lateral sclerosis. *American Journal of Human Genetics, 75*(5), 822–831. doi:10.1086/425287 PMID:15372378

O'Leary, J. C. III, Li, Q., Marinec, P., Blair, L. J., Congdon, E. E., Johnson, A. G., ... Dickey, C. A. (2010). Phenothiazine-mediated rescue of cognition in tau transgenic mice requires neuroprotection and reduced soluble tau burden. *Molecular Neurodegeneration, 5*(1), 45. doi:10.1186/1750-1326-5-45 PMID:21040568

Paxinou, E., Chen, Q., Weisse, M., Giasson, B. I., Norris, E. H., Rueter, S. M., ... Ischiropoulos, H. (2001). Induction of α-synuclein aggregation by intracellular nitrative insult. *The Journal of Neuroscience: The Official Journal of the Society for Neuroscience, 21*(20), 8053–8061. doi:10.1523/JNEUROSCI.21-20-08053.2001 PMID:11588178

Petrucelli, L., Dickson, D., Kehoe, K., Taylor, J., Snyder, H., Grover, A., & ... . (2004). CHIP and Hsp70 regulate tau ubiquitination, degradation and aggregation. *Human Molecular Genetics, 13*(7), 703–714. doi:10.1093/hmg/ddh083 PMID:14962978

Pierrot, N., Ghisdal, P., Caumont, A. S., & Octave, J. N. (2004). Intraneuronal amyloid-β1-42 production triggered by sustained increase of cytosolic calcium concentration induces neuronal death. *Journal of Neurochemistry, 88*(5), 1140–1150. doi:10.1046/j.1471-4159.2003.02227.x PMID:15009669

Pottoo, F. H., Alshayban, D. M., Joseph, R., Al-Musa, F., Al-Jabran, O., & Aljaafari, D. (2020). Impact of adherence to antiepileptic medications on quality of life of epileptic patients in the Eastern Province of Saudi Arabia: A cross-sectional study. *Imam. J. Appl. Sci., 5*, 1-8. Retrieved from http://www.e-ijas.org/text.asp?2020/5/1/1/274291

Pottoo, F. H., Bhowmik, M., & Vohora, D. (2014). Raloxifene protects against seizures and neurodegeneration in a mouse model mimicking epilepsy in postmenopausal woman. *European Journal of Pharmaceutical Sciences, 65*, 167–173. doi:10.1016/j.ejps.2014.09.002 PMID:25218046

Pottoo, F. H., Tabassum, N., Javed, M. N., Nigar, S., Rasheed, R., Khan, A., ... Ashraf, G. M. (2019). The Synergistic Effect of Raloxifene, Fluoxetine, and Bromocriptine Protects Against Pilocarpine-Induced Status Epilepticus and Temporal Lobe Epilepsy. *Molecular Neurobiology*, *56*(2), 1233–1247. doi:10.100712035-018-1121-x PMID:29881945

Power, J. H., Barnes, O. L., & Chegini, F. (2017). Lewy Bodies and the Mechanisms of Neuronal Cell Death in Parkinson's Disease and Dementia with Lewy Bodies. *Brain Pathology (Zurich, Switzerland)*, *27*(1), 3–12. doi:10.1111/bpa.12344 PMID:26667592

Pratt, W., Morishima, Y., Murphy, M., & Harrell, M. (2006). Chaperoning of glucocorticoid receptors. In *Molecular Chaperones in Health and Disease* (pp. 111–138). Springer. doi:10.1007/3-540-29717-0_5

Prusiner, S. B. (1982). Novel proteinaceous infectious particles cause scrapie. *Science*, *216*(4542), 136–144. doi:10.1126cience.6801762 PMID:6801762

Prusiner, S. B. (1998). Prions. *Proceedings of the National Academy of Sciences of the United States of America*, *95*(23), 13363–13383. doi:10.1073/pnas.95.23.13363 PMID:9811807

Prusiner, S. B. (2001). Neurodegenerative Diseases and Prions. *The New England Journal of Medicine*, *344*(20), 1516–1526. doi:10.1056/NEJM200105173442006 PMID:11357156

Qi, L., Zhang, X. D., Wu, J. C., Lin, F., Wang, J., DiFiglia, M., & Qin, Z.-H. (2012). The role of chaperone-mediated autophagy in huntingtin degradation. *PLoS One*, *7*(10), e46834. doi:10.1371/journal.pone.0046834 PMID:23071649

Richter-Landberg, C., & Goldbaum, O. (2003). Stress proteins in neural cells: Functional roles in health and disease. *Cellular and Molecular Life Sciences CMLS*, *60*(2), 337–349. doi:10.1007000180300028 PMID:12678498

Rose, C., Menzies, F. M., Renna, M., Acevedo-Arozena, A., Corrochano, S., Sadiq, O., ... Rubinsztein, D. C. (2010). Rilmenidine attenuates toxicity of polyglutamine expansions in a mouse model of Huntington's disease. *Human Molecular Genetics*, *19*(11), 2144–2153. doi:10.1093/hmg/ddq093 PMID:20190273

Rosen, D. R., Siddique, T., Patterson, D., Figlewicz, D. A., Sapp, P., Hentati, A., ... Brown, R. H. Jr. (1993). Mutations in Cu/Zn superoxide dismutase gene are associated with familial amyotrophic lateral sclerosis. *Nature*, *362*(6415), 59–62. doi:10.1038/362059a0 PMID:8446170

Rothenberg, C., Srinivasan, D., Mah, L., Kaushik, S., Peterhoff, C. M., Ugolino, J., ... Monteiro, M. J. (2010). Ubiquilin functions in autophagy and is degraded by chaperone-mediated autophagy. *Human Molecular Genetics*, *19*(16), 3219–3232. doi:10.1093/hmg/ddq231 PMID:20529957

Sarkar, S., Davies, J. E., Huang, Z., Tunnacliffe, A., & Rubinsztein, D. C. (2007). Trehalose, a novel mTOR-independent autophagy enhancer, accelerates the clearance of mutant huntingtin and alpha-synuclein. *The Journal of Biological Chemistry*, *282*(8), 5641–5652. doi:10.1074/jbc.M609532200 PMID:17182613

Sarkar, S., Floto, R. A., Berger, Z., Imarisio, S., Cordenier, A., Pasco, M., ... Rubinsztein, D. C. (2005). Lithium induces autophagy by inhibiting inositol monophosphatase. *The Journal of Cell Biology*, *170*(7), 1101–1111. doi:10.1083/jcb.200504035 PMID:16186256

Sarkar, S., & Rubinsztein, D. C. (2008). Huntington's disease: Degradation of mutant huntingtin by autophagy. *The FEBS Journal, 275*(17), 4263–4270. doi:10.1111/j.1742-4658.2008.06562.x PMID:18637946

Scheffner, M., Nuber, U., & Huibregtse, J. M. (1995). Protein ubiquitination involving an E1–E2–E3 enzyme ubiquitin thioester cascade. *Nature, 373*(6509), 81–83. doi:10.1038/373081a0 PMID:7800044

Schmitz, A., Schneider, A., Kummer, M. P., & Herzog, V. (2004). Endoplasmic reticulum-localized amyloid β-peptide is degraded in the cytosol by two distinct degradation pathways. *Traffic (Copenhagen, Denmark), 5*(2), 89–101. doi:10.1111/j.1600-0854.2004.00159.x PMID:14690498

Seidel, K., Schols, L., Nuber, S., Petrasch-Parwez, E., Gierga, K., Wszolek, Z., & … . (2010). First appraisal of brain pathology owing to A30P mutant alpha-synuclein. *Annals of Neurology, 67*, 684–689. doi:10.1002/ana.21966 PMID:20437567

Serrano-Pozo, A., Frosch, M. P., Masliah, E., & Hyman, B. T. (2011). Neuropathological alterations in Alzheimer disease. *Cold Spring Harbor Perspectives in Medicine, 1*(1), a006189. doi:10.1101/cshperspect.a006189 PMID:22229116

Sheng, Y., Chattopadhyay, M., Whitelegge, J., & Valentine, J. S. (2012). SOD1 aggregation and ALS: Role of metallation states and disulfide status. *Current Topics in Medicinal Chemistry, 12*(22), 2560–2572. doi:10.2174/1568026611212220010 PMID:23339308

Shpilka, T., & Elazar, Z. (2011). Shedding light on mammalian microautophagy. *Developmental Cell, 20*(1), 1–2. doi:10.1016/j.devcel.2010.12.010 PMID:21238917

Sikorska, B., Liberski, P. P., & Brown, P. (2007). Neuronal autophagy and aggresomes constitute a consistent part of neurodegeneration in experimental scrapie. *Folia Neuropathologica, 45*, 170–178. PMID:18176890

Singleton, A., Farrer, M., Johnson, J., Singleton, A., Hague, S., & Kachergus, J. (2003). α-Synuclein locus triplication causes Parkinson's disease. *Science, 302*, 841-841.

Sittler, A., Lurz, R., Lueder, G., Priller, J., Lehrach, H., & Hayer-Hartl, M. K. (2001). Geldanamycin activates a heat shock response and inhibits huntingtin aggregation in a cell culture model of Huntington's disease. *Human Molecular Genetics, 10*(12), 1307–1315. doi:10.1093/hmg/10.12.1307 PMID:11406612

Spilman, P., Podlutskaya, N., Hart, M. J., Debnath, J., Gorostiza, O., Bredesen, D., ... Galvan, V. (2010). Inhibition of mTOR by rapamycin abolishes cognitive deficits and reduces amyloid-beta levels in a mouse model of Alzheimer's disease. *PLoS One, 5*(4), e9979. doi:10.1371/journal.pone.0009979 PMID:20376313

Stefanis, L. (2012). α-Synuclein in Parkinson's disease. *Cold Spring Harbor Perspectives in Medicine, 2*(2), a009399–a009399. doi:10.1101/cshperspect.a009399 PMID:22355802

Sweeney, P., Park, H., Baumann, M., Dunlop, J., Frydman, J., Kopito, R., ... Hodgson, R. (2017). Protein misfolding in neurodegenerative diseases: Implications and strategies. *Translational Neurodegeneration, 6*(1), 6–6. doi:10.118640035-017-0077-5 PMID:28293421

Thangavelu, S. R., Tripathi, P. P., Arya, U., Mishra, H. K., & Subramaniam, J. R. (2011). ALS associated mutant SOD1 impairs the motor neurons and astrocytes and wild type astrocyte secreted-factors reverse the impaired motor neurons. *Annals of Neurosciences, 18*, 48–55. PMID:25205921

Thomas, B., & Beal, M.F. (2007). Parkinson's disease. *Human Molecular Genetics, 16*(2), R183-194.

Tofaris, G. K., Kim, H. T., Hourez, R., Jung, J. W., Kim, K. P., & Goldberg, A. L. (2011). Ubiquitin ligase Nedd4 promotes alpha-synuclein degradation by the endosomal-lysosomal pathway. *Proceedings of the National Academy of Sciences of the United States of America, 108*(41), 17004–17009. doi:10.1073/pnas.1109356108 PMID:21953697

Varshavsky, A. (2005). Regulated protein degradation. *Trends in Biochemical Sciences, 30*(6), 283–286. doi:10.1016/j.tibs.2005.04.005 PMID:15950869

Vogiatzi, T., Xilouri, M., Vekrellis, K., & Stefanis, L. (2008). Wild type α-synuclein is degraded by chaperone-mediated autophagy and macroautophagy in neuronal cells. *The Journal of Biological Chemistry, 283*(35), 23542–23556. doi:10.1074/jbc.M801992200 PMID:18566453

Waelter, S., Boeddrich, A., Lurz, R., Scherzinger, E., Lueder, G., Lehrach, H., & Wanker, E. E. (2001). Accumulation of mutant huntingtin fragments in aggresome-like inclusion bodies as a result of insufficient protein degradation. *Molecular Biology of the Cell, 12*(5), 1393–1407. doi:10.1091/mbc.12.5.1393 PMID:11359930

Wang, C., Tan, Z., Niu, B., Tsang, K. Y., Tai, A., Chan, W. C. W., & … . (2018). Inhibiting the integrated stress response pathway prevents aberrant chondrocyte differentiation thereby alleviating chondrodysplasia. *eLife*, 7. PMID:30024379

Wang, F., Wang, X., Yuan, C. G., & Ma, J. (2010). Generating a prion with bacterially expressed recombinant prion protein. *Science, 327*(5969), 1132–1135. doi:10.1126cience.1183748 PMID:20110469

Wang, Y., Martinez-Vicente, M., Krüger, U., Kaushik, S., Wong, E., Mandelkow, E.-M., … Mandelkow, E. (2010). Synergy and antagonism of macroautophagy and chaperone-mediated autophagy in a cell model of pathological tau aggregation. *Autophagy, 6*(1), 182–183. doi:10.4161/auto.6.1.10815 PMID:20023429

Warby, S. C., Visscher, H., Collins, J. A., Doty, C. N., Carter, C., Butland, S. L., … Hayden, M. R. (2011). HTT haplotypes contribute to differences in Huntington disease prevalence between Europe and East Asia. European journal of human genetics. *European Journal of Human Genetics, 19*(5), 561–566. doi:10.1038/ejhg.2010.229 PMID:21248742

Webb, J. L., Ravikumar, B., Atkins, J., Skepper, J. N., & Rubinsztein, D. C. (2003). Alpha-Synuclein is degraded by both autophagy and the proteasome. *The Journal of Biological Chemistry, 278*(27), 25009–25013. doi:10.1074/jbc.M300227200 PMID:12719433

Welzel, A. T., & Walsh, D. M. (2011). Aberrant protein structure and diseases of the brain. *Irish Journal of Medical Science, 180*(1), 15–22. doi:10.100711845-010-0606-z PMID:20960238

Who Has Parkinson's? (n.d.). Available from https://www.parkinson.org/Understanding-Parkinsons/Statistics

Williams, A. J., & Paulson, H. L. (2008). Polyglutamine neurodegeneration: Protein misfolding revisited. *Trends in Neurosciences, 31*(10), 521–528. doi:10.1016/j.tins.2008.07.004 PMID:18778858

Willingham, S., Outeiro, T. F., DeVit, M. J., Lindquist, S. L., & Muchowski, P. J. (2003). Yeast genes that enhance the toxicity of a mutant huntingtin fragment or α-synuclein. *Science, 302*(5651), 1769–1772. doi:10.1126cience.1090389 PMID:14657499

World Alzheimer Report. (2018). Available from https://www.alz.co.uk/research/WorldAlzheimerReport2018.pdf

Yang, Z., & Klionsky, D. J. (2010). Eaten alive: A history of macroautophagy. *Nature Cell Biology*, *12*(9), 814–822. doi:10.1038/ncb0910-814 PMID:20811353

Ye, Y., & Rape, M. (2009). Building ubiquitin chains: E2 enzymes at work. *Nature Reviews. Molecular Cell Biology*, *10*(11), 755–764. doi:10.1038/nrm2780 PMID:19851334

Zarei, S., Carr, K., Reiley, L., Diaz, K., Guerra, O., Altamirano, P. F., ... Chinea, A. (2015). A comprehensive review of amyotrophic lateral sclerosis. *Surgical Neurology International*, *6*(1), 171–171. doi:10.4103/2152-7806.169561 PMID:26629397

Zarow, C., Lyness, S. A., Mortimer, J. A., & Chui, H. C. (2003). Neuronal loss is greater in the locus coeruleus than nucleus basalis and substantia nigra in Alzheimer and Parkinson diseases. *Archives of Neurology*, *60*(3), 337–341. doi:10.1001/archneur.60.3.337 PMID:12633144

Zhao, L., Longo-Guess, C., Harris, B. S., Lee, J.-W., & Ackerman, S. L. (2005). Protein accumulation and neurodegeneration in the woozy mutant mouse is caused by disruption of SIL1, a cochaperone of BiP. *Nature Genetics*, *37*(9), 974–979. doi:10.1038/ng1620 PMID:16116427

Chapter 2
Autophagic Dysfunction in Neurodegeneration:
Pathogenic Cellular and Molecular Mechanistic Approaches

Meenakshi Kaira
Department of Pharmaceutical Sciences, Maharshi Dayanand University, Rohtak, India

Abhilasha Ahlawat
Department of Pharmaceutical Sciences, Maharshi Dayanand University, Rohtak, India

Vaibhav Walia
Faculty of Pharmacy, DIT University, Dehradun, India

Munish Garg
Department of Pharmaceutical Sciences, Maharshi Dayanand University, Rohtak, India

ABSTRACT

Autophagy is a normal physiological process characterized by the degradation of complex cellular contents into a simpler one and reutilized them in biosynthetic pathways. Lysosomes are the cell organelle that participates in the process of autophagy. The brain is the most vulnerable organ in most lysosome disorders because neurons are inefficient in removing impaired organelles and waste materials. In the brain, autophagy suppresses the accumulation of ubiquitinated proteins that results in further damage to the neurons responsible for neurodegeneration. Autophagy mediates protective effects in age-related diseases. In the chapter, the authors describe the process of autophagy, the mechanism involved, and the implication of the autophagic pathways in the various neurodegenerative disorders.

DOI: 10.4018/978-1-7998-1317-0.ch002

INTRODUCTION

Autophagy is a self-degradative process of cellular housekeeping that plays a crucial role in multiple physiological processes. Autophagy is mainly responsible for catabolism in the body (Yu et al., 2010; Rong et al., 2011). Autophagy can be selective or non-selective. Selective autophagy is a receptor-mediated process and therefore the molecules to be degraded recognized by receptors before degradation whereas non-selective autophagy is the non-receptor-mediated process that works randomly without any recognition. Autophagy is not only responsible for the degradation of the damaged cellular constituents or organelles but also breaks the complex molecule simpler constituents so that they can be reutilized further. Various conditions result in the induction of autophagy including nutrient deprivation, oxidative stress and ultraviolet radiation (Levine and Kroemer, 2008).

Lysosomes are the cell organelle that participates in the process of autophagy (Huotari and Helenius, 2011; Luzio et al., 2007). Lysosomes consisted of several degradative enzymes (Kolter and Sandhoff, 2005). After lysosomal mediated degradation, products are transported out of lysosomes (Ruivo et al., 2009; Saftig and Klumperman, 2009). The brain is the most vulnerable organ in most lysosome disorders because neurons are inefficient in removing the impaired organelles, waste materials and disperse harmful substances (Boland et al., 2008; Lee et al., 2011). The ability of the neurons to remove the wastes by the process of autophagy decreases with the increases in the age of the individual suggesting the accumulation of the wastes and altered proteins in the brain of the older individual responsible for the generation of the various pathologies. In the brain, autophagy suppresses the accumulation of ubiquitinated proteins that results in further damage to the neurons responsible for neurodegeneration (Komatsu et al., 2006). Autophagy not only suppresses the aggregation of proteins and lipid, oxidative stress, chronic cell death, inflammation and cancer (Mathew et al., 2009) but also mediates protective effects in age-related diseases (Kapahi et al., 2010). Thus, it is relevant for further investigation in the therapeutics of the various disorders (Sarkar et al., 2007; White and DiPaola, 2009). Here, the authors describe the process of autophagy and its impact on neurodegenerative disorders.

TYPES OF AUTOPHAGY

In the case of mammalian cells, there are mainly three types of autophagy. These are microautophagy, macroautophagy, and chaperone-mediated autophagy (CMA).

Macroautophagy

Macroautophagy includes a series of steps i.e., membrane initiation, nucleation, elongation, autophagosome maturation and lysosomal fusion (Nakatogawa et al., 2009).

Microautophagy

It involves a process of membrane invagination or deformation to engulf cytoplasmic contents into the lysosome by forming a vesicle (Marzella et al., 1981) and thus also known as endosomal microautophagy (Sahu et al., 2011). Microautophagy participates in maintaining energy levels under stress conditions (Dubouloz et al., 2005).

Chaperone-Mediated Autophagy

Chaperone-mediated autophagy (CMA) recognizes the specific cargo and translocates it into the lumen of the lysosome by various mechanisms for degradation (Kaushik and Cuervo, 2018). The process is initiated with binding of hsc-70 (a cytosolic molecular chaperone) to cargo followed by its translocation into the lumen of the lysosome (Chiang et al., 1989; Bandyopadhyay et al., 2008). It has been reported that under the stress conditions, macroautophagy gets inhibited and CMA gets activated for maintaining the energy level and quality of cells (Schneider et al., 2014; Chava et al., 2017).

MECHANISM OF AUTOPHAGY

The process of autophagy is characterized by the influx of molecules, formation of autophagosome, fusion with the lysosome and the degradation of molecules inside the lysosomes (Cecconi and Levine, 2008; Mizushima and Komatsu, 2011). In the first step, the molecules to be degraded are enwrapped in the membrane and the structure so formed is known as termed as phagophore (Rubinsztein et al., 2012; Yen et al., 2010). Phagophore (Figure 1) then sequesters the cargoes to form autophagosomes. Autophagosome then moves to the lysosome and fuses with the lysosomes to form a new structure known as autolysosome in which the cargoes are degraded by the lysosomal enzymes and the products so formed might be excreted out or reuse further (Agarwal et al., 2015; Klionsky et al., 2012; Korolchuk et al., 2010; Loos et al., 2014).

Autophagy is initiated by two major complexes such as ULK and phosphatidylinositol-3-kinase (PI3K) complex pathway (Jung et al., 2009; Fan et al., 2011; Agarwal et al., 2015; Ohsumi and Mizushima, 2004). PI3K activity is regulated by the anti-apoptotic proteins BCL-2 and BCL-XL. Beclin1 then dissociated from BCL-2 to coordinate with Vps34 (He and Levine, 2010; Martyniszyn et al., 2011; Pattingre et al., 2005) and concentrate on the surface of phagophore (Obara and Ohsumi, 2008; Puri et al., 2013). Phagophore is then converted into autophagosome and the process includes the interaction of Atg7, Atg5 links with Atg12 to form the covalent complex which further links with the Atg16 responsible for the elongation of phagophore (Shao et al., 2007). The next step is the formation of autophagosome which requires Atg9 binding with Atg2 and Atg18. Atg4B is required for the generation of LC3-I from LC3 (microtubule-associated protein 1 light chain 3) (Fujita et al., 2008). LC3-I is then conjugated with the phosphatidylethanolamine (PE) to form phosphatidylethanolamine (PE)-conjugated LC3-II by the Atg5-Atg12-Atg16 complex (Kabeya et al., 2000). Autophagosome fuses with the lysosome to form autolysosome in which the cargoes are degraded by the enzymatic action (Itakura et al., 2012).

ATG —THE CORE MACHINERY

The autophagy process is divided into several steps, comprising the induction of autophagy, the formation of autophagic bodies, the fusion of vesicles, as well as the decomposition of autophagic bodies followed by the release of the degradation products back into the cytosol. Different groups of Atg proteins are intricate in these phases. Autophagy is regulated by the 6 groups of autophagy-related genes (ATG) and namely:

Figure 1. Mechanism of autophagy.

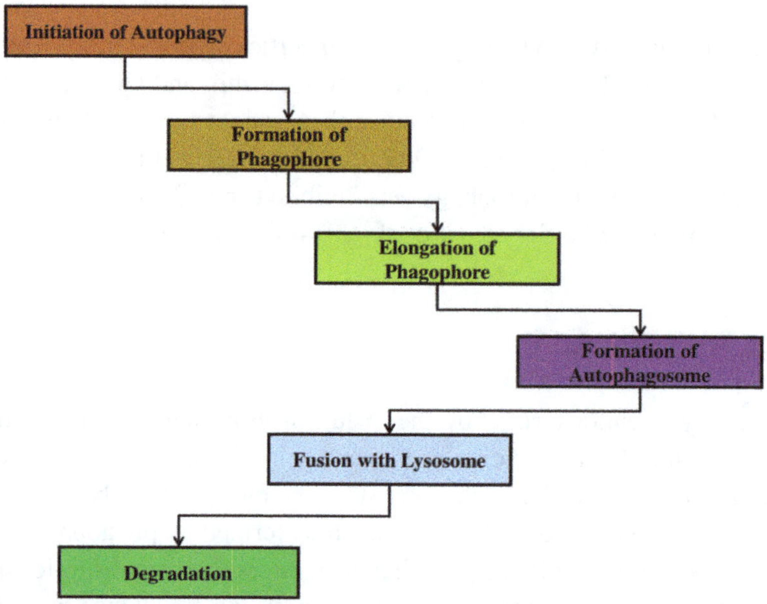

- **ATG1/ULK1 Kinase Complex**: ATG1 complex initiates autophagy either by forming a lipid bilayer (Tooze and Yoshimori, 2010) and is considered as the vital component of autophagy (Ganley et al., 2009; Kim et al., 2011).
- **ATG9**: ATG9 is the autophagy-specific phosphatidylinositol 3-kinase (AS-PI3K) complex which binds with ATG2-ATG18 to form ATG12-ATG5-ATG16 and is involved in the elongation of phagophore and causes the conversion of LC3-I to LC3-II (He et al., 2006).
- **PI3K Complex (ATG6-ATG14-Vps15-Vps34)**: Activation of autophagy is followed by the dissociation of PI3K complex from BCL-2 complex, coordination with Vps34 (He and Levine, 2010; Martyniszyn et al., 2011; Pattingre et al., 2005) and concentration on the surface of phagophore (Obara and Ohsumi, 2008; Puri et al., 2013).
- **PI3P-Binding ATG2-ATG18 Complex**: The ATG2-ATG18 complex is required for nucleation and to localize to autophagic membranes, endosomes or vacuolar membranes (Obara et al., 2008).
- **ATG12 Conjugation System (ATG12-ATG5-ATG16)**: ATG12 activating process consists of an E1 and E2 enzymes. E1-ubiquitin-activating enzyme (ATG7) which catalyzes the activation of ATG12. E2-ubiquitin-conjugating enzyme (ATG10) which eases the binding of activated ATG12 to ATG5 forming ATG12-ATG5 complex which further binds with ATG16 to form the ultimate complex i.e., ATG12-ATG5-ATG16 necessary for autophagy and mediation of ATG8-PE conjugation (Kuma et al., 2002).
- **ATG8 Conjugation System ATG8-Phosphatidylethanolamine**: ATG8-phosphatidylethanolamine (PE) conjugation can be done by two methods, either via ATG12-ATG5-ATG16 complex or via ATG8/LC3 (Hanada et al., 2007). ATG4 cleaves the newly synthesized LC3 to form its cytosolic form LC3 I, which is further converted toLC3 II by the molecules ATG7 and ATG3 (Table 1) (Komatsu et al., 2005; Hanada et al., 2009). This conjugation is required for the expansion and maturation of the autophagosome in the autophagic process (Weidberg et al., 2010; Mizushima and Yoshimori, 2007).

Table 1. Role of ATG in the process of autophagy

Type	Functions
ATG-1	Initiates autophagy.
ATG-2	ATG-2 interacts with ATG-18 responsible for ATG-2-ATG-18 complex and elongation of the phagophore.
ATG-3 and ATG-7	Converts LC3-I to LC3-II.
ATG-8	Forms conjugates by ATG-12-ATG-5-ATG-16 complex and ATG-8/LC-3 complex.
ATG-4	Cleaves LC-3 to LC3-I.
ATG-9	Forms complex with ATG-2-ATG-18 responsible for elongation of the phagophore.
ATG-10	Responsible for the binding of ATG-2 to ATG-5.
ATG-16	Forms ATG-5-ATG-12-ATG-16 complex with ATG-12-ATG-5 which is necessary for autophagy and activation of ATG-8 conjugation system.

AUTOPHAGY AND CELL DEATH

The process of cell death is characterized by the presence of the large number of AP in the cytoplasm then it is known as autophagic cell death but if the process of the cell death involved the number of caspases then it is known as apoptosis (Maiuri et al., 2007; Bursch, 2001). Normally both autophagy and apoptosis inhibit each other and stays inactive but under stressful conditions, mitochondria can lead to apoptosis by releasing cytochrome C whereas nutrient starvation may induce autophagy (Gonzalez-Polo et al., 2005). However, under nutrient starvation, autophagy generally acts as pro-survival but over increased autophagy results in autophagic cell death. Many kinds of research have suggested that autophagy precedes apoptosis. Apoptosis and autophagy also linked with each other as there are several pro-apoptotic signals which may induce autophagy. Further, the pro-survival signals suppress autophagy (Gustafsson and Gottlieb, 2008). Atg5 is an important link between autophagy and apoptosis and promotes autophagy-mediated cell death (Pyo et al., 2005). Calpain cleaves Atg5 and the fragment so formed then is translocated inside the mitochondria responsible for the release of the cytochrome c and activation of caspase (Yousefiand Simon, 2009).

It is suggested that the PI3K and Akt/PKB pathways activate the mammalian target of rapamycin (mTOR) to inhibit autophagy (Cong et al., 2016) (Figure 2). Phosphoinositide 3-kinases (PI3K) acts by phosphorylating 3′-hydroxyl group of the inositol ring of phosphatidylinositol 4,5-bisphosphate (PIP2) to generate phosphatidylinositol 3,4,5-trisphosphate (PIP3) to promote cell survival and growth (Whitman et al., 1988; Auger et al., 1989; Hanahan and Weinberg, 2011; Engelman, 2009; Fruman and Rommel, 2014). PIP3 binds to Akt and Akt is activated by sequential phosphorylation at T308 and S473 residues (Hay, 2005). AKT plays an important role in promoting glucose metabolism and cancer and functions as a critical regulator of cell survival and proliferation (Bellacosa et al., 2005; Manning and Cantley, 2007). The mammalian target of rapamycin (mTOR) is the best-characterized substrate for Akt. Akt activates mTOR through at least two mechanisms, either directly by phosphorylating mTORC1 at S2448 or indirectly through TSC2 (Nave et al., 1999; Inoki et al., 2002; Manning et al., 2002; Potter et al., 2002). mTORC1 further regulates cell proliferation, survival, and angiogenesis (De Benedetti and Graff, 2004; Soni et al., 2008; Culjkovic et al., 2008) while mTORC2 regulates growth factors such as insulin

Figure 2. Role of PI3K pathway in cell death, survival, and autophagy.

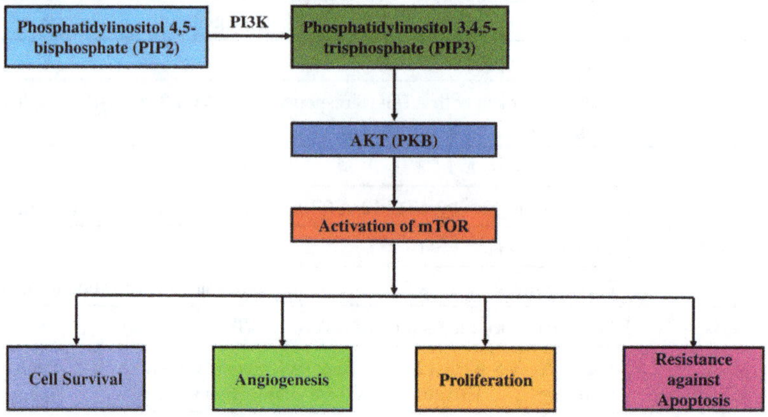

(Zinzalla et al., 2011). Thus, the activation of the PI3K/Akt/mTOR pathway increases cell survival and provides apoptotic resistance (Plastaras et al., 2008; Hay, 2005).

AUTOPHAGY IN NEURODEGENERATIVE DISORDERS

Autophagy in Alzheimer's Disease/Type-3 Diabetes Mellitus

Alzheimer's disease (AD) is the most common neurodegenerative disease characterized by the deposition of the intracellular tau tangles and extracellular amyloid-β (Aβ) plaques (Yoon and Kim, 2016). Aβ is formed by the cleavage of AAP (amyloid precursor protein) and is degraded by the process of autophagy (Iqbal et al., 2010; Boland et al., 2008; Spilman et al., 2010; Tian et al., 2011). Impaired clearance of contributing to elevated Aβ levels (Yu et al., 2005; Nixon et al., 2005; Boland et al., 2008) and the increased clearance of Aβ by autophagy improve cognition in the AD (Yang et al., 2010) as shown in Figure 3. The suppression of autophagy by the treatment of 3-methyl adenine results in an increased level of Aβ (Zheng et al., 2011). Autophagy results in the clearance of the tau (Berger et al., 2005; Kruger et al., 2012) and therefore the loss of autophagy results in the accumulation of the tau (Luna-Munoz et al., 2007) responsible for the synaptic dysfunction and neuronal degeneration in AD pathology (Irvine et al., 2008; Bloom, 2014). Further, the AD pathology is characterized by the loss of mTORC1 activity and the hyperactivation of the mTORC1 is correlated with cognitive decline in AD patients (Caccamo et al., 2011; Sun et al., 2014). Therefore, PI3K/Akt/mTORC1 pathways inhibit autophagy and reduced clearance of Aβ responsible for the pathology progression.

Insulin is responsible for the stimulation of AKT and insulin desensitization and interrupted Akt activation has been reported in post-mortem AD brain (Young et al., 2009; Morales-Corraliza et al., 2016). Further, insulin desensitization is the characteristic of type 2 diabetes mellitus (T2DM) and T2DM accelerates the risk of AD development (Biessels et al., 2006; Domínguez et al., 2014; Exalto et al., 2014). It is suggested that insulin resistance is accompanied by levels of Aβ oligomers (Moloney et al., 2010; O'Neill et al., 2012; Tramutola et al., 2015). Additionally, insulin-degrading enzyme (IDE),

Figure 3. Role of Autophagic dysfunction in the pathogenesis of Alzheimer's disease.

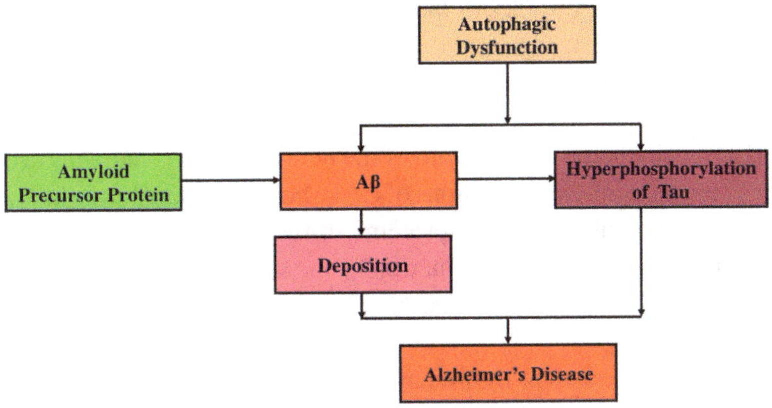

which degrades Aβ, showed reduced expression in AD (Vekrellis et al., 2000; van der Heide et al., 2006; Zhao et al., 2007).

Brain diabetes or type 3 diabetes mellitus (T3DM) is characterized by the chronic insulin resistance confined to the brain (de la Monte and Wands, 2008; Leszek et al., 2017). The chronic insulin resistance confined to the brain represents a pathogenic mechanism of neurodegeneration analogous to AD (de la Monte and Wands, 2008). AD is thus recognized as a metabolic disorder characterized by the impairment of brain insulin responsiveness and glucose utilization (Suzanne, 2014). It has been reported that the presence of diabetes increases the cognitive decline in older adults (Yaffe et al., 2004, 2006; Cheng et al., 2012; Biessels et al., 2014) and insulin resistance is a key factor involved in the development of dementia in older adults (Hassing et al., 2004; Yaffe et al., 2004; Strachan, 2011; Cheng et al., 2012). Therefore, individuals with insulin resistance are more prone to the development of AD and similar conditions, such as vascular dementia (Craft, 2009; Di Paolo and Kim, 2011). Thus, AD share common molecular and cellular pathways with diabetes associated with memory deficits and cognitive decline in older people (Virkamaki et al., 1999; Reiman et al., 1996; Small et al., 2000; Sperling et al., 2011). Also, both AD and DM are amyloidogenic conditions and amyloidosis occurs due to increased insulin resistance (Cooper et al., 1989; Hardy and Higgins, 1992; Lim et al., 2010). Augmentation of amylin initiates in the pancreas which may result in T2DM progression leading to the death of pancreatic beta cells which exacerbate the state of insulin resistance (Cooper et al., 1989; Jackson et al., 2013). However, in cerebral vasculature and brain parenchyma, augmentation of amylin (Despa et al., 2012; Jackson et al., 2013; Srodulski et al., 2014) has been reported in the patients of AD (Jackson et al., 2013) suggesting the pathogenic role of the amylin in AD (Yan et al., 2014; Oskarsson et al., 2015). It has been found that APP and beta-secretase 1 are involved in Aβ formation and thus, their upregulation occurs in brain insulin deficiency results in more Aβ formation (Devi et al., 2012). Aβ has a high propensity to form various types of oligomers and amyloid fibrils that disrupt communications between neurons and cause cell death (Chiti and Dobson, 2006). Insulin degrading enzymes (IDE) has been found to degrade the Aβ (Gilbert, 2013) thus prevent the formation of oligomers and aggregates of Aβ (Farris et al., 2003; Miller et al., 2003). Several single nucleotide polymorphisms (SNPs) in non-coding regions of the human IDE gene on chromosome 10q are associated with T2DM (Ghosh et al., 2000; Kim et al., 2007; Sladek et al., 2007; Zeggini et al., 2007). There is increasing evidence that improper IDE function, regulation, or trafficking might contribute to the etiology of metabolic diseases.

Autophagy in Parkinson's Disease

Parkinson's disease (PD) is characterized by death of the neurons of substantia nigra (SN) pars compacta (pc) and the presence of the intracellular inclusions of Lewy body (LB) and Lewy neurites (LN) composed of α-synuclein (α-syn) and polyubiquitinated proteins in the various regions of the brain (Dauer and Przedborski, 2003; Kalia et al., 2013). Accumulation of the misfolded α-syn and the impaired removal of the misfolded proteins due to the defected degradation pathway led to the neuronal cell death (Tong et al., 2010).

In PD, NF-κB (nuclear factor kappa-light-chain-enhancer of activated B cells) activity, regulated by LRRK2 (leucine-rich repeat serine-threonine protein kinase-2) (Gardet et al, 2010), is involved in initiation of cell death mediated by tumor necrosis factor (TNF) extrinsically hence, its inhibition may decrease neuronal death (Dauer and Przedborski, 2003; Hayley et al., 2004; McCoy et al., 2006). In neurons, expression of LRRK2 increases reactive oxygen species (ROS) level by inducing fragmentation of mitochondria, where it co-exists with DLP1 (dynamin like protein 1) (Li et al, 2014). LRRK2 also extinguishes autophagy which may elevate the augmentation of misfolded proteins by regulating their clearance which leads to neurodegeneration suggesting that inhibition of kinase activity of LRRK2 may induce autophagy (Alegre-Abarrategui et al., 2009). In older people, the mutation in the LRRK2 gene induces autosomal dominant PD more frequently (Klein and Westenberger, 2012; Vekrellis et al., 2011). LRRK2 also regulates various other cellular functions such as proteolysis, neuritic morphology, and outgrowth and vesicular trafficking (Li et al, 2014).

Also, the mutation of the Pink1-Parkin pathways is associated with the early-onset familial and sporadic PD cases (Valente et al., 2004; Kitada et al., 1998). In brief, Pink1 is localized on the mitochondrial membrane of the damaged mitochondria followed by the recruitment and activation of ubiquitin ligase Parkin followed by the sequestration of the damaged mitochondria into autophagosomes for degradation (Matsuda et al., 2010; Narendra et al., 2008). Thus, the mutations Pink1-Parkin pathways are the main causing factors for autosomal recessive forms of PD (Kazlauskaite and Muqit, 2015).

Autophagy degrades α-syn (Cuervo et al., 2004; Vogiatzi et al., 2008) and the PD brains are characterized by the presence of dysfunctional lysosomes and accumulation of autophagosomes (Dehay et al., 2010). CMA inhibition leads to the accumulation of insoluble and high molecular weight components of α-syn which states that CMA is mainly responsible for the degradation of wild type α-syn (Cuervo et al., 2004; Vogiatzi et al., 2008). Further, the mutants of α-synuclein (A53T and A30P) inhibit CMA and CMA linked degradation of α-syn leading to its accumulation causing PD (Cuervo et al., 2004). α-syn mutant inhibits CMA pathways by interfering with its main components such as LAMP2A and Hsc70 as it has a higher affinity for them (Kabuta et al., 2008). Thus, the inhibition of the lysosomal process results in the accumulation of α-syn further makes the neurons prone to damage (Dehay et al., 2010; Lee et al., 2004; Webb et al., 2003).

Autophagy in Huntington's Disease

Huntington's disease (HD) is another most common polyglutamine disease characterized by CAG repeat which leads to polyglutamine (polyQ) expansions and pathogenic aggregation (Imarisio et al., 2008; Jimenez-sanchez et al., 2017). Huntingtin plays a key role in autophagosome transport and the depletion of huntingtin results in abnormal accumulation of autophagosomes and impaired cargo degradation (Zheng et al., 2010). Mutant huntingtin (Htt) in HD is responsible for the formation of toxic oligomers. Increased levels of autophagosomes and impaired autolysosomes, accompanied by elevated augmentation of mutant Htt (Ravikumar et al., 2005). In HD cells, due to inadequate interaction of p62 and mutant Htt, impaired cargo recognition has been reported resulting in feeble deterioration of mutant Htt (Fu et al., 2017). Rapamycin induced autophagy may decrease the formation of aggregates and cytotoxicity (Ravikumar et al., 2002, 2004).

Autophagy in Amyotrophic Lateral Sclerosis

Amyotrophic lateral sclerosis (ALS) is a life-threatening, age-related neurodegenerative disorder indicated by progressive loss of both upper and lower motor neurons in the brain and spinal cord (Rowland and Shneider, 2001; Pasinelli and Brown, 2006). In the pathology of ALS, insoluble aggregates of misfolded protein and surrounding oligodendrocytes are augmented in degenerating motor neurons and in the spinal cord, hippocampus, frontal and temporal cortices respectively (Balch et al., 2008). Misfolded proteins also formed in the normal body but get eliminated continuously by the cells either by degrading them before accumulation or by removing them after accumulation, this phenomenon is known as autophagy. In disease neurons, aggregated proteins have been found suggesting an alteration in autophagy of cells failing which results in misfolded protein accumulation. These aggregates further attract more proteins and thus may lead to cytotoxicity.

ALS is broadly classified into two types, sporadic ALS (sALS) and familial ALS (fALS). The former one is more common as it is non-hereditary than the latter one which is hereditary (Zarei et al., 2015). At first, the genetic link of fALS was found by mutation in SOD1 (mSOD1) (Turner et al., 2013), with time, many more genes were discovered in ALS such as TDP-43 (TAR DNA-binding protein 43), OPTN (optineurin), FUS (fused in sarcoma), C9orf72, VCP (valosin-containing protein), sequestosome 1 (SQSTM1), DCTN1 (dynatin), MATR3 and UBQLN2 (ubiquilin 2) (Renton et al., 2014).

In most of the neurodegenerative disorders, brains are the most susceptible organ which reveals that autophagy is more necessary in neurons than other cells to regulate homeostasis of protein. In neurons, there are some rare structures such as large dendritic and axonal cytoplasm which cause hindrance in removing abnormal organelles in time (Boland et al., 2008). Autophagic vacuoles and lysosomes are two important components of autophagy located at two different places in neurons. The former one is formed in axons which have to travel to the latter one which is located near the cell body. Also, neurons are incapable in mitosis, unlike other mitotic cells which make them more prone to damage as they are unable to disperse harmful substances by division (Lee et al., 2011) due to which on aging, the situation gets worse resulting in more augmentation of abnormal autophagic substrates (cargo).

Autophagy in Tauopathies

Tauopathies are the heterogeneous group of dementias and movement disorders characterized neuropathologically by the accumulation of intracellular abnormal tau filaments and neurofibrillary tangles. It is suggested that the abnormalities in tau protein are caused by the microtubule-associated protein tau (MAPT) which has been implicated in the neurodegenerative disease (Ghetti et al., 2015). Further, most of the tauopathies are associated with the deposition of at least one amyloidogenic protein, such as α-synuclein or huntingtin (Hashiguchi et al., 2000; Jensen et al., 1999).

Human tau is encoded by the gene present on chromosome 17 (Andreadis, 2006) and consists of domains including N-terminal acidic projection domain does not bind to microtubules directly and projects away from microtubules but is involved in the regulation of microtubule dynamics, attachment, etc (Chen et al., 1992). The microtubule-binding domain binds and stabilizes the microtubules (Mukrasch et al., 2009). The proline-rich domain serves as the site for the binding of the kinase belonging to Src family, for example, PI3K (Morris et al., 2011; Mandelkow and Mandelkow, 2012). PI3K is responsible for the phosphorylation of tau and is important for tau modification (Hanger et al., 2009). Tau phosphorylation is regulated by glycogen synthase kinase (GSK) and protein phosphatase 2A (PP2A) (Hanger et al., 2009). Glycogen synthase kinase (GSK)-3 plays a prominent role in the process of neuronal degeneration and the over-activation of GSK3β significantly contributes to tau phosphorylation (Pei et al., 1997) and induces tau aggregation (Rankin et al., 2007). Increase activity of GSK3 is responsible for the increase tau phosphorylation and tangle formation (Meske et al., 2008) and contributes to the neurofibrillary tangle burden in AD (Leroy et al., 2002; Hanger et al., 1992). Therefore, the inhibition of GSK3 results in the inhibition of tau phosphorylation and prevents neuronal loss in AD pathology (Caccamo et al., 2007; Leroy et al., 2010; Sereno et al., 2009). Protein phosphatase 2A (PP2A) is responsible for the dephosphorylation of GSK3β at Ser9 (Lee et al., 2005). PP2A performs dephosphorylation of tau (Gong et al., 2000) and the reduced activity of PP2A is responsible for the pathogenesis of AD (Liu et al., 2005).

Tau is important in regulating synaptic plasticity, synaptic maturation and neurogenesis in hippocampal neurons (Chen et al., 2012; Pallas-Bazarra et al., 2016; Pallas-Bazarra et al., 2016). Further, the removal of tau led to impaired neuronal migration and neurogenesis (Fuster-Matanzo et al., 2009; Hong et al., 2010). Tau interacts with microtubules and stabilizes the microtubules (Aamodt and Williams, 1984) but the phosphorylation of tau, reduces its affinity for microtubules, resulting in cytoskeleton destabilization (Drewes et al., 1995) and increases tau aggregation (Eidenmuller et al., 2001; Von Berger et al., 2000; Kopke et al., 1993) contributing to neurofibrillary burden in AD (Leroy et al., 2002; Hanger et al., 1992). The extent of the neurofibrillary tangle and amyloid plaque deposited in different brain regions is linked with the severity of AD (Braak and Braak, 1995). The proliferation of AD pathology follows a steady trend - Braak stage I i.e., transentorhinal/peripheral cortex; Braak stage II i.e., CA1 region of the hippocampus, Braak stage III i.e., limbic structures; Braak stage IV i.e., amygdala; Braak Stage IV i.e., thalamus and claustrum; Braak stage V i.e., isocortical areas, and at last, Braak stage VI i.e., primary sensory, motor and visual regions (Braak and Braak, 1991). However, the neuronal death exceeds as compared to the degree of tangle pathology (Gomez-Isla et al., 1997).

Mutations in the *MAPT* gene also results in the development of parkinsonism independently of α-syn (Iijima et al., 1999; Ittner et al., 2008; Wade-Martins, 2012). Furthermore, tau proteins and α-synuclein co-aggregated together in tauopathies and synucleinopathies (Rousseaux et al., 2016; Irwin et al., 2013). Also, the sporadic PD and PD dementia shows the presence of neurofibrillary tangles containing tau (Joachim et al., 1987; Schneider et al., 2006) suggesting the involvement of both tau and α-syn (Muntane et al., 2008; Poulopoulos et al., 2012; Yamaguchi et al., 2005). Tau and α-syn co-localize in the neurons (Esposito et al., 2007; Arima et al., 2000; Ishizawa et al., 2003). α-Syn enhances GSK3β-mediated tau phosphorylation (Ciaccioli et al., 2013; Wills et al., 2011) and the presence of phosphorylated tau increased rate of cognitive decline in PD (Liu et al., 2015) suggesting the synergistic relationship between tau and α-synuclein in PD (Rousseaux et al., 2016).

Further, the increased tau phosphorylation has been observed with increased GSK3 activity and reduced amounts of protein phosphatases in HD (Blum et al., 2014; Gratuze et al., 2015; L'Episcopo et al., 2016; Fernandez-Nogales et al., 2014; Vuono et al., 2015). It is suggested that the htt plays a prominent role in the peculiar splicing of tau and related microtubule-associated protein MAP2 (Fernandez-Nogales et al., 2016).

Also, many studies proved the involvement of tau in HD by using animal models, for example, R6/2 mouse shows memory and learning impairment, motor dysfunction and intraneuronal inclusions of mutant htt (Davies et al., 1997; Lione et al., 1999) by excessively expressing htt exon 1 with an expanded polyglutamine repeat (Vuono et al., 2015; Carter et al., 1999).

RECENT DEVELOPMENTS AND FUTURE RESEARCH DIRECTIONS

Several studies have been revealed that lithium chloride (Sarkar et al., 2005) and trehalose induced autophagy increase degradation of cargo such as A30P and A53T, α-synuclein (α-syn) and mutant htt which may found to be effective in the treatment of AD (Kruger et al., 2012) and PD (Sarkar et al., 2007). In the tauopathy exhibiting mice model, long term oral treatment with lithium increases autophagy which further reduces p-tau-induced motor impairment (Shimada et al., 2012). Penitrem A induce autophagy by blocking Ca^{2+} channels as it is an irreversible inhibitor of high conductance Ca^{2+}-activated K^+ channels (Sarkar et al., 2009). Similarly, cycloheximide may inhibit segregational steps occurring before the actual formation of autolysosomes and hence may act as an effective autophagy inhibitor (Oliva et al., 1992). Some broadly used anti-malarial and anti-rheumatoid agents like chloroquine and hydroxychloroquine have been reported as clinical autophagy inhibitors. Also, lysosomal dysfunction mediated by chloroquine increases its antineoplastic effects (Harhaji-Trajkovic et al., 2012). Leupeptin, naturally occurring protease inhibitor, inhibits autophagy at the step of cytoplasmic degradation enclosed in lysosomes and causes the acquisition of autolysosomes and/or many cytoplasmic inclusions in the central vacuoles (Moriyasu and Inoue, 2008). Thus the drugs which induce the autophagic pathways in the brain might be the potential treatment of the neurodegenerative disorders.

CONCLUSION

Autophagy is crucial for the healthy aging of neurons. Neuronal autophagy is the main process for the degradation of atypical protein aggregate, which is the foremost cause of neurodegeneration. Furthermore, autophagic clearance mechanism reduces with the aging and therefore the ability of the neurons to remove the wastes decreases and the waste products accumulated in the brain of older individuals responsible for the development of various neurodegenerative disorders. Therefore, autophagy can be targeted in the treatment of these neurodegenerative disorders.

REFERENCES

Aamodt, E. J., & Williams, R. C. Jr. (1984). Microtubule-associated proteins connect microtubules and neurofilaments in vitro. *Biochemistry*, *23*(25), 6023–6031. doi:10.1021/bi00320a019 PMID:6543144

Agarwal, S., Tiwari, S. K., Seth, B., Yadav, A., Singh, A., Mudawal, A., ... Kumar, A. (2015). Activation of autophagic flux against xenoestrogen bisphenol-A-induced hippocampal neurodegeneration via AMP kinase (AMPK)/mammalian target of rapamycin (mTOR) pathways. *The Journal of Biological Chemistry*, *290*(34), 21163–21184. doi:10.1074/jbc.M115.648998 PMID:26139607

Alegre-Abarrategui, J., Christian, H., Lufino, M. M., Mutihac, R., Venda, L. L., Ansorge, O., & Wade-Martins, R. (2009). LRRK2 regulates autophagic activity and localizes to specific membrane microdomains in a novel human genomic reporter cellular model. *Human Molecular Genetics*, *18*(21), 4022–4034. doi:10.1093/hmg/ddp346 PMID:19640926

Andreadis, A. (2006). Misregulation of tau alternative splicing in neurodegeneration and dementia. In *Alternative Splicing and Disease* (pp. 89–107). Berlin: Springer. doi:10.1007/978-3-540-34449-0_5

Arima, K., Mizutani, T., Alim, M. A., Tonozuka-Uehara, H., Izumiyama, Y., Hirai, S., & Uéda, K. (2000). NACP/α-synuclein and tau constitute two distinctive subsets of filaments in the same neuronal inclusions in brains from a family of parkinsonism and dementia with Lewy bodies: Double-immunolabeling fluorescence and electron microscopic studies. *Acta Neuropathologica*, *100*(2), 115–121. doi:10.1007004010050002 PMID:10963357

Auger, K. R., Serunian, L. A., Soltoff, S. P., Libby, P., & Cantley, L. C. (1989). PDGF-dependent tyrosine phosphorylation stimulates production of novel polyphosphoinositides in intact cells. *Cell*, *57*(1), 167–175. doi:10.1016/0092-8674(89)90182-7 PMID:2467744

Bandyopadhyay, U., Kaushik, S., Varticovski, L., & Cuervo, A. M. (2008). The chaperone-mediated autophagy receptor organizes in dynamic protein complexes at the lysosomal membrane. *Molecular and Cellular Biology*, *28*(18), 5747–5763. doi:10.1128/MCB.02070-07 PMID:18644871

Bellacosa, A., Kumar, C. C., Di Cristofano, A., & Testa, J. R. (2005). Activation of AKT kinases in cancer: Implications for therapeutic targeting. *Advances in Cancer Research*, *94*, 29–86. doi:10.1016/S0065-230X(05)94002-5 PMID:16095999

Berger, Z., Ravikumar, B., Menzies, F. M., Oroz, L. G., Underwood, B. R., Pangalos, M. N., ... Rubinsztein, D. C. (2005). Rapamycin alleviates toxicity of different aggregate-prone proteins. *Human Molecular Genetics*, *15*(3), 433–442. doi:10.1093/hmg/ddi458 PMID:16368705

Biessels, G. J., De Leeuw, F. E., Lindeboom, J., Barkhof, F., & Scheltens, P. (2006). Increased cortical atrophy in patients with Alzheimer's disease and type 2 diabetes mellitus. *Journal of Neurology, Neurosurgery, and Psychiatry*, *77*(3), 304–307. doi:10.1136/jnnp.2005.069583 PMID:16484636

Biessels, G. J., Strachan, M. W., Visseren, F. L., Kappelle, L. J., & Whitmer, R. A. (2014). Dementia and cognitive decline in type 2 diabetes and prediabetic stages: Towards targeted interventions. *The Lancet. Diabetes & Endocrinology*, *2*(3), 246–255. doi:10.1016/S2213-8587(13)70088-3 PMID:24622755

Bloom, G. S. (2014). Amyloid-β and tau: The trigger and bullet in Alzheimer disease pathogenesis. *JAMA Neurology*, *71*(4), 505–508. doi:10.1001/jamaneurol.2013.5847 PMID:24493463

Blum, D., Herrera, F., Francelle, L., Mendes, T., Basquin, M., Obriot, H., ... Buée, L. (2014). Mutant huntingtin alters Tau phosphorylation and subcellular distribution. *Human Molecular Genetics*, *24*(1), 76–85. doi:10.1093/hmg/ddu421 PMID:25143394

Boland, B., Kumar, A., Lee, S., Platt, F. M., Wegiel, J., Yu, W. H., & Nixon, R. A. (2008). Autophagy Induction and Autophagosome Clearance in Neurons:Relationship to Autophagic Pathology in Alzheimer's Disease. *The Journal of Neuroscience: The Official Journal of the Society for Neuroscience*, *28*(27), 6926–6937. doi:10.1523/JNEUROSCI.0800-08.2008 PMID:18596167

Boland, B., Kumar, A., Lee, S., Platt, F. M., Wegiel, J., Yu, W. H., & Nixon, R. A. (2008). Autophagy induction and autophagosome clearance in neurons: Relationship to autophagic pathology in Alzheimer's disease. *The Journal of Neuroscience: The Official Journal of the Society for Neuroscience*, *28*(27), 6926–6937. doi:10.1523/JNEUROSCI.0800-08.2008 PMID:18596167

Bosco, D., Fava, A., Plastino, M., Montalcini, T., & Pujia, A. (2011). Possible implications of insulin resistance and glucose metabolism in Alzheimer's disease pathogenesis. *Journal of Cellular and Molecular Medicine*, *15*(9), 1807–1821. doi:10.1111/j.1582-4934.2011.01318.x PMID:21435176

Braak, H., & Braak, E. (1991). Neuropathological stageing of Alzheimer-related changes. *Acta Neuropathologica*, *82*(4), 239–259. doi:10.1007/BF00308809 PMID:1759558

Braak, H., & Braak, E. V. A. (1995). Staging of Alzheimer's disease-related neurofibrillary changes. *Neurobiology of Aging*, *16*(3), 271–278. doi:10.1016/0197-4580(95)00021-6 PMID:7566337

Bursch, W. (2001). The autophagosomal–lysosomal compartment in programmed cell death. *Cell Death and Differentiation*, *8*(6), 569–581. doi:10.1038j.cdd.4400852 PMID:11536007

Caccamo, A., Maldonado, M. A., Majumder, S., Medina, D. X., Holbein, W., Magrí, A., & Oddo, S. (2011). Naturally secreted amyloid-β increases mammalian target of rapamycin (mTOR) activity via a PRAS40-mediated mechanism. *The Journal of Biological Chemistry*, *286*(11), 8924–8932. doi:10.1074/jbc.M110.180638 PMID:21266573

Caccamo, A., Oddo, S., Tran, L. X., & LaFerla, F. M. (2007). Lithium reduces tau phosphorylation but not A beta or working memory deficits in a transgenic model with both plaques and tangles. *American Journal of Pathology*, *170*(5), 1669–1675. doi:10.2353/ajpath.2007.061178 PMID:17456772

Carter, R. J., Lione, L. A., Humby, T., Mangiarini, L., Mahal, A., Bates, G. P., ... Morton, A. J. (1999). Characterization of progressive motor deficits in mice transgenic for the human Huntington's disease mutation. *The Journal of Neuroscience: The Official Journal of the Society for Neuroscience*, *19*(8), 3248–3257. doi:10.1523/JNEUROSCI.19-08-03248.1999 PMID:10191337

Cecconi, F., & Levine, B. (2008). The role of autophagy in mammalian development: Cell makeover rather than cell death. *Developmental Cell*, *15*(3), 344–357. doi:10.1016/j.devcel.2008.08.012 PMID:18804433

Chava, S., Lee, C., Aydin, Y., Chandra, P. K., Dash, A., Chedid, M., ... Dash, S. (2017). Chaperone-mediated autophagy compensates for impaired macroautophagy in the cirrhotic liver to promote hepatocellular carcinoma. *Oncotarget*, *8*(25), 40019. doi:10.18632/oncotarget.16685 PMID:28402954

Chen, J., Kanai, Y., Cowan, N. J., & Hirokawa, N. (1992). Projection domains of MAP2 and tau determine spacings between microtubules in dendrites and axons. *Nature*, *360*(6405), 674–677. doi:10.1038/360674a0 PMID:1465130

Chen, Q., Zhou, Z., Zhang, L., Wang, Y., Zhang, Y. W., Zhong, M., ... Yu, Z. P. (2012). Tau protein is involved in morphological plasticity in hippocampal neurons in response to BDNF. *Neurochemistry International*, *60*(3), 233–242. doi:10.1016/j.neuint.2011.12.013 PMID:22226842

Cheng, G., Huang, C., Deng, H., & Wang, H. (2012). Diabetes as a risk factor for dementia and mild cognitive impairment: A meta-analysis of longitudinal studies. *Internal Medicine Journal*, *42*(5), 484–491. doi:10.1111/j.1445-5994.2012.02758.x PMID:22372522

Chiang, H. L., Plant, C. P., & Dice, J. F. (1989). A role for a 70-kilodalton heat shock protein in lysosomal degradation of intracellular proteins. *Science*, *246*(4928), 382–385. doi:10.1126cience.2799391 PMID:2799391

Chiti, F., & Dobson, C. M. (2006). Protein misfolding, functional amyloid, and human disease. *Annual Review of Biochemistry*, *75*(1), 333–366. doi:10.1146/annurev.biochem.75.101304.123901 PMID:16756495

Ciaccioli, G., Martins, A., Rodrigues, C., Vieira, H., & Calado, P. (2013). A powerful yeast model to investigate the synergistic interaction of α-synuclein and tau in neurodegeneration. *PLoS One*, *8*(2), e55848. doi:10.1371/journal.pone.0055848 PMID:23393603

Cong, J., Liu, R., Wang, X., Jiang, H., & Zhang, Y. (2016). MiR-634 decreases cell proliferation and induces apoptosis by targeting mTOR signaling pathway in cervical cancer cells. *Artificial Cells, Nanomedicine, and Biotechnology*, *44*(7), 1694–1701. doi:10.3109/21691401.2015.1080171 PMID:26367112

Cooper, G. J., Day, A. J., Willis, A. C., Roberts, A. N., Reid, K. B., & Leighton, B. (1989). Amylin and the amylin gene: Structure, function and relationship to islet amyloid and to diabetes mellitus. *Biochimica et Biophysica Acta (BBA)-. Molecular Cell Research*, *1014*(3), 247–258. PMID:2690958

Craft, S. (2009). The role of metabolic disorders in Alzheimer disease and vascular dementia: Two roads converged. *Archives of Neurology*, *66*(3), 300–305. doi:10.1001/archneurol.2009.27 PMID:19273747

Cuervo, A. M., Stefanis, L., Fredenburg, R., Lansbury, P. T., & Sulzer, D. (2004). Impaired degradation of mutant α-synuclein by chaperone-mediated autophagy. *Science*, *305*(5688), 1292–1295. doi:10.1126cience.1101738 PMID:15333840

Culjkovic, B., Tan, K., Orolicki, S., Amri, A., Meloche, S., & Borden, K. L. (2008). The eIF4E RNA regulon promotes the Akt signaling pathway. *The Journal of Cell Biology*, *181*(1), 51–63. doi:10.1083/jcb.200707018 PMID:18391071

Dauer, W., & Przedborski, S. (2003). Parkinson's disease: mechanisms and models. *Neuron, 39*(6), 889-909.

Davies, S. W., Turmaine, M., Cozens, B. A., DiFiglia, M., Sharp, A. H., Ross, C. A., ... Bates, G. P. (1997). Formation of neuronal intranuclear inclusions underlies the neurological dysfunction in mice transgenic for the HD mutation. *Cell, 90*(3), 537–548. doi:10.1016/S0092-8674(00)80513-9 PMID:9267033

De Benedetti, A., & Graff, J. R. (2004). eIF-4E expression and its role in malignancies and metastases. *Oncogene, 23*(18), 3189–3199. doi:10.1038j.onc.1207545 PMID:15094768

De la Monte, S. M., & Wands, J. R. (2008). Alzheimer's disease is type 3 diabetes—Evidence reviewed. *Journal of Diabetes Science and Technology, 2*(6), 1101–1113. doi:10.1177/193229680800200619 PMID:19885299

Dehay, B., Bové, J., Rodríguez-Muela, N., Perier, C., Recasens, A., Boya, P., & Vila, M. (2010). Pathogenic lysosomal depletion in Parkinson's disease. *The Journal of Neuroscience: The Official Journal of the Society for Neuroscience, 30*(37), 12535–12544. doi:10.1523/JNEUROSCI.1920-10.2010 PMID:20844148

Despa, S., Margulies, K. B., Chen, L., Knowlton, A. A., Havel, P. J., Taegtmeyer, H., ... Despa, F. (2012). Hyperamylinemia contributes to cardiac dysfunction in obesity and diabetes: A study in humans and rats. *Circulation Research, 110*(4), 598–608. doi:10.1161/CIRCRESAHA.111.258285 PMID:22275486

Devi, L., Alldred, M. J., Ginsberg, S. D., & Ohno, M. (2012). Mechanisms underlying insulin deficiency-induced acceleration of β-amyloidosis in a mouse model of Alzheimer's disease. *PLoS One, 7*(3), e32792. doi:10.1371/journal.pone.0032792 PMID:22403710

Di Paolo, G., & Kim, T. W. (2011). Linking lipids to Alzheimer's disease: Cholesterol and beyond. *Nature Reviews. Neuroscience, 12*(5), 284–296. doi:10.1038/nrn3012 PMID:21448224

Domínguez, R. O., Pagano, M. A., Marschoff, E. R., González, S. E., Repetto, M. G., & Serra, J. A. (2014). Alzheimer disease and cognitive impairment associated with diabetes mellitus type 2: Associations and a hypothesis [English Edition]. *Neurologia (Barcelona, Spain), 29*(9), 567–572. PMID:24140159

Drewes, G., Trinczek, B., Illenberger, S., Biernat, J., Schmitt-Ulms, G., Meyer, H. E., ... Mandelkow, E. (1995). Microtubule-associated protein/microtubule affinity-regulating kinase (p110mark) a novel protein kinase that regulates tau-microtubule interactions and dynamic instability by phosphorylation at the Alzheimer-specific site serine 262. *The Journal of Biological Chemistry, 270*(13), 7679–7688. doi:10.1074/jbc.270.13.7679 PMID:7706316

Dubouloz, F., Deloche, O., Wanke, V., Cameroni, E., & De Virgilio, C. (2005). The TOR and EGO protein complexes orchestrate microautophagy in yeast. *Molecular Cell, 19*(1), 15–26. doi:10.1016/j.molcel.2005.05.020 PMID:15989961

Eidenmuller, J., Thomas, F. A. T. H., Thorsten, M. A. A. S., & Madeline, P. O. O. L. (2001). Phosphorylation-mimicking glutamate clusters in the proline-rich region are sufficient to simulate the functional deficiencies of hyperphosphorylated tau protein. *The Biochemical Journal, 357*(3), 759–767. doi:10.1042/bj3570759 PMID:11463346

Engelman, J. A. (2009). Targeting PI3K signalling in cancer: Opportunities, challenges and limitations. *Nature Reviews. Cancer, 9*(8), 550–562. doi:10.1038/nrc2664 PMID:19629070

Esposito, A., Dohm, C. P., Kermer, P., Bähr, M., & Wouters, F. S. (2007). α-Synuclein and its disease-related mutants interact differentially with the microtubule protein tau and associate with the actin cytoskeleton. *Neurobiology of Disease*, *26*(3), 521–531. doi:10.1016/j.nbd.2007.01.014 PMID:17408955

Exalto, L. G., van der Flier, W. M., Scheltens, P., Vrenken, H., & Biessels, G. J. (2014). Dysglycemia, brain volume and vascular lesions on MRI in a memory clinic population. *Journal of Diabetes and Its Complications*, *28*(1), 85–90. doi:10.1016/j.jdiacomp.2012.12.003 PMID:23352495

Fan, W., Nassiri, A., & Zhong, Q. (2011). Autophagosome targeting and membrane curvature sensing by Barkor/Atg14 (L). *Proceedings of the National Academy of Sciences of the United States of America*, *108*(19), 7769–7774. doi:10.1073/pnas.1016472108 PMID:21518905

Farris, W., Mansourian, S., Chang, Y., Lindsley, L., Eckman, E. A., Frosch, M. P., ... Guénette, S. (2003). Insulin-degrading enzyme regulates the levels of insulin, amyloid β-protein, and the β-amyloid precursor protein intracellular domain in vivo. *Proceedings of the National Academy of Sciences of the United States of America*, *100*(7), 4162–4167. doi:10.1073/pnas.0230450100 PMID:12634421

Fernández-Nogales, M., Cabrera, J. R., Santos-Galindo, M., Hoozemans, J. J., Ferrer, I., Rozemuller, A. J., ... Lucas, J. J. (2014). Huntington's disease is a four-repeat tauopathy with tau nuclear rods. *Nature Medicine*, *20*(8), 881–885. doi:10.1038/nm.3617 PMID:25038828

Fernández-Nogales, M., Santos-Galindo, M., Hernández, I. H., Cabrera, J. R., & Lucas, J. J. (2016). Faulty splicing and cytoskeleton abnormalities in H untington's disease. *Brain Pathology (Zurich, Switzerland)*, *26*(6), 772–778. doi:10.1111/bpa.12430 PMID:27529534

Fruman, D. A., & Rommel, C. (2014). PI3K and cancer: Lessons, challenges and opportunities. *Nature Reviews. Drug Discovery*, *13*(2), 140–156. doi:10.1038/nrd4204 PMID:24481312

Fu, Y., Wu, P., Pan, Y., Sun, X., Yang, H., Difiglia, M., & Lu, B. (2017). A toxic mutant huntingtin species is resistant to selective autophagy. *Nature Chemical Biology*, *13*(11), 1152–1154. doi:10.1038/nchembio.2461 PMID:28869595

Fujita, N., Hayashi-Nishino, M., Fukumoto, H., Omori, H., Yamamoto, A., Noda, T., & Yoshimori, T. (2008). An Atg4B mutant hampers the lipidation of LC3 paralogues and causes defects in autophagosome closure. *Molecular Biology of the Cell*, *19*(11), 4651–4659. doi:10.1091/mbc.e08-03-0312 PMID:18768752

Fuster-Matanzo, A., de Barreda, E. G., Dawson, H. N., Vitek, M. P., Avila, J., & Hernández, F. (2009). Function of tau protein in adult newborn neurons. *FEBS Letters*, *583*(18), 3063–3068. doi:10.1016/j.febslet.2009.08.017 PMID:19695252

Ganley, I. G., Lam, D. H., Wang, J., Ding, X., Chen, S., & Jiang, X. (2009). ULK1· ATG13· FIP200 complex mediates mTOR signaling and is essential for autophagy. *The Journal of Biological Chemistry*, *284*(18), 12297–12305. doi:10.1074/jbc.M900573200 PMID:19258318

Gardet, A., Benita, Y., Li, C., Sands, B. E., Ballester, I., Stevens, C., ... Podolsky, D. K. (2010). LRRK2 is involved in the IFN-γ response and host response to pathogens. *Journal of Immunology (Baltimore, Md.: 1950)*, *185*(9), 5577–5585. doi:10.4049/jimmunol.1000548 PMID:20921534

Ghetti, B., Oblak, A. L., Boeve, B. F., Johnson, K. A., Dickerson, B. C., & Goedert, M. (2015). Invited review: frontotemporal dementia caused by microtubule-associated protein tau gene (MAPT) mutations: a chameleon for neuropathology and neuroimaging. *Neuropathology and Applied Neurobiology*, *41*(1), 24–46. doi:10.1111/nan.12213 PMID:25556536

Ghosh, S., Watanabe, R. M., Valle, T. T., Hauser, E. R., Magnuson, V. L., Langefeld, C. D., ... Chines, P. (2000). The Finland–United States investigation of non–insulin-dependent diabetes mellitus genetics (FUSION) study. I. An autosomal genome scan for genes that predispose to type 2 diabetes. *American Journal of Human Genetics*, *67*(5), 1174–1185. PMID:11032783

Gilbert, B. J. (2013). The role of amyloid β in the pathogenesis of Alzheimer's disease. *Journal of Clinical Pathology*, *66*(5), 362–366. doi:10.1136/jclinpath-2013-201515 PMID:23526599

Goldwurm, S., Di Fonzo, A., Simons, E. J., Rohe, C. F., Zini, M., Canesi, M., ... Meucci, N. (2005). The G6055A (G2019S) mutation in LRRK2 is frequent in both early and late onset Parkinson's disease and originates from a common ancestor. *Journal of Medical Genetics*, *42*(11), e65–e65. doi:10.1136/jmg.2005.035568 PMID:16272257

Gómez-Isla, T., Hollister, R., West, H., Mui, S., Growdon, J. H., Petersen, R. C., ... Hyman, B. T. (1997). Neuronal loss correlates with but exceeds neurofibrillary tangles in Alzheimer's disease. *Annals of Neurology: Official Journal of the American Neurological Association and the Child Neurology Society*, *41*(1), 17–24. doi:10.1002/ana.410410106 PMID:9005861

Gong, C. X., Lidsky, T., Wegiel, J., Zuck, L., Grundke-Iqbal, I., & Iqbal, K. (2000). Phosphorylation of microtubule-associated protein tau is regulated by protein phosphatase 2A in mammalian brain implications for neurofibrillary degeneration in Alzheimer's disease. *The Journal of Biological Chemistry*, *275*(8), 5535–5544. doi:10.1074/jbc.275.8.5535 PMID:10681533

González-Polo, R. A., Boya, P., Pauleau, A. L., Jalil, A., Larochette, N., Souquère, S., ... Kroemer, G. (2005). The apoptosis/autophagy paradox: Autophagic vacuolization before apoptotic death. *Journal of Cell Science*, *118*(14), 3091–3102. doi:10.1242/jcs.02447 PMID:15985464

Gratuze, M., Noël, A., Julien, C., Cisbani, G., Milot-Rousseau, P., Morin, F., ... Bissonnette, S. (2014). Tau hyperphosphorylation and deregulation of calcineurin in mouse models of Huntington's disease. *Human Molecular Genetics*, *24*(1), 86–99. doi:10.1093/hmg/ddu456 PMID:25205109

Gustafsson, Å. B., & Gottlieb, R. A. (2008). Recycle or die: The role of autophagy in cardioprotection. *Journal of Molecular and Cellular Cardiology*, *44*(4), 654–661. doi:10.1016/j.yjmcc.2008.01.010 PMID:18353358

Hanada, T., Noda, N. N., Satomi, Y., Ichimura, Y., Fujioka, Y., Takao, T., ... Ohsumi, Y. (2007). The Atg12-Atg5 conjugate has a novel E3-like activity for protein lipidation in autophagy. *The Journal of Biological Chemistry*, *282*(52), 37298–37302. doi:10.1074/jbc.C700195200 PMID:17986448

Hanada, T., Satomi, Y., Takao, T., & Ohsumi, Y. (2009). The amino-terminal region of Atg3 is essential for association with phosphatidylethanolamine in Atg8 lipidation. *FEBS Letters*, *583*(7), 1078–1083. doi:10.1016/j.febslet.2009.03.009 PMID:19285500

Hanahan, D., & Weinberg, R. A. (2011). Hallmarks of cancer: the next generation. *Cell, 144*(5), 646-674.

Hanger, D. P., Anderton, B. H., & Noble, W. (2009). Tau phosphorylation: The therapeutic challenge for neurodegenerative disease. *Trends in Molecular Medicine*, *15*(3), 112–119. doi:10.1016/j.molmed.2009.01.003 PMID:19246243

Hanger, D. P., Hughes, K., Woodgett, J. R., Brion, J. P., & Anderton, B. H. (1992). Glycogen synthase kinase-3 induces Alzheimer's disease-like phosphorylation of tau: Generation of paired helical filament epitopes and neuronal localisation of the kinase. *Neuroscience Letters*, *147*(1), 58–62. doi:10.1016/0304-3940(92)90774-2 PMID:1336152

Hardy, J. A., & Higgins, G. A. (1992). Alzheimer's disease: The amyloid cascade hypothesis. *Science*, *256*(5054), 184–186. doi:10.1126cience.1566067 PMID:1566067

Harhaji-Trajkovic, L., Arsikin, K., Kravic-Stevovic, T., Petricevic, S., Tovilovic, G., Pantovic, A., ... Trajkovic, V. (2012). Chloroquine-mediated lysosomal dysfunction enhances the anticancer effect of nutrient deprivation. *Pharmaceutical Research*, *29*(8), 2249–2263. doi:10.100711095-012-0753-1 PMID:22538436

Hashiguchi, M., Sobue, K., & Paudel, H. K. (2000). 14-3-3ζ is an effector of tau protein phosphorylation. *The Journal of Biological Chemistry*, *275*(33), 25247–25254. doi:10.1074/jbc.M003738200 PMID:10840038

Hassing, L. B., Grant, M. D., Hofer, S. M., Pedersen, N. L., Nilsson, S. E., Berg, S., ... Johansson, B. (2004). Type 2 diabetes mellitus contributes to cognitive decline in old age: A longitudinal population-based study. *Journal of the International Neuropsychological Society*, *10*(4), 599–607. doi:10.1017/S1355617704104165 PMID:15327738

Hay, N. (2005). The Akt-mTOR tango and its relevance to cancer. *Cancer Cell*, *8*(3), 179–183. doi:10.1016/j.ccr.2005.08.008 PMID:16169463

Hayley, S., Crocker, S. J., Smith, P. D., Shree, T., Jackson-Lewis, V., Przedborski, S., ... Park, D. S. (2004). Regulation of dopaminergic loss by Fas in a 1-methyl-4-phenyl-1, 2, 3, 6-tetrahydropyridine model of Parkinson's disease. *The Journal of Neuroscience: The Official Journal of the Society for Neuroscience*, *24*(8), 2045–2053. doi:10.1523/JNEUROSCI.4564-03.2004 PMID:14985447

He, C., Song, H., Yorimitsu, T., Monastyrska, I., Yen, W. L., Legakis, J. E., & Klionsky, D. J. (2006). Recruitment of Atg9 to the preautophagosomal structure by Atg11 is essential for selective autophagy in budding yeast. *The Journal of Cell Biology*, *175*(6), 925–935. doi:10.1083/jcb.200606084 PMID:17178909

Hong, X. P., Peng, C. X., Wei, W., Tian, Q., Liu, Y. H., Yao, X. Q., ... Wang, J. Z. (2010). Essential role of tau phosphorylation in adult hippocampal neurogenesis. *Hippocampus*, *20*(12), 1339–1349. doi:10.1002/hipo.20712 PMID:19816983

Huotari, J., & Helenius, A. (2011). Endosome maturation. *The EMBO Journal*, *30*(17), 3481–3500. doi:10.1038/emboj.2011.286 PMID:21878991

Iijima, M., Tabira, T., Poorkaj, P., Schellenberg, G. D., Trojanowski, J. Q., Lee, V. M. Y., ... Yamashita, Y. (1999). A distinct familial presenile dementia with a novel missense mutation in the tau gene. *Neuroreport*, *10*(3), 497–501. doi:10.1097/00001756-199902250-00010 PMID:10208578

Imarisio, S., Carmichael, J., Korolchuk, V., Chen, C. W., Saiki, S., Rose, C., ... Rubinsztein, D. C. (2008). Huntington's disease: From pathology and genetics to potential therapies. *The Biochemical Journal, 412*(2), 191–209. doi:10.1042/BJ20071619 PMID:18466116

Inoki, K., Li, Y., Zhu, T., Wu, J., & Guan, K. L. (2002). TSC2 is phosphorylated and inhibited by Akt and suppresses mTOR signalling. *Nature Cell Biology, 4*(9), 648–657. doi:10.1038/ncb839 PMID:12172553

Iqbal, K., Liu, F., Gong, C. X., & Grundke-Iqbal, I. (2010). Tau in Alzheimer disease and related tauopathies. *Current Alzheimer Research, 7*(8), 656–664. doi:10.2174/156720510793611592 PMID:20678074

Irvine, G. B., El-Agnaf, O. M., Shankar, G. M., & Walsh, D. M. (2008). Protein aggregation in the brain: The molecular basis for Alzheimer's and Parkinson's diseases. *Molecular Medicine (Cambridge, Mass.), 14*(7-8), 451–464. doi:10.2119/2007-00100.Irvine PMID:18368143

Irwin, D. J., Lee, V. M. Y., & Trojanowski, J. Q. (2013). Parkinson's disease dementia: Convergence of α-synuclein, tau and amyloid-β pathologies. *Nature Reviews. Neuroscience, 14*(9), 626–636. doi:10.1038/nrn3549 PMID:23900411

Ishizawa, T., Mattila, P., Davies, P., Wang, D., & Dickson, D. W. (2003). Colocalization of tau and alpha-synuclein epitopes in Lewy bodies. *Journal of Neuropathology and Experimental Neurology, 62*(4), 389–397. doi:10.1093/jnen/62.4.389 PMID:12722831

Itakura, E., Kishi-Itakura, C., & Mizushima, N. (2012). The hairpin-type tail-anchored SNARE syntaxin 17 targets to autophagosomes for fusion with endosomes/lysosomes. *Cell, 151*(6), 1256–1269. doi:10.1016/j.cell.2012.11.001 PMID:23217709

Ittner, L. M., Fath, T., Ke, Y. D., Bi, M., van Eersel, J., Li, K. M., ... Götz, J. (2008). Parkinsonism and impaired axonal transport in a mouse model of frontotemporal dementia. *Proceedings of the National Academy of Sciences of the United States of America, 105*(41), 15997–16002. doi:10.1073/pnas.0808084105 PMID:18832465

Jackson, K., Barisone, G. A., Diaz, E., Jin, L. W., DeCarli, C., & Despa, F. (2013). Amylin deposition in the brain: A second amyloid in Alzheimer disease? *Annals of Neurology, 74*(4), 517–526. doi:10.1002/ana.23956 PMID:23794448

Jensen, P. H., Hager, H., Nielsen, M. S., Højrup, P., Gliemann, J., & Jakes, R. (1999). α-Synuclein binds to tau and stimulates the protein kinase A-catalyzed tau phosphorylation of serine residues 262 and 356. *The Journal of Biological Chemistry, 274*(36), 25481–25489. doi:10.1074/jbc.274.36.25481 PMID:10464279

Jimenez-Sanchez, M., Licitra, F., Underwood, B. R., & Rubinsztein, D. C. (2017). Huntington's disease: Mechanisms of pathogenesis and therapeutic strategies. *Cold Spring Harbor Perspectives in Medicine, 7*(7), a024240. doi:10.1101/cshperspect.a024240 PMID:27940602

Joachim, C. L., Morris, J. H., Kosik, K. S., & Selkoe, D. J. (1987). Tau antisera recognize neurofibrillary tangles in a range of neurodegenerative disorders. *Annals of Neurology, 22*(4), 514–520. doi:10.1002/ana.410220411 PMID:2963585

Jung, C. H., Jun, C. B., Ro, S. H., Kim, Y. M., Otto, N. M., Cao, J., ... Kim, D. H. (2009). ULK-Atg13-FIP200 complexes mediate mTOR signaling to the autophagy machinery. *Molecular Biology of the Cell, 20*(7), 1992–2003. doi:10.1091/mbc.e08-12-1249 PMID:19225151

Kabeya, Y., Mizushima, N., Ueno, T., Yamamoto, A., Kirisako, T., Noda, T., ... Yoshimori, T. (2000). LC3, a mammalian homologue of yeast Apg8p, is localized in autophagosome membranes after processing. *The EMBO Journal, 19*(21), 5720–5728. doi:10.1093/emboj/19.21.5720 PMID:11060023

Kabuta, T., Furuta, A., Aoki, S., Furuta, K., & Wada, K. (2008). Aberrant interaction between Parkinson disease-associated mutant UCH-L1 and the lysosomal receptor for chaperone-mediated autophagy. *The Journal of Biological Chemistry, 283*(35), 23731–23738. doi:10.1074/jbc.M801918200 PMID:18550537

Kalia, L. V., Kalia, S. K., McLean, P. J., Lozano, A. M., & Lang, A. E. (2013). α-Synuclein oligomers and clinical implications for Parkinson disease. *Annals of Neurology, 73*(2), 155–169. doi:10.1002/ana.23746 PMID:23225525

Kapahi, P., Chen, D., Rogers, A. N., Katewa, S. D., Li, P. W. L., Thomas, E. L., & Kockel, L. (2010). With TOR, less is more: A key role for the conserved nutrient-sensing TOR pathway in aging. *Cell Metabolism, 11*(6), 453–465. doi:10.1016/j.cmet.2010.05.001 PMID:20519118

Kaushik, S., & Cuervo, A. M. (2018). The coming of age of chaperone-mediated autophagy. *Nature Reviews. Molecular Cell Biology, 19*(6), 1. doi:10.103841580-018-0001-6 PMID:29626215

Kazlauskaite, A., & Muqit, M. M. (2015). PINK1 and Parkin–mitochondrial interplay between phosphorylation and ubiquitylation in Parkinson's disease. *The FEBS Journal, 282*(2), 215–223. doi:10.1111/febs.13127 PMID:25345844

Kim, J., Kundu, M., Viollet, B., & Guan, K. L. (2011). AMPK and mTOR regulate autophagy through direct phosphorylation of Ulk1. *Nature Cell Biology, 13*(2), 132–141. doi:10.1038/ncb2152 PMID:21258367

Kim, M., Hersh, L. B., Leissring, M. A., Ingelsson, M., Matsui, T., Farris, W., ... Tanzi, R. E. (2007). Decreased catalytic activity of the insulin-degrading enzyme in chromosome 10-linked Alzheimer disease families. *The Journal of Biological Chemistry, 282*(11), 7825–7832. doi:10.1074/jbc.M609168200 PMID:17244626

Kitada, T., Asakawa, S., Hattori, N., Matsumine, H., Yamamura, Y., Minoshima, S., ... and Shimizu, N. (1998). Mutations in the parkin gene cause autosomal recessive juvenile parkinsonism. *Nature, 392*(6676), 605.

Klein, C., & Westenberger, A. (2012). Genetics of Parkinson's Disease. *Cold Spring Harbor Perspectives in Medicine, 2*(1), a008888. doi:10.1101/cshperspect.a008888 PMID:22315721

Klionsky, D. J., Abdalla, F. C., Abeliovich, H., Abraham, R. T., Acevedo-Arozena, A., Adeli, K., ... Zuckerbraun, B. (2012). Guidelines for the use and interpretation of assays for monitoring autophagy. *Autophagy, 8*(8), 445–544. doi:10.4161/auto.19496 PMID:22966490

Komatsu, M., Waguri, S., Chiba, T., Murata, S., Iwata, J. I., Tanida, I., ... Tanaka, K. (2006). Loss of autophagy in the central nervous system causes neurodegeneration in mice. *Nature, 441*(7095), 880–884. doi:10.1038/nature04723 PMID:16625205

Komatsu, M., Waguri, S., Ueno, T., Iwata, J., Murata, S., Tanida, I., ... Kominami, E. (2005). Impairment of starvation-induced and constitutive autophagy in Atg7-deficient mice. *The Journal of Cell Biology, 169*(3), 425–434. doi:10.1083/jcb.200412022 PMID:15866887

Kopke, E., Tung, Y. C., Shaikh, S., Alonso, A. D. C., Iqbal, K., & Grundke-Iqbal, I. (1993). Microtubule-associated protein tau. Abnormal phosphorylation of a non-paired helical filament pool in Alzheimer disease. *The Journal of Biological Chemistry, 268*(32), 24374–24384. PMID:8226987

Korolchuk, V. I., Menzies, F. M., & Rubinsztein, D. C. (2010). Mechanisms of cross-talk between the ubiquitin-proteasome and autophagy-lysosome systems. *FEBS Letters, 584*(7), 1393–1398. doi:10.1016/j.febslet.2009.12.047 PMID:20040365

Kruger, U., Wang, Y., Kumar, S., & Mandelkow, E. M. (2012). Autophagic degradation of tau in primary neurons and its enhancement by trehalose. *Neurobiology of Aging, 33*(10), 2291–2305. doi:10.1016/j.neurobiolaging.2011.11.009 PMID:22169203

Krüger, U., Wang, Y., Kumar, S., & Mandelkow, E. M. (2012). Autophagic degradation of tau in primary neurons and its enhancement by trehalose. *Neurobiology of Aging, 33*(10), 2291–2305. doi:10.1016/j.neurobiolaging.2011.11.009 PMID:22169203

Ksiezak-Reding, H., Pyo, H. K., Feinstein, B., & Pasinetti, G. M. (2003). Akt/PKB kinase phosphorylates separately Thr212 and Ser214 of tau protein in vitro. *Biochimica et Biophysica Acta (BBA)-. Molecular Basis of Disease, 1639*(3), 159–168. doi:10.1016/j.bbadis.2003.09.001 PMID:14636947

Kuma, A., Mizushima, N., Ishihara, N., & Ohsumi, Y. (2002). Formation of the~ 350-kDa Apg12-Apg5·Apg16 multimeric complex, mediated by Apg16 oligomerization, is essential for autophagy in yeast. *The Journal of Biological Chemistry, 277*(21), 18619–18625. doi:10.1074/jbc.M111889200 PMID:11897782

L'episcopo, F., Drouin-Ouellet, J., Tirolo, C., Pulvirenti, A., Giugno, R., Testa, N., ... Cicchetti, F. (2016). GSK-3β-induced Tau pathology drives hippocampal neuronal cell death in Huntington's disease: Involvement of astrocyte–neuron interactions. *Cell Death & Disease, 7*(4), e2206. doi:10.1038/cddis.2016.104 PMID:27124580

Labbadia, J., & Morimoto, R. I. (2013). Huntington's disease: Underlying molecular mechanisms and emerging concepts. *Trends in Biochemical Sciences, 38*(8), 378–385. doi:10.1016/j.tibs.2013.05.003 PMID:23768628

Larson, M. E., Sherman, M. A., Greimel, S., Kuskowski, M., Schneider, J. A., Bennett, D. A., & Lesné, S. E. (2012). Soluble α-synuclein is a novel modulator of Alzheimer's disease pathophysiology. *The Journal of Neuroscience: The Official Journal of the Society for Neuroscience, 32*(30), 10253–10266. doi:10.1523/JNEUROSCI.0581-12.2012 PMID:22836259

Lau, D. H., Hogseth, M., Phillips, E. C., O'Neill, M. J., Pooler, A. M., Noble, W., & Hanger, D. P. (2016). Critical residues involved in tau binding to fyn: Implications for tau phosphorylation in Alzheimer's disease. *Acta Neuropathologica Communications, 4*(1), 49. doi:10.118640478-016-0317-4 PMID:27193083

Le, N. T., Chang, L., Kovlyagina, I., Georgiou, P., Safren, N., Braunstein, K. E., ... Morrison, B. M. (2016). Motor neuron disease, TDP-43 pathology, and memory deficits in mice expressing ALS–FTD-linked UBQLN2 mutations. *Proceedings of the National Academy of Sciences of the United States of America, 113*(47), E7580–E7589. doi:10.1073/pnas.1608432113 PMID:27834214

Lee, G., Newman, S. T., Gard, D. L., Band, H., & Panchamoorthy, G. (1998). Tau interacts with src-family non-receptor tyrosine kinases. *Journal of Cell Science, 111*(21), 3167–3177. PMID:9763511

Lee, H. J., Khoshaghideh, F., Patel, S., & Lee, S. J. (2004). Clearance of α-synuclein oligomeric intermediates via the lysosomal degradation pathway. *The Journal of Neuroscience: The Official Journal of the Society for Neuroscience, 24*(8), 1888–1896. doi:10.1523/JNEUROSCI.3809-03.2004 PMID:14985429

Lee, H. K., & Marzella, L. (1994). Regulation of intracellular protein degradation with special reference to lysosomes: Role in cell physiology and pathology. *International Review of Experimental Pathology, 35*, 39–147. PMID:7860191

Lee, J. H., Yu, W. H., Kumar, A., Lee, S., Mohan, P. S., Peterhoff, C. M., ... Uchiyama, Y. (2010). Lysosomal proteolysis and autophagy require presenilin 1 and are disrupted by Alzheimer-related PS1 mutations. *Cell, 141*(7), 1146–1158. doi:10.1016/j.cell.2010.05.008 PMID:20541250

Lee, S., Sato, Y., & Nixon, R. A. (2011). Lysosomal Proteolysis Inhibition Selectively Disrupts Axonal Transport of Degradative Organelles and Causes an Alzheimer's-Like Axonal Dystrophy. *The Journal of Neuroscience: The Official Journal of the Society for Neuroscience, 31*(21), 7817–7830. doi:10.1523/JNEUROSCI.6412-10.2011 PMID:21613495

Lee, S., Sato, Y., & Nixon, R. A. (2011). Lysosomal proteolysis inhibition selectively disrupts axonal transport of degradative organelles and causes an Alzheimer's-like axonal dystrophy. *The Journal of Neuroscience, 31*(21), 7817–7830.

Lee, Y. I., Seo, M., Kim, Y., Kim, S. Y., Kang, U. G., Kim, Y. S., & Juhnn, Y. S. (2005). Membrane depolarization induces the undulating phosphorylation/dephosphorylation of glycogen synthase kinase 3β, and this dephosphorylation involves protein phosphatases 2A and 2B in SH-SY5Y human neuroblastoma cells. *The Journal of Biological Chemistry, 280*(23), 22044–22052. doi:10.1074/jbc.M413987200 PMID:15799972

Leissring, M. A., Farris, W., Chang, A. Y., Walsh, D. M., Wu, X., Sun, X., ... Selkoe, D. J. (2003). Enhanced proteolysis of β-amyloid in APP transgenic mice prevents plaque formation, secondary pathology, and premature death. *Neuron, 40*(6), 1087–1093. doi:10.1016/S0896-6273(03)00787-6 PMID:14687544

Lemmon, M. A. (2007, January). Pleckstrin homology (PH) domains and phosphoinositides. [). Portland Press Limited.]. *Biochemical Society Symposium, 74*, 81–93. doi:10.1042/BSS2007c08 PMID:17233582

Leroy, K., Ando, K., Héraud, C., Yilmaz, Z., Authelet, M., Boeynaems, J. M., ... Brion, J. P. (2010). Lithium treatment arrests the development of neurofibrillary tangles in mutant tau transgenic mice with advanced neurofibrillary pathology. *Journal of Alzheimer's Disease, 19*(2), 705–719. doi:10.3233/JAD-2010-1276 PMID:20110614

Leroy, K., Boutajangout, A., Authelet, M., Woodgett, J. R., Anderton, B. H., & Brion, J. P. (2002). The active form of glycogen synthase kinase-3β is associated with granulovacuolar degeneration in neurons in Alzheimer's disease. *Acta Neuropathologica*, *103*(2), 91–99. doi:10.1007004010100435 PMID:11810173

Leszek, J., Trypka, E., Tarasov, V., Md, V., Ashraf, G., & Aliev, G. (2017). Type 3 diabetes mellitus: A novel implication of Alzheimers disease. *Current Topics in Medicinal Chemistry*, *17*(12), 1331–1335. doi:10.2174/1568026617666170103163403 PMID:28049395

Leverenz, J. B., Umar, I., Wang, Q., Montine, T. J., McMillan, P. J., Tsuang, D. W., ... Zhang, J. (2007). Proteomic identification of novel proteins in cortical lewy bodies. *Brain Pathology (Zurich, Switzerland)*, *17*(2), 139–145. doi:10.1111/j.1750-3639.2007.00048.x PMID:17388944

Levine, B., Sinha, S., & Kroemer, G. (2008). Bcl-2 family members: Dual regulators of apoptosis and autophagy. *Autophagy*, *4*, 600–606. doi:10.4161/auto.6260

Levy, O. A., Malagelada, C., & Greene, L. A. (2009). Cell death pathways in Parkinson's disease: Proximal triggers, distal effectors, and final steps. *Apoptosis*, *14*(4), 478–500. doi:10.100710495-008-0309-3 PMID:19165601

Levy-Lahad, E., Wasco, W., Poorkaj, P., Romano, D. M., Oshima, J., Pettingell, W. H., ... Wang, K. (1995). Candidate gene for the chromosome 1 familial Alzheimer's disease locus. *Science*, *269*(5226), 973–977. doi:10.1126cience.7638622 PMID:7638622

Li, J.-Q., Tan, L., & Yu, J.-T. (2014). The role of the LRRK2 gene in Parkinsonism. *Molecular Neurodegeneration*, *9*(1), 47. doi:10.1186/1750-1326-9-47 PMID:25391693

Li, X., Alafuzoff, I., Soininen, H., Winblad, B., & Pei, J. J. (2005). Levels of mTOR and its downstream targets 4E-BP1, eEF2, and eEF2 kinase in relationships with tau in Alzheimer's disease brain. *The FEBS Journal*, *272*(16), 4211–4220. doi:10.1111/j.1742-4658.2005.04833.x PMID:16098202

Li, X. C., Hu, Y., Wang, Z. H., Luo, Y., Zhang, Y., Liu, X. P., ... Wang, J. Z. (2016). Human wild-type full-length tau accumulation disrupts mitochondrial dynamics and the functions via increasing mitofusins. *Scientific Reports*, *6*(1), 24756. doi:10.1038rep24756 PMID:27099072

Lim, Y. A., Rhein, V., Baysang, G., Meier, F., Poljak, A., & Raftery, J., M., ... and Götz, J. (. (2010). Aβ and human amylin share a common toxicity pathway via mitochondrial dysfunction. *Proteomics*, *10*(8), 1621–1633. doi:10.1002/pmic.200900651 PMID:20186753

Lione, L. A., Carter, R. J., Hunt, M. J., Bates, G. P., Morton, A. J., & Dunnett, S. B. (1999). Selective discrimination learning impairments in mice expressing the human Huntington's disease mutation. *The Journal of Neuroscience: The Official Journal of the Society for Neuroscience*, *19*(23), 10428–10437. doi:10.1523/JNEUROSCI.19-23-10428.1999 PMID:10575040

Liu, C., Cholerton, B., Shi, M., Ginghina, C., Cain, K. C., Auinger, P., & Zhang, J. (2015). CSF tau and tau/Aβ42 predict cognitive decline in Parkinson's disease. *Parkinsonism & Related Disorders*, *21*(3), 271–276. doi:10.1016/j.parkreldis.2014.12.027 PMID:25596881

Liu, F., Grundke-Iqbal, I., Iqbal, K., & Gong, C. X. (2005). Contributions of protein phosphatases PP1, PP2A, PP2B and PP5 to the regulation of tau phosphorylation. *The European Journal of Neuroscience*, *22*(8), 1942–1950. doi:10.1111/j.1460-9568.2005.04391.x PMID:16262633

Liu, F., Li, B., Tung, E. J., Grundke-Iqbal, I., Iqbal, K., & Gong, C. X. (2007). Site-specific effects of tau phosphorylation on its microtubule assembly activity and self-aggregation. *The European Journal of Neuroscience*, *26*(12), 3429–3436. doi:10.1111/j.1460-9568.2007.05955.x PMID:18052981

Long, Y. C., & Zierath, J. R. (2006). AMP-activated protein kinase signaling in metabolic regulation. *The Journal of Clinical Investigation*, *116*(7), 1776–1783. doi:10.1172/JCI29044 PMID:16823475

Lu, J., Xie, G., Jia, W., & Jia, W. (2013). Insulin resistance and the metabolism of branched-chain amino acids. *Frontiers of Medicine*, *7*(1), 53–59. doi:10.100711684-013-0255-5 PMID:23385611

Luna-Munoz, J., Chavez-Macias, L., Garcia-Sierra, F., & Mena, R. (2007). Earliest Stages of Tau Conformational Changes are Related to the Appearance of a Sequence of Specific Phospho-Dependent Tau Epitopes in Alzheimer's Disease 1. *Journal of Alzheimer's Disease*, *12*(4), 365–375. doi:10.3233/JAD-2007-12410 PMID:18198423

Luzio, J. P., Pryor, P. R., & Bright, N. A. (2007). Lysosomes: Fusion and function. *Nature Reviews. Molecular Cell Biology*, *8*(8), 622–632. doi:10.1038/nrm2217 PMID:17637737

Maiuri, M. C., Le Toumelin, G., Criollo, A., Rain, J. C., Gautier, F., Juin, P., ... Hickman, J. A. (2007). Functional and physical interaction between Bcl-XL and a BH3-like domain in Beclin-1. *The EMBO Journal*, *26*(10), 2527–2539. doi:10.1038j.emboj.7601689 PMID:17446862

Majid, T., Ali, Y. O., Venkitaramani, D. V., Jang, M. K., Lu, H. C., & Pautler, R. G. (2014). In vivo axonal transport deficits in a mouse model of fronto-temporal dementia. *NeuroImage. Clinical*, *4*, 711–717. doi:10.1016/j.nicl.2014.02.005 PMID:24936422

Mandelkow, E. M., & Mandelkow, E. (2012). Biochemistry and cell biology of tau protein in neurofibrillary degeneration. *Cold Spring Harbor Perspectives in Medicine*, *2*(7), a006247. doi:10.1101/cshperspect.a006247 PMID:22762014

Manning, B. D., & Cantley, L. C. (2007). AKT/PKB signaling: Navigating downstream. *Cell*, *129*(7), 1261–1274. doi:10.1016/j.cell.2007.06.009 PMID:17604717

Manning, B. D., Tee, A. R., Logsdon, M. N., Blenis, J., & Cantley, L. C. (2002). Identification of the tuberous sclerosis complex-2 tumor suppressor gene product tuberin as a target of the phosphoinositide 3-kinase/akt pathway. *Molecular Cell*, *10*(1), 151–162. doi:10.1016/S1097-2765(02)00568-3 PMID:12150915

Martin, L., Latypova, X., & Terro, F. (2011). Post-translational modifications of tau protein: Implications for Alzheimer's disease. *Neurochemistry International*, *58*(4), 458–471. doi:10.1016/j.neuint.2010.12.023 PMID:21215781

Martinez-Vicente, M., Talloczy, Z., Wong, E., Tang, G., Koga, H., Kaushik, S., ... Cuervo, A. M. (2010). Cargo recognition failure is responsible for inefficient autophagy in Huntington's disease. *Nature Neuroscience*, *13*(5), 567–576. doi:10.1038/nn.2528 PMID:20383138

Martyniszyn, L., Szulc, L., Boratyńska, A., & Niemiałtowski, M. G. (2011). Beclin 1 is involved in regulation of apoptosis and autophagy during replication of ectromelia virus in permissive L929 cells. *Archivum Immunologiae et Therapiae Experimentalis, 59*(6), 463–471. doi:10.100700005-011-0149-7 PMID:21972018

Marui, W., Iseki, E., Uéda, K., & Kosaka, K. (2000). Occurrence of human α-synuclein immunoreactive neurons with neurofibrillary tangle formation in the limbic areas of patients with Alzheimer's disease. *Journal of the Neurological Sciences, 174*(2), 81–84. doi:10.1016/S0022-510X(99)00327-5 PMID:10727692

Marzella, L., Ahlberg, J., & Glaumann, H. (1981). Autophagy, heterophagy, microautophagy and crinophagy as the means for intracellular degradation. *Virchows Archiv B Cell Pathology Zell-pathologie, 36*(1), 219-234.

Masliah, E. (1995). Mechanisms of synaptic dysfunction in Alzheimer's disease. *Histology and Histopathology, 10*(2), 509–519. PMID:7599445

Mathew, R., Karp, C. M., Beaudoin, B., Vuong, N., Chen, G., Chen, H. Y., ... DiPaola, R. S. (2009). Autophagy suppresses tumorigenesis through elimination of p62. *Cell, 137*(6), 1062–1075. doi:10.1016/j.cell.2009.03.048 PMID:19524509

Matsuda, N., Sato, S., Shiba, K., Okatsu, K., Saisho, K., Gautier, C. A., ... Kimura, M. (2010). PINK1 stabilized by mitochondrial depolarization recruits Parkin to damaged mitochondria and activates latent Parkin for mitophagy. *The Journal of Cell Biology, 189*(2), 211–221. doi:10.1083/jcb.200910140 PMID:20404107

McCoy, M. K., Martinez, T. N., Ruhn, K. A., Szymkowski, D. E., Smith, C. G., Botterman, B. R., ... Tansey, M. G. (2006). Blocking soluble tumor necrosis factor signaling with dominant-negative tumor necrosis factor inhibitor attenuates loss of dopaminergic neurons in models of Parkinson's disease. *The Journal of Neuroscience: The Official Journal of the Society for Neuroscience, 26*(37), 9365–9375. doi:10.1523/JNEUROSCI.1504-06.2006 PMID:16971520

Meske, V., Albert, F., & Ohm, T. G. (2008). Coupling of mammalian target of rapamycin with phosphoinositide 3-kinase signaling pathway regulates protein phosphatase 2a-and glycogen synthase kinase-3β-dependent phosphorylation of tau. *The Journal of Biological Chemistry, 283*(1), 100–109. doi:10.1074/jbc.M704292200 PMID:17971449

Miller, B. C., Eckman, E. A., Sambamurti, K., Dobbs, N., Chow, K. M., Eckman, C. B., ... Thiele, D. L. (2003). Amyloid-β peptide levels in brain are inversely correlated with insulysin activity levels in vivo. *Proceedings of the National Academy of Sciences of the United States of America, 100*(10), 6221–6226. doi:10.1073/pnas.1031520100 PMID:12732730

Mizushima, N., & Komatsu, M. (2011). Autophagy: Renovation of cells and tissues. *Cell, 147*(4), 728–741. doi:10.1016/j.cell.2011.10.026 PMID:22078875

Mizushima, N., & Yoshimori, T. (2007). How to interpret LC3 immunoblotting. *Autophagy, 3*(6), 542–545. doi:10.4161/auto.4600 PMID:17611390

Moloney, A. M., Griffin, R. J., Timmons, S., O'Connor, R., Ravid, R., & O'Neill, C. (2010). Defects in IGF-1 receptor, insulin receptor and IRS-1/2 in Alzheimer's disease indicate possible resistance to IGF-1 and insulin signalling. *Neurobiology of Aging*, *31*(2), 224–243. doi:10.1016/j.neurobiolaging.2008.04.002 PMID:18479783

Morales-Corraliza, J., Wong, H., Mazzella, M. J., Che, S., Lee, S. H., Petkova, E., ... Mathews, P. M. (2016). Brain-wide insulin resistance, tau phosphorylation changes, and hippocampal neprilysin and amyloid-β alterations in a monkey model of type 1 diabetes. *The Journal of Neuroscience: The Official Journal of the Society for Neuroscience*, *36*(15), 4248–4258. doi:10.1523/JNEUROSCI.4640-14.2016 PMID:27076423

Moreno, H., Morfini, G., Buitrago, L., Ujlaki, G., Choi, S., Yu, E., ... Sugimori, M. (2016). Tau pathology-mediated presynaptic dysfunction. *Neuroscience*, *325*, 30–38. doi:10.1016/j.neuroscience.2016.03.044 PMID:27012611

Moriyasu, Y., & Inoue, Y. (2008). Chapter Thirty-Two Use of Protease Inhibitors for Detecting Autophagy in Plants. *Methods in Enzymology*, *451*, 557–580. doi:10.1016/S0076-6879(08)03232-1 PMID:19185740

Morris, M., Maeda, S., Vossel, K., & Mucke, L. (2011). The many faces of tau. *Neuron*, *70*(3), 410–426. doi:10.1016/j.neuron.2011.04.009 PMID:21555069

Mosconi, L. (2005). Brain glucose metabolism in the early and specific diagnosis of Alzheimer's disease. *European Journal of Nuclear Medicine and Molecular Imaging*, *32*(4), 486–510. doi:10.100700259-005-1762-7 PMID:15747152

Moussaud, S., Jones, D. R., Moussaud-Lamodière, E. L., Delenclos, M., Ross, O. A., & McLean, P. J. (2014). Alpha-synuclein and tau: Teammates in neurodegeneration? *Molecular Neurodegeneration*, *9*(1), 43. doi:10.1186/1750-1326-9-43 PMID:25352339

Mukrasch, M. D., Bibow, S., Korukottu, J., Jeganathan, S., Biernat, J., Griesinger, C., ... Zweckstetter, M. (2009). Structural polymorphism of 441-residue tau at single residue resolution. *PLoS Biology*, *7*(2), e1000034. doi:10.1371/journal.pbio.1000034 PMID:19226187

Müller, R., Heinrich, M., Heck, S., Blohm, D., & Richter-Landsberg, C. (1997). Expression of microtubule-associated proteins MAP2 and tau in cultured rat brain oligodendrocytes. *Cell and Tissue Research*, *288*(2), 239–249. doi:10.1007004410050809 PMID:9082959

Muntane, G., Dalfo, E., Martinez, A., & Ferrer, I. (2008). Phosphorylation of tau and α-synuclein in synaptic-enriched fractions of the frontal cortex in Alzheimer's disease, and in Parkinson's disease and related α-synucleinopathies. *Neuroscience*, *152*(4), 913–923. doi:10.1016/j.neuroscience.2008.01.030 PMID:18343584

Nakatogawa, H., Suzuki, K., Kamada, Y., & Ohsumi, Y. (2009). Dynamics and diversity in autophagy mechanisms: Lessons from yeast. *Nature Reviews. Molecular Cell Biology*, *10*(7), 458–467. doi:10.1038/nrm2708 PMID:19491929

Narendra, D., Tanaka, A., Suen, D. F., & Youle, R. J. (2008). Parkin is recruited selectively to impaired mitochondria and promotes their autophagy. *The Journal of Cell Biology*, *183*(5), 795–803. doi:10.1083/jcb.200809125 PMID:19029340

Nave, B. T., Ouwens, D. M., Withers, D. J., Alessi, D. R., & Shepherd, P. R. (1999). Mammalian target of rapamycin is a direct target for protein kinase B: Identification of a convergence point for opposing effects of insulin and amino-acid deficiency on protein translation. *The Biochemical Journal*, *344*(2), 427–431. doi:10.1042/bj3440427 PMID:10567225

Nixon, R. A. (2007). Autophagy, amyloidogenesis and Alzheimer disease. *Journal of Cell Science*, *120*(23), 4081–4091. doi:10.1242/jcs.019265 PMID:18032783

Nixon, R. A., Wegiel, J., Kumar, A., Yu, W. H., Peterhoff, C., Cataldo, A., & Cuervo, A. M. (2005). Extensive involvement of autophagy in Alzheimer disease: An immuno-electron microscopy study. *Journal of Neuropathology and Experimental Neurology*, *64*(2), 113–122. doi:10.1093/jnen/64.2.113 PMID:15751225

Nixon, R. A., Wegiel, J., Kumar, A., Yu, W. H., Peterhoff, C., Cataldo, A., & Cuervo, A. M. (2005). Extensive involvement of autophagy in Alzheimer disease: An immuno-electron microscopy study. *Journal of Neuropathology and Experimental Neurology*, *64*(2), 113–122. doi:10.1093/jnen/64.2.113 PMID:15751225

Nowacek, A., Kosloski, L. M., & Gendelman, H. E. (2009). Neurodegenerative disorders and nanoformulated drug development. *Nanomedicine (London)*, *4*(5), 541–555. doi:10.2217/nnm.09.37 PMID:19572820

Nübling, G., Bader, B., Levin, J., Hildebrandt, J., Kretzschmar, H., & Giese, A. (2012). Synergistic influence of phosphorylation and metal ions on tau oligomer formation and coaggregation with α-synuclein at the single molecule level. *Molecular Neurodegeneration*, *7*(1), 35. doi:10.1186/1750-1326-7-35 PMID:22824345

O'Neill, C., Kiely, A. P., Coakley, M. F., Manning, S., & Long-Smith, C. M. (2012). *Insulin and IGF-1 signalling: longevity, protein homoeostasis and Alzheimer's disease*. Academic Press.

Oaks, A. W., Frankfurt, M., Finkelstein, D. I., & Sidhu, A. (2013). Age-dependent effects of A53T alpha-synuclein on behavior and dopaminergic function. *PLoS One*, *8*(4), e60378. doi:10.1371/journal.pone.0060378 PMID:23560093

Obara, K., & Ohsumi, Y. (2008). Dynamics and function of PtdIns (3) P in autophagy. *Autophagy*, *4*(7), 952–954. doi:10.4161/auto.6790 PMID:18769109

Obara, K., & Ohsumi, Y. (2008). Dynamics and function of PtdIns (3) P in autophagy. *Autophagy*, *4*(7), 952–954. doi:10.4161/auto.6790 PMID:18769109

Ohsumi, Y., & Mizushima, N. (2004). Two ubiquitin-like conjugation systems essential for autophagy. *Seminars in Cell & Developmental Biology*, *2*(15), 231–236. doi:10.1016/j.semcdb.2003.12.004 PMID:15209383

Oliva, O., Réz, G., Pálfia, Z., & Fellinger, E. (1992). Dynamics of vinblastine-induced autophagocytosis in murine pancreatic acinar cells: Influence of cycloheximide post-treatments. *Experimental and Molecular Pathology*, *56*(1), 76–86. doi:10.1016/0014-4800(92)90025-7 PMID:1547871

Oskarsson, M. E., Paulsson, J. F., Schultz, S. W., Ingelsson, M., Westermark, P., & Westermark, G. T. (2015). In vivo seeding and cross-seeding of localized amyloidosis: A molecular link between type 2 diabetes and Alzheimer disease. *American Journal of Pathology, 185*(3), 834–846. doi:10.1016/j.ajpath.2014.11.016 PMID:25700985

Oz-Levi, D., Ben-Zeev, B., Ruzzo, E. K., Hitomi, Y., Gelman, A., Pelak, K., ... Alkelai, A. (2012). Mutation in TECPR2 reveals a role for autophagy in hereditary spastic paraparesis. *American Journal of Human Genetics, 91*(6), 1065–1072. doi:10.1016/j.ajhg.2012.09.015 PMID:23176824

Ozcelik, S., Fraser, G., Castets, P., Schaeffer, V., Skachokova, Z., Breu, K., ... Tolnay, M. (2013). Rapamycin attenuates the progression of tau pathology in P301S tau transgenic mice. *PLoS One, 8*(5), e62459. doi:10.1371/journal.pone.0062459 PMID:23667480

Pallas-Bazarra, N., Jurado-Arjona, J., Navarrete, M., Esteban, J. A., Hernández, F., Ávila, J., & Llorens-Martín, M. (2016). Novel function of Tau in regulating the effects of external stimuli on adult hippocampal neurogenesis. *The EMBO Journal, 35*(13), 1417–1436. doi:10.15252/embj.201593518 PMID:27198172

Panda, P. K., Fahrner, A., Vats, S., Seranova, E., Sharma, V., Chipara, M., ... Sarkar, S. (2019). Chemical screening approaches enabling drug discovery of autophagy modulators for biomedical applications in human diseases. *Frontiers in Cell and Developmental Biology, 7*, 7. doi:10.3389/fcell.2019.00038 PMID:30949479

Papasozomenos, S. C., & Binder, L. I. (1987). Phosphorylation determines two distinct species of tau in the central nervous system. *Cell Motility and the Cytoskeleton, 8*(3), 210–226. doi:10.1002/cm.970080303 PMID:2446784

Park, W. S., Do Heo, W., Whalen, J. H., O'Rourke, N. A., Bryan, H. M., Meyer, T., & Teruel, M. N. (2008). Comprehensive identification of PIP3-regulated PH domains from C. elegans to H. sapiens by model prediction and live imaging. *Molecular Cell, 30*(3), 381–392. doi:10.1016/j.molcel.2008.04.008 PMID:18471983

Pattingre, S., Tassa, A., Qu, X., Garuti, R., Liang, X. H., Mizushima, N., ... Levine, B. (2005). Bcl-2 antiapoptotic proteins inhibit Beclin 1-dependent autophagy. *Cell, 122*(6), 927–939. doi:10.1016/j.cell.2005.07.002 PMID:16179260

Pei, J. J., Tanaka, T., Tung, Y. C., Braak, E. V. A., Iqbal, K., & Grundke-Iqbal, I. (1997). Distribution, levels, and activity of glycogen synthase kinase-3 in the Alzheimer disease brain. *Journal of Neuropathology and Experimental Neurology, 56*(1), 70–78. doi:10.1097/00005072-199701000-00007 PMID:8990130

Perez, S. E., He, B., Nadeem, M., Wuu, J., Scheff, S. W., Abrahamson, E. E., ... Mufson, E. J. (2015). Resilience of precuneus neurotrophic signaling pathways despite amyloid pathology in prodromal Alzheimer's disease. *Biological Psychiatry, 77*(8), 693–703. doi:10.1016/j.biopsych.2013.12.016 PMID:24529280

Perez-Nievas, B. G., Stein, T. D., Tai, H. C., Dols-Icardo, O., Scotton, T. C., Barroeta-Espar, I., ... Serrano-Pozo, A. (2013). Dissecting phenotypic traits linked to human resilience to Alzheimer's pathology. *Brain, 136*(8), 2510–2526. doi:10.1093/brain/awt171 PMID:23824488

Pickford, F., Masliah, E., Britschgi, M., Lucin, K., Narasimhan, R., Jaeger, P. A., ... Wyss-Coray, T. (2008). The autophagy-related protein beclin 1 shows reduced expression in early Alzheimer disease and regulates amyloid β accumulation in mice. *The Journal of Clinical Investigation, 118*(6), 2190–2199. doi:10.1172/JCI33585 PMID:18497889

Piras, A., Collin, L., Grüninger, F., Graff, C., & Rönnbäck, A. (2016). Autophagic and lysosomal defects in human tauopathies: Analysis of post-mortem brain from patients with familial Alzheimer disease, corticobasal degeneration and progressive supranuclear palsy. *Acta Neuropathologica Communications, 4*(1), 22. doi:10.118640478-016-0292-9 PMID:26936765

Plastaras, J. P., Dorsey, J. F., Carroll, K., Kim, S. H., Birnbaum, M. J., & El-Deiry, W. S. (2008). Role of PI3K/Akt signaling in TRAIL-and radiation-induced gastrointestinal apoptosis. *Cancer Biology & Therapy, 7*(12), 2047–2053. doi:10.4161/cbt.7.12.7570 PMID:19106634

Pooler, A. M., & Hanger, D. P. (2010). *Functional implications of the association of tau with the plasma membrane*. Academic Press.

Potter, C. J., Pedraza, L. G., & Xu, T. (2002). Akt regulates growth by directly phosphorylating Tsc2. *Nature Cell Biology, 4*(9), 658–665. doi:10.1038/ncb840 PMID:12172554

Poulopoulos, M., Levy, O. A., & Alcalay, R. N. (2012). The neuropathology of genetic Parkinson's disease. *Movement Disorders, 27*(7), 831–842. doi:10.1002/mds.24962 PMID:22451330

Puri, C., Renna, M., Bento, C. F., Moreau, K., & Rubinsztein, D. C. (2013). Diverse autophagosome membrane sources coalesce in recycling endosomes. *Cell, 154*(6), 1285–1299. doi:10.1016/j.cell.2013.08.044 PMID:24034251

Pyo, J. O., Jang, M. H., Kwon, Y. K., Lee, H. J., Jun, J. I., Woo, H. N., ... Mizushima, N. (2005). Essential roles of Atg5 and FADD in autophagic cell death dissection of autophagic cell death into vacuole formation and cell death. *The Journal of Biological Chemistry, 280*(21), 20722–20729. doi:10.1074/jbc.M413934200 PMID:15778222

Qiu, W. Q., Walsh, D. M., Ye, Z., Vekrellis, K., Zhang, J., Podlisny, M. B., ... Selkoe, D. J. (1998). Insulin-degrading enzyme regulates extracellular levels of amyloid β-protein by degradation. *The Journal of Biological Chemistry, 273*(49), 32730–32738. doi:10.1074/jbc.273.49.32730 PMID:9830016

Rankin, C. A., Sun, Q., & Gamblin, T. C. (2007). Tau phosphorylation by GSK-3β promotes tangle-like filament morphology. *Molecular Neurodegeneration, 2*(1), 12. doi:10.1186/1750-1326-2-12 PMID:17598919

Ravikumar, B., Acevedo-Arozena, A., Imarisio, S., Berger, Z., Vacher, C., O'Kane, C. J., ... Rubinsztein, D. C. (2005). Dynein mutations impair autophagic clearance of aggregate-prone proteins. *Nature Genetics, 37*(7), 771–776. doi:10.1038/ng1591 PMID:15980862

Ravikumar, B., Duden, R., & Rubinsztein, D. C. (2002). Aggregate-prone proteins with polyglutamine and polyalanine expansions are degraded by autophagy. *Human Molecular Genetics, 11*(9), 1107–1117. doi:10.1093/hmg/11.9.1107 PMID:11978769

Ravikumar, B., Vacher, C., Berger, Z., Davies, J. E., Luo, S., Oroz, L. G., ... Rubinsztein, D. C. (2004). Inhibition of mTOR induces autophagy and reduces toxicity of polyglutamine expansions in fly and mouse models of Huntington disease. *Nature Genetics*, *36*(6), 585–595. doi:10.1038/ng1362 PMID:15146184

Reiman, E. M., Caselli, R. J., Yun, L. S., Chen, K., Bandy, D., Minoshima, S., ... Osborne, D. (1996). Preclinical evidence of Alzheimer's disease in persons homozygous for the ε4 allele for apolipoprotein E. *The New England Journal of Medicine*, *334*(12), 752–758. doi:10.1056/NEJM199603213341202 PMID:8592548

Renvoisé, B., Chang, J., Singh, R., Yonekawa, S., FitzGibbon, E. J., Mankodi, A., ... Mahuran, D. J. (2014). Lysosomal abnormalities in hereditary spastic paraplegia types SPG 15 and SPG 11. *Annals of Clinical and Translational Neurology*, *1*(6), 379–389. doi:10.1002/acn3.64 PMID:24999486

Rodríguez-Martín, T., Cuchillo-Ibáñez, I., Noble, W., Nyenya, F., Anderton, B. H., & Hanger, D. P. (2013). Tau phosphorylation affects its axonal transport and degradation. *Neurobiology of Aging*, *34*(9), 2146–2157. doi:10.1016/j.neurobiolaging.2013.03.015 PMID:23601672

Rodríguez-Muela, N., Koga, H., García-Ledo, L., de la Villa, P., De la Rosa, E. J., Cuervo, A. M., & Boya, P. (2013). Balance between autophagic pathways preserves retinal homeostasis. *Aging Cell*, *12*(3), 478–488. doi:10.1111/acel.12072 PMID:23521856

Rohn, T. T., Wirawan, E., Brown, R. J., Harris, J. R., Masliah, E., & Vandenabeele, P. (2011). Depletion of Beclin-1 due to proteolytic cleavage by caspases in the Alzheimer's disease brain. *Neurobiology of Disease*, *43*(1), 68–78. doi:10.1016/j.nbd.2010.11.003 PMID:21081164

Rong, Y., McPhee, C. K., Deng, S., Huang, L., Chen, L., Liu, M., ... Lenardo, M. J. (2011). Spinster is required for autophagic lysosome reformation and mTOR reactivation following starvation. *Proceedings of the National Academy of Sciences of the United States of America*, *108*(19), 7826–7831. doi:10.1073/pnas.1013800108 PMID:21518918

Roriz-Filho, J. S., Sa-Roriz, T. M., Rosset, I., Camozzato, A. L., Santos, A. C., Chaves, M. L., ... Roriz-Cruz, M. (2009). (Pre) diabetes, brain aging, and cognition. *Biochimica et Biophysica Acta (BBA)-. Molecular Basis of Disease*, *1792*(5), 432–443. doi:10.1016/j.bbadis.2008.12.003

Rose, C., Menzies, F. M., Renna, M., Acevedo-Arozena, A., Corrochano, S., Sadiq, O., ... Rubinsztein, D. C. (2010). Rilmenidine attenuates toxicity of polyglutamine expansions in a mouse model of Huntington's disease. *Human Molecular Genetics*, *19*(11), 2144–2153. doi:10.1093/hmg/ddq093 PMID:20190273

Rousseaux, M. W., de Haro, M., Lasagna-Reeves, C. A., De Maio, A., Park, J., Jafar-Nejad, P., ... Vilanova-Velez, L. (2016). TRIM28 regulates the nuclear accumulation and toxicity of both alpha-synuclein and tau. *eLife*, *5*, e19809. doi:10.7554/eLife.19809 PMID:27779468

Rubinsztein, D. C., Shpilka, T., & Elazar, Z. (2012). Mechanisms of autophagosome biogenesis. *Current Biology*, *22*(1), R29–R34. doi:10.1016/j.cub.2011.11.034 PMID:22240478

Ruivo, R., Anne, C., Sagné, C., & Gasnier, B. (2009). Molecular and cellular basis of lysosomal transmembrane protein dysfunction. *Biochimica et Biophysica Acta (BBA)-. Molecular Cell Research*, *1793*(4), 636–649. PMID:19146888

Saftig, P., & Klumperman, J. (2009). Lysosome biogenesis and lysosomal membrane proteins: Trafficking meets function. *Nature Reviews. Molecular Cell Biology*, *10*(9), 623–635. doi:10.1038/nrm2745 PMID:19672277

Sahu, R., Kaushik, S., Clement, C. C., Cannizzo, E. S., Scharf, B., Follenzi, A., ... Santambrogio, L. (2011). Microautophagy of cytosolic proteins by late endosomes. *Developmental Cell*, *20*(1), 131–139. doi:10.1016/j.devcel.2010.12.003 PMID:21238931

Sarbassov, D. D., Ali, S. M., Kim, D. H., Guertin, D. A., Latek, R. R., Erdjument-Bromage, H., ... Sabatini, D. M. (2004). Rictor, a novel binding partner of mTOR, defines a rapamycin-insensitive and raptor-independent pathway that regulates the cytoskeleton. *Current Biology*, *14*(14), 1296–1302. doi:10.1016/j.cub.2004.06.054 PMID:15268862

Sarkar, S., Davies, J. E., Huang, Z., Tunnacliffe, A., & Rubinsztein, D. C. (2007). Trehalose, a novel mTOR-independent autophagy enhancer, accelerates the clearance of mutant huntingtin and α-synuclein. *The Journal of Biological Chemistry*, *282*(8), 5641–5652. doi:10.1074/jbc.M609532200 PMID:17182613

Sarkar, S., Davies, J. E., Huang, Z., Tunnacliffe, A., & Rubinsztein, D. C. (2007). Trehalose, a novel mTOR-independent autophagy enhancer, accelerates the clearance of mutant huntingtin and α-synuclein. *The Journal of Biological Chemistry*, *282*(8), 5641–5652. doi:10.1074/jbc.M609532200 PMID:17182613

Sarkar, S., Floto, R. A., Berger, Z., Imarisio, S., Cordenier, A., Pasco, M., ... Rubinsztein, D. C. (2005). Lithium induces autophagy by inhibiting inositol monophosphatase. *The Journal of Cell Biology*, *170*(7), 1101–1111. doi:10.1083/jcb.200504035 PMID:16186256

Sarkar, S., Ravikumar, B., Floto, R. A., & Rubinsztein, D. C. (2009). Rapamycin and mTOR-independent autophagy inducers ameliorate toxicity of polyglutamine-expanded huntingtin and related proteinopathies. *Cell Death and Differentiation*, *16*(1), 46–56. doi:10.1038/cdd.2008.110 PMID:18636076

Sarkar, S., Ravikumar, B., Floto, R. A., & Rubinsztein, D. C. (2009). Rapamycin and mTOR-independent autophagy inducers ameliorate toxicity of polyglutamine-expanded huntingtin and related proteinopathies. *Cell Death and Differentiation*, *16*(1), 46–56. doi:10.1038/cdd.2008.110 PMID:18636076

Schneider, J. A., Aggarwal, N. T., Barnes, L., Boyle, P., & Bennett, D. A. (2009). The neuropathology of older persons with and without dementia from community versus clinic cohorts. *Journal of Alzheimer's Disease*, *18*(3), 691–701. doi:10.3233/JAD-2009-1227 PMID:19749406

Schneider, J. A., Li, J. L., Li, Y., Wilson, R. S., Kordower, J. H., & Bennett, D. A. (2006). Substantia nigra tangles are related to gait impairment in older persons. *Annals of Neurology*, *59*(1), 166–173. doi:10.1002/ana.20723 PMID:16374822

Schneider, J. L., Suh, Y., & Cuervo, A. M. (2014). Deficient chaperone-mediated autophagy in liver leads to metabolic dysregulation. *Cell Metabolism*, *20*(3), 417–432. doi:10.1016/j.cmet.2014.06.009 PMID:25043815

Sengupta, A., Kabat, J., Novak, M., Wu, Q., Grundke-Iqbal, I., & Iqbal, K. (1998). Phosphorylation of tau at both Thr 231 and Ser 262 is required for maximal inhibition of its binding to microtubules. *Archives of Biochemistry and Biophysics*, *357*(2), 299–309. doi:10.1006/abbi.1998.0813 PMID:9735171

Sereno, L., Coma, M., Rodriguez, M., Sanchez-Ferrer, P., Sanchez, M. B., Gich, I., ... Clarimon, J. (2009). A novel GSK-3β inhibitor reduces Alzheimer's pathology and rescues neuronal loss in vivo. *Neurobiology of Disease*, *35*(3), 359–367. doi:10.1016/j.nbd.2009.05.025 PMID:19523516

Shao, Y., Gao, Z., Feldman, T., & Jiang, X. (2007). Stimulation of ATG12-ATG5 conjugation by ribonucleic acid. *Autophagy*, *3*(1), 10–16. doi:10.4161/auto.3270 PMID:16963840

Sherrington, R., Rogaev, E. I., Liang, Y. A., Rogaeva, E. A., Levesque, G., Ikeda, M., ... Tsuda, T. (1995). Cloning of a gene bearing missense mutations in early-onset familial Alzheimer's disease. *Nature*, *375*(6534), 754–760. doi:10.1038/375754a0 PMID:7596406

Shimada, K., Motoi, Y., Ishiguro, K., Kambe, T., Matsumoto, S. E., Itaya, M., ... Mizuno, Y. (2012). Long-term oral lithium treatment attenuates motor disturbance in tauopathy model mice: Implications of autophagy promotion. *Neurobiology of Disease*, *46*(1), 101–108. doi:10.1016/j.nbd.2011.12.050 PMID:22249108

Silvestri, L., Caputo, V., Bellacchio, E., Atorino, L., Dallapiccola, B., Valente, E. M., & Casari, G. (2005). Mitochondrial import and enzymatic activity of PINK1 mutants associated to recessive parkinsonism. *Human Molecular Genetics*, *14*(22), 3477–3492. doi:10.1093/hmg/ddi377 PMID:16207731

Sinha, S. C., Colbert, C. L., Becker, N., Wei, Y., & Levine, B. (2008). Molecular basis of the regulation of Beclin 1-dependent autophagy by the γ-herpesvirus 68 Bcl-2 homolog M11. *Autophagy*, *4*(8), 989–997. doi:10.4161/auto.6803 PMID:18797192

Sladek, R., Rocheleau, G., Rung, J., Dina, C., Shen, L., Serre, D., ... Balkau, B. (2007). A genome-wide association study identifies novel risk loci for type 2 diabetes. *Nature*, *445*(7130), 881–885. doi:10.1038/nature05616 PMID:17293876

Small, G. W., Ercoli, L. M., Silverman, D. H., Huang, S. C., Komo, S., Bookheimer, S. Y., ... Mazziotta, J. C. (2000). Cerebral metabolic and cognitive decline in persons at genetic risk for Alzheimer's disease. *Proceedings of the National Academy of Sciences of the United States of America*, *97*(11), 6037–6042. doi:10.1073/pnas.090106797 PMID:10811879

Soni, A., Akcakanat, A., Singh, G., Luyimbazi, D., Zheng, Y., Kim, D., ... Meric-Bernstam, F. (2008). eIF4E knockdown decreases breast cancer cell growth without activating Akt signaling. *Molecular Cancer Therapeutics*, *7*(7), 1782–1788. doi:10.1158/1535-7163.MCT-07-2357 PMID:18644990

Sottejeau, Y., Bretteville, A., Cantrelle, F. X., Malmanche, N., Demiaute, F., Mendes, T., ... Dourlen, P. (2015). Tau phosphorylation regulates the interaction between BIN1's SH3 domain and Tau's proline-rich domain. *Acta Neuropathologica Communications*, *3*(1), 58. doi:10.118640478-015-0237-8 PMID:26395440

Sperbera, B. R., Leight, S., Goedert, M., & Lee, V. Y. (1995). Glycogen synthase kinase-3β phosphorylates tau protein at multiple sites in intact cells. *Neuroscience Letters*, *197*(2), 149–153. doi:10.1016/0304-3940(95)11902-9 PMID:8552282

Sperling, R. A., Aisen, P. S., Beckett, L. A., Bennett, D. A., Craft, S., Fagan, A. M., ... Park, D. C. (2011). Toward defining the preclinical stages of Alzheimer's disease: Recommendations from the National Institute on Aging-Alzheimer's Association workgroups on diagnostic guidelines for Alzheimer's disease. *Alzheimer's & Dementia*, *7*(3), 280–292. doi:10.1016/j.jalz.2011.03.003 PMID:21514248

Spillantini, M. G., Crowther, R. A., Jakes, R., Cairns, N. J., Lantos, P. L., & Goedert, M. (1998). Filamentous α-synuclein inclusions link multiple system atrophy with Parkinson's disease and dementia with Lewy bodies. *Neuroscience Letters*, *251*(3), 205–208. doi:10.1016/S0304-3940(98)00504-7 PMID:9726379

Spillantini, M. G., Schmidt, M. L., Lee, V. M. Y., Trojanowski, J. Q., Jakes, R., & Goedert, M. (1997). α-Synuclein in Lewy bodies. *Nature*, *388*(6645), 839–840. doi:10.1038/42166 PMID:9278044

Spilman, P., Podlutskaya, N., Hart, M. J., Debnath, J., Gorostiza, O., Bredesen, D., ... Galvan, V. (2010). Inhibition of mTOR by rapamycin abolishes cognitive deficits and reduces amyloid-β levels in a mouse model of Alzheimer's disease. *PLoS One*, *5*(4), e9979. doi:10.1371/journal.pone.0009979 PMID:20376313

Srodulski, S., Sharma, S., Bachstetter, A. B., Brelsfoard, J. M., Pascual, C., Xie, X. S., ... Despa, F. (2014). Neuroinflammation and neurologic deficits in diabetes linked to brain accumulation of amylin. *Molecular Neurodegeneration*, *9*(1), 30. doi:10.1186/1750-1326-9-30 PMID:25149184

Stadel, D., Millarte, V., Tillmann, K. D., Huber, J., Tamin-Yecheskel, B. C., Akutsu, M., ... Dötsch, V. (2015). TECPR2 cooperates with LC3C to regulate COPII-dependent ER export. *Molecular Cell*, *60*(1), 89–104. doi:10.1016/j.molcel.2015.09.010 PMID:26431026

Steen, E., Terry, B. M., & Rivera, E., Cannon, J. L., Neely, T. R., Tavares, R., ... de la Monte, S. M. (2005). Impaired insulin and insulin-like growth factor expression and signaling mechanisms in Alzheimer's disease–is this type 3 diabetes? *Journal of Alzheimer's Disease*, *7*(1), 63–80. doi:10.3233/JAD-2005-7107 PMID:15750215

Strachan, M. W. J. (2011). RD Lawrence Lecture 2010^. The brain as a target organ in Type 2 diabetes: Exploring the links with cognitive impairment and dementia. *Diabetic Medicine*, *28*(2), 141–147. doi:10.1111/j.1464-5491.2010.03199.x PMID:21219420

Strachan, M. W. J. (2011). RD Lawrence Lecture 2010^. The brain as a target organ in Type 2 diabetes: Exploring the links with cognitive impairment and dementia. *Diabetic Medicine*, *28*(2), 141–147. doi:10.1111/j.1464-5491.2010.03199.x PMID:21219420

Sun, Y. X., Ji, X., Mao, X., Xie, L., Jia, J., Galvan, V., ... Jin, K. (2014). Differential activation of mTOR complex 1 signaling in human brain with mild to severe Alzheimer's disease. *Journal of Alzheimer's Disease*, *38*(2), 437–444. doi:10.3233/JAD-131124 PMID:23979023

Surridge, C. D., & Burns, R. G. (1994). The difference in the binding of phosphatidylinositol distinguishes MAP2 from MAP2C and Tau. *Biochemistry*, *33*(26), 8051–8057. doi:10.1021/bi00192a009 PMID:8025110

Tanaka, K., Suzuki, T., Hattori, N., & Mizuno, Y. (2004). Ubiquitin, proteasome and parkin. *Biochimica et Biophysica Acta (BBA)-. Molecular Cell Research*, *1695*(1-3), 235–247.

Tashiro, K., Hasegawa, M., Ihara, Y., & Iwatsubo, T. (1997). Somatodendritic localization of phosphorylated tau in neonatal and adult rat cerebral cortex. *Neuroreport*, *8*(12), 2797–2801. doi:10.1097/00001756-199708180-00029 PMID:9295120

Tian, Y., Bustos, V., Flajolet, M., & Greengard, P. (2011). A small-molecule enhancer of autophagy decreases levels of Aβ and APP-CTF via Atg5-dependent autophagy pathway. *The FASEB Journal*, *25*(6), 1934–1942. doi:10.1096/fj.10-175158 PMID:21368103

Toker, A. (2012). Phosphoinositide 3-kinases—A historical perspective. In I. Phosphoinositides (Ed.), *Enzymes of Synthesis and Degradation* (pp. 95–110). Dordrecht: Springer.

Toker, A., & Cantley, L. C. (1997). Signalling through the lipid products of phosphoinositide-3-OH kinase. *Nature*, *387*(6634), 673–676. doi:10.1038/42648 PMID:9192891

Tong, Y., Yamaguchi, H., Giaime, E., Boyle, S., Kopan, R., Kelleher, R. J., & Shen, J. (2010). Loss of leucine-rich repeat kinase 2 causes impairment of protein degradation pathways, accumulation of α-synuclein, and apoptotic cell death in aged mice. *Proceedings of the National Academy of Sciences of the United States of America*, *107*(21), 9879–9884. doi:10.1073/pnas.1004676107 PMID:20457918

Tooze, S. A., & Yoshimori, T. (2010). The origin of the autophagosomal membrane. *Nature Cell Biology*, *12*(9), 831–835. doi:10.1038/ncb0910-831 PMID:20811355

Tramutola, A., Triplett, J. C., Di Domenico, F., Niedowicz, D. M., Murphy, M. P., Coccia, R., ... Butterfield, D. A. (2015). Alteration of mTOR signaling occurs early in the progression of Alzheimer disease (AD): Analysis of brain from subjects with pre-clinical AD, amnestic mild cognitive impairment and late-stage AD. *Journal of Neurochemistry*, *133*(5), 739–749. doi:10.1111/jnc.13037 PMID:25645581

Uddin, M., Stachowiak, A., Mamun, A. A., Tzvetkov, N. T., Takeda, S., Atanasov, A. G., ... Stankiewicz, A. M. (2018). Autophagy and Alzheimer's disease: From molecular mechanisms to therapeutic implications. *Frontiers in Aging Neuroscience*, *10*, 4. doi:10.3389/fnagi.2018.00004 PMID:29441009

Usardi, A., Pooler, A. M., Seereeram, A., Reynolds, C. H., Derkinderen, P., Anderton, B., ... Williamson, R. (2011). Tyrosine phosphorylation of tau regulates its interactions with Fyn SH2 domains, but not SH3 domains, altering the cellular localization of tau. *The FEBS Journal*, *278*(16), 2927–2937. doi:10.1111/j.1742-4658.2011.08218.x PMID:21692989

Valente, E. M., Abou-Sleiman, P. M., Caputo, V., Muqit, M. M., Harvey, K., Gispert, S., ... Albanese, A. (2004). Hereditary early-onset Parkinson's disease caused by mutations in PINK1. *Science*, *304*(5674), 1158–1160. doi:10.1126cience.1096284 PMID:15087508

van der Heide, L. P., Ramakers, G. M., & Smidt, M. P. (2006). Insulin signaling in the central nervous system: Learning to survive. *Progress in Neurobiology*, *79*(4), 205–221. doi:10.1016/j.pneurobio.2006.06.003 PMID:16916571

Vanderweyde, T., Apicco, D. J., Youmans-Kidder, K., Ash, P. E., Cook, C., da Rocha, E. L., ... Ivanov, P. (2016). Interaction of tau with the RNA-binding protein TIA1 regulates tau pathophysiology and toxicity. *Cell Reports*, *15*(7), 1455–1466. doi:10.1016/j.celrep.2016.04.045 PMID:27160897

Vanhaesebroeck, B., & Waterfield, M. D. (1999). Signaling by distinct classes of phosphoinositide 3-kinases. *Experimental Cell Research, 253*(1), 239–254. doi:10.1006/excr.1999.4701 PMID:10579926

Vantaggiato, C., Crimella, C., Airoldi, G., Polishchuk, R., Bonato, S., Brighina, E., ... Santorelli, F. M. (2013). Defective autophagy in spastizin mutated patients with hereditary spastic paraparesis type 15. *Brain, 136*(10), 3119–3139. doi:10.1093/brain/awt227 PMID:24030950

Varga, R. E., Khundadze, M., Damme, M., Nietzsche, S., Hoffmann, B., Stauber, T., ... Kessels, M. M. (2015). In vivo evidence for lysosome depletion and impaired autophagic clearance in hereditary spastic paraplegia type SPG11. *PLOS Genetics, 11*(8), e1005454. doi:10.1371/journal.pgen.1005454 PMID:26284655

Vekrellis, K., Xilouri, M., Emmanouilidou, E., Rideout, H. J., & Stefanis, L. (2011). Pathological roles of α-synuclein in neurological disorders. *Lancet Neurology, 10*(11), 1015–1025. doi:10.1016/S1474-4422(11)70213-7 PMID:22014436

Vekrellis, K., Ye, Z., Qiu, W. Q., Walsh, D., Hartley, D., Chesneau, V., ... Selkoe, D. J. (2000). Neurons regulate extracellular levels of amyloid β-protein via proteolysis by insulin-degrading enzyme. *The Journal of Neuroscience: The Official Journal of the Society for Neuroscience, 20*(5), 1657–1665. doi:10.1523/JNEUROSCI.20-05-01657.2000 PMID:10684867

Vingtdeux, V., Chandakkar, P., Zhao, H., d'Abramo, C., Davies, P., & Marambaud, P. (2011). Novel synthetic small-molecule activators of AMPK as enhancers of autophagy and amyloid-β peptide degradation. *The FASEB Journal, 25*(1), 219–231. doi:10.1096/fj.10-167361 PMID:20852062

Virkamäki, A., Ueki, K., & Kahn, C. R. (1999). Protein–protein interaction in insulin signaling and the molecular mechanisms of insulin resistance. *The Journal of Clinical Investigation, 103*(7), 931–943. doi:10.1172/JCI6609 PMID:10194465

Vogiatzi, T., Xilouri, M., Vekrellis, K., & Stefanis, L. (2008). Wild type α-synuclein is degraded by chaperone-mediated autophagy and macroautophagy in neuronal cells. *The Journal of Biological Chemistry, 283*(35), 23542–23556. doi:10.1074/jbc.M801992200 PMID:18566453

Von Bergen, M., Friedhoff, P., Biernat, J., Heberle, J., Mandelkow, E. M., & Mandelkow, E. (2000). Assembly of τ protein into Alzheimer paired helical filaments depends on a local sequence motif (306VQIVYK311) forming β structure. *Proceedings of the National Academy of Sciences of the United States of America, 97*(10), 5129–5134. doi:10.1073/pnas.97.10.5129 PMID:10805776

Vonsattel, J. P. G., & DiFiglia, M. (1998). Huntington disease. *Journal of Neuropathology and Experimental Neurology, 57*(5), 369. doi:10.1097/00005072-199805000-00001 PMID:9596408

Vuono, R., Winder-Rhodes, S., de Silva, R., Cisbani, G., Drouin-Ouellet, J., Spillantini, M. G., ... Barker, R. A. (2015). The role of tau in the pathological process and clinical expression of Huntington's disease. *Brain, 138*(7), 1907–1918. doi:10.1093/brain/awv107 PMID:25953777

Wade-Martins, R. (2012). Genetics: The MAPT locus—a genetic paradigm in disease susceptibility. *Nature Reviews. Neurology, 8*(9), 477–478. doi:10.1038/nrneurol.2012.169 PMID:22940644

Wang, J. Z., Grundke-Iqbal, I., & Iqbal, K. (2007). Kinases and phosphatases and tau sites involved in Alzheimer neurofibrillary degeneration. *The European Journal of Neuroscience*, *25*(1), 59–68. doi:10.1111/j.1460-9568.2006.05226.x PMID:17241267

Wang, L., Harris, T. E., & Lawrence, J. C. Jr. (2008). Regulation of proline-rich Akt substrate of 40 kDa (PRAS40) function by mammalian target of rapamycin complex 1 (mTORC1)-mediated phosphorylation. *The Journal of Biological Chemistry*, *283*(23), 15619–15627. doi:10.1074/jbc.M800723200 PMID:18372248

Wang, L., Harris, T. E., & Lawrence, J. C. Jr. (2008). Regulation of proline-rich Akt substrate of 40 kDa (PRAS40) function by mammalian target of rapamycin complex 1 (mTORC1)-mediated phosphorylation. *The Journal of Biological Chemistry*, *283*(23), 15619–15627. doi:10.1074/jbc.M800723200 PMID:18372248

Wang, X., Campbell, L. E., Miller, C. M., & Proud, C. G. (1998). Amino acid availability regulates p70 S6 kinase and multiple translation factors. *The Biochemical Journal*, *335*(3), 711–711. doi:10.1042/bj3350711w PMID:9693128

Webb, J. L., Ravikumar, B., Atkins, J., Skepper, J. N., & Rubinsztein, D. C. (2003). α-Synuclein is degraded by both autophagy and the proteasome. *The Journal of Biological Chemistry*, *278*(27), 25009–25013. doi:10.1074/jbc.M300227200 PMID:12719433

Weidberg, H., Shvets, E., Shpilka, T., Shimron, F., Shinder, V., & Elazar, Z. (2010). LC3 and GATE-16/GABARAP subfamilies are both essential yet act differently in autophagosome biogenesis. *The EMBO Journal*, *29*(11), 1792–1802. doi:10.1038/emboj.2010.74 PMID:20418806

White, E., & DiPaola, R. S. (2009). The double-edged sword of autophagy modulation in cancer. *Clinical Cancer Research*, *15*(17), 5308–5316. doi:10.1158/1078-0432.CCR-07-5023 PMID:19706824

Whitman, M., Downes, C. P., Keeler, M., Keller, T., & Cantley, L. (1988). Type I phosphatidylinositol kinase makes a novel inositol phospholipid, phosphatidylinositol-3-phosphate. *Nature*, *332*(6165), 644–646. doi:10.1038/332644a0 PMID:2833705

Wills, J., Credle, J., Haggerty, T., Lee, J. H., Oaks, A. W., & Sidhu, A. (2011). Tauopathic changes in the striatum of A53T α-synuclein mutant mouse model of Parkinson's disease. *PLoS One*, *6*(3), e17953. doi:10.1371/journal.pone.0017953 PMID:21445308

Yaffe, K., Blackwell, T., Kanaya, A. M., Davidowitz, N., Barrett-Connor, E., & Krueger, K. (2004). Diabetes, impaired fasting glucose, and development of cognitive impairment in older women. *Neurology*, *63*(4), 658–663. doi:10.1212/01.WNL.0000134666.64593.BA PMID:15326238

Yaffe, K., Blackwell, T., Whitmer, R. A., Krueger, K., & Barrett-Connor, E. (2006). Glycosylated hemoglobin level and development of mild cognitive impairment or dementia in older women. *The Journal of Nutrition, Health & Aging*, *10*(4), 293. PMID:16886099

Yamaguchi, K., Cochran, E. J., Murrell, J. R., Polymeropoulos, M. H., Shannon, K. M., Crowther, R. A., ... Ghetti, B. (2005). Abundant neuritic inclusions and microvacuolar changes in a case of diffuse Lewy body disease with the A53T mutation in the α-synuclein gene. *Acta Neuropathologica*, *110*(3), 298–305. doi:10.100700401-005-1042-4 PMID:15981014

Yan, L. M., Velkova, A., & Kapurniotu, A. (2014). Molecular characterization of the hetero-assembly of β-amyloid peptide with islet amyloid polypeptide. *Current Pharmaceutical Design*, *20*(8), 1182–1191. doi:10.2174/13816128113199990064 PMID:23713771

Yang, A. J., Chandswangbhuvana, D., Margol, L., & Glabe, C. G. (1998). Loss of endosomal/lysosomal membrane impermeability is an early event in amyloid Aβ1-42 pathogenesis. *Journal of Neuroscience Research*, *52*(6), 691–698. doi:10.1002/(SICI)1097-4547(19980615)52:6<691::AID-JNR8>3.0.CO;2-3 PMID:9669318

Yang, D. S., Stavrides, P., Mohan, P. S., Kaushik, S., Kumar, A., Ohno, M., ... Pawlik, M. (2010). Reversal of autophagy dysfunction in the TgCRND8 mouse model of Alzheimer's disease ameliorates amyloid pathologies and memory deficits. *Brain*, *134*(1), 258–277. doi:10.1093/brain/awq341 PMID:21186265

Yao, X. Q., Zhang, X. X., Yin, Y. Y., Liu, B., Luo, D. J., Liu, D., ... Wang, J. Z. (2011). Glycogen synthase kinase-3β regulates Tyr307 phosphorylation of protein phosphatase-2A via protein tyrosine phosphatase 1B but not Src. *The Biochemical Journal*, *437*(2), 335–344. doi:10.1042/BJ20110347 PMID:21554241

Yen, W. L., Shintani, T., Nair, U., Cao, Y., Richardson, B. C., Li, Z., ... Klionsky, D. J. (2010). The conserved oligomeric Golgi complex is involved in double-membrane vesicle formation during autophagy. *The Journal of Cell Biology*, *188*(1), 101–114. doi:10.1083/jcb.200904075 PMID:20065092

Yoon, S. Y., & Kim, D. H. (2016). Alzheimer's disease genes and autophagy. *Brain Research*, *1649*, 201–209. doi:10.1016/j.brainres.2016.03.018 PMID:27016058

Young, K. F., Pasternak, S. H., & Rylett, R. J. (2009). Oligomeric aggregates of amyloid β peptide 1–42 activate ERK/MAPK in SH-SY5Y cells via the α7 nicotinic receptor. *Neurochemistry International*, *55*(8), 796–801. doi:10.1016/j.neuint.2009.08.002 PMID:19666073

Yousefi, S., & Simon, H. U. (2009). Autophagy in cancer and chemotherapy. In *Death Receptors and Cognate Ligands in Cancer* (pp. 183–190). Berlin: Springer. doi:10.1007/400_2008_25

Yu, L., McPhee, C. K., Zheng, L., Mardones, G. A., Rong, Y., Peng, J., ... Hailey, D. W. (2010). Termination of autophagy and reformation of lysosomes regulated by mTOR. *Nature*, *465*(7300), 942–946. doi:10.1038/nature09076 PMID:20526321

Yu, W. H., Cuervo, A. M., Kumar, A., Peterhoff, C. M., Schmidt, S. D., Lee, J. H., ... Jiang, Y. (2005). Macroautophagy—A novel β-amyloid peptide-generating pathway activated in Alzheimer's disease. *The Journal of Cell Biology*, *171*(1), 87–98. doi:10.1083/jcb.200505082 PMID:16203860

Yue, Z., Jin, S., Yang, C., Levine, A. J., & Heintz, N. (2003). Beclin 1, an autophagy gene essential for early embryonic development, is a haploinsufficient tumor suppressor. *Proceedings of the National Academy of Sciences of the United States of America*, *100*(25), 15077–15082. doi:10.1073/pnas.2436255100 PMID:14657337

Zeggini, E., Weedon, M. N., Lindgren, C. M., Frayling, T. M., Elliott, K. S., Lango, H., ... Barrett, J. C. (2007). Replication of genome-wide association signals in UK samples reveals risk loci for type 2 diabetes. *Science*, *316*(5829), 1336–1341. doi:10.1126cience.1142364 PMID:17463249

Zhao, Z., Xiang, Z., Haroutunian, V., Buxbaum, J. D., Stetka, B., & Pasinetti, G. M. (2007). Insulin degrading enzyme activity selectively decreases in the hippocampal formation of cases at high risk to develop Alzheimer's disease. *Neurobiology of Aging*, *28*(6), 824–830. doi:10.1016/j.neurobiolaging.2006.05.001 PMID:16769157

Zheng, L., Terman, A., Hallbeck, M., Dehvari, N., Cowburn, R. F., Benedikz, E., ... Marcusson, J. (2011). Macroautophagy-generated increase of lysosomal amyloid β-protein mediates oxidant-induced apoptosis of cultured neuroblastoma cells. *Autophagy*, *7*(12), 1528–1545. doi:10.4161/auto.7.12.18051 PMID:22108004

Zheng, S., Clabough, E. B., Sarkar, S., Futter, M., Rubinsztein, D. C., & Zeitlin, S. O. (2010). Deletion of the huntingtin polyglutamine stretch enhances neuronal autophagy and longevity in mice. *PLOS Genetics*, *6*(2), e1000838. doi:10.1371/journal.pgen.1000838 PMID:20140187

Zhong, Y., Wang, Q. J., Li, X., Yan, Y., Backer, J. M., Chait, B. T., ... Yue, Z. (2009). Distinct regulation of autophagic activity by Atg14L and Rubicon associated with Beclin 1–phosphatidylinositol-3-kinase complex. *Nature Cell Biology*, *11*(4), 468–476. doi:10.1038/ncb1854 PMID:19270693

Zinzalla, V., Stracka, D., Oppliger, W., & Hall, M. N. (2011). Activation of mTORC2 by association with the ribosome. *Cell*, *144*(5), 757–768. doi:10.1016/j.cell.2011.02.014 PMID:21376236

Chapter 3
Molecular Mechanisms Underlying the Role of Mitophagy in Neurodegeneration

Abhilasha Ahlawat
Department of Pharmaceutical Sciences, Maharshi Dayanand University, Rohtak, India

Meenakshi Kaira
Department of Pharmaceutical Sciences, Maharshi Dayanand University, Rohtak, India

Vaibhav Walia
Faculty of Pharmacy, DIT University, Dehradun, India

Munish Garg
Department of Pharmaceutical Sciences, Maharshi Dayanand University, Rohtak, India

ABSTRACT

Mitophagy is a selective autophagy process in which damaged or surplus mitochondria are removed to sustain normal homeostasis. The efficient removal of damaged or stressed mitochondria is crucial for cellular health. Recent literature emphasizes the role of PINK1-Parkin pathways in the pathogenesis process of various neurodegenerative disorders. Further, mitophagy has shown potential therapeutic activity in treating neurodegenerative diseases. Thus, mitophagy might be important in the field of pharmacotherapeutics. In the present chapter, the authors explain mitophagy, mitophagy signaling pathways, as well as mechanisms and roles of mitophagy in various neurodegenerative disorders.

DOI: 10.4018/978-1-7998-1317-0.ch003

INTRODUCTION

Mitochondria are the organelle mainly involved in the production of energy, regulation of metabolism and cell death (Ding and Yin, 2012). The process of removal of defected/damaged mitochondria from the cell is known as mitophagy. Mitophagy is related to the autophagy of mitochondria and was first termed by Lemasters (Lemasters, 2005). Mitophagy plays a crucial role in mitochondrial disruption and neurodegenerative disorders (Ding and Yin, 2012). In mitophagy, the mitochondria to be removed are packed into autophagosomes followed by the fusion with lysosomes and leading to complete mitochondrial degradation (Xie et al., 2011). Pathways involved in the regulation of mitophagy are classified either as ubiquitin-dependent or ubiquitin-independent (Khaminets et al., 2016). Mitophagy is further classified as basal (progressive mitochondrial housekeeping that recycles old and damaged organelles), stress-induced (extracellular stress signals affecting mitochondrial physiology which may trigger mitochondrial clearance) or programmed (gets activated in various cell types during development) (Sekine and Youle, 2018). Mitophagy impairment alters mitochondrial function and leads to aggregation of nonfunctioning organelles further causing cellular and tissue damage (Durcan and Fon, 2015). Various stimuli contribute to mitophagy through various signaling pathways in different cellular aspects (Palikaras et al., 2017). Basal mitophagy is known to be independent of PINK1 in metabolically active tissues (McWilliams et al., 2018). Mitophagy mediated by starvation or hypoxia is partially independent of the typical macro-autophagy process (Hirota et al., 2015).

Neurons are found to be more vulnerable to mitophagic impairment as seen in various neurodegenerative disorders (Martinez-Vicente, 2017). Alzheimer's disease (AD), Parkinson's disease (PD), Huntington's disease (HD) and amyotrophic lateral sclerosis (ALS) share common pathology marked by the neuronal loss which may further lead to various physical and cognitive dysfunction in patients. These disorders have aggregated neurotoxic proteins and dysfunctional mitochondria in common that alter the cellular homeostasis and neuronal function. Removal of these neurotoxic proteins and dysfunctional organelles is important for neuronal homeostasis. Any impairment in the removal process alters neuronal homeostasis and plays a role in the development of these neurodegenerative disorders (Rodolfo et al., 2018).

Mitochondrial dysregulation and defective mitophagy have been associated with neurodegenerative disorders. Mutations in the PINK1 and Parkin genes are responsible for PD (Valente et al., 2004). In AD, amyloid beta-derived diffusible ligands cause mitochondrial splitting and mitophagy (Ryu et al., 2015). In HD, the mutant huntingtin is known to induce mitophagy (Khalil et al., 2015). In ALS, mitochondrial dysfunction is seen where ubiquitinated mitochondria's targeting to autophagosomes is reduced further contributing to the development of ALS (Benatar, 2007).

So, the proper degradation of dysfunctional mitochondria through mitophagy is important for maintaining mitochondrial quality and quantity in neurons. Thus, mitophagy has found to play a crucial role in aforementioned disorders (Martinez-Vicente, 2017). So, in the present chapter, authors describe the multifaced role of mitophagy in neurodegenerative disorders.

REGULATION OF MITOPHAGY

Mitophagy removes the defected mitochondria and these defected mitochondria need tagging to get recognized by lysosomes. This function is performed by some special classes of proteins including Uth1, Aup1p, and Atg32 (Kanki et al., 2009) (Figure 1). It is suggested that the protein Atg32 aggregates on

Figure 1. Signaling pathways and mitophagy regulation.
PINK 1: phosphatase and tensin homolog (PTEN)-induced kinase 1; USP: universal stress protein; Smurf 1: SMAD-specific E3 ubiquitin ligase protein 1; Mul 1: mitochondrial E3 ubiquitin protein ligase 1; March 5: membrane associated ring-CH-type finger 5; Ambra 1: Activating molecule in Beclin 1 regulated autophagy protein 1; Bnip 3: BCL2 Interacting Protein 3; Fundc 1: FUN14 domain containing protein 1.

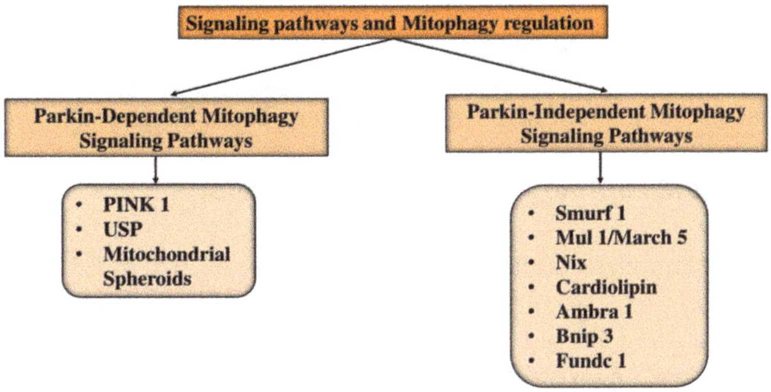

the outer mitochondrial membrane in deprived conditions and interacts with Atg8 (Kanki et al., 2009). Atg32 gets phosphorylated at Ser 114 and 119 position by casein kinase 2, then N-terminal of Atg32 binds to the C-terminal of Atg11 and Atg11 further connects to Dnm1, which is a mitochondrial fission protein recruiting fission complex to mitochondria (Kanki et al., 2013).

Parkin Dependent Mitophagy Signaling Pathways

Phosphatase and tensin homolog (PTEN)-induced kinase 1 (Pink1) activates and recruits Parkin on mitochondria. Overexpressed Pink1 is responsible for the translocation of Parkin to mitochondria (Kawajiri et al., 2010). Pink1 enters the mitochondria with the help of a transporter present in the inner mitochondrial membrane. Pink1 is degraded in the mitochondria (Greene et al., 2003). Truncated Pink1 gets liberated into the cytosol and further undergoes degradation by proteasomes (Yamano and Youle, 2013). Pink1 fragments present in the cytosol block Parkin's translocation to mitochondria (Fedorowicz et al., 2014).

Parkin, an E3-ubiquitin ligase encoded by the *PARK2* gene, is found in the spleen, brain, thymus, kidney, liver, heart and muscle (Ding and Yin 2012). Mitophagy induced by Parkin is regulated by deubiquitination (Narendra et al., 2008). Pink1 phosphorylates Parkin and Ser65 which results in recruiting and binding Parkin to mitochondria, increase in Parkin action and ubiquitination of Parkin-induced outer mitochondrial membrane protein. Parkin incorporates ubiquitin chains on outer mitochondrial membrane proteins after activation and Pink1 phosphorylates these chains and results in Parkin recruitment and activation (Shiba-Fukushima et al., 2014).

Ubiquitin-specific peptidase (USP) 30 and USP 15 removes ubiquitin from outer mitochondrial membrane proteins that were earlier ubiquitinated by Parkin (Cornelissen et al., 2014). Further, Parkin gets ubiquitinated to undergo its deterioration (Zhang et al., 2000). Parkin causes the ubiquitination of mitofusin (Mfn) 1/2 responsible for their proteasomal deterioration and the splitting of mitochondria (Gegg et al., 2010). The ubiquitination of Miro1 separates damaged and healthy mitochondria to promote the enveloping of distorted mitochondria by an autophagosome (Wang et al., 2011). So, both, Pink1 and

Parkin, stimulate the removal of damaged mitochondria and Pink1 causes outer mitochondrial membrane protein ubiquitination and induction of mitophagy.

Mitochondrial spheroids are ring or cup-like structured mitochondria having the same appearance as that of autophagosomes and are produced in oxidative stress conditions. Their production is dependent on Mfn1 and Mfn2 fusion proteins. They are known to enwrap various cellular contents including endoplasmic reticulum, lipid droplets or mitochondria. They are also known to degrade lysosomal protein contents (Ding et al., 2012 a,b). Parkin is known to inhibit mitochondrial spheroid generation by commencing proteasomal deterioration of Mfn1 and Mfn2 (Ding et al., 2012 a,b).

Parkin-Independent Mitophagy Signaling Pathways

Mitophagy is also induced through Parkin-independent pathways exerted by E3 ubiquitin ligases including Smurf1, Mul1/March5, Nix, cardiolipin, Ambra1, Bcl2/adenovirus E1B 19 kDa interacting protein 3 (Bnip3), Fun 14 domain containing 1 (Fundc1). Bnip3 interacts with Pink1 to block its proteolytic action and promotes its aggregation on the mitochondrial membrane for recruiting Parkin on distorted mitochondria (Zhang et al., 2016).

Nix causes induction of mitophagy in cells undergoing higher oxidative phosphorylation leading to raised reactive oxygen species (ROS) production, dysfunction of mitochondria and cell death. Nix promotes the ROS generation which is further known to activate autophagy and may also act as a substrate of Parkin in Parkin-dependent mitophagy (Gao et al., 2015).

Hypoxia promotes Fundc1 and March5 interaction responsible for Fundc1 ubiquitination and degeneration and thereby protects mitochondria from mitophagic destruction. Hypoxic conditions dephosphorylate Fundc1 and increase its interaction with microtubule-associated protein 1 light chain 3 (LC3) and promote mitophagy (Chen et al., 2017). Fundc1 also promotes mitochondrial fragmentation by interacting with Drp1. Phosphorylation of Fundc1 mediates mitochondrial fission, fusion along with mitophagy (Chen et al., 2016).

Smurf1, an E3 ubiquitin ligase, does not exert a direct action on mitophagy directly but C2 domain of Smurf1 is necessary for engulfing degraded mitochondria through autophagosomes. In various studies, it is demonstrated that Smurf1 deficient mice have shown the presence of aggregated distorted mitochondria in the liver, brain, and heart (Orvedahl et al., 2011).

Mul1, an E3 ligase, ubiquitinate Mfn proteins. Mul1 action is found to be independent of Parkin as overexpression or knockdown in Mul1 or Parkin-expressing HeLa cells does not affect the translocation of Parkin to depolarized mitochondria (Yun et al., 2014).

Cardiolipin is a phospholipid of the inner mitochondrial membrane (Ren et al., 2014). Upon depolarization of mitochondria, cardiolipin translocates to the outer mitochondrial membrane (Chu et al., 2013). Translocation of cardiolipin results in the activation of pathways responsible for apoptosis and mitophagy (Ren et al., 2014).

Ambra1 activates mitophagy, independent of Parkin. Ambra1 interacts with LC3 and is involved in both Parkin-dependent and independent mitophagy (Williams and Ding, 2018). Thus, the mutations in Ambra1 signaling block mitophagy (Strappazzon et al., 2015).

Table 1. Proteins associated with neurodegenerative disorders which play role in mitophagy.

Neurodegenerative Disorders	Proteins Involved in Mitophagy
Alzheimer's disease	Tau, PINK 1/Parkin, Sirtuins
Parkinson's disease	PINK 1/Parkin, α-Syn, DJ-1, LRRK 2, p62
Huntington's disease	PINK 1/Parkin, GAPDH, TG 2, VCP, HAP 1, DRP 1
Amyotrophic lateral sclerosis	OPTN, TBK 1, SOD 1, VCP, TDP-43

PATHOPHYSIOLOGICAL ROLES OF MITOPHAGY

Mitochondria play a crucial role in adult neuroplasticity (Mattson et al., 2008). Brain-derived neurotrophic factor (BDNF) plays a critical role in hippocampal synaptic plasticity, learning, memory and stress resistance in neurons. BDNF also stimulates PGC-1α expression and mitochondrial biogeny (Cheng et al., 2012). Defective BDNF signaling has shown to be involved in synaptic dysfunction and neuronal degradation in AD (Marosi and Mattson, 2014). Mitochondrial protein causes deacetylation of sirtuin 3 (SIRT3) which is upregulated in a neuronal activity-dependent manner in response to exercise and further regulates Ca^{2+} levels at glutamatergic synapses to enhance neuroplasticity and cellular stress resistance (Cheng et al., 2016). Mitochondria are also known to regulate Ca^{2+} levels, as disturbed neuronal Ca^{2+} levels cause neuronal death in various neurodegenerative disorders (Bezprozvanny and Mattson, 2008). Damaged mitochondria may also trigger caspase-9-dependent neuronal cell death by releasing cytochrome c. Thus, maintaining a healthy mitochondrial level is necessary for neuronal health and this, various mitochondrial quality control pathways including misfolded protein degradation, fission, fusion and engulfment and deterioration of degraded mitochondria exists (Cai and Tammineni, 2016). Therefore, mitochondria serve a crucial part in neuron development, functioning, maintenance and survival (Kerr et al., 2017). A number of protein and protein complexes are linked with neurodegeneration which play role in mitophagy as shown in Table 1.

Mitophagy and Aging

Mitochondria are the primary source of ROS (Sena and Chandel, 2012). Mitochondrial functionality and integrity are important for neuronal survival and activity (Rugarli and Langer, 2012). Neuronal cells require more mitochondria to meet higher energy demand. Mitochondria constantly undergo recurrent fission and fusion episodes. Various constituents of the fission/fusion events are associated with numerous neurological disorders that underlie the importance of mitochondrial dynamics in neuronal homeostasis as these events remove damaged mitochondrial constituents by diffusing in the mitochondrial network and also cause fragmentation and isolation of effective mitochondria before mitophagy (Waterham et al., 2007). Excessive fusion alters mitochondrial dynamics and mitophagy processes thereby preventing autophagial degradation of mitochondrial by starvation-induced autophagy (Rambold et al., 2011). Increased fission or decreased fusion by modulating mitochondrial dynamics promote the isolation of disrupted mitochondria and subsequently eliminates through mitophagy. Therefore, mitochondrial damage due to dysregulation in mitophagy is associated with the onset and progression of various age-related neurodegenerative disorders (Batlevi and La Spada, 2011). Mitochondrial defects in aging cause a decrease in the ability to survive stress, functional impairment, and death with time (Navarro and Boveris,

Figure 2. Effect of aging on mitophagy.
MFN: mitofusin; Drp 1: dynamin-related protein 1; Fis 1: mitochondrial fission 1 protein; Mgm 1: mitochondrial genome maintainance 1 protein.

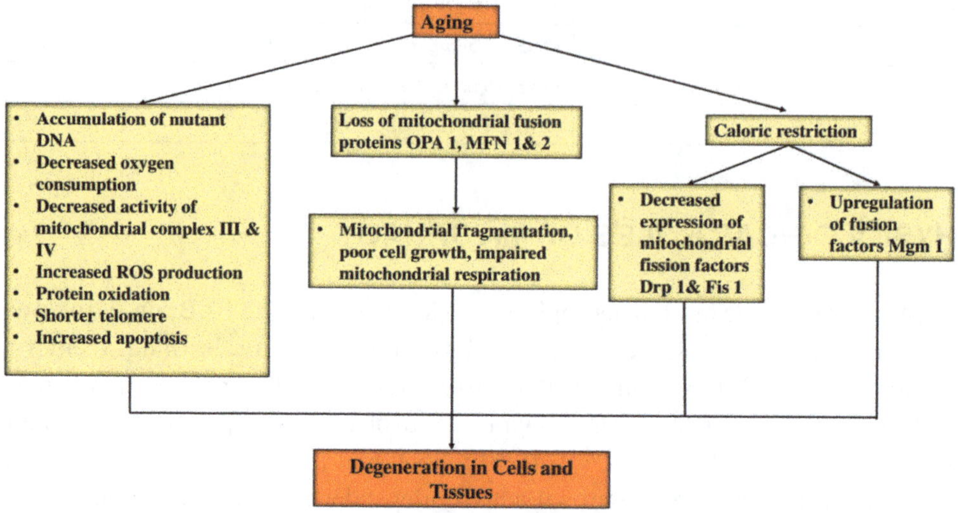

2004). Further, the mutations in the mtDNA during aging are characterized by reduced autophagic activity (Hubbard et al., 2012). Caloric restriction may lead to longevity due to increased removal of defective mitochondria by inducing autophagy (Yen and Klionsky, 2008). Aging also increases the risk of neurodegeneration and thus, neurodegenerative disorders are more prominent in older age individuals. Figure 2 represents the impact of aging on mitophagy.

Mitophagy and Alzheimer's Disease

Alzheimer's disease (AD) is a chronic degenerative disorder of central nervous system (CNS) characterized by impaired cognition and behavior and pathological changes involving formation and accumulation of senile plaques-containing beta-amyloid (Aβ) and neurofibrillary tangles (NFTs), neuronal and synaptic loss (Chen, 2018). The disoriented fusion between autophagosomes and lysosomes causes mitophagy defects in AD (Nixon, 2013). Aggregation in autophagosomes in cortical neurons is due to oxidative stress associated with AD (Boland et al., 2008). Sirtuins are nicotinamide adenine dinucleotide (NAD+)-dependent histone deacetylases that regulate various metabolic pathways involved in survival, differentiation, metabolism and cell death (Carafa et al, 2016).

Sirtuins have been implicated in the neuronal degeneration process through the regulation of the enzyme poly (ADP-ribose) polymerase (PARP) (Alano et al, 2004) and the activation of PARP1 results in the neuronal death as seen in preclinical models of the PD (Outeiro et al, 2007). There are seven mammalian SIRTs, out of which nuclear SIRT1 and mitochondrial SIRT3 are responsible for neuroprotection and the reduced level of these proteins causes neurodegenerative disorders (Fang et al., 2016).

SIRT1, a nuclear protein, is widely expressed in neurons (Ramadori et al, 2008). SIRT1 deacetylates autophagy genes and stimulates autophagy (Lee et al, 2008) required for the elimination of toxic altered protein deposits accumulated in neurodegenerative disorders. It is reported that the overexpression of SIRT1 suppresses the production of α-synuclein deposits (Donmez et al, 2012). Further SIRT1 interacts with p53 and deacetylates p53 and thus reduces the transcriptional activation of p53 (Vaziri et al, 2001) and thus inhibits the apoptosis induced by DNA damage or oxidative stress. Vitamin E is known to stimulate SIRT1 and thus reduce oxidative damage caused by a high-fat and high-sugar diet (Wu et al, 2006).

SIRT2 regulates the deacetylation of microtubules (North et al, 2003), condensation of chromatin during metaphase (Vaquero et al, 2006) and cell-cycle progression (Dryden et al, 2003). SIRT2 also interact with p65 of NF-κB family member, deacetylate it at lysine 310 (Rothgiesser et al, 2010). Thus, the antagonism of the SIRT1 through the deacetylation of p65 results in the inhibition of NF-κB activity (Yeung et al, 2004). Further, SIRT2 is known to promote neuronal death and therefore the inhibition of SIRT2 has shown neuronal protection against α-synuclein toxicity both in vitro and in flies (Outeiro et al, 2007). Pharmacological inhibition of SIRT2 has shown to promote α-synuclein Lewy body-like inclusions (Outeiro et al, 2007). Thus, SIRT2 protects against neurodegenerative disorders (de Oliveira et al, 2012).

SIRT3, localized in inner mitochondrial membrane cristae and matrix (He et al, 2012), regulates β-oxidation of fatty acid, ATP generation, ROS formation (Giblin et al, 2014). SIRT3 reacts with NAD+ to produce nicotinamide (Onyango et al, 2002) which is further converted to NAD+ by nicotinamide phosphoribosyltransferase (Rongvaux et al, 2002). SIRT3 regulates various metabolic enzymes through deacetylation. SIRT3 deacetylate and activate acetyl-CoA synthetase 2 which converts acetate to acetyl-CoA, which further participates in the tricarboxylic acid cycle (TCA) cycle to produce energy.

SIRT3 activates the enzyme glutamate dehydrogenase (GDH) which catalyzes the formation of α-ketoglutarate from glutamate and thus the loss of SIRT3 function affect TCA cycle and mitochondrial oxidation mechanism which further results in the increased production of the ROS and its downstream signaling pathways (Shih and Donmez, 2013). Increased mitochondrial ROS production has been implicated in age-related neuronal degeneration (Singh, 2006). ROS mediates neuronal death via the opening of the mitochondrial permeability transition pore (mtPTP) leading to neuronal cell death (Du and Yan, 2010). It is further suggested that the formation of mtPTP results in the rapid outflow of NAD+ from the mitochondria (Lemasters et al, 2009). SIRT3 suppresses mtPTP formation (Hafner et al, 2010) and thus significantly enhances neuronal lifespan by inhibiting neuronal cell death (Weir et al, 2012). SIRT3 has been shown to provide neuronal protection in HD (Fu et al, 2012).

SIRT4 can affect metabolic function in the cell (Shih and Donmez, 2013). SIRT4 inhibits mitochondrial glutamine metabolism (Miyo et al, 2015) by inhibiting glutamate dehydrogenase (GDH) activity (Haigis et al, 2006) and uptake of glutamine (Csibi et al, 2013). Further, the inhibition of glutamine metabolism leads to the phosphorylation and activation of p53 responsible for preventing genomic instability and SIRT4 inhibit mitochondrial glutamine metabolism followed by cell cycle arrest making it a critical regulator of genome fidelity (Reid et al, 2013).

Brain tissues of AD patients show decreased SIRT1 protein levels (Figure 3) in the parietal cortex and it is related to aggregation of Aβ and tau tangles (Julien et al., 2009). SIRT1 is known to upregulate PGC-1α and exert a critical role in the induction of mitophagy through deacetylation and activation of some key proteins involved in autophagy including ATG5, ATG7, ATG8/LC3, PINK1 stabilization, and upregulation of mitophagy receptors Nix/BNIP3L and LC3 (Fang et al., 2016). SIRT3 activates FOXO3 to generate an autophagy p62 protein, accumulating on ubiquitinated mitochondrial substrates thereby forming autolysosomes (Tseng et al., 2013). SIRT3 levels are found to be decreased in APP/PS1 double mutant AD mice neurons (Yang et al., 2015). SIRT3 protects mitochondria and neurons from excitotoxicity, metabolic stress and cell death caused by SOD2 and cyclophilin D deacetylation-dependent manner (Cheng et al., 2016). So, impairment in SIRT activity may cause abnormal mitochondrial function and

Figure 3. Mitophagic dysfunction in the pathogenesis of neurodegenerative disorders.
PINK1: phosphatase and tensin homolog (PTEN)-induced kinase 1; LRRK2: leucine rich repeat kinase 2; DJ-1: protein deglycase 1; SIRT: sirtuin 1; Aβ: amyloid beta; VCP: valosin-containing protein; TBK1: TANK binding kinase 1; SOD1: superoxide dismutase; TDP-43: transactive response DNA binding protein-43; GADPH: glyceraldehyde-3-phosphate dehydrogenase.

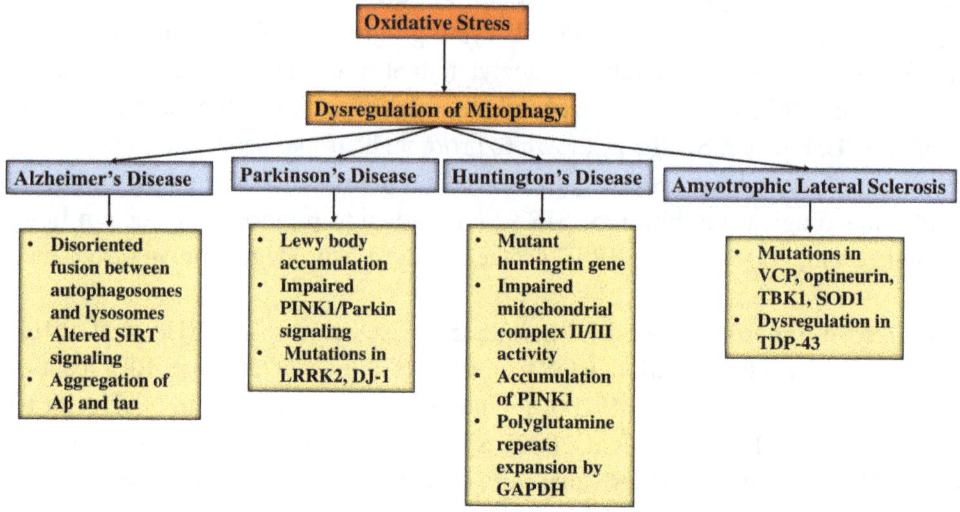

blockade of mitophagy and further leads to aggregation of disrupted mitochondria, Aβ plaques and tau tangles (Kerr et al., 2017).

Mitophagy and Parkinson's Disease

Mitochondria are a dynamic organelle that undergoes fusion, fission, and mitophagy (Ni and Williams, 2015) and these processes are important for normal neuronal functions (McInnes, 2013). Greater number of enlarged, swollen mitochondria denotes mitochondrial dysfunction in Parkinson's disease (PD) (Poole et al, 2008).

PINK1-Parkin signaling has been found altered in mitochondrial impairment (Sterky et al, 2011). Under normal conditions, PINK1 is transferred to the inner mitochondrial membrane from the outer membrane. PINK1 is degraded by presenilin-associated rhomboid-like protein (PARL, a protease of the inner membrane) and the ubiquitin-proteasome system. Thus, a very small undetectable amount of PINK1 is found to be present in mitochondria. However, the loss of potential of mitochondrial membranes and increased aggregation of the misfolded proteins had led to the stabilization of PINK1 on the outer mitochondrial membrane. The accumulation of PINK1 in the outer membrane is the sign of elimination of the mitochondria (Pickrell and Youle, 2015). PINK1 may either act directly or indirectly. The indirect action of the PINK1 involves the activation of Omi/HtrA2, a serine protease that activates pro-apoptotic proteins. Inhibition of this protease prevents cell death in response to oxidative stress (Dagda and Chu, 2009). PINK1 activates the Parkin present in the cytoplasm. PINK1 phosphorylate mitofusin2 (Mfn2) at Thr111 and Ser442 for binding of Parkin to the mitochondria for its removal through mitophagy (Pickrell and Youle, 2015). Parkin is a ubiquitin E3 ligase that causes degradation of misfolded proteins and prevents cell apoptosis (Yoshii et al, 2011). Parkin acts as a protective agent, prevents mitochondrial swelling, kainic acid-induced excitotoxicity in neurons, manganese-induced cell death and protects against dopaminergic and α-synuclein toxicity. Parkin ubiquitinates the synphilin-1(α-synuclein interacting protein) and this led to the formation of protein aggregate and the Parkin mediated ubiquitination initiates the degradation of the ubiquitinated substance by the ubiquitin-proteasome system, thereby, suggesting that Parkin is involved in inclusion formation and autophagic clearance of targeted proteins (Dawson and Dawson, 2010). Further, Parkin is responsible for maintaining normal functioning of mitochondria and mit-DNA (Kuroda et al, 2006), decreased ROS formation (Kuroda et al, 2006), protection of mtDNA from damage (Rothfuss et al, 2009), regulation of mitochondrial fission and fusion mechanism (Narendra et al., 2008) and mutations in the gene encodes for Parkin responsible for PD (Miklya et al, 2014). After activation of Parkin, PINK1 phosphorylates Parkin at Ser65 in the N-terminal UBL domain (Truban et al, 2016). Further, PINK1 phosphorylates the substrate i.e. ubiquitin forms phospho-ubiquitin that confers Parkin to regulate distinct downstream targets. The downstream targets of Parkin are VDAC1 ubiquitination and SQSTM1/p62 (p62) involved in the process of mitophagy. Parkin appears to ubiquitinate VDAC1 that plays a minor role in mitophagy (Narendra et al, 2010). Parkin mediates the recruitment of p62, a shuttle protein that transports ubiquitinated proteins for degradation (Narendra et al, 2010). Dysregulation of p62 results in the accumulation of protein aggregates that further produce neuronal damage (Song et al, 2016). UPS and autophagy are the major pathways involved in the degradation of the damaged proteins and the increase in the age reduced the activities and efficiency of these pathways (Rubinsztein et al, 2011). It is reported that the PD is characterized by the reduced proteasome activity in SN region (McNaught et al, 2003).

Parkin mutations have been linked with the PD (Ebrahimi-Fakhari et al, 2012), early onset of PD (Klein and Westenberger, 2012), oxidative stress-mediated by α-synuclein aggregation (Bendor et al. 2013), structural changes in the mitochondria (Botella et al, 2008) decreased activity of complex-I of electron transport chain (ETC) and ROS production (Parihar et al, 2009) increased mitochondrial membrane permeability and leakage of pro-apoptotic molecules (Bandopadhyay and de Belleroche, 2010), finally resulting in death of neurons.

In the neurons, LRRK2 colocalizes with dynamin like protein 1 (DLP1) and induces mitochondrial fragmentation and enhanced ROS levels (Niu et al, 2012). LRRK2 also regulates several cellular functions such as regulation of neuritic outgrowth and morphology (Li et al, 2014). LRRK2 regulates nuclear factor kappa-light-chain-enhancer of activated B cells (NF-κB) activity responsible for the activation of tumor necrosis factor (TNF) mediated extrinsic pathway of cell death and inhibition of this pathway may decrease neuronal death in PD (Gardet et al, 2010). LRRK2 genetic mutations are linked with the onset of autosomal dominant PD in later ages (Klein and Westenberger, 2012).

DJ-1, a cytosolic protein in the brain, is known to provide neuroprotection against ROS (Björkblom et al, 2013). Any alterations in the normal functioning of cysteine residue (particularly C106) of DJ-1 leads to the complete loss of DJ1 function. DJ-1 quenches ROS (Ariga et al, 2013), inhibits p53 and stimulates superoxide dismutase (SOD) expression (Ariga et al, 2013). Mutations in DJ-1 are known to develop an autosomal recessive form of PD in rare cases (Klein and Westenberger, 2012).

Mitophagy and Huntington's Disease

Huntington's disease (HD), an autosomal-dominant disorder, responsible for motor and cognitive impairment, psychiatric disorders, progressive dementia and death even after 15–20 years after the outset of the disease (Bates et al., 2002). Mutant huntingtin (mHtt) gene causes an alteration in various cellular pathways particularly it impairs mitochondrial metabolism which ultimately leads to neuronal dysfunction and death (Khalil et al., 2015). mHtt causes disruption of mitochondrial functions and leads to energetic defect, increased ROS and release of pro-apoptotic molecules. Impaired mitochondrial complex II/III activity is found in the brain samples of HD patients (Browne et al., 1997).

Functional mitochondria need biogenesis and mitochondrial fusion/fission dynamics to replenish the damaged component pool. PGC-1α is the main transcriptional coactivator that maintains mitochondrial biogenesis and energy metabolism. mHtt downregulates PGC-1α by interfering CREB/TAF4-dependent transcriptional pathway (Cui et al., 2006). Mitochondrial fragments due to enhanced dynamin-related protein 1 (DRP1) action are linked to HD (Shirendeb et al., 2012; Song et al., 2011). The aggregation of damaged mitochondria in neurons of the postmitotic phase is considered an important process in the pathogenesis of HD (Khalil et al., 2015).

PINK1 accumulates in the outer membrane of deleterious mitochondria to recruit E3 ubiquitin ligase Parkin and commence mitophagy (Narendra et al., 2010). Another molecular mechanism involved in mitophagy in HD is glyceraldehyde-3-phosphate dehydrogenase (GAPDH) which causes the expansion of polyglutamine repeats. Expanded polyglutamine tract expression catalytically inactivates GAPDH (iGAPDH), which accelerates its association with distorted mitochondria in HD. Therefore, iGAPDH acts as a signaling molecule that directly engulfs deleterious mitochondria into lysosomes. Further, mitochondrial GAPDH interacts abnormally with long polyglutamine tracts and mediate GAPDH-mediated mitophagy and leads to aggregated distorted mitochondria and enhances cellular death (Hwang et al., 2015).

Mitophagy and Amyotrophic Lateral Sclerosis

Amyotrophic lateral sclerosis (ALS) is a neurodegenerative disorder of adulthood that causes progressive motor neuronal dysfunction, sensory, lingual, behavioral and cognitive impairment and may also cause significant dementia (Wang et al., 2019). In ALS, collective pathological changes lead to axonal accumulation of damaged mitochondria (Lin et al., 2017). Mutations in genes regulating the mitophagy process such as valosin-containing protein (VCP), optineurin, and TANK-binding kinase-1 (TBK-1) are directly linked to ALS (Majcher, 2015). In ALS, the mutant VCP is unable to migrate to damaged mitochondria and segregate ubiquitinylated proteins, therefore causing abnormal mitochondrial accumulation in mouse embryonic fibroblasts (Kim et al., 2013). Various mutations in genes (SOD1, OPTN, TBK1, VCP, C9ORF72) are associated with ALS. Mitochondrial quality control can be retained by inhibiting OPTN or TBK1 mutations, pharmacological inhibition or genetic knockdown of PINK1 or Parkin. Any kind of alteration in TBK1/OPTN may improve neuronal function and prevent further progression of the disease (Wong et al., 2015). OPTN mutations play a role in the pathogenesis of ALS (Moore and Holzbaur, 2016). OPTN can trigger the formation of autophagosomes around damaged mitochondria. Further, OPTN mutations have been shown to enhance NF-κB activity and disrupt intracellular transport. These mutations may activate the inflammatory factors further contributing to disease progression (Li et al., 2018). SOD1 is extensively studied ALS-related genes. Mutations in SOD1 cause accumulation of hydroxyl radicals, trigger cytotoxicity and contribute to ALS. SOD1 may indirectly participate in the regulation of mitochondrial homeostasis in ALS (Kaur et al., 2016). SOD1 relies on Parkin to reduce the mitochondrial Rho-GTPase 1 (Miro1), whereas overexpression of Miro1 and inhibition of PINK1 may reverse the autophagosome trafficking defect resulted from the SOD1 mutations. Mutant SOD1 may also impair axonal transport in PINK1/Parkin dependent manner (Moller et al., 2017). Also, VCP exhibits similar properties as that of OPTN; such as it relies on the Parkin-mediated ubiquitination to be recruited to the mitochondria and is involved in mitophagy. The ALS-related VCP mutations will disrupt the mitophagy balance through the PINK1/Parkin pathway, thereby affecting the clearance of abnormal mitochondria (Evans and Holzbaur, 2019).

TDP-43 also leads to aggregation of abnormal mitochondria (Magrané et al., 2014). TDP-43 is known to disturb the anterograde transport of mitochondria leading to accumulation in the neurons and neuromuscular junctions (Shan et al., 2010). Sequestration in kinesin-associated proteins seen in TDP-43-positive cytoplasmic aggregates in mouse motor neurons might indicate dysregulation in mitochondrial axonal transport (Stoica et al., 2016). Thus, improving mitophagy and restoring mitochondrial homeostasis may offer a potential treatment approach for ALS (Wang et al., 2019).

RECENT DEVELOPMENTS AND FUTURE RESEARCH DIRECTIONS

Recently, PD associated genes are also found to be involved in mitophagy. Regarding the role of these proteins in mitophagy, various questions remain unaddressed. By which mechanism parkin get recruited to the mitochondria and under which conditions remains to be elucidated. Various recent studies describe that PINK1, NIX (a pro-apoptotic gene) or environmental stress may be responsible for the recruitment of Parkin to the outer mitochondrial membrane as Parkin does not possess mitochondrial targeting sequence. Further studies showing the interaction of PINK1/NIX with Parkin near mitochondria or particularly at the mitochondria and translocation of Parkin to mitochondria in NIX deficient neurons are further needed. PINK1 is also known to phosphorylate Parkin on Thr175 and favor Parkin localization to mitochondria (Kim et al., 2008). NIX is also important for Parkin mitochondrial translocation. It is also important to find out whether the interaction of Parkin with NIX is reliant on Parkin's phosphorylation or whether NIX act as an adaptor to assist in Parkin phosphorylation (Pridgeon et al., 2007).

Another question that has been raised recently is whether mitophagy has a defensive role or its dysregulation is the underlying cause of neurodegeneration (Ravikumar et al., 2004). Dysfunction in mitochondria is directly linked to neurodegenerative disorders (Burchell et al., 2010). Neurons survive on mitochondrial respiration for energy and failure to remove mitochondria due to respiratory defects and enhanced ROS levels will lead to the oxidative burden. Factors influencing the signaling of mitophagy are still needed to be recognized. Mechanisms underlying both, mitophagy and mitochondrial biogenesis, and methods to regulate both these processes are also required to be elucidated as regulation of these processes may serve as potential therapeutic intervention to act on various cell populations without having deleterious effects on other organs and brain/body tissues (Deas et al., 2011).

There is a persistent demand for the drugs or natural compounds responsible for activating mitophagy without having undesirable side effects. Various secondary metabolites such as polyphenols obtained from plants are being studied for protecting against ROS (Pallauf and Rimbach, 2013). They are known to protect mitochondria and may facilitate mitophagy. Various transcription factors such as forkhead transcription factor (FOXO) and transcription factor EB (TFEB) also serve as potential therapeutic targets for regulating mitophagy and polyphenols are also considered to affect TFEB signaling to upregulate mitophagy (Zhang et al., 2016; Fang et al., 2016). From these studies, it is clear that polyphenols modulate mitophagy transcription to withstand mitochondrial stress (Cagin and Enriquez, 2016). Resveratrol may improve mitochondrial bioenergetics and re-establish mitochondrial oxidative phosphorylation mechanisms (Valenti et al., 2016). Resveratrol is known to protect from respiratory dysfunction caused by mitochondrial mutations in fibroblast cells (Mizuguchi et al., 2017). Rosmarinic acid is known to enhance mitochondrial proliferation by enhancing mitochondrial synthesis factors such as PGC-1α, SIRT1, and transcription factor A mitochondria (TFAM) (Jayanthy et al., 2017). Epicatechin also increases the level of important mitochondrial respiratory and biogenesis factors such as PGC-1α, TFAM, and SIRT1 to boost mitochondrial respiratory function (Lee et al., 2015). Alma also protects from oxidative stress by stimulating mitochondrial biogenesis and respiration by activating AMPK (Yamamoto et al., 2016).

Fortunately, recent technological advancements have highlighted mitophagy research including development of animal models and techniques for directly measuring mitophagy activity in living tissues (Sun et al., 2015), identification of molecular mechanism involved in mitophagy followed by its utilization to develop strategies for disease treatment in near future (Um and Yun, 2017).

CONCLUSION

Mitochondria have a critical role in energy metabolism, signal transduction, cellular survival. Proliferation in defective organelles plays a central role in various neurodegenerative disorders. Maintenance or rejuvenation of mitochondrial function or network is crucial in these disorders. Therefore, mitophagy is a realistic target for the prevention of these pathologies. Various attempts are being made to improve neurodegenerative and age-related disorders including antioxidants and mitophagy inducers or modulators. Besides these advancements, further mechanistic studies of mitophagy are needed to improve the understanding of the mechanisms involved in neurodegeneration and to suggest novel therapeutic interventions to combat neurodegenerative disorders.

REFERENCES

Alano, C. C., Ying, W., & Swanson, R. A. (2004). Poly (ADP-ribose) polymerase-1-mediated cell death in astrocytes requires NAD+ depletion and mitochondrial permeability transition. *The Journal of Biological Chemistry, 279*(18), 18895–18902. doi:10.1074/jbc.M313329200 PMID:14960594

Ariga, H., Takahashi-Niki, K., Kato, I., Maita, H., Niki, T., & Iguchi-Ariga, S. M. M. (2013). Neuroprotective Function of DJ-1 in Parkinson's Disease. *Oxidative Medicine and Cellular Longevity, 2013*, 683920. doi:10.1155/2013/683920 PMID:23766857

Bandopadhyay, R., & de Belleroche, J. (2010). Pathogenesis of Parkinson's disease: Emerging role of molecular chaperones. *Trends in Molecular Medicine, 16*(1), 27–36. doi:10.1016/j.molmed.2009.11.004 PMID:20036196

Bates, G. P., Harper, P. S., & Jones, A. L. (Eds.). (2002). *Huntington's Disease*. Oxford, UK: Oxford University Press.

Batlevi, Y., & La Spada, A. R. (2011). Mitochondrial autophagy in neural function, neurodegenerative disease, neuron cell death, and aging. *Neurobiology of Disease, 43*(1), 46–51. doi:10.1016/j.nbd.2010.09.009 PMID:20887789

Benatar, M. (2007). Lost in translation: Treatment trials in the SOD1 mouse and in human ALS. *Neurobiology of Disease, 26*(1), 1–13. doi:10.1016/j.nbd.2006.12.015 PMID:17300945

Bendor, J. T., Logan, T. P., & Edwards, R. H. (2013). The function of alpha-synuclein. *Neuron, 79*(6), 1044–1066. doi:10.1016/j.neuron.2013.09.004 PMID:24050397

Bezprozvanny, I., & Mattson, M. P. (2008). Neuronal calcium mishandling and the pathogenesis of Alzheimer's disease. *Trends in Neurosciences, 31*(9), 454–463. doi:10.1016/j.tins.2008.06.005 PMID:18675468

Björkblom, B., Adilbayeva, A., Maple-Grødem, J., Piston, D., Ökvist, M., Xu, X. M., ... Møller, S. G. (2013). Parkinson Disease Protein DJ-1 Binds Metals and Protects against Metal-induced Cytotoxicity. *The Journal of Biological Chemistry, 288*(31), 22809–22820. doi:10.1074/jbc.M113.482091 PMID:23792957

Boland, B., Kumar, A., Lee, S., Platt, F. M., Wegiel, J., Yu, W. H., & Nixon, R. A. (2008). Autophagy induction and autophagosome clearance in neurons: Relationship to autophagic pathology in Alzheimer's disease. *The Journal of Neuroscience: The Official Journal of the Society for Neuroscience, 28*(27), 6926–6937. doi:10.1523/JNEUROSCI.0800-08.2008 PMID:18596167

Botella, J. A., Bayersdorfer, F., & Schneuwly, S. (2008). Superoxide dismutase overexpression protects dopaminergic neurons in a Drosophila model of Parkinson's disease. *Neurobiology of Disease, 30*(1), 65–73. doi:10.1016/j.nbd.2007.11.013 PMID:18243716

Browne, S. E., Bowling, A. C., Macgarvey, U., Baik, M. J., Berger, S. C., Muquit, M. M., ... Beal, M. F. (1997). Oxidative damage and metabolic dysfunction in Huntington's disease: Selective vulnerability of the basal ganglia. *Annals of Neurology: Official Journal of the American Neurological Association and the Child Neurology Society, 41*(5), 646–653. doi:10.1002/ana.410410514 PMID:9153527

Burchell, V. S., Gandhi, S., Deas, E., Wood, N. W., Abramov, A. Y., & Plun-Favreau, H. (2010). Targeting mitochondrial dysfunction in neurodegenerative disease. *Part II. Expert Opinion on Therapeutic Targets, 14*(5), 497–511. doi:10.1517/14728221003730434 PMID:20334487

Cagin, U., & Enriquez, J. A. (2015). The complex crosstalk between mitochondria and the nucleus: What goes in between? *The International Journal of Biochemistry & Cell Biology, 63*, 10–15. doi:10.1016/j.biocel.2015.01.026 PMID:25666554

Cai, Q., & Tammineni, P. (2016). Alterations in mitochondrial quality control in Alzheimer's disease. *Frontiers in Cellular Neuroscience, 10*, 24. doi:10.3389/fncel.2016.00024 PMID:26903809

Carafa, V., Rotili, D., Forgione, M., Cuomo, F., Serretiello, E., Hailu, G. S., ... Altucci, L. (2016). Sirtuin functions and modulation: From chemistry to the clinic. *Clinical Epigenetics, 8*(1), 61. doi:10.118613148-016-0224-3 PMID:27226812

Chan, N. C., Salazar, A. M., Pham, A. H., Sweredoski, M. J., Kolawa, N. J., Graham, R. L., ... Chan, D. C. (2011). Broad activation of the ubiquitin–proteasome system by Parkin is critical for mitophagy. *Human Molecular Genetics, 20*(9), 1726–1737. doi:10.1093/hmg/ddr048 PMID:21296869

Chen, M., Chen, Z., Wang, Y., Tan, Z., Zhu, C., Li, Y., ... Chen, Q. (2016). Mitophagy receptor FUNDC1 regulates mitochondrial dynamics and mitophagy. *Autophagy, 12*(4), 689–702. doi:10.1080/15548627.2016.1151580 PMID:27050458

Chen, Y. G. (2018). Research Progress in the Pathogenesis of Alzheimer's Disease. *Chinese Medical Journal, 131*(13), 1618–1624. doi:10.4103/0366-6999.235112 PMID:29941717

Chen, Z., Liu, L., Cheng, Q., Li, Y., Wu, H., Zhang, W., ... Wang, J. (2017). Mitochondrial E3 ligase MARCH5 regulates FUNDC1 to fine-tune hypoxic mitophagy. *EMBO Reports*, *18*(3), 495–509. doi:10.15252/embr.201643309 PMID:28104734

Cheng, A., Wan, R., Yang, J.-L., Kamimura, N., Son, T. G., Ouyang, X., ... Mattson, M. P. (2012). Involvement of PGC-1alpha in the formation and maintenance of neuronal dendritic spines. *Nature Communications*, *3*(1), 1250. doi:10.1038/ncomms2238 PMID:23212379

Cheng, A., Yang, Y., Zhou, Y., Maharana, C., Lu, D., Peng, W., ... Bohr, V. A. (2016). Mitochondrial SIRT3 mediates adaptive responses of neurons to exercise and metabolic and excitatory challenges. *Cell Metabolism*, *23*(1), 128–142. doi:10.1016/j.cmet.2015.10.013 PMID:26698917

Chu, C. T., Ji, J., Dagda, R. K., Jiang, J. F., Tyurina, Y. Y., Kapralov, A. A., ... Wang, K. Z. Q. (2013). Cardiolipin externalization to the outer mitochondrial membrane acts as an elimination signal for mitophagy in neuronal cells. *Nature Cell Biology*, *15*(10), 1197–1205. doi:10.1038/ncb2837 PMID:24036476

Cornelissen, T., Haddad, D., Wauters, F., Van Humbeeck, C., Mandemakers, W., Koentjoro, B., ... Vandenberghe, W. (2014). The deubiquitinase USP15 antagonizes Parkin-mediated mitochondrial ubiquitination and mitophagy. *Human Molecular Genetics*, *23*(19), 5227–5242. doi:10.1093/hmg/ddu244 PMID:24852371

Csibi, A., Fendt, S. M., Li, C., Poulogiannis, G., Choo, A. Y., Chapski, D. J., ... Henske, E. P. (2013). The mTORC1 pathway stimulates glutamine metabolism and cell proliferation by repressing SIRT4. *Cell*, *153*(4), 840–854. doi:10.1016/j.cell.2013.04.023 PMID:23663782

Cui, L., Jeong, H., Borovecki, F., Parkhurst, C. N., Tanese, N., & Krainc, D. (2006). Transcriptional repression of PGC-1α by mutant huntingtin leads to mitochondrial dysfunction and neurodegeneration. *Cell*, *127*(1), 59–69. doi:10.1016/j.cell.2006.09.015 PMID:17018277

Dagda, R. K., & Chu, C. T. (2009). Mitochondrial quality control: Insights on how Parkinson's disease related genes PINK1, Parkin, and Omi/HtrA2 interact to maintain mitochondrial homeostasis. *Journal of Bioenergetics and Biomembranes*, *41*(6), 473–479. doi:10.100710863-009-9255-1 PMID:20012177

Dawson, T. M., and Dawson, V. L. (2010). The Role of Parkin in Familial and Sporadic Parkinson's Disease. *Movement Disorders: Official Journal of the Movement Disorder Society, 25*(1), S32-S39.

de Oliveira, R. M., Sarkander, J., Kazantsev, A. G., & Outeiro, T. F. (2012). SIRT2 as a therapeutic target for age-related disorders. *Frontiers in Pharmacology*, *3*, 3. doi:10.3389/fphar.2012.00082 PMID:22563317

Deas, E., Wood, N. W., & Plun-Favreau, H. (2011). Mitophagy and Parkinson's disease: The PINK1-parkin link. *Biochimica et Biophysica Acta*, *1813*(4), 623–633. doi:10.1016/j.bbamcr.2010.08.007 PMID:20736035

Ding, W. X., Guo, F., Ni, H. M., Bockus, A., Manley, S., Stolz, D. B., ... Yin, X. M. (2012a). Parkin and mitofusins reciprocally regulate mitophagy and mitochondrial spheroid formation. *The Journal of Biological Chemistry*, *287*(50), 42379–42388. doi:10.1074/jbc.M112.413682 PMID:23095748

Ding, W. X., Li, M., Biazik, J. M., Morgan, D. G., Guo, F., Ni, H. M., ... Yin, X. M. (2012b). Electron microscopic analysis of a spherical mitochondrial structure. *The Journal of Biological Chemistry*, *287*(50), 42373–42378. doi:10.1074/jbc.M112.413674 PMID:23093403

Ding, W. X., & Yin, X. M. (2012). Mitophagy: Mechanisms, pathophysiological roles, and analysis. *Biological Chemistry*, *393*(7), 547–564. doi:10.1515/hsz-2012-0119 PMID:22944659

Donmez, G., Arun, A., Chung, C. Y., McLean, P. J., Lindquist, S., & Guarente, L. (2012). SIRT1 protects against α-synuclein aggregation by activating molecular chaperones. *The Journal of Neuroscience: The Official Journal of the Society for Neuroscience*, *32*(1), 124–132. doi:10.1523/JNEUROSCI.3442-11.2012 PMID:22219275

Dryden, S. C., Nahhas, F. A., Nowak, J. E., Goustin, A. S., & Tainsky, M. A. (2003). Role for human SIRT2 NAD-dependent deacetylase activity in control of mitotic exit in the cell cycle. *Molecular and Cellular Biology*, *23*(9), 3173–3185. doi:10.1128/MCB.23.9.3173-3185.2003 PMID:12697818

Du, H., & Yan, S. S. (2010). Mitochondrial permeability transition pore in Alzheimer's disease: Cyclophilin D and amyloid beta. Biochimica et Biophysica Acta (BBA)-. *Molecular Basis of Disease*, *1802*(1), 198–204.

Durcan, T. M., Tang, M. Y., Pérusse, J. R., Dashti, E. A., Aguileta, M. A., McLelland, G. L., ... Fon, E. A. (2014). USP8 regulates mitophagy by removing K6-linked ubiquitin conjugates from parkin. *The EMBO Journal*, *33*(21), 2473–2491.

Ebrahimi-Fakhari, D., Wahlster, L., & McLean, P. J. (2012). Protein degradation pathways in Parkinson's disease: Curse or blessing. *Acta Neuropathologica*, *124*(2), 153–172. doi:10.100700401-012-1004-6 PMID:22744791

Evans, C. S., & Holzbaur, E. (2019). Autophagy and mitophagy in ALS. *Neurobiology of Disease*, *122*, 35–40. doi:10.1016/j.nbd.2018.07.005 PMID:29981842

Fang, E. F., Kassahun, H., Croteau, D. L., Scheibye-Knudsen, M., Marosi, K., Lu, H., ... Iser, W. B. (2016). NAD+ replenishment improves lifespan and healthspan in ataxia telangiectasia models via mitophagy and DNA repair. *Cell Metabolism*, *24*(4), 566–581. doi:10.1016/j.cmet.2016.09.004 PMID:27732836

Fang, E. F., Scheibye-Knudsen, M., Chua, K. F., Mattson, M. P., Croteau, D. L., & Bohr, V. A. (2016). Nuclear DNA damage signalling to mitochondria in ageing. *Nature Reviews. Molecular Cell Biology*, *17*(5), 308–321. doi:10.1038/nrm.2016.14 PMID:26956196

Fu, J., Jin, J., Cichewicz, R. H., Hageman, S. A., Ellis, T. K., Xiang, L., ... Ross, C. A. (2012). trans-(−)-ϵ-Viniferin increases mitochondrial sirtuin 3 (SIRT3), activates AMP-activated protein kinase (AMPK), and protects cells in models of Huntington Disease. *The Journal of Biological Chemistry*, *287*(29), 24460–24472. doi:10.1074/jbc.M112.382226 PMID:22648412

Gardet, A., Benita, Y., Li, C., Sands, B. E., Ballester, I., Stevens, C., ... Podolsky, D. K. (2010). LRRK2 is involved in the IFN-γ response and host response to pathogens. *Journal of Immunology (Baltimore, Md.: 1950)*, *185*(9), 5577–5585. doi:10.4049/jimmunol.1000548 PMID:20921534

Gegg, M. E., Cooper, J. M., Chau, K. Y., Rojo, M., Schapira, A. H., & Taanman, J. W. (2010). Mitofusin 1 and mitofusin 2 are ubiquitinated in a PINK1/parkin-dependent manner upon induction of mitophagy. *Human Molecular Genetics, 19*(24), 4861–4870. doi:10.1093/hmg/ddq419 PMID:20871098

Giblin, W., Skinner, M. E., & Lombard, D. B. (2014). Sirtuins: Guardians of mammalian healthspan. *Trends in Genetics, 30*(7), 271–286. doi:10.1016/j.tig.2014.04.007 PMID:24877878

Greene, J. C., Whitworth, A. J., Kuo, I., Andrews, L. A., Feany, M. B., & Pallanck, L. J. (2003). Mitochondrial pathology and apoptotic muscle degeneration in Drosophila parkin mutants. *Proceedings of the National Academy of Sciences of the United States of America, 100*(7), 4078–4083. doi:10.1073/pnas.0737556100 PMID:12642658

Hafner, A. V., Dai, J., Gomes, A. P., Xiao, C. Y., Palmeira, C. M., Rosenzweig, A., & Sinclair, D. A. (2010). Regulation of the mPTP by SIRT3-mediated deacetylation of CypD at lysine 166 suppresses age-related cardiac hypertrophy. *Aging (Albany NY), 2*(12), 914–923. doi:10.18632/aging.100252 PMID:21212461

Haigis, M. C., Mostoslavsky, R., Haigis, K. M., Fahie, K., Christodoulou, D. C., Murphy, A. J., ... Wolberger, C. (2006). SIRT4 inhibits glutamate dehydrogenase and opposes the effects of calorie restriction in pancreatic β cells. *Cell, 126*(5), 941–954. doi:10.1016/j.cell.2006.06.057 PMID:16959573

He, W., Newman, J. C., Wang, M. Z., Ho, L., & Verdin, E. (2012). Mitochondrial sirtuins: Regulators of protein acylation and metabolism. *Trends in Endocrinology and Metabolism, 23*(9), 467–476. doi:10.1016/j.tem.2012.07.004 PMID:22902903

Hirota, Y., Yamashita, S. I., Kurihara, Y., Jin, X., Aihara, M., Saigusa, T., ... Kanki, T. (2015). Mitophagy is primarily due to alternative autophagy and requires the MAPK1 and MAPK14 signaling pathways. *Autophagy, 11*(2), 332–343. doi:10.1080/15548627.2015.1023047 PMID:25831013

Hubbard, V. M., Valdor, R., Macian, F., & Cuervo, A. M. (2012). Selective autophagy in the maintenance of cellular homeostasis in aging organisms. *Biogerontology, 13*(1), 21–35. doi:10.100710522-011-9331-x PMID:21461872

Hwang, S., Disatnik, M. H., & Mochly-Rosen, D. (2015). Impaired GAPDH-induced mitophagy contributes to the pathology of Huntington's disease. *EMBO Molecular Medicine, 7*(10), 1307–1326. doi:10.15252/emmm.201505256 PMID:26268247

Jayanthy, G., Roshana Devi, V., Ilango, K., & Subramanian, S. P. (2017). Rosmarinic acid mediates mitochondrial biogenesis in insulin resistant skeletal muscle through activation of AMPK. *Journal of Cellular Biochemistry, 118*(7), 1839–1848. doi:10.1002/jcb.25869 PMID:28059465

Jeong, S. M., Xiao, C., Finley, L. W., Lahusen, T., Souza, A. L., Pierce, K., ... Xu, X. (2013). SIRT4 has tumor-suppressive activity and regulates the cellular metabolic response to DNA damage by inhibiting mitochondrial glutamine metabolism. *Cancer Cell, 23*(4), 450–463. doi:10.1016/j.ccr.2013.02.024 PMID:23562301

Julien, C., Tremblay, C., Emond, V., Lebbadi, M., Salem, N. Jr, Bennett, D. A., & Calon, F. (2009). Sirtuin 1 reduction parallels the accumulation of tau in Alzheimer disease. *Journal of Neuropathology and Experimental Neurology, 68*(1), 48–58. doi:10.1097/NEN.0b013e3181922348 PMID:19104446

Kagan, V. E., Jiang, J., Huang, Z., Tyurina, Y. Y., Desbourdes, C., Cottet-Rousselle, C., ... Cheikhi, A. (2016). NDPK-D (NM23-H4)-mediated externalization of cardiolipin enables elimination of depolarized mitochondria by mitophagy. *Cell Death and Differentiation*, *23*(7), 1140. doi:10.1038/cdd.2015.160 PMID:26742431

Kanki, T., Kurihara, Y., Jin, X., Goda, T., Ono, Y., Aihara, M., ... Kang, D. (2013). Casein kinase 2 is essential for mitophagy. *EMBO Reports*, *14*(9), 788–794. doi:10.1038/embor.2013.114 PMID:23897086

Kanki, T., Wang, K., Cao, Y., Baba, M., & Klionsky, D. J. (2009). Atg32 is a mitochondrial protein that confers selectivity during mitophagy. *Developmental Cell*, *17*(1), 98–109. doi:10.1016/j.devcel.2009.06.014 PMID:19619495

Kaur, S. J., McKeown, S. R., & Rashid, S. (2016). Mutant SOD1 mediated pathogenesis of amyotrophic lateral sclerosis. *Gene*, *577*(2), 109–118. doi:10.1016/j.gene.2015.11.049 PMID:26657039

Kawajiri, S., Saiki, S., Sato, S., Sato, F., Hatano, T., Eguchi, H., & Hattori, N. (2010). PINK1 is recruited to mitochondria with parkin and associates with LC3 in mitophagy. *FEBS Letters*, *584*(6), 1073–1079. doi:10.1016/j.febslet.2010.02.016 PMID:20153330

Kerr, J. S., Adriaanse, B. A., Greig, N. H., Mattson, M. P., Cader, M. Z., Bohr, V. A., & Fang, E. F. (2017). Mitophagy and Alzheimer's disease: Cellular and molecular mechanisms. *Trends in Neurosciences*, *40*(3), 151–166. doi:10.1016/j.tins.2017.01.002 PMID:28190529

Khalil, B., El Fissi, N., Aouane, A., Cabirol-Pol, M. J., Rival, T., & Liévens, J. C. (2015). PINK1-induced mitophagy promotes neuroprotection in Huntington's disease. *Cell Death & Disease*, *6*(1), e1617. doi:10.1038/cddis.2014.581 PMID:25611391

Khalil, B., El Fissi, N., Aouane, A., Cabirol-Pol, M.-J., Rival, T., & Liévens, J.-C. (2015). PINK1-induced mitophagy promotes neuroprotection in Huntington's disease. *Cell Death & Disease*, *6*(1), e1617. doi:10.1038/cddis.2014.581 PMID:25611391

Khaminets, A., Behl, C., & Dikic, I. (2016). Ubiquitin-dependent and independent signals in selective autophagy. *Trends in Cell Biology*, *26*(1), 6–16. doi:10.1016/j.tcb.2015.08.010 PMID:26437584

Kim, N. C., Tresse, E., Kolaitis, R. M., Molliex, A., Thomas, R. E., Alami, N. H., ... Winborn, B. J. (2013). VCP is essential for mitochondrial quality control by PINK1/Parkin and this function is impaired by VCP mutations. *Neuron*, *78*(1), 65–80. doi:10.1016/j.neuron.2013.02.029 PMID:23498974

Kim, Y., Park, J., Kim, S., Song, S., Kwon, S. K., Lee, S. H., ... Chung, J. (2008). PINK1 controls mitochondrial localization of Parkin through direct phosphorylation. *Biochemical and Biophysical Research Communications*, *377*(3), 975–980. doi:10.1016/j.bbrc.2008.10.104 PMID:18957282

Klein, C., & Westenberger, A. (2012). Genetics of Parkinson's Disease. *Cold Spring Harbor Perspectives in Medicine*, *2*(1), a008888. doi:10.1101/cshperspect.a008888 PMID:22315721

Kuroda, Y., Mitsui, T., Kunishige, M., Shono, M., Akaike, M., Azuma, H., & Matsumoto, T. (2006). Parkin enhances mitochondrial biogenesis in proliferating cells. *Human Molecular Genetics*, *15*(6), 883–895. doi:10.1093/hmg/ddl006 PMID:16449237

Lee, I., Hüttemann, M., Kruger, A., Bollig-Fischer, A., & Malek, M. H. (2015). (–)–Epicatechin combined with 8 weeks of treadmill exercise is associated with increased angiogenic and mitochondrial signaling in mice. *Frontiers in Pharmacology*, *6*, 43. doi:10.3389/fphar.2015.00043 PMID:25821434

Lee, I. H., Cao, L., Mostoslavsky, R., Lombard, D. B., Liu, J., Bruns, N. E., ... Finkel, T. (2008). A role for the NAD-dependent deacetylase Sirt1 in the regulation of autophagy. *Proceedings of the National Academy of Sciences of the United States of America*, *105*(9), 3374–3379. doi:10.1073/pnas.0712145105 PMID:18296641

Lemasters, J. J. (2005). Selective mitochondrial autophagy, or mitophagy, as a targeted defense against oxidative stress, mitochondrial dysfunction, and aging. *Rejuvenation Research*, *8*(1), 3–5. doi:10.1089/rej.2005.8.3 PMID:15798367

Lemasters, J. J., Theruvath, T. P., Zhong, Z., & Nieminen, A. L. (2009). Mitochondrial calcium and the permeability transition in cell death. *Biochimica et Biophysica Acta (BBA)-. Bioenergetics*, *1787*(11), 1395–1401. doi:10.1016/j.bbabio.2009.06.009 PMID:19576166

Li, J.-Q., Tan, L., & Yu, J.-T. (2014). The role of the LRRK2 gene in Parkinsonism. *Molecular Neurodegeneration*, *9*(1), 47. doi:10.1186/1750-1326-9-47 PMID:25391693

Lin, M. Y., Cheng, X. T., Xie, Y., Cai, Q., & Sheng, Z. H. (2017). Removing dysfunctional mitochondria from axons independent of mitophagy under pathophysiological conditions. *Autophagy*, *13*(10), 1792–1794. doi:10.1080/15548627.2017.1356552 PMID:28812939

Liu, L., Feng, D., Chen, G., Chen, M., Zheng, Q., Song, P., ... Huang, L. (2012). Mitochondrial outer-membrane protein FUNDC1 mediates hypoxia-induced mitophagy in mammalian cells. *Nature Cell Biology*, *14*(2), 177–185. doi:10.1038/ncb2422 PMID:22267086

Liu, Z., Li, H., Hong, C., Yue, T., Chen, C., Wang, Z., ... Hu, R. (2018). ALS-associated E478G mutation in human OPTN (optineurin) promotes inflammation and induces neuronal cell death. *Frontiers in Immunology*, *9*, 2647. doi:10.3389/fimmu.2018.02647 PMID:30519240

Magrane, J., Cortez, C., Gan, W. B., & Manfredi, G. (2013). Abnormal mitochondrial transport and morphology are common pathological denominators in SOD1 and TDP43 ALS mouse models. *Human Molecular Genetics*, *23*(6), 1413–1424. doi:10.1093/hmg/ddt528 PMID:24154542

Majcher, V., Goode, A., James, V., & Layfield, R. (2015). Autophagy receptor defects and ALS-FTLD. *Molecular and Cellular Neurosciences*, *66*, 43–52. doi:10.1016/j.mcn.2015.01.002 PMID:25683489

Marosi, K., & Mattson, M. P. (2014). BDNF mediates adaptive brain and body responses to energetic challenges. *Trends in Endocrinology and Metabolism*, *25*(2), 89–98. doi:10.1016/j.tem.2013.10.006 PMID:24361004

Martinez-Vicente, M. (2017). Neuronal mitophagy in neurodegenerative diseases. *Frontiers in Molecular Neuroscience*, *10*, 64. doi:10.3389/fnmol.2017.00064 PMID:28337125

Mattson, M. P., Gleichmann, M., & Cheng, A. (2008). Mitochondria in neuroplasticity and neurological disorders. *Neuron*, *60*(5), 748–766. doi:10.1016/j.neuron.2008.10.010 PMID:19081372

McCoy, M. K., Martinez, T. N., Ruhn, K. A., Szymkowski, D. E., Smith, C. G., Botterman, B. R., ... Tansey, M. G. (2006). Blocking soluble tumor necrosis factor signaling with dominant-negative tumor necrosis factor inhibitor attenuates loss of dopaminergic neurons in models of Parkinson's disease. *The Journal of Neuroscience: The Official Journal of the Society for Neuroscience*, *26*(37), 9365–9375. doi:10.1523/JNEUROSCI.1504-06.2006 PMID:16971520

McInnes, J. (2013). Insights on altered mitochondrial function and dynamics in the pathogenesis of neurodegeneration. *Translational Neurodegeneration*, *2*(1), 12. doi:10.1186/2047-9158-2-12 PMID:23711354

McNaught, K. S. P., Belizaire, R., Isacson, O., Jenner, P., & Olanow, C. W. (2003). Altered proteasomal function in sporadic Parkinson's disease. *Experimental Neurology*, *179*(1), 38–46. doi:10.1006/exnr.2002.8050 PMID:12504866

McWilliams, T. G., Prescott, A. R., Allen, G. F., Tamjar, J., Munson, M. J., Thomson, C., ... Ganley, I. G. (2016). mito-QC illuminates mitophagy and mitochondrial architecture in vivo. *The Journal of Cell Biology*, *214*(3), 333–345. doi:10.1083/jcb.201603039 PMID:27458135

McWilliams, T. G., Prescott, A. R., Montava-Garriga, L., Ball, G., Singh, F., Barini, E., ... Ganley, I. G. (2018). Basal mitophagy occurs independently of PINK1 in mouse tissues of high metabolic demand. *Cell Metabolism*, *27*(2), 439–449. doi:10.1016/j.cmet.2017.12.008 PMID:29337137

Miklya, I., Göltl, P., Hafenscher, F., & Pencz, N. (2014). The role of parkin in Parkinson's disease. *Neuropsychopharmacologia Hungarica: a Magyar Pszichofarmakologiai Egyesulet lapja= official journal of the Hungarian Association of Psychopharmacology*, *16*(2), 67-76.

Mizuguchi, Y., Hatakeyama, H., Sueoka, K., Tanaka, M., & Goto, Y. I. (2017). Low dose resveratrol ameliorates mitochondrial respiratory dysfunction and enhances cellular reprogramming. *Mitochondrion*, *34*, 43–48. doi:10.1016/j.mito.2016.12.006 PMID:28093354

Moller, A., Bauer, C. S., Cohen, R. N., Webster, C. P., & De Vos, K. J. (2017). Amyotrophic lateral sclerosis-associated mutant SOD1 inhibits anterograde axonal transport of mitochondria by reducing Miro1 levels. *Human Molecular Genetics*, *26*(23), 4668–4679. doi:10.1093/hmg/ddx348 PMID:28973175

Moore, A. S., & Holzbaur, E. L. (2016). Spatiotemporal dynamics of autophagy receptors in selective mitophagy. *Autophagy*, *12*(10), 1956–1957. doi:10.1080/15548627.2016.1212788 PMID:27484150

Narendra, D., Kane, L. A., Hauser, D. N., Fearnley, I. M., & Youle, R. J. (2010). p62/SQSTM1 is required for Parkin-induced mitochondrial clustering but not mitophagy; VDAC1 is dispensable for both. *Autophagy*, *6*(8), 1090–1106. doi:10.4161/auto.6.8.13426 PMID:20890124

Narendra, D., Tanaka, A., Suen, D. F., & Youle, R. J. (2008). Parkin is recruited selectively to impaired mitochondria and promotes their autophagy. *The Journal of Cell Biology*, *183*(5), 795–803. doi:10.1083/jcb.200809125 PMID:19029340

Narendra, D. P., Jin, S. M., Tanaka, A., Suen, D. F., Gautier, C. A., Shen, J., ... Youle, R. J. (2010). PINK1 is selectively stabilized on impaired mitochondria to activate Parkin. *PLoS Biology*, *8*(1), e1000298. doi:10.1371/journal.pbio.1000298 PMID:20126261

Navarro, A., & Boveris, A. (2004). Rat brain and liver mitochondria develop oxidative stress and lose enzymatic activities on aging. *American Journal of Physiology. Regulatory, Integrative and Comparative Physiology*, *287*(5), R1244–R1249. doi:10.1152/ajpregu.00226.2004 PMID:15271654

Ni, H. M., Williams, J. A., & Ding, W. X. (2015). Mitochondrial dynamics and mitochondrial quality control. *Redox Biology*, *4*, 6–13. doi:10.1016/j.redox.2014.11.006 PMID:25479550

Niu, J., Yu, M., Wang, C., & Xu, Z. (2012). Leucine-rich repeat kinase 2 disturbs mitochondrial dynamics via Dynamin-like protein. *Journal of Neurochemistry*, *122*(3), 650–658. doi:10.1111/j.1471-4159.2012.07809.x PMID:22639965

Nixon, R. A. (2013). The role of autophagy in neurodegenerative disease. *Nature Medicine*, *19*(8), 983–997. doi:10.1038/nm.3232 PMID:23921753

North, B. J., Marshall, B. L., Borra, M. T., Denu, J. M., & Verdin, E. (2003). The human Sir2 ortholog, SIRT2, is an NAD+-dependent tubulin deacetylase. *Molecular Cell*, *11*(2), 437–444. doi:10.1016/S1097-2765(03)00038-8 PMID:12620231

Okamoto, K., Kondo-Okamoto, N., & Ohsumi, Y. (2009). A landmark protein essential for mitophagy: Atg32 recruits the autophagic machinery to mitochondria. *Autophagy*, *5*(8), 1203–1205. doi:10.4161/auto.5.8.9830 PMID:19770589

Onyango, P., Celic, I., McCaffery, J. M., Boeke, J. D., & Feinberg, A. P. (2002). SIRT3, a human SIR2 homologue, is an NAD-dependent deacetylase localized to mitochondria. *Proceedings of the National Academy of Sciences of the United States of America*, *99*(21), 13653–13658. doi:10.1073/pnas.222538099 PMID:12374852

Orvedahl, A., Sumpter, R. Jr, Xiao, G., Ng, A., Zou, Z., Tang, Y., ... Forst, C. V. (2011). Image-based genome-wide siRNA screen identifies selective autophagy factors. *Nature*, *480*(7375), 113–117. doi:10.1038/nature10546 PMID:22020285

Outeiro, T. F., Grammatopoulos, T. N., Altmann, S., Amore, A., Standaert, D. G., Hyman, B. T., & Kazantsev, A. G. (2007). Pharmacological inhibition of PARP-1 reduces α-synuclein-and MPP+-induced cytotoxicity in Parkinson's disease in vitro models. *Biochemical and Biophysical Research Communications*, *357*(3), 596–602. doi:10.1016/j.bbrc.2007.03.163 PMID:17449015

Palikaras, K., Daskalaki, I., Markaki, M., & Tavernarakis, N. (2017). Mitophagy and age-related pathologies: Development of new therapeutics by targeting mitochondrial turnover. *Pharmacology & Therapeutics*, *178*, 157–174. doi:10.1016/j.pharmthera.2017.04.005 PMID:28461251

Pallauf, K., & Rimbach, G. (2013). Autophagy, polyphenols and healthy ageing. *Ageing Research Reviews*, *12*(1), 237–252. doi:10.1016/j.arr.2012.03.008 PMID:22504405

Parihar, M. S., Parihar, A., Fujita, M., Hashimoto, M., & Ghafourifar, P. (2009). Alpha-synuclein overexpression and aggregation exacerbates impairment of mitochondrial functions by augmenting oxidative stress in human neuroblastoma cells. *The International Journal of Biochemistry & Cell Biology*, *41*(10), 2015–2024. doi:10.1016/j.biocel.2009.05.008 PMID:19460457

Pickrell, A. M., & Youle, R. J. (2015). The roles of PINK1, parkin, and mitochondrial fidelity in Parkinson's disease. *Neuron, 85*(2), 257–273. doi:10.1016/j.neuron.2014.12.007 PMID:25611507

Poole, A. C., Thomas, R. E., Andrews, L. A., McBride, H. M., Whitworth, A. J., & Pallanck, L. J. (2008). The PINK1/Parkin pathway regulates mitochondrial morphology. *Proceedings of the National Academy of Sciences of the United States of America*, *105*(5), 1638–1643. doi:10.1073/pnas.0709336105 PMID:18230723

Pridgeon, J. W., Olzmann, J. A., Chin, L. S., & Li, L. (2007). PINK1 protects against oxidative stress by phosphorylating mitochondrial chaperone TRAP1. *PLoS Biology*, *5*(7), e172. doi:10.1371/journal.pbio.0050172 PMID:17579517

Quiros, P. M., Mottis, A., & Auwerx, J. (2016). Mitonuclear communication in homeostasis and stress. *Nature Reviews. Molecular Cell Biology*, *17*(4), 213–226. doi:10.1038/nrm.2016.23 PMID:26956194

Ramadori, G., Lee, C. E., Bookout, A. L., Lee, S., Williams, K. W., Anderson, J., ... Coppari, R. (2008). Brain SIRT1: Anatomical distribution and regulation by energy availability. *The Journal of Neuroscience: The Official Journal of the Society for Neuroscience*, *28*(40), 9989–9996. doi:10.1523/JNEUROSCI.3257-08.2008 PMID:18829956

Rambold, A. S., Kostelecky, B., Elia, N., & Lippincott-Schwartz, J. (2011). Tubular network formation protects mitochondria from autophagosomal degradation during nutrient starvation. *Proceedings of the National Academy of Sciences of the United States of America*, *108*(25), 10190–10195. doi:10.1073/pnas.1107402108 PMID:21646527

Ravikumar, B., Vacher, C., Berger, Z., Davies, J. E., Luo, S., Oroz, L. G., ... Rubinsztein, D. C. (2004). Inhibition of mTOR induces autophagy and reduces toxicity of polyglutamine expansions in fly and mouse models of Huntington disease. *Nature Genetics*, *36*(6), 585–595. doi:10.1038/ng1362 PMID:15146184

Reid, M. A., Wang, W. I., Rosales, K. R., Welliver, M. X., Pan, M., & Kong, M. (2013). The B55α subunit of PP2A drives a p53-dependent metabolic adaptation to glutamine deprivation. *Molecular Cell*, *50*(2), 200–211. doi:10.1016/j.molcel.2013.02.008 PMID:23499005

Ren, M., Phoon, C. K., & Schlame, M. (2014). Metabolism and function of mitochondrial cardiolipin. *Progress in Lipid Research*, 55, 1-16.

Rodolfo, C., Campello, S., & Cecconi, F. (2018). Mitophagy in neurodegenerative diseases. *Neurochemistry International*, *117*, 156–166. doi:10.1016/j.neuint.2017.08.004 PMID:28797885

Rongvaux, A., Shea, R. J., Mulks, M. H., Gigot, D., Urbain, J., Leo, O., & Andris, F. (2002). Pre-B-cell colony-enhancing factor, whose expression is up-regulated in activated lymphocytes, is a nicotinamide phosphoribosyl transferase, a cytosolic enzyme involved in NAD biosynthesis. *European Journal of Immunology*, *32*(11), 3225–3234. doi:10.1002/1521-4141(200211)32:11<3225::AID-IMMU3225>3.0.CO;2-L PMID:12555668

Rothfuss, O., Fischer, H., Hasegawa, T., Maisel, M., Leitner, P., Miesel, F., ... Patenge, N. (2009). Parkin protects mitochondrial genome integrity and supports mitochondrial DNA repair. *Human Molecular Genetics*, *18*(20), 3832–3850. doi:10.1093/hmg/ddp327 PMID:19617636

Rothgiesser, K. M., Erener, S., Waibel, S., Lüscher, B., & Hottiger, M. O. (2010). SIRT2 regulates NF-κB-dependent gene expression through deacetylation of p65 Lys310. *Journal of Cell Science, 123*(24), 4251–4258. doi:10.1242/jcs.073783 PMID:21081649

Rubinsztein, D. C., Mariño, G., & Kroemer, G. (2011). Autophagy and aging. *Cell, 146*(5), 682–695. doi:10.1016/j.cell.2011.07.030 PMID:21884931

Rugarli, E. I., & Langer, T. (2012). Mitochondrial quality control: A matter of life and death for neurons. *The EMBO Journal, 31*(6), 1336–1349. doi:10.1038/emboj.2012.38 PMID:22354038

Ryu, S. W., Han, E. C., Yoon, J., & Choi, C. (2015). The mitochondrial fusion-related proteins Mfn2 and OPA1 are transcriptionally induced during differentiation of bone marrow progenitors to immature dendritic cells. *Molecules and Cells, 38*(1), 89. PMID:25387754

Schapira, A. H. V. (2012). Targeting mitochondria for neuroprotection in Parkinson's disease. *Antioxidants & Redox Signalling, 16*(9), 965–973. doi:10.1089/ars.2011.4419 PMID:22229791

Seirafi, M., Kozlov, G., & Gehring, K. (2015). Parkin structure and function. *The FEBS Journal, 282*(11), 2076–2088. doi:10.1111/febs.13249 PMID:25712550

Sekine, S., & Youle, R. J. (2018). PINK1 import regulation; a fine system to convey mitochondrial stress to the cytosol. *BMC Biology, 16*(1), 2. doi:10.118612915-017-0470-7 PMID:29325568

Sena, L. A., & Chandel, N. S. (2012). Physiological roles of mitochondrial reactive oxygen species. *Molecular Cell, 48*(2), 158–167. doi:10.1016/j.molcel.2012.09.025 PMID:23102266

Shan, X., Chiang, P. M., Price, D. L., & Wong, P. C. (2010). Altered distributions of Gemini of coiled bodies and mitochondria in motor neurons of TDP-43 transgenic mice. *Proceedings of the National Academy of Sciences of the United States of America, 107*(37), 16325–16330. doi:10.1073/pnas.1003459107 PMID:20736350

Shiba-Fukushima, K., Arano, T., Matsumoto, G., Inoshita, T., Yoshida, S., Ishihama, Y., ... Imai, Y. (2014). Phosphorylation of mitochondrial polyubiquitin by PINK1 promotes Parkin mitochondrial tethering. *PLOS Genetics, 10*(12), e1004861. doi:10.1371/journal.pgen.1004861 PMID:25474007

Shih, J., & Donmez, G. (2013). Mitochondrial Sirtuins as Therapeutic Targets for Age-Related Disorders. *Genes & Cancer, 4*(3-4), 91–96. doi:10.1177/1947601912474931 PMID:24019999

Shirendeb, U. P., Calkins, M. J., Manczak, M., Anekonda, V., Dufour, B., McBride, J. L., ... Reddy, P. H. (2011). Mutant huntingtin's interaction with mitochondrial protein Drp1 impairs mitochondrial biogenesis and causes defective axonal transport and synaptic degeneration in Huntington's disease. *Human Molecular Genetics, 21*(2), 406–420. doi:10.1093/hmg/ddr475 PMID:21997870

Singh, K. K. (2006). Mitochondria damage checkpoint, aging, and cancer. *Annals of the New York Academy of Sciences, 1067*(1), 182–190. doi:10.1196/annals.1354.022 PMID:16803984

Song, P., Li, S., Wu, H., Gao, R., Rao, G., Wang, D., ... Chen, Q. (2016). Parkin promotes proteasomal degradation of p62: Implication of selective vulnerability of neuronal cells in the pathogenesis of Parkinson's disease. *Protein & Cell, 7*(2), 114–129. doi:10.100713238-015-0230-9 PMID:26746706

Song, W., Chen, J., Petrilli, A., Liot, G., Klinglmayr, E., Zhou, Y., ... Masliah, E. (2011). Mutant huntingtin binds the mitochondrial fission GTPase dynamin-related protein-1 and increases its enzymatic activity. *Nature Medicine*, *17*(3), 377–382. doi:10.1038/nm.2313 PMID:21336284

Sterky, F. H., Lee, S., Wibom, R., Olson, L., & Larsson, N. G. (2011). Impaired mitochondrial transport and Parkin-independent degeneration of respiratory chain-deficient dopamine neurons in vivo. *Proceedings of the National Academy of Sciences of the United States of America*, *108*(31), 12937–12942. doi:10.1073/pnas.1103295108 PMID:21768369

Stoica, R., Paillusson, S., Gomez-Suaga, P., Mitchell, J. C., Lau, D. H., Gray, E. H., ... Hanger, D. P. (2016). ALS/FTD-associated FUS activates GSK-3β to disrupt the VAPB–PTPIP51 interaction and ER–mitochondria associations. *EMBO Reports*, *17*(9), 1326–1342. doi:10.15252/embr.201541726 PMID:27418313

Strappazzon, F., Nazio, F., Corrado, M., Cianfanelli, V., Romagnoli, A., Fimia, G. M., ... Cecconi, F. (2015). AMBRA1 is able to induce mitophagy via LC3 binding, regardless of PARKIN and p62/SQSTM1. *Cell Death and Differentiation*, *22*(3), 419–432. doi:10.1038/cdd.2014.139 PMID:25215947

Sun, N., Yun, J., Liu, J., Malide, D., Liu, C., Rovira, I. I., ... Finkel, T. (2015). Measuring in vivo mitophagy. *Molecular Cell*, *60*(4), 685–696. doi:10.1016/j.molcel.2015.10.009 PMID:26549682

Temme, C., Weissbach, R., Lilie, H., Wilson, C., Meinhart, A., Meyer, S., ... Wahle, E. (2009). The Drosophila melanogaster gene cg4930 encodes a high affinity inhibitor for endonuclease G. *The Journal of Biological Chemistry*, *284*(13), 8337–8348. doi:10.1074/jbc.M808319200 PMID:19129189

Truban, D., Hou, X., Caulfield, T. R., Fiesel, F. C., & Springer, W. (2016). PINK1, Parkin, and Mitochondrial Quality Control: What can we Learn about Parkinson's Disease Pathobiology? *Journal of Parkinson's Disease*, *7*(1), 13–29. doi:10.3233/JPD-160989 PMID:27911343

Tseng, A. H., Shieh, S. S., & Wang, D. L. (2013). SIRT3 deacetylates FOXO3 to protect mitochondria against oxidative damage. *Free Radical Biology & Medicine*, *63*, 222–234. doi:10.1016/j.freeradbiomed.2013.05.002 PMID:23665396

Um, J. H., & Yun, J. (2017). Emerging role of mitophagy in human diseases and physiology. *BMB Reports*, *50*(6), 299–307. doi:10.5483/BMBRep.2017.50.6.056 PMID:28366191

Valente, E. M., Abou-Sleiman, P. M., & Caputo, V. (2004). Hereditary early-onset Parkinson's disease caused by mutations in PINK1. *Science*, *304*(5674), 1158–1160. doi:10.1126cience.1096284 PMID:15087508

Valenti, D., de Bari, L., de Rasmo, D., Signorile, A., Henrion-Caude, A., Contestabile, A., & Vacca, R. A. (2016). The polyphenols resveratrol and epigallocatechin-3-gallate restore the severe impairment of mitochondria in hippocampal progenitor cells from a Down syndrome mouse model. *Biochimica et Biophysica Acta (BBA)-. Molecular Basis of Disease*, *1862*(6), 1093–1104. doi:10.1016/j.bbadis.2016.03.003 PMID:26964795

Vaziri, H., Dessain, S. K., Eaton, E. N., Imai, S. I., Frye, R. A., Pandita, T. K., ... Weinberg, R. A. (2001). hSIR2SIRT1 functions as an NAD-dependent p53 deacetylase. *Cell*, *107*(2), 149–159. doi:10.1016/S0092-8674(01)00527-X PMID:11672523

Wang, X., Winter, D., Ashrafi, G., Schlehe, J., Wong, Y. L., Selkoe, D., ... Schwarz, T. L. (2011). PINK1 and Parkin target Miro for phosphorylation and degradation to arrest mitochondrial motility. *Cell, 147*(4), 893–906. doi:10.1016/j.cell.2011.10.018 PMID:22078885

Wang, Y., Liu, N., & Lu, B. (2019). Mechanisms and roles of mitophagy in neurodegenerative diseases. *CNS Neuroscience & Therapeutics, 25*(7), 859–875. doi:10.1111/cns.13140 PMID:31050206

Waterham, H. R., Koster, J., van Roermund, C. W., Mooyer, P. A., Wanders, R. J., & Leonard, J. V. (2007). A lethal defect of mitochondrial and peroxisomal fission. *The New England Journal of Medicine, 356*(17), 1736–1741. doi:10.1056/NEJMoa064436 PMID:17460227

Weir, H. J., Murray, T. K., Kehoe, P. G., Love, S., Verdin, E. M., O'Neill, M. J., ... Balthasar, N. (2012). CNS SIRT3 expression is altered by reactive oxygen species and in Alzheimer's disease. *PLoS One, 7*(11), e48225. doi:10.1371/journal.pone.0048225 PMID:23139766

Williams, J. A., & Ding, W.-X. (2018). Mechanisms, pathophysiological roles and methods for analyzing mitophagy – recent insights. *Biological Chemistry, 399*(2), 147–178. doi:10.1515/hsz-2017-0228 PMID:28976892

Wong, Y. C., & Holzbaur, E. L. (2015). Temporal dynamics of PARK2/parkin and OPTN/optineurin recruitment during the mitophagy of damaged mitochondria. *Autophagy, 11*(2), 422–424. doi:10.1080/15548627.2015.1009792 PMID:25801386

Wu, A., Ying, Z., & Gomez-Pinilla, F. (2006). Oxidative stress modulates Sir2α in rat hippocampus and cerebral cortex. *The European Journal of Neuroscience, 23*(10), 2573–2580. doi:10.1111/j.1460-9568.2006.04807.x PMID:16817860

Xie, H., Li, Y. J., & Chen, Y. Y. (2011). Mitophagy. *Chinese Journal of Biochemistry and Molecular Biology, 27*(12), 1081–1087.

Yamamoto, H., Morino, K., Mengistu, L., Ishibashi, T., Kiriyama, K., Ikami, T., & Maegawa, H. (2016). Amla enhances mitochondrial spare respiratory capacity by increasing mitochondrial biogenesis and antioxidant systems in a murine skeletal muscle cell line. *Oxidative Medicine and Cellular Longevity, 2016*, 2016. doi:10.1155/2016/1735841 PMID:27340504

Yamano, K., & Youle, R. J. (2013). PINK1 is degraded through the N-end rule pathway. *Autophagy, 9*(11), 1758–1769. doi:10.4161/auto.24633 PMID:24121706

Yang, W., Zou, Y., Zhang, M., Zhao, N., Tian, Q., Gu, M., ... Yu, W. (2015). Mitochondrial Sirt3 expression is decreased in APP/PS1 double transgenic mouse model of Alzheimer's disease. *Neurochemical Research, 40*(8), 1576–1582. doi:10.100711064-015-1630-1 PMID:26045440

Yen, W. L., & Klionsky, D. J. (2008). How to live long and prosper: Autophagy, mitochondria, and aging. *Physiology (Bethesda, MD), 23*(5), 248–262. doi:10.1152/physiol.00013.2008 PMID:18927201

Yeung, F., Hoberg, J. E., Ramsey, C. S., Keller, M. D., Jones, D. R., Frye, R. A., & Mayo, M. W. (2004). Modulation of NF-κB-dependent transcription and cell survival by the SIRT1 deacetylase. *The EMBO Journal, 23*(12), 2369–2380. doi:10.1038j.emboj.7600244 PMID:15152190

Yin, X. M., & Ding, W. X. (2013). The reciprocal roles of PARK2 and mitofusins in mitophagy and mitochondrial spheroid formation. *Autophagy*, *9*(11), 1687–1692. doi:10.4161/auto.24871 PMID:24162069

Yoshii, S. R., Kishi, C., Ishihara, N., & Mizushima, N. (2011). Parkin mediates proteasome-dependent protein degradation and rupture of the outer mitochondrial membrane. *The Journal of Biological Chemistry*, *286*(22), 19630–19640. doi:10.1074/jbc.M110.209338 PMID:21454557

Yun, J., Puri, R., Yang, H., Lizzio, M. A., Wu, C., Sheng, Z. H., & Guo, M. (2014). MUL1 acts in parallel to the PINK1/parkin pathway in regulating mitofusin and compensates for loss of PINK1/parkin. *eLife*, *3*, e01958. doi:10.7554/eLife.01958 PMID:24898855

Zhang, T., Xue, L., Li, L., Tang, C., Wan, Z., Wang, R., ... Billiar, T. R. (2016). BNIP3 protein suppresses PINK1 kinase proteolytic cleavage to promote mitophagy. *The Journal of Biological Chemistry*, *291*(41), 21616–21629. doi:10.1074/jbc.M116.733410 PMID:27528605

Zhang, X., Cheng, X., Yu, L., Yang, J., Calvo, R., Patnaik, S., ... Delling, M. (2016). MCOLN1 is a ROS sensor in lysosomes that regulates autophagy. *Nature Communications*, *7*(1), 12109. doi:10.1038/ncomms12109 PMID:27357649

Zhang, Y., Gao, J., Chung, K. K., Huang, H., Dawson, V. L., & Dawson, T. M. (2000). Parkin functions as an E2-dependent ubiquitin–protein ligase and promotes the degradation of the synaptic vesicle-associated protein, CDCrel-1. *Proceedings of the National Academy of Sciences of the United States of America*, *97*(24), 13354–13359. doi:10.1073/pnas.240347797 PMID:11078524

Chapter 4
Microglial Mitophagy and Neurodegenerative Disorders

Eyitayo Adeyemi Oyindamola
Kwame Nkrumah University of Science and Technology, Ghana

Maxwell Kwadwo Agyemang
Kwame Nkrumah University of Science and Technology, Ghana

Joseph Owusu-Sarfo
Kwame Nkrumah University of Science and Technology, Ghana

Oduro Kofi Yeboah
https://orcid.org/0000-0001-7080-0730
Kwame Nkrumah University of Science and Technology, Ghana

Newman Osafo
https://orcid.org/0000-0001-8142-2368
Kwame Nkrumah University of Science and Technology, Ghana

ABSTRACT

Microglia are important in the regulation of the inflammatory response in regulating the release of proinflammatory mediators in the brain. Through their phagocytic actions, microglia are significant in the CNS when it comes to the body's response to physiological insults by promoting repair of impaired brain function. They do so by engulfing and degrading microbes as well as brain-derived debris and proteins such as myelin and axonal fragments, amyloid-beta, and apoptotic cells. This mitophagic activity of microglia is of importance in neurodegeneration. In most neurodegenerative disorders, mitophagy is impaired with resultant accumulation of dysfunctional mitochondria as well as processes such as lysosomal fusion and autophagosomes. In Parkinson's and Alzheimer's for example, impaired mitophagy accounts for the build-up of α-synuclein and amyloid respectively in affected individuals. The chapter discusses extensively the link between microglia mitophagy and neurodegeration and how dysfunctional mitophagy increases the likelihood of their occurrence.

DOI: 10.4018/978-1-7998-1317-0.ch004

INTRODUCTION

The mitochondrion is a membrane-bound organelle with several important roles in cellular function, including the production of adenosine triphosphate (ATP), the main energy currency of the cell, via oxidative phosphorylation, calcium homeostasis, and the metabolism of fatty and amino acids, and steroids (Wager and Russell, 2013). Notwithstanding, the mitochondrion is also the primary source of potentially damaging endogenous reactive oxygen species (ROS), which have been associated with a number of pathological pathways such as neurodegeneration, and the induction of lipid peroxidation, protein carbonyls and DNA damage (Santos et al., 2012; Murphy et al., 2011). It has been shown that the release of the cytochrome c, a hemeprotein relevant in the mitochondrial electron transport system and apoptosis, from mitochondria, triggers apoptosis under the regulation of several regulators, the most prominent being members of the B-cell lymphoma protein-2 (BCL2) family (Ow et al., 2008). ROS can induce mitochondrial permeability transition pore (mPTP) and also increase the release of cytochrome c from mitochondria, both of which results in programmed cell death (Murphy, 2008).

The removal of damaged mitochondria is essential for cell survival. Neurons, being highly specialized cells, are peculiarly liable to defects in autophagic mechanisms. These impairments in mitochondrial function and their dynamics have been identified in many neurodegenerative disorders, and modulators of both mitochondrial physiology and autophagy have presented themselves as promising therapeutic targets (Wager and Russell, 2013). Studies have demonstrated that deletion of certain pivotal autophagic genes such as *ATG-7* and *ATG-5* in successive post-mitotic cells enhances the formation and accumulation of cytoplasmic inclusions and induces neurodegeneration in the absence of any other pathological pathway that can also contribute to neural tissue death (Plaza-Zabala et al., 2017; Hara et al., 2006; Komatsu et al., 2006). This selective degradation of mitochondria by highly specialized autophagic mechanisms is what is termed mitophagy, and represents an important quality control mechanism in protein folding (Wager and Russell, 2013).

Microglia are the brain's resident macrophages and contribute to a major part of the brain's innate immune system. By orchestrating an inflammatory response, regulating the release of proinflammatory mediators within the brain and through phagocytosis, microglia respond to physiological insults to the central nervous system (CNS) and promote the correction and repair of the brain function following CNS damage (Plaza-Zabala et al., 2017). In addition to engulfing and degrading microbes, microglia also phagocytose different types of brain-derived cargo such as myelin and axonal fragments, synaptic materials, apoptotic cells, and protein deposits such as amyloid-β (Aβ), among other things (Sierra, 2013). Autophagy and phagocytosis are strikingly similar in morphology and mechanisms, including the formation of the transient vesicular structures, autophagosomes, and phagosomes, respectively. These vesicular structures aid in the delivery of cargo to lysosomes for digestion and degradation (Martinez et al., 2011). Both processes, as such, play a crucial role in maintaining cellular homeostasis through the degradation of harmful substrates of both intracellular and extracellular origins. However, in contrast to phagocytosis which occurs in a selective population of immune cells including macrophages and microglia, neutrophils and dendritic cells, autophagy occurs in almost all cell types in mammals (Sierra, 2013). It is believed that, the two processes are not mutually exclusive, and that functional cross-talk may exist between them during innate immune response in peripheral macrophages. The potential regulatory action of autophagy over phagocytosis has been suggested to unfold at different steps along the phagocytic cascade, which may affect the engulfment and degradation of the phagocytic cargo (Plaza-Zabala et al., 2017).

Three distinct autophagic mechanisms have been described based on mechanisms by which substrates are delivered into lysosomes for degradation, that is, microautophagy, chaperone-mediated autophagy and macroautophagy (Lahiri et al., 2019; Yu et al., 2018; Badadani, 2012). Macroautophagy, a tightly regulated process that guards against the inappropriate removal of cytosolic components was for many years thought to e a nonselective bulk degradation process. However, it is now evident that there exist very distinct subtypes of this autophagic mechanism, including mitophagy, mediated by autophagic adaptor proteins (Youle and Narendra, 2011; Johansen and Lamark, 2011). In this chapter, authors focus on the studies of mitophagy in microglial cells, and its role in the accumulation of misfolded protein aggregates implicated in neurodegeneration.

PATHWAYS OF MITOPHAGY IN MAMMALS

In the mammalian cell, the activation of mitophagy occurs via two major pathways; receptor-mediated and PINK1/Parkin-mediated, also known as ubiquitin-independent and ubiquitin-dependent pathways, respectively (Figure 1) (Wang et al., 2019). Although these two pathways have been relatively well studied, the exact nature of the difference between the two mechanisms in terms of expression in different tissues or in the output of mitochondria degradation remains unknown (Wei et al., 2015). Two receptors, NIX (BNIP3L) and SQSTM1/62, in mammalian cells have been recognized to link mitochondria with autophagy mechanism in various cell types. The transient vesicular structures formed during autophagy recognize target mitochondria through LC3 adapters either in a ubiquitin-dependent or ubiquitin-independent manner and also via direct interaction of LC3 with its receptors (Li et al., 2019). Although molecular mechanisms of mitophagy in mammals differ from that of yeast, some functional counterparts of the yeast Atg32 have been proposed as mitophagy receptors in mammalian cells. These functional counterparts include BNIP3 and its homologous protein BNIP3L (also known as NIX), and FUN14 domain-containing protein 1 (FUNDC1), all of which can be recruited to the outer mitochondrial membrane (Shi et al., 2014). These receptors also possess the LC3-interacting regions (LIRs), thus can interact with LC3, the mammalian homolog of Atg8, an interaction crucial in the selection of mitochondria as cargo. Reversible protein phosphorylation has also been suggested in the effective controlling of receptor-mediated mitophagy. A perfect illustration is the phosphorylation of FUNDC1 by Src and CK2 under conditions of normal oxygen pressures (pO_2, resulting in suppression of mitophagy. Under hypoxic conditions, however, FUNDC1 is dephosphorylated by phosphoglycerate mutase 5 (PGAM5) resulting in the upregulation of mitophagy (Liu et al., 2014; Chen et al., 2014).

Receptor-Mediated Mitophagy

BNIP3 and BNIP3L/NIX are BH3-only proteins and belong to the BCL2 family. These receptors function in cell death and together with the LC3 family proteins, they remove mitochondria (Kang and Saif, 2008; Wang et al., 2019). It is believed that NIX is essential in reticulocyte maturation and in addition to Atg7 and ULK1, functions to remove mitochondria in reticulocytes (Schweers et al., 2007; Wang et al., 2019). In addition to the enhancement of mitophagy by upregulation of BNIP3 or its homologous protein, the interplay between BNIP3 and LC3 through phosphorylation of Ser17 and Ser24, adjacent to the LIRs of BNIP3, also enhances mitophagy. Moreover, the interaction between NIX and a small GTPase Rheb has also been demonstrated to initiate mitophagy (Li et al., 2007).

Figure 1. Mechanisms of Mitophagy in mammals.
Some OMM proteins, such as BNIP3, NIX, and FUNDC1, possess the LC3-interacting regions (LIRs) that interact with LC3, serving as a crucial link in the selection of mitochondria as cargo. In the damaged mitochondria, the loss of mitochondrial membrane potential leads to the accumulation of PINK1 on the membrane. PINK1 then recruits and phosphorylate Parkin which in turn stimulates the ubiquitination of its substrates on OMM via K48 or K63 linkage, followed by protein quality control and subsequent mitochondrial quality control. NBR1, p62, OPTN, NDP52, and TAX1BP1 are autophagy adaptor proteins involved in this process. Ambra1, activating molecule in beclin 1-regulated autophagy; BCL2L13, BCL2-like 13; BNIP3, Bcl2/adenovirus E1B 19-kDa protein-interacting protein 3; Drp1, dynamin-related protein 1; FKBP8, FK506-binding protein 8; FUNDC1, FUN14 domain-containing protein 1; LC3, microtubule-associated protein 1 light chain 3; Mfn 2, mitofusin 2; NDP52 (CALCOCO2), calcium-binding and coiled-coil domain 2; NIX (BNIP3L), BCL2/adenovirus E1B 19-kDa interacting protein 3-like; OMM, outer mitochondrial membrane; OPTN, optineurin; P, phosphate; p62 (SQSTM1), sequestosome 1; PGAM5, phosphoglycerate mutase family member 5; PHB2, Prohibitin 2; PINK1, PTEN-induced putative kinase 1; TOM70, translocase of outer mitochondrial membrane 70; TRAF, tumor necrosis factor receptor-associated factor; Ub, ubiquitin; ULK1, UNC51-like kinase 1.

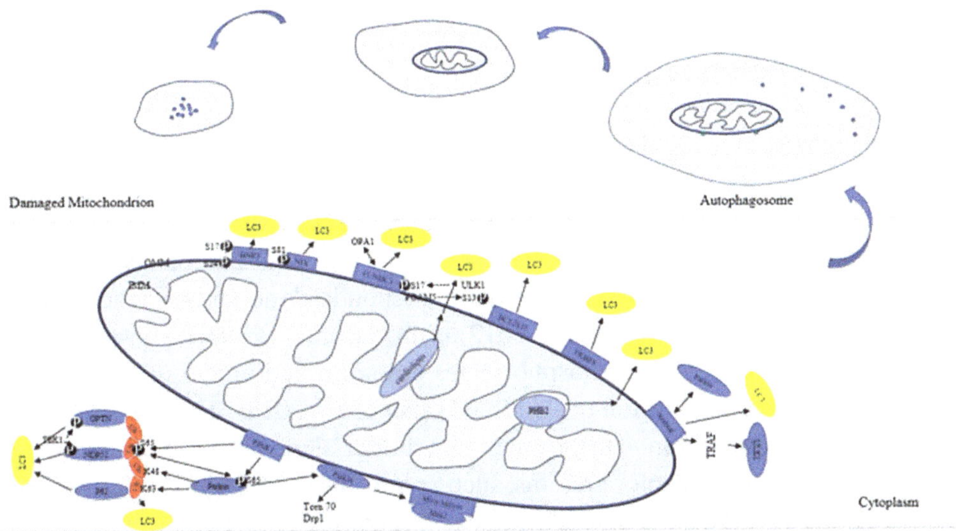

The outer mitochondrial membrane protein, FUNDC1, has been recognized to effectively activate autophagy mechanisms and promote mitophagy under hypoxic conditions. The mechanism is thought to involve the dramatic downregulation of FUNDC1 mRNA and protein levels in the condition of low oxygen tension. Under these hypoxic conditions, FUNDC1 is phosphorylated at Ser17 by ULK1 and dephosphorylated at Ser13 by PGAM5. The latter mechanism can be blocked by BCL2L1/B-cell lymphoma-extra large (Bcl-xL). Phosphorylation of Ser17 by ULK1, under conditions of low oxygen tension, promotes interaction between FUNDC1 and LC3 (Wu et al., 2014). In this interaction, FUNDC1 via its LIR motif can recruit LC3, resulting in the activation of mitophagy. This outer mitochondrial membrane protein can also interact with DNM1L/DRP1 and OPA1, to regulate mitochondrial fission or fusion and mediate mitophagy. FUNDC1 can, therefore, coordinate the dynamics and quality control of mitochondria (Chen et al., 2016) (Figure 1).

Other outer mitochondrial membrane proteins that mediate mitophagy include FK506-binding protein 8 (FKBP8) and Bcl2-like 13 (BCL2L13). The latter interacts with LC3 via its LIR motif located on its N-terminal helix facing the cytoplasm. FKBP8 also interacts with LC3 (Bhujabal et al., 2017). When treated with the chemical ionophore, CCCP, FKBP8 translocate into the endoplasmic reticulum. However, the FKBP8 N412K mutant cannot migrate from the mitochondrion into the endoplasmic reticulum and is defective in the suppression of apoptosis during mitophagy. This implies that, during mitophagy, not all mitochondrial-derived proteins are degraded, and that the subcellular localization of FKBP8 can coordinate the dynamics of cell survival during mitophagy (Saita et al., 2013).

Following Parkin-mediated degradation of the outer mitochondrial membrane, proteins such as prohibitin 2 (PHB2), located on the inner mitochondrial membrane can be exposed. The exposed PHB2 then interacts with LC3 to induce mitophagy (Wang et al., 2019). Cardiolipin, another inner mitochondrial membrane receptor can be externalized during mitochondrial depolarization. The externalized receptor can then bind to the N-terminal helix of LC3 to regulate mitophagy (Lee et al., 2017). It has been shown that Parkin translocation can trigger mitophagy through the activation of a Beclin 1 interactor, known as Ambra1 (van Humbeeck, 2011). The mechanism is thought to involve the slow translocation of Ambra1 from the cytoplasm into the endoplasmic reticulum and its subsequent regulation of autophagosome nucleation (van Humbeeck et al., 2011).

Ubiquitin-Mediated Mitophagy

The PINK1-Parkin pathway is recognized to be crucial in regulating mitochondrial homeostasis in cells. The pathway promotes mitochondrial health through several mitochondrial quality control mechanisms, including the turnover of outer membrane mitochondrial proteins by the proteasome, formation of mitochondria-derived autophagosomes, and degradation of organelles via mitophagy. In a mitochondrion, PINK1 is highly expressed and stabilized on the outer mitochondrial membrane, where it recruits Parkin to the damaged mitochondrion (Greene et al., 2012). Subsequently, PINK1 phosphorylates Parkin at Ser65 and at the same position PINK1 phosphorylates ubiquitin and ubiquitin-like domain of Parkin to allosterically relieve the autoinhibition of Parkin (Okatsu et al., 2018). Thereafter, Parkin directs the ubiquitylation of different proteins found in the outer mitochondrial membrane. Subsequent interaction of these ubiquitylated proteins with adaptor proteins, such as mitofusin, Miro, TOM70, and Drp1, mediates mitochondrial sequestration (Okatsu et al., 2010). These findings suggest that the mitochondria-targeted serine/threonine kinase, PINK1, and the cytoplasmic E3 ubiquitin ligase, Parkin, plays multifaceted roles in the dynamics and survival homeostasis of the mitochondria.

Autophagy adaptor proteins possess ubiquitin-binding domains that bind to ubiquitinated mitochondrial proteins and LIR motifs to induce mitophagy (Johansen and Lamark, 2011; Lazarou et al., 2015). Of the several adaptor proteins, optineurin (OPTN) is the most studied, for PINK1/Parkin-mediated mitochondrial phagophore recruitment (Wong and Holzbaur, 2014). By interacting with ubiquitylated proteins on the outer mitochondrial membrane, OPTN is recruited to the damaged mitochondria where it interacts with LC3 to induce mitochondrial isolation by an autophagosome (Chu, 2019). Other adaptor proteins include NBR1, NDP52 (CALCOCO2), SQSTM1 (p62) and TAX1BP1. The roles of these proteins in mitophagy are summarised in Figure 1. Subsequent sections in this chapter will define the roles and relevance of these mitophagy receptors and ubiquitylated proteins in the regulation of dynamics of brain function by microglial cells in neurodegenerative disorders such as Alzheimer's disease (AD), Parkinson's disease (PD), Huntington's disease (HD) and amyotrophic lateral sclerosis (ALS).

MITOPHAGY AND ALZHEIMER'S DISEASE

AD is said to account for 70% of dementia (Kametani and Hasegawa, 2018). It is an irreversible and progressive disease of the brain that slowly destroys thinking skills and memory. The capacity to perform simple tasks is even impaired. The disease usually occurs in people within the age group of 65 and above (De, Bala, and DasGupta, 2011).

AD is characterized by the appearance of senile plaques and neurofibrillary tangles. The former is reported to develop first in brain regions that are associated with cognition after which it spreads to other cortical regions with the progression of the disease. The senile plaques consist of several other components of insoluble deposits of Aβ-peptides which is a fragment of amyloid precursor protein (APP). The generation of Aβ-peptide occurs via two consecutive cleavage processes. First, β-secretase generates one end of the Aβ-peptide, while γ-secretase generates the other end. Two forms of Aβ are known to exist. Aβ42 is a longer species and Aβ40 is the shorter species. The longer species are reported to be deposited first and thus may partake in the initiation of events which eventually lead to the deposition of amyloid (De et al., 2011). As to whether senile plaques are the cause or product of AD remains unclear. There is however a lot of evidence that dysfunction in APP metabolism with a consequent increase in Aβ is responsible for AD. Neurofibrillary tangles, which is the second distinguishing characteristic of AD are mostly formed via chemically modified tau protein; the formation of tangle is limited to severity of disease such that, the more advanced the disease is, the more tau angles are formed on the brain.

The loss of neurons and synapses in the cerebral cortex and subcortical region represents the main pathophysiology of AD. The loss leads to atrophy of the affected brain areas – degeneration of the temporal and parietal lobes as well as parts of the frontal cortex and cingulate gyrus (Wenk, 2003). In the brains of individuals affected by AD, microscopy shows clearly visible neurofibrillary tangles and amyloid plaques (Tiraboschi et al., 2004). Even though some individuals may develop some tangles and plaques due to aging, the number of tangles and plaques present in the brains of Alzheimer's patients is far greater.

Amyloid Protein

For decades, the amyloid hypothesis has been the main explanation for the pathogenesis of AD (Hardy and Allsop, 1991; Hardy and Selkoe, 2002; Hardy and Higgins, 1992; Selkoe, 1991). Unlike normal individuals, the metabolism of APP and the ability to degrade Aβ is reduced, leading to the accumulation of Aβ-peptides. As mentioned earlier, Aβ42 initiates amyloid fibril formation. Elevated levels of Aβ42 causes an induction of the formation of amyloid fibrils. The accumulated Aβ amyloid fibrils subsequently develop into senile plaques, which causes toxicity of neurons. There is also the induction of tau pathology, which results in the death of neurons.

The amyloid hypothesis, however, faces some setbacks. A study by Bryan et al., using genetically modified mice models, in which Aβ amyloid is deposited in the brain found out that although Aβ amyloid fibrils accumulation were observed, neurofibrillary tangles (resulting from the accumulation of tau) and neuronal cell death were not observed (Bryan et al., 2009). This finding suggests that Aβ fibrils accumulation is not intrinsically neurotoxic and also, that Aβ does not induce tau accumulation. Due to the fact that AB is a normal product of APP metabolism, and is not toxic under normal physiological conditions, there is an idea that Aβ oligomers are the key toxic agents (Kametani and Hasegawa, 2018). Due to the advancement in amyloid imaging, observation of Aβ accumulation in the patient's brain has been made possible. These have been several reports of normal individuals with amyloid deposits; very few amyloid deposits in AD patients have also been observed (Edison et al., 2007; Li et al., 2008). There have also been reported that in the brains of some elderly patients without dementia, the distributions of senile plaques are sometimes as extensive as dementia patients (Chételat et al., 2013; Davies et al., 1987; Fagan et al., 2009; Karpuj et al., 1999). These findings suggest the deposition of Aβ is an aging phenomenon and that it has no direct relation with the onset of AD. With these facts on hand, it appears that neuronal loss and amyloid deposition are independent phenomena (Chételat et al., 2013).

Tau is a microtubule-associated protein that is involved in the regulation of the stability of tubulin assemblies. Isoforms of the tau proteins exist – 3R tau and 4R tau (Kametani and Hasegawa, 2018). It is reported that in the AD brains, both isoforms are accumulated in a hyperphosphorylated state in the pathological inclusions (Goedert, 1993; Goedert et al., 1996; Iqbal, Liu, and Gong, 2016; Serrano-Pozo et al., 2011). These inclusions have different names depending on where they are formed. If formed in the axons or dendrites, they are referred to as threads, while they are called neurofibrillary tangles. The spreading of tau is reported to correlate strongly with the extent of chemical presentation and cognitive symptoms (Braak and Braak, 1991).

The involvement of Aβ in the pathogenesis of AD has necessitated research into therapeutic agents that would affect the production of Aβ. According to Golde et al., there are currently 3 distinct strategies targeting Aβ production and these have all moved to clinical trials (Golde et al., 2010). These include β-secretase inhibitors, β and γ- secretase inhibitors are known to block the production of all Aβ species (Vassar, 2002; Wolfe, 2008). The modulators, on the other hand, shift the cleavage by γ-secretase, the consequence of which is the alteration, that is, the profile of Aβ peptides. Aβ42 lowering γ-secretase modulators reduce the production of Aβ-42 while increasing the levels of Aβ-peptide shorter than Aβ1-40 (Weggen et al., 2001).

Therapeutic agents that target tau, however, remain in the conceptual stage despite the fact that some preclinical studies provide evidence of that reduction in the production of tau can be beneficial in the disease (Golde et al., 2010). The molecular mechanisms of impaired mitochondrial homeostasis in AD are still under investigation. The study by Fang et al., reported that in AD, there is impairment of mitophagy in the hippocampus of the patient in with AD (Fang et al., 2019). The impairment was also observed in induced pluripotent stem cell cultured human AD neurons and in animal AD models.

In individuals who develop the condition of AD, there is impairment of glucose utilization, which seems to be a widely documented abnormality in neuronal physiology that occurs before the onset of the discernible cognitive deficit in such individuals. This has been demonstrated by using 2-deoxy-D-glucose positron emission tomography brain imaging (Kapogiannis and Mattson, 2011). As early as 1991, the proposal that disruptions in the health of mitochondria and neuronal metabolism are early features of AD was put forward by Gibson and Blass (Blass and Gibson, 1991). Afterwards, there have been several studies reporting in abnormalities in mitochondria in AD. Results from studies in living AD patients and the post-mortem brain tissues show that neuronal cells in the affected brain region suffer from mitochondrial dysfunction (Kerr et al., 2017). The positron emission tomography (PET) brain scans have shown that there is decreased radiolabelled glucose uptake into neurons. Biochemical studies have also reported that there is decreased activity of the mitochondrial enzyme involved in oxidative phosphorylation and the citric acid cycle (CAC) (Kapogiannis and Mattson, 2011). The consequence of the accumulation of dysfunctional mitochondria is the reduced cellular ATP and excessive production of ROS, which can worsen mitochondrial damage, resulting in the anomalous amyloidogenic processing of APP and pTau and the subsequent forming of AD-defining Aβ plaques and neurofibrillary tangles (Mattson et al., 2008).

Mitophagy, being a center to the survival of mitochondria, has been reported to be impaired in AD (Kerr et al., 2017). Neurons that exhibit abnormal accumulation of autophagosomal vacuoles are a very distinct characteristic of AD. Their accumulation may stem from lysosomal dysfunction which may be due to dysregulation of calcium homeostasis in the neurons (Nixon, 2013). Abnormal mitophagy in AD may be attributed to a dysfunctional fusion between lysosomes and autophagosomes. For instance, the lysosomal dysfunction in healthy cells has been shown to result in neuronal phenotype like those observed in AD (Nixon, 2013; Nixon et al., 2008).

The stimulation of mitophagy has been demonstrated to ameliorate the cognitive deficits in AD. In the study by Fang et al., using Aβ and tau *Caenorhabditis elegans* models of AD, mitophagy was stimulated via supplementation of nicotinamide adenine dinucleotide (NAD^+), acitonin and urolithin A. The result was a reversal of memory impairment. The reversal of memory impairment was observed to be mediated via PINK1, DCT-1 or PDR-1 pathways. Mitophagy was shown to reduce the levels of Aβ42 and Aβ40 and prevents impairment of cognition in an APP/PSI mouse model. This was attributed to the microglial phagocytosis of extracellular Aβ plaques and suppression of neuroinflammation. The stimulation of mitophagy was shown to stop AD-related tau hyperphosphorylation in human neuronal cells and cause a reversal of memory impairment in transgenic mice and nematodes (Fang et al., 2019). Another study showed that treating of 3xTgAD mice with nicotinamide (a precursor of NAD^+) led to an improvement in tau and AB pathologies as well as amelioration of memory and learning deficits (Liu et al., 2013). Using rapamycin, the inhibitor of mammalian target of rapamycin (mTOR), resulted in the amelioration of cognitive deficits and decreased Aβ pathologies in an APP mutant AD mouse model (Spilman et al., 2010).

These findings suggest that impairment mitophagy plays a pivotal role in the pathogenesis of AD and that using mitophagy inducing agents like acitonin, spermidine and urolithins (Morselli et al., 2011; Ryu et al., 2016; Sun et al., 2015) represent an important therapeutic potential.

Sirtuins

Sirtuins represent a class of NAD+ dependent enzymes and are known for their beneficial effects in age-related disorders. A number of substrates have been uncovered for sirtuins, typically for SIRT1, which appears to be the most widely studied sirtuins in the brain (Haigis and Sinclair, 2010). This particular sirtuin has been studied in many neurodegenerative disorders like AD and HD.

Sirtuins were initially identified in yeast and termed as silent information regulators (SIRs). They perform their functions via deacetylation of lysines through the consumption of NAD+ (Sinclair and Guarente, 1997). The seven human homologs of sirtuins perform different enzymatic roles. The NAD+ dependent ability to deacetylate nonhistone and histone substrate is crucial to the regulation of cellular function (Anekonda and Reddy, 2006; Gan and Mucke, 2008).

Because SIRT1 might play a role in the development of AD, via regulation of aging and other metabolic processes, this enzyme has obtained tremendous research attention (Anekonda and Reddy, 2006; Gan and Mucke, 2008; Nunomura et al., 2007; Outeiro, Marques, and Kazantsev, 2008). Studies have reported that SIRT1 suppresses gamma-secretase activity in a number of *in vitro* models. This serves to decrease the production of Aβ (Qin et al., 2006a). The decrease in the activity of gamma-secretase has been duplicated *in vivo* using transgenic mice that overexpress SIRT1 (Qin et al., 2006b). This provides the first evidence that links SIRTs, calorie restriction, and AD. A correlation between the load of AB and SIRT1 has also been observed in the brain of nonhuman primates that underwent calorie restriction (Qin et al., 2006a).

Attenuation of amyloidogenic processing of APP by SIRT1 has been demonstrated using AD transgenic mouse models and cell cultures (Bonda et al., 2011). The mechanism is via SIRT1 mediated increase in the alpha-secretase production and activity through alpha-secretase gene activation. Because alpha-secretase is responsible for non-amyloidogenic cleavage of APP and increase in its activity tilts the balance of the APP processing to decrease the accumulation of toxic Aβ, that would otherwise result from the activities of beta-secretase and gamma-secretase. The overexpression of SIRT1 may be mediated by NAD+ or the antioxidant resveratrol and has been shown to result in a decrease in oligomerized Aβ-peptides (Wang et al., 2010), as well as to attenuate oxidative stress (Albani et al., 2009; Feige et al., 2008). Chen et al., report that SIRT1 expression prevents microglial-dependent Aβ toxicity via inhibition of NF-κB signaling (Chen et al., 2005).

Studies using PET imaging have revealed that the brain regions that are prone to the deposition of Aβ plaques and neurodegeneration in AD are spatially similar to areas that metabolize glucose through glycolysis in the brains of young normal individuals (Vlassenko et al., 2010). Aerobic glycolysis an immediate source of ATP through the metabolism of glucose-6-phosphate to pyruvate and lactic acid despite the availability of oxygen and is important particularly for fast-growing cells (Vander Heiden, Cantley, and Thompson, 2009). Aerobic glycolysis incidentally depletes the reserves of NAD+. This results in an inhibition of NAD+ through enhanced NADH production and reduced regeneration of NAD+. This results in an inhibition of NAD+ dependent histone deacetylase activity of SIRT1 (Nakahata et al., 2008). This serves to suggest that there may be a link between the utilization of aerobic glycolysis in the mediotemporal lobe of the brain and the coincidental distribution of Aβ plaques in AD. That is to say that, in those brain regions. The dependence aerobic glycolysis by neurons possibly inhibits the activity of SIRT1 (via depletion of NAD+ reserves), resulting in a tilt in the balance of APP processing towards the amyloidogenic pathway.

The critical role of sirtuins, particularly SIRT1 in age-related disorders has also been studied extensively, with a lot of research still going on. The sirtuin pathway possibly has a role in the pathogenesis of AD. However, whether or not that role is causal still remains unknown. Overexpression of SIRT1 has shown tremendous benefits as demonstrated in animal and cell culture models. Agents that enhance SIRT1 overexpression, thus represent therapeutic potential in the management of AD.

The PINK/Parkin Pathway

The PINK1/Parkin pathway represents one of the most studied ways of clearing damaged mitochondria in mammalian cells. The PTEN-induced kinase 1 (PINK1) is a mitochondrial serine/threonine-protein kinase that has been shown to protect cells via attenuation of stress-induced mitochondrial dysfunction. The activity of this enzyme leads to the binding of Parkin protein to depolarised mitochondria. Binding serves to result in autophagy of those mitochondria (Lazarou et al., 2013; Narendra et al., 2010). In normal, healthy cells, mitochondria maintain a membrane potential. This potential can be used to bring in PINK1 to the inner mitochondrial membrane. There, PINK1 is cleaved by presenilins-associated rhomboid-like protein (PARL), after which it is removed from the outer membrane. In severely damaged mitochondria, the ability to import PINK1 is lacking due to the insufficient potential. Thus, PINK1 is accumulated on the outer membrane. The consequence of the accumulation of PINK1 is the recruitment of E3 ubiquitin ligase Parkin which targets the damaged mitochondrial via polyubiquitination of mitochondrial proteins (Lazarou et al., 2015; Youle and van der Bliek, 2012) for removal via mitophagy.

Alternatively, PINK1 may cause the recruitment of autophagy receptors in a direct manner via Parkin-independent fashion. This pathway results in low levels of mitophagy (Lazarou et al., 2015). The critical understanding of the mechanisms through which AD-linked pathological processes affect mitophagy mediated by Parkin is however limited.

Aβ has been identified in mitochondrial membranes and has been shown to have interactions with mitochondrial proteins. Through the interaction, Aβ may affect mitochondrial dynamics and cause changes in the motility of the mitochondria (Calkins et al., 2011; Devi et al., 2006; Du et al., 2010; Manczak et al., 2011; X. Wang et al., 2008). The interaction may cause disruptions in the electron transfer chain and increase the production of ROS. Impairment of mitochondrial function via interaction has also been reported (Devi et al., 2006; Du et al., 2008; Lustbader et al., 2004; Mattson et al., 1998). Whether mitophagy is efficiently induced under these pathophysiological conditions however remains unknown.

APP transgenic mouse models revealed that mitophagy mediated by Parkin is heavily induced in mutant human APP transgenic neurons and AD patients' brains, which is closely linked to an increased mitophagic flux. The mutant human APP neurons showed increased recruitment of cytosolic Parkin to depolarised mitochondria. The translocation of Parkin primarily takes place in somatodendritic regions; the effects of this compartmentalization being a decrease in anterograde mitochondrial transport in axons. Increased mitophagy in the brains of AD patients was coupled with a reduction of cytosolic Parkin as the disease progressed, suggesting that aberrant accumulation of damaged mitochondria may be linked to inadequate mitophagy in the removal of defective mitochondrial (Ye et al., 2015). The findings in the aforementioned study actually provide the first *in vivo* indication that there is progressive depletion of Parkin in the course of the development of AD.

Future investigations are needed to probe further into the role of the PINK1/Parkin pathway in the development of AD. With available evidence showing that an increase in Parkin-mediated mitophagy enhances the removal of defective mitochondria, approaches that target the modulation of mitophagy may ameliorate the mitochondrial pathology in AD and may represent therapeutic importance.

MITOPHAGY IN PARKINSON'S DISEASE

PD is the second most common neurodegenerative disorder after AD and the most common movement disorder (Martinez-Vicente, 2017). PD is pathologically defined by progressive loss of dopaminergic neurons in substantia nigra resulting in a decrease of dopamine (DA) in the striatum and the development of intraneuronal cytoplasmic inclusions called Lewy bodies (LBs) composed primarily of aggregated alpha-synuclein proteins (Ferreira and Romero-Ramos, 2018). The disease affects nearly 2% of the world's population over 50 years (Gao et al., 2017). Generally, the cause of PD is unknown, but it is believed to involve both genetic and environmental factors (Gan-Or et al., 2015). The cardinal features of PD are bradykinesia, resting tremor, rigidity and postural instability (Gazewood et al., 2013). Despite great advances in biomedical research, available treatments are aimed at relieving symptoms; without stopping or slowing down the progressive death of neurons (Martinez-Vicente, 2017). From the last two decades, rigorous studies have provided a better understanding of the pathogenesis of the disease and revealed new pathways of which impaired mitophagy is implicated (Gao et al., 2017).

Mitophagy and Alpha-Synuclein

Mitophagy is the specific autophagic elimination of mitochondria (De Duve, 1963). It is the only known metabolism via which whole mitochondria can be selectively eliminated (Martinez-Vicente, 2017). In PD, mitophagy is impaired resulting in the accumulation of dysfunctional mitochondria and bioenergetic deficits that occur early and promote the disease-related alpha-synucleinopathy (Martinez-Vicente, 2017; Gao et al., 2017).

Studies using genetic-modulated or toxin-induced animal and cellular models as well as postmortem human tissue indicate that compromised mitophagy might be a pivotal factor in the pathogenesis of synaptic dysfunction and the aggregation of misfolded proteins (LBs), which in turn impairs mitochondrial homeostasis.

The primary component of LBs is α-synuclein (Spillantini et al., 1997). Alpha-synuclein is a protein constitutively synthesized by neurons and is normally found in the synapses. It can also be found localized to mitochondria and connected to the endoplasmic reticulum (ER) through mitochondrial-associated ER membrane (Gao et al., 2017). Alpha-synuclein is suggested to play a central role in both the neuronal events and immune responses occurring in PD (Pique et al., 2016; Ferreira and Romero-Ramos, 2018). During PD, α-synuclein progressively transforms from soluble to insoluble complexes, most likely through intermediate highly diffusible small soluble oligomeric forms, which abnormally aggregate leading to neuronal dysfunction (Eschbach and Danzer, 2014; Gao et al., 2017). Dopaminergic neurons are the most vulnerable and will degenerate and die resulting in nigral cell loss, decrease in dopamine levels which will, in turn, impair basal ganglia circuitry (Obeso et al., 2008). This results in the four cardinal symptoms of PD as mentioned above.

Overexpression of α-synuclein inhibits the normal function of inner-mitochondrial membrane-anchored respiratory chain complexes in the whole brain of PD patient, but mostly in nigrostriatal neurons (Gao et al., 2017). α-Synuclein overexpression has been shown to increase the number of fragmented mitochondria *in vitro* (Plotegher et al., 2014). In addition, pre-fibrillar forms of α-synuclein reduce mitochondrial Ca^{2+} retention. Ca^{2+} is required by the mitochondria to generate ATP via the tricarboxylic acid cycle. The impaired mitochondrial function may lead to a reduction in cellular function and decreased ATP levels as well as excessive reactive oxygen species production in neurons, which further exacerbate mitochondrial damage (Mattson et al., 2008). α-Synuclein has a tendency to spread from one cell to another (Ferreira and Romero-Ramos, 2018). Accumulation of α-synuclein in different neuronal populations seems to be associated with all the symptoms of PD. The various neurodegenerative stages proposed by Braak show a correlation between an α-synuclein distribution and PD progression.

Most PD patients are diagnosed as sporadic but other less common forms of familial PD have been linked to different autosomal recessive and autosomal dominant genes. Two genes causing autosomal recessive forms of the disease are PARK 2 and PARK 6, which encode Parkin and PINK1 respectively. PINK1 is localized to mitochondria and Parkin resides in the cytosol. These two proteins mediate a type of mitophagy known as PINK1/Parkin-mediated mitophagy (Pickrell and Youle 2015).

PARK2, the first among the two to be identified, contains 12 exons that encode the 465 amino acid protein, Parkin (Kitada et al., 1998). Parkin is an E3 ubiquitin ligase with an amino-terminal ubiquitin-like domain and a carboxyl-terminal ubiquitin ligase domain (Hristova et al., 2009; Shimura et al., 2000). Parkin plays a key role in mitochondrial homeostasis. Cell biology studies reveal that Parkin is recruited from the cytosol to depolarized mitochondria to mediate mitophagy (Narendra et al., 2008). Mutations in the Parkin protein produce a parkinsonian syndrome which is the most frequent cause of juvenile PD

(Matsumine et al., 1997; Davie, 2008). Accordingly, it has been shown that Parkin facilities the binding of ubiquitin to other proteins such as the α-synuclein interacting protein synphilin-1 leading to the formation of Lewy bodies (Davie, 2008).

Phosphatase and tensin homolog (PTEN)-induced kinase 1 (PINK1) is made up of 581 amino acids. PINK1 and Parkin normally work together to govern mitochondrial quality control via PINK1/Parkin-mediated mitophagy (Pickrell and Youle 2015). PINK1 is the major player, essential for the phosphorylation of ubiquitin, while Parkin plays an important role by amplifying this signal (Martinez-Vicente, 2017). The fact that mutations in these two genes are linked to autosomal recessive forms of PD led to greater attention being focused on mitophagy, and the possibility that impaired mitochondrial turnover might be one of the main contributors to PD pathogenesis (Martinez-Vicente, 2017). Few studies using some particular mouse models (PINK1- and Parkin-knockout mice) could only show mitochondrial dysfunction, but probably not enough to trigger neuronal death arising from neuronal loss, altered DA metabolism, presence of LBs, or abnormality in motor behavior (Martinez-Vicente, 2017). However, Parkin- and PINK1-mutant fly models have shown a more clear-cut PD phenotype, including mitochondrial dysfunction as well as dopaminergic neuronal loss, significant motor disabilities and reduced lifespan (Martinez-Vicente, 2017).

Protein Deglycase DJ-1

A rare autosomal recessive form of PD caused by mutations in the *PARK7* gene which encode DJ-1 has been reported (Bonifati et al., 2003; Hague et al., 2003). DJ-1 often known as a redox sensor/reductase influences mitochondrial homeostasis and mitophagy (Canet-Aviles et al., 2004) As a transcriptional regulator, it has been shown to regulate the clearance of endogenous ROS through the modulation of scavenging systems by thioredoxin/apoptosis signal-regulating kinase 1 (Trx/Ask1) complex (Andres-Mateos et al., 2007). Deficiency of DJ-1 decreases brain mitochondria consumption of hydrogen peroxide (H_2O_2), leading to an increased level of oxidative stress, and eventually causes cell death in DA neurons. In addition, DJ-1 directly interacts with alpha-synuclein. Overexpression of DJ-1 reduces the dimerization of alpha-synuclein while the mutant form of DJ-1 in PD causes misfolded alpha-synuclein aggregates in DA neurons (Zondler et al., 2014). Remarkably, the deficiency of DJ-1 reduces the levels of membrane lipid rafts and simultaneously limits the internalization of extracellular alpha-synuclein. Furthermore, DJ-1 deficiency also leads to a decrease in the ability to degrade alpha-synuclein by autophagy (Nash et al., 2017)

NF-κB

Nuclear factor kappa-light-chain-enhancer of activated B cells (NF- κB) has been reported in studies involving alpha-synuclein pathology in rodent cell lines and human microglia (Ferreira and Romero-Ramos, 2018). Translocation of NF-κB resulting from a phosphorylation cascade which is initiated by the interaction of alpha-synuclein with Toll-like receptor (TLR) has been implicated in PD. Alpha-synuclein through the induction of pro-inflammatory mediators leads to NF- κB translocation and an increase in the level of TNF and IL-1β. This results in a cascade of events that relates to alpha-synuclein toxicity and inflammation.

MHC Class II

Major histocompatibility complex class II (MHC class II) is an immune marker that has a direct correlation with alpha-synuclein neuronal pathology. This is true not only in humans but also in rodents (Croisier et al., 2015). As mentioned earlier, alpha-synuclein play a role in the immune response during PD. MHC class II is involved in antigen presentation to T-cell, specifically CD4-T cells. There is speculation that upon uptake by microglia (antigen-presenting cell), the protein alpha-synuclein can be processed in endosomes and presented to T-cells via MHC class II (Ferreira and Romero-Ramos, 2018). As such, there is increased T-cell proliferation following aggregated alpha-synuclein in an MHC class II dependent manner *in vitro* (Harms et al., 2013). Failure in the adaptive immune signal (MHCII-CD4 T cell), may result in alpha-synuclein aggregation and that could be a contributing factor to the inhibition of mitochondrial function. In PD patients, the CD4 T cells seem to be more prone to apoptosis and so their population seems to be altered (Saunders et al., 2012; Stevens et al., 2012).

Nrf2

Oxidative stress has been implicated in alpha-synucleinopathy during PD. The antioxidant transcription factor, the nuclear factor erythroid 2-related factor 2 (Nrf2), reduces the oxidative stress posed by aggregated alpha-synuclein. In an *in vivo* study, lack of Nfr2 resulted in increased alpha-synuclein with a corresponding increase in pro-inflammatory transcription factors (IL-6, IL-1β, iNOS) and reduced microglia (Lastres-Becker et al., 2012). Consequently, overexpression of Nfr2 in the brain has been shown to be neuroprotective in alpha-synuclein based models (Gan et al., 2012).

Inflammasome

Substantial studies have reported the involvement of inflammasome in the alpha-synuclein immune response. In human monocytes, aggregated alpha-synuclein induced release of IL-1β in a phagocytosis dependent event, which required caspase-1 activation and involved the NLRP3 inflammasome (Codolo et al., 2013). In other studies, aggregated alpha-synuclein in a human monocytic cell line (LPS-primed THP-1), IL-1β release through a caspase-1 activation was observed, suggestive of the inflammasome (Ferreira and Romero-Ramos, 2018). Inflammasome related caspase-1 activation leads to the truncation of alpha-synuclein and generation of a pro-aggregatory form of truncated alpha-synuclein that promotes aggregation and neuronal toxicity.

MITOPHAGY AND HUNTINGTON'S DISEASE

HD is a neurodegenerative disorder that results in the progressive destruction of cells of the brain (Frank, 2014). It is also referred to as Huntington's chorea, because of the characteristic jerky involuntary movements of the face, shoulder and hips associated with the disease. Higher mental functions eventually degenerate into dementia. HD is an inherited disorder, however, it is reported that up to a tenth of HD cases are due to mutations (Dayalu and Albin, 2015). The condition is generally an autosomal dominant one that occurs as a result of an expansion of the CAG trinucleotide repeat that encodes a polyglutamine tract in the amino-terminal region of the huntingtin protein. This leads to an abnormal protein which eventually causes severe damage to the brain (Ross and Tabrizi, 2011). The mutant huntingtin (mtHtt) is widely expressed, however, in HD, there is preferential atrophy of the corpus striatum. This atrophy is caused by the destruction of GABAergic medium spiny neurons in the putamen as well as the caudate nucleus. Other regions of the brain such as the cortex are also affected (Bates et al., 2015; Rosas et al., 2003).

Several studies have been carried out to assess the functions that the huntingtin protein plays in the human body. Some studies have reported their interaction with important proteins involved in intracellular transport, gene transcription and cell signaling (Glajch and Sadri-Vakili, 2015; Van et al., 2007). Another study also showed that the protein is very essential for the development of the embryo (Duyao et al., 1995; Harjes and Wanker, 2003; Nasir et al., 1995; Zeitlin et al., 1995). Since the loss of Htt in adults results in neurodegeneration, Htt is said to play a crucial role in the maintenance and survival of neurons.

For proper cellular homeostasis to be achieved, there is a need for all the quality control mechanisms to function well. These quality control mechanisms are responsible for degrading and recycling cellular components as well as ensuring continuous turnover. Through autophagy, intracellular components like organelles and proteins are sent to the lysosome for degradation (Cuervo and Dice, 1998; He and Klionsky, 2009). Neurons, being post-mitotic cells need a very good autophagic activity to prevent the accumulation of improperly folded proteins. Thus, the essential role autophagy plays in neuronal homeostasis cannot be overemphasized (Hara et al., 2006; Komatsu et al., 2006).

Two systems, namely the ubiquitin-proteasome system (UPS) and the autophagy-lysosome system are responsible for getting rid of improperly folded proteins. After the identification of unfolded or misfolded protein, the UPS targets them for degradation. This system is involved in the removal of short-lived proteins through ubiquitination (Altuntas et al., 2014). Long-lived abnormal proteins, on the other hand, are cleared by autophagy. Damaged organelles and other components of the cytoplasm are cleared by this system. The autophagy-lysosome system works by sequestration of cellular components in double-membrane vesicles known as autophagosomes, which are fused with lysosomes for degradation. Studies have shown that in neurodegenerative conditions like HD, this quality control system is impaired. Thus, with the inability to degrade abnormal proteins into aggregates. These aggregates interfere with the metabolism of the cell. They cause neurotoxicity and lead to the death of the neurons when they get in brain inclusions. The gradual loss of neurons forms the basis of several neurodegenerative diseases like HD (Martinez-Vicente and Cuervo, 2007).

It has been found out that different forms of autophagy are affected in HD. Some scientific works have reported defective macroautophagy in neurons that have been affected by HD (Heng et al., 2010; Martinez-Vicente et al., 2010). Mitophagy is a key process in maintaining cellular health as it ensures the proper mitochondrial turnover and helps prevents the accumulation of defective mitochondria which can result in degeneration of the cell. Basal mitophagy accounts for the continuous turnover of the mitochondrial pool, nonetheless, mitophagy may also be induced following mitochondrial damage, stress or hypoxia (Martinez-Vicente, 2017).

GAPDH

Several studies have assessed the role of glyceraldehyde-3-phosphate dehydrogenase (GAPDH) in mitophagy. This is a new molecular mechanism that has been uncovered. Yogalingam et al., found out that under conditions of oxidative stress resulting from ischemia-reperfusion injury in cardiomyocytes, GAPDH associated with the damaged mitochondria leading to uptake into the lysosomes (Yogalingam et al., 2013). An earlier study in 1996 found an interaction between the enzyme and huntingtin proteins (Li and Li, 2004; Wu et al., 2007). The activity of GAPDH in glycolysis has been reported to decline in the fibroblast cells obtained patients with HD (Mazzola and Sirover, 2002). Physiological and biochemical effects following the association of GAPDH and mtHtt have however not been evaluated. Several other researchers have reported on other non-glycolytic roles that this enzyme is involved in. Some of these include gene transcription regulation (Zheng et al., 2003), vesicular transport (Bryksin and Laktionov, 2008), the dynamics of the microtubule (Tisdale, 2002) and several interactions with molecules like nitric oxide (Hara and Snyder, 2006) and glutathione (Puder and Soberman, 1997). This diversity in functions with respect to GAPDH has been attributed to its oligomerization status, modifications it undergoes post-translation and subcellular localization.

GAPDH has been reported to play a signaling role in the initiation of mitophagy under conditions of oxidative stress (Hwang et al., 2015). Expanded polyQ (polyglutamine) tracts have been shown to selectively associate with GAPDH and catalytically inactivate the enzyme (iGAPDH). The iGAPDH is then able to interact with damaged mitochondria, serving as a signaling molecule to induce the engulfment of damaged mitochondria into the lysosome (Hwang et al., 2015).

Studies have shown that an abnormal association between the long polyQ tract and mitochondrial GAPDH stalls GAPDH-mediated mitophagy. Such an abnormal association is what occurs between the mtHtt and iGAPDH. The rippling effect of this interaction is the accumulation of dysfunctional mitochondria in the cells, elevated levels of reactive oxygen species (ROS) and its consequent effects of oxidative stress and eventually death of the cell (Hwang et al., 2015). This cause of dysfunction of the mitochondria is especially detrimental to neurons since healthy mitochondria are very crucial for cellular activity. It is thus, not a surprise that when mitochondrial dysfunction occurs, several neurodegenerative diseases can result.

GAPDH has been demonstrated to have a high sensitivity to a number of oxidative post-translation modifications, like S-nitrosylation and glutathionyaltion. Cysteine residues that are present at the active site of the enzyme are especially susceptible to these modifications, the consequence of these being inactivation of the enzyme iGAPDH (Hwang et al., 2015). Hwang et al., have confirmed that with the oxidation of the -SH group of the catalytically important Cys152 of the enzyme to an -SO_3 group, a change in the activity of the enzyme occurs (Hwang et al., 2015). The modified GAPDH, iGAPDH has been shown to selectively interact with damaged mitochondria in several cell culture models of HD, serving as a signal for microautophagy. Hwang et al., in their study demonstrated that in HD, an abnormal interaction between the polyglutamine tract and mitochondrial GAPDH occurs with the end result being inhibition of GAPDH-induced mitophagy (Hwang et al., 2015).

Under normal cellular conditions, microautophagy occurs to get rid of damaged mitochondria which ensures the health of the cell. According to Hwang et al., in cells with mtHtt however, the expanded polyglutamine tracts associated with the mitochondria preventing their engulfment by lysosomes – abnormal mitophagy, leading to cell death by apoptosis (Hwang et al., 2015). Oxidized inactive GAPDH has been shown to induce mitophagy by selectively associating with the damaged mitochondria. In cells expressing the mutant huntingtin however, this process is blunted by the expanded polyQ tracts resulting from the mutant huntingtin. As such, there is an accumulation of damaged mitochondria and eventual cell death. The area of GAPDH-induced mitophagy has not been studied to a great extent. This could be attributed to the minimal levels of the enzyme on the mitochondria under normal cellular conditions. A study by Tarze et al., has, however, demonstrated that under stressed conditions, like treatment with mitochondriotoxic agents, the levels GAPDH increase (Tarze et al., 2007).

Further, Hwang et al., demonstrated that the impairment of mitophagy by mutant huntingtin can be resolved by overexpressed iGAPDH (Hwang et al., 2015). In their research, increased levels of iGAPDH enhanced the blunted mitophagy and resulted in improvement in the function of mitochondria in the cells that expressed expanded polyQ tracts. The ultimate result of this was an enhancement of cell survival. This finding presents a potential approach in the treatment of HD and thus, it is required that further studies are carried out to explore the potential therapeutic role that GAPDH may have in the treatment of HD.

Valosin Containing Protein

Available evidence shows that mutant huntingtin protein causes neurotoxicity by bringing about mitochondrial defects, leading to a bio-energetic failure, HD-associated neuronal dysfunction and eventual death of the cell (Bossy-Wetzel et al., 2008; Costa and Scorrano, 2012). Some studies have also demonstrated how mutant huntingtin protein triggers fragmentation of mitochondria by causing hyperactivation of the primary mitochondrial fission protein, Dynamin-related protein 1 (Drp 1) (Guo et al., 2013; Shirendeb et al., 2011; Song et al., 2011). Using pharmacological agents to inhibit Drp 1 resolved mutant huntingtin induced mitochondrial dysfunction and neuronal deaths in several models of HD (Guo et al., 2013; Song et al., 2011). These findings, while proving that mitochondrial damage is crucial in the etiology of HD, also show that when the mitochondrial injury is attenuated, there will be a reduction in the neuronal pathology that occurs in HD.

The mutant protein associates with the mitochondria. Upon association, it can do either of two things - recruitment of soluble cytosolic proteins or interaction with components of the mitochondria (Braun et al., 2006; Yano et al., 2014). Li et al., reported that certain alterations to the binding of Htt to target proteins can contribute to HD pathogenesis (Li and Li, 2004). One such protein, valosin containing protein (VCP) has been identified as a high abundance mt-Htt interacting protein in mitochondria.

The VCP is a widely expressed protein belonging to the AAA+ (ATPase associated with diverse cellular activities) protein family. The protein is highly conserved in all eukaryotes and it is made up of an N-terminal domain, a C-terminal domain and two ATPase domains namely D1 and D2 (DeLaBarre et al., 2003). Whereas the ATPase domains cause hydrolysis of ATP, the other two domains interact with substrates and other molecules and cofactors for the purpose of a wide range of cellular functions such as apoptosis, ubiquitin proteome system (UPS) mediated protein degradation and autophagosome maturation and initiation (Xia et al., 2016).

Mutations in VCP has been identified in patients with neurodegenerative disorders like amyotrophic lateral sclerosis (ALS) and HD (Guo et al., 2016; Johnson et al., 2010; Watts et al., 2004). The mechanisms as to how mutations in VCP contribute to the pathogenesis of such disease are however not clear.

The gene that codes for VCP has been identified on chromosome 9p133 in humans (Watts et al., 2004), comprising 17 exons. The VCP itself is made up of 806 amino acids and weighs 97kDa (Xia et al., 2016). VCP has been identified in the skeletal muscle, ovaries, brain, testes, liver, kidney, whole blood, lung, heart and lymph nodes (Meyer and Weihl, 2014). It is mostly found in the nucleus and cytosol (Ye, 2006). Other locations include the ER, Golgi body and mitochondria, accounting for its diverse cellular functions.

The role of VCP in autophagy is a new research area. Through autophagy, components of the cytosol are degraded via sequestration by autophagosomes which fuse with the lysosomes for degradation. The involvement of VCP in mammalian autophagy was first noticed during the study of inclusion body myopathy associated with Paget disease of bone and frontotemporal dementia (IBMPED) pathogenesis by mutations in VCP (Ju et al., 2009). It was observed that patients with this condition had light chain 3 (LC3) enriched vacuoles in muscle tissue. Accumulation of autophagosomes as a result of impairment of autophagosome fusion with lysosome and maturation. This was reproduced in the mouse models that expressed the mutated VCP, where LC3 and p62 accumulated because the fusion between lysosomes and autophagosomes failed to occur (Ju et al., 2009). These emphasized the critical role that VCP plays in autophagy initiation as well as autophagosome maturation.

Studies have reported that mutations in VCP are considered to underly the pathogenesis of a wide variety of neurodegenerative diseases such as IBMPFD, ALS, and HD (Guo et al., 2016; Johnson et al., 2010; Watts et al., 2004). A mutation, k524A, has been shown to particularly cause a disruption in protein degradation and induce HD (Hirabayashi et al., 2001; Poksay et al., 2011). Mice expressing mutant VCP show degeneration of mitochondria, enhanced autophagy degeneration of motor neurons and early death (Nalbandian et al., 2013; Nalbandian et al., 2012; Yin et al., 2012). In humans, VCP gene mutation has been demonstrated to underly the pathogenesis of ALS, bone and muscular deterioration and frontotemporal dementia, all of which are typical manifestations of dysfunctions of the mitochondria (Kakizuka, 2008; Meyer, Bug, and Bremer, 2012).

Endogenous VCP associates with polyQ containing aggregates in HD patients (Almeida et al., 2015; Hirabayashi et al., 2001; Mori et al., 2013) and it is able to bind directly with several polyQ disease proteins such as huntingtin (Fujita et al., 2013; Yang et al., 2014). Studies have confirmed that VCP is probably an effector of cell death in polyQ induced neurodegeneration. For instance, in a transgenic *Drosophila* model that expressed a fragment of the polyQ gene carrying 79 or 92 CAG repeats, there was an upregulation of the expression of VCP prior to cell death. Also, there was severely enhanced eye degeneration (Higashiyama et al., 2002; Hirabayashi et al., 2001). As to how VCP mediates neuronal pathology in HD and whether VCP manipulation can blunt the neuronal degeneration that occurs in HD is a question of research and further studies. Mutations in the gene that codes for VCP may lead to a disruption in the normal functions of VCP with the consequence being the widespread accumulation of intracellular ubiquitinated proteins that may add on to the disease pathogenesis.

A study reported that VCP recruitment to mitochondria was caused by impairment of mitophagy (Xing et al., 2016). The recruitment of VCP to mitochondria was demonstrated using HD mouse striatal HdhQ111 (mutant) and HdhQ7 (wild-type, wt) cells. They also observed that VIP binds mtHtt on mitochondria. They reported that binding does not only contribute to disease pathology but to severity as well. This assertion is supported by a study by of the Htt interactions in total brain lysates of BACHD transgenic mice (Shirasaki et al., 2012).

One remarkable finding is how HV-3, a novel peptide is able to ameliorate mitochondrial damage and cell death via interference with the HH/ VCP interaction. This interaction inhibited VCP translocation to mitochondria. The overall effect was a significant reduction in neuropathy of HD R6/2 and mice. HV-3 was shown to bind to VCP due to a relatively higher affinity for VCP. This finding may suggest that HV-3 offers competition to Htt. Another suggestion is that HV-3 targets and blocks the VCP binding site on HL. These findings set the ball rolling for further research into the development of inhibitors like HV-3. These agents may bring about a new therapeutic approach for HD and other diseases in which VCP translocation to mitochondria is characterized.

TG2

Tissue transglutaminase-2 (TG2) is one enzyme that has been implicated in the pathogenesis of HD. In recent times, several studies have been conducted on the role of the enzyme in the etiology of HD. TG2 is an enzyme that performs several functions. It is reported to the most ubiquitously expressed enzyme of the transglutaminase family (Altuntas et al., 2014; Tatsukawa et al., 2016). This family of the enzyme is designated as blood coagulator XIII and TG1-7 (Tatsukawa et al., 2016).

The primary function of TG2 is catalyzation of isopeptide bond formation between the gamma-carboxamide group of glutamine residue side chain and the epsilon-amino groups of the lysine reside chains with subsequent release of ammonia. The bonds that are formed by TGs are highly resistant to degradation by proteolysis. TG2 is involved in the catalysis of the post-translational modification of proteins via calcium-dependent crosslinking (Fesus and Piacentini, 2002). TG2 performs other several enzymatic processes apart from its transamidation activity.

In HD, the number of damaged mitochondria is elevated leading to the production of ROS which has been shown to play a role in neuronal death (Mastroberardino and Piacentini, 2010). It is reported that the amount of TG2 in cells is elevated during stressful conditions like hypoxia and nutrient deprivation, and calcium-dependent activity leads to the formation of insoluble protein aggregates (Ricotta et al., 2010).

Some studies have shown that this ubiquitously expressed enzyme is involved in the maturation of autophagosome and that, if it is present, there is an impairment in the clearance of damaged mitochondria (D'Eletto et al., 2009; Rossin et al., 2012). Also, it was shown that TG2 knockout mice exhibited impairment in autophagy and accumulated ubiquitinated protein aggregates on starvation. Further, the induction of autophagy has been reported to increase TG2 dependent post-translational modification of its substrate proteins. This post-translational process is also enhanced with proteasome inhibition.

Altuntas et al., (2014) also reported on the involvement of TG2 in mitophagy. Cells that lacked TG2 expressed an accumulation of dysfunctional mitochondria. They further demonstrated that cells that lacked the TG2 showed alterations on the functionality and morphology of mitochondria.

Research has proven TG2 to be an essential regulator of mitochondrial function and energy metabolism (Szondy et al., 2006). Piacentini et al., (2002) found out that apoptosis rapidly occurred in neural cells overexpressing TG2. In a study, the chemotherapeutic agent staurosporine caused apoptosis via mitochondria. This led to a rapid loss of mitochondrial potential in cells that overexpressed TG2 only. In diseases like HD, which is characterized by increased TG2 activity in the brain and impairment of mitochondrial function, this evidence may be worthwhile (Lesort et al., 1999). Other studies have reported that the increase in the activity of TG in the HD caudate may add to mitochondria dysfunction via incorporation of a substrate of TG2 - mitochondrial aconitase into inactive polymers (Kim et al., 2005). It is important to note that inhibition of TG2 has proven to be a great benefit in the treatment in HD in several animal models (Mastroberardino and Piacentini, 2010). Available evidence points to the fact that TG2 forms crosslinks with several proteins like Htt and that overexpression of TG2 are elevated in brain areas that are affected by HD (Mastroberardino and Piacentini, 2010). There is some compelling evidence that TG2 can contribute to abnormal protein aggregation, causing neurodegenerative disorders and eventually neuronal death, making the inhibition of TG2 a possible therapeutic target in the treatment of HD.

Several studies have reported that TG2 expression and TG2 activity are increased in both mouse models of HD and HD patients (Dedeoglu et al., 2002; Jeitner, Matson, Folk, Blass, and Cooper, 2008; Karpuj et al., 2002; Karpuj et al., 1999; Lesort et al., 1999). Again, a study showed that genetic deletion of TG2 in R6/1 and R6/2 mouse models of HD led to phenotypical improvements including reduced neuronal cell death and improved motor function and survival (Bailey and Johnson, 2005; Mastroberardino et al., 2002).

Another study by Menalled et al., (Menalled et al., 2014) has however found conflicting results. Contrary to other studies, this study reported that under rigorous conditions, using R6/1 and R6/2 mouse models, there was no effect on motor and cognitive deficits upon ablation of TG2. They also reported that the ablation of TG2 does not extend the lifespan of R6/2 under optimal husbandry conditions. Furthermore, the deletion of TG2 did not alter the Htt aggregate load in the striatum or cortex. Both mouse lines did not show a reduction in the brain atrophy. These findings make TG2 an unsuitable therapeutic target for HD. With these conflicting findings, it is necessary that more studies be carried out to assess the therapeutic potential of TG2 in the treatment of HD.

MITOPHAGY IN MICROGLIA: ROLE IN THE DEVELOPMENT OF AMYOTROPHIC LATERAL SCLEROSIS

ALS is a rare group of progressive neurological diseases that specifically affects neurons responsible for controlling voluntary muscle movement. Their neuropathologic findings are unique and include the degeneration of anterior horn cells resulting in muscle atrophy or amyotrophy, degeneration and sclerosis of the corticospinal tracts (Osafo et al., 2019). The spectrum of the disease is clinically and genetically heterogeneous, devastating and rapidly progressive, with cognitive and behavioral impairments as core features, in addition to motor impairments (Schmidt et al., 2016). Although research interest in the field has been heightened over the years, providing a great insight into the ultimate consequences of the disease on the human brain, the exact pathological mechanism underlying the sequelae of neurological deficits remains largely unknown. Nonetheless, several gene mutations in superoxide dismutase 1 (SOD1), OPTN, TANK binding kinase 1 (TBK1), valosin-containing protein (VCP), and chromosome 9 open reading frame 72 (C9ORF72) have been implicated in the development of the disease and the accumulation of damaged mitochondria in ALS has been observed with live-cell imaging (Wang et al., 2019). Studies have shown that inhibition of mutations in *OPTN* and the *TBK1* gene significantly improves neuronal function and blocks disease progression (Wong and Holzbaur, 2014; Moore and Holzbaur, 2016). This implies that mitophagy plays a potential role in the development and progression of ALS since *OPTN* and *TBK1* are important regulators of mitophagy.

Evidence from *in vivo* studies conducted in mice has demonstrated that resident microglia increase their number during disease progression, and their activation states represent a continuum between the neuroprotective M2 and the toxic M1 phenotype (Liao et al., 2012; Chiu et al., 2013). Microglia have consistently been shown to exhibit an anti-inflammatory profile with attenuated Toll-like receptor 2 responses to controlled immune challenge, at the pre-onset phase of SOD1-mediated disease. In addition, these brain macrophages also exhibit overexpression of the anti-inflammatory cytokine IL-10 (Gravel et al., 2016). Interestingly, microglia-mediated neuroinflammation is modulated by histamine, and that the administration of the antihistamine, clemastine, to SOD1^{G93A} mice at the asymptomatic phase to the onset of the disease delays disease onset and improves motor function and survival. It has also been observed that the administration of Clemastine activates autophagy in the SOD1^{G93A} mice (Apolloni et al., 2016). This suggests that mitophagy in microglia could be targeted in the development of new therapeutic strategies in the treatment of ALS.

Figure 2. Disruption of autophagy by OPTN at several stages.
By inhibiting ATG5/ATG12/ATG6L1 complex formation, the depletion of optineurin, as well as overexpression of ALS mutant OPTN, inhibits the formation of phagophore and autophagosome. OPTN links the ubiquitinated cargo to LC3, together with facilitating autophagosomal engulfment of the cargo, such as mitochondria. Subsequent degradation of the engulfed mitochondria is inhibited by ALS mutant OPTN. Furthermore, an imbalance of optineurin expressions, such as overexpression or depletion, also interferes with autophagy, resulting in the accumulation of damaged mitochondria. In addition, ALS mutant OPTN also inhibits the fusion of the autophagosome with the lysosome, further inhibiting autophagy flux. OPTN−, optineurin depletion; OPTN+, optineurin overexpression; OPTN ALS, ALS mutant optineurin.

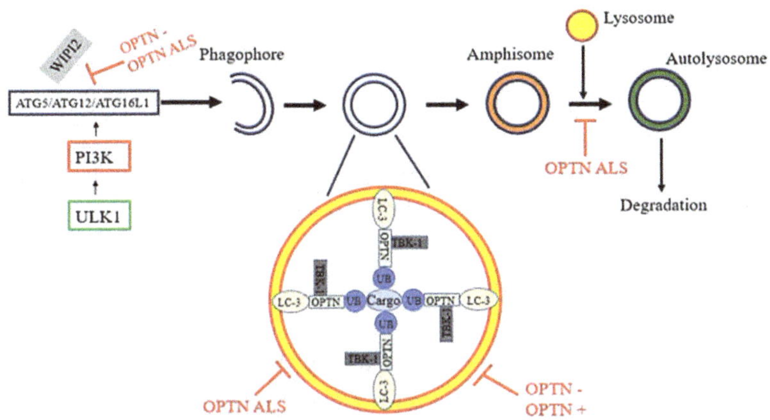

OPTN and TBK1 in Microglial Mitophagy

Optineurin, originally named FIP2, is a highly conserved, 64 kDa hexameric multifunctional protein involved in several cellular processes (Ying et al., 2012; Toth and Atkin, 2018). Importantly, it functions as a ubiquitin-LC3-binding autophagy receptor that is implicated in several forms of selective autophagy, including xenophagy, mitophagy, and aggrephagy (Wild et al., 2011; Moore and Holzbaur, 2016b). Moreover, OPTN interacts with several putative optineurin-binding partners to produce diverse functions, including influencing the innate immune response by negatively regulating the nuclear factor kappa B (NF-κB) pathway (Li et al., 1998). Mutations in OPTN, therefore, enhances NF-κB activity and impede intracellular transport processes contributing to the development of ALS (Liu et al., 2018). OPTN can bind with mitochondria in a Parkin-independent manner, however, the binding is greatly stabilized in the presence of Parkin. Following this interaction, double FYVE-containing protein 1 (DFCP1) translocates into the mitochondria and recruit LC. OPTN then interacts with LC via its LIR to mediate the formation of autophagosomes (Figure 2). This mechanism, therefore, indicates the importance of OPTN in Parkin-mediated mitophagy (Wang et al., 2019). In support of this, wild-type mice with OPTN siRNA resistance have been shown to withstand mitochondrial accumulation, whereas the induction of ALS-related OPTN mutation (E478G) in mice reduces the ability to withstand impaired mitophagy. In addition, ALS-related OPTN UBAN mutant has been utilized to block OPTN's mitochondria translocation, resulting in abnormal mitophagy. Thus linking OPTN with ALS through the mitochondrial degradation efficiency and demonstrating the significance of mitophagy in the development and progression of ALS (Wang et al., 2019).

Normally, OPTN undergoes post-translational modifications, including phosphorylation and ubiquitination by several enzymatic processes, impacting significantly on its function (Ying et al., 2010). One such major enzyme, capable of phosphorylating optineurin on at least nine serine and two threonine residues, is TBK1 (Richter et al., 2016). In a damaged mitochondrion, PINK1 accumulates on the outer mitochondrial membrane and phosphorylate ubiquitin chains on several other resident proteins with subsequent interaction with autophagy adaptors (Heo et al., 2015). PINK1 then recruits and phosphorylates Parkin resulting in its activation and allowing further autophagy adaptor binding. Subsequently, TBK1 is activated in a Parkin-, OPTN-, NDP52 and OPTN's ubiquitin-binding ability-dependent manner. The active TBK1 can then phosphorylate p62, OPTN and NDP52 to greatly improve their ability to bind to ubiquitin and LC3-II, leading to degradation of the mitochondria (Figure 2) (Lazarou et al., 2015). It is recognized that, TBK1 is essential for efficient mitophagy and that, loss of function in this kinase would result in impaired mitophagy with resultant accumulation of damaged mitochondria, thus may enhance the pathogenesis of ALS (Sasaki and Iwata, 1996; Oakes et al., 2017).

SOD1 and Mitophagy in Microglia

Although the exact molecular pathway involved in the degeneration of motor neurons in ALS is unknown, possible primary mechanisms may include the toxic effects of mutant SOD1. These effects include abnormal protein aggregation, disorganization of intermediate filaments, glutamate-mediated excitotoxicity, mitochondrial dysfunction and impaired apoptosis (Osafo et al., 2019). Studies have shown that, in the SOD1 model of ALS, the flux of mitophagy is decreased leading to the accumulation of impaired mitochondria. Subsequent restoration of mitophagy by mitochondria quality control reduce this accumulation, a mechanism thought to involve the blocking of reverse transport of autophagosomes in axons by SOD1. It is proposed that mutant SOD1 impairs axonal transport in a PINK1/Parkin pathway-dependent manner, where the mutation results in autophagosome trafficking defect (Moller et al., 2017). This defect has been shown to be reversible with the overexpression of mitochondrial Rho GTPase 1 and inhibition of PINK1 (Palomo et al., 2018). The discovery of an increasing list of genetic causes and risk factors of ALS, such as SOD1 mutations, has led to the realization that errors in RNA metabolism and quality control in protein folding are pivotal in the initiation of the disease.

NEUROPROTECTIVE STRATEGIES

The complexity of the pathogenesis of neurodegenerative diseases had led to extensive investigations of novel compounds and nonpharmacological approaches to modify the course of these conditions while reducing adverse drug effects (Monteiro et al., 2017). Current pharmacological and non-pharmacological approaches have limitations as treatments are majorly focused on alleviating symptoms rather than preventing or slowing down the progression of the disease. In the past decade, in-depth research into neurodegenerative diseases has led to a better understanding of the complex cascade of events involved. Therefore, the development of drugs to offer neuroprotection in neurodegenerative diseases might logically evolve from an improved understanding of the etiology and pathogenesis of these diseases (Monteiro et al., 2017).

Neuroprotection is a complex mechanism that involves preventing cell death and restoring function to damaged neurons as well as regeneration of neuronal numbers (Monteiro et al., 2017). Although the origin, progression, and heterogeneity of neurodegenerative diseases differ, these disorders are characterized by common features at the molecular level. These include accumulation of aggregated misfolded proteins; impairment of degradative processes including autophagy; oxidative stress; neuroinflammation; and impaired mitochondrial function (Martinez-Vincente, 2017). For this reason, some neuroprotective strategies may be beneficial to more than one neurodegenerative disorder. In some instances, target pathways may differ and so therapeutic approach may be specific for a particular type of neurodegenerative disease. In this section, neuroprotective strategies have described generally for aforementioned neurodegenerative diseases.

As noted, one common feature in neurodegenerative diseases is neuroinflammation. Inflammation has always played an integral part in many diseases and neurodegenerative diseases are no exceptions. Recently, there have been growing concerns that neuroinflammation mechanisms might contribute to the cascade of events leading to neuronal degeneration observed in AD, PD and other neurodegenerative diseases. Experimental models and postmortem studies in many neurodegenerative diseases revealed increased concentrations of activated glial cells as well as high levels of certain types of cytokines and interleukins in the brain (IL-6, IL-1β) (Monteiro et al., 2017).

Several neuroprotective agents have been demonstrated in either an animal model or in the *in vitro* model to target neuroinflammation and are considered as anti-inflammatory agents. These include non-steroidal anti-inflammatory drugs (NSAIDs), statins and minocycline. In-vitro studies in animal models showed that NSAIDs can prevent the degeneration of dopaminergic neurons in PD (Esposito et al., 2007). Statins, commonly used clinically to treat dyslipidemia, have been demonstrated to significantly mitigate the risk of having PD by inhibiting pro-inflammatory cytokines and free oxy radical production (Di Napoli et al., 2002). In a recent study, minocycline a second-generation tetracycline well known for its antibacterial property has been shown to have a potential property to protect cells from inflammation-induced injuries in the brain (Sakar et al., 2016). Furthermore, minocycline, has been demonstrated to inhibit the activation of macrophage in the brain and also prevent apoptosis in cell culture (Duan et al., 2002). Minocycline exhibits inhibitory effects on caspase activity. The caspase inhibiting function of minocycline reduces proteolysis of mutant huntingtin and the formation of cytotoxic N-terminal fragments and extends survival in HD mice. Minocycline is well tolerated and safe in HD patients. A recent clinical trial using minocycline in a small cohort of HD patients for 24 months showed stabilization of motor and neuropsychological functions, amelioration of psychiatric symptoms, and improved outcomes (Ferrante et al., 2009). Minocycline's potential as an anti-inflammatory agent is currently being investigated in humans (NINDS NET-PD Investigators, 2006). In ALS, the trophic effect of micrologia that is induced by T cells is suggested to be a novel mechanism to alter disease progression by manipulating inflammatory pathways (Morrison et al., 2012).

Oxidative stress has been well documented to play a vital role in the pathology of neurodegenerative diseases. Reactive oxygen species helps in regulating cell function as they trigger subsequent important signaling event but when produced in excess can overwhelm the body's antioxidant defenses and may cause oxidative stress (Wang et al., 2007). The presence of reactive oxygen species in the brain makes it vulnerable to oxidative damage (Valko et al., 2007).

A wide range of putative neuroprotective agents that target different pathways has been developed. They include monoamine oxidases (MAO) inhibitors that inhibit dopamine metabolism, agents like coenzyme Q (CoQ10) that can increase the electron transport, and chemicals like selenium which promotes glutathione, an antioxidant protein, and thus confer protection against free radicals in the cell. Low levels of glutathione, is found in the substantia nigra of human PD suggesting that the anti-oxidative system is compromised (Sakar et al., 2016). CoQ10 has also been shown to be effective in slowing the progression of neurological damage in the animal model (Beal, 1998). CoQ10 is essential for complexes I and II electron transfer activities during oxidative phosphorylation and thus plays a vital role in ATP production. CoQ10 also has membrane-stabilizing properties and acts as an antioxidant. The therapeutic rationale includes targeting improved cellular energy production and reducing oxidative stress (Ferrante et al., 2009). A small study involving PD patients reveals that Q10 administration resulted in a reduction in UDPRS (Unified Parkinson Disease Rating Scale) score. CoQ10 is also efficacious in animal models of HD (Ferrante et al., 2009). Treatment with CoQ10 in HD patients showed a trend toward slowing functional decline and a trend toward a beneficial effect on cognitive tests and behavior. It has recently been suggested that higher doses of CoQ10 may provide greater efficacy in treating neurodegenerative diseases (Ferrante et al., 2009).

Besides its regulatory effects in the light-dark cycle, melatonin is a hormone with neuroprotective antioxidant and anti-inflammatory properties. Available evidence indicates that PD is associated with impaired brain expression of melatonin and its receptors presented a neuroprotective effect in animal models of PD induced by different toxins. Non-motor symptoms commonly experienced by PD patients such as sleep and anxiety disorders, depression and memory function may be improved by melatonin (Monteiro et al., 2017).

Natural products have always been a good source of lead compounds, which might ameliorate central nervous system neurodegeneration. Green coffee, a non-toxic small molecule has been shown to improve cognitive and motor performance in mouse models with tau pathology in AD studies. Green coffee is an inhibitor of protein phosphate A2 methylesterase and in effect helps in the modulation of tau phosphorylation, a crucial step in AD pathogenesis (Varghese et al., 2014). The antioxidant properties of the bee-derived plants' resin propolis have been extensively studied. Propolis was shown to restore neuronal function, increase antioxidant enzyme activity, and decreased lipid peroxidation which in turn decreased oxidative stress. Yellow propolis was shown to elicit antioxidant activity associated with anxiolytic, antidepressant and cognitive enhancer properties (Kocot et al., 2018). Derivatives of oleanic acid compounds have been shown to activate Nrf2/ARE, a pivotal system in ALS pathogenesis. In ALS, regulation of Nrf2/ARE is suboptimal resulting in oxidative stress, neuroinflammation and mitochondrial dysfunction (Morrison et al., 2012). Another potential treatment for ALS is focused on alterations of the muscles and motor axons that occur before apparent pathology in motor neuron cell bodies (Morrison et al., 2012). Targeting peripheral axons, neuromuscular junction and possibly muscles may provide better alternatives for the current treatment of ALS.

Aside from western medicine, integrative medicine has been reported to improve the conditions of patients with neuroprotective disorders. Traditional chinese medicine (TCM) together with western medicine has been shown to treat dyskinesias, improve sleep disorders and quality of life of PD patients (Chen et al., 2014). Also, a remarkable improvement in behavioral and psychological symptoms of dementia in patients with AD was observed following therapy with TCM (Chen et al., 2014). Another potential treatment for ALS is focused on alterations of the muscles and motor axons that occur before apparent pathology in motor neuron cell bodies (Morrison et al., 2012). Targeting peripheral axons, neuromuscular junction and possibly muscles may provide better alternatives for the current treatment of ALS.

RECENT DEVELOPMENTS AND FUTURE RESEARCH DIRECTIONS

Mitophagy plays an important role in maintaining mitochondrial quality. It is essential in maintaining homeostasis in the mitochondrial network, sustaining cell survival, and the health of the organisms as a whole (Roger et al., 2017). Mitochondrial quality is an indicator of chronic diseases such as neurological disorders and degenerative diseases. Through extensive work in *in vitro* biochemistry, biomedical sciences, molecular components required for mitophagy have been identified, but the evaluation of their physiological importance has largely remained unexplored. (Roger et al., 2017). This is because the necessary tools required for the detection of mitophagy *in vivo* were not readily available.

With the advent of biomedical research, the means by which the occurrence of *mitophagy in vivo* can be analyzed may be elicited. This would also help boost the knowledge of the significance of mitophagy in mammals and roles currently undiscovered that mitophagy may play in disease progression or mitophagic pathways currently unexplored. Electron microscopy is the gold standard method of detecting and monitoring mitophagy *in vivo*, as it allows visualization of autophagosome membranes and other cellular compartments. Electron microscopy, however, requires a high level of expertise and experience and is difficult to execute consistently (McWilliams, 2019). In recent times, mouse models have been developed which employ fluorescent reporter systems to detect mitophagy *in vivo*. Endpoint readouts of mitophagy *in vivo* are elicited using the acid-labile properties of fluorescent reporter proteins. The endpoint readouts of mitophagy depends on the mitochondrial delivery to the acidic microenvironment of the lysosome.

A recently developed model is the PMitoTimer reporter gene, developed from a fluorescent reporter gene, pTimer. The MitoTimer protein targets the mitochondria as a mitochondria targeting sequence has been added to the N-terminus of the Mito-Timer coding sequence. This model is employed in the monitoring of mitochondrial biogenesis and mitophagy with time. Another mouse model is the Mt-kiema, which is a newly developed mouse model, which expresses a mitochondria-targeted, pH-sensitive fluorophore. With the occurrence of mitophagy, the acidity of the lysosome shifts the excitation wavelength, changing the fluorescence of the fluorophore from green to red, allowing the detection and assessment of mitophagy (Sun et al., 2015). The fluorescent reporter system is utilized by another mouse model-Mito-QC, which is designed to detect and monitor mitochondrial quality control (QC). This system is based on a tandem-fusion protein (mCherry-GFP) whose target is the outer mitochondrial membrane, using the mitochondrial targeting sequence (MTS) of the outer mitochondrial membrane protein FIS1. Mitochondria fluoresce both red and green colors, and when combined, they give off a yellow fluorescence. With the occurrence of mitophagy, there is the exposure of the mitochondria to the acidity of the lysosome, resulting in the quenching of the GFP fluorescence of the protein, providing an endpoint readout which can be visualized (McWilliams et al., 2019)

Another method that is used is the colocalization of the fluorescent tracker, which is used by another novel mouse model – RedMIT/GFP-LC3 mouse model which has red fluorescent mitochondria and green fluorescent autophagosomes. Mitophagy is known to have taken place when the red fluorescence and green fluorescence colocalize, and this serves as a readout to detect mitophagy (Diot et al., 2018)

These models provide a starting point for the analysis of the physiological significance of mitophagy, and they will prove useful for further studies in degenerative diseases and the roles played by mitophagy in neurodegeneration.

CONCLUSION

Mitophagy is essential for cell survival. In neurons, imperfect mitophagy results in the accretion of damaged mitochondria and lastly causes various neurodegenerative disorders. Neurons, being highly specialized cells, are peculiarly liable to defects in autophagic mechanisms. These impairments in mitochondrial function and their dynamics have been identified in many neurodegenerative diseases such as AD, PD, HD and ALS and modulators of both mitochondrial physiology and autophagy have presented themselves as promising therapeutic targets. However, further studies are necessary to comprehend exactly how defective mitophagy contributes to neurodegeneration.

REFERENCES

Albani, D., Polito, L., Batelli, S., De Mauro, S., Fracasso, C., Martelli, G., ... Forloni, G. (2009). The SIRT1 activator resveratrol protects SK-N-BE cells from oxidative stress and against toxicity caused by alpha-synuclein or amyloid-beta (1-42) peptide. *Journal of Neurochemistry, 110*(5), 1445–1456. doi:10.1111/j.1471-4159.2009.06228.x PMID:19558452

Almeida, B., Abreu, I. A., Matos, C. A., Fraga, J. S., Fernandes, S., Macedo, M. G., . . . MacedoRibeiro, S. (2015). *SUMOylation of the brain-predominant Ataxin-3 isoform modulates its interaction with p97*. Academic Press.

Altuntas, S., D'Eletto, M., Rossin, F., Hidalgo, L. D., Farrace, M. G., Falasca, L., . . . Piacentini, M. (2014). *Type 2 Transglutaminase, mitochondria and Huntington's disease: menage a trois*. Academic Press.

Andres-Mateos, E., Perier, C., Zhang, L., Blanchard-Fillion, B., Greco, T. M., Thomas, B., ... Dawson, V. L. (2007). DJ-1 gene deletion reveals that DJ-1 is an atypical peroxiredoxin-like peroxidase. *Proceedings of the National Academy of Sciences of the United States of America, 104*(37), 14807–14812. doi:10.1073/pnas.0703219104 PMID:17766438

Anekonda, T. S., & Reddy, P. H. (2006). Neuronal protection by sirtuins in Alzheimer's disease. *Journal of Neurochemistry, 96*(2), 305–313. doi:10.1111/j.1471-4159.2005.03492.x PMID:16219030

Badadani, M. (2012). Autophagy Mechanism, Regulation, Functions, and Disorders. ISRN Cell Biology, Article ID 927064.

Bailey, C. D., & Johnson, G. V. (2005). *Tissue transglutaminase contributes to disease progression in the R6/2 Huntington's disease mouse model via aggregate-independent mechanisms*. Academic Press.

Bates, G. P., Dorsey, R., Gusella, J. F., Hayden, M. R., Kay, C., Leavitt, B. R., . . . Wetzel, R. (2015). *Huntington disease*. Academic Press.

Beal, M. F., Matthews, R. T., Tieleman, A., & Shults, C. W. (1998). Coenzyme Q10 attenuates the 1-Methyl-4-Phenyl-1,2,3,Tetrahydropyridine (Mptp) induced loss of striatal dopamine and dopaminergic axons in aged mice. *Brain Research, 783*(1), 109–114. doi:10.1016/S0006-8993(97)01192-X PMID:9479058

Bhujabal, Z., Birgisdottir, Å. B., Sjøttem, E., Brenne, H. B., Øvervatn, A., Habisov, S., ... Johansen, T. (2017). FKBP8 recruits LC3A to mediate Parkin-independent mitophagy. *EMBO Reports, 18*(6), 947–961. doi:10.15252/embr.201643147 PMID:28381481

Blass, J., & Gibson, G. (1991). *The role of oxidative abnormalities in the pathophysiology of Alzheimer's disease*. Academic Press.

Bonda, D. J., Lee, H. G., Camins, A., Pallas, M., Casadesus, G., Smith, M. A., & Zhu, X. (2011). The sirtuin pathway in ageing and Alzheimer disease: Mechanistic and therapeutic considerations. *Lancet Neurology, 10*(3), 275–279. doi:10.1016/S1474-4422(11)70013-8 PMID:21349442

Bonifati, V., Rizzu, P., Squitieri, F., Krieger, E., Vanacore, N., ... Van Swieten, J. C. (2003). DJ-1(PARK7), a novel gene for autosomal recessive, early onset parkinsonism. *Neurological Sciences, 24*(3), 159–160. doi:10.100710072-003-0108-0 PMID:14598065

Bossy-Wetzel, E., Petrilli, A., & Knott, A. B. (2008). *Mutant huntingtin and mitochondrial dysfunction*. Academic Press.

Braak, H., & Braak, E. (1991). *Neuropathological stageing of Alzheimer-related changes*. Academic Press.

Braun, R. J., Zischka, H., Madeo, F., Eisenberg, T., Wissing, S., Büttner, S., . . . Ueffing, M. (2006). *Crucial mitochondrial impairment upon CDC48 mutation in apoptotic yeast*. Academic Press.

Bryan, K. J., Lee, H.-g., Perry, G., Smith, M. A., & Casadesus, G. (2009). Transgenic mouse models of Alzheimer's disease: behavioral testing and considerations. In *Methods of Behavior Analysis in Neuroscience* (2nd ed.). CRC Press/Taylor and Francis.

Bryksin, A., & Laktionov, P. (2008). *Role of glyceraldehyde-3-phosphate dehydrogenase in vesicular transport from golgi apparatus to endoplasmic reticulum*. Academic Press.

Calkins, M. J., Manczak, M., Mao, P., Shirendeb, U., & Reddy, P. H. (2011). Impaired mitochondrial biogenesis, defective axonal transport of mitochondria, abnormal mitochondrial dynamics and synaptic degeneration in a mouse model of Alzheimer's disease. *Human Molecular Genetics, 20*(23), 4515–4529. doi:10.1093/hmg/ddr381 PMID:21873260

Canet-Avilés, R. M., Wilson, M. A., Miller, D. W., Ahmad, R., McLendon, C., ... Bandyopadhyay, S. (2004). The Parkinson's disease protein DJ-1 is neuroprotective due to cysteine-sulfnic acid-driven mitochondrial localization. *Proceedings of the National Academy of Sciences of the United States of America, 101*(24), 9103–9108. doi:10.1073/pnas.0402959101 PMID:15181200

Chen, G., Han, Z., Feng, D. U., Chen, Y., Chen, L., Wu, H., ... Chen, Q. (2014). A regulatory signaling loop comprising the PGAM5 phosphatase and CK2 controls receptor-mediated mitophagy. *Molecular Cell, 54*(3), 362–377. doi:10.1016/j.molcel.2014.02.034 PMID:24746696

Chen, J., Zhou, Y., Mueller-Steiner, S., Chen, L. F., Kwon, H., Yi, S., ... Gan, L. (2005). SIRT1 protects against microglia-dependent amyloid-beta toxicity through inhibiting NF-kappaB signaling. *The Journal of Biological Chemistry, 280*(48), 40364–40374. doi:10.1074/jbc.M509329200 PMID:16183991

Chen, M., Chen, Z., Wang, Y., Tan, Z., Zhu, C., Li, Y., ... Chen, Q. (2016). Mitophagy receptor FUNDC1 regulates mitochondrial dynamics and mitophagy. *Autophagy, 12*(4), 689–702. doi:10.1080/15548627.2016.1151580 PMID:27050458

Chen, X., & Pan, W. (2014). The Strategies for neurodegenerative diseases by Integrative Medicine. *Integrative Medicine (Encinitas, Calif.), 1*, 223–225.

Chételat, G. (2013). *Alzheimer disease: Aβ-independent processes—rethinking preclinical AD*. Academic Press.

Chételat, G., La Joie, R., Villain, N., Perrotin, A., de La Sayette, V., Eustache, F., and Vandenberghe, R. (2013). *Amyloid imaging in cognitively normal individuals, at-risk populations and preclinical Alzheimer's disease*. Academic Press.

Chu, C. T. (2019). Mechanisms of selective autophagy and mitophagy: Implications for neurodegenerative diseases. *Neurobiology of Disease, 122*, 23–34. doi:10.1016/j.nbd.2018.07.015 PMID:30030024

Codolo, G., Plotegher, N., Pozzobon, T., Brucale, M., Tessari, I., ... Bubacco, L. (2013). Triggering of inflammasome by aggregated alpha-synuclein, an inflammatory response in synucleinopathies. *PLoS One, 8*(1), e55375. doi:10.1371/journal.pone.0055375 PMID:23383169

Costa, V., & Scorrano, L. (2012). *Shaping the role of mitochondria in the pathogenesis of Huntington's disease*. Academic Press.

Croisier, E., Moran, L. B., Dexter, D. T., Pearce, R. K., & Graeber, M. B. (2005). Microglial inflammation in the parkinsonian substantia nigra: Relationship to alpha-synuclein deposition. *Journal of Neuroinflammation, 2*(1), 14. doi:10.1186/1742-2094-2-14 PMID:15935098

Cuervo, A. M., & Dice, J. F. (1998). *Lysosomes, a meeting point of proteins, chaperones, and proteases*. Academic Press.

D'Eletto, M., Grazia Farrace, M., Falasca, L., Reali, V., Oliverio, S., Melino, G., . . . Piacentini, M. (2009). *Transglutaminase 2 is involved in autophagosome maturation*. Academic Press.

Davie, C. A. (2008). A review of Parkinson's disease. *British Medical Bulletin, 86*(1), 109–127. doi:10.1093/bmb/ldn013 PMID:18398010

Davies, C., Mann, D., Sumpter, P., & Yates, P. (1987). *A quantitative morphometric analysis of the neuronal and synaptic content of the frontal and temporal cortex in patients with Alzheimer's disease*. Academic Press.

Dayalu, P., & Albin, R. L. (2015). *Huntington disease: pathogenesis and treatment*. Academic Press.

De, A., Bala, N., & DasGupta, P. (2011). Alzheimer's Disease and its. *Management, 2*(4), 1439–1443.

De Duve, C. (1963). The lysosome. *Scientific American, 208*(5), 64–72. doi:10.1038cientificamerican0563-64 PMID:14025755

Dedeoglu, A., Kubilus, J. K., Jeitner, T. M., Matson, S. A., Bogdanov, M., Kowall, N. W., . . . Beal, M. F. (2002). *Therapeutic effects of cystamine in a murine model of Huntington's disease*. Academic Press.

DeLaBarre, B., Brunger, A. T., & Biology, M. (2003). *Complete structure of p97/valosin-containing protein reveals communication between nucleotide domains*. Academic Press.

Devi, L., Prabhu, B. M., Galati, D. F., Avadhani, N. G., & Anandatheerthavarada, H. K. (2006). Accumulation of amyloid precursor protein in the mitochondrial import channels of human Alzheimer's disease brain is associated with mitochondrial dysfunction. *The Journal of Neuroscience: The Official Journal of the Society for Neuroscience*, *26*(35), 9057–9068. doi:10.1523/JNEUROSCI.1469-06.2006 PMID:16943564

Di Napoli, P., Taccardi, A. A., Oliver, M., & de Caterina, R. (2002). Statins and stroke: Evidence for cholesterol-independent effects. *European Heart Journal*, *23*(24), 1908–1921. doi:10.1053/euhj.2002.3236 PMID:12473253

Diot, A., Agnew, T., Sanderson, J., Liao, C., Carver, J., Neves, R. P., ... Poulton, J. (2018). Validating the RedMIT/GFP-LC3 Mouse Model by studying mitophagy in autosomal dominant optic atrophy due to OPA1Q285STOP mutation. *Frontiers in Cell and Developmental Biology*, *6*, 103. doi:10.3389/fcell.2018.00103 PMID:30283778

Du, H., Guo, L., Fang, F., Chen, D., Sosunov, A. A., McKhann, G. M., ... Yan, S. D. (2008). Cyclophilin D deficiency attenuates mitochondrial and neuronal perturbation and ameliorates learning and memory in Alzheimer's disease. *Nature Medicine*, *14*(10), 1097–1105. doi:10.1038/nm.1868 PMID:18806802

Du, H., Guo, L., Yan, S., Sosunov, A. A., McKhann, G. M., & Yan, S. S. (2010). Early deficits in synaptic mitochondria in an Alzheimer's disease mouse model. *Proceedings of the National Academy of Sciences of the United States of America*, *107*(43), 18670–18675. doi:10.1073/pnas.1006586107 PMID:20937894

Duan, W., Ladenheim, B., Cutler, R. G., Kruman, I. I., Cadet, J. L., & Mattson, M. P. (2002). Dietary folate deficiency and elevated homocysteine levels endanger dopaminergic neurons in models of Parkinson's Disease. *Journal of Neurochemistry*, *80*(1), 101–110. doi:10.1046/j.0022-3042.2001.00676.x PMID:11796748

Duyao, M. P., Auerbach, A. B., Ryan, A., Persichetti, F., Barnes, G. T., McNeil, S. M., . . . Joyner, A. (1995). *Inactivation of the mouse Huntington's disease gene homolog Hdh*. Academic Press.

Edison, P., Archer, H., Hinz, R., Hammers, A., Pavese, N., Tai, Y., . . . Kennedy, A. (2007). *Amyloid, hypometabolism, and cognition in Alzheimer disease: an [11C] PIB and [18F] FDG PET study*. Academic Press.

Eschbach, J., & Danzer, K. M. (2014). Alpha-Synuclein in Parkinson's disease: Pathogenic function and translation into animal models. *Neurodegenerative Diseases*, *14*(1), 1–17. doi:10.1159/000354615 PMID:24080741

Esposito, E., Di Matteo, V., Benigno, A., Pierucci, M., Crescimanno, G., & Di Giovanni, G. (2007). Non-Steroidal Anti-Inflammatory Drugs in Parkinson's Disease. *Experimental Neurology*, *205*(2), 295–312. doi:10.1016/j.expneurol.2007.02.008 PMID:17433296

Fagan, A. M., Mintun, M. A., Shah, A. R., Aldea, P., Roe, C. M., Mach, R. H., . . . Holtzman, D. M. (2009). *Cerebrospinal fluid tau and ptau181 increase with cortical amyloid deposition in cognitively normal individuals: Implications for future clinical trials of Alzheimer's disease*. Academic Press.

Fang, E. F., Hou, Y., Palikaras, K., Adriaanse, B. A., Kerr, J. S., Yang, B., . . . Dan, X. (2019). *Mitophagy inhibits amyloid-β and tau pathology and reverses cognitive deficits in models of Alzheimer's disease*. Academic Press.

Feige, J. N., Lagouge, M., Canto, C., Strehle, A., Houten, S. M., Milne, J. C., ... Auwerx, J. (2008). Specific SIRT1 activation mimics low energy levels and protects against diet-induced metabolic disorders by enhancing fat oxidation. *Cell Metabolism, 8*(5), 347–358. doi:10.1016/j.cmet.2008.08.017 PMID:19046567

Ferrante, R. J. (2009). Mouse models of Huntington's Disease and methodological consideration for therapeutic trials. *Biochimica et Biophysica Acta, 1792*(6), 506–520. doi:10.1016/j.bbadis.2009.04.001 PMID:19362590

Ferreira, S. A., & Romero-Ramos, M. (2018). Microglia response during Parkinson's disease: Alpha-synuclein intervention. *Frontiers in Cellular Neuroscience, 12*, 247. doi:10.3389/fncel.2018.00247 PMID:30127724

Fesus, L., & Piacentini, M. (2002). *Transglutaminase 2: an enigmatic enzyme with diverse functions*. Academic Press.

Frank, S. (2014). *Treatment of Huntington's disease*. Academic Press.

Fujita, K., Nakamura, Y., Oka, T., Ito, H., Tamura, T., Tagawa, K., . . . Shiwaku, H. (2013). *A functional deficiency of TERA/VCP/p97 contributes to impaired DNA repair in multiple polyglutamine diseases*. Academic Press.

Gan, L., & Mucke, L. (2008). Paths of convergence: Sirtuins in aging and neurodegeneration. *Neuron, 58*(1), 10–14. doi:10.1016/j.neuron.2008.03.015 PMID:18400158

Gan, L., Vargas, M. R., Johnson, D. A., & Johnson, J. A. (2012). Astrocyte-specific overexpression of Nrf2 delays motor pathology and synuclein aggregation throughout the CNS in the alpha-synuclein mutant (A53T) mouse model. *The Journal of Neuroscience: The Official Journal of the Society for Neuroscience, 32*(49), 17775–17787. doi:10.1523/JNEUROSCI.3049-12.2012 PMID:23223297

Gan-Or, Z., Dion, P. A., & Rouleau, G. A. (2015). Genetic perspective on the role of the autophagy-lysosome pathway in Parkinson disease. *Autophagy, 11*(9), 1443–1457. doi:10.1080/15548627.2015.1067364 PMID:26207393

Gao, F., Yang, J., Wang, D., Li, C., Fu, Y., Wang, H., ... Zhang, J. (2017). Mitophagy in Parkinson's Disease: Pathogenic and therapeutic implications. *Frontiers in Neurology, 8*, 527. doi:10.3389/fneur.2017.00527 PMID:29046661

Gazewood, D. J., Richards, R. D., & Clebak, K. (2013). Parkinson Disease: An update. *American Family Physician, 87*(4), 267–273. PMID:23418798

Glajch, K. E., & Sadri-Vakili, G. (2015). *Epigenetic mechanisms involved in Huntington's disease pathogenesis*. Academic Press.

Goedert, M. (1993). *Tau protein and the neurofibrillary pathology of Alzheimer's disease*. Academic Press.

Goedert, M., Spillantini, M., Hasegawa, M., Jakes, R., Crowther, R., & Klug, A. (1996). *Molecular dissection of the neurofibrillary lesions of Alzheimer's disease*. Paper presented at the Cold Spring Harbor symposia on quantitative biology.

Golde, T. E., Petrucelli, L., & Lewis, J. (2010). *Targeting Aβ and tau in Alzheimer's disease, an early interim report*. Academic Press.

Greene, A. W., Grenier, K., Aguileta, M. A., Muise, S., Farazifard, R., Haque, M. E., ... Fon, E. A. (2012). Mitochondrial processing peptidase regulates PINK1 processing, import and Parkin recruitment. *EMBO Reports*, *13*(4), 378–385. doi:10.1038/embor.2012.14 PMID:22354088

Guo, X., Disatnik, M.-H., Monbureau, M., Shamloo, M., Mochly-Rosen, D., & Qi, X. (2013). *Inhibition of mitochondrial fragmentation diminishes Huntington's disease–associated neurodegeneration*. Academic Press.

Guo, X., Sun, X., Hu, D., Wang, Y. J., Fujioka, H., Vyas, R., ... Qi, X. (2016). VCP recruitment to mitochondria causes mitophagy impairment and neurodegeneration in models of Huntington's disease. *Nature Communications*, *7*(1), 12646. doi:10.1038/ncomms12646 PMID:27561680

Hague, S., Rogaeva, E., Hernandez, D., Gulick, C., Singleton, A., ... Hanson, M. (2003). Early-onset Parkinson's disease caused by a compound heterozygous DJ-1 mutation. *Annals of Neurology*, *54*, 271–274. doi:10.1002/ana.10663 PMID:12891685

Haigis, M. C., & Sinclair, D. A. (2010). Mammalian sirtuins: Biological insights and disease relevance. *Annual Review of Pathology*, *5*(1), 253–295. doi:10.1146/annurev.pathol.4.110807.092250 PMID:20078221

Hara, M. R., & Snyder, S. H. (2006). *Nitric oxide–GAPDH–Siah: a novel cell death cascade*. Academic Press.

Hara, T., Nakamura, K., Matsui, M., Yamamoto, A., Nakahara, Y., Suzuki-Migishima, R., ... Mizushima, N. (2006). Suppression of basal autophagy in neural cells causes neurodegenerative disease in mice. *Nature*, *441*(7095), 885–889. doi:10.1038/nature04724 PMID:16625204

Hara, T., Nakamura, K., Matsui, M., Yamamoto, A., Nakahara, Y., Suzuki-Migishima, R., . . . Okano, H. (2006). *Suppression of basal autophagy in neural cells causes neurodegenerative disease in mice*. Academic Press.

Hardy, J., & Allsop, D. (1991). *Amyloid deposition as the central event in the aetiology of Alzheimer's disease*. Academic Press.

Hardy, J., & Selkoe, D. J. (2002). *The amyloid hypothesis of Alzheimer's disease: progress and problems on the road to therapeutics*. Academic Press.

Hardy, J. A., & Higgins, G. A. (1992). *Alzheimer's disease: the amyloid cascade hypothesis*. Academic Press.

Harjes, P., & Wanker, E. E. (2003). *The hunt for huntingtin function: interaction partners tell many different stories*. Academic Press.

He, C., & Klionsky, D. J. (2009). *Regulation mechanisms and signaling pathways of autophagy*. Academic Press.

Heng, M. Y., Duong, D. K., Albin, R. L., Tallaksen-Greene, S. J., Hunter, J. M., Lesort, M. J., . . . Detloff, P. J. (2010). *Early autophagic response in a novel knock-in model of Huntington disease*. Academic Press.

Higashiyama, H., Hirose, F., Yamaguchi, M., Inoue, Y., Fujikake, N., & Matsukage, A. (2002). *Identification of ter94, Drosophila VCP, as a modulator of polyglutamine-induced neurodegeneration*. Academic Press.

Hirabayashi, M., Inoue, K., Tanaka, K., Nakadate, K., Ohsawa, Y., Kamei, Y., . . . Kimura, Y. (2001). *VCP/p97 in abnormal protein aggregates, cytoplasmic vacuoles, and cell death, phenotypes relevant to neurodegeneration*. Academic Press.

Hristova, V. A., Beasley, S. A., Rylett, R. J., & Shaw, G. S. (2009). Identification of a novel Zn2+-binding domain in the autosomal recessive juvenile parkinson-related E3 ligase parkin. *The Journal of Biological Chemistry, 284*(22), 14978–14986. doi:10.1074/jbc.M808700200 PMID:19339245

Hwang, S., Disatnik, M. H., & Mochly-Rosen, D. (2015). Impaired GAPDH-induced mitophagy contributes to the pathology of Huntington's disease. *EMBO Molecular Medicine, 7*(10), 1307–1326. doi:10.15252/emmm.201505256 PMID:26268247

Iqbal, K., Liu, F., & Gong, C.-X. (2016). *Tau and neurodegenerative disease: the story so far*. Academic Press.

Jeitner, T. M., Matson, W. R., Folk, J. E., Blass, J. P., & Cooper, A. J. (2008). *Increased levels of γ-glutamylamines in Huntington disease CSF*. Academic Press.

Johansen, T., & Lamark, T. (2011). Selective autophagy mediated by autophagic adapter proteins. *Autophagy, 7*(3), 279–296. doi:10.4161/auto.7.3.14487 PMID:21189453

Johnson, J. O., Mandrioli, J., Benatar, M., Abramzon, Y., Van Deerlin, V. M., Trojanowski, J. Q., . . . Wuu, J. (2010). *Exome sequencing reveals VCP mutations as a cause of familial ALS*. Academic Press.

Ju, J.-S., Fuentealba, R. A., Miller, S. E., Jackson, E., Piwnica-Worms, D., Baloh, R. H., Weihl, C. C. (2009). *Valosin-containing protein (VCP) is required for autophagy and is disrupted in VCP disease*. Academic Press.

Kakizuka, A. (2008). *Roles of VCP in human neurodegenerative disorders*. Portland Press Limited. doi:10.1042/BST0360105

Kametani, F., & Hasegawa, M. (2018). *Reconsideration of amyloid hypothesis and tau hypothesis in Alzheimer's disease*. Academic Press.

Kang, S. P., & Saif, M. W. (2008). Pharmacogenomics and pancreatic cancer treatment. Optimizing current therapy and individualizing future therapy. *JOP, 9*, 251–266. PMID:18469437

Kapogiannis, D., & Mattson, M. P. (2011). *Disrupted energy metabolism and neuronal circuit dysfunction in cognitive impairment and Alzheimer's disease*. Academic Press.

Karpuj, M. V., Becher, M. W., Springer, J. E., Chabas, D., Youssef, S., Pedotti, R., . . . Steinman, L. (2002). *Prolonged survival and decreased abnormal movements in transgenic model of Huntington disease, with administration of the transglutaminase inhibitor cystamine*. Academic Press.

Karpuj, M. V., Garren, H., Slunt, H., Price, D. L., Gusella, J., Becher, M. W., & Steinman, L. (1999). *Transglutaminase aggregates huntingtin into nonamyloidogenic polymers, and its enzymatic activity increases in Huntington's disease brain nuclei*. Academic Press.

Kerr, J. S., Adriaanse, B. A., Greig, N. H., Mattson, M. P., Cader, M. Z., Bohr, V. A., & Fang, E. F. (2017). *Mitophagy and Alzheimer's disease: cellular and molecular mechanisms*. Academic Press.

Kim, S.-Y., Marekov, L., Bubber, P., Browne, S. E., Stavrovskaya, I., Lee, J., . . . Gibson, G. E. (2005). *Mitochondrial aconitase is a transglutaminase 2 substrate: transglutamination is a probable mechanism contributing to high-molecular-weight aggregates of aconitase and loss of aconitase activity in Huntington disease brain*. Academic Press.

Kitada, T., Tong, Y., Gautier, C. A., & Shen, J. (2009). Absence of nigral degeneration in aged parkin/DJ-1/PINK1 triple knockout mice. *Journal of Neurochemistry*, *111*(3), 696–702. doi:10.1111/j.1471-4159.2009.06350.x PMID:19694908

Komatsu, M., Waguri, S., Chiba, T., Murata, S., Iwata, J.-i., Tanida, I., . . . Kominami, E. (2006). *Loss of autophagy in the central nervous system causes neurodegeneration in mice*. Academic Press.

Komatsu, M., Waguri, S., Chiba, T., Murata, S., Iwata, J., Tanida, I., ... Tanaka, K. (2006). Loss of autophagy in the central nervous system causes neurodegeneration in mice. *Nature*, *441*(7095), 880–884. doi:10.1038/nature04723 PMID:16625205

Lahiri, V., Hawkins, W. D., & Klionsky, D. J. (2019). Watch what you (self-) eat: Autophagic mechanisms that modulate metabolism. *Cell Metabolism*, *29*(4), 803–826. doi:10.1016/j.cmet.2019.03.003 PMID:30943392

Lastres-Becker, I., Ulusoy, A., Innamorato, N. G., Sahin, G., Rabano, A., ... Kirik, D. (2012). Alpha-synuclein expression and Nrf2 deficiency cooperate to aggravate protein aggregation, neuronal death and inflammation in early-stage Parkinson's disease. *Human Molecular Genetics*, *21*(14), 3173–3192. doi:10.1093/hmg/dds143 PMID:22513881

Lazarou, M., Narendra, D. P., Jin, S. M., Tekle, E., Banerjee, S., & Youle, R. J. (2013). PINK1 drives Parkin self-association and HECT-like E3 activity upstream of mitochondrial binding. *The Journal of Cell Biology*, *200*(2), 163–172. doi:10.1083/jcb.201210111 PMID:23319602

Lazarou, M., Sliter, D. A., Kane, L. A., Sarraf, S. A., Wang, C., Burman, J. L., & Youle, R. J. (2015). The ubiquitin kinase PINK1 recruits autophagy receptors to induce mitophagy. *Nature*, *524*(7565), 309–314. doi:10.1038/nature14893 PMID:26266977

Lee, Y.-K., Jun, Y.-W., Choi, H.-E., Huh, Y. H., Kaang, B.-K., Jang, D.-J., & Lee, J.-A. (2017). Development of LC3/GABARAP sensors containing a LIR and a hydrophobic domain to monitor autophagy. *The EMBO Journal*, *36*(8), 1100–1116. doi:10.15252/embj.201696315 PMID:28320742

Lesort, M., Chun, W., Johnson, G., & Ferrante, R. (1999). *Tissue transglutaminase is increased in Huntington's disease brain*. Academic Press.

Li, H., Ham, A., Ma, T. C., Kuo, S.-H., Kanter, E., Kim, D., ... Tang, G. (2019). Mitochondrial dysfunction and mitophagy defect triggered by heterozygous GBA mutations. *Autophagy, 15*(1), 113–130. doi:10.1080/15548627.2018.1509818 PMID:30160596

Li, S.H., & Li, X.-J. (2004). *Huntingtin–protein interactions and the pathogenesis of Huntington's disease*. Academic Press.

Li, Y., Rinne, J. O., Mosconi, L., Pirraglia, E., Rusinek, H., DeSanti, S., . . . Tsui, W. (2008). *Regional analysis of FDG and PIB-PET images in normal aging, mild cognitive impairment, and Alzheimer's disease*. Academic Press.

Li, Y., Wang, Y., Kim, E., Beemiller, P., Wang, C.-Y., Swanson, J., ... Guan, K.-L. (2007). Bnip3 mediates the hypoxia-induced inhibition on mammalian target of rapamycin by interacting with Rheb. *The Journal of Biological Chemistry, 282*(49), 35803–35813. doi:10.1074/jbc.M705231200 PMID:17928295

Liu, D., Pitta, M., Jiang, H., Lee, J. H., Zhang, G., Chen, X., ... Mattson, M. P. (2013). Nicotinamide forestalls pathology and cognitive decline in Alzheimer mice: Evidence for improved neuronal bioenergetics and autophagy procession. *Neurobiology of Aging, 34*(6), 1564–1580. doi:10.1016/j.neurobiolaging.2012.11.020 PMID:23273573

Liu, L., Sakakibara, K., Chen, Q., & Okamoto, K. (2014). Receptor-mediated mitophagy in yeast and mammalian systems. *Cell Research, 24*(7), 787–795. doi:10.1038/cr.2014.75 PMID:24903109

Lopert, P., & Patel, M. (2014). Brain mitochondria from DJ-1 knockout mice show increased respiration-dependent hydrogen peroxide consumption. *Redox Biology, 2*, 667–672. doi:10.1016/j.redox.2014.04.010 PMID:24936441

Lustbader, J. W., Cirilli, M., Lin, C., Xu, H. W., Takuma, K., Wang, N., ... Wu, H. (2004). ABAD directly links Abeta to mitochondrial toxicity in Alzheimer's disease. *Science, 304*(5669), 448–452. doi:10.1126cience.1091230 PMID:15087549

Manczak, M., Calkins, M. J., & Reddy, P. H. (2011). Impaired mitochondrial dynamics and abnormal interaction of amyloid beta with mitochondrial protein Drp1 in neurons from patients with Alzheimer's disease: Implications for neuronal damage. *Human Molecular Genetics, 20*(13), 2495–2509. doi:10.1093/hmg/ddr139 PMID:21459773

Martinez, J., Almendinger, J., Oberst, A., Ness, R., Dillon, C. P., Fitzgerald, P., ... Green, D. R. (2011). for the efficient clearance of dead cells. *Proceedings of the National Academy of Sciences of the United States of America, 108*(42), 17396–17401. doi:10.1073/pnas.1113421108 PMID:21969579

Martinez-Vicente, M. (2017). Neuronal mitophagy in neurodegenerative diseases. *Frontiers in Molecular Neuroscience, 10*, 64. doi:10.3389/fnmol.2017.00064 PMID:28337125

Martinez-Vicente, M., & Cuervo, A. M. (2007). *Autophagy and neurodegeneration: when the cleaning crew goes on strike*. Academic Press.

Martinez-Vicente, M., Talloczy, Z., Wong, E., Tang, G., Koga, H., Kaushik, S., . . . Sulzer, D. (2010). *Cargo recognition failure is responsible for inefficient autophagy in Huntington's disease*. Academic Press.

Mastroberardino, P., Iannicola, C., Nardacci, R., Bernassola, F., De Laurenzi, V., Melino, G., . . . Fesus, L. (2002). *'Tissue' transglutaminase ablation reduces neuronal death and prolongs survival in a mouse model of Huntington's disease*. Academic Press.

Mastroberardino, P. G., & Piacentini, M. (2010). *Type 2 transglutaminase in Huntington's disease: a double-edged sword with clinical potential*. Academic Press.

Matsumine, H., Saito, M., Shimoda-Matsubayashi, S., Tanaka, H., Ishikawa, A., Nakagawa-Hattori, Y., ... Takano, H. (1997). Localization of a gene for an autosomal recessive form of juvenile Parkinsonism to chromosome 6q25.2-27. *American Journal of Human Genetics, 60*, 588–596. PMID:9042918

Mattson, M. P., Gleichmann, M., & Cheng, A. (2008). Mitochondria in neuroplasticity and neurological disorders. *Neuron, 60*(5), 748–766. doi:10.1016/j.neuron.2008.10.010 PMID:19081372

Mattson, M. P., Partin, J., & Begley, J. G. (1998). Amyloid beta-peptide induces apoptosis-related events in synapses and dendrites. *Brain Research, 807*(1-2), 167–176. doi:10.1016/S0006-8993(98)00763-X PMID:9757026

Mazzola, J. L., & Sirover, M. A. (2002). *Alteration of intracellular structure and function of glyceraldehyde-3-phosphate dehydrogenase: a common phenotype of neurodegenerative disorders*. Academic Press.

McWilliams T.G., & Ganley, I.G. (2019). *Investigating mitophagy and Mitochondrial morphology in vivo using mito-QC: A comprehensive guide*. Academic Press.

Menalled, L. B., Kudwa, A. E., Oakeshott, S., Farrar, A., Paterson, N., Filippov, I., . . . Beltran, J. (2014). *Genetic deletion of transglutaminase 2 does not rescue the phenotypic deficits observed in R6/2 and zQ175 mouse models of Huntington's disease*. Academic Press.

Meyer, H., Bug, M., & Bremer, S. (2012). *Emerging functions of the VCP/p97 AAA-ATPase in the ubiquitin system*. Academic Press.

Meyer, H., & Weihl, C. C. (2014). The VCP/p97 system at a glance: connecting cellular function to disease pathogenesis. The Company of Biologists Ltd.

Monteiro, M. C., Coleman, M. D., Hill, E. J., Prediger, R. D., & Maia, C. S. F. (2017). Neuroprotection in neurodegenerative disease from basic science to clinical applications. *Oxidative Medicine and Cellular Longevity, 2017*, 1–3. doi:10.1155/2017/2949102 PMID:28337247

Mori, F., Tanji, K., Toyoshima, Y., Sasaki, H., Yoshida, M., Kakita, A., . . . Wakabayashi, K. (2013). *Valosin-containing protein immunoreactivity in tauopathies, synucleinopathies, polyglutamine diseases and intranuclear inclusion body disease*. Academic Press.

Morrison, B., Hensley, K., Pioro, P. E., Petri, S., & Kiaei, M. (2012). Amyotrophic Lateral Sclerosis and novel therapeutic strategies. *Neurology Research International, 2012*, 1–3. doi:10.1155/2012/798028 PMID:23316357

Morselli, E., Marino, G., Bennetzen, M. V., Eisenberg, T., Megalou, E., Schroeder, S., ... Kroemer, G. (2011). Spermidine and resveratrol induce autophagy by distinct pathways converging on the acetylproteome. *The Journal of Cell Biology*, *192*(4), 615–629. doi:10.1083/jcb.201008167 PMID:21339330

Murphy, M. P. (2009). How mitochondria produce reactive oxygen species. *The Biochemical Journal*, *417*(1), 1–13. doi:10.1042/BJ20081386 PMID:19061483

Murphy, M. P., Holmgren, A., Larsson, N.-G., Halliwell, B., Chang, C. J., Kalyanaraman, B., ... Winterbourn, C. C. (2011). Unraveling the biological roles of reactive oxygen species. *Cell Metabolism*, *13*(4), 361–366. doi:10.1016/j.cmet.2011.03.010 PMID:21459321

Nakahata, Y., Kaluzova, M., Grimaldi, B., Sahar, S., Hirayama, J., Chen, D., ... Sassone-Corsi, P. (2008). The NAD+-dependent deacetylase SIRT1 modulates CLOCK-mediated chromatin remodeling and circadian control. *Cell*, *134*(2), 329–340. doi:10.1016/j.cell.2008.07.002 PMID:18662547

Nalbandian, A., Llewellyn, K. J., Badadani, M., Yin, H. Z., Nguyen, C., Katheria, V., . . . Caiozzo, V. (2013). *A progressive translational mouse model of human valosin-containing protein disease: The VCPR155H/+ mouse*. Academic Press.

Nalbandian, A., Llewellyn, K. J., Kitazawa, M., Yin, H. Z., Badadani, M., Khanlou, N., . . . Mozaffar, T. (2012). *The homozygote VCPR155H/R155H mouse model exhibits accelerated human VCP-associated disease pathology*. Academic Press.

Narendra, D., Tanaka, A., Suen, D. F., & Youle, R. J. (2008). Parkin is recruited selectively to impaired mitochondria and promotes their autophagy. *The Journal of Cell Biology*, *183*(5), 795–803. doi:10.1083/jcb.200809125 PMID:19029340

Narendra, D. P., Jin, S. M., Tanaka, A., Suen, D. F., Gautier, C. A., Shen, J., ... Youle, R. J. (2010). PINK1 is selectively stabilized on impaired mitochondria to activate Parkin. *PLoS Biology*, *8*(1), e1000298. doi:10.1371/journal.pbio.1000298 PMID:20126261

Nasir, J., Floresco, S. B., O'Kusky, J. R., Diewert, V. M., Richman, J. M., Zeisler, J., . . . Hayden, M. R. (1995). *Targeted disruption of the Huntington's disease gene results in embryonic lethality and behavioral and morphological changes in heterozygotes*. Academic Press.

Nixon, R. A. (2013). *The role of autophagy in neurodegenerative disease*. Academic Press.

Nixon, R. A., Yang, D.-S., & Lee, J.-H. (2008). *Neurodegenerative lysosomal disorders: a continuum from development to late age*. Academic Press.

Nunomura, A., Moreira, P. I., Lee, H. G., Zhu, X., Castellani, R. J., Smith, M. A., & Perry, G. (2007). Neuronal death and survival under oxidative stress in Alzheimer and Parkinson diseases. *CNS & Neurological Disorders - Drug Targets*, *6*(6), 411–423. doi:10.2174/187152707783399201 PMID:18220780

Obeso, J. A., Marin, C., Rodriguez-Oroz, C., Blesa, J., Benitez-Temino, B., Mena-Segovia, J., ... Olanow, C. W. (2008). The basal ganglia in Parkinson's disease: Current concepts and unexplained observations. *Annals of Neurology*, *64*(S2Suppl. 2), S30–S46. doi:10.1002/ana.21481 PMID:19127584

Okatsu, K., Saisho, K., Shimanuki, M., Nakada, K., Shitara, H., Sou, Y., ... Matsuda, N. (2010). p62/SQSTM1 cooperates with Parkin for perinuclear clustering of depolarized mitochondria. *Genes to Cells, 15*, 887–900. doi:10.1111/j.1365-2443.2010.01426.x PMID:20604804

Okatsu, K., Sato, Y., Yamano, K., Matsuda, N., Negishi, L., Takahashi, A., ... Fukai, S. (2018). Structural insights into ubiquitin phosphorylation by PINK1. *Scientific Reports, 8*(1), 10382. doi:10.103841598-018-28656-8 PMID:29991771

Osafo, N., Obiri, D. D., Yeboah, O. K., & Minkah, P. A. (2019). Amyotrophic Lateral Sclerosis: A Predominant Form of Degenerative Disease of the Motor Neuron System. In M. Uddin & M. Amran (Eds.), *Handbook of Research on Critical Examinations of Neurodegenerative Disorders* (pp. 322–346). Hershey, PA: IGI Global. doi:10.4018/978-1-5225-5282-6.ch015

Outeiro, T. F., Marques, O., & Kazantsev, A. (2008). Therapeutic role of sirtuins in neurodegenerative disease. *Biochimica et Biophysica Acta, 1782*(6), 363–369. doi:10.1016/j.bbadis.2008.02.010 PMID:18373985

Ow, Y. P., Green, D. R., Hao, Z., & Mark, T. W. (2008). Cytochrome c: Functions beyond respiration. *Nature Reviews. Molecular Cell Biology, 9*(7), 532–542. doi:10.1038/nrm2434 PMID:18568041

Piacentini, M., Grazia Farrace, M., Piredda, L., Matarrese, P., Ciccosanti, F., Falasca, L., . . . Griffin, M. (2002). *Transglutaminase overexpression sensitizes neuronal cell lines to apoptosis by increasing mitochondrial membrane potential and cellular oxidative stress*. Academic Press.

Pickrell, A. M., & Youle, R. J. (2015). The roles of PINK1, parkin, and mitochondrial fidelity in Parkinson's disease. *Neuron, 85*(2), 257–273. doi:10.1016/j.neuron.2014.12.007 PMID:25611507

Plaza-Zabala, A., Sierra-Torre, V., & Sierra, A. (2017). Autophagy and Microglia: Novel Partners in Neurodegeneration and Aging. *International Journal of Molecular Sciences, 18*(598), 1–28. doi:10.3390/ijms18030598 PMID:28282924

Plotegher, N., Gratton, E., & Bubacco, L. (2014). Number and brightness analysis of alpha-synuclein oligomerization and the associated mitochondrial morphology alterations in live cells. *Biochimica et Biophysica Acta, 1840*(6), 2014–2024. doi:10.1016/j.bbagen.2014.02.013 PMID:24561157

Poksay, K. S., Madden, D. T., Peter, A. K., Niazi, K., Banwait, S., Crippen, D., . . . Rao, R. V. (2011). *Valosin-containing protein gene mutations: cellular phenotypes relevant to neurodegeneration*. Academic Press.

Puder, M., & Soberman, R. J. (1997). *Glutathione conjugates recognize the Rossmann fold of glyceraldehyde-3-phosphate dehydrogenase*. Academic Press.

Qin, W., Chachich, M., Lane, M., Roth, G., Bryant, M., de Cabo, R., ... Pasinetti, G. M. (2006a). Calorie restriction attenuates Alzheimer's disease type brain amyloidosis in Squirrel monkeys (Saimiri sciureus). *Journal of Alzheimer's Disease, 10*(4), 417–422. doi:10.3233/JAD-2006-10411 PMID:17183154

Qin, W., Yang, T., Ho, L., Zhao, Z., Wang, J., Chen, L., ... Pasinetti, G. M. (2006b). Neuronal SIRT1 activation as a novel mechanism underlying the prevention of Alzheimer disease amyloid neuropathology by calorie restriction. *The Journal of Biological Chemistry, 281*(31), 21745–21754. doi:10.1074/jbc.M602909200 PMID:16751189

Ricotta, M., Iannuzzi, M., De Vivo, G., & Gentile, V. (2010). *Physio-pathological roles of transglutaminase-catalyzed reactions.* Academic Press.

Rodger, C.E., McWilliams, T.G., & Ganley, I.G. (2017). *Mammalian mitophagy – from in vitro molecules to in vivo models.* doi:10.1111/febs.14336

Rosas, H., Koroshetz, W., Chen, Y., Skeuse, C., Vangel, M., Cudkowicz, M., . . . Makris, N. (2003). *Evidence for more widespread cerebral pathology in early HD: an MRI-based morphometric analysis.* Academic Press.

Ross, C. A., & Tabrizi, S. J. (2011). *Huntington's disease: from molecular pathogenesis to clinical treatment.* Academic Press.

Rossin, F., D'Eletto, M., Macdonald, D., Farrace, M. G., & Piacentini, M. J. A. a. (2012). *TG2 transamidating activity acts as a reostat controlling the interplay between apoptosis and autophagy.* Academic Press.

Ryu, D., Mouchiroud, L., Andreux, P. A., Katsyuba, E., Moullan, N., Nicolet-Dit-Felix, A. A., ... Auwerx, J. (2016). Urolithin A induces mitophagy and prolongs lifespan in C. elegans and increases muscle function in rodents. *Nature Medicine, 22*(8), 879–888. doi:10.1038/nm.4132 PMID:27400265

Saita, S., Shirane, M., & Nakayama, K. I. (2013). Selective escape of proteins from the mitochondria during mitophagy. *Nature Communications, 4*(1), 1410. doi:10.1038/ncomms2400 PMID:23361001

Santos, D., & Cardoso, S. M. (2012). Mitochondrial dynamics and neuronal fate in Parkinson's disease. *Mitochondrion, 12*(4), 428–437. doi:10.1016/j.mito.2012.05.002 PMID:22609323

Sarkar, S., Raymick, J., & Imam, R. J. (2016). Neuroprotective and therapeutic strategies against Parkinson's Disease: Recent perspective. *International Journal of Molecular Sciences, 17*(6), 904. doi:10.3390/ijms17060904 PMID:27338353

Saunders, J. A., Estes, K. A., Kosloski, L. M., Allen, H. E., Dempsey, K. M., ... Torres-Russotto, D. R. (2012). CD4C regulatory and effector/memory T cell subsets profile motor dysfunction in Parkinson's disease. *Journal of Neuroimmune Pharmacology, 7*(4), 927–938. doi:10.100711481-012-9402-z PMID:23054369

Schmidt, R., de Reus, M. A., Scholtens, S. H., van den Berg, L. H., & van den Heuvel, M. P. (2016). Simulating disease propagation across white matter connectome reveals anatomical substrate for neuropathology staging in amyotrophic lateral sclerosis. *NeuroImage, 124*, 762–769. doi:10.1016/j.neuroimage.2015.04.005 PMID:25869856

Schweers, R. L., Zhang, J., Randall, M. S., Loyd, M. R., Li, W., Dorsey, F. C., ... Ney, P. A. (2007). NIX is required for programmed mitochondrial clearance during reticulocyte maturation. *Proceedings of the National Academy of Sciences of the United States of America, 104*(49), 19500–19505. doi:10.1073/pnas.0708818104 PMID:18048346

Selkoe, D. J. (1991). *The molecular pathology of Alzheimer's disease.* Academic Press.

Serrano-Pozo, A., Frosch, M. P., Masliah, E., & Hyman, B. T. (2011). *Neuropathological alterations in Alzheimer disease.* Academic Press.

Shi, R.-Y., Zhu, S.-H., Li, V., Gibson, S. B., Xu, X.-S., & Kong, J.-M. (2014). BNIP3 interacting with LC3 triggers excessive mitophagy in delayed neuronal death in stroke. *CNS Neuroscience & Therapeutics, 20*(12), 1045–1055. doi:10.1111/cns.12325 PMID:25230377

Shimura, H., Hattori, N., Kubo, S., Mizuno, Y., Asakawa, S., Minoshima, S., ... Suzuki, T. (2000). Familial Parkinson disease gene product, parkin, is a ubiquitin-protein ligase. *Nature Genetics, 25*(3), 302–305. doi:10.1038/77060 PMID:10888878

Shirasaki, D. I., Greiner, E. R., Al-Ramahi, I., Gray, M., Boontheung, P., Geschwind, D. H., . . . Loo, J. A. (2012). *Network organization of the huntingtin proteomic interactome in mammalian brain*. Academic Press.

Shirendeb, U. P., Calkins, M. J., Manczak, M., Anekonda, V., Dufour, B., McBride, J. L., . . . Reddy, P. H. (2011). *Mutant huntingtin's interaction with mitochondrial protein Drp1 impairs mitochondrial biogenesis and causes defective axonal transport and synaptic degeneration in Huntington's disease*. Academic Press.

Sierra, A., Abiega, O., Shahraz, A., & Neumann, H. (2013). Janus-faced microglia: Beneficial and detrimental consequences of microglial phagocytosis. *Frontiers in Cellular Neuroscience, 7*, 7. doi:10.3389/fncel.2013.00006 PMID:23386811

Sinclair, D. A., & Guarente, L. (1997). Extrachromosomal rDNA circles--a cause of aging in yeast. *Cell, 91*(7), 1033–1042. doi:10.1016/S0092-8674(00)80493-6 PMID:9428525

Song, W., Chen, J., Petrilli, A., Liot, G., Klinglmayr, E., Zhou, Y., . . . Hayden, M. R. (2011). *Mutant huntingtin binds the mitochondrial fission GTPase dynamin-related protein-1 and increases its enzymatic activity*. Academic Press.

Spillantini, M. G., Schmidt, M. L., Lee, V. M., Trojanowski, J. Q., Jakes, R., & Goedert, M. (1997). Alpha-synuclein in Lewy bodies. *Nature, 388*(6645), 839–840. doi:10.1038/42166 PMID:9278044

Spilman, P., Podlutskaya, N., Hart, M. J., Debnath, J., Gorostiza, O., Bredesen, D., ... Galvan, V. (2010). Inhibition of mTOR by rapamycin abolishes cognitive deficits and reduces amyloid-beta levels in a mouse model of Alzheimer's disease. *PLoS One, 5*(4), e9979. doi:10.1371/journal.pone.0009979 PMID:20376313

Stevens, C. H., Rowe, D., Morel-Kopp, M. C., Orr, C., Russell, T., ... Ranola, M. (2012). Reduced T helper and B lymphocytes in Parkinson's disease. *Journal of Neuroimmunology, 252*(1-2), 95–99. doi:10.1016/j.jneuroim.2012.07.015 PMID:22910543

Sun, N., Yun, J., Liu, J., Malide, D., Liu, C., Rovira, I. I., ... Finkel, T. (2015). Measuring *in vivo* mitophagy. *Molecular Cell, 60*(4), 685–696. doi:10.1016/j.molcel.2015.10.009 PMID:26549682

Szondy, Z., Mastroberardino, P., Váradi, J., Farrace, M., Nagy, N., Bak, I., . . . Rizzuto, R. (2006). *Tissue transglutaminase (TG2) protects cardiomyocytes against ischemia/reperfusion injury by regulating ATP synthesis*. Academic Press.

Tarze, A., Deniaud, A., Le Bras, M., Maillier, E., Mollé, D., Larochette, N., . . . Brenner, C. (2007). *GAPDH, a novel regulator of the pro-apoptotic mitochondrial membrane permeabilization*. Academic Press.

Tatsukawa, H., Furutani, Y., Hitomi, K., & Kojima, S. (2016). *Transglutaminase 2 has opposing roles in the regulation of cellular functions as well as cell growth and death*. Academic Press.

The NINDS NET-PD Investigators. (2006). A randomized, double-blind, futility clinical trial of creatine and minocycline in early Parkinson disease. *Neurology, 66*(5), 664–671. doi:10.1212/01.wnl.0000201252.57661.e1 PMID:16481597

Tiraboschi, P., Hansen, L., Thal, L., & Corey-Bloom, J. (2004). *The importance of neuritic plaques and tangles to the development and evolution of AD*. Academic Press.

Tisdale, E. J. (2002). *Glyceraldehyde-3-phosphate dehydrogenase is phosphorylated by protein kinase Cι/λ and plays a role in microtubule dynamics in the early secretory pathway*. Academic Press.

Van Duijn, E., Kingma, E., & Van der Mast, R. (2007). *Psychopathology in verified Huntington's disease gene carriers*. Academic Press.

van Humbeeck, C., Cornelissen, T., & Vandenberghe, W. (2011). Ambra1. *Autophagy, 7*(12), 1555–1556. doi:10.4161/auto.7.12.17893 PMID:21921694

van Humbeeck, C., Cornelissen, T., & Vandenberghe, W. (2011). Ambra1: A Parkin-binding protein involved in mitophagy. *Autophagy, 7*(12), 1555–1556. doi:10.4161/auto.7.12.17893 PMID:21921694

Vander Heiden, M. G., Cantley, L. C., & Thompson, C. B. (2009). Understanding the Warburg effect: The metabolic requirements of cell proliferation. *Science, 324*(5930), 1029–1033. doi:10.1126cience.1160809 PMID:19460998

Varghese, M., Ho, L., Wang, J., Zhao, W., Levine, S., Ono, K., ... Pasinetti, G. (2014). Green coffee as a novel agent for Alzheimer's disease prevention by attenuating diabetes. *Translational Neuroscience, 5*(2), 111–116. doi:10.247813380-014-0213-y

Vassar, R. (2002). *β-secretase (BACE) as a drug target for Alzheimer's disease*. Academic Press.

Villar-Pique, A., Lopes da Fonseca, T., & Outeiro, T. F. (2016). Structure, function and toxicity of alpha-synuclein: The Bermuda triangle in synucleinopathies. *Journal of Neurochemistry, 139*(Suppl. 1), 240–255. doi:10.1111/jnc.13249 PMID:26190401

Vlassenko, A. G., Vaishnavi, S. N., Couture, L., Sacco, D., Shannon, B. J., Mach, R. H., ... Mintun, M. A. (2010). Spatial correlation between brain aerobic glycolysis and amyloid-beta (Abeta) deposition. *Proceedings of the National Academy of Sciences of the United States of America, 107*(41), 17763–17767. doi:10.1073/pnas.1010461107 PMID:20837517

Wager, K., & Russell, C. (2013). Mitophagy and neurodegeneration: The zebrafish model system. *Autophagy, 9*(11), 1693–1709. doi:10.4161/auto.25082 PMID:23939015

Wang, J., Fivecoat, H., Ho, L., Pan, Y., Ling, E., & Pasinetti, G. M. (2010). The role of Sirt1: At the crossroad between promotion of longevity and protection against Alzheimer's disease neuropathology. *Biochimica et Biophysica Acta, 1804*(8), 1690–1694. doi:10.1016/j.bbapap.2009.11.015 PMID:19945548

Wang, X., Su, B., Siedlak, S. L., Moreira, P. I., Fujioka, H., Wang, Y., ... Zhu, X. (2008). Amyloid-beta overproduction causes abnormal mitochondrial dynamics via differential modulation of mitochondrial fission/fusion proteins. *Proceedings of the National Academy of Sciences of the United States of America*, *105*(49), 19318–19323. doi:10.1073/pnas.0804871105 PMID:19050078

Wang, Y., Liu, N., & Lu, B. (2019). Mechanisms and roles of mitophagy in neurodegenerative diseases. *CNS Neuroscience & Therapeutics*, *25*, 859–875. doi:10.1111/cns.13140 PMID:31050206

Watts, G. D., Wymer, J., Kovach, M. J., Mehta, S. G., Mumm, S., Darvish, D., . . . Kimonis, V. E. (2004). *Inclusion body myopathy associated with Paget disease of bone and frontotemporal dementia is caused by mutant valosin-containing protein*. Academic Press.

Weggen, S., Eriksen, J. L., Das, P., Sagi, S. A., Wang, R., Pietrzik, C. U., . . . Bulter, T. (2001). *A subset of NSAIDs lower amyloidogenic Aβ42 independently of cyclooxygenase activity*. Academic Press.

Wei, H., Liu, L., & Chen, Q. (2015). Selective removal of mitochondria via mitophagy: Distinct pathways for different mitochondrial stresses. *Biochimica et Biophysica Acta*, *1853*(10), 2784–2790. doi:10.1016/j.bbamcr.2015.03.013 PMID:25840011

Wenk, G. L. (2003). *Neuropathologic changes in Alzheimer's disease*. Academic Press.

Wilson, R.J., Drake, J.C., Cui, D., Zhang, M., Perry, H.M., Kashatus, J.A., … Yan, Z. (2017). *Conditional Mitotimer reporter mice for assessment of mitochondrial structure, oxidative stress, and mitophagy*. Academic Press.

Wolfe, M. S. (2008). *Inhibition and modulation of γ-secretase for Alzheimer's disease*. Academic Press.

Wong, Y. C., & Holzbaur, E. L. (2014). Optineurin is an autophagy receptor for damaged mitochondria in parkin-mediated mitophagy that is disrupted by an ALS-linked mutation. *Proceedings of the National Academy of Sciences of the United States of America*, *111*(42), E4439–E4448. doi:10.1073/pnas.1405752111 PMID:25294927

Wu, J., Lin, F., & Qin, Z. (2007). *Sequestration of glyceraldehyde-3-phosphate dehydrogenase to aggregates formed by mutant huntingtin*. Academic Press.

Wu, W., Tian, W., Hu, Z., Chen, G., Huang, L., Li, W., ... Feng, D. (2014). ULK1 translocates to mitochondria and phosphorylates FUNDC1 to regulate mitophagy. *EMBO Reports*, *15*(5), 566–575. doi:10.1002/embr.201438501 PMID:24671035

Xia, D., Tang, W. K., & Ye, Y. (2016). *Structure and function of the AAA+ ATPase p97/Cdc48p*. Academic Press.

Yang, H., Li, J.-J., Liu, S., Zhao, J., Jiang, Y.-J., Song, A.-X., & Hu, H.-Y. (2014). *Aggregation of polyglutamine-expanded ataxin-3 sequesters its specific interacting partners into inclusions: implication in a loss-of-function pathology*. Academic Press.

Yano, H., Baranov, S. V., Baranova, O. V., Kim, J., Pan, Y., Yablonska, S., . . . Friedlander, R. M. J. N. n. (2014). *Inhibition of mitochondrial protein import by mutant huntingtin*. Academic Press.

Ye, X., Sun, X., Starovoytov, V., & Cai, Q. (2015). Parkin-mediated mitophagy in mutant hAPP neurons and Alzheimer's disease patient brains. *Human Molecular Genetics, 24*(10), 2938–2951. doi:10.1093/hmg/ddv056 PMID:25678552

Ye, Y. (2006). *Diverse functions with a common regulator: ubiquitin takes command of an AAA ATPase*. Academic Press.

Yin, H., Nalbandian, A., Hsu, C., Li, S., Llewellyn, K., Mozaffar, T., . . . Weiss, J. (2012). *Slow development of ALS-like spinal cord pathology in mutant valosin-containing protein gene knock-in mice*. Academic Press.

Yogalingam, G., Hwang, S., Ferreira, J. C., & Mochly-Rosen, D. (2013). *Glyceraldehyde-3-phosphate dehydrogenase (GAPDH) phosphorylation by protein kinase Cδ (PKCδ) inhibits mitochondria elimination by lysosomal-like structures following ischemia and reoxygenation-induced injury*. Academic Press.

Youle, R. J., & Narendra, D. P. (2011). Mechanisms of mitophagy. *Nature Reviews. Molecular Cell Biology, 12*(1), 9–14. doi:10.1038/nrm3028 PMID:21179058

Youle, R. J., & Van der Bliek, A. M. (2012). Mitochondrial fission, fusion, and stress. *Science, 337*(6098), 1062–1065. doi:10.1126cience.1219855 PMID:22936770

Yu, L., Chen, Y., & Tooze, S. (2018). Autophagy pathway: Cellular and molecular mechanisms. *Autophagy, 14*(2), 207–215. doi:10.1080/15548627.2017.1378838 PMID:28933638

Zeitlin, S., Liu, J.-P., Chapman, D. L., Papaioannou, V. E., & Efstratiadis, A. J. N. g. (1995). *Increased apoptosis and early embryonic lethality in mice nullizygous for the Huntington's disease gene homologue*. Academic Press.

Zheng, L., Roeder, R. G., & Luo, Y. (2003). *S phase activation of the histone H2B promoter by OCA-S, a coactivator complex that contains GAPDH as a key component*. Academic Press.

Zondler, L., Miller-Fleming, L., Repici, M., Gonçalves, S., Tenreiro, S., Rosado-Ramos, R., ... Outeiro, T. F. (2014). DJ-1 interactions with α-synuclein attenuate aggregation and cellular toxicity in models of Parkinson's disease. *Cell Death, 5*(7), e1350. doi:10.1038/cddis.2014.307 PMID:25058424

Chapter 5
Post-Translational Modifications in Neurodegeneration:
Tools of Protein Quality Control System

Abdullah Al Mamun
https://orcid.org/0000-0002-8774-5616
Department of Pharmacy, Southeast University, Dhaka, Bangladesh

Md. Farhad Hossain
Department of Physical Therapy, Inje University, Gimhae, South Korea

ABSTRACT

Post-translational modifications (PTMs) increase proteome activity for controlling every feature of normal cell biology. PTMs such as phosphorylation, acetylation, glycosylation, fatty acylation, palmitoylation, myristoylation, ubiquitination, SUMOylation (small ubiquitin-like modifiers), methylation, deamidation, nitrosylation, etc. of proteins can regulate the properties of protein including intracellular distribution, functionality, stability, accumulation, as well as interactions. PTMs take place at any stage of the protein life cycle, regulating protein folding and activity in time and space, subcellular localization of the protein, and their activity. Hence, PTMs play a pivotal role in the regulation of numerous cellular processes. Abnormal PTMs of one or more culprit proteins might contribute to neurodegeneration, which is shown in some neurodegenerative disorders including Alzheimer's, Parkinson's, and prion disease. In this chapter, the authors focus on the most essential PTMs that are observed in neurodegenerative disorders and elucidating the pathogenesis wherein they are involved.

DOI: 10.4018/978-1-7998-1317-0.ch005

INTRODUCTION

Neurodegenerative disorders (NDDs) take place when nerve cells in the central nervous system (CNS) or peripheral nervous system lose their activity progressively and eventually die. There are several NDDs such as Alzheimer's disease (AD), prion diseases, Parkinson's disease (PD), motor neuron diseases, Huntington's disease (HD), polyglutamine (PolyQ) diseases, and lyme diseases (Dugger and Dickson, 2017) and these disorders are characterized by various behavioral, cognitive, and motor symptoms. Characteristically NDDs are late-onset and noticeable with gradually degenerating phenotypes. Besides, most of the disorders are sporadic, but they are familial for specified cases. Causative genes have been detected merely for some NDDs. A specific group of populations who contain neurons that become deteriorated and ultimately dies leading to the diverse motor and cognitive symptoms. NDDs are primarily connected with the mechanisms of toxic gain of function, which play an essential role in the modification of numerous cellular pathways. Furthermore, the aggregation of aberrantly folded proteins in the formations of inclusion bodies, as well as micro-accumulates both outside and inside the neurons, are the hallmarks of NDDs. These disorders are also called as proteinopathies or brain folding diseases owing to the existence of these insoluble species. Based on the causative agent that that leads to the accumulation of proteins, brain folding diseases can be categorized into four types of proteinopathies, such as α-synucleinopathies, amyloidoses, tauopathies, as well as transactivation response-DNA binding protein 43 proteinopathies (Dugger and Dickson 2017).

At the transcriptional level, differential splicing and the usage of alternative promoters/terminators can considerably raise the number of messenger ribonucleic acids that are encoded by a single gene, which leads to the human transcriptome to above 100 thousand probable transcripts (Yura et al. 2006). Furthermore, covalent alterations in specific amino acids increase the human proteome to 1,000,000 proteins (Jensen 2004). Generally, modifications occurring at specified amino acids are jointly called as post-translational modifications (PTMs), autonomously by their biochemical properties. Unlike genetic variations that take place on an evolutionary time scale, PTMs work as molecular switches with regard to intra-cellular as well as extra-cellular stimuli. Numerous PTMs can concurrently occur and co-operate to determine the protein's molecular state such as its structure, and activity of the enzyme, cellular localization, turn-over as well as interactions with other biomolecules. Moreover, above 90,000 different PTMs have been detected until now via biophysical and biochemical analyses (Khoury et al. 2011).

PTMs are strongly controlled in space and time and they also play a pivotal role in the protein functionality. Therefore, any dyshomeostasis of PTMs can possibly contribute to a pathological condition. Besides, numerous NDDs are featured by abnormal PTMs of one or more culprit proteins. The aim of this study is to emphasize the significant role of PTMs in NDDs as well as focus on the pathogenesis where they are involved.

PROTEIN POST-TRANSLATIONAL MODIFICATIONS

Proteins are the major component in living organisms that exert cellular and physiological function, and its physical and chemical features indicate their activities. Protein folding and final structure including biochemical activity, half-life and stability are determined by the primary sequence of a protein (Marks et al. 2012). However, during the life span, the proteome of a protein would be two or three folds more complex rather than encoding genomes (Yokoyama et al. 2017). PTM, a proteome expansion route, is present in eukaryotes and prokaryotes but the exposure of PTMs is more common in eukaryotic cells.

Either enzymatic attachment or nonenzymatic binding of precise chemical groups to the side chains of the amino acid is responsible for protein PTM and this modification is occurred due to following protein translation. PTM also plays a crucial role in the protein's structure and both cellular and physiological functions of the protein. Furthermore, some processes such as phosphorylation, glycosylation, acetylation, methylation, SUMOylation (small ubiquitin-like modifiers), palmitoylation, biotinylation, ubiquitination, nitration, chlorination, and oxidation/reduction are the example of enzymatic PTMs (Omenn et al. 2017). On the other hand, glycation, nitrosylation, oxidation/reduction, acetylation, and succination are known as nonenzymatic PTMs (Kakizawa 2013; Merkley et al. 2014; Kuhn et al. 2014; Nedić et al. 2015; Greifenhagen et al. 2016). Furthermore, unconventional PTMs, for example, glypiation, neddylation, siderophorylation, AMPylation, and cholesteroylation, are associated with the function and structure of the protein (Basak et al.).

Alteration of protein synthesis, accuracy and PTMs are the major risk factors behind aging. During aging, damaged proteins turn to the abnormal protein via proteasomal and lysosomal signaling pathways (Shiozawa-West et al. 2015). The sequential improvement in molecular heterogeneity and imbalanced functioning of proteins is associated with various NDDs and age-related pathologies such as cataracts and sarcopenia (Rattan 2008). As a result, the contribution of PTMs and their functional ability in aging will accelerate the effective intervention, prevention as well as therapy not only for aging but also for age-related disorders including NDDs.

TYPES OF POST-TRANSLATIONAL MODIFICATIONS

PTMs Proteins are divided into two groups. The first one is surrounded by covalent attachment of some chemical groups by enzymatic catalysis. An electrophilic portion of a cosubstrate is consistently attached to electron enriched protein side-chain that works as a nucleophile in the transfer. The other group of PTMs is surrounded by the presence of covalent cleavage of peptide backbones. This cleavage is occurred by either proteases or autocatalysis processes. Furthermore, phosphorylation, acylation, glycosylation, alkylation, and oxidation are the most common covalent protein PTMs (Figure 1). These PTMs are catalyzed through specific mechanisms, which play a crucial role in aging and age-related disorders. The description of PTMs, related to aging and age-related disorders including NDDs, is briefly given below:

- PTMs of proteins is done by covalent modification of nucleophilic amino acid side chain through the electrophilic fragmentation of cosubstrate and cleavage of protein backbone at a specific peptide bond (Santos and Lindner 2017).
- There are five types of covalent attachments to protein side chains such as acylation, glycosylation, oxidation, phosphorylation, and alkylation (Santos and Lindner 2017).

Figure 1. Outline of the post-translational modifications of proteins by different processes.

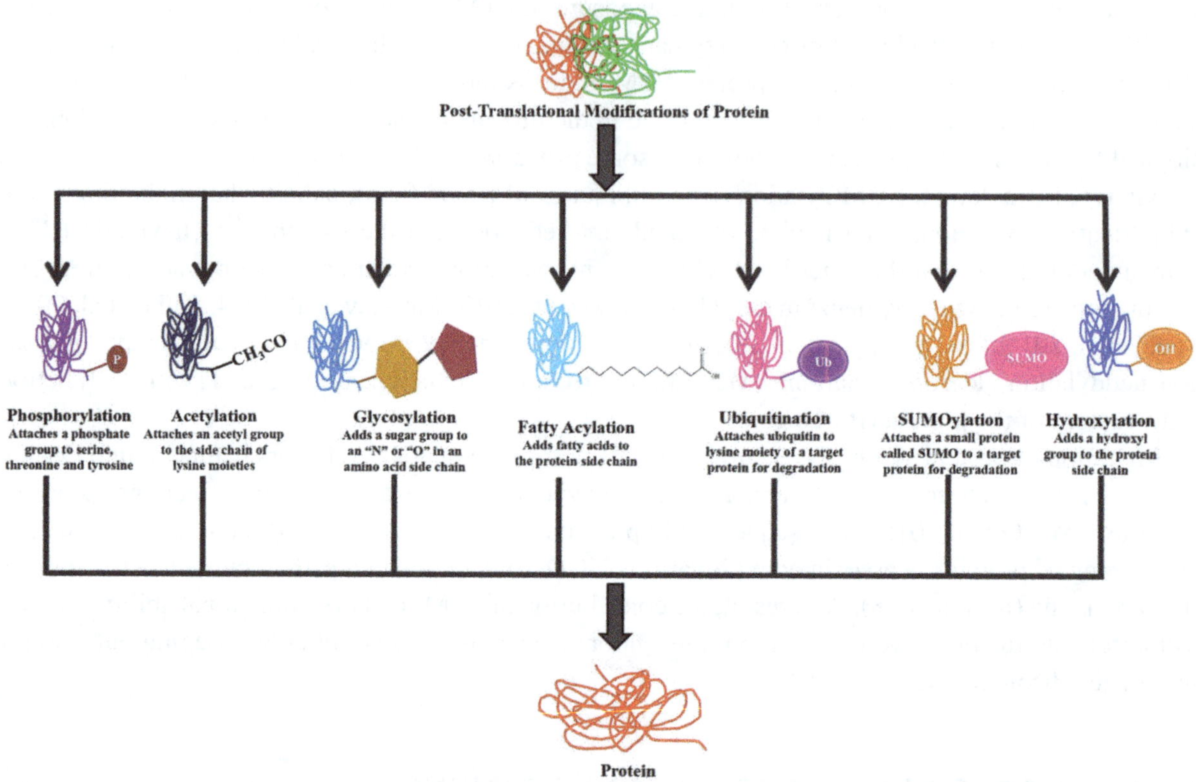

POST-TRANSLATIONAL MODIFICATIONS IN NEURODEGENERATION

Phosphorylation

Protein phosphorylation is denoted as the most extensive PTM. By regulating phosphorylation, almost all types of protein show their physiological action to the target cell. Phosphorylation is known as a reversible modification that is mainly occurred at serine residues **(Figure 2)** of protein substrate (95%), threonine (4%) and tyrosine (1%) (Nestler and Greengard 1999). Phosphorylation is mediated by protein kinases, which catalyze the shifting of the g-phosphate group of adenosine triphosphate to the hydroxyl group of target moieties. Protein phosphatases counterbalance their actions, which act antagonistically to promote the hydrolysis of phosphoester bonds. Therefore, the final pattern of phosphorylation is the outcome of these two groups of enzymes ' concerted activities. Any dyshomeostasis in this process generates the hypo- or hyper-phosphorylated of target proteins. Therefore, it can be said that abnormal phosphorylation leads to the pathogenesis of neurological diseases.

Figure 2. Phosphorylation of serine in the post-translational modification of proteins.

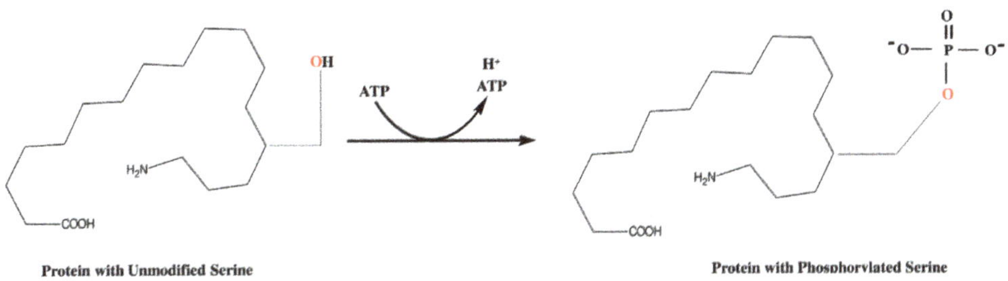

Tau Phosphorylation in Tauopathies

Tauopathies involve a group of neurodegenerative diseases such as AD, Pick's disease, progressive supranuclear palsy, corticobasal degeneration and post-encephalitic Parkinsonism and so on. At the molecular level, they are distinguished by the formation of insoluble aggregates called neurofibrillary tangles (NFTs) in neurons, consisting mainly of misfolded tau protein organized in paired helical filaments (PHFs) (Williams 2006). Tau, a naturally unfolded protein, exists in 6 isoforms because of splicing of 2, 3 and 10 exons in microtubule-associated protein tau gene and the most expression of tau is appeared in the CNS. A variety of N-terminal inserts including 0N, 1N, and 2N, and C-terminal containing (3R or 4R) microtubule-binding repeats are present in isoform. Tau is mainly located at the axons in matured neurons and by binding their surfaces it develops microtubule assembly (Mandelkow and Mandelkow 2011). 2N4R, containing 85 phosphorylation sites, is referred to as the longest isoform located inside the C-terminal repeats as well as the flanking regions. In the presence of different kinases including cyclin-dependent kinase 5 (CDK5), glycogen synthase kinase-3β and the mitogen-activated protein kinase family, about 30 of them are phosphorylated among 85 phosphorylation sites, during physiological conditions (Hanger et al. 2009). During disease progression, abnormal hyper-phosphorylation occurs at specific pathological sites although the etiology of this process is not completely elucidated until now. The level of CDK5 is upregulated in the AD brain which has been questioned by initial research evidence (Gong and Iqbal 2008).

The activity of protein phosphatase 2A (PP2A), a robust candidate, is inhibited in the AD brain due to the progression of endogenous inhibitors I_1^{PP2A} and I_2^{PP2A} (Tanimukai et al. 2005). Other substrates of PPA2 are also hyperphosphorylated during AD progression (Ulloa et al. 1994; Wang et al. 2001). Tau hyper-phosphorylation is the key factor for the enhancement of tau aggregation that results in NFTs formation. Hyper-phosphorylation could trigger to separate tau from the microtubule network and consequently leads to PHFs, whereas tau attaches microtubules more competently in its less phosphorylated form (Lindwall and Cole 1984). Moreover, epitopes' phosphorylation can directly impact on tau conformation, making it more vulnerable to accumulate into PHFs. However, the molecular mechanism regarding NFTs neurotoxicity is unclear. According to scientific evidence, it has been demonstrated that neuronal survival might be affected by tau aggregations via a toxic gain-of-function by sequestering essential cellular factors. Recently it is proved that larger tau aggregates remove oligomers to show its neuroprotective effect while soluble tau oligomers have been experienced as real toxic species (Jeganathan et al. 2008; Tenreiro et al. 2014).

Figure 3. Post-translational modification of proteins by attaching an acetyl group to the side chain of lysine residues.

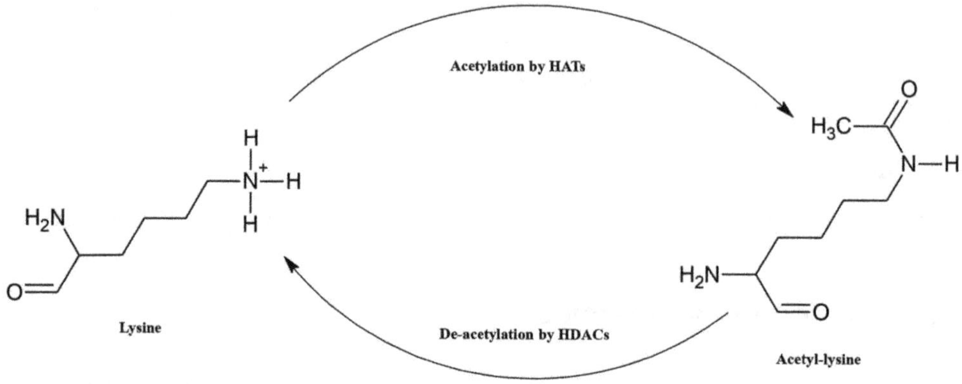

α-Synuclein Phosphorylation in Parkinson's Disease

PD is a neurological disorder that is characterized by the excessive loss of dopaminergic neurons in the substantia nigra. Several symptoms such as bradykinesia, rigidity, resting tremor and loss of postural reflex are associated with PD. Dying dopaminergic neurons exert intracytoplasmic inclusions of α- synuclein, called Lewy bodies (LBs), at a cellular level (Thomas and Flint Beal 2007). α- synuclein, an unfolded protein, which abundantly found in the presynaptic terminals of neurons in the CNS. The physiological role of α- synuclein is not completely understood but some evidence refers that it plays a pivotal role in the release of neurotransmitters (Bendor et al. 2013). There are 5 phosphorylation sites such as Ser87, Ser129, Tyr125, Tyr133, and Tyr136 present in its primary structure that are phosphorylated by several kinases like casein kinases 1 and 2, G-protein coupled receptor kinases, and the polo-like kinases (Waxman and Giasson 2011). Near about 4% and 90% of α- synuclein is phosphorylated in Ser129 and LBs respectively during physiological conditions (Anderson et al. 2006). During PD pathogenesis, the level of phospho-Ser87 is increased and Tyr125 is decreased in cortical tissue rather than healthy individuals (Paleologou et al. 2010; Cavallarin et al. 2012). This evidence strongly proves that the presence of α-synuclein phosphorylation is related to PD etiology, although the role of α- synuclein in LBs formation is not elucidated yet. On the other hand, the Ser87 and Ser129 phosphorylation plays an essential role in triggering the aggregation of α- synuclein (Gorbatyuk et al. 2008; da Silveira et al. 2009; Kragh et al. 2009). Excessive oxidative stress or inhibition of proteasome has been suggested for the explanation of α- synuclein hyperphosphorylation (Xu et al. 2015). On the contrary, α- synuclein aggregation is one of the causes of hyperphosphorylation because aggregates might be attached to protein kinases but not phosphatases (Waxman and Giasson 2008). Similarly, LBs and α- synuclein oligomers lead to tau pathogenesis resulting in neurodegeneration.

Acetylation

Acetylation is one of the most common forms of PTMs which consists of the transfer of the acetyl group (Figure 3) reversibly from acetyl-coenzyme A to the e-amino group of lysine residues. Histone acetyltransferases (HATs) and histone deacetylases (HDACs) catalyze PTM. Due to the imbalance of HAT/

HDAC, the risk factors of NDDs are increased in the nervous system. The effects of atypical acetylation on histone and non-histone substrates will be discussed below.

Histone Proteins

Histone acetylation is the most important mechanism and valuable epigenetic mark in CNS (Lilja et al. 2013). This PTM alleviates histones' positive charge also enhance chromatin conformation by reducing the reciprocal action of having negative charging phosphate groups of the deoxyribonucleic acid. As a result, chromatin topology permits gene transcription in the genome. Furthermore, acetyl tags then exert their functional docking sites for transcriptional activators as well as an epigenetic reader (Cheung et al. 2000). The antagonistic activity of HDACs and HATs in the physiological state regulates chromatin acetylation by maintaining gene expression. Abnormal histone acetylation is associated with the etiology of various neurodegenerative diseases by two probable approaches. Firstly, hypoacetylation at a precise genetic position decreases protein expression by a loss-of-function mechanism. Secondly, histone hypo-acetylation occurs at several genetic loci and resulting in extensive transcriptional deficits.

Friedreich ataxia is known as the main example of the first scenario. Friedreich ataxia is caused due to GAA pathogenic expansion located in the frataxin gene (Campuzano et al. 1996). Fragile X syndrome, a symptom of neurological disorder which is caused by fragile X mental retardation 1 gene slicing via the analogous mechanism. Hypoacetylation is occurred because of the expansion of CGG trinucleotide located in the 5'-untranslated region (Pietrobono et al. 2005). PolyQ diseases are the other best example of the second scenario. Due to cytosine-adenine-guanine repeat expansion in the wildtype protein, there are 9 NDDs are including huntingtin (HTT) for HD; ataxins (ATXNs) for spinocerebellar ataxias (SCAs) 1, 2, 3, 6, 7 and TATA-binding protein (TBP) for spinocerebellar ataxia type 17 (Pietrobono et al. 2005). Through several mechanisms, PolyQ expansion alters chromatin acetylation by inducing transcription defects. Furthermore, PolyQ proteins suppress transcription by inhibiting the activity of HAT as if HTT can bind with CREB-binding protein (CBP). ATXN3, another PolyQ protein, was involved in silence transcription either binding with histone or blocking the acetylation sites (Li et al. 2002). Eventually, PolyQ proteins such as ATXN1 reduces histone acetylation through engaging HDACs of target genes (Cvetanovic et al. 2007; Venkatraman et al. 2014).

Non-Histone Proteins

α-tubulin, a substrate of HAT/HDAC that is acetylated at Lys4 residue and postulated to observe in the microtubules' luminal face (Hammond et al. 2008). The involvement of HDACs—HDAC6 and sirtuin 2 is clearly identified and it deactivates α- tubulin not only *in vitro* but also *in vivo* model (Hubbert et al. 2002; North et al. 2003) but the role of HAT is not well specified yet. Hyper-acetylation of tubulin through the inhibition of HDAC increases axonal trafficking of mitochondria in the primary hippocampal neurons (Chen et al. 2010). In contrast, alteration of α-tubulin acetylation contributes to the progression of NDDs such as AD, HD and Charcot-Marie-Tooth disorders (Hempen and Brion 1996; Dompierre et al. 2007). Recently, it has been demonstrated that pharmacologically inhibited HDAC6 rescues axonal trafficking and returns the tubulin acetylation in the CTM mouse model (D'Ydewalle et al. 2011).

Glycosylation

Glycosylation is a complex PTMs that is occurred in more than 50% of the human proteome. It occurs in the endoplasmic reticulum (ER)/Golgi compartment via multi-step enzymatic reaction, resulting in the composition of some protein- bond oligosaccharides with diverse biological functions. There are mainly three PTMs covering carbohydrates are N-linked and O-linked glycosylation, and glypiation (Spiro 2002). Tremendously, at different levels, all these three types have been connected to neurodegeneration.

N-linked Glycosylation

N-linked glycosylation comprises the alteration of a precursor glycan from an isoprenoid lipid carrier to the side-chain among selected asparagine residues. Oligosaccharyltransferase in the ER is responsible for the catalysis of this reaction. Furthermore, these target moieties are identified by the consensus sequence Asp-X-Ser/Thr. In the Golgi apparatus, other glycosyltransferases and glycosidases are remodeled by the prime oligosaccharides and up-regulate the main glycans such as high mannose, hybrid and complex glycans (Aebi 2013).

In the case of neurodegeneration, the cellular prion protein (PrP^{C}) is the standard example of glycosylation. A cellular protein PrP^{C} that is transformed into a pathological conformer (PrP^{Sc} or prion), causing the so-called transmissible spongiform encephalopathies (TSEs) or prion disorders (Prusiner 1998). Because of the presence of its primary sequence of two asparagines (Asn181 and Asn187 in human PrP^{C}), PrP^{C} exists as an un-, mono- or di-glycosylated form which can undergo N-linked glycosylation under physiological conditions. Furthermore, glycosylation patterns of PrP^{Sc} may contribute to the molecular basis of TSE strain variation (Lawson et al. 2005) and show strain-specific properties, even though PrP^{C} glycosylation is not essential for conversion (Taraboulos et al. 1990). Undeniably, alterations in the glycoform ratio or composition have been found amid different prion diseases (Safar et al. 1998; Somerville 1999).

O-Linked Glycosylation

O-linked glycosylation begins in the Golgi apparatus without any obstacle where N-acetylgalactosamine moiety is transferred in the presence of N-acetyl galactosaminyltransferase to the hydroxyl-oxygen in the side chain of a serine or a threonine moiety. Thereafter, the sugar moiety is modified or elongated by acetylation, sialylation, fucosylation, polylactosamine, and sulfatation extension (Van Den Steen et al. 1998). Noticeably, this type of glycosylation was exerted to counteract phosphorylation at the same or other sites on protein's backbone (Comer and Hart 2001). This capability leads to a great impact on several neurodegenerative venues characterized by inappropriate protein phosphorylation such as tau progression in AD (Lefebvre et al. 2005). This hypothesis is supported by the previous research that tau-enriched AD brains exert a significant reduction in O-GlcNAc glycosylation rather than controls (Robertson et al. 2004). At the same time, the overexpression of O-GlcNAc transferase in cultured cells is enough to change tau phosphorylation which in turn to dephosphorylation of target sites such as Ser202, Ser396, and Ser404 (Robertson et al. 2004).

Glypiation

Glypiation is the attachment of the covalent bonding of a glycosylphosphatidylinositol (GPI) anchor as well as is a typical PTM that contains proteins to cell membranes. GPI-anchored proteins show a variety of functions such as signal transduction, antigen presentation and cell adhesion (Boonstra et al. 2016). PrP^{C} is an example of PTM that is linked to a GPI anchor during the maturation process. This linkage aims to specialize microdomains of the cellular membrane known as lipid rafts that are highly enriched in sphingolipids as well as cholesterol (Campana et al. 2005; Gasperini and Legname 2014). A few studies reveal that the conversion of PrP^{C} to PrP^{Sc} may play a crucial role in the GPI anchor in pathogenic progression (Taraboulos et al. 1995; Baron et al. 2002). On the other hand, anchorless PrP^{C} are susceptible to not only prion infection but also accumulate PrP^{Sc} in transgenic mice model (Chesebro et al. 2005). It is interesting to note that there were no clinical signs related to prior infection, reccommending that GPI anchor may be vital for mediating neurotoxicity in comparison to conversion itself (Chesebro et al. 2005). In addition, recent data demonstrate that GPI anchor could be also pivotal for establishing persistent prion infection (McNally et al. 2009).

Fatty Acylation

Fatty acylation comprises the covalent binding of fatty acids to proteins. PTMs have been observed in numerous eukaryotic membrane glycoproteins, playing an essential role in cellular conformation as well as function including signal transduction and the attachment of cells (Munday and López 2007). The modifications with myristate and palmitate (a 14-carbon saturated fatty acid and a 16-carbon saturated fatty acid respectively) are the two most common types of protein fatty acylation (Resh 1999). Both forms are vigorously controlled and their dysregulation is responsible for the involvement in various NDDs.

Palmitoylation

Palmitoylation is one of the fatty acylations that comprises of the reversible binding of a palmitoyl group to the –SH (sulfhydryl) group of a cysteine moiety through an unstable thioester bond, which has been catalyzed by S-palmitoyl transferases (PATs) inside the Golgi apparatus. For the development of neurons and the activity of synapses, protein palmitoylation is predominantly significant in the nervous system. Abnormal palmitoylation has played a crucial role in the development of various NDDs such as AD and HD. Furthermore, expansion of PolyQ in HTT was shown to decrease the protein's affinity for its precise PATs huntingtin interacting protein 14 (HIP14) as well as HIP14-like (DHHC13), which leads to the reduction of HTT palmitoylation at Cys214 that is necessary for its functions as well as trafficking (Resh 1999).

Moreover, palmitoylation-resistant HTT mutants demonstrate increased neurotoxicity and susceptibility to produce accumulates, which advocate the role of hypo-palmitoylation in examining the phenotype of HD (Resh 1999). On the other hand, a decline in palmitoylation may be favorable in terms of AD. Actually, palmitoylation in amyloid precursor protein (APP) at Cys^{186} as well as Cys^{187} increases its amyloidogenic processing by targeting APP to lipid rafts and improving its γ-secretase induced cleavage (Bhattacharyya et al. 2013). Captivatingly, the complex of γ-secretase itself is also gone through palmitoylation and its regulation may be actively linked with amyloid-beta (Aβ)-burden in the AD brain. Indeed, in an AD mouse model, the expression of a palmitoylation-lacking form of γ-secretase considerably decreased the quantity of insoluble Aβ and the accumulation of amyloid in the frontal cortex of the brain (Meckler et al. 2010). Overall, the function of palmitoylation in NDDs is ambivalent as opposing actions have been stated.

Myristoylation

Myristoylation involves the binding of a myristoyl residue to the alpha-amino group of an N-terminal glycine moiety through an amide bond. In contrast to palmitoylation, this PTM occurs in the cytosol via the function of the key cellular enzymes N-myristoyltransferases (NMTs) and this is also an irreversible process (Farazi et al. 2001). In addition, the exact substrate proteins specificity of these enzymes is specified by the consensus motif Met-Gly-X-X-X-Ser/Thr. The initial Met is detached with the help of a methionine aminopeptidase during translation, producing the glycine at position 2 in the N-terminal moiety (Resh 1999).

Current research recommends that myristoylation might play an essential role in HD by controlling the autophagy process. In a study by Martin et al., demonstrated that the internal Gly553 of HTT had undergone myristoylation after the cleavage of caspase at Asp552 and subsequently supported the development of autophagosomes (Martin et al. 2012). Significantly, this process was shown to be decreased in mutant HTT in comparison with the wild type protein, which might be owing to steric interference of the aberrant PolyQ stretch or to NMT suppression (Martin et al. 2014). Therefore, it is evident that the pathological PolyQ expansion in HTT might change the physiological dynamics of autophagy through reducing myristoylation at Gly553, leading to the development of cytotoxic materials found in HD (Martin et al. 2015).

Ubiquitination

Ubiquitination is one type of PTMs that affects the cellular processes prompting the degradation of the protein (Cichanover 2005), controlling cellular protein's localization, activation and deactivation of proteins, and intervening protein-protein interactions (Schnell and Hicke 2003; Mukhopadhyay and Riezman 2007). Especially, the ubiquitin (Ub) proteasome system (UPS) is the main proteolytic pathway, which is used by eukaryotic cells for the positioning of damaged or misfolded proteins (Cichanover 2005). Ubiquitination comprises of the covalent binding of single or many Ub molecules to a protein, producing mono- and poly-ubiquitination successively. Furthermore, Ub is a greatly preserved protein that consists of 76 amino acids (8 kDa), which is expressed in nearly all eukaryotic cells.

Ubiquitination occurs via 3 consecutive stages: (i) an Ub -activating enzyme (E1) activates an Ub molecule in its C-terminal glycine contributing to an intermediate thiolester producing E1-S~Ub. (ii) An Ub-conjugating enzyme (E2) triggers the transfer of the activated Ub to a member of the Ub-protein ligase (E3) family; (iii) E3 supports the shift of the Ub to a lysine moiety in the target protein (Figure 4). Poly-ubiquitination needs the shift of extra Ub residues to Lys48 of the formerly conjugated Ub molecule (Zeng et al. 2006). Mono-ubiquitination is played a key role in controlling cellular events, whereas poly-ubiquitination targets the substrate proteins for degradation with the help of 26S proteasome (Inc 2003). Any alterations in the ubiquitination process have been connected with the causes of many NDDs. Numerous NDDs are characterized by the aggregation of accumulated misfolded or damaged proteins as well as inclusion bodies. Additionally, these accumulate comprise Ub and proteasome subunits indicating a failure of the UPS in the clearance of misfolded or damaged proteins.

Figure 4. The ubiquitin-proteasome systemin in the post-translational modifications of proteins. Ubiquitination involves the covalent attachment of Ub to lysine moieties demonstrated at the surface of targeted substrates. Ub-activating enzyme E1 activates an Ub molecule in the presence of ATP. Then the Ub is shifted from E1 to the Ub conjugating enzyme (E2). Subsequently, a ubiquitin ligase (E3) interacts with the ubiquitin-bound E2 enzyme as well as a substrate to shift the Ub from E2 to the substrate. Multiple cycles of ubiquitylation can take place to form a lysine 48-linked polyubiquitin chain, which is the key signal for proteasomal degradation. When targeted to the 26S proteasome, substrates are degraded into short peptides that are subsequently broken down into amino acids in the presence of aminopeptidases (APPs). The polyubiquitin molecules are typically detached from the substrate by the proteasome with the help of DUBs prior to the degradation of the substrate. The free ubiquitin molecules are reused for a new round of ubiquitylation. Abbreviations used are Ub, ubiquitin; DUBs, deubiquitinating enzymes.

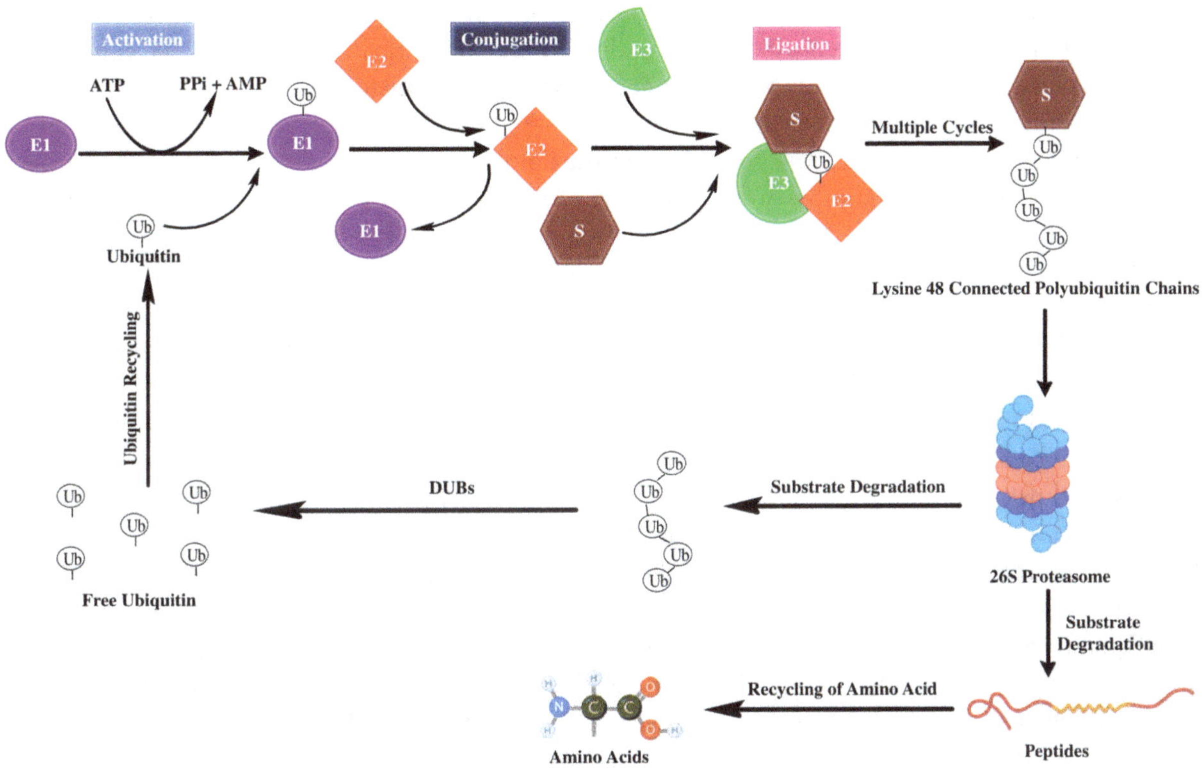

In addition, HD is characterized by the existence of intracellular accumulates of HTT protein with a prolonged PolyQ repeat (Mitra and Finkbeiner 2008; Schipper-Krom et al. 2012). Ubiquitination of HTT occurs at the position of Lys48 and Lys63, however, the proteasome may unable to digest prolonged PolyQ sequences (Kalchman et al. 1996; Venkatraman et al. 2004). Besides, mutants of HTT are exposed to sequester components of the proteasome within accumulates, probably improving the suppression of UPS (Suhr et al. 2001). Constantly with this model, the activation of proteasome enhances the phenotype of HD both *in vivo and in vitro*, (Seo et al. 2007; Jia et al. 2012).

Furthermore, Ub was also detected in NFTs as well as senile plaques from AD individuals like HD (Perry et al. 1987). Most importantly, Aβ was revealed to suppress the proteolytic activity of 26S proteasome, playing a crucial role in a sharp rise of ubiquitinated proteins in neuron cells (Lopez Salon et al. 2003; Almeida et al. 2006). These effects appear because of the aggregation of Aβ oligomers with the proteasomal substrates (Zhao and Yang 2010). Furthermore, mutations in the *Ub B* gene itself might lead to the phenotype of AD. The frameshift mutant Ub-B+1 (UbB+1) was identified in AD brain as well as

UbB(+1)-capped unanchored polyubiquitin chains were suggested to suppress the 26S proteasome through a dominant-negative mode of action (Van Leeuwen et al. 1998; Lam et al. 2000; Chadwick et al. 2012).

SUMOylation

SUMOylation comprises the covalent binding of SUMO proteins to lysine residues in precise target proteins like ubiquitination. Furthermore, this PTM is dependent on an enzymatic pathway similar to, however different from the ubiquitination process(Wilkinson and Henley 2010). SUMOylation does not target specific proteins for degradation; however, it controls numerous functions such as the stability of protein, transcription, and nuclear-cytosolic shuttling, unlike ubiquitination. SUMOylated protein has been identified in neuronal inclusion bodies connected with diverse misfolding diseases, recommending a crucial role in the development of NDDs.

As shown, SUMOylation has an effect on neurotoxicity via diverse pathways in PolyQ diseases. Besides, SUMO-induced alterations in the transport of protein through the nuclear membrane appear to be a causative factor for NDDs. Spinocerebellar ataxia type 1 is the best example of this condition. Actually, ATXN1 could be SUMOylated on as a minimum of 5 different positions of Lys moieties (Lys16, Lys194, Lys610, Lys697, and Lys746) as well as the process depends on phosphorylation at the position of Ser776 and a functional nuclear localization signal. The expanded PolyQ of ATXN1 was demonstrated to decline SUMOylation levels, probably hampering its accurate transferring between the cytosol and nucleus upon disease (Riley et al. 2005). Moreover, SUMOylation could also change the transcriptional process, in a Drosophila model for HD, it has been observed that the capability of a SUMOylated pathogenic portion of HTT in suppressing transcription and aggravating NDDs (Steffan et al. 2004). Furthermore, SUMOylation deleteriously controls the inherent transcriptional function of the androgen receptor (Poukka et al. 2000). The transcriptional activity of mutant androgen receptor-induced disease can be enhanced by interrupting the SUMOylation patterns(Chua et al. 2015). Finally, SUMOylation has also an effect on accumulate formation as demonstrated in dentatorubral pallidoluysian atrophy. The increased expression of SUMO1 in a neuronal disease model was exposed to trigger mutant ATN1 intranuclear accumulation as well as apoptosis (Terashima et al. 2002).

Deamidation

Unlike others, deamidation is the PTM that does not need enzymatic catalysis. This irreversible reaction comprises the cleavage of ammonia from the (Asn) moieties (amide group) and to a smaller amount of glutamines (Geiger and Clarke 1987). Deamidation of Asn continues via the formation of a 5 membered succinimide ring intermediate by the nucleophilic attack of the nitrogen atom in the subsequent peptide-bond on the carbonyl family of the Asn side chain, during physiological conditions. Subsequently, the intermediate ring goes through hydrolysis to produce a combination of aspartate (Asp) and isoaspartate (isoAsp) in a ratio of 1:3. *In vivo*, isoAsp could be transformed back to normal Asp with the help of enzyme protein L-isoaspartyl methyltransferase (PIMT) (Paper 2011). Deamidation is initially thought to be an only type of molecular damage linked with aging. In addition, a disparity in the physiological Asp: isoAsp ratio-probably because of PIMT scarcities, which has been connected with numerous pathological conditions such as multiple sclerosis (MS), AD, and vascular dementia.

In the case of AD, analysis of mass spectrometry in the accumulated tau protein from NFTs exposed the existence of isoAsp at the position of Asp193, Asp387, and Asn381 as a result of deamidation (Watanabe et al. 1999). Captivatingly, these isoAsp(s) are situated in nearby pathological hyper-phosphorylation sites, recommending that isoAsp transformation may trigger conformational rearrangements that can sequentially ease the process of phosphorylation. Currently, wide-ranging deamidation has been revealed to occur also at the position of Asn279 inside the microtubule-binding repeat domain. Particularly, this type of modification abolishes the immunoreactivity of tau to RD4 antibody that acknowledges the 4R isoform and has an effect on the capability of tau to attach microtubules (Dan et al. 2014). Overall, these outcomes focus on the significance of deamidation for the accumulation of tau into PHFs. Additionally, the repair enzyme PIMT is revealed to co-localize with NFTs (Shimizu et al. 2000), validating its key role in opposing the aggregation of isoAsp moieties in aged proteins. Constantly, PIMT-lacking mice demonstrate an abnormal rise of isoAsp in the brain at neonatal phases and die within 3 months of age because of lethal epileptic seizures (Yamamoto et al. 1998).

Oxidative/Nitrosative Stress and Post-Translational Modifications

Oxidative stress (OS) could alter various biological molecules through redox-induced reactions (Sayre et al. 2005). OS is capable to trigger several PTMs through direct oxidation of protein side-chains by reactive nitrogen and oxygen species (Stadtman 1992). Furthermore, the chief radical species that are responsible for such types of modifications are hydrogen peroxide (H_2O_2), the highly reactive hydroxyl radical (HO•), superoxide anion (O2−), and some nitrogen species including nitric oxide (NO). The characteristics of these PTMs could be either reversible or irreversible. The former comprise cysteine modification products such as S-nitrosylation, S-sulfenation, S-glutathionylation, and disulfides, whereas the latter contains carbonylation and tyrosine nitration (Cai and Yan 2013). OS is the result of a dyshomeostasis between biochemical processes that play a pivotal role in the generation of free radical species and these are accountable for their elimination and the cellular antioxidant pathway. A good example here is an imbalance of metal ions including iron, manganese, zinc, copper, and other traces of redox-active transition metals, which have an effect on metalloprotein activity and sequentially rise the generation of free radical (Sayre et al. 2000, 2012). Though radical species could exert controlling activities (Uttara et al. 2009; Gasperini et al. 2015), however, their generation is typically connected with aging as well as a copious pathological condition such as NDDs.

In addition, tyrosine nitration and protein carbonylation are frequently linked with oxidative damage as well as considered as the biomarkers for the determination of OS in aging and other diseases (Stadtman and Earl 2001). Conversely from the other PTMs, carbonylation could take place on some amino acid moieties, such as lysines, cysteines, prolines, arginines, threonines, and histidines. Additionally, carbonylated protein has been identified in numerous NDDs such as amyotrophic lateral sclerosis, AD, MS, and PD. Generally, cytoskeletal constituents including tubulin and neurofilaments appear to be the key targets for protein carbonylation (Bizzozero 2009). In addition to its harmful effects, protein carbonylation can play an essential role in signal transduction (Wong et al. 2008, 2012) as well as exerts defense against reoxygenation injury (Serviddio et al. 2005). Furthermore, tyrosine nitration (specifically 3-nitrotyrosine) is a very selective process because of merely distinct tyrosine moieties could be nitrated. This type of PTM is connected with diverse NDDs and acute or chronic inflammation (Lee et al. 2009). Tau nitration at the position of Tyr29 has been linked with the formation of NFT whereas nitration at the position of Tyr18 and Tyr394 has exposed to decrease the accumulation of tau *in vitro*, in AD (Reynolds et al. 2005, 2006). Besides, α-synuclein nitration appears to enhance the formation of LB through reducing its solubility in PD (Giasson et al. 2000; Takahashi et al. 2002; Hodara et al. 2004).

Furthermore, S-nitrosylation is one of the redox-induced PTMs that regulate the activity of the protein by covalent attachment of NO to thiol family of cysteine moieties. The formation of S-nitrosylation is balanced by denitrosylation enzymes including protein disulfide isomerase, the thioredoxin system, S-nitrosoglutathione reductase, and by protein-protein interaction of transnitrosylation during physiological conditions (Benhar et al. 2009; Nakamura et al. 2013). Abnormal S-nitrosylated proteins can contribute to protein misfolding/damage, synaptic damage, and neuronal loss, hence leading to the development of AD (Zhao et al. 2015) and PD (Chung et al. 2005). The change in the level of S-nitrosylation in these proteins might play a significant role in various cellular mechanisms such as mitochondrial function, protein quality control, and transcription factors, synaptic transmission, as well as homeostasis of iron, receptors and ion channels (Nakamura et al. 2013). Besides, OS and nitrosative appear to lead to the misfolding and accumulation of protein through chaperone and the dysfunction of the proteasome.

RECENT DEVELOPMENTS AND FUTURE RESEARCH DIRECTIONS

The recent improvements in the arena of proteomics and the quantification of PTMs, could deliver an enhanced mode of targeting various diseases including diabetes, cancer, NDDs, and heart diseases. Numerous cellular processes such as DNA replication, transcription, the progression of the cell cycle, dynamics of chromatin, gene silencing, apoptosis, neuronal repression, nuclear import, and DNA repair are principally regulated via histone acetylation and ultimately plays a pivotal role in the treatment of NDDs. It can be said that PTMs have a remarkable opportunity in every arena of science and technology; it can be recombinant DNA technology, genomics, proteomics, and the vital biological processes that control the incidence of all the lethal diseases. Effective and well-organized research is needed to characterize the significant role of PTMs with their clinical outlooks for the treatment of NDDs.

Understanding the role of PTMs in NDDs increases, thus it is high time for the advancement of methods to identify protein PTMs more quickly as well as precisely. Additionally, the current outcome of exceptional and irregular modifications in NDDs demands the advancement of more sensitive and precise methods to identify such types of PTMs (Basak et al.). In this case, the current quick improvement in large-scale genomics, as well as proteomics technologies, is likely to be an essential factor for such type of studies. Furthermore, drugs that target PTMs, including acetylation, ubiquitination, phosphorylation, and methylation will consider as valuable tools in examining the fundamental mechanism of PTMs regulation as well as deliver a pharmacological treatment option to fight the harmful effects of NDDs (Kaeberlein 2010).

CONCLUSION

PTMs provide a benefit in the arena of protein biology and the identification and quantification of these PTMs have played a crucial role in Proteomics, cell biology, genomics, and the prevention and treatment of many diseases especially NDDs. In this chapter, we have discussed the important roles of PTMs such as phosphorylation, acetylation, glycosylation, fatty acylation, palmitoylation, myristoylation, ubiquitination, SUMOylation, methylation, deamidation, and nitrosylation etc., of proteins that can regulate the proper functions of proteins. Furthermore, PTMs play an essential role in the functioning of chief homeostatic proteins, which sequentially control numerous cellular processes to further control pathogenesis in numerous diseases like NDDs. Therefore, exploring the roles of PTMs in NDDs pathways could provide us a better comprehending of the pathogenesis of NDDs as well as discover therapeutic targets for the treatment of these NDDs.

REFERENCES

Aebi, M. (2013). N-linked protein glycosylation in the ER. *BBA-Mol Cell Res*, *1833*, 2430–2437. PMID:23583305

Almeida, C. G., Takahashi, R. H., & Gouras, G. K. (2006). β-amyloid accumulation impairs multivesicular body sorting by inhibiting the ubiquitin-proteasome system. *The Journal of Neuroscience: The Official Journal of the Society for Neuroscience*, *26*(16), 4277–4288. doi:10.1523/JNEUROSCI.5078-05.2006 PMID:16624948

Anderson, J. P., Walker, D. E., Goldstein, J. M., de Laat, R., Banducci, K., Caccavello, R. J., ... Chilcote, T. J. (2006). Phosphorylation of Ser-129 is the dominant pathological modification of α-synuclein in familial and sporadic lewy body disease. *The Journal of Biological Chemistry*, *281*(40), 29739–29752. doi:10.1074/jbc.M600933200 PMID:16847063

Baron, G. S., Wehrly, K., & Dorward, D. W. (2002). Conversion of raft associated prion protein to the protease-resistant state requires insertion of PrP-res (PrPSc) into contiguous membranes. *The EMBO Journal*, *21*(5), 1031–1040. doi:10.1093/emboj/21.5.1031 PMID:11867531

Bendor, J. T., Logan, T. P., & Edwards, R. H. (2013). The function of α-synuclein. *Neuron*, *79*(6), 1044–1066. doi:10.1016/j.neuron.2013.09.004 PMID:24050397

Benhar, M., Forrester, M. T., & Stamler, J. S. (2009). Protein denitrosylation: Enzymatic mechanisms and cellular functions. *Nature Reviews. Molecular Cell Biology*, *10*(10), 721–732. doi:10.1038/nrm2764 PMID:19738628

Bhattacharyya, R., Barren, C., & Kovacs, D. M. (2013). Palmitoylation of amyloid precursor protein regulates amyloidogenic processing in lipid rafts. *The Journal of Neuroscience: The Official Journal of the Society for Neuroscience*, *33*(27), 11169–11183. doi:10.1523/JNEUROSCI.4704-12.2013 PMID:23825420

Bizzozero, O. A. (2009) Protein Carbonylation in Neurodegenerative and Demyelinating CNS Diseases. In Handbook of Neurochemistry and Molecular Neurobiology. Springer US. doi:10.1007/978-0-387-30375-8_23

Boonstra, M. C., De Geus, S. W. L., & Prevoo, H. A. J. M. (2016). Selecting Targets for Tumor Imaging: An Overview of Cancer-Associated Membrane Proteins. Biomark Cancer 8. *BIC*, *S38542*. doi:10.4137/bic.s38542 PMID:27721658

Cai, Z., & Yan, L.-J. (2013). Protein Oxidative Modifications: Beneficial Roles in Disease and Health. *Journal of Biochemical and Pharmacological Research*, *1*, 15–26. PMID:23662248

Campana, V., Sarnataro, D., & Zurzolo, C. (2005). The highways and byways of prion protein trafficking. *TreNDDs Cell Biol.*, *15*(2), 102–111. doi:10.1016/j.tcb.2004.12.002 PMID:15695097

Campuzano, V., Montermini, L., & Moltò, M.D. (1996). Friedreich's ataxia: Autosomal recessive disease caused by an intronic GAA triplet repeat expansion. *Science, 271*, 1423–1427. doi:10.1126cience.271.5254.1423

Cavallarin, N., Vicario, M., & Negro, A. (2012). The Role of Phosphorylation in Synucleinopathies: Focus on Parkinsons Disease. *CNS & Neurological Disorders - Drug Targets*, *9*(4), 471–481. doi:10.2174/187152710791556140 PMID:20522010

Chadwick, L., Gentle, L., Strachan, J., & Layfield, R. (2012). Review: Unchained maladie - a reassessment of the role of Ubb+1-capped polyubiquitin chains in Alzheimer's disease. *Neuropathology and Applied Neurobiology*, *38*(2), 118–131. doi:10.1111/j.1365-2990.2011.01236.x PMID:22082077

Chen, S., Owens, G. C., Makarenkova, H., & Edelman, D. B. (2010). HDAC6 Regulates Mitochondrial Transport in Hippocampal Neurons. *PLoS One*, *5*(5), e10848. doi:10.1371/journal.pone.0010848 PMID:20520769

Chesebro, B., Trifilo, M., & Race, R. (2005). Medicine: Anchorless prion protein results in infectious amyloid disease without clinical scrapie. *Science, 308*, 1435–1439. doi:10.1126cience.1110837

Cheung, P., Allis, C. D., & Sassone-Corsi, P. (2000). Signaling to chromatin through historic modifications. *Cell*, *103*(2), 263–271. doi:10.1016/S0092-8674(00)00118-5 PMID:11057899

Chua, J. P., Reddy, S. L., Yu, Z., Giorgetti, E., Montie, H. L., Mukherjee, S., ... Lieberman, A. P. (2015). Disrupting SUMOylation enhances transcriptional function and ameliorates polyglutamine androgen receptor-mediated disease. *The Journal of Clinical Investigation*, *125*(2), 831–845. doi:10.1172/JCI73214 PMID:25607844

Chung, K. K. K., Dawson, T. M., & Dawson, V. L. (2005). Nitric oxide, S-nitrosylation and neurodegeneration. *Cellular and Molecular Biology*, *51*, 247–254. PMID:16191392

Cichanover, A. (2005). Intracellular protein degradation: From a vague idea thru the lysosome and the ubiquitin-proteasome system and onto human diseases and drug targeting. *Cell Death and Differentiation*, *12*(9), 1178–1190. doi:10.1038j.cdd.4401692 PMID:16094394

Comer, F. I., & Hart, G. W. (2001). Reciprocity between O-GlcNAc and O-phosphate on the carboxyl terminal domain of RNA polymerase II. *Biochemistry*, *40*(26), 7845–7852. doi:10.1021/bi0027480 PMID:11425311

Cvetanovic, M., Rooney, R. J., Garcia, J. J., Toporovskaya, N., Zoghbi, H. Y., & Opal, P. (2007). The role of LANP and ataxin 1 in E4F-mediated transcriptional repression. *EMBO Reports*, *8*(7), 671–677. doi:10.1038j.embor.7400983 PMID:17557114

D'Ydewalle, C., Krishnan, J., Chiheb, D. M., Van Damme, P., Irobi, J., Kozikowski, A. P., ... Van Den Bosch, L. (2011). HDAC6 inhibitors reverse axonal loss in a mouse model of mutant HSPB1-induced Charcot-Marie-Tooth disease. *Nature Medicine*, *17*(8), 968–974. doi:10.1038/nm.2396 PMID:21785432 da Silveira, S. A., Schneider, B. L., & Cifuentes-Diaz, C. (2009). Phosphorylation does not prompt, nor prevent, the formation of α-synuclein toxic species in a rat model of Parkinson's disease. *Human Molecular Genetics, 18*, 872–887. doi:10.1093/hmg/ddn417 PMID:19074459

Dan, A., Takahashi, M., & Masuda-Suzukake, M. (2014). Extensive deamidation at asparagine residue 279 accounts for weak immunoreactivity of tau with RD4 antibody in Alzheimer's disease brain. *Acta Neuropathologica Communications, 2*. doi:10.1186/2051-5960-1-54 PMID:24252707

Dompierre, J. P., Godin, J. D., Charrin, B. C., Cordelieres, F. P., King, S. J., Humbert, S., & Saudou, F. (2007). Histone deacetylase 6 inhibition compensates for the transport deficit in Huntington's disease by increasing tubulin acetylation. *The Journal of Neuroscience: The Official Journal of the Society for Neuroscience, 27*(13), 3571–3583. doi:10.1523/JNEUROSCI.0037-07.2007 PMID:17392473

Dugger, B. N., & Dickson, D. W. (2017). Pathology of neurodegenerative diseases. *Cold Spring Harbor Perspectives in Biology*, 9. PMID:28062563

Farazi, T. A., Waksman, G., & Gordon, J. I. (2001). The Biology and Enzymology of Protein N-Myristoylation. *The Journal of Biological Chemistry, 276*(43), 39501–39504. doi:10.1074/jbc.R100042200 PMID:11527981

Gasperini, L., & Legname, G. (2014). Prion protein and aging. *Frontiers in Cell and Developmental Biology*, 2. PMID:25364751

Gasperini, L., Meneghetti, E., Pastore, B., Benetti, F., & Legname, G. (2015). Prion protein and copper cooperatively protect neurons by modulating NMDA receptor through S-nitrosylation. *Antioxidants & Redox Signalling, 22*(9), 772–784. doi:10.1089/ars.2014.6032 PMID:25490055

Geiger, T., & Clarke, S. (1987). Deamidation, isomerization, and racemization at asparaginyl and aspartyl residues in peptides. Succinimide-linked reactions that contribute to protein degradation. *The Journal of Biological Chemistry, 262*, 785–794. PMID:3805008

Giasson, B.I., Duda, J.E., & Murray, I.V.J. (2000). Oxidative damage linked to neurodegeneration by selective α-synuclein nitration in synucleinopathy lesions. *Science, 290*, 985–989. doi:10.1126cience.290.5493.985

Gong, C.-X., & Iqbal, K. (2008). Hyperphosphorylation of Microtubule-Associated Protein Tau: A Promising Therapeutic Target for Alzheimer Disease. *Current Medicinal Chemistry, 15*(23), 2321–2328. doi:10.2174/092986708785909111 PMID:18855662

Gorbatyuk, O. S., Li, S., Sullivan, L. F., Chen, W., Kondrikova, G., Manfredsson, F. P., ... Muzyczka, N. (2008). The phosphorylation state of Ser-129 in human α-synuclein determines neurodegeneration in a rat model of Parkinson disease. *Proceedings of the National Academy of Sciences of the United States of America, 105*(2), 763–768. doi:10.1073/pnas.0711053105 PMID:18178617

Greifenhagen, U., Frolov, A., Blüher, M., & Hoffmann, R. (2016). Site-specific analysis of advanced glycation end products in plasma proteins of type 2 diabetes mellitus patients. *Analytical and Bioanalytical Chemistry, 408*(20), 5557–5566. doi:10.100700216-016-9651-4 PMID:27236317

Hammond, J. W., Cai, D., & Verhey, K. J. (2008). Tubulin modifications and their cellular functions. *Current Opinion in Cell Biology, 20*(1), 71–76. doi:10.1016/j.ceb.2007.11.010 PMID:18226514

Hanger, D. P., Seereeram, A., & Noble, W. (2009). Mediators of tau phosphorylation in the pathogenesis of Alzheimer's disease. *Expert Review of Neurotherapeutics*, *9*(11), 1647–1666. doi:10.1586/ern.09.104 PMID:19903024

Hempen, B., & Brion, J. P. (1996). Reduction of Acetylated α-Tubulin Immunoreactivity in Neurofibrillary Tangle-bearing Neurons in Alzheimer's Disease. *Journal of Neuropathology and Experimental Neurology*, *55*(9), 964–972. doi:10.1097/00005072-199609000-00003 PMID:8800092

Hodara, R., Norris, E. H., Giasson, B. I., Mishizen-Eberz, A. J., Lynch, D. R., Lee, V. M.-Y., & Ischiropoulos, H. (2004). Functional consequences of α-synuclein tyrosine nitration: Diminished binding to lipid vesicles and increased fibril formation. *The Journal of Biological Chemistry*, *279*(46), 47746–47753. doi:10.1074/jbc.M408906200 PMID:15364911

Hubbert, C., Guardiola, A., Shao, R., Kawaguchi, Y., Ito, A., Nixon, A., ... Yao, T.-P. (2002). HDAC6 is a microtubule-associated deacetylase. *Nature*, *417*(6887), 455–458. doi:10.1038/417455a PMID:12024216

Jeganathan, S., Hascher, A., Chinnathambi, S., Biernat, J., Mandelkow, E.-M., & Mandelkow, E. (2008). Proline-directed pseudo-phosphorylation at AT8 and PHF1 epitopes induces a compaction of the paperclip folding of tau and generates a pathological (MC-1) conformation. *The Journal of Biological Chemistry*, *283*(46), 32066–32076. doi:10.1074/jbc.M805300200 PMID:18725412

Jensen, O. N. (2004). Modification-specific proteomics: Characterization of post-translational modifications by mass spectrometry. *Current Opinion in Chemical Biology*, *8*(1), 33–41. doi:10.1016/j.cbpa.2003.12.009 PMID:15036154

Jia, H., Kast, R. J., Steffan, J. S., & Thomas, E. A. (2012). Selective histone deacetylase (HDAC) inhibition imparts beneficial effects in Huntington's disease mice: Implications for the ubiquitin-proteasomal and autophagy systems. *Human Molecular Genetics*, *21*(24), 5280–5293. doi:10.1093/hmg/dds379 PMID:22965876

Kaeberlein, M. (2010). Resveratrol and rapamycin: Are they anti-aging drugs? *BioEssays*, *32*(2), 96–99. doi:10.1002/bies.200900171 PMID:20091754

Kakizawa, S. (2013). *Nitric oxide-induced calcium release: activation of type 1 ryanodine receptor, a calcium release channel, through non-enzymatic post-translational modification by nitric oxide*. doi:10.3389/fendo.2013.00142

Kalchman, M. A., Graham, R. K., Xia, G., Koide, H. B., Hodgson, J. G., Graham, K. C., ... Hayden, M. R. (1996). Huntingtin is ubiquitinated and interacts with a specific ubiquitin- conjugating enzyme. *The Journal of Biological Chemistry*, *271*(32), 19385–19394. doi:10.1074/jbc.271.32.19385 PMID:8702625

Khoury, G. A., Baliban, R. C., & Floudas, C. A. (2011). Proteome-wide post-translational modification statistics: Frequency analysis and curation of the swiss-prot database. *Scientific Reports*, *1*(1), 90. doi:10.1038rep00090 PMID:22034591

Kim, E., Napierala, M., & Dent, S. Y. R. (2011). Hyperexpansion of GAA repeats affects post-initiation steps of FXN transcription in Friedreich's ataxia. *Nucleic Acids Research*, *39*(19), 8366–8377. doi:10.1093/nar/gkr542 PMID:21745819

Kragh, C. L., Lund, L. B., Febbraro, F., Hansen, H. D., Gai, W.-P., El-Agnaf, O., ... Jensen, P. H. (2009). α-synuclein aggregation and Ser-129 phosphorylation-dependent cell death in oligodendroglial cells. *The Journal of Biological Chemistry*, *284*(15), 10211–10222. doi:10.1074/jbc.M809671200 PMID:19203998

Kuhn, M. L., Zemaitaitis, B., Hu, L. I., Sahu, A., Sorensen, D., Minasov, G., ... Wolfe, A. J. (2014). Structural, kinetic and proteomic characterization of acetyl phosphate-dependent bacterial protein acetylation. *PLoS One*, *9*(4), e94816. doi:10.1371/journal.pone.0094816 PMID:24756028

Kumari, D., Biacsi, R. E., & Usdin, K. (2011). Repeat expansion affects both transcription initiation and elongation in Friedreich ataxia cells. *The Journal of Biological Chemistry*, *286*(6), 4209–4215. doi:10.1074/jbc.M110.194035 PMID:21127046

Lam, Y. A., Pickart, C. M., Alban, A., Landon, M., Jamieson, C., Ramage, R., ... Layfield, R. (2000). Inhibition of the ubiquitin-proteasome system in Alzheimer's disease. *Proceedings of the National Academy of Sciences of the United States of America*, *97*(18), 9902–9906. doi:10.1073/pnas.170173897 PMID:10944193

Lawson, V. A., Collins, S. J., Masters, C. L., & Hill, A. F. (2005). Prion protein glycosylation. *Journal of Neurochemistry*, *93*(4), 793–801. doi:10.1111/j.1471-4159.2005.03104.x PMID:15857383

Lawson, V. A., Priola, S. A., Meade-White, K., Lawson, M. K., & Chesebro, B. (2004). Flexible N-terminal Region of Prion Protein Influences Conformation of Protease-resistant Prion Protein Isoforms Associated with Cross-species Scrapie Infection in Vivo and in Vitro. *The Journal of Biological Chemistry*, *279*(14), 13689–13695. doi:10.1074/jbc.M303697200 PMID:14736880

Lee, J. R., Kim, J. K., Lee, S. J., & Kim, K. P. (2009). Role of protein tyrosine nitration in neurodegenerative diseases and atherosclerosis. *Archives of Pharmacal Research*, *32*(8), 1109–1118. doi:10.100712272-009-1802-0 PMID:19727603

Lefebvre, T., Guinez, C., Dehennaut, V., Beseme-Dekeyser, O., Morelle, W., & Michalski, J.-C. (2005). Does O-GlcNAc play a role in neurodegenerative diseases? *Expert Review of Proteomics*, *2*(2), 265–275. doi:10.1586/14789450.2.2.265 PMID:15892570

Li, F., Macfarlan, T., Pittman, R. N., & Chakravarti, D. (2002). Ataxin-3 is a histone-binding protein with two independent transcriptional corepressor activities. *The Journal of Biological Chemistry*, *277*(47), 45004–45012. doi:10.1074/jbc.M205259200 PMID:12297501

Lilja, T., Heldring, N., & Hermanson, O. (2013). Like a rolling histone: Epigenetic regulation of neural stem cells and brain development by factors controlling histone acetylation and methylation. *Biochimica et Biophysica Acta. G, General Subjects*, *1830*(2), 2354–2360. doi:10.1016/j.bbagen.2012.08.011 PMID:22986149

Lindwall, G., & Cole, R. D. (1984). Phosphorylation affects the ability of tau protein to promote microtubule assembly. *The Journal of Biological Chemistry*, *259*, 5301–5305. PMID:6425287

Lopez Salon, M., Pasquini, L., Besio Moreno, M., Pasquini, J. M., & Soto, E. (2003). Relationship between beta-amyloid degradation and the 26S proteasome in neural cells. *Experimental Neurology*, *180*(2), 131–143. doi:10.1016/S0014-4886(02)00060-2 PMID:12684027

Mandelkow, E. M., & Mandelkow, E. (2011). Biochemistry and cell biology of Tau protein in neurofibrillary degeneration. *Cold Spring Harbor Perspectives in Biology, 3*, 1–25.

Marks, D. S., Hopf, T. A., & Sander, C. (2012). Protein structure prediction from sequence variation. *Nature Biotechnology, 30*(11), 1072–1080. doi:10.1038/nbt.2419 PMID:23138306

Martin, D. D. O., Ahpin, C. Y., Heit, R. J., Perinpanayagam, M. A., Yap, M. C., Veldhoen, R. A., ... Berthiaume, L. G. (2012). Tandem reporter assay for myristoylated proteins posttranslationally (TRAMPP) identifies novel substrates for post-translational myristoylation: PKCε, a case study. *The FASEB Journal, 26*(1), 13–28. doi:10.1096/fj.11-182360 PMID:21965604

Martin, D. D. O., Heit, R. J., Yap, M. C., Davidson, M. W., Hayden, M. R., & Berthiaume, L. G. (2014). Identification of a post-translationally myristoylated autophagy-inducing domain released by caspase cleavage of huntingtin. *Human Molecular Genetics, 23*(12), 3166–3179. doi:10.1093/hmg/ddu027 PMID:24459296

Martin, D. D. O., Ladha, S., Ehrnhoefer, D. E., & Hayden, M. R. (2015). Autophagy in Huntington disease and huntingtin in autophagy. *TreNDDs Neurosci., 38*(1), 26–35. doi:10.1016/j.tins.2014.09.003 PMID:25282404

Mayor, S., & Riezman, H. (2004). Sorting GPI-anchored proteins. *Nature Reviews. Molecular Cell Biology, 5*(2), 110–120. doi:10.1038/nrm1309 PMID:15040444

McNally, K. L., Ward, A. E., & Priola, S. A. (2009). Cells expressing anchorless prion protein are resistant to scrapie infection. *Journal of Virology, 83*(9), 4469–4475. doi:10.1128/JVI.02412-08 PMID:19225008

Meckler, X., Roseman, J., Das, P., Cheng, H., Pei, S., Keat, M., ... Thinakaran, G. (2010). Reduced Alzheimer's disease β-amyloid deposition in transgenic mice expressing S-palmitoylation-deficient APH1aL and nicastrin. *The Journal of Neuroscience: The Official Journal of the Society for Neuroscience, 30*(48), 16160–16169. doi:10.1523/JNEUROSCI.4436-10.2010 PMID:21123562

Merkley, E. D., Metz, T. O., Smith, R. D., Baynes, J. W., & Frizzell, N. (2014). The succinated proteome. *Mass Spectrometry Reviews, 33*(2), 98–109. doi:10.1002/mas.21382 PMID:24115015

Mitra, S., & Finkbeiner, S. (2008). The ubiquitin-proteasome pathway in Huntington's disease. *TheScientificWorldJournal, 8*, 421–433. doi:10.1100/tsw.2008.60 PMID:18454252

Mukhopadhyay, D., & Riezman, H. (2007). Proteasome-independent functions of ubiquitin in endocytosis and signaling. *Science, 315*, 201–205.

Munday, A. D., & López, J. A. (2007). Posttranslational protein palmitoylation: Promoting platelet purpose. *Arteriosclerosis, Thrombosis, and Vascular Biology, 27*(7), 1496–1499. doi:10.1161/ATVBAHA.106.136226 PMID:17581830

Nakamura, T., Tu, S., Akhtar, M. W., Sunico, C. R., Okamoto, S., & Lipton, S. A. (2013). Aberrant Protein S-nitrosylation in neurodegenerative diseases. *Neuron, 78*(4), 596–614. doi:10.1016/j.neuron.2013.05.005 PMID:23719160

Nedić, O., Rogowska-Wrzesinska, A., & Rattan, S. I. S. (2015). Standardization and quality control in quantifying non-enzymatic oxidative protein modifications in relation to ageing and disease: Why is it important and why is it hard? *Redox Biology, 5*, 91–100. doi:10.1016/j.redox.2015.04.001 PMID:25909343

Nestler, E.J., & Greengard, P. (1999). *Protein Phosphorylation is of Fundamental Importance in Biological Regulation*. Academic Press.

North, B. J., Marshall, B. L., Borra, M. T., Denu, J. M., & Verdin, E. (2003). The human Sir2 ortholog, SIRT2, is an NAD+-dependent tubulin deacetylase. *Molecular Cell, 11*(2), 437–444. doi:10.1016/S1097-2765(03)00038-8 PMID:12620231

Omenn, G. S., Lane, L., Lundberg, E. K., Overall, C. M., & Deutsch, E. W. (2017). Progress on the HUPO Draft Human Proteome: 2017 Metrics of the Human Proteome Project. *Journal of Proteome Research, 16*(12), 4281–4287. doi:10.1021/acs.jproteome.7b00375 PMID:28853897

Paleologou, K. E., Oueslati, A., Shakked, G., Rospigliosi, C. C., Kim, H.-Y., Lamberto, G. R., ... Lashuel, H. A. (2010). Phosphorylation at S87 is enhanced in synucleinopathies, inhibits α-synuclein oligomerization, and influences synuclein-membrane interactions. *The Journal of Neuroscience: The Official Journal of the Society for Neuroscience, 30*(9), 3184–3198. doi:10.1523/JNEUROSCI.5922-09.2010 PMID:20203178

Paper, W. (2011). Upgrading Analog Video Surveillance with IP Technology Panasonic Video surveillance systems. doi:10.1002/elps.201000027.Mass

Perry, G., Friedman, R., Shaw, G., & Chau, V. (1987). Ubiquitin is detected in neurofibrillary tangles and senile plaque neurites of Alzheimer disease brains. *Proceedings of the National Academy of Sciences of the United States of America, 84*(9), 3033–3036. doi:10.1073/pnas.84.9.3033 PMID:3033674

Pietrobono, R., Tabolacci, E., Zalfa, F., Zito, I., Terracciano, A., Moscato, U., ... Neri, G. (2005). Molecular dissection of the events leading to inactivation of the FMR1 gene. *Human Molecular Genetics, 14*(2), 267–277. doi:10.1093/hmg/ddi024 PMID:15563507

Poukka, H., Karvonen, U., Jänne, O. A., & Palvimo, J. J. (2000). Covalent modification of the androgen receptor by small ubiquitin-like modifier 1 (SUMO-1). *Proceedings of the National Academy of Sciences of the United States of America, 97*(26), 14145–14150. doi:10.1073/pnas.97.26.14145 PMID:11121022

Prusiner, S. B. (1998). Prions. *Proceedings of the National Academy of Sciences of the United States of America*, 13363–13383.

Rattan, S. I. S. (2008). Increased molecular damage and heterogeneity as the basis of aging. *Biological Chemistry, 389*(3), 267–272. doi:10.1515/BC.2008.030 PMID:18208348

Resh, M. D. (1999). Fatty acylation of proteins: New insights into membrane targeting of myristoylated and palmitoylated proteins. *Biochimica et Biophysica Acta. Molecular Cell Research, 1451*(1), 1–16. doi:10.1016/S0167-4889(99)00075-0 PMID:10446384

Reynolds, M. R., Berry, R. W., & Binder, L. I. (2005). Site-specific nitration differentially influences τ assembly in vitro. *Biochemistry, 44*(42), 13997–14009. doi:10.1021/bi051028w PMID:16229489

Reynolds, M. R., Reyes, J. F., Fu, Y., Bigio, E. H., Guillozet-Bongaarts, A. L., Berry, R. W., & Binder, L. I. (2006). Tau nitration occurs at tyrosine 29 in the fibrillar lesions of Alzheimer's disease and other tauopathies. *The Journal of Neuroscience: The Official Journal of the Society for Neuroscience*, *26*(42), 10636–10645. doi:10.1523/JNEUROSCI.2143-06.2006 PMID:17050703

Riley, B. E., Zoghbi, H. Y., & Orr, H. T. (2005). SUMOylation of the polyglutamine repeat protein, ataxin-1, is dependent on a functional nuclear localization signal. *The Journal of Biological Chemistry*, *280*(23), 21942–21948. doi:10.1074/jbc.M501677200 PMID:15824120

Robertson, L. A., Moya, K. L., & Breen, K. C. (2004). The potential role of tau protein O-glycosylation in Alzheimer's disease. *Journal of Alzheimer's Disease*, *6*(5), 489–495. doi:10.3233/JAD-2004-6505 PMID:15505370

Safar, J., Wille, H., Itri, V., Groth, D., Serban, H., Torchia, M., ... Prusiner, S. B. (1998). Eight prion strains have PrP(Sc) molecules with different conformations. *Nature Medicine*, *4*(10), 1157–1165. doi:10.1038/2654 PMID:9771749

Santos, A. L., & Lindner, A. B. (2017). Protein Posttranslational Modifications: Roles in Aging and Age-Related Disease. *Oxidative Medicine and Cellular Longevity*, *2017*, 2017. doi:10.1155/2017/5716409 PMID:28894508

Sayre, L., Smith, M., & Perry, G. (2012). Chemistry and Biochemistry of Oxidative Stress in Neurodegenerative Disease. *Current Medicinal Chemistry*, *8*(7), 721–738. doi:10.2174/0929867013372922 PMID:11375746

Sayre, L. M., Moreira, P. I., Smith, M. A., & Perry, G. (2005). Metal ions and oxidative protein modification in neurological disease. *Annali dell'Istituto Superiore di Sanita*, *41*, 143–164. PMID:16244388

Sayre, L. M., Perry, G., Atwood, C. S., & Smith, M. A. (2000). The role of metals in neurodegenerative diseases. *Cellular and Molecular Biology*, *46*, 731–741. PMID:10875436

Schipper-Krom, S., Juenemann, K., & Reits, E. A. J. (2012). The ubiquitin-proteasome system in huntington's disease: Are proteasomes impaired, initiators of disease, or coming to the rescue? *Biochemistry Research International*, *2012*, 1–12. doi:10.1155/2012/837015 PMID:23050151

Schnell, J. D., & Hicke, L. (2003). Non-traditional functions of ubiquitin and ubiquitin-binding proteins. *The Journal of Biological Chemistry*, *278*(38), 35857–35860. doi:10.1074/jbc.R300018200 PMID:12860974

Schreurs, S., Gerard, M., Derua, R., Waelkens, E., Taymans, J.-M., Baekelandt, V., & Engelborghs, Y. (2014). In vitro phosphorylation does not influence the aggregation kinetics of WT α-synuclein in contrast to its phosphorylation mutants. *International Journal of Molecular Sciences*, *15*(1), 1040–1067. doi:10.3390/ijms15011040 PMID:24434619

Seo, H., Sonntag, K. C., Kim, W., Cattaneo, E., & Isacson, O. (2007). Proteasome activator enhances survival of Huntington's disease neuronal model cells. *PLoS One*, *2*(2), e238. doi:10.1371/journal.pone.0000238 PMID:17327906

Serviddio, G., Di Venosa, N., Federici, A., D'Agostino, D., Rollo, T., Prigigallo, F., ... Vendemiale, G. (2005). Brief hypoxia before normoxic reperfusion (postconditioning) protects the heart against ischemia-reperfusion injury by preventing mitochondria peroxyde production and glutathione depletion. *The FASEB Journal, 19*(3), 354–361. doi:10.1096/fj.04-2338com PMID:15746178

Shimizu, T., Watanabe, A., Ogawara, M., Mori, H., & Shirasawa, T. (2000). Isoaspartate formation and neurodegeneration in Alzheimer's disease. *Archives of Biochemistry and Biophysics, 381*(2), 225–234. doi:10.1006/abbi.2000.1955 PMID:11032409

Shiozawa-West, N., Dunlop, R. A., & Rodgers, K. J. (2015). Using an in vitro model to study oxidised protein accumulation in ageing fibroblasts. *Biochimica et Biophysica Acta. G, General Subjects, 1850*(11), 2177–2184. doi:10.1016/j.bbagen.2015.07.002 PMID:26187876

Somerville, R. A. (1999). Host and transmissible spongiform encephalopathy agent strain control glycosylation of PrP. *The Journal of General Virology, 80*(7), 1865–1872. doi:10.1099/0022-1317-80-7-1865 PMID:10423157

Spiro, R. G. (2002). Protein glycosylation: Nature, distribution, enzymatic formation, and disease implications of glycopeptide boNDDs. *Glycobiology, 12*(4), 12. doi:10.1093/glycob/12.4.43R PMID:12042244

Stadtman, E.R. (1992). Protein oxidation and aging. *Science, 257*, 1220–1224.

Stadtman, E. R. (2001). Protein oxidation in aging and age-related diseases. *Annals of the New York Academy of Sciences, 928*(1), 22–38. doi:10.1111/j.1749-6632.2001.tb05632.x PMID:11795513

Steffan, J.S., Agrawal, N., & Pallos, J. (2004). SUMO Modification of Huntingtin and Huntington's Disease Pathology. *Science, 304*, 100–104. doi:10.1126cience.1092194

Suhr, S. T., Senut, M. C., Whitelegge, J. P., Faull, K. F., Cuizon, D. B., & Gage, F. H. (2001). Identities of sequestered proteins in aggregates from cells with induced polyglutamine expression. *The Journal of Cell Biology, 153*(2), 283–294. doi:10.1083/jcb.153.2.283 PMID:11309410

Takahashi, T., Yamashita, H., Nakamura, T., Nagano, Y., & Nakamura, S. (2002). Tyrosine 125 of alpha-synuclein plays a critical role for dimerization following nitrative stress. *Brain Research, 938*(1-2), 73–80. doi:10.1016/S0006-8993(02)02498-8 PMID:12031537

Tanimukai, H., Grundke-Iqbal, I., & Iqbal, K. (2005). Up-regulation of inhibitors of protein phosphatase-2A in Alzheimer's disease. *American Journal of Pathology, 166*(6), 1761–1771. doi:10.1016/S0002-9440(10)62486-8 PMID:15920161

Taraboulos, A., Rogers, M., Borchelt, D. R., McKinley, M. P., Scott, M., Serban, D., & Prusiner, S. B. (1990). Acquisition of protease resistance by prion proteins in scrapie-infected cells does not require asparagine-linked glycosylation. *Proceedings of the National Academy of Sciences of the United States of America, 87*(21), 8262–8266. doi:10.1073/pnas.87.21.8262 PMID:1978322

Taraboulos, A., Scott, M., & Semenov, A. (1995). Cholesterol depletion and modification of COOH-terminal targeting sequence of the prion protein inhibit formation of the scrapie isoform. *The Journal of Cell Biology, 129*(1), 121–132. doi:10.1083/jcb.129.1.121 PMID:7698979

Tenreiro, S., Eckermann, K., & Outeiro, T. F. (2014). Protein phosphorylation in neurodegeneration: Friend or foe? *Frontiers in Molecular Neuroscience*, *7*, 7. doi:10.3389/fnmol.2014.00042 PMID:24860424

Terashima, T., Kawai, H., Fujitani, M., Maeda, K., & Yasuda, H. (2002). SUMO-I co-localized with mutant atrophin-I with expanded polyglutamines accelerates intranuclear aggregation and cell death. *Neuroreport*, *13*(17), 2359–2364. doi:10.1097/00001756-200212030-00038 PMID:12488827

Thomas, B., & Flint Beal, M. (2007). Parkinson's disease. *Human Molecular Genetics*, 16. PMID:17911161

Ulloa, L., Montejo de Garcini, E., Gómez-Ramos, P., Morán, M. A., & Avila, J. (1994). Microtubule-associated protein MAP1B showing a fetal phosphorylation pattern is present in sites of neurofibrillary degeneration in brains of Alzheimer's disease patients. *Brain Research. Molecular Brain Research*, *26*(1-2), 113–122. doi:10.1016/0169-328X(94)90081-7 PMID:7854037

Uttara, B., Singh, A., Zamboni, P., & Mahajan, R. (2009). Oxidative Stress and Neurodegenerative Diseases: A Review of Upstream and Downstream Antioxidant Therapeutic Options. *Current Neuropharmacology*, *7*(1), 65–74. doi:10.2174/157015909787602823 PMID:19721819

Van Den Steen, P., Rudd, P. M., Dwek, R. A., & Opdenakker, G. (1998). Concepts and principles of O-linked glycosylation. *Critical Reviews in Biochemistry and Molecular Biology*, *33*(3), 151–208. doi:10.1080/10409239891204198 PMID:9673446

Van Leeuwen, F.W., De Kleijn, D.P.V., & Van Den Hurk, H.H. (1998). Frameshift mutants of β amyloid precursor protein and ubiquitin-B in Alzheimer's and Down patients. *Science*, *279*, 242–247. doi:10.1126cience.279.5348.242

Venkatraman, A., Hu, Y. S., Didonna, A., Cvetanovic, M., Krbanjevic, A., Bilesimo, P., & Opal, P. (2014). The histone deacetylase HDAC3 is essential for Purkinje cell function, potentially complicating the use of HDAC inhibitors in SCA1. *Human Molecular Genetics*, *23*(14), 3733–3745. doi:10.1093/hmg/ddu081 PMID:24594842

Venkatraman, P., Wetzel, R., Tanaka, M., Nukina, N., & Goldberg, A. L. (2004). Eukaryotic proteasomes cannot digest polyglutamine sequences and release them during degradation of polyglutamine-containing proteins. *Molecular Cell*, *14*(1), 95–104. doi:10.1016/S1097-2765(04)00151-0 PMID:15068806

Wang, J., Tung, Y. C., Wang, Y., Li, X. T., Iqbal, K., & Grundke-Iqbal, I. (2001). Hyperphosphorylation and accumulation of neurofilament proteins in Alzheimer disease brain and in okadaic acid-treated SY5Y cells. *FEBS Letters*, *507*(1), 81–87. doi:10.1016/S0014-5793(01)02944-1 PMID:11682063

Watanabe, A., Takio, K., & Ihara, Y. (1999). Deamidation and isoaspartate formation in smeared tau in paired helical filaments: Unusual properties of the microtubule-binding domain of tau. *The Journal of Biological Chemistry*, *274*(11), 7368–7378. doi:10.1074/jbc.274.11.7368 PMID:10066801

Waxman, E. A., & Giasson, B. I. (2008). Specificity and regulation of casein kinase-mediated phosphorylation of alpha-synuclein. *Journal of Neuropathology and Experimental Neurology*, *67*(5), 402–416. doi:10.1097/NEN.0b013e3186fc995 PMID:18451726

Waxman, E. A., & Giasson, B. I. (2011). Characterization of kinases involved in the phosphorylation of aggregated α-synuclein. *Journal of Neuroscience Research*, *89*(2), 231–247. doi:10.1002/jnr.22537 PMID:21162130

Wilkinson, K. A., & Henley, J. M. (2010). Mechanisms, regulation and consequences of protein SUMOylation. *The Biochemical Journal*, *428*(2), 133–145. doi:10.1042/BJ20100158 PMID:20462400

Williams, D. R. (2006). Tauopathies: Classification and clinical update on neurodegenerative diseases associated with microtubule-associated protein tau. *Internal Medicine Journal*, *36*(10), 652–660. doi:10.1111/j.1445-5994.2006.01153.x PMID:16958643

Wong, C. M., Bansal, G., Marcocci, L., & Suzuki, Y. J. (2012). Proposed role of primary protein carbonylation in cell signaling. *Redox Report*, *17*(2), 90–94. doi:10.1179/1351000212Y.0000000007 PMID:22564352

Wong, C. M., Cheema, A. K., Zhang, L., & Suzuki, Y. J. (2008). Protein carbonylation as a novel mechanism in redox signaling. *Circulation Research*, *102*(3), 310–318. doi:10.1161/CIRCRESAHA.107.159814 PMID:18079412

Xu, Y., Deng, Y., & Qing, H. (2015). The phosphorylation of α-synuclein: Development and implication for the mechanism and therapy of the Parkinson's disease. *Journal of Neurochemistry*, *135*(1), 4–18. doi:10.1111/jnc.13234 PMID:26134497

Yamamoto, A., Takagi, H., Kitamura, D., Tatsuoka, H., Nakano, H., Kawano, H., ... Shirasawa, T. (1998). Deficiency in protein L-isoaspartyl methyltransferase results in a fatal progressive epilepsy. *The Journal of Neuroscience: The Official Journal of the Society for Neuroscience*, *18*(6), 2063–2074. doi:10.1523/JNEUROSCI.18-06-02063.1998 PMID:9482793

Yokoyama, A., Katsura, S., & Sugawara, A. (2017). Biochemical analysis of histone succinylation. *Biochemistry Research International*, *2017*, 1–7. doi:10.1155/2017/8529404 PMID:29225971

Yura, K., Shionyu, M., Hagino, K., Hijikata, A., Hirashima, Y., Nakahara, T., ... Go, M. (2006). Alternative splicing in human transcriptome: Functional and structural influence on proteins. *Gene*, *380*(2), 63–71. doi:10.1016/j.gene.2006.05.015 PMID:16872759

Zeng, L. R., Vega-Sánchez, M. E., Zhu, T., & Wang, G. L. (2006). Ubiquitination-mediated protein degradation and modification: An emerging theme in plant-microbe interactions. Cell Research, 413–426

Zhao, Q.-F., Yu, J.-T., & Tan, L. (2015). S-Nitrosylation in Alzheimer's disease. *Molecular Neurobiology*, *51*(1), 268–280. doi:10.100712035-014-8672-2 PMID:24664522

Zhao, X., & Yang, J. (2010). Amyloid-B peptide is a substrate of the human 20S proteasome. *ACS Chemical Neuroscience*, *1*(10), 655–660. doi:10.1021/cn100067e PMID:21116456

Chapter 6
Proteostasis and Neurodegeneration:
Perspectives in the Pathogenesis of Molecular and Cellular Mechanisms

Md. Ataur Rahman
https://orcid.org/0000-0001-6649-3694
Korea Institute of Science and Technology, South Korea

Md Saidur Rahman
https://orcid.org/0000-0003-4583-8731
Chung-Ang University, South Korea

Md Jamal Uddin
https://orcid.org/0000-0003-2911-3255
Ewha Womans University, South Korea

A. N. M. Mamun-Or-Rashid
Doshisha University, Japan

Md. Rezanur Rahman
https://orcid.org/0000-0002-8739-8714
Khwaja Yunus Ali University, Bangladesh

Myung-Geol Pang
Chung-Ang University, South Korea

Hyewhon Rhim
Korea Institute of Science and Technology, South Korea

ABSTRACT

Proteostasis or protein homeostasis consists of a complex interrelated cellular system that controls several steps of protein quality and function from the initial step of synthesis as well as folding, and eventually degradation over enormous biochemical pathways. Proteostasis involves controlling protein folding, modification of the post-translational protein, and degradation of misfolded protein. However, the failure of proteostasis has resulted to produce a toxic protein that leads to disrupt aging and neurodegeneration. Additionally, endoplasmic reticulum degradation and autophagy dysfunction may outcome in cellular additional stress that is responsible for cell death. Consequently, proteostasis targets provide an element of a promising neuronal protective therapeutic method to improve the development of these diseases as well. In this chapter, the authors represent the current knowledge regarding how cellular proteostasis interruption contributes to progress neurodegenerative disorders.

DOI: 10.4018/978-1-7998-1317-0.ch006

INTRODUCTION

Protein production, quality control, as well as degradation of entire proteins are required firmly to regulate and maintain cellular function and health (Newton et al., 2019). To employ its biological activity and function they must require to appropriate fold with the conformation of three-dimensional structure (Kurtishi et al., 2019). Neuronal cell death and neurodegenerative disorders such as Alzheimer's disease (AD), Parkinson's disease (PD), amyotrophic lateral sclerosis (ALS), Huntington's disease (HD), as well as prion disease (pD) are sometimes occurred by protein misfolding (Coppede and Migliore, 2015; Shiwaku and Okazawa, 2015).

The common risk factors of neurodegenerative diseases are caused by environmental factors, oxidative stress, protein dysfunction, and aging which ultimately produce distinctive pathologies (Coppede and Migliore, 2015). Among them, protein aggregation and misfolding is a general feather in neurodegenerative disorders. By balancing protein misfolding as well as foldings are major in protecting the proteome functionality (Hipp et al., 2014). However, the accumulation of inactive misfolded proteins may result in stress responses in cells as well as organelles (Rao and Bredesen, 2004). Unfolded protein response (UPR) and proteostasis is maintained by endoplasmic reticulum (ER) that is triggered via ER stress and protein accumulation. UPR stimulates protein folding as well as decreases protein levels of ER through translation mitigation, proteasomal degradation, and autophagy (Plate and Wiseman, 2017). Autophagy furthermore plays an important function in proteostasis as a result of its capability to destroy aggregate proteins that cannot be degraded through the proteasome system (Dong and Cui, 2019). For instance, in the autophagy system autophagosome binds to lysosome as a consequence of degrading as well as recycle whole organelles which encourage cell survival factors to control neurodegeneration (Rahman and Rhim, 2017). Proteins are degraded by autophagy and the proteasome use ubiquitination to recruit target proteins as well (Rahman and Rhim, 2017; Zientara-Rytter and Subramani, 2019). Ubiquitination and additional covalent attachments, for example, SUMOylation, phosphorylation, and oxidation control regular proteome function (Colignon et al., 2019).

Additionally, post-translational modifications (PTMs) of proteins also shown to stimulates protein aggregation in numerous neurodegenerative syndromes (Owen and Shewmaker, 2019). Disease-associated proteins accumulation cause in the aggregation of protein which leads to proteotoxicity that causes a problem in post-mitotic neurons emphasizing the strong connection between neurodegeneration and aging (McAlary et al., 2019). Moreover, mitochondrial proteostasis is similarly crucial for the survival of the cell, however, its dysfunction may also lead to the increase of reactive oxygen species (ROS) that could be disrupted to general cellular proteostasis (Veeresh et al., 2019). Certainly, a dysfunction of mitochondrial proteostasis can lead to permanent apoptosis induction (Moehle et al., 2019). In this chapter, the author summarizes essential factors that disturb proteostasis as well as additionally describe how dysfunction of proteostasis eventually affects neurodegenerative disorders.

PROTEOSTASIS AND POST-TRANSLATIONAL MODIFICATIONS IN NEURODEGENERATIVE DISORDERS

Protein structure and function are regulated by post-translational modifications (PTMs) that modify protein tertiary structures and thereby increase the complexity and diversity of the protein function (Lin, 2018). Phosphorylation, acetylation, glycosylation, ubiquitination, NEDDylation, and SUMOylating are the most common forms of PTMs currently well-known. Disruption of PTMs can cause numerous pathological events that disrupt cellular proteostasis and ultimately lead to the progression of various diseases (Gao et al., 2019). PTMs must be strictly regulated because they play a central role in countless cellular functions (Taipale, 2019). PTM regulatory pathways are reported to be compromised most often in neurodegenerative diseases, which showed toxicity on cultured cells and thus kill neuronal cells (Kurtishi et al., 2019). The relationship between unusual PTMs and their role in causing neurodegenerative diseases are discussed in this section.

Proteostasis in Alzheimer's Disease

AD is measured by the existence of aggregates of tau proteins that are abnormally phosphorylated along with amyloid β-peptide plaques usually known as neurofibrillary tangles (NFTs). Glycogen synthase kinase-3β (GSK3β) is an AD-associated kinase that phosphorylates tau at T231. It has been shown that the reduction of phosphorylation of tau protein by inhibiting GSK3β in mice blocks NFTs formation (Ochalek et al., 2017). The phosphatase activity, mainly protein phosphatase 2A (PP2A) responsible for 70% of total cellular phosphatase activity, is reported to decrease in AD patients as a result of the upregulation of inhibitors PP2A, I1, and I2 (Rahman et al., 2016). This dysregulation cascade of phosphorylation events is greatly responsible for the progression of AD (Maqbool et al., 2016).

Tau protein is microtubule-associated that largely expressed in neurons where it stabilizes the cytoskeleton of neuronal cells (Hromadkova and Ovsepian, 2019). Tau shows six isoforms, each of which contains N-terminal inserts and C-terminal repeats at a varying number. Among tau protein PTMs, phosphorylation is the most common as it comprises 85 potential phosphorylation sites among those, 45 are of the serine (53%) residue, 35 are threonine (41%), and rest 5 are tyrosine (6%) residue (Alonso et al., 2018; Zhang et al., 2009). Tau hyper-phosphorylates at different sites within the protein, and ultimately forms NFTs in high amount. Hyperphosphorylation interrupts the capability of the protein tau to bind effectively to microtubules, which can cause neurotoxicity (Zhang et al., 2009). NEDDylation modifies APP on its lysine residues acting like the ubiquitin pathway (Stastna and Van Eyk, 2015). These positions are playing an important role as they are located in the binding protein-1 (BP-1) positions of APP, and the NEDDylation-conjugation pathway comprises BP-1 as an important participant (Chow et al., 2010). The reduction of BP-1 prevents NEDDylation, and later results in the buildup of APP β-secretase fragment and APP, suggesting APP NEDDylation can signal for the APP degradation as well (Kurtishi et al., 2019). Tau and amyloid beta (Aβ) aggregation are facilitated by the ubiquitination of these two proteins along with some other proteins (Zhang et al., 2017).

The Aβ plaques form by amyloid precursor protein (APP) excision in different cellular compartments. APP is formed in the ER and it has been demonstrated that it interacted with HRD1 which is an E3 ligase that clears freshly produced proteins in the organelle ER (Zheng et al., 2016). In AD patients, a decline in the expression of HRD1, which causes ER stress, ultimately leads to the APP and Aβ accumulation and consequently apoptosis (Cui et al., 2017). The APP is then adopted in the trans-Golgi system in which it is divided and produce Aβ plaques. It has been suggested that these Aβ plaques stimulate the GSK3β, which will lead to the tau hyperphosphorylation (Cui et al., 2017; Nomura et al., 2016).

Another PTM is SUMOylation that is also known to be linked with the pathogenesis of AD. Even though the small ubiquitin-like modifier is analogous to the ubiquitin, but it covalently binds with target proteins on their lysine residues and does not follow the pathway of ubiquitin. In contrast to the ubiquitination process, SUMOylation involves the accomplishment of E1, E2, and E3 to assign SUMO to bind with the target proteins, and later, the SUMOylation pathway can modify these target proteins. SUMOylation has been known to play a vital role in the case of oxidative stress (Feligioni et al., 2011), and any regulatory mistake of it is linked with the progression of neurodegenerative diseases. Studies have revealed that covalent binding of SUMO-1 and SUMO-2 with the 587 and 595 lysines of APP, respectively regulating Aβ aggregates levels (Zhang et al., 2009). Current data also recommends a significant contribution of calcium homeostasis in ER of AD patients precisely, ryanodine receptors (RyR) dysfunction into the ER triggering Ca^{2+} leakage that has been linked in AD patients (Korecka et al., 2019). The RyR complex hyperphosphorylation and hypernitrosylation result in calstabin reduction and Ca^{2+} leakage from the ER RyR channel (Zugel et al., 2019). Therefore, it is reasonable to recommend that pointing these "leaky" RyR channels may open new therapeutic approaches for AD treatment.

Proteostasis in Parkinson's Disease

The generation of Lewy bodies (LBs) in dopaminergic neurons is one of the characteristics of the PD progression (Olfati et al., 2019). LBs are mainly large deposits of α-synuclein-enriched proteins, a naturally propagated protein, present as an aggregate in neurons of patients with PD. α-Synuclein is a 14-kDa soluble protein with a total of five phosphorylation sites, two in serine (Ser87, Ser129) and three in tyrosine residue (Tyr125, Tyr133, and Tyr136) (McFarland et al., 2008; Waxman and Giasson, 2011). The frequency of Ser129 phosphorylation is about 4% in case of usual physiological conditions. But recent studies demonstrated that about 90% α-synuclein of LBs are phosphorylated at the Ser129 residue (Anderson et al., 2006). This discovery has driven to the hypothesis according to which Ser129 hyperphosphorylation is closely linked to the aggregated α-synuclein formation. It has also been demonstrated that Ser129 phosphorylation plays a toxic effect by enabling the transferring between the nucleus and the cytosol (Goncalves and Outeiro, 2013; Schaser et al., 2019). Inside of the nucleus, the variant α-synuclein phospho-Ser129 may interact with histone proteins that mask them against histone deacetylases, which induce neurotoxicity. Even though the status of phosphorylation of α-synuclein at the rest of the five phosphorylation sites (Ser87, and Tyr125, 133, and 136) are also changed in the patients with PD, the PTMs contribution to the aggregation of proteins is not evidenced (Benskey et al., 2016). Moreover, acetylation of histone is regulated by two enzymes, the histone acetyltransferases (HATs) and the histone deacetylases (HDACs) (DesJarlais and Tummino, 2016). The histone proteins are linked with the acetyl group by HATs and thereby promote the transcription of target genes. On the other hand, HDACs eliminate the connected acetyl group from histone proteins. Histone acetylation was reported to increase in PD patients may be linked with the pathogenesis of PD (Sharma et al., 2019; Sharma and Taliyan, 2015).

Proteins degradation processes are essential to maintain cellular proteostasis (Noormohammadi et al., 2018). Ubiquitination is a PTM catalyzed by an enzyme that targets and marks the proteins because of their degradation by binding to ubiquitin (Ub) (Hegarty et al., 2016). The ubiquitination process comprises of three different classes of catalytic enzymes, including the enzyme activation Ub (E1), the enzyme of conjugation to Ub (E2) and the ligase Ub (E3) (Buetow and Huang, 2016). The target proteins that labeled for degradation may be ubiquitinated once or poly-ubiquitinated when the latter directs the proteins toward the 26S proteasome for further degradation. The parkin, a PD associated protein is a ligase ubiquitin E3 that will mark the proteins poorly folded for ubiquitination (Myeku and Duff, 2018). Besides, PINK1 (putative kinase 1 induced by PTEN) is generally a mitochondrial Ser/Thr kinase that triggers and activates the Parkin via phosphorylation. Ubiquitin is phosphorylated in serine

65 by PINK1, then binds with Parkin and activates Parkin in the mitochondria to ubiquitinate the proteins badly folded, this leads eventually to the deterioration of proteins by the proteasome (Kane et al., 2014; Kazlauskaite et al., 2015). Mutations and malfunctions of Parkin and PINK1 are associated with the early-onset familial and sporadic PD (Pickrell and Youle, 2015). It has been demonstrated that this malfunction caused extreme ROS production into the mitochondria that led to an unacceptable cellular proteostasis (D'Amico et al., 2017). The protein kinase c-Abl was reported to regulate the activity of Parkin negatively by Tyr143 phosphorylation, which influences the activity of ubiquitin E3 ligase. PD post-mortem brains have also been reported this Parkin inactivation by c-Abl (Abushouk et al., 2018). It has also been demonstrated that Parkin ubiquitinates p62, an important shuttle protein that regulates the quality of the protein (Song et al., 2016). For proteasomal degradation of polyubiquitinated proteins, p62 plays a vital role in the transportation and Parkin tags p62 in response to mitochondrial stress for the degradation by the proteasome (Komatsu et al., 2007). When this equilibrium is disturbed, p62 can lead to the aggregated proteins accumulation and alter mitochondrial dynamics, and thereby disrupt cellular homeostasis. Additionally, α-synuclein phosphorylation, nitration, and SUMOylation are found to change in patients with PD, therefore, they may assist as effective biomarkers of PD (Krumova et al., 2011). In effect, the phosphorylation of Tyr125 and nitration of Y39 have found to increase significantly in patients with PD comparing to the healthy subjects, in contrast, SUMO-1 was significantly reduced in patients with PD (Guerra de Souza et al., 2016; Krumova et al., 2011). SUMOylation reduces aggregation of α-synuclein both *in vitro* model and *in vivo* conditions, which could clarify the reduced level of SUMO-1 in the patients with PD (Krumova et al., 2011). These findings recommend PTMs alterations as promising diagnostic and predictive biomarkers for PD pathogenesis even though a more detailed study still required for further confirmation.

Proteostasis in Amyotrophic Lateral Sclerosis

ALS also recognized as Lou Gehrig's disease is another lethal neurodegenerative disorder that disrupts the motor neurons (Zarei et al., 2015). The pathogenesis of ALS is considered by the damage of motor neurons not only in the brain but also in the spinal cord, which can result in the progression of paralysis and eventually cause death (Kaur et al., 2016; Rosen et al., 1993). The crucial influences that are responsible for the ALS pathogenesis are gene mutations in the region of superoxide dismutase (SOD1) and the loss of function of the astroglial glutamate transporter, EAAT2 (Fei et al., 2006; Rosenblum and Trotti, 2017). It has been shown that SOD1 and EAAT2 ubiquitination, SUMOylation, and acetylation trigger to the late onset of ALS (Ford et al., 2019).

SOD1 plays an important role in the elimination of free superoxide species and SOD1 is associated with familial ALS (Sirangelo and Iannuzzi, 2017). SOD1 gene is observed as mutated and in sporadic ALS, the PTMs appear to play a significant role for the late onset of ALS (Ayers et al., 2016; Rakhit et al., 2002). The E3 ligase E6-AP ubiquitinate SOD1 and studies on mice have also shown to reduce the E6-AP levels and to increase SOD1 levels before the ALS onset (Bett, 2016; Mishra et al., 2013). Current studies have also recommended that lysine 123 (K123) of SOD1 acetylation, a region for protein folding and copper-binding, may play a vital function in ALS progression; however, the connection between dysregulation of the acetylation of SOD1 and ALS progression is not completely revealed yet (Banks et al., 2016; Kaliszewski et al., 2016). Lysine 75 of SOD1 SUMOylation occurs by the action of SUMO-1, SUMO-2, and SUMO-3 (Banks and Andersen, 2019). EAAT2, sliced by the caspase 3 and later accumulated in the spinal cord astrocytes, is found as SUMOylated. Studies demonstrated that the

SOD1 mutant's modification in SUMO-3 plays a significant role in the aggregation of SOD1, which leads to the pathogenesis of ALS (Rosenblum and Trotti, 2017). As a result, it causes failure of SOD1 to eliminate ROS causing oxidative stress to the cells and thereby disturbs cellular proteostasis, and as a result, higher oxidative impairment is detected in patients with ALS (Labbadia and Morimoto, 2015; Nakamura et al., 2012). It has further been revealed that SUMO-1 is capable of SUMOylate Lys9 of SOD1, and this alteration rises the susceptibility of the aggregation of SOD1 (Kurtishi et al., 2019).

EAAT2 is expressed in the spinal cord that is a glutamate transporter and its expression reduced in patients with ALS (Takahashi et al., 2015). EAAT2 SUMOylation occurs by the action of SUMO-1 in the intracellular C-terminal domain of it. The SUMOylated domains accumulation has been discovered at the final stages of ALS in astrocytes and a loss of generic function of astrocytic physiology, causing loss of motor neurons, has been recommended (Lee et al., 2016). Additionally, one of the mutated genes is UBQLN2 in familial cases of ALS that is found on the X chromosome which plays important roles in proteostasis (Deng et al., 2011). In fact, it seems that the mutations in UBQLN2 affect proteasomal degradation (Deng et al., 2011; Navone et al., 2015). Ubiquitinated proteins less effective binding to the proteasome was reported by the mutation of UBQLN2 gene which codes for the protein ubiquilin-2, resulting in the accumulation of target proteins showing the failure of the proteasomal degradation process (Ajroud-Driss and Siddique, 2015).

PROTEOSTASIS AND AUTOPHAGY IN NEURODEGENERATIVE DISORDERS

Autophagy is defined as a cellular process which comprises the effective as well as selective degradation of aggregated/misfolded proteins molecules and maintaining cellular homeostasis (Rahman and Rhim, 2017; Uddin et al., 2019b). It is a lysosomal-associated degrading process that is triggered through stress conditions and removes injured/damaged proteins or organelles (**Figure 1**) (Anding and Baehrecke, 2017). Particularly, the process initiates by the development of the autophagosome formation, a double-membrane vesicle, which degrades and engulfs damaged/injured intra-cellular components (Glick et al., 2010). A group of autophagy-related genes (ATGs) is actively associated to initiate the formation of autophagosome, expansion of vesicle, and fusion with the lysosome (Nakamura and Yoshimori, 2017). The failure of autophagy is highly related to the pathogenesis of numerous neurodegenerative disorders (Polajnar and Zerovnik, 2014). However, the modulation of autophagy might be shown very beneficial in the development of several therapeutic intervention approaches (Rubinsztein et al., 2012). Moreover, quality control of autophagy is important for cellular proteostasis maintenance (Jackson and Hewitt, 2016). Therefore, in neurodegenerative diseases, autophagy dysfunction leads to an accumulation of damaged organelles formation as well as pathogenic proteins (Figure 1).

Proteostasis and Autophagy in Alzheimer's Disease

In AD, damaged mitochondria accumulate in neuronal cells which may translocation of misfolded proteins molecules into mitochondria, therefore leads to distraction in the oxidative phosphorylation cycle and activate cellular autophagy (Rhein et al., 2009). This response is termed as mitophagy, type of macro-autophagy, in were injured/damaged mitochondria are usually engulfed through the autophagosome and eventually degradation. Certainly, if this process is delayed, it might lead to accumulating protein aggregation such as Aβ oligomers. Mitophagy protects cells from program cell death via degrading injured/

Figure 1. Diverse steps of the autophagy pathway in a diversity of neurodegenerative disorders promising relations to neuronal cell death.

Various modifications interrelated to neurodegeneration distressing autophagic flux comprising decreased autophagy initiation or improved autophagy suppression; changed cargo recognition; incompetent autophagosome/lysosome fusion; autophagosome clearance is inefficient; and autophagic cargo degradation is inefficient to lysosomes. Autophagy modification might stimulate neuronal cell death through two promising mechanisms: (a) impairment in cargo degradation by lysosome prominent to lysosomal membrane permeabilization, LMP, as well as release of cathepsin protein into cytosol, therefore encouraging to release both necrotic or apoptotic cell death; (b) dysfunction in mitophagy subsequent in addition to impaired mitochondrial membrane permeability (MMP) as well as mitochondria leading to release cytochrome c to stimulate apoptotic cell death.

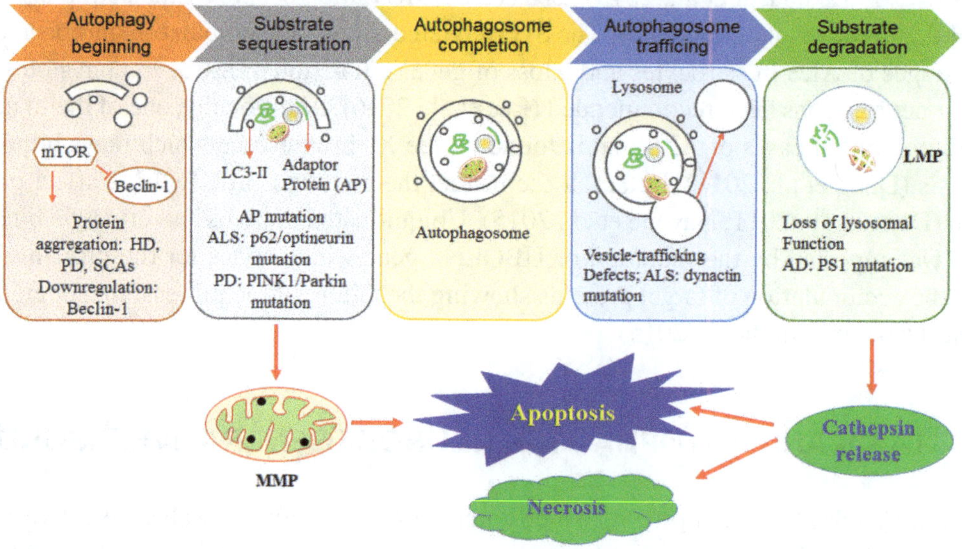

damaged mitochondria in the cells (Boya et al., 2005). It has been found that disruption of sporadic AD in the early stages is involved in autophagosome production (Ferro et al., 2019).

Along with some multiple proteins and microtubule-associated light chain 3-ll (LC3-ll), is associated in the early stage of autophagosome formation. Increasing LC3-ll is usually seen in AD early stages that indicate autophagy stimulation is an initial response to the progression of AD which impacts neuronal cell death (Liu and Li, 2019). Besides, lysosomal activity reduction can impair the autophagic pathway which also witnessed in AD pathogenesis. Reduced acidification of lysosomal activity may be directly connected to mutations of vacuolar ATPase, v-ATPase, in an early-onset AD, EOAD, patients (Uddin et al., 2019a). A common cause of EOAD, mutations of presenilin 1 (PS1), cause defective removal of several AD-associated proteins (Martin-Maestro et al., 2019). PS1 mutations result in the upregulation of Aβ accumulation via cleavage of APP.

Proteostasis and Autophagy in Parkinson's Disease

In PD, autophagy may also be repressed by pathogenic A30P and A53T α-synuclein mutants which inhibit endorsement of injured/damaged proteins and substances via lysosome (Tripathi et al., 2019). Inhibition of these aspects is the main progression and pathogenesis of PD. However, mitochondrial dysfunction also another cause where distended mitochondria not getting removed through the mitophagy process (Niu et al., 2019). PINK1, Parkin, and Htra2 (Park13) have been displayed to be related to a mitochondrial dysfunctional protein in PD. Particularly, PINK1 mutation causes mitochondrial dysfunction which eventually effects mitophagy (Miller and Muqit, 2019). Decreased lysosomal-autophagic degradation, as well as elevated ROS levels, caused mitophagy dysfunction and defects mitophagy or autophagy can be adequate to cause PD pathogenesis (Miller and Muqit, 2019; Shefa et al., 2019).

Proteostasis and Autophagy in Huntington's Disease

In HD, mitochondrial dysfunction has been assumed as a key role in pathogenesis (Intihar et al., 2019). Neuronal cell death may occur when autophagy and mitophagy initiation combats injured mitochondria as well as organelles. Several studies postulate that autophagosome cannot trap the huntingtin (Htt) aggregate protein (Rai et al., 2019). These results are an influx of autophagosomes and accumulate in cytoplasm performing as potential sources of damage to neuronal cells. Numerous studies have also revealed that lysosomal cathepsins overexpression aids Htt aggregate clearance from HEK cells by promising properties of autophagy to maintain cellular proteostasis (Liang et al., 2011; Valionyte et al., 2019).

Proteostasis and Autophagy in Amyotrophic Lateral Sclerosis

ER stress, oxidative stress, genetic factors, and mitochondrial dysfunctions are the main factors contribute to ALS pathogenesis (Klemann et al., 2018). Aggregated proteins which cause UPS impairment and mitochondrial dysfunction. It has been exposed that valsin-containing protein (VCP) mutations in a chaperon related autophagosome maturation, may be connected with ALS of autosomal dominant (Dols-Icardo et al., 2018). Accordingly, this could hamper mitochondrial quality control as well as autophagy. Autophagy stimulation is useful to decelerating ALS progression, autophagy induction by trehalose and resveratrol have been displayed to reduce aggregation, in addition, to promote cell survival and finally it regulates cellular proteostasis (Crippa et al., 2010; Silberman, 2018).

AGGREGATION AND PROTEOTOXICITY IN PROTEOSTASIS

The accumulated protein aggregates are the key feature of many neurodegenerative diseases. The toxic protein aggregates characterize PD, AD, HD, ALS, and pD as proteinopathies as shown in Figure 2 (Kurtishi et al., 2019). The accumulated proteins in specific neurons that are affected define the pathogenesis of the diseases. Although protein accumulation in neurons is the hallmarks of proteinopathies, but it is not clear the toxic effect or neuroprotective roles of these protein aggregates. It is considered that the accumulation of misfolded protein aggregates is one of the defense mechanisms to circumvent the proteotoxicity and maintenance of the proteostasis (Khan et al., 2018). These protein aggregates are thought to be proteotoxic. The loss-of-function of the aggregate proteins cannot perform normal physiological functions. The gain-of-functions where the protein inclusions gain new functions that may lead to cell death (Winklhofer et al., 2008).

Figure 2. In neurodegenerative pathways changing protein conformations and proteostatic lead to show pathological condition.

Excessive and injure protein straightly causes mitochondrial dysfunction, ER stress, oxidative stress, and aggregation. In addition, feedback loops (I, II) highlight the critical nature of injured/damaged proteins propagate mitochondrial dysfunction and ER stress causes the damage of normal functionality which additionally changing proteostasis. Cellular protective response to these pathways comprises proteasome degradation via apoptosis and autophagy in several neurodegenerative disorders.

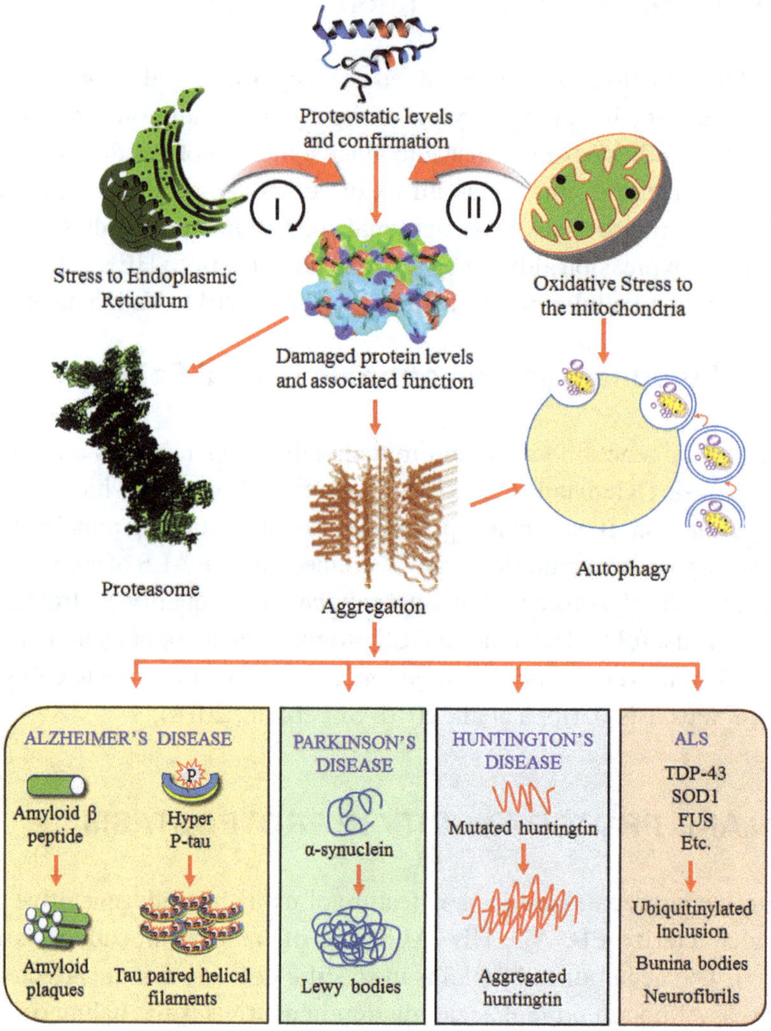

Aggregation and Proteotoxicity in Alzheimer's Disease

The mutations of amyloid precursor metabolic protein-encoding genes have been implicated in the plaque formation, but it is not entirely understood whether non-fibrillar Aβ multimers or amyloid plaques drives the pathogenesis of AD (Lesne et al., 2006). The disruption of two forms of amyloid plaques Aβ42 and Aβ40 in the mesenchyme in AD patients. An increase in the Aβ42/Aβ40 ratio results in inhibition of synaptic function leading to neurotoxicity and increased aggregation in AD neurons. Non-toxic protein aggregation has been found which is independent of AD, for instance; the increased Aβ40 aggregates have been observed in neuroprotection. It is thought that the inhibition of toxic Aβ42 into aggregates rich in Aβ40 is a mechanism of neuroprotection. Aβ has been associated with cytotoxicity with tau, but the only dysfunctional tau dysfunction cannot lead to neurodegeneration (Acquarone et al., 2019). This notion suggests the hypothesis that it is involved in plaque generation later stage in AD pathogenesis. Moreover, tau can also enhance Aβ toxicity via a feedback loop (Frost et al., 2015). The Tau aggregation into NFTs or paired helical filaments (PHFs) is implicated in neuroprotection (Flores-Rodriguez et al., 2015). The inhibition of gain-of-function of PHFs is implicated in AD (Flores-Rodriguez et al., 2015). In contrast, due to the instability of microtubules and dysfunction of cellular transport associated with loss-of- function toxicity by PHFs does not seem to be important (Venkatramani and Panda, 2019).

Aggregation and Proteotoxicity in Parkinson's Disease

The α-synuclein proteotoxicity has been considered responsible for the formation of LB in brain cells. Recent emerging evidence suggesting non-fibrillar various species of α-synuclein are very important to the pathogenesis of PD (Delenclos et al., 2019). For instance, emerging evidence proposes the α-synuclein multimers can disrupt cell membranes. LB formation can be a neuroprotective mechanism if the aggregation of α-synuclein multimers decreases their toxicity (Xu et al., 2019). Thus, it is considered the α-synuclein aggregation and LB formation in PD is a multidimensional process. It is considered the α-synuclein aggregation is a toxic mechanism, and its functional inactivation might detriment the normal physiological functions of neurons (Rodriguez-Nogales et al., 2016). Evidence also indicates the α-synuclein has been associated with glucose level maintenance, inhibition of apoptosis, alter the activity of calmodulin, increase the assembly of SNARE activity, molecular chaperone functions, etc., (Das and Eliezer, 2019). These functions may be reduced in response to protein aggregation.

Aggregation and Proteotoxicity in Huntington's Disease

The presence of 5′CAG repeat in the huntingtin encoding gene IT15 is responsible for the formation of Htt aggregation in HD. The amount of polyQ correlates with HD development and Htt aggregation (Wanker et al., 2019). The formation of inclusion bodies is not related to neurotoxicity and accumulated toxic misfolded Htt is observed to be involved in neuroprotection (Miller et al., 2010). Cell survival, autophagy, transcription, endocytosis, vesicle trafficking are the main physiological roles of Htt (Newcombe et al., 2015; Saudou and Humbert, 2016). By decreasing the level of Htt aggregates the reversal of the HD-like pathology in mice was observed of HTT a loss-of-function which may lead to HD development (McAdam et al., 2019).

Aggregation and Proteotoxicity in Amyotrophic Lateral Sclerosis

In ALS, TAR DNA-binding protein (TDP-43) most commonly develops disordered aggregates, while in some subtypes of ALS skein-like bodies have been observed also (Prasad et al., 2019). Loss-of-function and gain-of-function of toxicity induced by upregulation and partial knockdown of TDP-43 revealed its association of ALS pathogenesis (Ludolph and Brettschneider, 2015). The familial ALS is caused by mutant SOD1, and amyloid fibrils in various forms in wild type (WT) and mutant SOD1 that may show the varying level of neurotoxicity (Kaur et al., 2016). The gain-of-toxic neurotoxicity is shown by SOD1 mutants when localized near mitochondria, while loss-of-function neurotoxicity is observed due to reduced defense against oxidative stress (Carri et al., 2015; Sau et al., 2007). The amyloid fibrils are formed in a cascade that needs the formation of a nucleus leading to subsequent aggregation. Recent evidence has demonstrated that low complexity regions (LCRs) of proteins that act like a prion. Due to weak intermolecular forces, LCRs are capable to produce phase separation into liquid droplets which is responsible for the conversion of fibrils into solid protein aggregates (Patel et al., 2015). An ALS-associated protein FUS was used for these early-stage experiments but is desirable that other proteins namely α-synuclein, mutant HTT, PrPSC, TDP-43, and Aβ will be studied in future studies of LCR phase separation (Kurtishi et al., 2019). Much research and promise are generating by LCRs to identify a potential common event in neurodegenerative disease pathogenesis.

Aggregation and Proteotoxicity in Prion Diseases

Different proteins such as α-synuclein, tau, beta-amyloid, Htt, and SOD1 are regarded as prions because of their ability to transmission. Toxic protein accumulation in prion diseases is occurred due to genetic mutations or environmental factors (You and Ikezu, 2019). Toxic protein PrPSC is involved in prion diseases as explained by pD model. In response to infection, the normal PrSC protein is converted into toxic PrPSC (Forloni et al., 2019). It is controversial whether loss of function of PrSC is responsible for proteotoxicity (Westergard et al., 2016). Evidence proposes that aggregate PrPSC may halt or hinder the transmission of infection of proteins leading to the conversion of toxic PrPC to neuroprotection.

ER STRESS AND PROTEOSTASIS

ER serves multiple functions such as protein folding, degradation of misfolded proteins, and maintenance of calcium homeostasis (Zheng et al., 2019). In the ER, a continuous membrane system plays a key role in the maintenance of protein quality as well as proteostasis (Morimoto, 2019). ER stress is regarded as the buildup of misfolded or unfolded protein. The UPR maintains the normal function of the cell such as inhibiting protein translation, misfolded protein degradation, and activation of the signaling pathways. In cases where these are not maintained, cells undergo apoptosis (Figure 3). ER stress and UPR are becoming an interesting area of study neurodegenerative disorders (Poplawski et al., 2019). Since misfolded proteins gather during neurodegeneration, the ER gets stress leading to proteotoxicity. Under physiological conditions, the ER controls proteostasis cells; however, the chronic misfolding of protein overruns the capability of the ER and activates pathways which responsible for cytotoxicity in neurodegenerative diseases (Iurlaro and Munoz-Pinedo, 2016).

ER Stress and Proteostasis in Alzheimer's Disease

The occurrence of UPR in the brain indicates the link between AD and ER stress. Aβ42 is reported to activate AMPK, which suppresses the mTOR pathway while Aβ42 promotes AD (Yoon et al., 2012). UPR pathway is directly associated with AD pathology. For instance, XBP1, regulates γ-secretase complex, CDK5, and other APP processing factors. In addition, PERK-eIF2-α pathway activates β-secretase 1 (BACE1), leading to APP and β-amyloid (O'Connor et al., 2008). Fascinatingly, deficiency of PERK, GCN2, and PKR reduced the pathologies of AD (Venkatramani and Panda, 2019).

Figure 3. Protein synthesis pathway throughout the ER.
The entire protein produces in the ER through ribosomes. Innate protein is principally exported, although some others may aggregate. Deregulation and misfolded protein are aggregated and sent to the proteasome for clearance or degradation. Aggregates and misfolded proteins are encouraged to UPR that comprise UPR related gene regulation, chaperone response, ER-associated protein degradation (ERAD), and apoptosis to mitigate the adverse toxic effects of aggregates proteins.

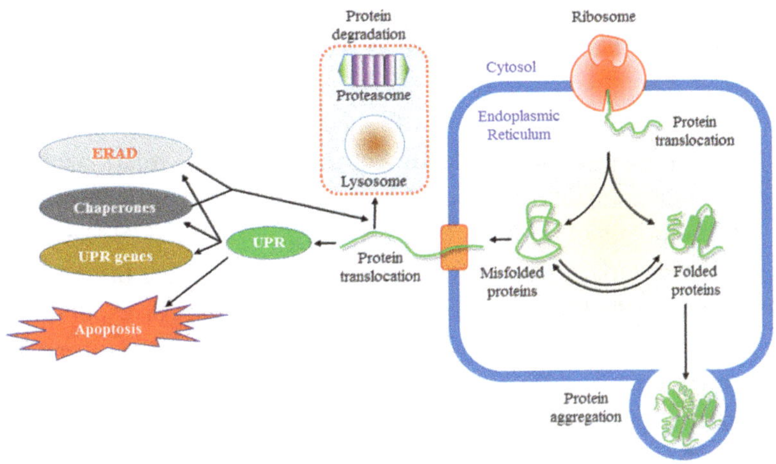

ER Stress and Proteostasis in Parkinson's Disease

PD pathology is also involved with the induction of ER stress. α-Synuclein misfolded is accumulated in a variety of cellular compartments such as the ER (Wang and Kaufman, 2016). In PD brain, ERAD exports the target proteins to the ubiquitin-proteosome system (UPS) and ERAD E3-ligase, HRD1, is elevated (Mahdi et al., 2016). In addition, α-synuclein also suppresses the ERAD pathway and promotes pathogenesis, while PD protein, Parkin, also inhibits the UPS. Rotenone is known to induce PD conditions *in vitro* and *in vivo* and it increases the activation of ER stress responses (Goswami et al., 2016, 2019).

ER Stress and Proteostasis in Huntington's Disease

HD pathology is linked with induction ER stress responses such as BiP, CHOP, and XBP1 (Carnemolla et al., 2009) and this takes place in response to destruction of ERAD by mutant huntingtin (mHTT) inhibiting the ERAD proteins Npl4, Ufd1, p97, and gp78 (Duennwald and Lindquist, 2008; Shacham et al., 2019). The UPS is overloaded by mHTT proteotoxicity and accumulates misfolded proteins (Harding and Tong, 2018; Wang and Kaufman, 2016). However, the association between XBP1 and autophagy has been observed in HD (Jiang et al., 2016).

ER Stress and Proteostasis in Prion Diseases

Along with others, pDs are involved with activation of ER stress and UPR; but the exact mechanism is not clear (Hetz and Mollereau, 2014; Xu and Zhu, 2012). PERK-eIF2-α- UPR pathway inhibition using inhibitors is involved in neuroprotection and decreased disease progression. However, the treatment of salubrinal increases the PERK-eIF2-α stress activation leading to PD progression (Gupta et al., 2019). Neurodegenerative disorders induce ER stress and hinder proteostasis. For example, AD and HD involved with activation of XBP1 where HD and PD are linked with activation of PERK (Morimoto, 2019). It is necessary to clearly define the molecular mechanisms that may involve these distinct neurodegenerative diseases.

RECENT DEVELOPMENTS AND FUTURE RESEARCH DIRECTIONS

As there is a cellular and molecular association with the aggregation of the protein, formation of inclusion, as well as dysfunction of neuronal cell death, but the systematic relationship between all of them remains one of the major questions in this field. However, proteostasis breakdown is a serious problem that can be affected by pathogenically misfolded peptides or proteins. Nevertheless, the cumulative ability of the proteostasis network system will be an essential part of the future research area in neurodegenerative disorders. Increasing the aging population as well as the corresponding anticipated growth in people suffering since neurodegeneration, therefore understanding the potential of the network of proteostasis in the central nervous system is an essential task for research and innovation. Improvements in height through-put single-cell analyses will initiate the systematic understanding of exactly how and why neurons are diverse to other cell categories, besides what creates them so vulnerable to defects the system of proteostasis network in neurodegenerative disorders. To recognize important players in this system will be more important for conveying novel therapeutic approaches. However, reductionist methods have recognized some components of proteostasis, but to improve a complete opinion there is an essential requirement to move concerning the system in addition to consider the impression of the diverse tissue and cell types. Therefore, an additional holistic method needs to consider the understanding of proteostasis in perspectives of neurodegenerative disorders will consequently be important for future research and development.

CONCLUSION

In neurodegenerative disorders, impaired proteostasis is an important contributor to pathogenesis. However, modifications in cellular proteostasis caused by aggregation of protein misfolded, mitochondrial dysfunction, ER stress, and autophagy, are also main contributors of neuron cell death. Thus, sustaining cell survival and longevity, it is required to ensure cellular proteostasis. The control of protein quality to offset the development of age-associated diseases has been the focus of several recent research programs. This collective approach appears logical, as the traditional neurodegenerative associated proteins have supported to be challenging as targets for the intervention of therapeutic approaches. Therefore, extending the possibility of research to comprise proteostatic pathways manipulation might be lead to favorable ways to diminish the neurodegenerative disorders.

ACKNOWLEDGMENT

This work was supported by the Korea Research Fellowship (KRF) Program (2016H1D3A1908615, 2017H1D3A1A02013844, and 2015H1D3A1062189) through the National Research Foundation of Korea (NRF) funded by the Ministry of Science and ICT, Republic of Korea.

REFERENCES

Abushouk, A. I., Negida, A., Elshenawy, R. A., Zein, H., Hammad, A. M., Menshawy, A., & Mohamed, W. M. Y. (2018). C-Abl Inhibition; A Novel Therapeutic Target for Parkinson's Disease. *CNS & Neurological Disorders - Drug Targets*, *17*(1), 14–21. doi:10.2174/1871527316666170602101538 PMID:28571531

Acquarone, E., Argyrousi, E. K., van den Berg, M., Gulisano, W., Fa, M., Staniszewski, A., ... Fiorito, J. (2019). Synaptic and memory dysfunction induced by tau oligomers is rescued by up-regulation of the nitric oxide cascade. *Molecular Neurodegeneration*, *14*(1), 14. doi:10.118613024-019-0326-4 PMID:31248451

Ajroud-Driss, S., & Siddique, T. (2015). Sporadic and hereditary amyotrophic lateral sclerosis (ALS). *Bba-Mol Basis Dis*, *1852*(4), 679–684. doi:10.1016/j.bbadis.2014.08.010 PMID:25193032

Alonso, A. D., Cohen, L. S., Corbo, C., Morozova, V., ElIdrissi, A., Phillips, G., & Kleiman, F. E. (2018). Hyperphosphorylation of Tau Associates With Changes in Its Function Beyond Microtubule Stability. *Frontiers in Cellular Neuroscience*, *12*, 338. doi:10.3389/fncel.2018.00338 PMID:30356756

Anderson, J. P., Walker, D. E., Goldstein, J. M., de Laat, R., Banducci, K., Caccavello, R. J., ... Chilcote, T. J. (2006). Phosphorylation of Ser-129 is the dominant pathological modification of alpha-synuclein in familial and sporadic Lewy body disease. *The Journal of Biological Chemistry*, *281*(40), 29739–29752. doi:10.1074/jbc.M600933200 PMID:16847063

Anding, A. L., & Baehrecke, E. H. (2017). Cleaning House: Selective Autophagy of Organelles. *Developmental Cell*, *41*(1), 10–22. doi:10.1016/j.devcel.2017.02.016 PMID:28399394

Ayers, J. I., Diamond, J., Sari, A., Fromholt, S., Galaleldeen, A., Ostrow, L. W., ... Borchelt, D. R. (2016). Distinct conformers of transmissible misfolded SOD1 distinguish human SOD1-FALS from other forms of familial and sporadic ALS. *Acta Neuropathologica*, *132*(6), 827–840. doi:10.100700401-016-1623-4 PMID:27704280

Banks, C., Rodriguez, N., Gashler, K., Bikman, B., & Andersen, J. (2016). Acylation of SOD1 Provides Tool to determine if Mitochondrial Aggregation of SOD1 is Main Driver of ALS. *The FASEB Journal*, 30.

Banks, C. J., & Andersen, J. L. (2019). Mechanisms of SOD1 regulation by post-translational modifications. *Redox Biology*, *26*, 26. doi:10.1016/j.redox.2019.101270 PMID:31344643

Benskey, M. J., Perez, R. G., & Manfredsson, F. P. (2016). The contribution of alpha synuclein to neuronal survival and function - Implications for Parkinson's disease. *Journal of Neurochemistry*, *137*(3), 331–359. doi:10.1111/jnc.13570 PMID:26852372

Bett, J. S. (2016). Proteostasis regulation by the ubiquitin system. *Essays in Biochemistry*, *60*(2), 143–151. doi:10.1042/EBC20160001 PMID:27744330

Boya, P., Gonzalez-Polo, R. A., Casares, N., Perfettini, J. L., Dessen, P., Larochette, N., ... Kroemer, G. (2005). Inhibition of macroautophagy triggers apoptosis. *Molecular and Cellular Biology*, *25*(3), 1025–1040. doi:10.1128/MCB.25.3.1025-1040.2005 PMID:15657430

Buetow, L., & Huang, D. T. (2016). Structural insights into the catalysis and regulation of E3 ubiquitin ligases. *Nature Reviews. Molecular Cell Biology*, *17*(10), 626–642. doi:10.1038/nrm.2016.91 PMID:27485899

Carnemolla, A., Fossale, E., Agostoni, E., Michelazzi, S., Calligaris, R., De Maso, L., ... Persichetti, F. (2009). Rrs1 Is Involved in Endoplasmic Reticulum Stress Response in Huntington Disease. *The Journal of Biological Chemistry*, *284*(27), 18167–18173. doi:10.1074/jbc.M109.018325 PMID:19433866

Carri, M. T., Valle, C., Bozzo, F., & Cozzolino, M. (2015). Oxidative stress and mitochondrial damage: Importance in non-SOD1 ALS. *Frontiers in Cellular Neuroscience*, *9*, 41. doi:10.3389/fncel.2015.00041 PMID:25741238

Chow, V. W., Mattson, M. P., Wong, P. C., & Gleichmann, M. (2010). An Overview of APP Processing Enzymes and Products. *Neuromol Med*, *12*(1), 1–12. doi:10.100712017-009-8104-z PMID:20232515

Colignon, B., Delaive, E., Dieu, M., Demazy, C., Muhovski, Y., Antoine, A., ... Mauro, S. (2019). Dual coordination of the SUMOylation and phosphorylation pathways during the response to heat stress in Solanum tuberosum. *Environmental and Experimental Botany*, *162*, 192–200. doi:10.1016/j.envexpbot.2019.02.024

Coppede, F., & Migliore, L. (2015). DNA damage in neurodegenerative diseases. *Mutation Research*, *776*, 84–97. doi:10.1016/j.mrfmmm.2014.11.010 PMID:26255941

Crippa, V., Sau, D., Rusmini, P., Boncoraglio, A., Onesto, E., Bolzoni, E., ... Poletti, A. (2010). The small heat shock protein B8 (HspB8) promotes autophagic removal of misfolded proteins involved in amyotrophic lateral sclerosis (ALS). *Human Molecular Genetics*, *19*(17), 3440–3456. doi:10.1093/hmg/ddq257 PMID:20570967

Cui, W. G., Wang, S. T., Wang, Z. P., Wang, Z. Y., Sun, C. L., & Zhang, Y. H. (2017). Inhibition of PTEN Attenuates Endoplasmic Reticulum Stress and Apoptosis via Activation of PI3K/AKT Pathway in Alzheimer's Disease. *Neurochemical Research*, *42*(11), 3052–3060. doi:10.100711064-017-2338-1 PMID:28819903

D'Amico, D., Sorrentino, V., & Auwerx, J. (2017). Cytosolic Proteostasis Networks of the Mitochondrial Stress Response. *Trends in Biochemical Sciences*, *42*(9), 712–725. doi:10.1016/j.tibs.2017.05.002 PMID:28579074

Das, T., & Eliezer, D. (2019). Membrane interactions of intrinsically disordered proteins: The example of alpha-synuclein. *Bba-Proteins Proteom*, *1867*(10), 879–889. doi:10.1016/j.bbapap.2019.05.001 PMID:31096049

Delenclos, M., Burgess, J. D., Lamprokostopoulou, A., Outeiro, T. F., Vekrellis, K., & McLean, P. J. (2019). Cellular models of alpha-synuclein toxicity and aggregation. *Journal of Neurochemistry*, *150*(5), 566–576. doi:10.1111/jnc.14806 PMID:31265132

Deng, H. X., Chen, W. J., Hong, S. T., Boycott, K. M., Gorrie, G. H., Siddique, N., ... Siddique, T. (2011). Mutations in UBQLN2 cause dominant X-linked juvenile and adult-onset ALS and ALS/dementia. *Nature*, *477*(7363), 211–U113. doi:10.1038/nature10353 PMID:21857683

DesJarlais, R., & Tummino, P. J. (2016). Role of Histone-Modifying Enzymes and Their Complexes in Regulation of Chromatin Biology. *Biochemistry-Us*, *55*(11), 1584–1599. doi:10.1021/acs.biochem.5b01210 PMID:26745824

Dols-Icardo, O., Garcia-Redondo, A., Rojas-Garcia, R., Borrego-Hernandez, D., Illan-Gala, I., Munoz-Blanco, J. L., ... Clarimón, J. (2018). Analysis of known amyotrophic lateral sclerosis and frontotemporal dementia genes reveals a substantial genetic burden in patients manifesting both diseases not carrying the C9orf72 expansion mutation. *J Neurol Neurosur Ps*, *89*(2), 162–168. doi:10.1136/jnnp-2017-316820 PMID:28889094

Dong, Z., & Cui, H.J. (2019). The Autophagy-Lysosomal Pathways and Their Emerging Roles in Modulating Proteostasis in Tumors. *Cells*, *8*.

Duennwald, M. L., & Lindquist, S. (2008). Impaired ERAD and ER stress are early and specific events in polyglutamine toxicity. *Genes & Development*, *22*(23), 3308–3319. doi:10.1101/gad.1673408 PMID:19015277

Fei, E., Jia, N., Yan, M., Ying, Z., Sun, Q., Wang, H., ... Wang, G. (2006). SUMO-1 modification increases human SOD1 stability and aggregation. *Biochemical and Biophysical Research Communications*, *347*(2), 406–412. doi:10.1016/j.bbrc.2006.06.092 PMID:16828461

Feligioni, M., Brambilla, E., Camassa, A., Sclip, A., Arnaboldi, A., Morelli, F., ... Borsello, T. (2011). Crosstalk between JNK and SUMO Signaling Pathways: deSUMOylation Is Protective against H2O2-Induced Cell Injury. *PLoS One*, *6*(12), 6. doi:10.1371/journal.pone.0028185 PMID:22164242

Ferro, F., Servais, S., Besson, P., Roger, S., Dumas, J. F., & Brisson, L. (2019). Autophagy and mitophagy in cancer metabolic remodelling. *Seminars in Cell & Developmental Biology*. PMID:31154012

Flores-Rodriguez, P., Ontiveros-Torres, M.A., Cardenas-Aguayo, M.C., Luna-Arias, J.P., Meraz-Rios, M.A., Viramontes-Pintos, A., … Floran-Garduno, B. (2015). The relationship between truncation and phosphorylation at the C-terminus of tau protein in the paired helical filaments of Alzheimer's disease. *Front Neurosci*, *9*.

Ford, L., Fioriti, L., & Kandel, E. R. (2019). Ubiquitination and SUMOylation of Amyloid and Amyloid-like Proteins in Health and Disease. *Current Issues in Molecular Biology*, *35*, 195–230. PMID:31422940

Forloni, G., Roiter, I., & Tagliavini, F. (2019). Clinical trials of prion disease therapeutics. *Current Opinion in Pharmacology*, *44*, 53–60. doi:10.1016/j.coph.2019.04.019 PMID:31108459

Frost, B., Gotz, J., & Feany, M. B. (2015). Connecting the dots between tau dysfunction and neurodegeneration. *Trends in Cell Biology*, *25*(1), 46–53. doi:10.1016/j.tcb.2014.07.005 PMID:25172552

Gao, J., Shao, K., Chen, X., Li, Z., Liu, Z., Yu, Z., ... Li, P. (2019). The involvement of post-translational modifications in cardiovascular pathologies: Focus on SUMOylation, neddylation, succinylation, and prenylation. *Journal of Molecular and Cellular Cardiology*, *138*, 49–58. doi:10.1016/j.yjmcc.2019.11.146 PMID:31751566

Glick, D., Barth, S., & Macleod, K. F. (2010). Autophagy: Cellular and molecular mechanisms. *The Journal of Pathology*, *221*(1), 3–12. doi:10.1002/path.2697 PMID:20225336

Goncalves, S., & Outeiro, T. F. (2013). Assessing the Subcellular Dynamics of Alpha-synuclein Using Photoactivation Microscopy. *Molecular Neurobiology, 47*(3), 1081–1092. doi:10.100712035-013-8406-x PMID:23389286

Goswami, P., Gupta, S., Biswas, J., Joshi, N., Swarnkar, S., Nath, C., & Singh, S. (2016). Endoplasmic Reticulum Stress Plays a Key Role in Rotenone-Induced Apoptotic Death of Neurons. *Molecular Neurobiology, 53*(1), 285–298. doi:10.100712035-014-9001-5 PMID:25428620

Goswami, P., Gupta, S., Biswas, J., Joshi, N., Swarnkar, S., Nath, C., & Singh, S. (2019). Correction to: Endoplasmic Reticulum Stress Plays a Key Role in Rotenone-Induced Apoptotic Death of Neurons. *Molecular Neurobiology, 56*(9), 6654–6655. doi:10.100712035-019-01674-x PMID:31230259

Guerra de Souza, A. C., Prediger, R. D., & Cimarosti, H. (2016). SUMO-regulated mitochondrial function in Parkinson's disease. *Journal of Neurochemistry, 137*(5), 673–686. doi:10.1111/jnc.13599 PMID:26932327

Gupta, S., Biswas, J., Gupta, P., Singh, A., Tiwari, S., Mishra, A., & Singh, S. (2019). Salubrinal attenuates nitric oxide mediated PERK:IRE1alpha: ATF-6 signaling and DNA damage in neuronal cells. *Neurochemistry International, 131*, 104581. doi:10.1016/j.neuint.2019.104581 PMID:31639405

Harding, R. J., & Tong, Y. F. (2018). Proteostasis in Huntington's disease: Disease mechanisms and therapeutic opportunities. *Acta Pharmacologica Sinica, 39*(5), 754–769. doi:10.1038/aps.2018.11 PMID:29620053

Hegarty, S. V., Sullivan, A. M., & O'Keeffe, G. W. (2016). The Epigenome as a therapeutic target for Parkinson's disease. *Neural Regeneration Research, 11*(11), 1735–1738. doi:10.4103/1673-5374.194803 PMID:28123403

Hetz, C., & Mollereau, B. (2014). Disturbance of endoplasmic reticulum proteostasis in neurodegenerative diseases. *Nature Reviews. Neuroscience, 15*(4), 233–249. doi:10.1038/nrn3689 PMID:24619348

Hipp, M. S., Park, S. H., & Hartl, F. U. (2014). Proteostasis impairment in protein-misfolding and -aggregation diseases. *Trends in Cell Biology, 24*(9), 506–514. doi:10.1016/j.tcb.2014.05.003 PMID:24946960

Hromadkova, L., & Ovsepian, S. V. (2019). Tau-Reactive Endogenous Antibodies: Origin, Functionality, and Implications for the Pathophysiology of Alzheimer's Disease. *Journal of Immunology Research, 2019*, 7406810. doi:10.1155/2019/7406810 PMID:31687413

Intihar, T. A., Martinez, E. A., & Gomez-Pastor, R. (2019). Mitochondrial Dysfunction in Huntington's Disease; Interplay Between HSF1, p53 and PGC-1alpha Transcription Factors. *Frontiers in Cellular Neuroscience, 13*, 103. doi:10.3389/fncel.2019.00103 PMID:30941017

Iurlaro, R., & Munoz-Pinedo, C. (2016). Cell death induced by endoplasmic reticulum stress. *The FEBS Journal, 283*(14), 2640–2652. doi:10.1111/febs.13598 PMID:26587781

Jackson, M. P., & Hewitt, E. W. (2016). Cellular proteostasis: Degradation of misfolded proteins by lysosomes. *Proteostasis, 60*, 173–180. PMID:27744333

Jiang, Y., Chadwick, S. R., & Lajoie, P. (2016). Endoplasmic reticulum stress: The cause and solution to Huntington's disease? *Brain Research*, *1648*, 650–657. doi:10.1016/j.brainres.2016.03.034 PMID:27040914

Kaliszewski, M., Kennedy, A. K., Blaes, S. L., Shaffer, R. S., Knott, A. B., Song, W., ... Bossy-Wetzel, E. (2016). SOD1 Lysine 123 Acetylation in the Adult Central Nervous System. *Frontiers in Cellular Neuroscience*, *10*, 287. doi:10.3389/fncel.2016.00287 PMID:28066183

Kane, L. A., Lazarou, M., Fogel, A. I., Li, Y., Yamano, K., Sarraf, S. A., ... Youle, R. J. (2014). PINK1 phosphorylates ubiquitin to activate Parkin E3 ubiquitin ligase activity. *The Journal of Cell Biology*, *205*(2), 143–153. doi:10.1083/jcb.201402104 PMID:24751536

Kaur, S. J., McKeown, S. R., & Rashid, S. (2016). Mutant SOD1 mediated pathogenesis of Amyotrophic Lateral Sclerosis. *Gene*, *577*(2), 109–118. doi:10.1016/j.gene.2015.11.049 PMID:26657039

Kazlauskaite, A., Martinez-Torres, R. J., Wilkie, S., Kumar, A., Peltier, J., Gonzalez, A., ... Muqit, M. M. K. (2015). Binding to serine 65-phosphorylated ubiquitin primes Parkin for optimal PINK1-dependent phosphorylation and activation. *EMBO Reports*, *16*(8), 939–954. doi:10.15252/embr.201540352 PMID:26116755

Khan, S. S., LaCroix, M., Boyle, G., Sherman, M. A., Brown, J. L., Amar, F., ... Lesne, S. E. (2018). Bidirectional modulation of Alzheimer phenotype by alpha-synuclein in mice and primary neurons. *Acta Neuropathologica*, *136*(4), 589–605. doi:10.100700401-018-1886-z PMID:29995210

Klemann, C. J. H. M., Visser, J. E., Van den Bosch, L., Martens, G. J. M., & Poelmans, G. (2018). Integrated molecular landscape of amyotrophic lateral sclerosis provides insights into disease etiology. *Brain Pathology (Zurich, Switzerland)*, *28*(2), 203–211. doi:10.1111/bpa.12485 PMID:28035716

Komatsu, M., Ueno, T., Waguri, S., Uchiyama, Y., Kominami, E., & Tanaka, K. (2007). Constitutive autophagy: Vital role in clearance of unfavorable proteins in neurons. *Cell Death and Differentiation*, *14*(5), 887–894. doi:10.1038j.cdd.4402120 PMID:17332773

Korecka, J. A., Talbot, R., Osborn, T. M., de Leeuw, S. M., Levy, S. A., Ferrari, E. J., ... Isacson, O. (2019). Neurite Collapse and Altered ER Ca2+ Control in Human Parkinson Disease Patient iPSC-Derived Neurons with LRRK2 G2019S Mutation. *Stem Cell Reports*, *12*(1), 29–41. doi:10.1016/j.stemcr.2018.11.021 PMID:30595548

Krumova, P., Meulmeester, E., Garrido, M., Tirard, M., Hsiao, H. H., Bossis, G., ... Weishaupt, J. H. (2011). Sumoylation inhibits alpha-synuclein aggregation and toxicity. *The Journal of Cell Biology*, *194*(1), 49–60. doi:10.1083/jcb.201010117 PMID:21746851

Kurtishi, A., Rosen, B., Patil, K. S., Alves, G. W., & Moller, S. G. (2019). Cellular Proteostasis in Neurodegeneration. *Molecular Neurobiology*, *56*(5), 3676–3689. doi:10.100712035-018-1334-z PMID:30182337

Labbadia, J., & Morimoto, R. I. (2015). The Biology of Proteostasis in Aging and Disease. *Annual Review of Biochemistry*, *84*(1), 435–464. doi:10.1146/annurev-biochem-060614-033955 PMID:25784053

Lee, J., Hyeon, S. J., Im, H., Ryu, H., Kim, Y., & Ryu, H. (2016). Astrocytes and Microglia as Non-cell Autonomous Players in the Pathogenesis of ALS. *Experimental Neurobiology*, *25*(5), 233–240. doi:10.5607/en.2016.25.5.233 PMID:27790057

Lesne, S., Koh, M. T., Kotilinek, L., Kayed, R., Glabe, C. G., Yang, A., ... Ashe, K. H. (2006). A specific amyloid-beta protein assembly in the brain impairs memory. *Nature*, *440*(7082), 352–357. doi:10.1038/nature04533 PMID:16541076

Liang, Q. L., Ouyang, X. S., Schneider, L., & Zhang, J. H. (2011). Reduction of mutant huntingtin accumulation and toxicity by lysosomal cathepsins D and B in neurons. *Molecular Neurodegeneration*, *6*(1), 6. doi:10.1186/1750-1326-6-37 PMID:21631942

Lin, Y. W. (2018). Structure and function of heme proteins regulated by diverse post-translational modifications. *Archives of Biochemistry and Biophysics*, *641*, 1–30. doi:10.1016/j.abb.2018.01.009 PMID:29407792

Liu, J., & Li, L. (2019). Targeting Autophagy for the Treatment of Alzheimer's Disease: Challenges and Opportunities. *Frontiers in Molecular Neuroscience*, *12*, 12. doi:10.3389/fnmol.2019.00203 PMID:31507373

Ludolph, A. C., & Brettschneider, J. (2015). TDP-43 in amyotrophic lateral sclerosis - is it a prion disease? *European Journal of Neurology*, *22*(5), 753–761. doi:10.1111/ene.12706 PMID:25846565

Mahdi, A. A., Rizvi, S. H., & Parveen, A. (2016). Role of Endoplasmic Reticulum Stress and Unfolded Protein Responses in Health and Diseases. *Indian Journal of Clinical Biochemistry*, *31*(2), 127–137. doi:10.100712291-015-0502-4 PMID:27069320

Maqbool, M., Mobashir, M., & Hoda, N. (2016). Pivotal role of glycogen synthase kinase-3: A therapeutic target for Alzheimer's disease. *European Journal of Medicinal Chemistry*, *107*, 63–81. doi:10.1016/j.ejmech.2015.10.018 PMID:26562543

Martin-Maestro, P., Sproul, A., Martinez, H., Paquet, D., Gerges, M., Noggle, S., & Starkov, A. A. (2019). Autophagy Induction by Bexarotene Promotes Mitophagy in Presenilin 1 Familial Alzheimer's Disease iPSC-Derived Neural Stem Cells. *Molecular Neurobiology*, *56*(12), 8220–8236. doi:10.100712035-019-01665-y PMID:31203573

McAdam, R. L., Morton, A., Gordon, S. L., Alterman, J. F., Khvorova, A., Cousin, M. A., & Smillie, K. J. (2019). Loss of huntingtin function slows synaptic vesicle endocytosis in striatal neurons from the htt(Q140/Q140) mouse model of Huntington's disease. *Neurobiology of Disease*, *134*, 104637. doi:10.1016/j.nbd.2019.104637 PMID:31614197

McAlary, L., Plotkin, S. S., & Cashman, N. R. (2019). Emerging Developments in Targeting Proteotoxicity in Neurodegenerative Diseases. *CNS Drugs*, *33*(9), 883–904. doi:10.100740263-019-00657-9 PMID:31414322

McFarland, M. A., Ellis, C. E., Markey, S. P., & Nussbaum, R. L. (2008). Proteomics Analysis Identifies Phosphorylation-dependent alpha-Synuclein Protein Interactions. *Molecular & Cellular Proteomics*, *7*(11), 2123–2137. doi:10.1074/mcp.M800116-MCP200 PMID:18614564

Miller, J., Arrasate, M., Shaby, B. A., Mitra, S., Masliah, E., & Finkbeiner, S. (2010). Quantitative Relationships between Huntingtin Levels, Polyglutamine Length, Inclusion Body Formation, and Neuronal Death Provide Novel Insight into Huntington's Disease Molecular Pathogenesis. *The Journal of Neuroscience: The Official Journal of the Society for Neuroscience, 30*(31), 10541–10550. doi:10.1523/JNEUROSCI.0146-10.2010 PMID:20685997

Miller, S., & Muqit, M. M. K. (2019). Therapeutic approaches to enhance PINK1/Parkin mediated mitophagy for the treatment of Parkinson's disease. *Neuroscience Letters, 705*, 7–13. doi:10.1016/j.neulet.2019.04.029 PMID:30995519

Mishra, A., Maheshwari, M., Chhangani, D., Fujimori-Tonou, N., Endo, F., Joshi, A.P., Jana, N.R., & Yamanaka, K. (2013). E6-AP association promotes SOD1 aggresomes degradation and suppresses toxicity. *Neurobiol Aging, 34*, 1311-1323.

Moehle, E. A., Shen, K., & Dillin, A. (2019). Mitochondrial proteostasis in the context of cellular and organismal health and aging. *The Journal of Biological Chemistry, 294*(14), 5396–5407. doi:10.1074/jbc.TM117.000893 PMID:29622680

Morimoto, R. I. (2019). Cell-Nonautonomous Regulation of Proteostasis in Aging and Disease. *Cold Spring Harbor Perspectives in Biology*, a034074. doi:10.1101/cshperspect.a034074 PMID:30962274

Myeku, N., & Duff, K. E. (2018). Targeting the 26S Proteasome To Protect Against Proteotoxic Diseases. *Trends in Molecular Medicine, 24*(1), 18–29. doi:10.1016/j.molmed.2017.11.006 PMID:29233753

Nakamura, S., & Yoshimori, T. (2017). New insights into autophagosome-lysosome fusion. *Journal of Cell Science, 130*(7), 1209–1216. doi:10.1242/jcs.196352 PMID:28302910

Nakamura, T., Cho, D. H., & Lipton, S. A. (2012). Redox regulation of protein misfolding, mitochondrial dysfunction, synaptic damage, and cell death in neurodegenerative diseases. *Experimental Neurology, 238*(1), 12–21. doi:10.1016/j.expneurol.2012.06.032 PMID:22771760

Navone, F., Genevini, P., & Borgese, N. (2015). Autophagy and Neurodegeneration: Insights from a Cultured Cell Model of ALS. *Cells-Basel, 4*(3), 354–386. doi:10.3390/cells4030354 PMID:26287246

Newcombe, E., Ramdzan, Y., Sethi, A., Lee, M., Loo, D., Graham, B., ... Hatters, D. (2015). Understanding the biology of Huntington's disease via the pathogenic huntingtin monomer. *Protein Science, 24*, 86–87.

Newton, T. M., Duce, J. A., & Bayle, E. D. (2019). The proteostasis network provides targets for neurodegeneration. *British Journal of Pharmacology, 176*(18), 3508–3514. doi:10.1111/bph.14643 PMID:30820936

Niu, Y. J., Nie, Z. W., Shin, K. T., Zhou, W., & Cui, X. S. (2019). PINK1 regulates mitochondrial morphology via promoting mitochondrial fission in porcine preimplantation embryos. *The FASEB Journal, 33*(7), 7882–7895. doi:10.1096/fj.201802473R PMID:30897005

Nomura, J., Hosoi, T., Kaneko, M., Ozawa, K., Nishi, A., & Nomura, Y. (2016). Neuroprotection by Endoplasmic Reticulum Stress-Induced HRD1 and Chaperones: Possible Therapeutic Targets for Alzheimer's and Parkinson's Disease. *Medical Sciences: Open Access Journal, 4*(3), 4. doi:10.3390/medsci4030014 PMID:29083378

Noormohammadi, A., Calculli, G., Gutierrez-Garcia, R., Khodakarami, A., Koyuncu, S., & Vilchez, D. (2018). Mechanisms of protein homeostasis (proteostasis) maintain stem cell identity in mammalian pluripotent stem cells. *Cellular and Molecular Life Sciences*, *75*(2), 275–290. doi:10.100700018-017-2602-1 PMID:28748323

O'Connor, T., Sadleir, K. R., Maus, E., Velliquette, R. A., Zhao, J., Cole, S. L., ... Vassar, R. (2008). Phosphorylation of the translation initiation factor eIF2alpha increases BACE1 levels and promotes amyloidogenesis. *Neuron*, *60*(6), 988–1009. doi:10.1016/j.neuron.2008.10.047 PMID:19109907

Ochalek, A., Mihalik, B., Avci, H. X., Chandrasekaran, A., Teglasi, A., Bock, I., ... Dinnyés, A. (2017). Neurons derived from sporadic Alzheimer's disease iPSCs reveal elevated TAU hyperphosphorylation, increased amyloid levels, and GSK3B activation. *Alzheimer's Research & Therapy*, *9*(1), 90. doi:10.118613195-017-0317-z PMID:29191219

Olfati, N., Shoeibi, A., & Litvan, I. (2019). Progress in the treatment of Parkinson-Plus syndromes. *Parkinsonism & Related Disorders*, *59*, 101–110. doi:10.1016/j.parkreldis.2018.10.006 PMID:30314846

Owen, I., & Shewmaker, F. (2019). The Role of Post-Translational Modifications in the Phase Transitions of Intrinsically Disordered Proteins. *International Journal of Molecular Sciences*, *20*(21), 20. doi:10.3390/ijms20215501 PMID:31694155

Patel, A., Lee, H. O., Jawerth, L., Maharana, S., Jahnel, M., Hein, M. Y., ... Alberti, S. (2015). A Liquid-to-Solid Phase Transition of the ALS Protein FUS Accelerated by Disease Mutation. *Cell*, *162*(5), 1066–1077. doi:10.1016/j.cell.2015.07.047 PMID:26317470

Pickrell, A. M., & Youle, R. J. (2015). The roles of PINK1, parkin, and mitochondrial fidelity in Parkinson's disease. *Neuron*, *85*(2), 257–273. doi:10.1016/j.neuron.2014.12.007 PMID:25611507

Plate, L., & Wiseman, R. L. (2017). Regulating Secretory Proteostasis through the Unfolded Protein Response: From Function to Therapy. *Trends in Cell Biology*, *27*(10), 722–737. doi:10.1016/j.tcb.2017.05.006 PMID:28647092

Polajnar, M., & Zerovnik, E. (2014). Impaired autophagy: A link between neurodegenerative and neuropsychiatric diseases. *Journal of Cellular and Molecular Medicine*, *18*(9), 1705–1711. doi:10.1111/jcmm.12349 PMID:25139375

Poplawski, T., Pytel, D., Dziadek, J., & Majsterek, I. (2019). Interplay between Redox Signaling, Oxidative Stress, and Unfolded Protein Response (UPR) in Pathogenesis of Human Diseases. *Oxidative Medicine and Cellular Longevity*, *2019*, 1–2. doi:10.1155/2019/6949347 PMID:31089415

Prasad, A., Bharathi, V., Sivalingam, V., Girdhar, A., & Patel, B. K. (2019). Molecular Mechanisms of TDP-43 Misfolding and Pathology in Amyotrophic Lateral Sclerosis. *Frontiers in Molecular Neuroscience*, *12*, 25. doi:10.3389/fnmol.2019.00025 PMID:30837838

Rahman, M. A., & Rhim, H. (2017). Therapeutic implication of autophagy in neurodegenerative diseases. *BMB Reports*, *50*(7), 345–354. doi:10.5483/BMBRep.2017.50.7.069 PMID:28454606

Rahman, M. M., Rumzhum, N. N., Hansbro, P. M., Morris, J. C., Clark, A. R., Verrills, N. M., & Ammit, A. J. (2016). Activating protein phosphatase 2A (PP2A) enhances tristetraprolin (TTP) anti-inflammatory function in A549 lung epithelial cells. *Cellular Signalling*, *28*(4), 325–334. doi:10.1016/j.cellsig.2016.01.009 PMID:26820662

Rai, S. N., Singh, B. K., Rathore, A. S., Zahra, W., Keswani, C., Birla, H., ... Singh, S. P. (2019). Quality Control in Huntington's Disease: A Therapeutic Target. *Neurotoxicity Research*, *36*(3), 612–626. doi:10.100712640-019-00087-x PMID:31297710

Rakhit, R., Cunningham, P., Furtos-Matei, A., Dahan, S., Qi, X. F., Crow, J. P., ... Chakrabartty, A. (2002). Oxidation-induced misfolding and aggregation of superoxide dismutase and its implications for amyotrophic lateral sclerosis. *The Journal of Biological Chemistry*, *277*(49), 47551–47556. doi:10.1074/jbc.M207356200 PMID:12356748

Rao, R. V., & Bredesen, D. E. (2004). Misfolded proteins, endoplasmic reticulum stress and neurodegeneration. *Current Opinion in Cell Biology*, *16*(6), 653–662. doi:10.1016/j.ceb.2004.09.012 PMID:15530777

Rhein, V., Song, X., Wiesner, A., Ittner, L. M., Baysang, G., Meier, F., ... Eckert, A. (2009). Amyloid-beta and tau synergistically impair the oxidative phosphorylation system in triple transgenic Alzheimer's disease mice. *Proceedings of the National Academy of Sciences of the United States of America*, *106*(47), 20057–20062. doi:10.1073/pnas.0905529106 PMID:19897719

Rodriguez-Nogales, C., Garbayo, E., Carmona-Abellan, M. M., Luquin, M. R., & Blanco-Prieto, M. J. (2016). Brain aging and Parkinson's disease: New therapeutic approaches using drug delivery systems. *Maturitas*, *84*, 25–31. doi:10.1016/j.maturitas.2015.11.009 PMID:26653838

Rosen, D. R., Siddique, T., Patterson, D., Figlewicz, D. A., Sapp, P., Hentati, A., ... Brown, R. H. Jr. (1993). Mutations in Cu/Zn superoxide dismutase gene are associated with familial amyotrophic lateral sclerosis. *Nature*, *362*(6415), 59–62. doi:10.1038/362059a0 PMID:8446170

Rosenblum, L. T., & Trotti, D. (2017). EAAT2 and the Molecular Signature of Amyotrophic Lateral Sclerosis. *Advances in Neurobiology*, *16*, 117–136. doi:10.1007/978-3-319-55769-4_6 PMID:28828608

Rubinsztein, D. C., Codogno, P., & Levine, B. (2012). Autophagy modulation as a potential therapeutic target for diverse diseases. *Nature Reviews. Drug Discovery*, *11*(9), 709–U784. doi:10.1038/nrd3802 PMID:22935804

Sau, D., De Biasi, S., Vitellaro-Zuccarello, L., Riso, P., Guarnieri, S., Porrini, M., ... Poletti, A. (2007). Mutation of SOD1 in ALS: A gain of a loss of function. *Human Molecular Genetics*, *16*(13), 1604–1618. doi:10.1093/hmg/ddm110 PMID:17504823

Saudou, F., & Humbert, S. (2016). The Biology of Huntingtin. *Neuron*, *89*(5), 910–926. doi:10.1016/j.neuron.2016.02.003 PMID:26938440

Schaser, A.J., Osterberg, V.R., Dent, S.E., Stackhouse, T.L., Wakeham, C.M., Boutros, S., … Luna, E. (2019). Alpha-synuclein is a DNA binding protein that modulates DNA repair with implications for Lewy body disorders. *Sci Rep*, *9*.

Shacham, T., Sharma, N., & Lederkremer, G. Z. (2019). Protein Misfolding and ER Stress in Huntington's Disease. *Frontiers in Molecular Biosciences*, *6*, 6. doi:10.3389/fmolb.2019.00020 PMID:31001537

Sharma, S., Sarathlal, K. C., & Taliyan, R. (2019). Epigenetics in Neurodegenerative Diseases: The Role of Histone Deacetylases. *Cns Neurol Disord-Dr*, *18*(1), 11–18. doi:10.2174/1871527317666181004155136 PMID:30289079

Sharma, S., & Taliyan, R. (2015). *Targeting Histone Deacetylases: A Novel Approach in Parkinson's Disease*. Parkinsons Dis.

Shefa, U., Jeong, N. Y., Song, I. O., Chung, H. J., Kim, D., Jung, J., & Huh, Y. (2019). Mitophagy links oxidative stress conditions and neurodegenerative diseases. *Neural Regeneration Research*, *14*(5), 749–756. doi:10.4103/1673-5374.249218 PMID:30688256

Shiwaku, H., & Okazawa, H. (2015). Impaired DNA damage repair as a common feature of neurodegenerative diseases and psychiatric disorders. *Current Molecular Medicine*, *15*(2), 119–128. doi:10.2174/1566524015666150303002556 PMID:25732151

Silberman, D. M. (2018). Metabolism, neurodegeneration and epigenetics: Emerging role of Sirtuins. *Neural Regeneration Research*, *13*(3), 417–418. doi:10.4103/1673-5374.228719 PMID:29623921

Sirangelo, I., & Iannuzzi, C. (2017). The Role of Metal Binding in the Amyotrophic Lateral Sclerosis-Related Aggregation of Copper-Zinc Superoxide Dismutase. *Molecules (Basel, Switzerland)*, *22*(9), 22. doi:10.3390/molecules22091429 PMID:28850080

Song, P. P., Li, S. S., Wu, H., Gao, R. Z., Rao, G. H., Wang, D. M., ... Chen, Q. (2016). Parkin promotes proteasomal degradation of p62: Implication of selective vulnerability of neuronal cells in the pathogenesis of Parkinson's disease. *Protein & Cell*, *7*(2), 114–129. doi:10.100713238-015-0230-9 PMID:26746706

Stastna, M., & Van Eyk, J. E. (2015). Posttranslational modifications of lysine and evolving role in heart pathologies-Recent developments. *Proteomics*, *15*(5-6), 1164–1180. doi:10.1002/pmic.201400312 PMID:25430483

Taipale, M. (2019). Disruption of protein function by pathogenic mutations: Common and uncommon mechanisms (1). *Biochemistry and Cell Biology*, *97*(1), 46–57. doi:10.1139/bcb-2018-0007 PMID:29693415

Takahashi, K., Foster, J. B., & Lin, C. L. G. (2015). Glutamate transporter EAAT2: Regulation, function, and potential as a therapeutic target for neurological and psychiatric disease. *Cellular and Molecular Life Sciences*, *72*(18), 3489–3506. doi:10.100700018-015-1937-8 PMID:26033496

Tripathi, M. K., Rajput, C., Mishra, S., Rasheed, M. S. U., & Singh, M. P. (2019). Malfunctioning of Chaperone-Mediated Autophagy in Parkinson's Disease: Feats, Constraints, and Flaws of Modulators. *Neurotoxicity Research*, *35*(1), 260–270. doi:10.100712640-018-9917-z PMID:29949106

Uddin, M. S., Al Mamun, A., Labu, Z. K., Hidalgo-Lanussa, O., Barreto, G. E., & Ashraf, G. M. (2019a). Autophagic dysfunction in Alzheimer's disease: Cellular and molecular mechanistic approaches to halt Alzheimer's pathogenesis. *Journal of Cellular Physiology*, *234*(6), 8094–8112. doi:10.1002/jcp.27588 PMID:30362531

Uddin, M. S., Mamun, A. A., Jakaria, M., Thangapandiyan, S., Ahmad, J., Rahman, M. A., ... Aleya, L. (2019b). Emerging promise of sulforaphane-mediated Nrf2 signaling cascade against neurological disorders. *The Science of the Total Environment*, 135624. PMID:31784171

Valionyte, E., Yang, Y., Roberts, S. L., Kelly, J., Lu, B., & Luo, S. (2019). Lowering mutant huntingtin levels and toxicity: Autophagy-endolysosome pathways in Huntington's disease. *Journal of Molecular Biology*. doi:10.1016/j.jmb.2019.11.012 PMID:31786267

Veeresh, P., Kaur, H., Sarmah, D., Mounica, L., Verma, G., Kotian, V., ... Bhattacharya, P. (2019). Endoplasmic reticulum-mitochondria crosstalk: From junction to function across neurological disorders. *Annals of the New York Academy of Sciences*, *1457*(1), 41–60. doi:10.1111/nyas.14212 PMID:31460675

Venkatramani, A., & Panda, D. (2019). Regulation of neuronal microtubule dynamics by tau: Implications for tauopathies. *International Journal of Biological Macromolecules*, *133*, 473–483. doi:10.1016/j.ijbiomac.2019.04.120 PMID:31004638

Wang, M., & Kaufman, R. J. (2016). Protein misfolding in the endoplasmic reticulum as a conduit to human disease. *Nature*, *529*(7586), 326–335. doi:10.1038/nature17041 PMID:26791723

Wanker, E. E., Ast, A., Schindler, F., Trepte, P., & Schnoegl, S. (2019). The pathobiology of perturbed mutant huntingtin protein-protein interactions in Huntington's disease. *Journal of Neurochemistry*, *151*(4), 507–519. doi:10.1111/jnc.14853 PMID:31418858

Waxman, E. A., & Giasson, B. I. (2011). Characterization of kinases involved in the phosphorylation of aggregated alpha-synuclein. *Journal of Neuroscience Research*, *89*(2), 231–247. doi:10.1002/jnr.22537 PMID:21162130

Westergard, T., Jensen, B. K., Wen, X., Cai, J., Kropf, E., Iacovitti, L., ... Trotti, D. (2016). Cell-to-Cell Transmission of Dipeptide Repeat Proteins Linked to C9orf72-ALS/FTD. *Cell Reports*, *17*(3), 645–652. doi:10.1016/j.celrep.2016.09.032 PMID:27732842

Winklhofer, K. F., Tatzelt, J., & Haass, C. (2008). The two faces of protein misfolding: Gain- and loss-of-function in neurodegenerative diseases. *The EMBO Journal*, *27*(2), 336–349. doi:10.1038j.emboj.7601930 PMID:18216876

Xu, K., & Zhu, X. P. (2012). Endoplasmic reticulum stress and prion diseases. *Reviews in the Neurosciences*, *23*(1), 79–84. doi:10.1515/rns.2011.062 PMID:22718614

Xu, M. M., Ryan, P., Rudrawar, S., Quinn, R. J., Zhang, H. Y., & Mellick, G. D. (2019). Advances in the development of imaging probes and aggregation inhibitors for alpha-synuclein. *Acta Pharmacologica Sinica*. doi:10.103841401-019-0304-y PMID:31586134

Yoon, S. O., Park, D. J., Ryu, J. C., Ozer, H. G., Tep, C., Shin, Y. J., ... Huang, K. (2012). JNK3 perpetuates metabolic stress induced by Abeta peptides. *Neuron*, *75*(5), 824–837. doi:10.1016/j.neuron.2012.06.024 PMID:22958823

You, Y., & Ikezu, T. (2019). Emerging roles of extracellular vesicles in neurodegenerative disorders. *Neurobiology of Disease*, *130*, 104512. doi:10.1016/j.nbd.2019.104512 PMID:31229685

Zarei, S., Carr, K., Reiley, L., Diaz, K., Guerra, O., Altamirano, P. F., ... Chinea, A. (2015). A comprehensive review of amyotrophic lateral sclerosis. *Surgical Neurology International*, *6*(1), 171. doi:10.4103/2152-7806.169561 PMID:26629397

Zhang, Y., Chen, X., Zhao, Y. F., Ponnusamy, M., & Liu, Y. (2017). The role of ubiquitin proteasomal system and autophagy-lysosome pathway in Alzheimer's disease. *Reviews in the Neurosciences*, *28*(8), 861–868. doi:10.1515/revneuro-2017-0013 PMID:28704199

Zhang, Y., Tian, Q., Zhang, Q., Zhou, X., Liu, S., & Wang, J. Z. (2009). Hyperphosphorylation of microtubule-associated tau protein plays dual role in neurodegeneration and neuroprotection. *Pathophysiology*, *16*(4), 311–316. doi:10.1016/j.pathophys.2009.02.003 PMID:19410438

Zheng, Q. Y., Huang, T., Zhang, L. S., Zhou, Y., Luo, H., Xu, H. X., & Wang, X. (2016). Dysregulation of Ubiquitin-Proteasome System in Neurodegenerative Diseases. *Frontiers in Aging Neuroscience*, *8*, 8. doi:10.3389/fnagi.2016.00303 PMID:28018215

Zheng, Z., Shang, Y., Tao, J., Zhang, J., & Sha, B. (2019). Endoplasmic Reticulum Stress Signaling Pathways: Activation and Diseases. *Current Protein & Peptide Science*, *20*(9), 935–943. doi:10.2174/1389203720666190621103145 PMID:31223084

Zientara-Rytter, K., & Subramani, S. (2019). The Roles of Ubiquitin-Binding Protein Shuttles in the Degradative Fate of Ubiquitinated Proteins in the Ubiquitin-Proteasome System and Autophagy. *Cells*, *8*.

Zugel, M., Wehrstein, F., Qiu, S., Diel, P., Steinacker, M., & Schumann, U. (2019). Moderate intensity continuous training reverses the detrimental effects of ovariectomy on RyR1 phosphorylation in rat skeletal muscle. *Molecular and Cellular Endocrinology*, *481*, 1–7. doi:10.1016/j.mce.2018.11.003 PMID:30465874

Section 2

Protein Quality Control in Progressive Neurodegenerative Disorders

Chapter 7
ER Stress Signaling in Alzheimer's Disease:
Molecular Mechanisms and Therapeutic Implications

Md. Motiar Rahman
Shenzhen Institute of Advanced Technology (SIAT), Chinese Academy of Sciences (CAS), China

Looniva Shrestha
Shenzhen Institute of Advanced Technology (SIAT), Chinese Academy of Sciences (CAS), China

Mst Ara Gulshan
Graduate School of Innovative Life Science, University of Toyama, Japan

ABSTRACT

Alzheimer's disease (AD) is the most common etiology of dementia amongst aged individuals and a principal public health-related abnormality. It is considered as a multifactorial disorder, with no particular origin identified, and also some modifiable, as well as non-modifiable threats are correlated with its progression and development. The endoplasmic reticulum (ER) stress response is considered as a key process in the pathogenesis of AD. In this chapter, the authors present a summary of related transmembrane kinase proteins responsible for the onset of AD as well as show the interrelationship between ER stress and AD. Finally, the authors demonstrate the therapeutics intervention for AD diagnosis by highlighting the current practices to advance novel therapies.

INTRODUCTION

Alzheimer's disease (AD) is an extremely severe or irreversible, and radical disease of the brain that progressively devastates memory as well as thinking abilities and, ultimately, the capability to perform the easiest works (Mann, 1996; Norfray and Provenzale, 2004; Strassnig and Ganguli, 2005). It is known that AD is the utmost basis for developing dementia amongst elderly peoples (Korolev, 2014). AD is

DOI: 10.4018/978-1-7998-1317-0.ch007

currently known as the 5th prominent cause of mortality in the US among 65 years or elderly individuals (Heron, 2013), and around $200 billion are exhausted every year on uninterrupted care of patients with progressive dementia. Worldwide, it is projected that about 35 million of populations possess either the symptoms of AD or any forms of dementia, and around 65 million are anticipated to develop dementia by 2030 and 115 million by 2050, respectively (Prince et al., 2013).

AD was coined after the name of Dr. Alois Alzheimer, a German psychiatrist, and neurologist. In 1906, he observed alterations in the tissue of the brain of a woman who had just died of abnormal brain disorder along with indications involved memory damage, vocal complications, and unusual manner (Alzheimer, 1907). Dr. Alzheimer then investigated her brain and uncovered several abnormal clumps (more familiar as amyloid plaques) and tangled bundles of fibers (familiar as tau or neurofibrillary tangles), which are still considered as some of the prime features of AD (Korolev, 2014). Besides, impairment of contacts between nerve cells observes in the AD brain. Generally, neurons pass a signal between various portions of the brain, and conversely, from the brain tissues to various organs of the body. Several other intricated brain alterations are assumed to govern a substantial function in AD, too. This loss primarily looks to develop in the brain hippocampus, responsible for growing memories. Since neurons die, other portions of the brain are agitated (Devanand et al., 2007; Jack et al., 1997). At the last step of AD, destruction is extensive, and brain tissue has disappeared drastically.

Endoplasmic reticulum (ER) is the major component for efficient protein folding and quality control that are prerequisites to support cellular biochemical reactions. But, ER stress and UPR are indulged in the incidence of synaptic dysfunction in prion-related abnormalities and AD. In normal conditions, ER maintains protein maturation, calcium balance and expression of certain altered proteins or vesicles transport (ER-Golgi apparatus). However, under ER stress, cells stimulate a passive response, the unfolded protein response (UPR), are responsible to enhance protein folding ability, in addition to activating the capacity of quality control as well as protein degeneration pathway to diminish the unfolded protein (UP) load (Walter and Ron, 2011). Under extreme ER stress conditions, the UPR turns its pathway toward apoptosis by initiating intricate pro-apoptotic pathways (Urra et al., 2013).

The UPR is controlled by several stress sensors present at the envelope of ER including inositol-requiring enzyme 1 (IRE1), protein kinase RNA-like ER kinase (PERK) and activating transcription factor 6 (ATF6). These sensors maintain a transcription factor (TF) which mediates a subclass of partly overlapping target genes expression involving apoptosis or stress adaptation. The endonuclease activity of IRE1 mediates the abnormal splicing of the mRNA that encodes the transcription factor X-box binding protein (XBP1) and activates those genes linked to folding and degradation of protein, biosynthesis of lipid, etc. (Acosta-Alvear et al., 2007; Lee et al., 2003). Moreover, IRE1 induces the breakdown of selective mRNAs through RIDD (IRE1-dependent decay) (Hollien et al., 2009), and initiates several kinases, which includes the apoptosis signal-regulating kinase 1 (ASK1) and JUN amino-terminal kinase (JNK) (Urano et al., 2000). Under ER dysfunction, ATF6 is degraded by protease after translocation to the Golgi body and liberates a cytosolic part which translocates to the cell nucleus and maintains as a UPR transcription factor (Ron and Walter, 2007). Soluble ATF6 can complex with XBP1 to yield heterodimers and stimulates certain gene expression (Shoulders et al., 2013). Stimulation of PERK directly phosphorylates the ubiquitous eukaryotic translation initiation factor 2α (eIF2α) to promptly inhibit translation and diminish the protein overload in the ER lumen. This pathway further mediates ATF4 translation, an important transcription factor that maintains the expression of gene clusters included in autophagy, apoptosis, metabolism of amino acid, and antioxidant reactions (Walter and Ron, 2011). This chapter presents a concise pathway of UPR-mediated AD pathogenesis followed by the interrelation between ER stress and AD. Finally, the authors give an overview of how to control AD with special emphasize on UPR-mediated therapeutics intervention.

ER AND ITS BIOLOGICAL FUNCTION

ER is a kind of intracellular eukaryotic organelle and that creates a collective network of the compacted, membrane-bound sac or a tube-like structure called cisternae. This organelle shares a part of its membranes with the outer part of the nuclear membrane. The outer membrane of the rough ER (also called cytosolic membrane) is linked with ribosomes, the protein synthetic machinery. ER involves in many biological roles including the folding and quality control of protein in cisternae and the passage of the newly formed proteins into vesicles to the Golgi body. ER is also included in the maintenance of Ca^{2+} balance, and cholesterol biosynthesis in the cell.

From the rough ER, only the perfectly folded proteins biomolecules are translocated to the Golgi body. The UPs that are not folded properly as the 3D (three-dimensional) network, are stuck in the ER lumen since they achieve their active structures. If these proteins can't achieve the final tertiary structure, they are returned back to the cytoplasm and undergo to ubiquitination as well as proteasomal degradation by ER-associated degradation (ERAD) (Smith et al., 2011; Walter and Ron, 2011). In this pathway, ATP provides energy to identify an undesirable protein with a ubiquitin chain and design it for demolition. Finally, the protein is subjected to proteasomal-dependent fragmentation as shown in Figure 1.

ER STRESS AND ALZHEIMER'S DISEASE

Neurons of the human brain are susceptible to various hereditary and environmental factors that disturb the functional stability of the ER via the deposition of extended proteins and disorders in redox and calcium equilibriums. Hence, it is usual that various reports have reported that ER dysfunction is existing in numerous forms of neurodegenerations (Lindholm et al., 2006; Paschen and Mengesdorf, 2005 and Scheper and Hoozemans, 2009). Indication of stimulated unfolded protein response pathway has been identified in Parkinson's, Alzheimer's, Huntington's, and also in amyotrophic lateral sclerosis (ALS) diseases (Hoozemans et al., 2009; Kanekura et al., 2009; Lindholm et al., 2006; Matus et al., 2008; Paschen and Mengesdorf, 2005 and Scheper and Hoozemans, 2009). In addition, cerebral ischemia can stimulate the unfolded protein response, even though a simultaneous drastic reduction in protein translation obviously reduces the UPR levels (DeGracia and Montie, 2004). Viral toxicities such as the Borna virus can stimulate obvious ER dysfunction in the brain hippocampus and consequently induce UPR cascade (Williams and Lipkin, 2006). In neurons, the cisternae and tubules of ER can extrude from the nuclear envelope into dendritic spines and dendrons or dendrites, and also near the axons so far presynaptic ends. This indicates that neuronal ER is exceptionally a unique organ having functionally diverse subcompartments (Bánhegyi et al., 2008; Verkhratsky, 2005). For instance, Murakami group (2007) shown that ER stress response might be restricted to dendrons. This heterogeneous neuronal ER linkage might be linked to the damage in synapsis as well as axons (Raff et al., 2002), predominantly in the incident of redox-based abnormalities (Bánhegyi et al., 2008). Kudo group (2008) reported that the expression of a biochemical stimulator of binding immunoglobulin protein (BiP) can inhibit neuronal decease in both laboratories as well as *in vivo* settings, accentuating a function for ER dysfunction in neural cell death.

Figure 1. The normal physiological function of the endoplasmic reticulum.
Properly folded proteins are translocated to the Golgi complex. The UPs are transported back to the cytosol, undergoes ubiquitination and proteolytic degradation. Redrawn with little modification from the reference (Jie et al., 2015).

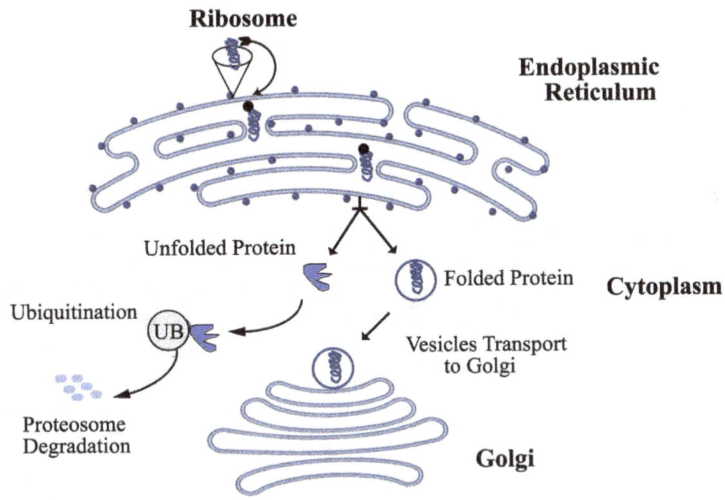

UPR, ER STRESS AND ALZHEIMER'S DISEASE

The UPR is a cell-mediated stress signaling connected to the ER dysfunction/stress. This response is stimulated in reply to the deposition of misfolded or unfolded proteins in the ER lumen. In this regard, the UPR has established three objectives:

- 1. Primarily to reestablish the usual task of the cell by inhibiting translation of protein
- 2. Destroying misfolded proteins
- 3. Finally, triggering the related-signaling that induce to upturn the fabrication of molecular chaperones indulged in the folding process of protein.

When these aims are obscure within a definite period of time or degeneration is delayed, the UPR undergoes towards apoptosis (Hetz and Papa, 2018). Continued UPR overactivation has been associated with prion and neurodegenerative disorders, and blocking the UPR can act as a therapeutic trial for those disorders. Ailments are responsive to UPR blockage, for instance, Creutzfeldt-Jakob disease, AD, and Huntington's disease (Moreno et al., 2013).

PATHWAY OF UPR-MEDIATED ALZHEIMER'S PATHOGENESIS

UPR belongs to a type of self-defense system induced by ER dysfunction in the event of being injured from several biological or physical insults. These responses comprise the increase of molecular chaperone, attenuation of translation, and ERAD (ER-associated protein degradation) stimulation in eukaryotes. Immunoglobulin binding protein binding immunoglobulin protein (BiP) is one of the instances of ER-resident chaperones that is believed to inhibit the emission of partially accumulated immunoglobulin (Gerakis and Hetz, 2018; Haas and Wabl, 1983; Santos and Ferreira, 2018). PERK, IRE1, and ATF6 serve as transducers of the transmembrane signal, which act as UPR sensors (Figure 2).

Figure 2. Role of UPR in Alzheimer's disease pathogenesis.
GRP78 detaches from PERK, IRE1, and ATF6 during ER stress and help them to become active. PERK induces phosphorylation of eIF2α that short-lived weakens protein synthesis in order to diminish in protein packing into the endoplasmic reticulum. But, phosphorylation of eIF2α particularly translates ATF4 and mediates nuclear import of Nrf2. IRE1 mediates a frameshift of XBP-1 protein, by inducing the splicing of XBP-1 mRNA, and its nuclear import maintains UPR target genes. COPII vesicle mediates stimulated ATF6 translocation to the Golgi body where it is sliced by S1P and S2P, and the cytoplasmic tail of ATF6 acts as a TF for the regulation UPR target genes. Redrawn with little modification from the reference (Jie et al., 2015).

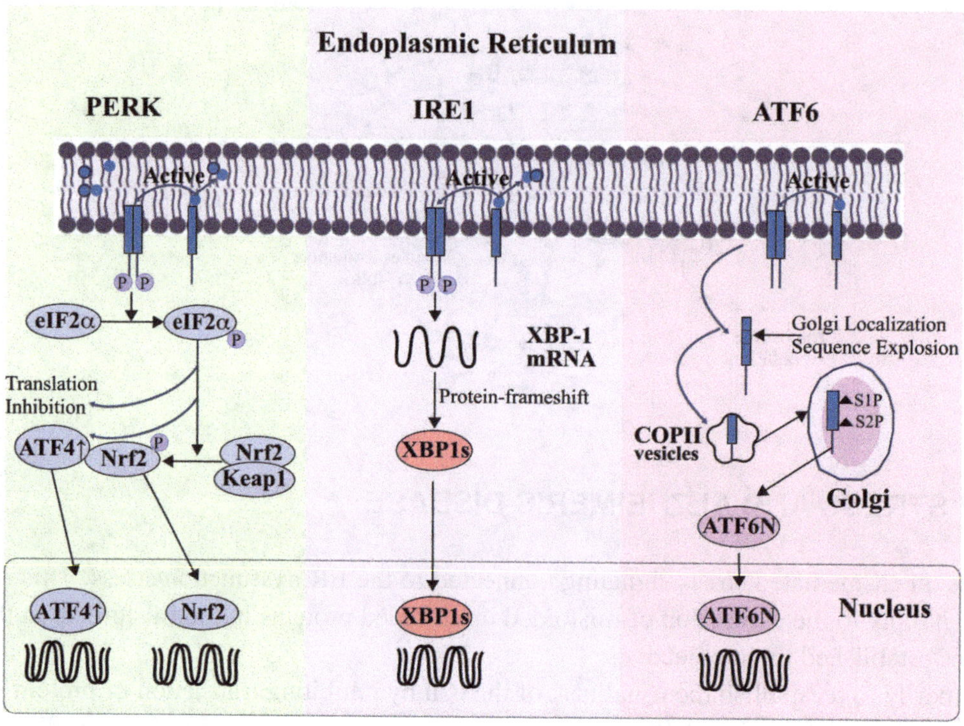

PERK

PERK belongs to a form of stress kinase in ER that governs an essential function in UPR control. Generally, PERK is inactivated by attaching the GRP78 protein and afterward, PERK and GRP78 form a composite under typical circumstances (Sanderson et al., 2010). On the other hand, in stressed ER cells, a rise of extended proteins in the lumen of ER dissociates GRP78 away from PERK and thus stimulates PERK to form oligomerization as well as self-phosphorylation.

While PERK is initiated, it triggers phosphorylation of α-subunit of eukaryotic eIF2α (translation initiation factor-2) at Ser51 (Marciniak et al., 2006). In this approach, PERK assists to decrease the protein fluctuation that is directed to enter the previously stressed lumen of the endoplasmic reticulum to relieve the stress of the ER (Walter and Ron, 2011). Since the proteins are hindered universally, an activating transcription factor 4 (ATF4) is skipped. The desired genes of activating transcription factor 4 include transcription factor C/EBP homologous protein (CHOP), activating transcription factor 3 (ATF3), DNA damage-inducible 34 (GADD34), and growth arrest.

CHOP is a well-known competitor in the activation of an organ or apoptotic regeneration that triggers the transcription of various genes, which in turn may enhance cell death. But, in highly ER stress, CHOP upregulation enhances the level of stress and therefore prompts apoptosis. Nonetheless, the opposite result has also been reported by another milder model. This signifies that the level and period of the ER stress should be stressed; because the deleterious or beneficial result depends on them (Roussel et al., 2013). GADD34 (DNA damage-inducible 34) protein attaches to the active site of protein phosphatase 1 (PP1c) and negatively stimulates PERK function by eIF2α dephosphorylation, and thus brings about the rescue of translation just after the early inhibition of protein synthesis (Park and Ozcan, 2013). ATF3 belongs to a member of the ATF/CREB subfamily of basic-region leucine zipper (bZIP) proteins that helps to mediate CHOP and GADD34 expression.

An alternative type of effector protein, which being PERK mediated phosphorylation, is nuclear factor erythroid2-related factor 2 (Nrf2), a constituent of Nrf2/Keap1 complex. This type of phosphorylation dissociates Nrf2/Keap1 complexes, and thus permits Nrf2 nuclear translocation. Furthermore, the nuclear import of Nrf2 is not reliant on the phosphorylation of eIF2α. Ideally, targeted removal of Nrf2 decreases the survival of cells after ER stress. These outcomes propose that the roles of Nrf2 might be a mediator of the PERK cell survival pathway (Cullinan et al., 2003).

IRE1

IRE1 also belongs to a group of type I transmembrane kinases like PERK. IRE1 has two subtypes (IRE1α and IRE1β) in mammals. The former type is known as ubiquitous, whilst IRE1β is limitedly expressed in gut cells. Moreover, IRE1α knockout mice reveal initial embryonic lethality, however; IRE1β knockout mice exhibit phenotypic normally (Iwawaki et al., 2009; Tsuru et al., 2013). Nonetheless, both are crucial for the UPR initiation. IRE1-bound GRP78 becomes freed from IRE1, during the period of ER stress. When this event happens, the assembly of IRE1 lets it bind with unfolded proteins, IRE1 becomes functional and defends cells from additional injury. In contrast, XBP1 (X-box binding protein 1) mRNA splicing could be prompted by IRE1α while ER dysfunction happens that brings about the deletion of a 26nt intron from XBP1 mRNA as well as a protein-frameshift which makes the active structure of this protein (Calfon et al., 2002). It would be broadly recognized that IRE1α cascade was excluded during irreversible ER dysfunction to stimulate cell-death (Chen and Brandizzi, 2013). However, numerous reports reveal that IRE1α insistently regulates the power of proper protein folding as well as guides UPR signaling (Han et al., 2009; Upton et al., 2012). The latest studies demonstrate that the role of IRE1α on the biogenesis of miRNA is uninterrupted and that it agonizes DICER-mediated conventional pathway to continue a translational inhibition to access into cell apoptosis (Upton et al., 2012).

ATF6

This is also a transmembrane protein similar to the former two kinases that shows no activity in the normal tissues for its binding to binding immunoglobulin protein. Under ER dysfunction, however; dissociation of BiP allows Golgi localization sequence sites of ATF6 explosion as well as transformation to the Golgi through COP II vesicles, where ATF6 subjected to cleave by two diverse proteases, known as site-1 and site-2 proteases, to form an N-terminal cytosolic fragment, ATF6(N) that enters to the cell nucleus to stimulate UPR-responsive genes transcription (Malhotra and Kaufman, 2007). But, it is not clear if the dissociation of BiP is only the mediator for the translocation of ATF6. The S1P (serine protease site-1 protease) triggers the cleavage of ATF6 in the lumen and the N-terminal site cleavage is then triggered by the S2P (metalloprotease site-2 protease) (Ye et al., 2000). However, ATF6 cleavage does not appear while cells are devoid of sterols, or while the cleavage is reliant on the SREBP (sterol-responsive element-binding protein) cleavage activating protein (SCAP) that guides SREBP to the Golgi for proteolysis (Patil and Walter, 2001).

Numerous analyses show that deposition of intracellular β−amyloid (Aβ) as well as phosphorylated tau, along with the agitation of calcium balance, govern a significant function in the Alzheimer's disease pathogenesis (LaFerla, 2007; Mattson and Chan, 2003; Selkoe, 2001). Newly, Hoozemans et al., (2009) reported the alterations in neuronal unfolded protein response in the patient's brains with AD in detail. They investigated that immunohistochemical analysis of stimulated UPR kinases, pPERK, peIF2α, and pIRE1α, was raised in the brain hippocampus of AD, particularly in neurons possessing granulovacuolar disintegration. Remarkably, staining of pPERK level was superior in those neurons which exhibited dispersed staining for a protein of phosphorylated tau. This staining, however; was less eminent compared to those neurons possessing neurofibrillary tangles and more specifically those tangles were not stained with pPERK. In the brain hippocampus, the CA1, CA4, and subiculum sections consist of plenty of pPERK (+ve) neurons. Unfolded protein response markers revealed a granulated staining array which was not colocalized with p62 or ubiquitin, representing that they did not form aggresomes. Neurons with pPERK (+ve) also exhibited ample staining for β−glycogen synthase kinase-3β (GSK-3) which is considered as a significant remark since it designated that endoplasmic reticulum stress can trigger GSK-3β expression, an eminent tau kinase, and by the way, it can increase neurofibrillary tangle deposition (Resende et al., 2008; Takashima, 2006). Unterberger et al., (2006) similarly showed that the initiations of PERK, eIF2α, and p38 MAPK are interrelated with the incidence of unusual tau protein in AD neurons. These findings have reported the intimate association between the tau pathology and ER dysfunction in neurons. Furthermore, they showed that unfolded protein response (+ve) staining is identified in neurons but not in glial cells.

Many data have explained demonstrations of ER dysfunction in samples of postmortem brain from AD peoples; repeatedly, some articles describe the existence of ER stress in *vitro* as well as animal model experiments. As an example, in human tests, BiP intensities enhanced in those neurons that coupled with amyloid amasses in AD patient's brain (Hoozemans et al., 2006). Also, phosphorylation of PERK and eIF2α was improved in brains of AD (Nijholt et al., 2011; O'Connor et al., 2008), and CHOP is overexpressed in the temporal cortex of AD patient's brains (Lee et al., 2010). There is also existence of XBP1 mRNA splicing in hippocampal tissue and temporal cortex of AD patients (Lee et al., 2010). Besides, heat shock protein (HSP72) was enhanced neighboring neurofibrillary tangles and neuritic plaques from AD neurons (Hamos et al., 1991). In both laboratory and animal models, using gene knockout or drugs as well as gene silencing model also reported the link between endoplasmic reticulum dysfunction and AD.

LINKS BETWEEN ER STRESS AND ALZHEIMER'S DISEASE

ER Stress and Aβ

Aβ belongs to the major constituent of extracellular senile plaques that synthesized from successive degradations of class I APP (transmembrane amyloid precursor protein) by β- and γ-secretases and its neurotoxicity govern an important function in the progression of AD. It stimulates disintegration and apoptosis of neurons that closely interrelate to the cognitive loss of AD individuals. A novel postulation proposed that failure of AD memory is triggered by tiny soluble Aβ proteins and toxins which target as well as to interrupt specific synapses. Currently, this hypothesis is buttressed by large body researches, with more than 1,000 articles focusing on the oligomer pustulosis (Ferreira and Klein, 2011).

Under standard conditions, GRP78/BiP could complex with APP to prevent Aβ deposition. However, while ER stress behaves unusual, such as mutation of presenilin 1 (*PSEN1*) and abnormal splicing of presenilin 2 (*PSEN2*), molecular chaperone expression will be halted triggering to the elevated deposition of β-amyloid as well as the susceptibility to ER dysfunction (Sato et al., 2001). APP thought to be translocated to the Golgi apparatus after glycosylation at its amino-terminal and further proceed with its C-terminus glycosylation. Due to lack of IREl in cells, this procedure will be halted inducing unfolded as well as an unusual breakdown of APP in the ER and thus, increase the concentration of Aβ. However, it has been anticipated that β-amyloid can openly trigger ER stress responses and cell death. Moreover, ER calcium discharge which in turn included in oligomer-mediated GSK-3β initiation and tau phosphorylation can trigger AD (Resende et al., 2008). Furthermore, Aβ-mediated continued initiation of the ER dysfunction that can be identified by the momentous elevations in the intensities of ATF4, active ATF6α, spliced XBP1, non-spliced XBP1, and the ER chaperone GRP78/BiP in cultivated cells can instigate apoptosis of brain endothelial cell that amasses in the cerebral vessels in various AD individuals and modified mice model. Elaborately, the development of rat brain endothelial cells incubated with β-amyloid exhibit substantial decline in cell viability as well as lifespan but the upturn in the number of dead cells. Also, resulting from incubation with β-amyloid, the Ca^{2+} balance outside or in the ER might be diminished. More calcium efflux from the ER to the cytoplasm, while before initiation of cell apoptosis, also considerably influences the pathological progression of AD (Fonseca et al., 2013). Concerning programmed cell death, Aβ1-42 stimulates caspase-12 in critical neurons via calpain stimulation and caspase-12 knockout neurons are weakly susceptible to Aβ-mediated apoptosis (Takuma et al., 2005). The existence of efficient mitochondria is needed for ER stress-prompted apoptosis and there is a crosstalk between mitochondria and ER in Aβ-treated cells (Costa et al., 2010). In contrast, mitochondrial disorder stimulates the response of ER dysfunction by increasing the neuronal vulnerability to those that ER stress-mediated by β-amyloid (Costa et al., 2013).

Besides, Aβ could disturb the attachment between microtubules (MT) and ER for affecting the design of the ER and thus triggers the failure of ER that may govern a significant effect in Aβ peptide-mediated neurodegenerative syndrome. Extraordinarily, this progression is not dependent on the unfolded protein response. When ER structure being distorted, autophagy is initiated and lysosomal breakdown increased, as displayed by EM (electron microscopy) as well as live-cell imaging (Lai et al., 2009). Using medications to alleviate microtubules could partly prevent the breakdown of the endoplasmic reticulum and stimulation of autophagy that might be a method of AD therapy.

Now the question what is the impact of Aβ on the UPR? Elevated expression of BiP discovered on the supplication of exogenous Aβ to major cortical brain neurons proposes that deposition of the peptide can initiate ER stress pathways (Lee et al., 2010; Resende et al., 2008). Truly, phosphorylation of kinases such as p-eIF2a, PERK, and breakdown of ATF6 could also be indicators of the presence of unfolded protein response initiation in β-amyloid treated neurons. In contrast, short-term care of β-amyloid (within 6 h) preferentially improved stimulation of the PERK signaling in neurons (Lee et al., 2010). In amyloid β-mediated ER stress, both eIF2α and PERK were phosphorylated to facilitate the stimulation of ER chaperones and thereby provides stability to accumulated protein toxicity of neuronal tissues. On the other hand, PERK silenced restricted the phosphorylation of eIF2α and increased apoptosis (Lee et al., 2010; Resende et al., 2008). Moreover, amyloid-β42 trial mediates expression of CHOP (also known as GADD153) in both cultivated cells as well as the rabbit hippocampus. Also, pretreatment with CHOP antisense RNA progresses survival following the Aβ exposure that proposes a function for CHOP in Aβ-induced apoptosis. That is the region of CHOP to transcriptionally block defensive cellular molecules like glutathione and Bcl-2. IP3-induced calcium discharge from ER, triggered by β-amyloid exposure, mediated the expression of CHOP (Schapansky et al., 2007). Besides, Aβ could initiate the response factor XBP1 of ER stress in mammalian reared neurons and in genetically modified flies, and its active structure XBP1s exhibits neuroprotection on the two diverse AD prototypes (inhibits the accumulation of free-floating Ca^{2+} in the cytoplasm), while knockdown of XBP1 factor aggravates Aβ toxicity (Casas-Tinto et al., 2011). Intraneuronal collection of Aβ might be related to ER dysfunction in AD individual's brains at an initial period.

ER Stress and Tau

Similar to senile plaques, hyperphosphorylated tangling in intracellular neurofibril amasses of microtubule-coupled protein tau directing to NFTs production which is another known pathological symptom of AD. As stated earlier, the UPR initiation can be seen in AD patient's brains after postmortem analysis, in turn, indications for unfolded protein response stimulation are ample in neurons with phosphorylated tau which advocate an intimate relationship between tau pathology and ER-stress.

UPR stimulation is witnessed in the postmortem brains of AD patients and is linked with the premature phase of tau phosphorylation and neurofibrillary collapse (Hoozemans et al., 2009). The immunoreactivity of p-PERK, reported to occur together with the phosphorylation of tau, was significantly elevated in the TgTau^{P301L} mice hippocampus in comparison to age-matched controls group signifying that ER dysfunction was elevated in experimental mice (Ho et al., 2012). OA that could decrease the activity of protein phosphatase 2A and prompts tau hyperphosphorylation (Kins et al., 2001) could elicit the UPR (Ho et al., 2012). The evidence might be the elevation of the immunoreactivity of p-eIF2α, p-PERK, increased concentrations of mRNA for GADD153, and splicing of mRNA for xbp-1. UPR stimulation (phosphorylation of pPERK as well as pIRE1) is existing matters which can be categorized neuropathologically as FTLD-tau but is not noticeable in other types of FTLD (Diana et al., 2012). These analyses showed that stimulation of the UPR is closely linked with the collection as well as the deposition of tau. A report on brains from rTg4510 tau modified animals and AD patients show unusual deposition of CD3∂ that is a substrate of ERAD in endoplasmic reticulum and stimulation of PERK. The pathway which shows how tau deposition assisting its detrimental is cooperating with the ER membrane and its related peptides that are crucial for ERAD, involving Hrd1 and VCP. Fascinatingly, this procedure could be inverted if soluble tau becomes exhausted, advocating that approaches targeted at decreasing soluble tau might be helpful for tauopathies involving Alzheimer's disease (Abisambra et al., 2013). Thapsigargin has exposed to prevent intracellular Ca^{2+}-transport ATPase and stimulate agitation of intracellular Ca^{2+} balance for triggering ER stress. It was found that the band intensity at various positions of phosphorylated tau (Thr231, Ser262, and Ser396) enhanced by Thapsigargin usage. All these reports signify that ER stress might mediate tau phosphorylated, which further advocated that tau phosphorylation and ER-stress could be coupled to induce a brutal cycle in Alzheimer's syndrome (Ho et al., 2012).

Further in *vitro* reports propose that $A\beta_{1-42}$ peptides can stimulate ER stress and accelerate the phosphorylation of tau and reduce cell apoptosis through a pathway facilitated by GSK-3β initiation (Resende et al., 2008). Moreover, GSK-3β is shown to be elevated co-localizing with pPERK in AD brain's neurons (Hoozemans et al., 2009), which demonstrate that UPR initiation mediated by β-amyloid is a primary occasion during pathology of tau and position to a dynamic crosstalk between these molecular pathways in tauopathies (Ferreiro and Pereira, 2012).

ER Stress and Autophagy

Autophagy is a self-devouring pathway, which performs a crucial function in cell removal and regulating cell metabolic homeostasis (Eisenberg et al., 2009). Neurons particularly, due to their size, severe polarization, and post-mitotic criteria, might be vulnerable to the deposition of injured or accumulated cytosolic membranes, or constituents, and reliant on autophagy for existence (Tooze and Schiavo, 2008). Thus, the noteworthy functions of autophagy in the brain are correlated with regulating the equilibrium between the synthesis and deprivation of proteins when faults in the autophagy mechanism have been linked to neurodegeneration including AD. Autophagy's mutation has also been coupled with AD, for example, some structures like autophagosomes were shown in DN (dystrophic neurites) mice models as well as AD individuals. This perhaps due to the damage of the maturation of autophagosomes into autolysosomes. There has also an assumption that weakened autophagic flux postulates a unique location for the production of Aβ peptide (Haung et al., 2005). As discussed earlier, ERAD is identified as the leading cellular pathway for elimination of UP in the event of cellular ER dysfunction; nevertheless, the latest investigations have demonstrated that the emergence of autophagosome is enhanced in the case of ER abnormalities (Ogata et al., 2006).

It has been reported that stimulated IRE1 employs TNF receptor-associated factor 2 (TRAF2) on the membrane of the ER and then triggers JNK (c-Jun N-terminal kinase) signaling. In IRE1-impaired cells or cells entertained with JNK agonist, the ER dysfunction mediated autophagy was hindered. Those reports designate that stimulation of JNK triggered by the IRE1-TRAF2 mechanism is necessary and governs a vital function in autophagy initiation in the eve of ER abnormality (Lerner et al., 2009). But, lack of adequate IRE1 cells or tissues, the autophagy might be stimulated, designating that rather than the IRE1-JNK mechanism, another signaling cascade may govern important functions in autophagy stimulation under ER dysfunction. Previously, it has been revealed that the PERK-eIF2α mechanism might involve autophagy stimulation under ER dysfunction. In contrast, new findings stated that the PERK-eIF2α signaling is unnecessary to initiate ER stress-mediated autophagy. ATF6 knockdown and PERK-inadequate cells exhibited that autophagy was initiated following ER dysfunction in a related fashion to that of standard control (Lerner et al., 2009). The disorder of autophagy delivered cells susceptible to ER dysfunction, advocating that autophagy involves the protection against ER stress-induced death. Thus, it is rational to think that autophagy might become a crucial point for therapeutic intervention of AD treatment. For instance, autophagy-induced upregulation by rapamycin regulator or its analog CCI-

779 that are the targets of rapamycin (TOR) antagonists, to defend against neurodegenerative disorder found in experiments in mice and Drosophila (Jiang et al., 2014). Also, ER dysfunction triggered by thapsigargin can be relieved by 4-phenyl butyric acid (4-PBA) through the proteins associated with the UPR such as C/EBP homologous protein, GRP78, GRP94, eIF-2α, phospho- JNK1, phospho-eIF-2α, and phosphor-JNK2/3, JNK1, PERK, IRE-1α, and sXBP-1. This proposes that 4-PBA could be an alternative clinical target against ER stress-related pathologic states (Kim et al., 2012).

ER Stress and *PSEN*

Presenilin 1 (PS1) and Presenilin 2 (PS2) are extremely homologous and both of the proteins (encoded by *PSEN*) are essential constituents of the multi-subunit protease complex (γ-secretase). γ-secretase enclosing mutationally-distorted presenilin can still induce the breakdown of APP, but the site of proteolysis is changed. Typical γ-secretase produces mainly $A\beta_{40}$ with a lower extent of $A\beta_{42}$, however; mutant γ-secretase yields a larger amount of $A\beta_{42}$. But, $A\beta_{42}$ is found greater amyloid genic as well as more susceptible to deposited than $A\beta_{40}$. There's more, Ca^{2+} imbalance has been supposed as an additional mechanism by which the homologous protein presenilins enhance to the AD pathogenesis and could cause AD. Alternatively, presenilin has an intimate relation with ER stress. In the following section, the authors present the link between ER dysfunction to PS1 as well as PS2 individually.

PS1 belongs to a constituent of the γ-secretase that plays a pivotal role in the breakdown of APP. An analysis of a family with AD indicates that *PSEN1* gene mutation was exposed among the peoples and their ancestors are also the bearer of this gene alteration. Altered presenilin transfected into cell lines of human kidneys and murine neuroblastoma activated β-amyloid production with elevated $A\beta_{42/40}$ ratio (Kulczycki et al., 2001). Modified PS1 may particularly upturn $A\beta_{42}$ secretion while N-glycosylation is defective.

ER stress is initiated by *PSEN1* gene alterations by blocking the signaling cascade of the UPR. Also, the reduction of GRP78 mRNA in altered PS1 brain neurons specifies that this alteration could disturb the stress response by halting the UPR and thus enhance cell vulnerability to ER dysfunction. In contrast, another research group reports that neither the initiation of PERK and IRE1α, nor the synchronize stimulation of CHOP mRNA and BiP, and protein is damaged in tissues deficient PS1 role. Moreover, contradicting with the aforementioned report that demonstrated reduced quantity of BiP in patient's neurons with PS1-coupled FAD, they were unable to discover momentous reductions in BiP concentrations in the individual's brains with sporadic Alzheimer's disease, or individuals with FAD hosting PS1 gene mutations (Katayama et al., 1999). These divergences are feasibly due to the impacts of PS1 modifications that could be disguised by dealing with higher amounts of ER stress stimulators or by sustained induction. The alternative probability may be the cells that used latter are less subtle to endoplasmic reticulum stress. Also, the cells require to be controlled carefully under similar experimental settings to identify those sensitive deficiencies of BiP/GRP78 mRNA in PS1 mutation-expressing cells. Presently, this area of research is still debated. It similarly designates that the BiP expression alone is insufficient to evaluate the initiation of the UPR pathway. The eIF2α phosphorylation is recognized to be enhanced in PS1 distorted knock-in mice, to be specific, PS1 could block eIF2α phosphorylation (Milhavet et al., 2002). Indeed, FAD-mutant PS1 interrupts the unfolded protein response by reducing both eIF2α phosphorylation and initiation of PERK. From the above analysis, FAD-mutant PS1 stimulates the unfolded protein response in several ways.

Impaired ER orientation, as well as alterations of overall ER activities, were also seen in relation to mutation of PS-1. Moreover, loss of ER balance, not induced by the pathways of UPR, autophagy might result (Høyer-Hansen and Jäättelä, 2007 and Nijholt et al., 2011). There are considerable data that the stability of autophagocytosis is distressed in AD that shows the accumulation of protein (Nixon and Yang, 2011). Aging, a key threat of Alzheimer's disease, is also correlated with faults in autophagocytosis (Salminen and Kaarniranta, 2009). Similarly, new analyses have reported that autophagy stimulation can decrease Aβ deposition and lessen memory loss in the modified AD mice (Yang et al., 2011).

Several other cell culture analyses demonstrate that the mutation of PS1 bothers subcellular Ca^{2+} balance and elevated synthesis of free radicals in mutated cells. Mutations in the PS1 gene induce neurons to DNA damage-mediated decease by inducing ER-facilitated apoptotic proteolytic pathways such as stimulation of calpains as well as caspase-12 (Takuma et al., 2005). Genetic alterations in PS1 elevate cellular vulnerability to cell death prompted by several upsets such as removal of trophic influences and exposure to β-amyloid.

PS2 is an alternative γ-secretase. A new analysis demonstrates that PS2 gene alteration has an intimate association with the cause of cognitive loss in APP genetically modified mice in associative trace eyeblink conditioning. This seems to be the maiden report to interpret the impact of PS2 gene alteration on mouse eyeblink conditioning (Kishimoto and Kirino, 2013). Another spliced event of the PS2 (PS2V) deficient exon 5 was described to be expressed previously in humans with sporadic AD. Gene lacking exon 5 shows frameshift mutations in 6 exons and encodes stop codon inducing to immature inhibition of translation. PS2V-encoding protein expresses mostly in the temporal cortex and CA1 hippocampus in AD individuals.

Under ER dysfunction, various genes including GRP94 and GRP78/BiP are broadly identified to be improved to refold back the UP and defend cells from damage. But, the procedure looks to be decreased in PS2 expressed cells. The reason following the reduction of GRP78 mRNA is the decreased phosphorylation of IRE1 is believed to be phosphorylated along with the UP in the ER (Sato et al., 2001). Moreover, PS2V protein can notably excite the deposition of amyloid β_{40} as well as β_{42} and alters tau protein organization that is an essential constituent of neurofibrillary tangles (Manabe et al., 2007). To confirm the compromised phosphorylation, investigators carried out some experimental analysis, which showed that PS2V attaches to IRE1 directly on the ER membrane. Nonetheless, the reason for the PS2V-mediated decline in phosphorylated IRE1 is yet to be considered (Sato et al., 2001). Investigators refine high mobility group A protein 1α that can attach to DNA as a TF and nearby to the 5′splice site of the PS2 pre-mRNA. The HMGA1 protein expression is found to become improved through investigations of patients with sporadic AD hippocampus (Manabe et al., 2003). Also, patients with sporadic Alzheimer's disease hippocampus exist unusual mRNA clip that mediates the loss of ER stress response pathway as a significant factor of the neurodegenerative syndrome.

ER Stress and Neuronal Cell Death

Neuronal death is the utmost and crucial characteristic of neurodegenerative disease including AD. Since autophagy governs defensive functions in AD pathogenesis, apoptotic govern just the reverse. Though many kinds of researches would stress on mitochondria as a central stimulator of apoptosis in Alzheimer's disease, numerous new analyses proposed that neuron's death in Alzheimer's disease has its basis in the ER (Jie-Qiong et al., 2015). Oxidative dysfunction and deposited misfolded proteins provoke cellular response including ER stress response which might defend cells from the lethally deposited misfolded proteins. Irreversible ER dysfunction and excessive misfolded protein deposition, however, devastate the cellular 'quality control' and impede the defense systems responsible for proper folding as well as degradation of defective proteins, ultimately leading to the stress of organelle and finally neuronal cell death (Shanker et al., 2008).

ER Stress and Apoptosis

ER dysfunction elicits a modification system called UPR, that targets to eliminate UP and reestablish endoplasmic reticulum homeostasis. Nevertheless, while the ER stress is very extreme, it cannot be altered; usually, apoptosis will take over. (Roussel et al., 2013). Amongst the UPR cascade, IRE1α is considered as a crucial protein that regulates cell fate in the stimulation of cell death (Upton et al., 2012). IRE1α stimulates the initiation of the apoptotic-signaling kinase-1 (ASK1) cascade, which successively activates the downstream genes of Jun-N-terminal kinase (JNK) and p38 MAPK pathways. These genes, in turn, progress apoptosis as part of the IRE1-TRAF2-ASK1 pathway (Song et al., 2014). The substrates of JNK are Bcl-2 and Bim, that are suppressed and initiated through JNK phosphorylation. Likewise, ASK1-dependent JNK initiations are competent to activate Alzheimer's disease pathogenesis by escalating β-amyloid synthesis, augmenting inflammation, or even stimulating NFTs accumulation. Additionally, p38 MAPK leads to the phosphorylation and stimulates CHOP, which induces the alterations in gene expression and thus promotes cell death, involving the upregulation of DR5 and Bim, while downregulation of Bcl-2 (Sano and Reed, 2013). However, β-amyloid could initiate the ASK1 pathway that leads to SH-SY5Y cell's death (Akterin et al., 2006). These data verify that ASK1 could be a novel approach in the prevention or therapeutic intervention of AD. Furthermore, PERK/eIF2α-dependent cascade activates the pro-apoptotic TF CHOP. CHOP-dependent initiation of GADD34 progresses dephosphorylation of elF2α overturning the inhibition of protein synthesis that permits the pro-apoptotic proteins synthesis encoding mRNAs (Sano and Reed, 2013). GADD34 expression accelerates ROS (reactive oxygen species) formation conceivably by escalating translation, which might induce proteotoxicity and finally apoptosis. BCL-2 family grabs the most attention in cell apoptosis triggered by CHOP. The BCL-2 expression is decreased by CHOP during ER dysfunction triggering cell death. Besides, a few members of the BCL-2 family-like BH3-only proteins NOXA, PUMA, and BIM similarly showed increased expression under endoplasmic reticulum dysfunction. Remarkably, the ER stress oversensitivity of PERK-defective cells is overturned by suppressing of NOXA, indicating a significant function of the proteins of the BCL-2 family in apoptosis (Gupta et al., 2012).

As stated, it has been confirmed that the altered PS1 and the anomalous splicing of PS2V enhance the cellular susceptibility to ER dysfunction and lead to cell death in FAD and SAD. Nonetheless, the particular molecules participating in AD-related cell death yet remains unknown (Katayama et al., 2004). It is now well-established that caspase-12 knockout mice exhibited endurance to death triggered by apoptosis stimulated Aβ protein that has a very critical part to play in the development of neuronal apoptosis in AD (Nakagawa et al., 2000). Caspase-12 is typically placed on the external membrane of the endoplasmic reticulum that requires to be cleaved for initiation at the N-terminal site by calpain. GRP78 has been shown to link to caspase-7 and caspase-12 and blocks the escape of caspase-12 from the endoplasmic reticulum. Under ER stress, caspase-7 slices caspase-12 into pieces by binding to and convert it into functional form (Martinez et al., 2010). Nevertheless, unfavorably caspase-12 expression was only discovered in rats and mice.

Figure 3. Links between ER stress and Alzheimer's pathogenesis.
APP is cleaved by secretases and deposit beta-amyloid in the brain and cause AD. Similarly, tau, autophagy, presenilin, and apoptosis induces the events related to the cause of AD.

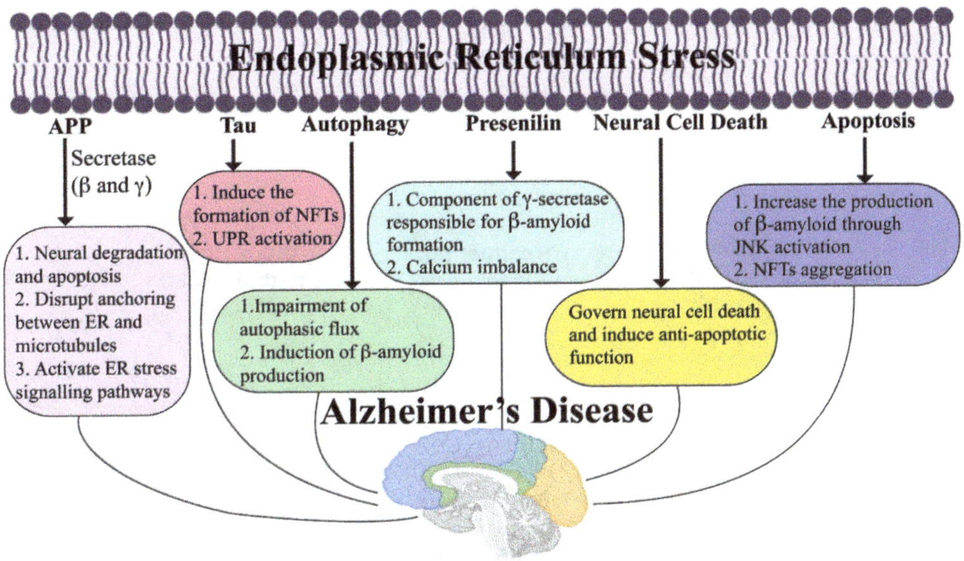

Afterward, researchers utilizing the mouse caspase-12 sequence as a molecular probe, found that caspase-4 had a higher structural uniformity to caspase-12 and thus might be regarded as human caspase-12 (Katayama et al., 2004). As caspase-4 occur only in human cells, it is therefore not feasible to perform the caspase-4 knockout tests in mice to confirm its link with apoptosis. On the other hand, immunohistochemical staining on the membrane of endoplasmic reticulum was reduced by utilizing the RNAi-based knockdown procedure, and cell apoptosis due to Tg induction was decreased. On the contrary, non-ER stress (Etop) related cell death remained unaffected. In SK-N-SH cell, immunohistochemical staining of caspase-4 showed increased expression in the pyramidal cell's cytoplasm from the hippocampal brains of AD individuals (Katayama et al., 2004). These results indicated that caspase-4 could be the specific caspase in humans and is closely related to cell lysis in patients with AD.

Furthermore, the disturbance of calcium homeostasis under endoplasmic reticulum might be another significant connection in ER dysfunction and cell death. Acute and immense discharge of calcium from the endoplasmic reticulum can activate different signaling pathways that cause cell apoptosis whereas the reduction of calcium ion in endoplasmic reticulum generates ER stress. Therefore, these processes are mutually dependent on each other. As stated earlier, ER stress reaction could induce the overproduction of Aβ that could precisely initiate the mitochondrial-mediated apoptotic pathway by releasing calcium ion from the endoplasmic reticulum via channels related to ryanodine receptors (RyR) and inositol 1,4,5-trisphosphate receptors (IP3R) (Ferreiro et al., 2008). When a substantial amount of free Ca^{2+} in the cytoplasm has reached, mitochondria can arrest them and disrupt the mitochondrial envelope leading to the secretion of proapoptotic factors that consist of Smac/Diablo, cytochrome-c, HtrA2/Omi, and so on (Ferreiro et al., 2008; Vaux, 2011). According to the current research, apoptosis does have a strong connection with ER stress and augments the susceptibility to AD. The major interaction between ER stress and Alzheimer's disease has been depicted in Figure 3.

Figure 4. Novel therapeutic approaches to Alzheimer's disease.

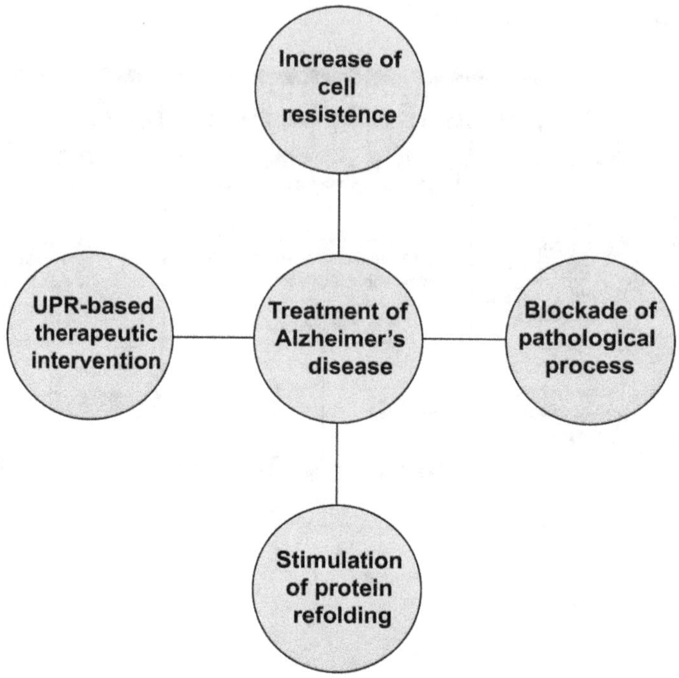

CONTROL OF ER STRESS AS PART OF ALZHEIMER'S TREATMENT

Several procedures could be alternatives for the therapeutic pathways, figure 4. Firstly, procedures must be designed to improve the endurance of cells to ER stress-related conditions. Moreover, to prevent the pathological activity, that is the core reason for the functional endoplasmic reticulum damage, should be scrutinized into therapeutic interference design. Lastly, inducing the refolding of the UP and to safeguarding cells from an unalterable impairment that is triggered by the buildup of the UP in ER stress situations seems to be necessary. Currently, based on lab data, three methods exhibited therapeutic significance in inhibiting or hindering the pathological pathway that triggers ER stress (Figure 4) (Paschen and Mengesdorf, 2005).

INCREASE OF CELL RESISTANCE

The first therapeutic approach is to enhance the cell's resistance. Since previously mentioned, the onset of AD is deeply related to ER stress. It has been assumed that increasing cell endurance could protect ER stress conditions. A research confirmed that HCT116 cells with a DICER hypomorphic mutation (Exn5/Exn5) or DICER or DROSHA knockdown were resilient to endoplasmic reticulum stress-mediated cell death, indicating that absence of miRNA biogenesis enhanced persistence to endoplasmic reticulum stress-mediated apoptosis (Cawley et al., 2013). This implies that miRNA biogenesis interference might be a tool for a future medicinal cure.

BLOCKADE OF PATHOLOGICAL PROCESS

The second approach is to impede the etiological processes. The main pathological activity resulting in endoplasmic reticulum malfunctioning in several acute and degenerative neural disorders is oxidative dysfunction triggered by an increase in ROS to the levels surpassing antioxidant activity. It has been shown that rats highly expressing SOD1, ischemia mediated ER stress is indeed noticeably inhibited. Thus, it can be predicted that medications with antioxidant properties are outstanding candidates for blocking ROS-stimulated damage of ER function (Paschen and Mengesdorf, 2005). The analyses executed with the SOD1 transgenic mice model stated above reveal that ROS governs a principal function in the chain of reactions causing in ischemia-mediated ER stress. Edaravone, a novel oxygen-free radicals' scavenger, has been reported to inhibit ischemia-mediated ER dysfunction and to inhibit post-ischemic development of brain edema as well as decrease infarct volume in focal cerebral ischemia (Qi et al., 2004). In brain neurodegeneration, it is still unestablished at which stage of the pathological process ROS interferes.

STIMULATION OF PROTEIN REFOLDING

The final approach aids in initiating protein refolding. Chemical chaperones (CC) could be the major tool to manage this issue. CC is a cluster of low-molecular-weight proteins that support proper protein folding and control anomalous protein accretion. CC such as DMSO and TMAO have been tested in the laboratory and they exhibited lowered cytotoxicity and apoptosis, which has been testified as a decent remedial approach (Yoshida et al., 2002). Medicines such as geldanamycin can regulate and boost chaperone levels (Hahn, 2009; Verhoef, 2002; Winklhofer et al., 2001). One research demonstrated that Geldanamycin that selectively links between hsp90 and GRP94, is an effective modulator of the cell response to ER dysfunction, causing the upregulation in the transcription of ER chaperones and higher expression of gadd153/CHOP transcription factor. Moreover, many shreds of evidence strongly propose that the rise in mRNA concentrations is caused by geldanamycin impacts on GRP94 but not hsp90. However, Hsp90 is a key chaperone in the cell that functions to establish a protein's quality control and aid in protein deprivation. Nevertheless, a lot of data specify that as aging and increased oxidative damage, allow this system to be less effective which further causes nitration and oxidation of proteins, in addition to the chaperones themselves, leading to the buildup of additional misfolded proteins (Cuervo and Dice, 2000; Lund et al., 2002; Tonoki et al., 2009). This shows that the Hsp90 antagonist would likely offer remedial benefits in AD. For instance, phenylbutyrate, an inhibitor of ATP attachment to Hsp90 protein (Bali et al., 2005), suspends cognitive loss and moderates tau pathology in the mouse of AD (Ricobaraza et al., 2009). Terracciano et al., (2017) described that dimeric and trimeric triazole-based compounds emerge as the latest category of Hsp90 molecular chaperone antagonists that might offer an innovative strategy for the treatment of AD.

UPR IN THERAPEUTIC INTERVENTION

As ER stress is an etiological feature of several disorders including diabetes, cancer malignancies, and PMDs, pharmacological regulation of the unfolded protein response is recently being developed as a feasible therapeutic approach. Evaluating the growing data that showed the UPR regulates the metabolism of APP, neuroplasticity, and phosphorylation of tau, the intervention of the UPR pathway in Alzheimer's disease may not only impact on proteostasis of endoplasmic reticulum but also additional key attributes of the disorder. Comprehensively, merely a small number of medications affecting the unfolded protein response have been investigated in preclinical studies of AD (Hetz et al., 2013). The advancement of a UPR-based pharmacological drug confronts a huge challenge owing to the important function of this signaling in several organs, with a greater possibility of causing negative impacts in the pancreas and liver with other tissues (Dufey et al., 2014). Besides, the researches in other PMD models revealed that the involvement of the unfolded protein response to the pathology may vary on the ailment condition, the affected neurons and the signaling branch examined (Hetz and Mollereau, 2014; Scheper and Hoozemans, 2015; H. L. Smith and Mallucci, 2016), making specificity concerns a major aspect to be outlined for therapeutic development (Hetz et al., 2013; Maly and Papa, 2014). UPR-targeting medications could be classified into two main categories:

- The compounds which specifically regulate definite UPR machinery and
- The compounds that unspecifically regulate proteostasis of ER by moderating processes such as folding and degradation of proteins, ERAD (Kraskiewicz and Fitzgerald, 2012).

Amongst the direct regulators, compounds targeting the PERK proteins of the UPR are the highly researched drugs in the models of neurodegenerative disorder. The majority of the existing drugs, for example, salubrinal does not focus PERK itself, instead eIF2α phosphorylation by inhibiting its phosphatase compounds, retaining the protein translation blockade (Boyce et al., 2005). As stated, the dispensation of salubrinal to AD models augmented the assembly of Aβ and BACE1 expression (Devi and Ohno, 2014). Prominently, other eIF2α inhibitors like sephin 1 and guanabenz have been studied in numerous PMD models presenting exceptional neuroprotective properties (Mercado and Hetz, 2017). Other tiny molecules affecting PERK are yet to be examined in AD experiments. Remarkably, the oral dispensation of GSK2606414 improved neurodegenerative disorders in tau transgenic mice (Radford et al., 2015). On the other hand, a new report proposed that GSK2606414 has a greater affinity to RIP kinases, besides producing severe reactions caused by pancreatic toxicity (Rojas-Rivera et al., 2017). On the contrary, the delivery of ISRIB provided neuroprotective effects in prion-related disease models whilst not inducing pancreatic infection at all (Halliday et al., 2015). On the other hand, the ISRIB administration of an AD mouse model has not altered the pathogenesis of the disease (Johnson and Kang, 2016). Moreover, ISRIB has less solubility, which may cause a decrease in its translational capacity (Halliday et al., 2017). The recent redirecting of the use of dibenzoyl-methane and trazodone as ISRIB counterfeits may further open significant routes for future examination in Alzheimer's disease models as these drugs showed remarkable protective outcomes on oral administration of tau transgenic mice (Halliday et al., 2017).

The second highly researched UPR pathway for the production of novel therapeutic drugs is the IRE1–XBP1s signaling axis, mainly for the cure of disorders related to metabolism as well as cancer. MKC-3946 or STF-083010 induced a decrease in IRE1 RNAse activity which specifically lowered the proliferation of myeloma tumors in xenograft models that are often linked with minimal toxicity (Hetz et al., 2013; Mimura et al., 2012). Some more compounds that could efficiently inhibit IRE1 yet need a complete pharmacological analysis and whether they can surpass the blood-brain barrier is still unknown (Maly and Papa, 2014). The kinase inhibiting RNase attenuators (KIRAs) are the other compounds that might allosterically combine and inhibit IRE1, lowering the detrimental impacts of its signaling in diabetes and obesity models (Morita et al., 2017; Wang et al., 2012).

CC are tiny proteins that stabilize the other proteins and are known to decrease the levels of ER dysfunction. Three major CC that has been broadly researched and is accepted by the FDA are specifically 4-phenyl butyric acid, tauroursodeoxycholic acid, and trehalose. Treatment of cells and organisms with chaperones have shown some protective effects against ER dysfunction in several cell types along with some disease models (Lindquist and Kelly, 2011). For example, 4-phenyl butyric acid augments cognitive failure in tau transgenic mice (Ricobaraza et al., 2009). Tauroursodeoxycholic acid supplementation moderates amyloid accumulation in AD mice and progresses cognitive ability (Dionísio et al., 2015). Trehalose has also been known to defend against AD pathogenesis (Du et al., 2013). Although CC require further classification in Alzheimer's disease models, they show a key option for the advancement of the novel treatment.

Gene therapy is evolving as a viable approach to particularly target specific brains and could be applicable to control ER proteostasis (Valenzuela et al., 2016). It has been contemplated that other precise means to decrease ER dysfunction and enhance the function and endurance of targeted neurons in Alzheimer's syndrome might be via supplementation of vectors coding for functional ATF6, ER chaperones, or XBP1s into the brain. As appraised, the inoculation of XBP1s into the brain hippocampus with lentivirus has stated to recover neuronal performance in the neurons of AD (Cissé et al., 2017). Additionally, the curative efficacy of the UPR might induce other features of the disorder which are beyond synaptic protein expression and ER stress buffering. Besides the neuronal plasticity, several neuroprotective properties of XBP1s-induced therapy are being published, progressing peripheral axonal regeneration (Oñate et al., 2016), improving motor recovery in spinal cord damage (Valenzuela et al., 2012), moderating the scarcity of PD dopaminergic neuron models (Valdés et al., 2014), and decreasing mutant huntingtin protein aggregation (Zuleta et al., 2012).

RECENT DEVELOPMENTS AND FUTURE RESEARCH DIRECTIONS

The involvement of untranslated protein response in AD was identified primarily as a downstream target of cellular response to the deposition of tau or amyloid-β rather than a key characteristic of the syndrome. Current analyses reported the direct association of the UPR in AD and placing the mechanism as a crucial player in both AD pathogenesis as well as plausible therapeutic invention. Numerous experimental analyses propose that ER stress signaling might stimulate the tau phosphorylation, amyloid pathway, and synaptic loss via divergent mechanisms. New outcomes show that UPR instability in Alzheimer's disease may affect neuron's role autonomously from β-amyloid or tau adversity. Outside the maintenance of cellular proteostasis, UPR has diverse functions that might be linked to other related characteristics of Alzheimer's disease. For example, insulin resistance, metabolic shortcomings, and dysregulation of lipid homeostasis are thought of as crucial constituents of AD etiology (Haughey et al., 1999; Yuyama et al., 2014). The chief organelle included in protein metabolism is ER and the UPR has been revealed to control lipid biosynthesis, especially through various TFs such as RIDD, XBP1s, and ATF4 (Hotamisligil, 2010; So et al., 2012; Volmer and Ron, 2015; Wang and Kaufman, 2016). In other words, changing the composition of membrane lipids has been stated to be a promising stimulator of the UPR (Volmer and Ron, 2015; Volmer et al., 2013). Therefore, the presence of long-persisting ER dysfunction in Alzheimer's disease can directly influence to mutated brain lipid constituent, insulin endurance and deposition of neurotoxic levels of lipids including ceramides. Moreover, the role of the UPR is important for continuing the regular activity of immune response cells including dendritic cells (DCs), macrophages, and B lymphocytes, stimulating inflammation as well as cytokine production (Todd et al., 2008). In general, there are still various experimental questions that require to fulfill to realize the effects and importance of the UPR pathway to various forms of AD.

During recent years, numerous progression and development of approaches for gene therapy have been seen, which might involve an outstanding option to pharmacology. Gene therapy, in the perspective of neurodegenerative disorders, brings multiple benefits such as local delivery and high specificity to particular brain sites and no universal health effects (Maguire et al., 2014; Valenzuela et al., 2016). Currently, adeno-associated viruses are an excellent alternative for gene therapy in the CNS that exhibit an exceptional safety profile as described in various therapeutic inventions (Maguire et al., 2014). The abundant viral serotypes would further permit an additional specific targeting of particular neuronal populations. Various pre-clinical analyses have reported widespread defensive impacts of UPR-mediated gene therapy in different PMDs models (Valenzuela et al., 2016). The current optimistic result to insert the XBP1s cDNA into the mouse models of AD brain potentiate fascinating opportunities for future clinical intervention (Cissé et al., 2017).

As XBP1s might collaborate with oncogenes to improve malignant tumors in the brain (Urra et al., 2016), the plausible negative impacts of prolonged utilization of UPR-mediated gene therapy for brain abnormalities necessary to be sincerely solved. Significantly, genetically modified mice highly expressing XBP1s in brain neurons for two years showing no symptoms of tumorigenicity, signifying that high XBP1s expression may be safe in the brain. More studies would be necessary to entirely solve the hypothesis of the ER proteostasis arrangement as Alzheimer's disease target. The point that UPR stimulates cardinal hallmark of AD such as synaptic function, cellular stress, and aggregation of abnormal protein, it sets the mechanism as a joining point that incorporates various cellular events indulged in the pathogenesis of AD.

CONCLUSION

It is clearly indicated that ER dysfunction is coupled in the AD pathogenesis. Some characteristics including impaired Ca^{2+} balance, oxidative stress, intracellular accumulation of tau as well as β-amyloid proteins, might be triggered by stress of ER in brain, but, this type of pathology could induce ER stress and hence increase AD pathology. The ER stress markers in AD are placed in neurons of brain rather than in glial tissues. This indication is in support with the amyloid postulation of AD because accumulated and misfolded peptides are problematic in term of quality in neurons and can initiate cellular protection through UPR. The objective of this protein response is to under stimulate protein production and rise protein folding capability by mounting ER as well as cellular chaperones levels. Moreover, through the autophagy process, the UPR can support the redox defense, buffering capacity of calcium, and stimulate cellular cleansing. ER dysfunction also motivate neurons to induce cell death by apoptosis. Remarkably, new reports have showed that ER dysfunction can also influence inflammation to protect brain cells from necrotic injuries. This response looks to be an apprehended form of response including chemokines and cytokines to stimulate glial cells. Prolonged and irreversible ER dysfunction might be harmful to neurons as a delayed defense reduces the neurons viability and can change the UPR response to control an apoptotic platform. But, the ER is extremely specific in neurons and extent of ER dysfunction greatly differ among various subsections, for example, in axonal synapses and dendrites. Initial indication designates that ER stress can cause synaptic damage and axonal relapse. In summary, ER stress implicates all the components that can promote AD pathogenesis.

ACKNOWLEDGMENT

The work was supported by Shenzhen Peacock project (KQTD2015033117210153) and Shenzhen Science and Technology Innovation Committee Basic Science Research Grant (JCYJ20150629151046896).

REFERENCES

Abisambra, J. F., Jinwal, U. K., Blair, L. J., O'Leary, J. C., Li, Q., Brady, S., ... Dickey, C. A. (2013). Tau accumulation activates the unfolded protein response by impairing endoplasmic reticulum-associated degradation. *The Journal of Neuroscience: The Official Journal of the Society for Neuroscience*, *33*(22), 9498–9507. doi:10.1523/JNEUROSCI.5397-12.2013 PMID:23719816

Acosta-Alvear, D., Zhou, Y., Blais, A., Tsikitis, M., Lents, N. H., Arias, C., ... Dynlacht, B. D. (2007). XBP1 controls diverse cell type- and condition- specific transcriptional regulatory networks. *Molecular Cell*, *27*(1), 53–66. doi:10.1016/j.molcel.2007.06.011 PMID:17612490

Akterin, S., Cowburn, R. F., Miranda-Vizuete, A., Jiménez, A., Bogdanovic, N., Winblad, B., & Cedazo-Minguez, A. (2006). Involvement of glutaredoxin-1 and thioredoxin-1 in β-amyloid toxicity and Alzheimer's disease. *Cell Death and Differentiation*, *13*(9), 1454–1465. doi:10.1038j.cdd.4401818 PMID:16311508

Alzheimer, A. (1907). Uber eine eigenartige Erkrankung der Hirnrinde. *Allg Zeitschr f Psychiatrie u Psych-Gerichtl Med*, *64*, 146–148. doi:10.1002/ca.980080612 PMID:8713166

Bali, P., Pranpat, M., Bradner, J., Balasis, M., Fiskus, W., Guo, F., ... Bhalla, K. (2005). Inhibition of histone deacetylase 6 acetylates and disrupts the chaperone function of heat shock protein 90: A novel basis for antileukemia activity of histone deacetylase inhibitors. *The Journal of Biological Chemistry*, *280*(29), 26729–26734. doi:10.1074/jbc.C500186200 PMID:15937340

Bánhegyi, G., Mandl, J., & Csala, M. (2008). Redox-based endoplasmic reticulum dysfunction in neurological diseases. *Journal of Neurochemistry*, *107*(1), 20–34. doi:10.1111/j.1471-4159.2008.05571.x PMID:18643792

Boyce, M., Bryant, K. F., Jousse, C., Long, K., Harding, H. P., Scheuner, D., ... Yuan, J. (2005). A selective inhibitor of eIF2α dephosphorylation protects cells from ER stress. *Science*, *307*(5711), 935–939. doi:10.1126cience.1101902 PMID:15705855

Calfon, M., Zeng, H., Urano, F., Till, J. H., Hubbard, S. R., Harding, H. P., ... Ron, D. (2002). IRE1 couples endoplasmic reticulum load to secretory capacity by processing the XBP-1 mRNA. *Nature*, *415*(6867), 92–96. doi:10.1038/415092a PMID:11780124

Casas-Tinto, S., Zhang, Y., Sanchez-Garcia, J., Gomez-Velazquez, M., Rincon-Limas, D. E., & Fernandez-Funez, P. (2011). The ER stress factor XBP1s prevents amyloid-β neurotoxicity. *Human Molecular Genetics*, *20*(11), 2144–2160. doi:10.1093/hmg/ddr100 PMID:21389082

Cawley, K., Logue, S. E., Gorman, A. M., Zeng, Q., Patterson, J., Gupta, S., & Samali, A. (2013). Disruption of microRNA biogenesis confers resistance to ER stress-induced cell death upstream of the mitochondrion. *PLoS One*, *8*(8), e73870. doi:10.1371/journal.pone.0073870 PMID:23977393

Chen, Y., & Brandizzi, F. (2013). IRE1: ER stress sensor and cell fate executor. *Trends in Cell Biology*, *23*(11), 547–555. doi:10.1016/j.tcb.2013.06.005 PMID:23880584

Cissé, M., Duplan, E., Lorivel, T., Dunys, J., Bauer, C., Meckler, X., ... Checler, F. (2017). The transcription factor XBP1s restores hippocampal synaptic plasticity and memory by control of the Kalirin-7 pathway in Alzheimer model. *Molecular Psychiatry*, *22*(11), 905–917. doi:10.1038/mp.2016.152 PMID:27646263

Costa, R. O., Ferreiro, E., Cardoso, S. M., Oliveira, C. R., & Pereira, C. M. F. (2010). ER stress-mediated apoptotic pathway induced by aβ peptide requires the presence of functional mitochondria. *Journal of Alzheimer's Disease*, *20*(2), 625–636. doi:10.3233/JAD-2010-091369 PMID:20182029

Costa, R. O., Ferreiro, E., Oliveira, C. R., & Pereira, C. M. F. (2013). Inhibition of mitochondrial cytochrome c oxidase potentiates Aβ-induced ER stress and cell death in cortical neurons. *Molecular and Cellular Neurosciences*, *52*, 1–8. doi:10.1016/j.mcn.2012.09.005 PMID:23026561

Cuervo, A. M., & Dice, J. F. (2000). Age-related decline in chaperone-mediated autophagy. *The Journal of Biological Chemistry*, *275*(40), 31505–31513. doi:10.1074/jbc.M002102200 PMID:10806201

Cullinan, S. B., Zhang, D., Hannink, M., Arvisais, E., Kaufman, R. J., & Diehl, J. A. (2003). Nrf2 Is a Direct PERK Substrate and Effector of PERK-Dependent Cell Survival. *Molecular and Cellular Biology*, *23*(20), 7198–7209. doi:10.1128/MCB.23.20.7198-7209.2003 PMID:14517290

DeGracia, D. J., & Montie, H. L. (2004). Cerebral ischemia and the unfolded protein response. *Journal of Neurochemistry*, *91*(1), 1–8. doi:10.1111/j.1471-4159.2004.02703.x PMID:15379881

Devanand, D. P., Pradhaban, G., Liu, X., Khandji, A., De Santi, S., Segal, S., ... De Leon, M. J. (2007). Hippocampal and entorhinal atrophy in mild cognitive impairment: Prediction of Alzheimer disease. *Neurology*, *68*(11), 828–836. doi:10.1212/01.wnl.0000256697.20968.d7 PMID:17353470

Devi, L., & Ohno, M. (2014). PERK mediates eIF2α phosphorylation responsible for BACE1 elevation, CREB dysfunction and neurodegeneration in a mouse model of Alzheimer's disease. *Neurobiology of Aging*, *35*(10), 2272–2281. doi:10.1016/j.neurobiolaging.2014.04.031 PMID:24889041

Dionísio, P. A., Amaral, J. D., Ribeiro, M. F., Lo, A. C., D'Hooge, R., & Rodrigues, C. M. P. (2015). Amyloid-β pathology is attenuated by tauroursodeoxycholic acid treatment in APP/PS1 mice after disease onset. *Neurobiology of Aging*, *36*(1), 228–240. doi:10.1016/j.neurobiolaging.2014.08.034 PMID:25443293

Du, J., Liang, Y., Xu, F., Sun, B., & Wang, Z. (2013). Trehalose rescues Alzheimer's disease phenotypes in APP/PS1 transgenic mice. *The Journal of Pharmacy and Pharmacology*, *65*(12), 1753–1756. doi:10.1111/jphp.12108 PMID:24236985

Dufey, E., Sepúlveda, D., Rojas-Rivera, D., & Hetz, C. (2014). Cellular mechanisms of endoplasmic reticulum stress signaling in health and disease. 1. An overview. *American Journal of Physiology. Cell Physiology, 307*(7), C582–C594. doi:10.1152/ajpcell.00258.2014 PMID:25143348

Eisenberg-Lerner, A., Bialik, S., Simon, H. U., & Kimchi, A. (2009). Life and death partners: Apoptosis, autophagy and the cross-talk between them. *Cell Death and Differentiation, 16*(7), 966–975. doi:10.1038/cdd.2009.33 PMID:19325568

Ferreira, S. T., & Klein, W. L. (2011). The Aβ oligomer hypothesis for synapse failure and memory loss in Alzheimer's disease. *Neurobiology of Learning and Memory, 96*(4), 529–543. doi:10.1016/j.nlm.2011.08.003 PMID:21914486

Ferreiro, E., Oliveira, C. R., & Pereira, C. M. F. (2008). The release of calcium from the endoplasmic reticulum induced by amyloid-beta and prion peptides activates the mitochondrial apoptotic pathway. *Neurobiology of Disease, 30*(3), 331–342. doi:10.1016/j.nbd.2008.02.003 PMID:18420416

Ferreiro, E., & Pereira, C. M. F. (2012). Endoplasmic reticulum stress: A new playER in tauopathies. *The Journal of Pathology, 226*(5), 687–692. doi:10.1002/path.3977 PMID:22190226

Fonseca, A. C. R. G., Ferreiro, E., Oliveira, C. R., Cardoso, S. M., & Pereira, C. F. (2013). Activation of the endoplasmic reticulum stress response by the amyloid-beta 1-40 peptide in brain endothelial cells. *Biochimica et Biophysica Acta. Molecular Basis of Disease, 1832*(12), 2191–2203. doi:10.1016/j.bbadis.2013.08.007 PMID:23994613

Frank, M. (2007). Intracellular amyloid-beta in Alzheimer's disease. *Nature Reviews. Neuroscience, 8*(7), 499–509. doi:10.1038/nrn2168 PMID:17551515

Gerakis, Y., & Hetz, C. (2018). Emerging roles of ER stress in the etiology and pathogenesis of Alzheimer's disease. *The FEBS Journal, 285*(6), 995–1011. doi:10.1111/febs.14332 PMID:29148236

Gupta, S., Giricz, Z., Natoni, A., Donnelly, N., Deegan, S., Szegezdi, E., & Samali, A. (2012). NOXA contributes to the sensitivity of PERK-deficient cells to ER stress. *FEBS Letters, 586*(22), 4023–4030. doi:10.1016/j.febslet.2012.10.002 PMID:23068609

Haas, I. G., & Wabl, M. (1983). Immunoglobulin heavy chain binding protein. *Nature, 306*(5941), 387–389. doi:10.1038/306387a0 PMID:6417546

Hahn, J. S. (2009). The Hsp90 chaperone machinery: From structure to drug development. *BMB Reports, 42*(10), 623–630. doi:10.5483/BMBRep.2009.42.10.623 PMID:19874705

Halliday, M., Radford, H., Sekine, Y., Moreno, J., Verity, N., Le Quesne, J., ... Mallucci, G. R. (2015). Partial restoration of protein synthesis rates by the small molecule ISRIB prevents neurodegeneration without pancreatic toxicity. *Cell Death & Disease, 6*(3), e1672. doi:10.1038/cddis.2015.49 PMID:25741597

Halliday, M., Radford, H., Zents, K. A. M., Molloy, C., Moreno, J. A., Verity, N. C., ... Mallucci, G. R. (2017). Repurposed drugs targeting eIF2α-P-mediated translational repression prevent neurodegeneration in mice. *Brain, 140*(6), 1768–1783. doi:10.1093/brain/awx074 PMID:28430857

Hamos, J. E., Oblas, B., Pulaski-Salo, D., Welch, W. J., Bole, D. G., & Drachman, D. A. (1991). Expression of heat shock proteins in Alzheimer's disease. *Neurology, 41*(3), 345–350. doi:10.1212/WNL.41.3.345 PMID:2005999

Han, D., Lerner, A. G., Vande Walle, L., Upton, J. P., Xu, W., Hagen, A., ... Papa, F. R. (2009). IRE1α Kinase Activation Modes Control Alternate Endoribonuclease Outputs to Determine Divergent Cell Fates. *Cell, 138*(3), 562–575. doi:10.1016/j.cell.2009.07.017 PMID:19665977

Haughey, N. J., Bandaru, V. V. R., Bae, M., & Mattson, M. P. (2010). Roles for dysfunctional sphingolipid metabolism in Alzheimer's disease neuropathogenesis. *Biochimica et Biophysica Acta. Molecular and Cell Biology of Lipids, 1801*(8), 878–886. doi:10.1016/j.bbalip.2010.05.003 PMID:20452460

Haung Yu, W., Cuervo, A. M., Kumar, A., Peterhoff, C. M., Schmidt, S. D., Lee, J. H., ... Nixon, R. A. (2005). Macroautophagy - A novel β-amyloid peptide-generating pathway activated in Alzheimer's disease. *The Journal of Cell Biology, 171*(1), 87–98. doi:10.1083/jcb.200505082 PMID:16203860

Heron, M. (2013). Death: Leading causes for 2010. *National Vital Statistics Reports, 62*(6). PMID:24364902

Hetz, C., Chevet, E., & Harding, H. P. (2013). Targeting the unfolded protein response in disease. *Nature Reviews. Drug Discovery, 12*(9), 703–719. doi:10.1038/nrd3976 PMID:23989796

Hetz, C., & Mollereau, B. (2014). Disturbance of endoplasmic reticulum proteostasis in neurodegenerative diseases. *Nature Reviews. Neuroscience, 15*(4), 233–249. doi:10.1038/nrn3689 PMID:24619348

Hetz, C., & Papa, F. R. (2018). The Unfolded Protein Response and Cell Fate Control. *Molecular Cell, 69*(2), 169–181. doi:10.1016/j.molcel.2017.06.017 PMID:29107536

Ho, Y. S., Yang, X., Lau, J. C. F., Hung, C. H. L., Wuwongse, S., Zhang, Q., ... Chang, R. C. C. (2012). Endoplasmic reticulum stress induces tau pathology and forms a vicious cycle: Implication in Alzheimer's disease pathogenesis. *Journal of Alzheimer's Disease, 28*(4), 839–854. doi:10.3233/JAD-2011-111037 PMID:22101233

Hollien, J., Lin, J. H., Li, H., Stevens, N., Walter, P., & Weissman, J. S. (2009). Regulated Ire1-dependent decay of messenger RNAs in mammalian cells. *The Journal of Cell Biology, 186*(3), 323–331. doi:10.1083/jcb.200903014 PMID:19651891

Hoozemans, J. J. M., Stieler, J., van Haastert, E. S., Veerhuis, R., Rozemuller, A. J. M., Baas, F., ... Scheper, W. (2006). The unfolded protein response affects neuronal cell cycle protein expression: Implications for Alzheimer's disease pathogenesis. *Experimental Gerontology, 41*(4), 380–386. doi:10.1016/j.exger.2006.01.013 PMID:16564150

Hoozemans, J. J. M., Van Haastert, E. S., Nijholt, D. A. T., Rozemuller, A. J. M., Eikelenboom, P., & Scheper, W. (2009). The unfolded protein response is activated in pretangle neurons in alzheimer's disease hippocampus. *American Journal of Pathology, 174*(4), 1241–1251. doi:10.2353/ajpath.2009.080814 PMID:19264902

Hotamisligil, G. S. (2010). Endoplasmic Reticulum Stress and the Inflammatory Basis of Metabolic Disease. *Cell, 140*(6), 900–917. doi:10.1016/j.cell.2010.02.034 PMID:20303879

Høyer-Hansen, M., & Jäättelä, M. (2007). Connecting endoplasmic reticulum stress to autophagy by unfolded protein response and calcium. *Cell Death and Differentiation*, *14*(9), 1576–1582. doi:10.1038j.cdd.4402200 PMID:17612585

Iwawaki, T., Akai, R., Yamanaka, S., & Kohno, K. (2009). Function of IRE1 alpha in the placenta is essential for placental development and embryonic viability. *Proceedings of the National Academy of Sciences of the United States of America*, *106*(39), 16657–16662. doi:10.1073/pnas.0903775106 PMID:19805353

Jack, C. R., Petersen, R. C., Xu, Y. C., Waring, S. C., O'Brien, P. C., Tangalos, E. G., ... Kokmen, E. (1997). Medial temporal atrophy on MRI in normal aging and very mild Alzheimer's disease. *Neurology*, *49*(3), 786–794. doi:10.1212/WNL.49.3.786 PMID:9305341

Jiang, T., Yu, J. T., Zhu, X. C., Tan, M. S., Wang, H. F., Cao, L., ... Tan, L. (2014). Temsirolimus promotes autophagic clearance of amyloid-β and provides protective effects in cellular and animal models of Alzheimer's disease. *Pharmacological Research*, *81*, 54–63. doi:10.1016/j.phrs.2014.02.008 PMID:24602800

Johnson, E. C. B., & Kang, J. (2016). A small molecule targeting protein translation does not rescue spatial learning and memory deficits in the hAPP-J20 mouse model of Alzheimer's disease. *PeerJ*, *4*, e2565. doi:10.7717/peerj.2565 PMID:27781164

Kanekura, K., Suzuki, H., Aiso, S., & Matsuoka, M. (2009). ER stress and unfolded protein response in amyotrophic lateral sclerosis. *Molecular Neurobiology*, *39*(2), 81–89. doi:10.100712035-009-8054-3 PMID:19184563

Katayama, T., Imaizumi, K., Manabe, T., Hitomi, J., Kudo, T., & Tohyama, M. (2004). Induction of neuronal death by ER stress in Alzheimer's disease. *Journal of Chemical Neuroanatomy*, *28*(1–2), 67–78. doi:10.1016/j.jchemneu.2003.12.004 PMID:15363492

Katayama, T., Imaizumi, K., Sato, N., Miyoshi, K., Kudo, T., Hitomi, J., ... Tohyama, M. (1999). Presenilin-1 mutations downregulate the signalling pathway of the unfolded-protein response. *Nature Cell Biology*, *1*(8), 479–485. doi:10.1038/70265 PMID:10587643

Kim, D. S., Li, B., Rhew, K. Y., Oh, H. W., Lim, H. D., Lee, W., ... Kim, H. R. (2012). The regulatory mechanism of 4-phenylbutyric acid against ER stress-induced autophagy in human gingival fibroblasts. *Archives of Pharmacal Research*, *35*(7), 1269–1278. doi:10.100712272-012-0718-2 PMID:22864750

Kins, S., Crameri, A., Evans, D. R. H., Hemmings, B. A., Nitsch, R. M., & Götz, J. (2001). Reduced Protein Phosphatase 2A Activity Induces Hyperphosphorylation and Altered Compartmentalization of Tau in Transgenic Mice. *The Journal of Biological Chemistry*, *276*(41), 38193–38200. PMID:11473109

Kishimoto, Y., & Kirino, Y. (2013). Presenilin 2 mutation accelerates the onset of impairment in trace eyeblink conditioning in a mouse model of Alzheimer's disease overexpressing human mutant amyloid precursor protein. *Neuroscience Letters*, *538*, 15–19. doi:10.1016/j.neulet.2013.01.025 PMID:23370287

Korolev, I. O. (2014). Alzheimer 's Disease : A Clinical and Basic Science Review. *Medical Student Research Journal*, *4*, 24–33.

Kraskiewicz, H., & Fitzgerald, U. (2012). InterfERing with endoplasmic reticulum stress. *Trends in Pharmacological Sciences*, *33*(2), 53–63. doi:10.1016/j.tips.2011.10.002 PMID:22112465

Kudo, T., Kanemoto, S., Hara, H., Morimoto, N., Morihara, T., Kimura, R., ... Takeda, M. (2008). A molecular chaperone inducer protects neurons from ER stress. *Cell Death and Differentiation*, *15*(2), 364–375. doi:10.1038j.cdd.4402276 PMID:18049481

Kulczycki, J, Bertrand, E, Lojkowska, W, Dowjat, W, & Wisniewski, T, and L.-Z. M. (2001). Familial Alzheimer's disease connected with mutation in presenilin gene. *Neurologia i Neurochirurgia Polska*, *35*(2), 213–224. PMID:11599221

Lai, C. S. W., Preisler, J., Baum, L., Lee, D. H. S., Ng, H. K., Hugon, J., ... Chang, R. C. C. (2009). Low molecular weight Aβ induces collapse of endoplasmic reticulum. *Molecular and Cellular Neurosciences*, *41*(1), 31–43. doi:10.1016/j.mcn.2009.01.006 PMID:19386229

Lee, A. H., Iwakoshi, N. N., & Glimcher, L. H. (2003). XBP-1 regulates a subset of endoplasmic reticulum resident chaperone genes in the unfolded protein response. *Molecular and Cellular Biology*, *23*(21), 7448–7459. doi:10.1128/MCB.23.21.7448-7459.2003 PMID:14559994

Lee, D. Y., Lee, K. S., Lee, H. J., Kim, D. H., Noh, Y. H., Yu, K., ... Kim, S. S. (2010). Activation of PERK signaling attenuates Aβ-mediated ER stress. *PLoS One*, *5*(5), e10489. doi:10.1371/journal.pone.0010489 PMID:20463975

Lee, J. H., Won, S. M., Suh, J., Son, S. J., Moon, G. J., Park, U. J., & Gwag, B. J. (2010). Induction of the unfolded protein response and cell death pathway in alzheimer's disease, but not in aged Tg2576 mice. *Experimental & Molecular Medicine*, *42*(5), 386–394. doi:10.3858/emm.2010.42.5.040 PMID:20368688

Lindholm, D., Wootz, H., & Korhonen, L. (2006). ER stress and neurodegenerative diseases. *Cell Death and Differentiation*, *13*(3), 385–392. doi:10.1038j.cdd.4401778 PMID:16397584

Lindquist, S. L., & Kelly, J. W. (2011). Chemical and biological approaches for adapting proteostasis to ameliorate protein misfolding and aggregation diseases-progress and prognosis. *Cold Spring Harbor Perspectives in Biology*, *3*(12), a004507. doi:10.1101/cshperspect.a004507 PMID:21900404

Lund, J., Tedesco, P., Duke, K., Wang, J., Kim, S. K., & Johnson, T. E. (2002). Transcriptional profile of aging in C. elegans. *Current Biology*, *12*(18), 1566–1573. doi:10.1016/S0960-9822(02)01146-6 PMID:12372248

Maguire, C. A., Ramirez, S. H., Merkel, S. F., Sena-Esteves, M., & Breakefield, X. O. (2014). Gene Therapy for the Nervous System: Challenges and New Strategies. *Neurotherapeutics; the Journal of the American Society for Experimental NeuroTherapeutics*, *11*(4), 817–839. doi:10.100713311-014-0299-5 PMID:25159276

Malhotra, J. D., & Kaufman, R. J. (2007). The endoplasmic reticulum and the unfolded protein response. *Seminars in Cell & Developmental Biology*, *18*(8), 716–731. doi:10.1016/j.semcdb.2007.09.003 PMID:18023214

Maly, D. J., & Papa, F. R. (2014). Druggable sensors of the unfolded protein response. *Nature Chemical Biology*, *10*(11), 892–901. doi:10.1038/nchembio.1664 PMID:25325700

Manabe, T., Katayama, T., Sato, N., Gomi, F., Hitomi, J., Yanagita, T., ... Tohyama, M. (2003). Induced HMGA 1a expression causes aberrant splicing a Presenilin-2 pre-mRNA in sporadic Alzhiemer's disease. *Cell Death and Differentiation, 10*(6), 698–708. doi:10.1038j.cdd.4401221 PMID:12761578

Manabe, T., Ohe, K., Katayama, T., Matsuzaki, S., Yanagita, T., Okuda, H., ... Mayeda, A. (2007). HMGA1a: Sequence-specific RNA-binding factor causing sporadic Alzheimer's disease-linked exon skipping of presenilin-2 pre-mRNA. *Genes to Cells, 12*(10), 1179–1191. doi:10.1111/j.1365-2443.2007.01123.x PMID:17903177

Mann, D. M. A. (1996). Pyramidal nerve cell loss in Alzheimer's disease. *Neurodegeneration, 5*(4), 423–427. doi:10.1006/neur.1996.0057 PMID:9117557

Marciniak, S. J., Garcia-Bonilla, L., Hu, J., Harding, H. P., & Ron, D. (2006). Activation-dependent substrate recruitment by the eukaryotic translation initiation factor 2 kinase PERK. *The Journal of Cell Biology, 172*(2), 201–209. doi:10.1083/jcb.200508099 PMID:16418533

Martinez, J. A., Zhang, Z., Svetlov, S. I., Hayes, R. L., Wang, K. K., & Larner, S. F. (2010). Calpain and caspase processing of caspase-12 contribute to the ER stress-induced cell death pathway in differentiated PC12 cells. *Apoptosis, 15*(12), 1480–1493. doi:10.100710495-010-0526-4 PMID:20640600

Martínez, G., Duran-Aniotz, C., and Hetz, C. (2016). Gene therapy to target ER stress in brain diseases. *Brain Research*, 561–570. doi:10.1016/j.brainres.2016.04.064 PMID:27131987

Mattson, M. P., & Chan, S. L. (2003). Neuronal and glial calcium signaling in Alzheimer's disease. *Cell Calcium, 34*(4-5), 385–397. doi:10.1016/S0143-4160(03)00128-3 PMID:12909083

Mattson, M. P., Pedersen, W. A., Duan, W., Culmsee, C., & Camandola, S. (1999). Cellular and molecular mechanisms underlying perturbed energy metabolism and neuronal degeneration in Alzheimer's and Parkinson's diseases. *Annals of the New York Academy of Sciences, 893*(1 OXIDATIVE/ENE), 154–175. doi:10.1111/j.1749-6632.1999.tb07824.x PMID:10672236

Matus, S., Lisbona, F., Torres, M., Leon, C., Thielen, P., & Hetz, C. (2008). The Stress Rheostat: An Interplay Between the Unfolded Protein Response (UPR) and Autophagy in Neurodegeneration. *Current Molecular Medicine, 8*(3), 157–172. doi:10.2174/156652408784221324 PMID:18473817

Mercado, G., & Hetz, C. (2017). Drug repurposing to target proteostasis and prevent neurodegeneration: Accelerating translational efforts. *Brain, 140*(6), 1544–1547. doi:10.1093/brain/awx107 PMID:28549133

Milhavet, O., Martindale, J. L., Camandola, S., Chan, S. L., Gary, D. S., Cheng, A., ... Mattson, M. P. (2002). Involvement of Gadd153 in the pathogenic action of presenilin-1 mutations. *Journal of Neurochemistry, 83*(3), 673–681. doi:10.1046/j.1471-4159.2002.01165.x PMID:12390529

Mimura, N., Fulciniti, M., Gorgun, G., Tai, Y. T., Cirstea, D., Santo, L., ... Anderson, K. C. (2012). Blockade of XBP1 splicing by inhibition of IRE1α is a promising therapeutic option in multiple myeloma. *Blood, 119*(24), 5772–5781. doi:10.1182/blood-2011-07-366633 PMID:22538852

Moreno, J. A., Halliday, M., Molloy, C., Radford, H., Verity, N., Axten, J. M., ... Mallucci, G. R. (2013). Oral treatment targeting the unfolded protein response prevents neurodegeneration and clinical disease in prion-infected mice. *Science Translational Medicine*, *5*(206), 206ra138. doi:10.1126citranslmed.3006767 PMID:24107777

Morita, S., Villalta, S. A., Feldman, H. C., Register, A. C., Rosenthal, W., Hoffmann-Petersen, I. T., ... Papa, F. R. (2017). Targeting ABL-IRE1α Signaling Spares ER-Stressed Pancreatic β Cells to Reverse Autoimmune Diabetes. *Cell Metabolism*, *25*(4), 883–897. doi:10.1016/j.cmet.2017.03.018 PMID:28380378

Murakami, T., Hino, S. I., Saito, A., & Imaizumi, K. (2007). Endoplasmic reticulum stress response in dendrites of cultured primary neurons. *Neuroscience*, *146*(1), 1–8. doi:10.1016/j.neuroscience.2007.01.069 PMID:17367944

Nakagawa, T., Zhu, H., Morishima, N., Li, E., Xu, J., Yankner, B. A., & Yuan, J. (2000). Caspase-12 mediates endoplasmic-reticulum-specific apoptosis and cytotoxicity by amyloid-β. *Nature*, *403*(6765), 98–103. doi:10.1038/47513 PMID:10638761

Nijholt, D. A. T., De Graaf, T. R., Van Haastert, E. S., Oliveira, A. O., Berkers, C. R., Zwart, R., ... Scheper, W. (2011). Endoplasmic reticulum stress activates autophagy but not the proteasome in neuronal cells: Implications for Alzheimer's disease. *Cell Death and Differentiation*, *18*(6), 1071–1081. doi:10.1038/cdd.2010.176 PMID:21252911

Nijholt, D. A. T., van Haastert, E. S., Rozemuller, A. J. M., Scheper, W., & Hoozemans, J. J. M. (2012). The unfolded protein response is associated with early tau pathology in the hippocampus of tauopathies. *The Journal of Pathology*, *226*(5), 693–702. doi:10.1002/path.3969 PMID:22102449

Nixon, R. A., & Yang, D. (2011). Autophagy failure in Alzheimer ' s disease-locating the primary defect. *Neurobiology of Disease*, *43*(1), 38–45. doi:10.1016/j.nbd.2011.01.021 PMID:21296668

Norfray, J. F., & Provenzale, J. M. (2004). Alzheimer's Disease: Neuropathologic Findings and Recent Advances in Imaging. *AJR. American Journal of Roentgenology*, *182*(1), 3–13. doi:10.2214/ajr.182.1.1820003 PMID:14684506

O'Connor, T., Sadleir, K. R., Maus, E., Velliquette, R. A., Zhao, J., Cole, S. L., ... Vassar, R. (2008). Phosphorylation of the Translation Initiation Factor eIF2α Increases BACE1 Levels and Promotes Amyloidogenesis. *Neuron*, *60*(6), 988–1009. doi:10.1016/j.neuron.2008.10.047 PMID:19109907

Ogata, M., Hino, S.-i., Saito, A., Morikawa, K., Kondo, S., Kanemoto, S., ... Imaizumi, K. (2006). Autophagy Is Activated for Cell Survival after Endoplasmic Reticulum Stress. *Molecular and Cellular Biology*, *26*(24), 9220–9231. doi:10.1128/MCB.01453-06 PMID:17030611

Oñate, M., Catenaccio, A., Martínez, G., Armentano, D., Parsons, G., Kerr, B., ... Court, F. A. (2016). Activation of the unfolded protein response promotes axonal regeneration after peripheral nerve injury. *Scientific Reports*, *6*(1), 21709. doi:10.1038rep21709 PMID:26906090

Park, S. W., & Ozcan, U. (2013). Potential for therapeutic manipulation of the UPR in disease. *Seminars in Immunopathology*, *35*(3), 351–373. doi:10.100700281-013-0370-z PMID:23572207

Paschen, W., & Mengesdorf, T. (2005). Cellular abnormalities linked to endoplasmic reticulum dysfunction in cerebrovascular disease-therapeutic potential. *Pharmacology & Therapeutics*, *108*(3), 362–375. doi:10.1016/j.pharmthera.2005.05.008 PMID:16140387

Patil, C., & Walter, P. (2001). Intracellular signaling from the endoplasmic reticulum to the nucleus: The unfolded protein response in yeast and mammals. *Current Opinion in Cell Biology*, *13*(3), 349–355. doi:10.1016/S0955-0674(00)00219-2 PMID:11343907

Prince, M., Bryce, R., Albanese, E., Wimo, A., Ribeiro, W., & Ferri, C. P. (2013). The global prevalence of dementia: A systematic review and metaanalysis. *Alzheimer's & Dementia*, *9*(1), 63–75.e2. doi:10.1016/j.jalz.2012.11.007 PMID:23305823

Qi, X., Okuma, Y., Hosoi, T., & Nomura, Y. (2004c). Edaravone protects against hypoxia/ischemia-induced endoplasmic reticulum dysfunction. *Journal de Pharmacologie*, *311*(1), 388–393. doi:10.1124/jpet.104.069088 PMID:15178695

Radford, H., Moreno, J. A., Verity, N., Halliday, M., & Mallucci, G. R. (2015). PERK inhibition prevents tau-mediated neurodegeneration in a mouse model of frontotemporal dementia. *Acta Neuropathologica*, *130*(5), 633–642. doi:10.100700401-015-1487-z PMID:26450683

Raff, M. C., Whitmore, A. V., & Finn, J. T. (2002). Axonal self-destruction and neurodegeneration. *Science*, *296*(5569), 868–871. doi:10.1126cience.1068613 PMID:11988563

Resende, R., Ferreiro, E., Pereira, C., & Oliveira, C. R. (2008). ER stress is involved in Aβ-induced GSK-3β activation and tau phosphorylation. *Journal of Neuroscience Research*, *86*(9), 2091–2099. doi:10.1002/jnr.21648 PMID:18335524

Ricobaraza, A., Cuadrado-Tejedor, M., Pérez-Mediavilla, A., Frechilla, D., Del Río, J., & García-Osta, A. (2009). Phenylbutyrate ameliorates cognitive deficit and reduces tau pathology in an alzheimer's disease mouse model. *Neuropsychopharmacology*, *34*(7), 1721–1732. doi:10.1038/npp.2008.229 PMID:19145227

Rojas-Rivera, D., Delvaeye, T., Roelandt, R., Nerinckx, W., Augustyns, K., Vandenabeele, P., & Bertrand, M. J. M. (2017). When PERK inhibitors turn out to be new potent RIPK1 inhibitors: Critical issues on the specificity and use of GSK2606414 and GSK2656157. *Cell Death and Differentiation*, *24*(6), 1100–1110. doi:10.1038/cdd.2017.58 PMID:28452996

Ron, D., & Walter, P. (2007). Signal integration in the endoplasmic reticulum unfolded protein response. *Nature Reviews. Molecular Cell Biology*, *8*(7), 519–529. doi:10.1038/nrm2199 PMID:17565364

Roussel, B. D., Kruppa, A. J., Miranda, E., Crowther, D. C., Lomas, D. A., & Marciniak, S. J. (2013). Endoplasmic reticulum dysfunction in neurological disease. *Lancet Neurology*, *12*(1), 105–118. doi:10.1016/S1474-4422(12)70238-7 PMID:23237905

Salminen, A., & Kaarniranta, K. (2009). Regulation of the aging process by autophagy. *Trends in Molecular Medicine*, *15*(5), 217–224. doi:10.1016/j.molmed.2009.03.004 PMID:19380253

Sanderson, T. H., Deogracias, M. P., Nangia, K. K., Wang, J., Krause, G. S., & Kumar, R. (2010). PKR-like endoplasmic reticulum kinase (PERK) activation following brain ischemia is independent of unfolded nascent proteins. *Neuroscience*, *169*(3), 1307–1314. doi:10.1016/j.neuroscience.2010.05.076 PMID:20538047

Sano, R., & Reed, J. C. (2013). ER stress-induced cell death mechanisms. *Biochimica et Biophysica Acta. Molecular Cell Research*, *1833*(12), 3460–3470. doi:10.1016/j.bbamcr.2013.06.028 PMID:23850759

Santos, L. E., & Ferreira, S. T. (2018). Crosstalk between endoplasmic reticulum stress and brain inflammation in Alzheimer's disease. *Neuropharmacology*, *136*, 35–360. doi:10.1016/j.neuropharm.2017.11.016 PMID:29129774

Sato, N., Imaizumi, K., Manabe, T., Taniguchi, M., Hitomi, J., Katayama, T., ... Tohyama, M. (2001). Increased production of β-amyloid and vulnerability to endoplasmic reticulum stress by an aberrant spliced form of presenilin 2*. *The Journal of Biological Chemistry*, *276*(3), 2108–2114. doi:10.1074/jbc.M006886200 PMID:11031265

Schapansky, J., Olson, K., Van Der Ploeg, R., & Glazner, G. (2007). NF-κB activated by ER calcium release inhibits Aβ-mediated expression of CHOP protein: Enhancement by AD-linked mutant presenilin 1. *Experimental Neurology*, *208*(2), 169–176. doi:10.1016/j.expneurol.2007.04.009 PMID:17927985

Scheper, W., & Hoozemans, J. (2009). Endoplasmic Reticulum Protein Quality Control in Neurodegenerative Disease: The Good, the Bad and the Therapy. *Current Medicinal Chemistry*, *16*(5), 615–626. doi:10.2174/092986709787458506 PMID:19199926

Scheper, W., & Hoozemans, J. J. M. (2015). The unfolded protein response in neurodegenerative diseases: A neuropathological perspective. *Acta Neuropathologica*, *130*(3), 315–331. doi:10.100700401-015-1462-8 PMID:26210990

Selkoe, D. J. (2001). Alzheimer's disease: Genes, proteins, and therapy. *Physiological Reviews*, *81*(2), 741–766. doi:10.1152/physrev.2001.81.2.741 PMID:11274343

Shankar, J. (2008). Coupling Endoplasmic Reticulum Stress to the Cell Death Program in Dopaminergic Cells: Effect of Paraquat. *Neuromolecular Medicine*, *10*(4), 333–342. doi:10.100712017-008-8047-9 PMID:18773310

Shoulders, M. D., Ryno, L. M., Genereux, J. C., Moresco, J. J., Tu, P. G., Wu, C., ... Wiseman, R. L. (2013). Stress-independent activation of XBP1s and/or ATF6 reveals three functionally diverse ER proteostasis environments. *Cell Reports*, *3*(4), 1279–1292. doi:10.1016/j.celrep.2013.03.024 PMID:23583182

Smith, H. L., & Mallucci, G. R. (2016). The unfolded protein response: Mechanisms and therapy of neurodegeneration. *Brain*, *139*(8), 2113–2121. doi:10.1093/brain/aww101 PMID:27190028

Smith, M. H., Ploegh, H. L., & Weissman, J. S. (2011). Road to ruin: Targeting proteins for degradation in the endoplasmic reticulum. *Science*, *334*(6059), 1086–1090. doi:10.1126cience.1209235 PMID:22116878

So, J. S., Hur, K. Y., Tarrio, M., Ruda, V., Frank-Kamenetsky, M., Fitzgerald, K., ... Lee, A. H. (2012). Silencing of lipid metabolism genes through ire1α-mediated Mrna decay lowers plasma lipids in mice. *Cell Metabolism*, *16*(4), 487–499. doi:10.1016/j.cmet.2012.09.004 PMID:23040070

Song, J., Park, K. A., Lee, W. T., & Lee, J. E. (2014). Apoptosis signal regulating kinase 1 (ASK1): Potential as a therapeutic target for Alzheimer's disease. *International Journal of Molecular Sciences, 15*(2), 2119–2129. doi:10.3390/ijms15022119 PMID:24481061

Strassnig, M., and Ganguli, M. (2005). About a peculiar disease of the cerebral cortex: Alzheimer's original case revisited. *Psychiatry, 1*, 3–8.

Takashima, A. (2006). GSK-3 is essential in the pathogenesis of Alzheimer's disease. *Journal of Alzheimer's Disease, 9*(3), 309–317. doi:10.3233/JAD-2006-9S335 PMID:16914869

Takuma, K., Du Yan, S. S., Stern, D. M., & Yamada, K. (2005). Mitochondrial dysfunction, endoplasmic reticulum stress, and apoptosis in Alzheimer's disease. *Journal of Pharmacological Sciences, 97*(3), 312–316. doi:10.1254/jphs.CPJ04006X PMID:15750290

Terracciano, A., An, Y., Sutin, A. R., Thambisetty, M., & Resnick, S. M. (2017). Personality change in the preclinical phase of Alzheimer disease. *JAMA Physiatry, 74*(12), 1259–1265. doi:10.1001/jamapsychiatry.2017.2816 PMID:28975188

Todd, D. J., Lee, A. H., & Glimcher, L. H. (2008). The endoplasmic reticulum stress response in immunity and autoimmunity. *Nature Reviews. Immunology, 8*(9), 663–674. doi:10.1038/nri2359 PMID:18670423

Tonoki, A., Kuranaga, E., Tomioka, T., Hamazaki, J., Murata, S., Tanaka, K., & Miura, M. (2009). Genetic Evidence Linking Age-Dependent Attenuation of the 26S Proteasome with the Aging Process. *Molecular and Cellular Biology, 29*(4), 1095–1106. doi:10.1128/MCB.01227-08 PMID:19075009

Tooze, S. A., & Schiavo, G. (2008). Liaisons dangereuses: Autophagy, neuronal survival and neurodegeneration. *Current Opinion in Neurobiology, 18*(5), 504–515. doi:10.1016/j.conb.2008.09.015 PMID:18840524

Tsuru, A., Fujimoto, N., Takahashi, S., Saito, M., Nakamura, D., & Iwano, M., … Kohno, K. (2013). Negative feedback by IRE1β optimizes mucin production in goblet cells. *Proceedings of the National Academy of Sciences of the United States of America, 110*(8), 2864–2869. 10.1073/pnas.1212484110

Unterberger, U., Höftberger, R., Gelpi, E., Flicker, H., Budka, H., & Voigtländer, T. (2006). Endoplasmic reticulum stress features are prominent in Alzheimer disease but not in prion diseases in vivo. *Journal of Neuropathology and Experimental Neurology, 65*(4), 348–357. doi:10.1097/01.jnen.0000218445.30535.6f PMID:16691116

Upton, J. P., Wang, L., Han, D., Wang, E. S., Huskey, N. E., Lim, L., ... Oakes, S. A. (2012). IRE1α cleaves select microRNAs during ER stress to derepress translation of proapoptotic caspase-2. *Science, 338*(6108), 818–822. doi:10.1126cience.1226191 PMID:23042294

Urano, F., Wang, X., Bertolotti, A., Zhang, Y., Chung, P., Harding, H. P., & Ron, D. (2000). Coupling of stress in the ER to activation of JNK protein kinases by transmembrane protein kinase IRE1. *Science, 287*(5453), 664–666. doi:10.1126cience.287.5453.664 PMID:10650002

Urra, H., Dufey, E., Avril, T., Chevet, E., & Hetz, C. (2016). Endoplasmic Reticulum Stress and the Hallmarks of Cancer. *Trends in Cancer, 2*(5), 252–262. doi:10.1016/j.trecan.2016.03.007 PMID:28741511

Urra, H., Dufey, E., Lisbona, F., Rojas-Rivera, D., & Hetz, C. (2013). When ER stress reaches a dead end. *Biochimica et Biophysica Acta, 1833*(12), 3507–3517. doi:10.1016/j.bbamcr.2013.07.024 PMID:23988738

Valdés, P., Mercado, G., Vidal, R. L., Molina, C., Parsons, G., & Court, F. A., ... Hetz, C. (2014). Control of dopaminergic neuron survival by the unfolded protein response transcription factor XBP1. *Proceedings of the National Academy of Sciences of the United States of America, 111*, 6804–6809. 10.1073/pnas.1321845111

Valenzuela, V., Collyer, E., Armentano, D., Parsons, G. B., Court, F. A., & Hetz, C. (2012). Activation of the unfolded protein response enhances motor recovery after spinal cord injury. *Cell Death & Disease, 3*(2), e272. doi:10.1038/cddis.2012.8 PMID:22337234

Vaux, D. L. (2011). Apoptogenic factors released from mitochondria. *Biochimica et Biophysica Acta. Molecular Cell Research, 1813*(4), 546–550. doi:10.1016/j.bbamcr.2010.08.002 PMID:20713095

Verhoef, L. G. G. C. (2002). Aggregate formation inhibits proteasomal degradation of polyglutamine proteins. *Human Molecular Genetics, 11*(22), 2689–2700. doi:10.1093/hmg/11.22.2689 PMID:12374759

Verkhratsky, A. (2005). Physiology and pathophysiology of the calcium store in the endoplasmic reticulum of neurons. *Physiological Reviews, 85*(1), 201–279. doi:10.1152/physrev.00004.2004 PMID:15618481

Volmer, R., & Ron, D. (2015). Lipid-dependent regulation of the unfolded protein response. *Current Opinion in Cell Biology, 33*, 67–73. doi:10.1016/j.ceb.2014.12.002 PMID:25543896

Volmer, R., Van Der Ploeg, K., & Ron, D. (2013). Membrane lipid saturation activates endoplasmic reticulum unfolded protein response transducers through their transmembrane domains. *Proceedings of the National Academy of Sciences of the United States of America, 110*(12), 4628–4633. doi:10.1073/pnas.1217611110 PMID:23487760

Walter, P., & Ron, D. (2011). The unfolded protein response: From stress pathway to homeostatic regulation. *Science, 334*(6059), 1081–1086. doi:10.1126cience.1209038 PMID:22116877

Wang, L., Perera, B. G. K., Hari, S. B., Bhhatarai, B., Backes, B. J., Seeliger, M. A., ... Maly, D. J. (2012). Divergent allosteric control of the IRE1α endoribonuclease using kinase inhibitors. *Nature Chemical Biology, 8*(12), 982–989. doi:10.1038/nchembio.1094 PMID:23086298

Wang, M., & Kaufman, R. J. (2016). Protein misfolding in the endoplasmic reticulum as a conduit to human disease. *Nature, 529*(7586), 326–335. doi:10.1038/nature17041 PMID:26791723

Williams, B. L., & Lipkin, W. I. (2006). Endoplasmic Reticulum Stress and Neurodegeneration in Rats Neonatally Infected with Borna Disease Virus. *Journal of Virology, 80*(17), 8613–8626. doi:10.1128/JVI.00836-06 PMID:16912310

Winklhofer, K. F., Reintjes, A., Hoener, M. C., Voellmy, R., & Tatzelt, J. (2001). Geldanamycin Restores a Defective Heat Shock Response in Vivo. *The Journal of Biological Chemistry, 276*(48), 45160–45167. doi:10.1074/jbc.M104873200 PMID:11574536

Yang, D. S., Stavrides, P., Mohan, P. S., Kaushik, S., Kumar, A., Ohno, M., ... Nixon, R. A. (2011). Reversal of autophagy dysfunction in the TgCRND8 mouse model of Alzheimer's disease ameliorates amyloid pathologies and memory deficits. *Brain, 134*(Pt 1), 258–277. doi:10.1093/brain/awq341

Ye, J., Rawson, R. B., Komuro, R., Chen, X., Davé, U. P., Prywes, R., ... Goldstein, J. L. (2000). ER stress induces cleavage of membrane-bound ATF6 by the same proteases that process SREBPs. *Molecular Cell*, *6*(6), 1355–1364. doi:10.1016/S1097-2765(00)00133-7 PMID:11163209

Yoshida, H., Yoshizawa, T., Shibasaki, F., Shoji, S., & Kanazawa, I. (2002). Chemical chaperones reduce aggregate formation and cell death caused by the truncated Machado-Joseph disease gene product with an expanded polyglutamine stretch. *Neurobiology of Disease*, *10*(2), 88–99. doi:10.1006/nbdi.2002.0502 PMID:12127147

Yuyama, K., Mitsutake, S., & Igarashi, Y. (2014). Pathological roles of ceramide and its metabolites in metabolic syndrome and Alzheimer's disease. *Biochimica et Biophysica Acta. Molecular and Cell Biology of Lipids*, *1841*(5), 793–798. doi:10.1016/j.bbalip.2013.08.002 PMID:23948264

Zuleta, A., Vidal, R. L., Armentano, D., Parsons, G., & Hetz, C. (2012). AAV-mediated delivery of the transcription factor XBP1s into the striatum reduces mutant Huntingtin aggregation in a mouse model of Huntington's disease. *Biochemical and Biophysical Research Communications*, *420*(3), 558–563. doi:10.1016/j.bbrc.2012.03.033 PMID:22445760

Chapter 8
Molecular Interactions of α-Synuclein, Mitochondria, and Cellular Degradation Pathways in Parkinson's Disease

Mansi Verma
Jaypee Institute of Information Technology, India

Sujata Basu
Enago-Publication Support, Crimson Interactive Pvt. Ltd., India

Manisha Singh
Jaypee Institute of Information Technology, India

Rachana R.
Jaypee Institute of Information Technology, India

Simrat Kaur
Jaypee Institute of Information Technology, India

Neha Goel
Jaypee Institute of Information Technology, India

ABSTRACT

Parkinson's disease (PD) has been reported to be the most common neurodegenerative diseases all over the world. Several proteins are associated and responsible for causing PD. One such protein is α-synuclein. This chapter discusses the role of α-synuclein in PD. Various genetic and epigenetic factors, which cause structural and functional changes for α-synuclein, have been described. Several molecular mechanisms, which are involved in regulating mitochondrial and lysosomal related pathways and are linked to α-synuclein, have been discussed in detail. The knowledge gathered is further discussed in terms of using α-synuclein as a diagnostic marker for PD and as a novel therapeutic target for the same.

DOI: 10.4018/978-1-7998-1317-0.ch008

INTRODUCTION

One of the most common age-associated neurodegenerative disorders is Parkinson's disease (PD) which, is a motor disease, that affects around 1% of people whose age is more than sixty years (Pagano et al, 2016). Features associated with the disease include motor deficits, slowness of movement, stiffness in the trunk or limbs, distinctive resting tremor, damaged coordination and balance (Magrinelli et al., 2016). Sporadic and familial PD are the two major classifications of the disease. Familial PD is caused by inheriting the genes with mutation(s) in DNA that are passed through generations (Schiesling et al., 2008). Idiopathic PD is another name for sporadic PD, makes around 85% of all PD cases. These people do not have a prior family history of PD. Even though the causes behind these cases are still not clear, sporadic cases also often result from a complex interaction between genetic factors with the environment (Del et al., 2016).

The decrease in dopamine neurotransmitter level has been attributed as the cause of motor symptoms observed in PD (Barone, 2010). Furthermore, inside the substantia nigra pars compacta region of the midbrain, there is a progressive loss of dopaminergic neurons which is the main reason behind the lowered dopamine neurotransmitter levels (Brichta and Greengard, 2014). Recognition of dopamine's role in PD has helped in the discovery of the drug levodopa's (L-DOPA, dopamine precursor). Despite soothing the symptoms associated with PD it does not help in stopping the disease progression. Not only has that, but prolonged use of levodopa also led to many side effects like hallucinations and involuntary movements (dyskinesias), etc (Shaw et al., 1980). Thus, it is high time to investigate the other components involved in PD and target them not only to appease the symptoms but also, to stop the disease progression.

There are several proteins, which have been found out to have disturbed level/function/structure in the neurons and lead to neurodegenerative disorders (NDDs). One of those kinds of protein is α-synuclein (α-syn), which is found to be aggregated in PD and is an integral part of Lewy's bodies – the major hallmark PD and it can lead to various dysfunctions of neurons (Smith and DeLong, 2012).

α-Syn is known to be overexpressed in various cases of PD. Several gene mutations are responsible for causing the accumulation of this protein. The level/structure/function of this protein is found to be linked to PD. α-Synuclein's accumulation in cytoplasm leads to dysfunction of lysosome and mitochondria. It also affects the processes of cleansing the cellular environment the autophagy and mitophagy. This background information has led the researchers to explore the role of this protein in various cellular processes, which are found to be affected in the case of PD. Besides neural cells, this protein has also been detected in various other cells/organs, and has been looked upon in two ways such as it indicates the beginning of this disease in a patient and it can also be utilized as a diagnostic marker for early detection or to monitor the progression of disease during treatment. It is also thought to be a target molecule for anti-PD drugs as it is found to interfere in various normal cellular processes. In this chapter, the authors have discussed the structure, function, mechanism of action of α-syn and its linkage with mitochondria and cellular degradation pathways in PD.

STRUCTURE AND FUNCTION OF α-SYNUCLEIN IN PARKINSON'S DISEASE

α-Syn was first recognized through an antibody used to purify cholinergic vesicles of the torpedo electric organ (Maroteaux and Scheller, 1991)and,α-synuclein involvement in the pre-synaptic functions was first identified. The expression of α-syn was detected by the antibody at the nuclear envelope, along with its localization at the synapse, accounting for its name synuclein (Maroteaux et al., 1988). The presence of α-syn in the nucleus has been accounted for in subsequent research (Gonçalves and Outeiro, 2013, McLean et al., 2000, Mori et al., 2002). Synuclein is a small protein (140 amino acid residues, as shown in Figure 1) which is less than the molecular weight cut off of the nuclear pore (~40 kDa).

The small size of α-syn has been thought to be aiding its entry to the nucleus (Bendor et al, 2013). Untagged and endogenous synuclein is hypothesized to enter the nucleus through simple diffusion, despite its localization being affected by nuclear and cytoplasmic proteins (Goers et al., 2003, Kontopoulos et al., 2006, Specht et al., 2005).

The membrane's interaction of α-syn and its presynaptic location strongly suggests the role of synuclein in the transmitter release. Many publications have reported that α-syn promotes release of neurotransmitters, but some others indicate it having an inhibitory role as well (Cabin et al., 2002, Murphy et al., 2000) as, α-syn knockout mice were found to be recovered faster due to dopamine release than their wildtype counterparts, a mild reduction in striatal dopamine is observed in the knockout mice (Abeliovich et al., 2000).

LINK BETWEEN α-SYNUCLEIN AND PARKINSON'S DISEASE

Genetic Factors Responsible for Parkinson's Disease

Autosomal dominant PD is caused mainly due to several point mutations in SNCA gene and multiplications of the *SNCA* gene, (Deng et al., 2015). The genetic analysis in the Contursi kindred case gave the first hint for the molecular basis of PD (Golbe, et al., 1996). 60 individuals in 5 generations from Contursi in Salerno province, Italy, were a part of this clinical genetic analysis study (Golbe et al., 1996). Mutation from alanine to threonine at residue 53 (A53T) and the substitution of alanine at position 53 by glutamate (A53E) residue was identified in the *SNCA* gene (Kang, et al., 2011; Picillo et al., 2018) as given in Table 1. Familial-linked point mutations like A53T and pathogenic mechanism of α-syn lead to fast fibrilization or oligomerization of the protein (Stojkovska et al., 2018), hinting at protein aggregation mediated gain-in-toxic function. Eventually, other point mutations in *SNCA* had also been revealed, such as H50Q, G51D, A30P, and E46K. Most of these are known to enhance the assembly kinetics of α-syn (Kiely et al., 2015).

Epigenetic Causes of Parkinson's Disease

While the genetic and molecular basis of the connection between α-synuclein and PD had been established years ago, the involvement of epigenetics in the progression and development of the disease and the influence of α-syn is also being explored. Epigenetic modifications are changes in the expression or functions of the gene which occur without a change in the deoxyribonucleic acid (DNA) sequence. It includes post-modifications of histone, DNA methylation, and regulation by non-coding ribonucleic acids (RNAs).

Figure 1. NMR structure of 1XQ8 human micelle-bound α-synuclein.

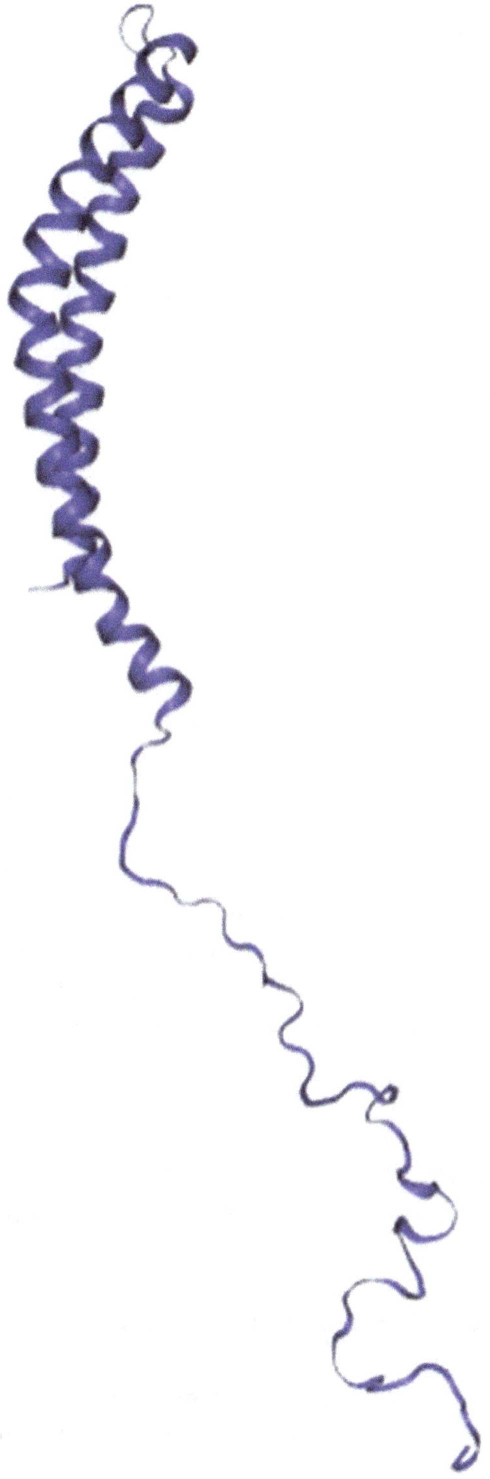

Table 1. Mutations in SNCA gene enhancing assembly kinetics of α-synuclein.

Mutation	Function	References
A53E	• Reduces α-synuclein aggregation; • Enhances proteasome activity affecting regular proteostasis.	Chen-Plotkin, 2014
A53T	• Induce postsynaptic deficits, tau phosphorylation-dependent postsynaptic dysfunction, which increases the chances ofα-synuclein aggregation.	Anderson et al., 2010; Conway et al., 2000; el-Agnaf and Irvine, 2002; Volles and Lansbury, 2003
H50Q	• Accelerated α-synuclein fibrillization, increases α-synuclein secretion and extracellular toxicity.	Khalaf, Ossama et al., 2014
G51D	• Phosphorylation, nuclear localization, and increase in the secretion of the G51D mutant in primary neurons and mammalian cells, in which α-syn-induced mitochondrial fragmentation, was further aggravated.	Fares, Mohamed-Bilal et al.,2014
A30P	• Loss of viability protection against oxidative stress; Modulation of dopamine vesicle trafficking by α-synuclein is affected; also influences α-synuclein's interaction with the cell membranes.	Jensen et al., 1998; Saha et al., 2004
E46K	• Most toxic induces robust PLK2-dependent α-synuclein phosphorylation at serine 129; • α-Syn's ability to bind to negatively charged liposome is enhanced greatly.	Choi et al., 2004; Inigo-Marco et al., 2017

DNA Methylation

DNA methylation is an extensively researched epigenetic modification. In eukaryotic cells, methyl from SAM (S-adenosyl methionine) is transferred to 5-C of cytosine through DNA methyltransferases (DNMT), which leads to 5-methylcytosine (5-mC) formation, this process is called DNA methylation. DNA methylation is known for regulating α-synuclein expression (Holliday 1990).

In some patients, SNCA intron 1 was found to be hypo-methylated (Ai et al., 2014). Furthermore, in HEK-293 cells hypo-methylation of CpG was shown to have a contribution to SNCA overexpression, eventually leading to the development of PD (Matsumoto et al., 2010). Interestingly, on comparing samples of DNA from peripheral blood leukocytes from PD patients and the controls, it was found that the level of methylation in CpG-2 decreases significantly in the patients (Song et al., 2014).

Histone Acetylation

The interaction of SNCA with histone alters its acetylation and can mediate its neurotoxicity. Through histone masking, it reduces H3 acetylation, this implies inhibition of SNCA gene expression and cell death (Harrison &Dexter, 2013).

Non-Coding RNAs Related to Parkinson's Disease

The microRNAs (miRNAs) are known for translational inhibition and/or degradation of mRNA. miRNA dysregulation is being explored as a key component for its contribution to neurodegeneration in PD. There are specific miRNAs that are known to regulate SNCA (Johnson et al., 2012). Upon binding to SNCA, gene mRNA, miR-7 represses SNCA protein level. Therefore, SNCA overexpression is observed when miR-7 decreases in 1-methyl-4-phenylpyridinium-positively charged organic molecule (MPP) induced PD cellular model. Furthermore, SNCA expression is regulated by miR-7 along with miR-153 (Doxakis, 2010) as well. Contrary to this, the protein level of SNCA in SH-SY5Y cells was reported to be increased by miR-106a.

There are many other examples of miRNA, which are indirectly found to be involved in the progression of PD, for example, the level of miR-133b was found to be decreased in the midbrain, which could have contributed to reducing the number of dopaminergic neurons there. Yet, the role of miR133-b's is still not clear *in vivo*. Another miRNA related to midbrain dopaminergic neuron differentiation is miR-132, upon its increase in rat model PD, its target protein Nurr 1 (nuclear receptor-related 1 protein) levels decreased (Kim et al., 2007)

During adult neurogenesis and central nervous system's neuronal differentiation, miR-124 was found to be decreased in the SNpC (substantia nigra pars compacta) of the MPTP (1-methyl-4-phenyl-1,2,3,6-tetrahydropyridine) induced PD mouse model, which lead to dopaminergic neuronal loss in PD (Kanagaraj et al., 2014). Additionally, downregulation of PI3K/AKT/ IGF-1 pathway due to an increase in miR-126 in dopaminergic neurons involved with pathogenesis is also being extensively researched (Kim et al., 2014). A list of miRNAs has been tabulated (Table 2) with their functions.

MECHANISM OF ACTION OF α-SYNUCLEIN IN PARKINSON'S DISEASE

The mechanism of action of α-syn is based upon interactions of various parts of this protein with other cellular components. α-syn–membrane interaction is modulated by amphipathic lysine-rich amino terminus (Barbour et al., 2008). The central region of the α-syn consists of a hydrophobic-rich motif that is made up of 65-69 amino acid residues which are known as the non-amyloid-beta component (NAC) (George, 2002). α-syn oligomerization and fibrillogenesis are decrease due to the deletion of large segments in this motif (Kahle, 2008). As mentioned above, the autosomal dominant PD occurs due to point mutations in *SNCA* gene (A53T, A30P, E46K, H50Q, G51D, and A53E) shown to cause familial forms of PD and dementia with Lewy bodies (Murphy, et.al, 2002). In fact α-syn goes through a helix-rich intermediate before forming a highly helical structure. The three major ways in which α-synuclein is involved in the initiation and/or progression of PD is through mitochondrial, lysosomal dysfunction, and nuclear dysfunction as discussed below.

Table 2. Role of miRNA in Parkinson's disease.

miRNA	Function	References
miR-7	Regulates SNCA expression	Doxakis, 2010
miR-106a	Protein level of SNCA in SH-SY5Y cells increased	Alvarez-Erviti et al., 2013
miR-132	Lowering the number of dopaminergic neurons in the midbrain	Kim et al., 2007
miR-133b	Midbrain dopaminergic neuron differentiation	Kanagaraj et al., 2014
miR-124	Dopaminergic neuronal loss in PD	Kim et al., 2014
miR-126	Down regulating IGF-1/PI3K/AKT pathway	Kim et al., 2014

Mitochondrial Dysfunction

Sporadic and familial both PDs are influenced by mitochondrial dysfunction. Strong evidence is available in the form of bioenergetics defects, increased reactive oxygen species (ROS) generation, disruptions in mitochondrial dynamics and complex I inhibition of the electron transport chain (ETC), in both experimental models and patients (Ryan et al.,2015; Winklhofer and Haass, 2010).

Many PD-associated genes regulate mitochondrial homeostasis. The mutations that link PD with mitochondrial dysfunction and also contribute to the progression of the disease have been studied in detail by Ryan et al., (2015). According to the recent data available, the interaction between α-synuclein and mitochondria is facilitated by α-synuclein binding to the outer mitochondrial membrane and in certain conditions; it can be imported, and also interacts with the F-type ATPase (Di Maio et al., 2016; Ludtmann et al., 2016). This indicates that there is some relationship between mitochondria and α-synuclein under pathological conditions and probably during normal physiological conditions.

Disturbance in Protein Import to Mitochondria

The localization of α-synuclein in the mitochondria is debatable (Guardia-Laguarta et al., 2014). Current research indicates that mitochondrial protein import mechanisms in PD can be disrupted by α-syn (Devi et al., 2008). The transportation of nuclear-encoded mitochondrial proteins into the mitochondria is mediated by the mitochondrial targeting sequence (MTS). The TOM (translocase of the outer membrane) receptors located on the outer membrane of the mitochondria, identify MTS. The TOM complex is involved in the translocation of these matrix targeted proteins to the TIM (translocase inner membrane) and eventually in the matrix. In models of PD and post-mortem PD brain tissue, the localization of α-synuclein into the outer mitochondrial membrane is observed along with its interaction with the TOM complex (Bender et al., 2013; Devi et al., 2008; Di Maio et al., 2016).

Similarly, some α-synuclein species like Ser-129E phosphomimetic are known for binding TOM20 receptor and prevent its interaction with co-receptor TOM22 *in vitro* experiments and inhibits mitochondrial protein import (Di Maio et al., 2016). This α-synuclein-TOM20 interaction was found to be linked with excessive ROS production and mitochondrial impairment (Di Maio et al., 2016). However, this remains unclear that whether blocking mitochondrial import is enough for driving nigrostriatal degeneration. In the case of HD model, neuronal death is caused by a lack of mitochondrial protein import (Yano et al., 2014). This suggests that there is potential for developing therapeutics involving mitochondrial protein import due to the converging mechanisms in neurodegenerative disorders.

Considering α-synuclein's affinity for detergent-resistant membranes rich in sphingolipids, acidic phospholipids, and cholesterol, its role in mitochondrial membrane is not even surprising (Jensen et al., 2011; Middleton and Rhoades, 2010). α-Syn adopts an amphipathic α-helical structure since that favors fibril formation when it binds to lipid vesicles and membranes and this is considered a crucial step when the oligomerization process is initiated (Beyer, 2007; Lee et al'., 2002; Tsigelnyet al., 2012). Mitochondrial homeostasis pathways and the protein import mechanisms are impaired due to the aggregation of α-syn at the outer membrane.

Disturbance in Mitochondrial Dynamics

Neuronal dysfunction and PD has been associated with dysregulation of mitochondrial dynamics (Van Laar and Berman, 2013). Mitochondrial dynamics is a key component to maintain cellular health. It includes autophagic degradation (mitophagy), mitochondrial fission, fusion, biogenesis, and transport of mitochondria. It is well known that α-synuclein alters the fusion process to hamper the mitochondrial dynamics. In the *C. elegans* model, mitochondria were fragmented as a result of the overexpression of α-syn (Kamp et al., 2010). Mitofusin (Mfn) 1 and 2 and Opa1 (optic atrophy type 1), mitochondrial fusion proteins, were not able to rescue the fusion deficit while, overexpression of many PD related proteins such as DJ-1 and Parkin (protein deglycase better known as PD protein 7) were successful in recovering the fusion deficit (Kamp et al., 2010). This domain can also be explored for developing therapeutics. Studies conducted on neuronal cells, both *in vivo* and *in vitro*, and mammalian cell lines suggests that α-syn binds to mitochondrial membranes, and mitochondrial fragmentation is induced by α-synuclein, independent of protein Drp1(dynamin-related protein 1), that mediates outer mitochondrial membrane fission, or Mfn2, another important component in the mitochondrial dynamics (Guardia-Laguarta et al., 2014; Kamp et al., 2010; Nakamura et al., 2011).

In addition, mitochondrial morphology in the CNS neurons is modulated by mutant A53T α-synuclein, in an age-dependent fashion, and mitochondrial transport is impaired during *in vitro* experiments (Xie and Chung, 2012). Mitochondrial transport and their position in neurons depend on microtubules, this microtubule transport is further based on molecular motors kinesins and dyneins(Vale, 2003). This demonstrates that mitochondrial transport and fusion/fission are impaired/altered by α-synuclein. Further characterization is required to understand the mechanism involved in how α-synuclein affects mitochondrial dynamics.

α-Syn can affect the clearance of dysfunctional mitochondria. During *in vivo* experiments in transgenic mice, where A53T α-syn was expressed in dopaminergic neurons, mitochondrial accumulation was observed which was linked with the rise in lysosome-mediated mitophagy respiratory dysfunction and increased lysosome-mediated mitophagy (Chinta et al., 2010). Interestingly, in both familial and sporadic PD, Miro, the outer mitochondrial protein, involved in mitochondrial transport is retained for prolonged duration and it delays the clearance of dysfunctional mitochondria (Hsieh et al., 2016). This indicates that when mitochondrial clearance is affected due to α-synuclein, which can also lead to PD pathogenesis.

LYSOSOME

Lysosomal Dysfunction

Dysfunctional clearance mechanism of lysosome promotes aggregation of soluble α-syn oligomers and this might be central to the progression of PD (Lee et al., 2004; Mak et al., 2010; Rideout et al., 2004). There is roughly a 7% probability of development of sporadic PD in people having a heterozygous mutation in the lysosomal hydrolase GBA1 and glucocerebrosidase (GCase) (Sidransky et al., 2009). Glycolipids are broken down by GCase. The activity of GCase protein decreases upon mutation in GBA1 gene, on the other hand, the level of α-synuclein protein and glycolipids increases. There is not much clarity about how the progression of PD is promoted by a modest reduction (30-50%) in glucocerebrosidase (GCase) due to GBA1. Based on genetics, a link between functional GCase and α-synuclein accumulation has been demonstrated by many research groups(Rocha and Sander, 2018). Due to the accumulation of glucosylceramide (GluCer) – the main substrate for GCase, an inverse relationship between α-synuclein and GCase is observed. Direct interaction of GluCer with α-synuclein promotes its conformational conversion into toxic oligomeric species and amyloidogenic fibrils upon the reduction of GCase (Mazzulli et al., 2011). Further, in the case of aged non-human primate brains, this aggregation of oligomeric and phosphorylated Ser-129 α-synuclein is linked with a reduction in PP2A (protein phosphatase 2 as well as, GCase in regions more vulnerable to neurodegeneration in PD (Chen et al., 2016; Liu et al., 2015). There is a lot more evidence that hints at an inverse relationship between PP2A) and GCase activity with α-synuclein phosphorylation.

The inhibition of GCase can increase phosphorylated Ser-129 α-synuclein and reduces PP2A activity (Chen et al., 2016; Du et al., 2015; Liu et al., 2015; Volpicelli-Daley et al., 2014). On the other hand, GCase knockdown leads to inhibition of autophagy-related proteins which further results in the inactivation of PP2A (Du et al., 2015).

Levels of the phosphorylated form of Ser-129 α-synuclein can be reduced by pharmacological induction of autophagy through metformin or rapamycin stimulated PP2A (Perez-Revuelta et al., 2014; Peterson et al., 1999). There's a possibility that PP2A activity is affected by a reduction in GCase activity due to overall lysosomal dysfunction hence, enhancing aggregation of α-synuclein. Or perhaps, PP2A is modulated by GCase by ceramide, a known activator of PP2A and autophagy (Chalfant et al., 1999; Dobrowsky et al., 1993). Thus, the α-synuclein load is diminished in GCase mutant mice by a rise in ceramide levels (Du et al., 2015). It can be said that in the relationship between α-synuclein and GCase, PP2A acts as an upstream mediator. However, the mechanism linking α-synuclein levels with PP2A and GBA is still unknown.

Autophagy-Lysosomal Pathway and Other Degradation Pathways

The degradation of α-synuclein involves the autophagy-lysosomal pathway (ALP) and the ubiquitin-proteasome system (UPS) (Tofaris et al., 2011). Short-lived soluble proteins are degraded by UPS (Goldberg, 2003), on the other hand, the degradation of dysfunctional organelles and cytosolic organelles is the responsibility of ALP (Klionsky and Emr, 2000). When these interconnected proteolytic systems fail to function properly, aggregated α-synuclein starts accumulating, which eventually hinders regular cell functions and promotes the pathogenesis of PD (Xilouri et al., 2013b). Also, there is a proposal that α-synuclein turnover and metabolism are being affected by chaperone-mediated autophagy (CMA). CMA is a selective type of autophagy that is found to be up-regulated during cellular stress (Cuervo and Dice, 2000a; Cuervo and Dice, 2000b) (Figure 2).

The complex of chaperons recognizes substrates having KFERQ motif which binds to LAMP-2A (lysosome-associated mitochondrial protein type 2A) (Cuervo et al., 2000). Levels of LAMP2A are directly correlated with CMA activity (Cuervo and Dice, 2000b), and in the SN (substantia nigra) of PD brains, the LAMP-2A expression is decreased (Alvarez-Erviti et al., 2010). Accumulations of α-synuclein and nigral cell death are reported to be a result of the reduction in LAMP-2A levels (Vogiatzi et al., 2008; Xilouri et al., 2016). However, in the case of rodents, overexpression of LAMP-2A protected them from α-synuclein-induced neuronal cell death (Xilourietal., 2013a) Further, CMA can be blocked by pathogenic α-synuclein mutants such as A30P and A53T etc (Cuervo et al., 2004.). The

Figure 2. The involvement of cellular degradation pathways in the α-synuclein accumulation.
ALP; autophagy-lysosomal pathway; CMA, chaperone-mediated autophagy; UPS, ubiquitin-proteasome system.

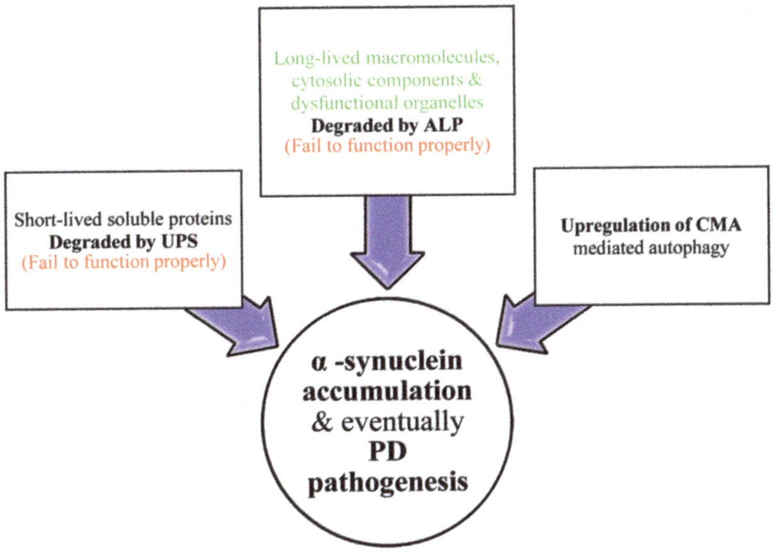

Figure 3. The interplay between chaperones, mitochondrial proteins and α-synuclein in cellular degradation pathways.

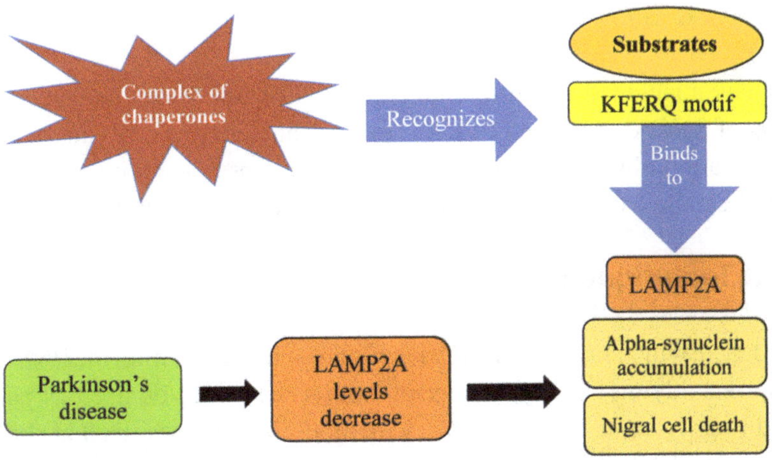

above mechanisms of action, as well as the interplay between chaperones, mitochondrial proteins, and α-synuclein are shown in Figure 3.

Despite a lot of research conducted in this field, an exact mechanism is yet to be relieved from controversies, for the degradation of α-synuclein. It is proposed that it varies in different systems. According to *in vitro* studies, in purified isolated systems, proteasomes, as well as lysosomes, seem to be involved in the degradation of α-synuclein (Cuervo et al., 2004Liu et al., 2003). However, the mechanism responsible for the degradation of α-synuclein – UPS or the autophagy pathway is still a mystery.

RECENT DEVELOPMENTS AND FUTURE RESEARCH DIRECTIONS

Diagnostic Marker

A candidate biomarker is chosen to investigate the pathophysiology of the disease, based on existing information. Cytotoxic role of α-synuclein in PD has been testified through several *in vitro* and *in vivo* studies (Steiner et al., 2011). Many research groups are using the α-synuclein level as PD biomarkers. This is based on the increasing evidence describing the role of α-synuclein in PD that eventually leads to the formation of Lewy's bodies – a hallmark of PD, often associated with neurodegeneration (Rosborough, 2017).

Most of the researches has concluded that the abnormal deposition of α-synuclein in the different parts of the neurons of the brain regions is the pathological hallmark of PD. The aggregates are found in the peripheral regions, outside central nervous system (CNS) as well. However, studies are still going on to understand whether peripheral pathology occurs at the same time as the central pathology and progresses predictably as it has been seen for Lewy pathology in CNS (Visanji et al., 2013). Some studies suggested that the accumulation of α-synuclein in CNS can represent a consequence of an abnormal effect of brain proteins such as Aβ and tau that promotes the fibrillation and aggregation of α-synuclein. The progression of the disease varies both in CNS and PNS. In CNS, the cases that have shorter survival, and are more aggressive, demonstrated higher loads of Lewy's bodies in comparison to cases with longer clinical courses (Kalia, 2018).

Molecular Biological Techniques

Several researchers have concluded that the pathogenicity of α-synuclein comes from missense mutations or the multiplication of SNCA, the gene encoding for α-synuclein protein, causes autosomal dominant forms of PD. Determination of these mutations and levels of expression with molecular biological techniques (RT-PCR etc.) can be used as a diagnostic marker (Hope et al., 2004).

Antibody-Based Techniques

Based on earlier evidence, which supports the presence of α-synuclein within the PNS (Beach et al., 2010), several researchers are studying if, α-syn in peripheral tissues could be used as pathological markers when it is detected through immune histochemistry methods. The hypothesis that aberrant accumulation of α-synuclein reaches the CNS starting from the periphery is what drives these investigators to explore this area. So, a detecting molecule measuring peripheral α-synuclein could detect the disease, long before the motor symptoms are observed by investigating the SnPC. the gastrointestinal tract, salivary glands, skin, and olfactory epithelium are the tissues included in peripheral tissues involved in the development of an α-synuclein-based biomarker (Beach et al., 2018; Witt et al., 2009).

Histochemical staining using antibodies against SNCA and anti-antibodies (AAbs) for α-synuclein can also be used to diagnose this disease. In the brains of PD patients, α-synuclein aggregates in SN co-localize with deposited immunoglobulin (Ig) G, showing that α-synuclein can potentially induce *in situ* auto-antibodies response (Horvath et al., 2018). There have been several methods deduced to quantify anti-antibodies to α-synuclein in fluid samples. Many researchers used ELISA-based approach to estimate anti-antibodies to α-synuclein in body fluid, including human cerebrospinal fluid and blood plasma of patients. Several studies demonstrated that α-syn AAbs levels were found to be higher in PD patients at their middle stage of motor impairment. Some studies showed that the levels of AAb were found to be decreasing with as the disease developed further (Roodveldt, et al. 2008).

Studies analyzed blood and cerebrospinal fluid (CSF) samples of patients and healthy subjects for α-synuclein AAbs. The levels were compared and found to be in a similar range however, there was a significant difference in CSF samples in comparison with males and females (who has higher). It was also found that the serum was rich in α-syn AAbs, relative to CSF. Using the hypothesis-based approach, α-synuclein AAb levels would reflect α-synucleinopathy, and this would rise with time. But, this correlation was not observed in the patients. However, the overall α-syn AAbs level of CSF in PD patients was higher than healthy people. This suggests that α-synuclein AAb levels can be used as a diagnostic biomarker for PD patients (Kalia, 2018).

The development and discovery of α-synuclein itself as biochemical markers is another area of research. Numerous assays have been developing to assess α-synuclein levels in biofluids like blood, saliva, and CSF, which are more readily accessible than the tissues (Goldman et al, 2018). Saliva is under investigation to determine the α-synuclein levels outside the CNS and the pathological process involved. Similarly, blood is interesting to study as α-synuclein could be transported from the CNS to blood. Studies conducted on rodents suggest it's possible to detect labeled α-synuclein which was injected in the brain, in the plasma with some CNS derived exosomes. In the exosome fraction of PD patients, the α-synuclein levels were detected to the higher than healthy controls, while the total plasma levels of α-synuclein were not very different (Shi et al., 2014). Yet there's still ambiguity about α-synuclein level in blood reflects aberrations of α-synuclein occurring in neurons of the CNS. For now, CNS abnormalities are best reflected in the biofluid CSF α-synuclein levels in these studies were measured using either enzyme-linked immunosorbent assay (ELISA), Luminex assays or mass spectrometry (Mollenhauer et al., 2012).

Brain Scanning

There is a need to develop a non-peptide positron emission tomography (PET) ligand for α-synuclein aggregates to assess Lewy body in PD patients with dementia (Velasco, et al., 2008). Functional imaging has advanced the level of diagnosis in neurodegenerative diseases. PET can examine the metabolic processes in the human body using nuclear medicine (Gambhir et al, 2001). A positron-emitting radionuclide emits pairs of gamma rays (indirectly) which are detected by the system, commonly fluorine-18, and then this is introduced through a radioactive tracer, a biologically active molecule. Computational analysis is used to construct 3D images of tracer concentration. In premotor cases of PD, the α-synuclein aggregates are described in different regions of the body including the esophagus, stomach, endocrine system and heart. Therefore, the extensive peripheral nervous system involvement in Lewy body disorders like PD has suggested that intensive tissue and functional studies are needed to be carried out for an accurate *in vivo* diagnosis of neurodegenerative diseases (Djaldetti et al., 2009).

Aggregation of α-synuclein spreading brainstem structures to the neocortex is a characteristic feature of α-synucleinopathies, like PD and dementia with Lewy bodies (DLB). While PD and DLB can be distinguished clinically, variation amongst PD dementia (PDD) and DLB is based on the temporal relationship between cognitive symptoms and motor and is usually subtle. The patterns of subcortical atrophy in DLB, PDD and PD, and analyze the key differences amongst DLB and PDD, and PD and PDD. An magnetic resonance imaging (MRI) based study conducted in PD, PDD and DLB patients involved segmentation of subcortical structures through the well-validated and fully-automated tool, and the shape and volume for each structure were also compared among different groups. Global subcortical atrophy was observed in case of PDD and DLB when compared to PD patients. In PDD and PD, greater hippocampal atrophy emerged as a distinguishing trait between them. Structural differences in the shape of both structures were identified through vertex analysis. This suggests that time-sparing, automated, subcortical volumetry can be used for deriving diagnostic information related to α-synucleinopathies (Gazzina et al., 2016).

Therapeutic Applications

Based on the above discussion, we suggest that one of the most attractive targets for drug development for PD is inhibition of α-synuclein aggregation. Many research groups have been exploring the disaggregation pathway for this reason. This pathway consists of heat-shock proteins (Hsp) 40, 70, and 104 and several other molecular chaperones. Hsp70 and Hsp40 are present in Lewy bodies, in case of yeast (Auluck et al., 2002). Hsp70's pharmacological activation or overexpression protects against α-synuclein toxicity in *in vitro* and *Drosophila melanogaster* models of PD, through the reduction in oligomer concentrations (Auluck and Bonini, 2002; Shin et al., 2004). The rat model of PD has revealed that Hsp104 overexpression decreases α-synuclein aggregation (Bianco et al., 2008). Formation of stable fibrils could be an interesting strategy to prevent cell death since oligomeric forms are toxic.

Identification of aggregation inhibitors developed based upon natural and synthetic compound libraries involved with α-synuclein available online may be screened. Not many inhibitors have been developed yet and the ones that have been developed are intended to offer neuroprotection, this includes compounds like EGCG (epigallocatechin-3-gallate), studied *in vitro*, (Bieschke et al.,2010) and those investigated *in vivo*, such as CLR01, (Prabhudesai et al., 2012) lysine-specific molecular tweezers that can inhibit aggregation as well as toxicity of amyloidogenic proteins, they bind to lysine and thus disrupt hydrophobic and electrostatic interactions which are required for nucleation, oligomerization, and fibril elongation. Anle 138b (3-[1,3-benzodioxol-5-yl]-5-[3-bromophenyl]-1H-pyrazole) inhibits α-synuclein oligomer formation, (Wagner et al., 2013) and the prolyl oligopeptidase inhibitor KYP-2047 (Savolainen et al., 2014). These therapeutics emphasize that there is a need to improve on how the mechanism behind α-syn aggregation has been defined so far and through that identifies better therapeutic targets.

TOM complex translocates α -syn to TIM and eventually in the matrix. In models of PD and post-mortem PD brain tissue, the localization of α-synuclein into the outer mitochondrial membrane is observed along with its interaction with the TOM complex (Bender et al., 2013; Devi et al., 2008; Di Maio et al., 2016). Similarly, some α-synuclein species like Ser-129E phosphomimetic are known for binding to TOM20 receptor and prevention of its interaction with co-receptor TOM22 and during in vitro conditions for inhibition of mitochondrial proteins like Mfn2 (Di Maio et al., 2016). This α-synuclein and TOM20 interaction was linked with excessive ROS production and mitochondrial impairment (Di Maio et al., 2016). Inhibition of α-synuclein translocation may also be targeted to stop the progression of PD. However, this remains unclear that whether blocking mitochondrial import is enough for driving nigrostriatal degeneration. It is important to note that, in the case of Huntington's disease (HD) model, neuronal death occurs due to impaired mitochondrial protein import (Yano et al., 2014).

Targeting α-synuclein by antibodies has emerged as a popular strategy, just like what was being tried last decade for Alzheimer's disease (AD). Neuroprotection has been reported by several groups after an active immunization-vaccination-based approach using short peptides or full-length protein and passive immunization-based approach using antibodies against the protein in transgenic mouse models of PD. As per monoclonal antibodies targeting α-synuclein, Roche and many other pharmaceutical companies have started phase I clinical trials. In a pilot study involving 32 PD patients, active immunization Affitope PD01 was proven to be safe. In the European SYMPATH consortium, PD03, from Affitope, will soon be evaluated in patients with PD. These studies involve numerous exploratory outcome measures. Despite the fact that active and passive immunization seems appealing many questions are still unanswered (Dehay et al., 2015). Firstly most of the studies were performed in *in vivo* conditions involving transgenic mice with restricted α-synuclein expression (PDGFß- and Thy1-A30P-synuclein transgenic mice), while the majority of α-synuclein is found on the membrane of red blood cells. It is therefore important to investigate it in early clinical studies to study the reaction of antibodies and peripheral α-synuclein and determine if the unbound antibodies are gaining access to the brain compartment at sufficient levels. Secondly, Lewy bodies are intraneuronal inclusions while α-synuclein is a cytosolic protein.

Antibodies recognition of intracellular protein, encouraging intracellular degradation has not been established yet but this could stop/halt transcellular α-syn propagation. Surprisingly, passive immunization against α-synuclein activates autophagy. Antibodies might thus trigger non-selective autophagy, eventually culminating in non-α-synuclein-related mechanism being responsible for α-synuclein clearance. This information suggests that there is a potential for developing therapeutics involving mitochondrial protein import due to the converging mechanisms in neurodegenerative disorders (Dehay, et al., 2015).

CONCLUSION

This chapter discussed the molecular mechanism of α-synuclein known so far, which is based on its effects in mitochondrial and lysosomal degradation pathways, and macroautophagy. These pathways are involved in the degrading α-synuclein in the neuronal system. Research publications support combating α-synuclein's toxicity with numerous therapeutic strategies till preclinical tests. Besides these, validation of using α-synuclein as a biomarker for the techniques like imaging, biochemical characterization and genetic link of PD has also been discussed. Targeting α-synuclein as a part of PD treatment seems to be promising but further experimentation and pre-clinical trials are needed to limit the risks involved with their clinical applications.

REFERENCES

Ai, S. X., Xu, Q., Hu, Y. C., Song, C. Y., Guo, J. F., Shen, L., ... Tang, B. S. (2014). Hypomethylation of SNCA in blood of patients with sporadic Parkinson's disease. *Journal of the Neurological Sciences*, *337*(1-2), 123–128. doi:10.1016/j.jns.2013.11.033 PMID:24326201

Alvarez-Erviti, L., Rodriguez-Oroz, M. C., Cooper, J. M., Caballero, C., Ferrer, I., Obeso, J. A., & Schapira, A. H. (2010). Chaperone-mediated autophagy markers in Parkinson disease brains. *Archives of Neurology*, *67*(12), 1464–1472. doi:10.1001/archneurol.2010.198 PMID:20697033

Atik, A., Stewart, T., & Zhang, J. (2016). Alpha-synuclein as a biomarker for Parkinson's disease. *Brain Pathology (Zurich, Switzerland)*, *26*(3), 410–418. doi:10.1111/bpa.12370 PMID:26940058

Auluck, P. K., & Bonini, N. M. (2002). Pharmacological prevention of Parkinson disease in Drosophila. *Nature Medicine*, *8*(11), 1185–1186. doi:10.1038/nm1102-1185 PMID:12411925

Barbour, R., Kling, K., Anderson, J. P., Banducci, K., Cole, T., Diep, L., & Chilcote, T. J. (2008). Red blood cells are the major source of alpha-synuclein in blood. *Neurodegenerative Diseases*, *5*(2), 55–59. doi:10.1159/000112832 PMID:18182779

Barone, P. (2010). Neurotransmission in Parkinson's disease: Beyond dopamine. *European Journal of Neurology*, *17*(3), 364–376. doi:10.1111/j.1468-1331.2009.02900.x PMID:20050885

Beach, T. G., Adler, C. H., Sue, L. I., Vedders, L., Lue, L., White, C. L. III, ... Walker, D. G. (2010). Multi-organ distribution of phosphorylated α-synuclein histopathology in subjects with Lewy body disorders. *Acta Neuropathologica*, *119*(6), 689–702. doi:10.100700401-010-0664-3 PMID:20306269

Beach, T. G., Serrano, G. E., Kremer, T., Canamero, M., Dziadek, S., Sade, H., ... Chahine, L. (2018). Immunohistochemical method and histopathology judging for the systemic synuclein sampling study (S4). *Journal of Neuropathology and Experimental Neurology*, *77*(9), 793–802. doi:10.1093/jnen/nly056 PMID:30107604

Bender, A., Desplats, P., Spencer, B., Rockenstein, E., Adame, A., Elstner, M., & Pham, E. (2013). TOM40 mediates mitochondrial dysfunction induced by α-synuclein accumulation in Parkinson's disease. *PLoS One*, *8*(4), e62277. doi:10.1371/journal.pone.0062277 PMID:23626796

Bendor, J. T., Logan, T. P., & Edwards, R. H. (2013). The function of α-synuclein. *Neuron*, *79*(6), 1044–1066. doi:10.1016/j.neuron.2013.09.004 PMID:24050397

Beyer, K. (2007). Mechanistic aspects of Parkinson's disease: A-synuclein and the biomembrane. *Cell Biochemistry and Biophysics*, *47*(2), 285–299. doi:10.100712013-007-0014-9 PMID:17652776

Bianco, C. L., Shorter, J., Régulier, E., Lashuel, H., Iwatsubo, T., Lindquist, S., & Aebischer, P. (2008). Hsp104 antagonizes α-synuclein aggregation and reduces dopaminergic degeneration in a rat model of Parkinson disease. *The Journal of Clinical Investigation*, *118*(9), 3087–3097. doi:10.1172/JCI35781 PMID:18704197

Bieschke, J., Russ, J., Friedrich, R. P., Ehrnhoefer, D. E., Wobst, H., Neugebauer, K., & Wanker, E. E. (2010). EGCG remodels mature α-synuclein and amyloid-β fibrils and reduces cellular toxicity. *Proceedings of the National Academy of Sciences of the United States of America*, *107*(17), 7710–7715. doi:10.1073/pnas.0910723107 PMID:20385841

Brichta, L., & Greengard, P. (2014). Molecular determinants of selective dopaminergic vulnerability in Parkinson's disease: An update. *Frontiers in Neuroanatomy*, *8*, 152–152. doi:10.3389/fnana.2014.00152 PMID:25565977

Cabin, D. E., Shimazu, K., Murphy, D., Cole, N. B., Gottschalk, W., McIlwain, K. L., ... Nussbaum, R. L. (2002). Synaptic vesicle depletion correlates with attenuated synaptic responses to prolonged repetitive stimulation in mice lacking α-synuclein. *The Journal of Neuroscience: The Official Journal of the Society for Neuroscience*, *22*(20), 8797–8807. doi:10.1523/JNEUROSCI.22-20-08797.2002 PMID:12388586

Chalfant, C. E., Kishikawa, K., Mumby, M. C., Kamibayashi, C., Bielawska, A., & Hannun, Y. A. (1999). Long chain ceramides activate protein phosphatase-1 and protein phosphatase-2A Activation is stereospecific and regulated by phosphatidic acid. *The Journal of Biological Chemistry*, *274*(29), 20313–20317. doi:10.1074/jbc.274.29.20313 PMID:10400653

Chen, M., Yang, W., Li, X., Li, X., Wang, P., Yue, F., & Yu, S. (2016). Age- and brain region-dependent α-synuclein oligomerization is attributed to alterations in intrinsic enzymes regulating α-synuclein phosphorylation in aging monkey brains. *Oncotarget*, *7*(8), 8466–8480. doi:10.18632/oncotarget.6445 PMID:27032368

Chen-Plotkin, A. S. (2014). Unbiased approaches to biomarker discovery in neurodegenerative diseases. *Neuron*, *84*(3), 594–607. doi:10.1016/j.neuron.2014.10.031 PMID:25442938

Chinta, S. J., Mallajosyula, J. K., Rane, A., & Andersen, J. K. (2010). Mitochondrial alpha-synuclein accumulation impairs complex I function in dopaminergic neurons and results in increased mitophagy in vivo. *Neuroscience Letters*, *486*(3), 235–239. doi:10.1016/j.neulet.2010.09.061 PMID:20887775

Choi, W., Zibaee, S., Jakes, R., Serpell, L. C., Davletov, B., Anthony Crowther, R., & Goedert, M. (2004). Mutation E46K increases phospholipid binding and assembly into filaments of human α-synuclein. *FEBS Letters*, *576*(3), 363–368. doi:10.1016/j.febslet.2004.09.038 PMID:15498564

Conway, K. A., Harper, J. D., & Lansbury, P. T. (1998). Accelerated in vitro fibril formation by a mutant α-synuclein linked to early-onset Parkinson disease. *Nature Medicine*, *4*(11), 1318–1320. doi:10.1038/3311 PMID:9809558

Cooper, A. A., Gitler, A. D., Cashikar, A., Haynes, C. M., Hill, K. J., Bhullar, B., & Cao, S. (2006). α-Synuclein blocks ER-Golgi traffic and Rab1 rescues neuron loss in Parkinson's models. *Science*, *313*(5785), 324–328. doi:10.1126cience.1129462 PMID:16794039

Cuervo, A. M., & Dice, J. F. (2000). Age-related decline in chaperone-mediated autophagy. *The Journal of Biological Chemistry*, *275*(40), 31505–31513. doi:10.1074/jbc.M002102200 PMID:10806201

Cullen, V., Sardi, S. P., Ng, J., Xu, Y. H., Sun, Y., Tomlinson, J. J., ... Schlossmacher, M. G. (2011). Acid β-glucosidase mutants linked to Gaucher disease, Parkinson disease, and Lewy body dementia alter α-synuclein processing. *Annals of Neurology*, *69*(6), 940–953. doi:10.1002/ana.22400 PMID:21472771

Deas, E., Cremades, N., Angelova, P. R., Ludtmann, M. H., Yao, Z., Chen, S., ... Gissen, P. (2016). Alpha-synuclein oligomers interact with metal ions to induce oxidative stress and neuronal death in Parkinson's disease. *Antioxidants & Redox Signalling*, *24*(7), 376–391. doi:10.1089/ars.2015.6343 PMID:26564470

Dehay, B., Bourdenx, M., Gorry, P., Przedborski, S., Vila, M., Hunot, S., ... Meissner, W. G. (2015). Targeting α-synuclein for treatment of Parkinson's disease: Mechanistic and therapeutic considerations. *Lancet Neurology*, *14*(8), 855–866. doi:10.1016/S1474-4422(15)00006-X PMID:26050140

Del Tredici, K., & Braak, H. (2016). Sporadic Parkinson's disease: Development and distribution of α-synuclein pathology. *Neuropathology and Applied Neurobiology*, *42*(1), 33–50. doi:10.1111/nan.12298 PMID:26662475

Deng, S., Deng, X., Yuan, L., Song, Z., Yang, Z., Xiong, W., & Deng, H. (2015). Genetic analysis of SNCA coding mutation in Chinese Han patients with Parkinson disease. *Acta Neurologica Belgica, 115*(3), 267–271. doi:10.100713760-014-0347-2 PMID:25092551

Di Maio, R., Barrett, P. J., Hoffman, E. K., Barrett, C. W., Zharikov, A., Borah, A., ... & Hastings, T. G. (2016). α-Synuclein binds to TOM20 and inhibits mitochondrial protein import in Parkinson's disease. *Science Translational Medicine, 8*(342), 342ra78-342ra78.

Dobrowsky, R. T., Kamibayashi, C., Mumby, M. C., & Hannun, Y. A. (1993). Ceramide activates heterotrimeric protein phosphatase 2A. *The Journal of Biological Chemistry, 268*(21), 15523–15530. PMID:8393446

Doxakis, E. (2010). Post-transcriptional regulation of α-synuclein expression by mir-7 and mir-153. *The Journal of Biological Chemistry, 285*(17), 12726–12734. doi:10.1074/jbc.M109.086827 PMID:20106983

Fares, M. B., Ait-Bouziad, N., Dikiy, I., Mbefo, M. K., Jovičić, A., Kiely, A., ... Lashuel, H. A. (2014). The novel Parkinson's disease linked mutation G51D attenuates in vitro aggregation and membrane binding of α-synuclein, and enhances its secretion and nuclear localization in cells. *Human Molecular Genetics, 23*(17), 4491–4509. doi:10.1093/hmg/ddu165 PMID:24728187

Gambhir, S., Czernin, J., Schwimmer, J., Silverman, D. H. S., Coleman, R. E., & Phelps, M. E. (2001). A Tabulated Summary of the FDG PET Literature. *Journal of Nuclear Medicine, 42*(S1), 1S–93S. PMID:11483694

Gazzina, S., Premi, E., Turrone, R., Acosta-Cabronero, J., Rizzetti, M. C., Cotelli, M. S., & Borroni, B. (2016). Subcortical matter in the α-synucleinopathies spectrum: An MRI pilot study. *Journal of Neurology, 263*(8), 1575–1582. doi:10.100700415-016-8173-5 PMID:27230856

George, J. M. (2001). The synucleins. *Genome Biology, 3*(1), reviews3002-1.

Ghosh, D., Mondal, M., Mohite, G. M., Singh, P. K., Ranjan, P., Anoop, A., ... Maji, S. K. (2013). The Parkinson's disease-associated H50Q mutation accelerates alpha-Synuclein aggregation *in vitro*. *Biochemistry, 52*(40), 6925–6927. doi:10.1021/bi400999d PMID:24047453

Goers, J., Manning-Bog, A. B., McCormack, A. L., Millett, I. S., Doniach, S., Di Monte, D. A., ... Fink, A. L. (2003). Nuclear localization of α-synuclein and its interaction with histones. *Biochemistry, 42*(28), 8465–8471. doi:10.1021/bi0341152 PMID:12859192

Golbe, L. I., Lazzarini, A. M., Duvoisin, R. C., Iorio, G. D., Sanges, G., Bonavita, V., & la Sala, S. (1996). Clinical genetic analysis of Parkinson's disease in the Contursi kindred. *Annals of Neurology: Official Journal of the American Neurological Association and the Child Neurology Society, 40*(5), 767–775. doi:10.1002/ana.410400513 PMID:8957018

Goldberg, M. S., Fleming, S. M., Palacino, J. J., Cepeda, C., Lam, H. A., Bhatnagar, A., ... Shen, J. (2003). Parkin-deficient mice exhibit nigrostriataldeficits but not loss of dopaminergic neurons. *The Journal of Biological Chemistry, 278*(44), 43628–43635. doi:10.1074/jbc.M308947200 PMID:12930822

Goldman, J. G., Andrews, H., Amara, A., Naito, A., Alcalay, R. N., Shaw, L. M., ... Kang, U. J. (2018). Cerebrospinal fluid, plasma, and saliva in the BioFIND study: Relationships among biomarkers and Parkinson's disease Features. *Movement Disorders, 33*(2), 282–288. doi:10.1002/mds.27232 PMID:29205509

Gonçalves, S., & Outeiro, T. F. (2013). Assessing the subcellular dynamics of alpha-synuclein using photoactivation microscopy. *Molecular Neurobiology, 47*(3), 1081–1092. doi:10.100712035-013-8406-x PMID:23389286

Greenbaum, E. A., Graves, C. L., Mishizen-Eberz, A. J., Lupoli, M. A., Lynch, D. R., Englander, S. W., ... Giasson, B. I. (2005). The E46K mutation in α-synuclein increases amyloid fibril formation. *The Journal of Biological Chemistry, 280*(9), 7800–7807. doi:10.1074/jbc.M411638200 PMID:15632170

Guardia-Laguarta, C., Area-Gomez, E., Rüb, C., Liu, Y., Magrané, J., Becker, D., & Przedborski, S. (2014). α-Synuclein is localized to mitochondria-associated ER membranes. *The Journal of Neuroscience: The Official Journal of the Society for Neuroscience, 34*(1), 249–259. doi:10.1523/JNEUROSCI.2507-13.2014 PMID:24381286

Harrison, I. F., & Dexter, D. T. (2013). Epigenetic targeting of histone deacetylase: Therapeutic potential in Parkinson's disease? *Pharmacology & Therapeutics, 140*(1), 34–52. doi:10.1016/j.pharmthera.2013.05.010 PMID:23711791

Hope, A. D., Myhre, R., Kachergus, J., Lincoln, S., Bisceglio, G., Hulihan, M., & Farrer, M. J. (2004). α-Synuclein missense and multiplication mutations in autosomal dominant Parkinson's disease. *Neuroscience Letters, 367*(1), 97–100. doi:10.1016/j.neulet.2004.05.100 PMID:15308306

Horvath, I., Iashchishyn, I. A., Moskalenko, R. A., Wang, C., Wärmländer, S. K., Wallin, C., & Morozova-Roche, L. A. (2018). Co-aggregation of pro-inflammatory S100A9 with α-synuclein in Parkinson's disease: Ex vivo and in vitro studies. *Journal of Neuroinflammation, 15*(1), 172. doi:10.118612974-018-1210-9 PMID:29866153

Hsieh, C. H., Shaltouki, A., Gonzalez, A. E., da Cruz, A. B., Burbulla, L. F., Lawrence, E. S., & Wang, X. (2016). Functional impairment in miro degradation and mitophagy is a shared feature in familial and sporadic Parkinson's disease. *Cell Stem Cell, 19*(6), 709–724. doi:10.1016/j.stem.2016.08.002 PMID:27618216

Íñigo-Marco, I., Valencia, M., Larrea, L., Bugallo, R., Martínez-Goikoetxea, M., Zuriguel, I., & Arrasate, M. (2017). E46K α-synuclein pathological mutation causes cell-autonomous toxicity without altering protein turnover or aggregation. *Proceedings of the National Academy of Sciences of the United States of America, 114*(39), E8274–E8283. doi:10.1073/pnas.1703420114 PMID:28900007

Jensen, T. S., Baron, R., Haanpää, M., Kalso, E., Loeser, J. D., Rice, A. S., & Treede, R. D. (2011). A new definition of neuropathic pain. *Pain, 152*(10), 2204–2205. doi:10.1016/j.pain.2011.06.017 PMID:21764514

Johnson, R., Noble, W., Tartaglia, G. G., & Buckley, N. J. (2012). Neurodegeneration as an RNA disorder. *Progress in Neurobiology, 99*(3), 293–315. doi:10.1016/j.pneurobio.2012.09.006 PMID:23063563

Kahle, P. J. (2008). α-synucleinopathy models and human neuropathology: Similarities and differences. *Acta Neuropathologica, 115*(1), 87–95. doi:10.100700401-007-0302-x PMID:17932682

Kamp, F., Exner, N., Lutz, A. K., Wender, N., Hegermann, J., Brunner, B., ... Eimer, S. (2010). Inhibition of mitochondrial fusion by α-synuclein is rescued by PINK1, Parkin and DJ-1. *The EMBO Journal, 29*(20), 3571–3589. doi:10.1038/emboj.2010.223 PMID:20842103

Kanagaraj, N., Beiping, H., Dheen, S. T., & Tay, S. S. W. (2014). Downregulation of miR-124 in MPTP-treated mouse model of Parkinson's disease and MPP iodide-treated MN9D cells modulates the expression of the calpain/cdk5 pathway proteins. *Neuroscience, 272*, 167–179. doi:10.1016/j.neuroscience.2014.04.039 PMID:24792712

Kang, L., Wu, K. P., Vendruscolo, M., & Baum, J. (2011). The A53T mutation is key in defining the differences in the aggregation kinetics of human and mouse α-synuclein. *Journal of the American Chemical Society, 133*(34), 13465–13470. doi:10.1021/ja203979j PMID:21721555

Khalaf, O., Fauvet, B., Oueslati, A., Dikiy, I., Mahul-Mellier, A. L., Ruggeri, F. S., ... Eliezer, D. (2014). The H50Q mutation enhances α-synuclein aggregation, secretion, and toxicity. *The Journal of Biological Chemistry, 289*(32), 21856–21876. doi:10.1074/jbc.M114.553297 PMID:24936070

Kiely, A. P., Ling, H., Asi, Y. T., Kara, E., Proukakis, C., Schapira, A. H., ... Holton, J. L. (2015). Distinct clinical and neuropathological features of G51D SNCA mutation cases compared with SNCA duplication and H50Q mutation. *Molecular Neurodegeneration, 10*(1), 41. doi:10.118613024-015-0038-3 PMID:26306801

Kim, J., Inoue, K., Ishii, J., Vanti, W. B., Voronov, S. V., Murchison, E., ... Abeliovich, A. (2007). AmicroRNA feedback circuit in midbrain dopamine neurons. *Science, 317*(5842), 1220–1224. doi:10.1126cience.1140481 PMID:17761882

Kim, W., Lee, Y., McKenna, N. D., Yi, M., Simunovic, F., Wang, Y., ... Sonntag, K. C. (2014). miR-126 contributes to Parkinson's disease by dysregulating the insulin-like growth factor/phosphoinositide 3-kinase signaling. *Neurobiology ofAging, 35*(7), 1712–1721. doi:10.1016/j.neurobiolaging.2014.01.021 PMID:24559646

Klein, C. J., & Benarroch, E. E. (2014). Epigenetic regulation: Basic concepts and relevance to neurologic disease. *Neurology, 82*(20), 1833–1840. doi:10.1212/WNL.0000000000000440 PMID:24759839

Klionsky, D. J., & Emr, S. D. (2000). Autophagy as a regulated pathway of cellular degradation. *Science, 290*(5497), 1717–1721. doi:10.1126cience.290.5497.1717 PMID:11099404

Klucken, J., Shin, Y., Masliah, E., Hyman, B. T., & McLean, P. J. (2004). Hsp70 reduces α-synuclein aggregation and toxicity. *The Journal of Biological Chemistry, 279*(24), 25497–25502. doi:10.1074/jbc.M400255200 PMID:15044495

Kontopoulos, E., Parvin, J. D., & Feany, M. B. (2006). α-synuclein acts in the nucleus to inhibit histone acetylation and promote neurotoxicity. *Human Molecular Genetics, 15*(20), 3012–3023. doi:10.1093/hmg/ddl243 PMID:16959795

Liu, C.W., Corboy, M.J., DeMartino, G.N., & Thomas, P.J. (2003). Endoproteolytic activity of the proteasome. *Science, 299*, 408–411.

Magrinelli, F., Picelli, A., Tocco, P., Federico, A., Roncari, L., Smania, N., & Tamburin, S. (2016). Pathophysiology of motor dysfunction in Parkinson's disease as the rationale for drug treatment and rehabilitation. *Parkinson's Disease*, *2016*, 9832839. doi:10.1155/2016/9832839 PMID:27366343

Maroteaux, L., & Scheller, R. H. (1991). The rat brain synucleins; family of proteins transiently associated with neuronal membrane. *Brain Research. Molecular Brain Research*, *11*(3-4), 335–343. doi:10.1016/0169-328X(91)90043-W PMID:1661825

Matsumoto, L., Takuma, H., Tamaoka, A., Kurisaki, H., Date, H., Tsuji, S., & Iwata, A. (2010). CpG demethylation enhances alpha-synuclein expression and affects the pathogenesis of Parkinson's disease. *PLoS One*, *5*(11), e15522. doi:10.1371/journal.pone.0015522 PMID:21124796

McCormack, A. L., Mak, S. K., Henderson, J. M., Bumcrot, D., Farrer, M. J., & Di Monte, D. A. (2010). α-synuclein suppression by targeted small interfering RNA in the primate substantianigra. *PLoS One*, *5*(8), e12122. doi:10.1371/journal.pone.0012122 PMID:20711464

Middleton, E. R., & Rhoades, E. (2010). Effects of curvature and composition on α-synuclein binding to lipid vesicles. *Biophysical Journal*, *99*(7), 2279–2288. doi:10.1016/j.bpj.2010.07.056 PMID:20923663

Mollenhauer, B., Trautmann, E., Otte, B., Ng, J., Spreer, A., Lange, P., & Trenkwalder, C. (2012). α-Synuclein in human cerebrospinal fluid is principally derived from neurons of the central nervous system. *Journal of Neural Transmission (Vienna, Austria)*, *119*(7), 739–746. doi:10.100700702-012-0784-0 PMID:22426833

Mori, H., Kondo, T., Yokochi, M., Matsumine, H., Nakagawa-Hattori, Y., Miyake, T., ... Mizuno, Y. (1998). Pathologic and biochemical studies of juvenile parkinsonism linked to chromosome 6q. *Neurology*, *51*(3), 890–892. doi:10.1212/WNL.51.3.890 PMID:9748052

Nakamura, K., Nemani, V. M., Azarbal, F., Skibinski, G., Levy, J. M., Egami, K., ... Sesaki, H. (2011). Direct membrane association drives mitochondrial fission by the Parkinson disease-associated protein α-synuclein. *The Journal of Biological Chemistry*, *286*(23), 20710–20726. doi:10.1074/jbc.M110.213538 PMID:21489994

Pagano, G., Ferrara, N., Brooks, D. J., & Pavese, N. (2016). Age at onset and Parkinson disease phenotype. *Neurology*, *86*(15), 1400–1407. doi:10.1212/WNL.0000000000002461 PMID:26865518

Pérez-Revuelta, B. I., Hettich, M. M., Ciociaro, A., Rotermund, C., Kahle, P. J., Krauss, S., & Di Monte, D. A. (2014). Metformin lowers Ser-129 phosphorylated α-synuclein levels via mTOR-dependent protein phosphatase 2A activation. *Cell Death & Disease*, *5*(5), e1209. doi:10.1038/cddis.2014.175 PMID:24810045

Peterson, R. T., Desai, B. N., Hardwick, J. S., & Schreiber, S. L. (1999). Protein phosphatase 2A interacts with the 70-kDa S6 kinase and is activated by inhibition of FKBP12–rapamycinassociated protein. *Proceedings of the National Academy of Sciences of the United States of America*, *96*(8), 4438–4442. doi:10.1073/pnas.96.8.4438 PMID:10200280

Picillo, M., Lizarraga, K. J., Friesen, E. L., Chau, H., Zhang, M., Sato, C., & Kalia, S. K. (2018). Parkinsonism due to A53E α-synuclein gene mutation: Clinical, genetic, epigenetic, and biochemical features. *Movement Disorders*, *33*(12), 1950–1955. doi:10.1002/mds.27506 PMID:30423204

Prabhudesai, S., Sinha, S., Attar, A., Kotagiri, A., Fitzmaurice, A. G., Lakshmanan, R., & Stahl, M. (2012). A novel "molecular tweezer" inhibitor of α-synuclein neurotoxicity in vitro and in vivo. *Neurotherapeutics; the Journal of the American Society for Experimental NeuroTherapeutics*, *9*(2), 464–476. doi:10.100713311-012-0105-1 PMID:22373667

Qureshi, I. A., & Mehler, M. F. (2011). Advances in epigenetics and epigenomics for neurodegenerative diseases. *Current Neurology and Neuroscience Reports*, *11*(5), 464–473. doi:10.100711910-011-0210-2 PMID:21671162

Rideout, H. J., Dietrich, P., Wang, Q., Dauer, W. T., & Stefanis, L. (2004). α-Synuclein is required for the fibrillar nature of ubiquitinated inclusions induced by proteasomal inhibition in primary neurons. *The Journal of Biological Chemistry*, *279*(45), 46915–46920. doi:10.1074/jbc.M405146200 PMID:15322100

Rocha, E. M., De Miranda, B., & Sanders, L. H. (2018). Alpha-synuclein: Pathology, mitochondrial dysfunction and neuroinflammation in Parkinson's disease. *Neurobiology of Disease*, *109*, 249–257. doi:10.1016/j.nbd.2017.04.004 PMID:28400134

Roodveldt, C., Christodoulou, J., & Dobson, C. M. (2008). Immunological features of α-synuclein in Parkinson's disease. *Journal of Cellular and Molecular Medicine*, *12*(5b), 1820–1829. doi:10.1111/j.1582-4934.2008.00450.x PMID:18671754

Rosborough, K., Patel, N., & Kalia, L. V. (2017). Alpha-synuclein and Parkinsonism: Updates and Future Perspectives. *Current Neurology and Neuroscience Reports*, *17*(4), 31. doi:10.100711910-017-0737-y PMID:28324300

Rutherford, N. J., Moore, B. D., Golde, T. E., & Giasson, B. I. (2014). Divergent effects of the H50Q and G51D SNCA mutations on the aggregation of α-synuclein. *Journal of Neurochemistry*, *131*(6), 859–867. doi:10.1111/jnc.12806 PMID:24984882

Ryan, B. J., Hoek, S., Fon, E. A., & Wade-Martins, R. (2015). Mitochondrial dysfunction and mitophagy in Parkinson's: From familial to sporadic disease. *Trends in Biochemical Sciences*, *40*(4), 200–210. doi:10.1016/j.tibs.2015.02.003 PMID:25757399

Savolainen, M. H., Richie, C. T., Harvey, B. K., Männistö, P. T., Maguire-Zeiss, K. A., & Myöhänen, T. T. (2014). The beneficial effect of a prolyloligopeptidase inhibitor, KYP-2047, on alpha-synuclein clearance and autophagy in A30P transgenic mouse. *Neurobiology of Disease*, *68*, 1–15. doi:10.1016/j.nbd.2014.04.003 PMID:24746855

Schiesling, C., Kieper, N., Seidel, K., & Krüger, R. (2008). Familial Parkinson's disease–genetics, clinical phenotype and neuropathology in relation to the common sporadic form of the disease. *Neuropathology and Applied Neurobiology*, *34*(3), 255–271. doi:10.1111/j.1365-2990.2008.00952.x PMID:18447897

Shaw, K., Lees, A., & Stern, G. M. (1980). The impact of treatment with levodopa on Parkinson's disease. *The Quarterly Journal of Medicine*, *49*(3), 283–293. PMID:7465763

Shi, M., Liu, C., Cook, T. J., Bullock, K. M., Zhao, Y., Ginghina, C., ... Zhang, J. (2014). Plasma exosomal α-synuclein is likely CNS-derived and increased in Parkinson's disease. *Acta Neuropathologica*, *128*(5), 639–650. doi:10.100700401-014-1314-y PMID:24997849

Sidransky, E., Nalls, M. A., Aasly, J. O., Aharon-Peretz, J., Annesi, G., Barbosa, E. R., ... Chen, C. M. (2009). Multicenter analysis of glucocerebrosidase mutations in Parkinson's disease. *The New England Journal of Medicine, 361*(17), 1651–1661. doi:10.1056/NEJMoa0901281 PMID:19846850

Singleton, A., Farrer, M., Johnson, J., Singleton, A., Hague, S., Kachergus, J., ... Gwinn-Hardy, K. (2003). α-Synuclein locus triplication causes Parkinson's disease. *Science, 302*(5646), 841–841. doi:10.1126cience.1090278 PMID:14593171

Smith, Y., Wichmann, T., Factor, S. A., & DeLong, M. R. (2012). Parkinson's disease therapeutics: new developments and challenges since the introduction of levodopa. *Neuropsychopharmacology, 37*(1), 213-246. doi:10.1038/npp.2011.212

Song, Y., Ding, H., Yang, J., Lin, Q., Xue, J., Zhang, Y., ... Cai, Y. (2014). Pyrosequencing analysis of SNCA methylation levels in leukocytes from Parkinson's disease patients. *Neuroscience Letters, 569*, 85–88. doi:10.1016/j.neulet.2014.03.076 PMID:24721670

Specht, C. G., Tigaret, C. M., Rast, G. F., Thalhammer, A., Rudhard, Y., & Schoepfer, R. (2005). Subcellular localisation of recombinant α-and γ-synuclein. *Molecular and Cellular Neurosciences, 28*(2), 326–334. doi:10.1016/j.mcn.2004.09.017 PMID:15691713

Steiner, J. A., Angot, E., & Brundin, P. (2011). A deadly spread: Cellular mechanisms of α-synuclein transfer. *Cell Death and Differentiation, 18*(9), 1425–1433. doi:10.1038/cdd.2011.53 PMID:21566660

Stojkovska, I., Krainc, D., & Mazzulli, J. R. (2018). Molecular mechanisms of α-synuclein and GBA1 in Parkinson's disease. *Cell and Tissue Research, 373*(1), 51–60. doi:10.100700441-017-2704-y PMID:29064079

Tofaris, G. K., Kim, H. T., Hourez, R., Jung, J. W., Kim, K. P., & Goldberg, A. L. (2011). Ubiquitin ligase Nedd4 promotes α-synuclein degradation by the endosomal–lysosomal pathway. *Proceedings of the National Academy of Sciences of the United States of America, 108*(41), 17004–17009. doi:10.1073/pnas.1109356108 PMID:21953697

Tsigelny, I. F., Sharikov, Y., Wrasidlo, W., Gonzalez, T., Desplats, P. A., Crews, L., & Masliah, E. (2012). Role of α-synuclein penetration into the membrane in the mechanisms of oligomer pore formation. *The FEBS Journal, 279*(6), 1000–1013. doi:10.1111/j.1742-4658.2012.08489.x PMID:22251432

Vale, R. D. (2003). The molecular motor toolbox for intracellular transport. *Cell, 112*(4), 467–480. doi:10.1016/S0092-8674(03)00111-9 PMID:12600311

Van Laar, V. S., & Berman, S. B. (2013). The interplay of neuronal mitochondrial dynamics and bioenergetics: Implications for Parkinson's disease. *Neurobiology of Disease, 51*, 43–55. doi:10.1016/j.nbd.2012.05.015 PMID:22668779

Vilarino-Guell, C., Wider, C., Ross, O. A., Dachsel, J. C., Kachergus, J. M., Lincoln, S. J., ... Farrer, M. J. (2011). VPS35 mutations in Parkinson disease. *American Journal of Human Genetics, 89*(1), 162–167. doi:10.1016/j.ajhg.2011.06.001 PMID:21763482

Visanji, N. P., Brooks, P. L., Hazrati, L. N., & Lang, A. E. (2013). The prion hypothesis in Parkinson's disease: Braak to the future. *Acta Neuropathologica Communications, 1*(1), 2. doi:10.1186/2051-5960-1-2 PMID:24252164

Vogiatzi, T., Xilouri, M., Vekrellis, K., & Stefanis, L. (2008). Wild type α-synuclein is degraded by chaperone-mediated autophagy and macroautophagy in neuronal cells. *The Journal of Biological Chemistry*, *283*(35), 23542–23556. doi:10.1074/jbc.M801992200 PMID:18566453

Volpicelli-Daley, L. A., Gamble, K. L., Schultheiss, C. E., Riddle, D. M., West, A. B., & Lee, V. M. Y. (2014). Formation of α-synuclein Lewy neurite–like aggregates in axons impedes the transport of distinct endosomes. *Molecular Biology of the Cell*, *25*(25), 4010–4023. doi:10.1091/mbc.e14-02-0741 PMID:25298402

Wagner, J., Ryazanov, S., Leonov, A., Levin, J., Shi, S., Schmidt, F., ... Giese, A. (2013). Anle138b: A novel oligomer modulator for disease-modifying therapy of neurodegenerative diseases such as prion and Parkinson's disease. *Acta Neuropathologica*, *125*(6), 795–813. doi:10.100700401-013-1114-9 PMID:23604588

Winklhofer, K. F., & Haass, C. (2010). Mitochondrial dysfunction in Parkinson's disease. *Biochimicaet Biophysica Acta (BBA)- Molecular Basis of Disease*, *1802*(1), 29–44. doi:10.1016/j.bbadis.2009.08.013

Witt, M., Bormann, K., Gudziol, V., Pehlke, K., Barth, K., Minovi, A., ... Hummel, T. (2009). Biopsies of olfactory epithelium in patients with Parkinson's disease. *Movement Disorders*, *24*(6), 906–914. doi:10.1002/mds.22464 PMID:19205070

Xie, W., & Chung, K. K. (2012). Alpha-synuclein impairs normal dynamics of mitochondria in cell and animal models of Parkinson's disease. *Journal of Neurochemistry*, *122*(2), 404–414. doi:10.1111/j.1471-4159.2012.07769.x PMID:22537068

Xilouri, M., Brekk, O. R., Kirik, D., & Stefanis, L. (2013). LAMP2A as a therapeutic target in Parkinson disease. *Autophagy*, *9*(12), 2166–2168. doi:10.4161/auto.26451 PMID:24145820

Xu, L., & Pu, J. (2016). Alpha-synuclein in Parkinson's Disease: From pathogenetic dysfunction to potential clinical application. *Parkinson's Disease*, *2016*, 1720621. doi:10.1155/2016/1720621 PMID:27610264

Yano, H., Baranov, S. V., Baranova, O. V., Kim, J., Pan, Y., Yablonska, S., ... Friedlander, R. M. (2014). Inhibition of mitochondrial protein import by mutant huntingtin. *Nature Neuroscience*, *17*(6), 822–831. doi:10.1038/nn.3721 PMID:24836077

Zimprich, A., Benet-Pages, A., Struhal, W., Graf, E., Eck, S. H., Offman, M. N., ... Strom, T. M. (2011). A mutation in VPS35, encoding a subunit of the retromer complex, causes late-onset Parkinson disease. *American Journal of Human Genetics*, *89*(1), 168–175. doi:10.1016/j.ajhg.2011.06.008 PMID:21763483

Section 3

Proteins and Protein Complexes in Neurodegeneration

Chapter 9
Beclin 1 Complex and Neurodegenerative Disorders

Ozaifa Kareem
Department of Pharmaceutical Sciences (Pharmacology Division), Faculty of Applied Sciences and Technology, University of Kashmir, Srinagar, India

Ghulam Nabi Bader
https://orcid.org/0000-0002-6820-0805
Department of Pharmaceutical Sciences (Pharmacology Division), Faculty of Applied Sciences and Technology, University of Kashmir, Srinagar, India

Faheem Hyder Pottoo
Department of Pharmacology, College of Clinical Pharmacy, Imam Abdulrahman Bin Faisal University, Saudi Arabia

Mohd. Amir
College of Clinical Pharmacy, Imam Abdulrahman Bin Faisal University, Saudi Arabia

Md. Abul Barkat
Department of Pharmaceutics, College of Pharmacy, University of Hafr Al-Batin, Al Jamiah, Hafr Al-Batin, Saudi Arabia

Mukesh Pandey
https://orcid.org/0000-0003-4474-0417
Department of Pharmaceutics, Delhi Institute of Pharmaceutical Sciences and Research, India

ABSTRACT

Beclin1 is the mammalian orthologue of yeast Atg6/vacuolar protein sorting-30 (VPS30). Beclin1 interacts with various biological macromolecules like ATG14, BIF-1, NRBF2, RUBICON, UVRAG, AMBRA1, HMGB1, PINK1, and PARKIN. Such interactions promote Beclin1-PI3KC3 complex formation. Autophagy is blocked in apoptosis owing to the breakdown of Beclin1 by caspase whereas autophagy induction inhibits effector caspase degradation, therefore, blocks apoptosis. Thus, the Beclin1 is an essential biomolecular species for cross-regulation between autophagy and apoptosis. Various studies carried out in neurodegenerative animal models associated with aggregated proteins have confirmed that multifunctional Beclin1 protein is necessary for neuronal integrity. The role of Beclin1 protein has been investigated and was reported in various human neurodegeneration disorders. This chapter aims to provide an insight into the role of Beclin1 in the development of neurodegenerative disorders.

DOI: 10.4018/978-1-7998-1317-0.ch009

INTRODUCTION

Several neurodegenerative disorders like Alzheimer's disease (AD), Parkinson's disease (PD), amyotrophic lateral sclerosis (ALS), epilepsy, etc., share a similar generalized pathology of acute or chronic loss of nerve cells (Pottoo et al., 2014; Nigar et al., 2016; Pottoo et al., 2019). This loss is triggered by different injuries and insults to neurons including the accumulation of misfolded proteins and damage to the cellular organelles. The removal of these abnormal and damaged cellular components is mandatory for maintaining normal cellular functions and activities of the nerve cells. This is done by the process named as autophagy. This process maintains and preserves the normal homeostasis of cells in the biological system through lysosomal degradation of the nonfunctional, defective or damaged organelles as well as biological macromolecules (Son et al., 2012). It is a tightly conserved process that starts with the combined operation of the Atg1 (autophagy-related 1)/ULK1 (Unc-51 like autophagy activating kinase-1) complex, and the PI3K-III (phosphatidylinositol 3-kinase, class III) complex (Bento et al., 2016).

Beclin1, a mammalian autophagy protein component of PI3K-IIIcomplex is crucial for membrane restructuring, endocytosis, cytokinesis, and phagocytosis. It is also required to start vesicular nucleation in autophagy (McKnight et al., 2014). This Beclin1 mediated autophagy has a prominent role and importance in various physiological and pathological conditions like neurodegeneration, tumorigenesis, and immunity. The dysregulation of Beclin1 affects other processes including apoptosis and cell death. This chapter highlights the basic knowledge of the Beclin1 functioning in autophagy and apoptosis and also focuses on its role in neurodegeneration.

MOLECULAR MECHANISMS OF AUTOPHAGY

Autophagy can be described as an intracellular biochemical process of catabolic nature that facilitates degradation and removal of dysfunctional proteins, damaged organelles, unutilized biological macromolecules, and invaded pathogens or foreign biological materials present within the cytoplasm of a cell through an autophagosome-lysosomal pathway. It functions as a survival mechanism against various kinds of stress such as starvation, during which unwanted cytoplasmic material is recycled and adenosine triphosphate (ATP) molecules are synthesized and also as a defense system against pathogenic microorganisms and cellular debris (Mizushima et al., 2011) as shown in Figure 1. Based on the mechanisms utilized for transporting cargo to lysosomes, three different autophagy processes were reported in the literature (Figure 2); macroautophagy, microautophagy and chaperone-mediated autophagy (CMA). Macroautophagy also called autophagy is the basic pathway utilized by cells for degrading cytoplasmic organelles and malformed proteins (Deretic et al., 2018).

The initiation of autophagy requires the sequestration of damaged cell organelles and the cellular proteins into an isolated membrane structure termed as phagophore. The structure expands and engulfs a small fragment of the cytosol to form an autophagosome (double-membrane vesicles). Subsequently, these autophagosomes undergo fusion with endosomal vesicles to form amphisomes. Amphisomes form autolysosomes by fusing with acidified lysosomes. Autolysosomes are the units responsible for the recycling and degradation of cytoplasmic contents. Thus, autophagy is simply a programmed cellular mechanism for cell survival, but if it becomes uncontrolled or unrestricted, it leads to cell component degeneration and ultimately cell death (Dikic et al., 2018).

Figure 1. Roles of autophagy in biological systems.

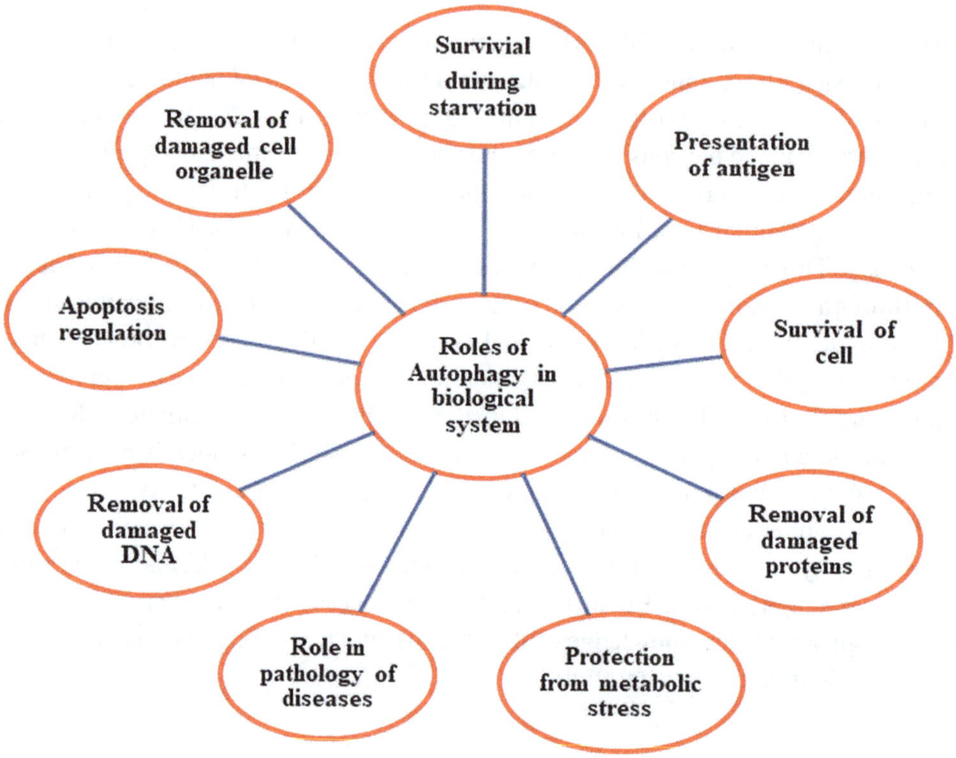

The mammalian cell autophagy process comprises of several steps (Figure 3). The first step is called initiation step, followed by the nucleation step (second step). The elongation is the third step in autophagy and the fourth step is called the closure step. Maturation and extrusion or degradation are the fifth and sixth steps respectively. Autophagy-related proteins (Atgs), regulate these steps. Thirty-six different Atgs are involved along with various other regulators that form different protein complexes (Xie et al., 2015). Most of the Atgs having a role in autophagy were first identified in the year 1999 in yeast cells. The first mammalian autophagy gene to be cloned was Beclin1, also called Atg6 in yeast (Nakatogawa et al., 2009).

The ULK1 Complex

This protein complex plays the role of initiator of autophagy. It consists of the serine/threonine kinase ULK1/2 (uncoordinated-51-like kinase 1 or 2 (Atg1 in yeast), FAK family kinase-interacting protein of 200 kDa; FIP200 (Atg17 in yeast), ATG13, and ATG101 subunits. The ULK1 complex acquires nutrient level related upstream signals and then accordingly initiates the process of autophagy (Mercer et al., 2009). Under the conditions of starvation, autophagy begins from modifications in cellular nutritional sensors AMK-activated protein kinase (AMPK) and mammalian target of rapamycin (mTOR); a kinase of the serine/threonine-protein kinase family. It is the principal component of mTOR complex 1 (mTOR1) and mTOR complex 2 (mTOR2) protein complexes. In normal cells, the autophagic process is negatively regulated by mTOR1 which phosphorylatesULK1 and inactivates it while mTOR2 regulates various

Figure 2. Different types of autophagy in mammalian cells.

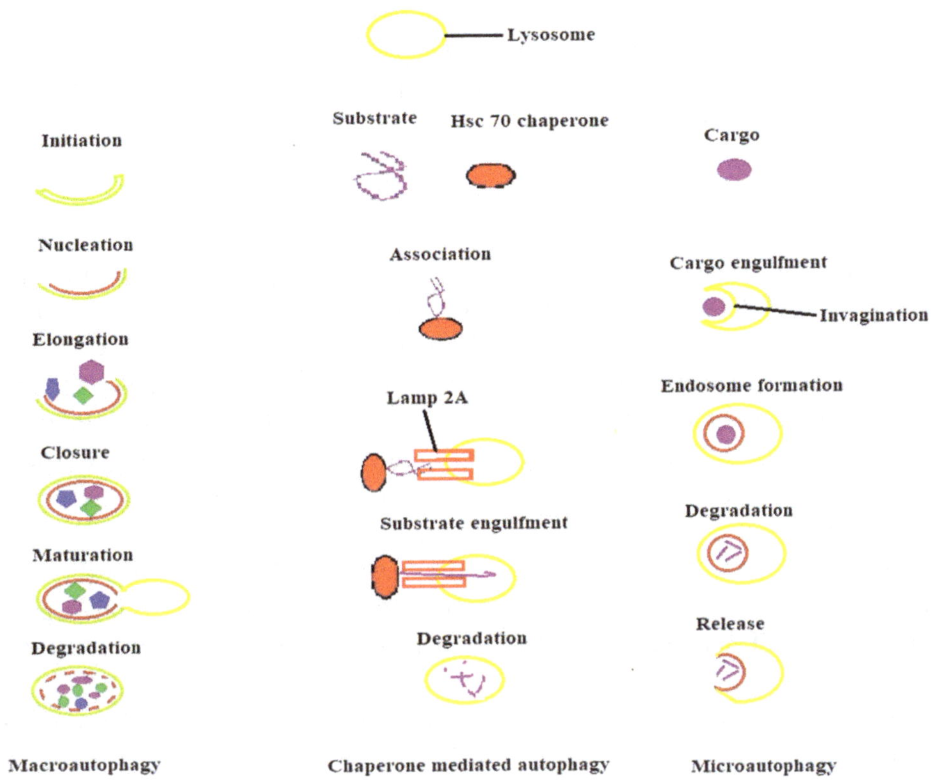

Figure 3. Steps involved in autophagy process

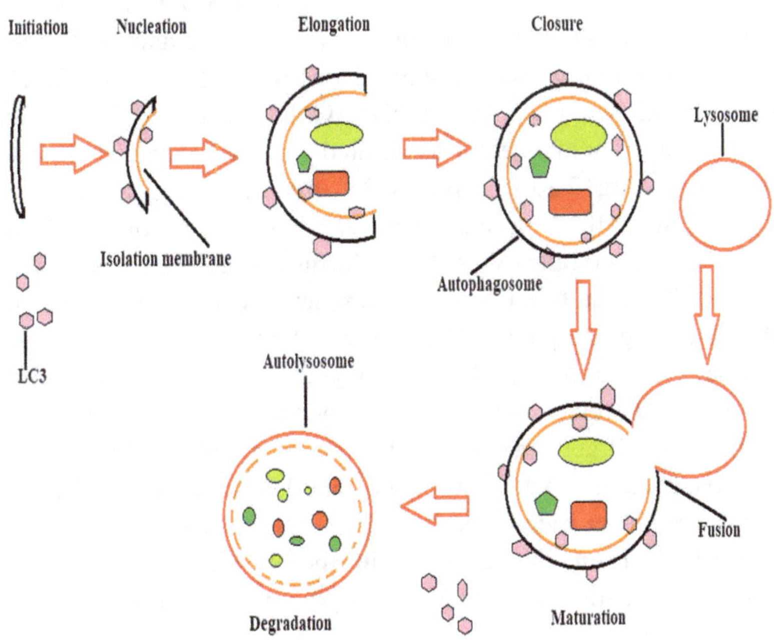

cellular processes. AMK-activated protein kinase (AMPK) phosphorylates different amino acid residues of ULK1 and thus suppresses autophagy. When the nutrient level gets depleted, the inhibitory effect of mTOR1 is released resulting in the dephosphorylation by protein phosphatase 2A (PP2A). ULK1/2 then autophosphorylates itself and other components of the protein complex (FIP200, Atg101, and Atg13) thereby inducing classic autophagy (Jung et al., 2009).

PI3K Complex

It is the second important autophagy protein complex and is very important for the formation of autophagosome and generation of PI3P, It facilitates the production of PIP3 by phosphorylating phosphatidylinositol at the 3'-hydroxyl position (Kihara et al., 2001). The autophagosome specific PI3 kinase complex consists of two subunit complexes; complex I (Beclin1, phosphoinositide 3-kinase [PI3K]C3, p150, and Atg14L subunits) and complex II (VPS34, VPS34-VPS15-Beclin 1-UVRAG and VPS34-VPS15-Beclin 1-UVRAG-Rubicon subunits). The former subunits are essential for the endocytic pathway while the latter one regulates the merging of the autophagosome with the lysosome. Beclin1 is regulated by B-cell leukemia/lymphoma-2 (Bcl-2) family proteins, such as Bcl-2, Bcl-XL, and interacts with PI3KC3 to generate PIP3 for the formation of autophagosome (Levine et al., 2008). The Atg14L subunit facilitates the localization of PI3KC3 complex into the endoplasmic reticulum (ER) for phagophore nucleation (Matsunaga et al., 2010). The ULK 1 complex also regulates the PI3KC3 complex to generate isolation membranes. Beclin1 undergoes phosphorylation to activate ULK1 through mTORC1 inhibition due to the withdrawal of amino acids. This results in increased lipid kinase activity of the complex and hence autophagy (Russell et al., 2013). Additionally, AMPK can directly phosphorylate Beclin1 under glucose starvation and low energy conditions to activate the complex (Kim et al., 2013).

ATG12-ATG8 and LC3 Pathways

There are two Atg7 dependent protein conjugating systems needed for autophagy (both selective and non-selective). These systems include ubiquitin-like proteins Atg12 and Atg8 (Ohsumi, 2001). They mediate the lipidation of three subfamilies of proteins in mammals – MAP1LC3 (microtubule-associated protein 1 light chain 3; also known as LC3), GATE16 (Golgi-associated ATPase enhancer of 16kDa) and GABARAP (γ-aminobutyric acid receptor-associated protein) (Tanida et al., 2003). Atg12 attaches covalently with Atg5 to form Atg12-Atg5 complex. E1 ubiquitin-conjugating enzyme (Atg7) and E2 ubiquitin-conjugating enzyme (Atg10) proteins catalyze the formation of Atg12-Atg5 complex. Non-covalent binding of Atg12-Atg5 complex with Atg16 forms a Atg12-Atg5-Atg16 multimeric structure (Mizushima et al., 1999). This multimeric Atg16 complex specifies the LC3 lipidation site for the biogenesis of membrane during autophagy (Fujita et al., 2008). Atg8/LC3 protein, is proteolytically cleaved by Atg4 and finally bound to Atg7 (E1-like enzyme) causing activation of Atg8. The activated Atg8 is transferred to Atg3 and eventually forms LC3II by conjugating with phosphatidylethanolamine (PE). Then newly formed LC3-II gets attached to the developing phagophore and to the outer and inner membranes of autophagosomes. After autophagosome-lysosome fusion, LC3-II attached to inner membrane degrades whereas outer membrane-bound LC3-II is recycled back by Atg4. Atg4 cleaves off the PE group. Since LC3-II remains attached with the phagophore, mature autophagosomes and autolysosomes, therefore, frequently used for monitoring the autophagosome and autophagy process as a biomarker (Kabeya et al., 2000).

BECLIN 1 IN AUTOPHAGY

BECN1 (autophagy gene), originally cloned by Levine et al., in 1988 (Liang et al., 1998) is a highly conserved mammalian orthologue of the Atg6/vacuolar protein sorting (VPS)-30 protein in yeast having a molecular weight of 600kDa (Aita et al., 1999). It is coiled-coil myosin-like BCL-2 interacting protein (Beclin) that has a potential role in the promotion and regulation of autophagy and genesis of tumors in mammals (Liang et al., 1998). Beclin1 contains three structural domains such as a BH3 domain, a central coiled-coil domain (CCD), and a C-terminal evolutionarily-conserved domain (ECD). Interaction of Beclin1 with Bcl-2 and Bcl-XL through BH-3 domain causes inhibition of the activity of Beclin1. Central coiled-coil domain (CCD) provides a platform for interaction of factors like, UV radiation resistance-associated gene (UVRAG), activating molecule in Beclin1-regulated autophagy (AMBRA1), and ATG14. Beclin1-PI3KC3 core complex is activated due to these interactions. Nuclear export signal (NES) of ECD has a vital contribution in Beclin1 mediated tumor suppression whose expression is reduced in many cancer types including breast, ovarian, and prostate. Both ECD and CCD have a contribution in Beclin1 interaction with PI3KC3 for inducing autophagy. The reduction in Beclin1 levels in the cellular environment inhibits autophagy and prevents the turnover of distorted or dysfunctional mitochondria leading to genotoxic stress due to free radical generation (Pattingre et al., 2005).

Beclin1-PI3KC3 Core Complex in Autophagy

Phosphatidylinositol 3-kinase (PI3K) is a group of enzymes that include class I, class II, and class III/ PI3KC3. These enzymes are involved in cell proliferation, differentiation, and migration. PI3KC3 plays a pleiotropic role in the autophagy induction and vacuolar protein sorting (VPS) pathways. Wortmannin and 3-methyladenine inhibit the activity of PI3KC3 thereby preventing omegasome formation and blocking autophagy (Axe et al., 2008). The core complex of PI3KC3 forms different complexes in yeast and mammals. In yeast cells, PI3KC3 kinase is VPS34 which forms two distinct types of PI3KC3 complexes. Type 1 complex consists of VPS34, VPS15, Atg6/VPS30, and Atg14 whereas type 2 complex consists of VPS34, VPS15, Atg6/VPS30, and VPS38. Both these complexes are important for autophagy and vacuolar protein sorting pathways respectively.

The PI3KC3 complex in mammalian cells consists of VPS34 along with a regulatory subunit, p150. Beclin1, interacts with UV irradiation resistance-associated gene (UVRAG) and a WD-40 domain-containing protein viz., AMBRA1 (activating molecule in Beclin1-regulated autophagy protein). Both UVRAG and AMBRA1 positively regulate Beclin1, while Bcl-2 is involved in negative regulation of the autophagy induction process (Fimia et al., 2007).UVRAG interacts with CCD domain of Beclin1 and activates Beclin1-PI3KC3 complex in mammalian cells that promotes the maturation of autophagosome by affixing fusion machinery on the late endosomes. UVRAG also contributes to apoptotic resistance and endosomal trafficking. Knockdown of Beclin1 and PI3KC3 diminishes the stability of UVRAG which suggests that it might be an autophagy generated degradation substrate. Thus, Beclin1 is necessary for both autophagy as well as for endocytosis. The functioning of the PI3KC3 complex also depends upon Beclin1-binding proteins (Itakura et al., 2008) (Figure 4).

Figure 4. Role of Beclin 1 in the process of autophagy.

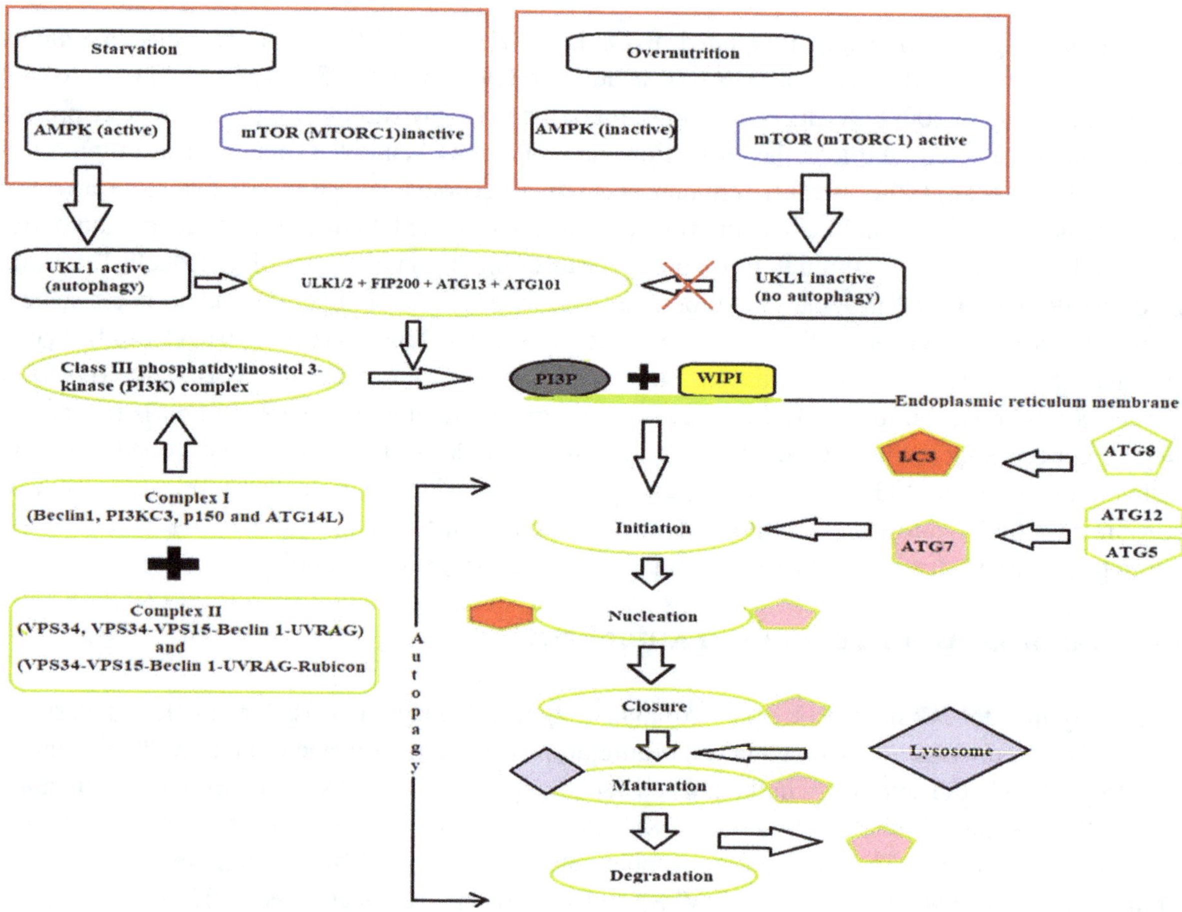

Beclin1-Bcl-2 Complex in Autophagy

Proteins of the Bcl-2 family are anti-apoptotic mediators and include Bcl-2, Bcl-XL, and myeloid cell leukemia 1 (MCL-1) proteins. They directly bind with the BH3 domain of Beclin1 to perform their activity. This binding inhibits autophagy by preventing the assembly of pro-autophagosomal structure by Beclin1. Thus, any mutation in the BH3 domain decreases the interaction and leads to the induction of autophagy. Further, Bcl-2 protein exerts its cytoprotective action by antagonizing BAX, BAK, and BIM proteins thereby preventing apoptosis. Beclin1 and Bcl-2/Bcl-XL proteins share a complex relationship. Anti-apoptotic action of Bcl-2 at mitochondria membrane cannot be inhibited by Beclin-1 whereas pro-autophagic activity of Beclin-1 can be inhibited by Bcl-2/Bcl-XL complex. Thus, the cell either undergoes apoptosis or autophagy depending on Beclin1 (Maiuri et al., 2007). The formation of Beclin1-Bcl-2 complex is affected by different kinases such as MAPK and DAPK. These kinases phosphorylate the respective domains of Beclin1 and Bcl-2 proteins.

Beclin1-Binding Proteins in Autophagy

Beclin1 protein, apart from making PI3KC3 and Bcl-2 protein complexes also interact with multiple protein complexes (Table 1) that affect the diverse functioning of the autophagic nucleation and maturation processes. Some of these proteins include:

- ***NRBF-2:*** Nuclear receptor binding factor-2 (NRBF-2) also called as co-modulator of PPAR and RXR, is an important constituent of Beclin1-PI3KC3 complex and interacts with its N-terminal to negatively regulate the autophagic flux. NBRF-2, also suppress the activity of VPS34 and genesis of autophagosomes through modulation of protein-protein interactions in the Atg14L-containing Beclin1-VPS34/PI3KC3 complex (Zhong et al., 2014).
- ***EGFR:*** Overexpression of epidermal growth factor receptor (EGFR) protein or its genetic mutations are known as a driving factor for various types of cancers, tumor cell proliferation, invasiveness, apoptosis suppression, and angiogenesis. EGFR inhibits autophagy either by enhancing mTOR's activity or by direct Beclin1inhibition. The suppression of this catabolic process is considered as a factor that contributes to the development of early phases of carcinogenesis, and therapy resistance in advanced neoplasms (Tini et al., 2015).
- ***HER2:*** Human epidermal growth factor receptor 2 (HER2) overexpression has been implicated in the pathogenesis of breast cancer. It confers resistance to the cancer cells against Lapatinib (FDA approved drug for breast cancer) by binding with ECD of Beclin1 in breast cancer cells (Han et al., 2013).
- ***Endophilin B1/BIF1:*** BAX interacting factor 1 (BIF1)/endophilin B1, is a multifunctional protein and is very important for regulating the process of apoptosis and autophagy. It also contributes in regulating mitochondrial morphology within the cellular environment. The loss of BIF1 inhibits apoptosis while its knockdown promotes apoptosis. Interaction of BIF1 with Beclin1 UVRAG positively mediates the actions of PI3KC3. Therefore, BIF1 contributes to both autophagy activation and tumor suppression (D. B. Wang et al., 2014; Takahashi et al., 2007).
- ***AMBRA1:*** Autophagy and Beclin 1 regulator 1 (AMBRA1) is a pro-autophagic gene that has a role in neurodevelopment. In the embryos of the mouse, neural tube defects were observed due to *AMBRA1* deficiency. These defects are related to impaired autophagy, exacerbated apoptosis, assembling of ubiquitinated proteins, and deranged cell proliferation. Both Beclin1 and *AMBRA1* are the components of PI3KC3 complex, which is involved in one of the early steps of autophagy induction, the biogenesis of autophagosome (Cianfanelli et al., 2015).
- ***HMGB1:*** High mobility group box 1 (HMGB1), chromatin-associated DNA-binding protein, has been reported for having a role in processes like autophagy, apoptosis, necrosis, and inflammation. HMGB1 and BCL-2 compete with each other to bind with Beclin1 to orient Beclin1 to the autophagosome. HMGB1-Beclin 1 complex formation is affected by various positive and negative regulators of autophagy. Release of HMGB1 during or along with extended autophagy, delayed apoptosis and cellular necrosis in mammalian cells suggests that it has an important role in human pathologies (Huang et al., 2017).
- ***Neuronal Isoform of Protein Interaction, Specifically with TC10:*** Initially identified as glutamate receptor δ2 (GluRδ2) by PDZ domain. It can synergize with CCD of Beclin1 to induce autophagy (Yue et al., 2002).

- ***VMP 1:*** A stress-induced transmembrane protein associated with pancreatitis. Its overexpression induces and promotes various autophagic human diseases ultimately causing cellular death. It binds with the BH3 motif of Beclin1 to form PI3KC3 complex (a key regulator of autophagy) at the site of autophagosome generation. Furthermore, the interaction of Beclin1with vacuole membrane protein 1 (VMP 1) results in the breakdown of Beclin1: BCL-2 complex Beclin1. This inhibits the activity of Beclin1. The Beclin1-VMP 1 interaction also strengthens the union of ATG16L1 and LC3 with the membrane of the autophagosome. Thus, VMP1 expression is essential for the activation of PI3KC3 complex during mammalian autophagy at the autophagosome formation site (Molejon et al., 2013).
- ***SLAMF1:*** Signaling lymphocyte activation molecule family 7(SLAMF1), a gram-negative microbial sensor also designated as CD150. It gets incorporated into a phagosome after the sequestration of the associated bacteria within the phagosome. Binding SLAMF1 with Beclin1 induces the maturation of phagosome and death of the engulfed bacteria within the phagosome. Therefore, is an important factor in cellular pathogen clearance mechanism or phagocytosis and is also important for autophagy induction (X Li et al., 2013).
- ***IP3R:*** Inositol trisphosphate receptor (IP3R) is a membrane glycoprotein that acts as a mediator of calcium ions released from endoplasmic reticulum which affects the mitochondrial processes like cellular bioenergetics and apoptosis. Thus, IP3R, have an impact on cell death and survival. Calcium (Ca^{2+}) ions are important signaling units for the onset of autophagy process but the contribution of reduced IP3R/Beclin1-complex formation in basal autophagy is not very clear. Its impact on tumorigenesis is also not very well understood (Kania et al., 2017).
- ***PTEN-Induced Putative Kinase 1/PARKIN:*** PINK1 is a serine/threonine-protein kinase that has a neuroprotective action. It selectively accumulates on the depolarized mitochondria and promotes translocation of PARK2/PARKIN to depolarized mitochondria making it important for mitophagy activation. In autophagy induced due to the starvation, localized accumulation of Beclin1 protein occurs at specific contact regions between mitochondria and endoplasmic reticulum. These contact regions are called mitochondria-associated membranes. The autophagosome originates from these regions (Gelmetti et al., 2017).
- ***Survivin:*** It is a protein expressed during the development of the fetus and in malignancy. It has antiapoptotic action and directly inhibits caspase 3 activity. Beclin1 overexpression inhibits survivin activity and its activity gets increased when the expression of Beclin1 is silenced (Niu et al., 2010).
- ***Pathogen-Derived Beclin1 Interaction Partners:*** Various proteins derived from pathogens e.g., vBCL-2 (KSHV and murine γHV68 M11, A179L), Nef (HIV), M2 (influenza A), orf16 (KSHV), ICP34.5 (HSV-1), and E1B19K (adenovirus) have been characterized which can bind with Beclin 1. Binding of such proteins with Beclin1 inhibits autophagosome formation (Pattingre et al., 2005). Viruses such as influenza A, herpesviruses, HIV also inhibits macroautophagy. This inhibition appears to be beneficial for viral replication by different mechanisms. Thus, in viral infection autophagy has a role in both antiviral defense as well as pro-viral survival (Münz et al., 2011).

Table 1. Role and actions of Beclin1 and various associated proteins and their effect on autophagy.

Protein	Action	Effect on Autophagy
Beclin1-PI3KC3 core complex; *BENC gene*	BENC 1 Regulates the activity of VPS34 promotes maturation of autophagosomes	Induction of autophagy
BCL-2 family of proteins	Binds with the BH3 domain of Beclin1 inhibits autophagy induction	Inhibition of autophagy
NBRF-2	suppresses the activity of VPS34 and genesis of autophagosomes	Inhibition of autophagy
EGFR	Increases the activity of mTOR and Inhibition of Beclin1	Inhibition of autophagy
(BIF1)/endophilin B1	Mediates action of PI3KC3 by binding with Beclin1-UVRAG	Induction of autophagy
AMBRA1	Biogenesis of autophagosome	Induction of autophagy
HMGB1	Competes with BCL-2 for binding with Beclin1 and promotes autophagosome formation	Induction of autophagy
Neuronal Isoform of Protein Interaction	synergize with CCD of Beclin1to induce autophagy	Induction of autophagy
Vacuole Membrane Protein 1	Activation of PI3KC3 complex, dissociation of BCL-2 with Beclin1, enhances the association of ATG16L1 and LC3 with autophagosome membranes, over the induction of autophagy may cause cell death	Induction of autophagy
SLAMF1	Maturation of autophagosomes, important for removal of pathogenic material	Induction of autophagy
PINK1	Translocation of PARKIN to damaged mitochondria	Induction of mitophagy
PARKIN	Localizes to mitochondria	Induction of mitophagy
Survivin	Inhibits caspase 3 activity	Inhibition of autophagy
Pathogen-Derived Beclin1 Interaction Partners	Binding of these pathogenic components with Beclin1 inhibits autophagosome formation	Inhibition of autophagy

PHYSIOLOGICAL ROLE OF AUTOPHAGY IN NEURODEGENERATION

Autophagy is a highly conserved intracellular pathway utilized by eukaryotic cells for recycling and decomposition of proteins (unutilized or long-lived) and cell organelles. Ubiquitin/proteasome system is responsible for the degradation of the short-lived protein. Autophagy, depending on the mechanism of induction is categorized into two types such as constitutive and inducible. The former regularly remove abnormal proteins to maintain homeostasis while the latter helps to reduce cellular injury against various noxious stimuli. Autophagy, based on the amount of substance degraded can be categorized into two types; selective and bulk. If small subcellular particles are digested in a nonselective manner, it is termed as bulk autophagy whereas degradation of specific organelles and molecules, by binding to selective receptors, is known as selective autophagy.

Neuronal autophagy has an important housekeeping role in the development of synapses (Zaffagnini et al., 2016). A study demonstrating the effect of autophagy in neurons on synapse reported that autophagy, induced by overexpression of Atg1, positively regulated synaptic development in the neuromuscular junction of *Drosophila*. On the other hand, the lowering of autophagic activity due to autophagy gene mutations resulted in a subsequent reduction in the size of synapses (Shen et al., 2009). A build-up of cytoplasmic inclusion bodies in neurons was reported along with the progressive loss of motoric functions of neurons in the study conducted in Atg5 deficient mice (Hara et al., 2006). These results indicated that the removal of abnormal proteins that can disrupt normal neuronal functions and cause neurodegeneration are very essential for the survival of neurons. Thus, autophagy is an important intracellular mechanism for the prevention of abnormal protein accumulation within the neurons.

NEURODEGENERATIVE DISORDERS AND BECLIN 1

Neurodegenerative disorders are increasingly being realized to occur due to the accumulation of aggregated, ubiquitinated and misfolded/unfolded proteins in neurons which eventually lead to impairment of synapses, cell organelles damage and ultimately nerve cell death (Ross et al., 2004). Autophagy is the mechanism responsible for the degeneration and elimination of the aggregated proteins from neuronal cells. Constitutive and basal autophagy inhibition cause nerve cell damage (Komatsu et al., 2006).

Normally, misfolded proteins are attached with ubiquitin (Ub) and digested by the proteasome. If the degenerative capacity of the proteasome (ubiquitin-proteasome pathway) becomes saturated then the autophagy system gets activated for removing damaged cell organelles and accumulated proteins. This capacity declines with the advancing age and the accumulation of these ubiquitylated protein aggregates damage the nerve cells which ultimately lead to neurodegeneration (Dikic et al., 2017). In addition, reduction in the activity of lysosomal hydrolytic enzymes inhibits autophagy and cause neurodegeneration (Shen et al., 20). Mutations in Atg genes or deletion of essential Atg7 and Atg5 genes are also considered responsible for the formation of cytoplasmic inclusions and neurodegeneration (Lipinski et al., 2010). Beclin1 has a direct multifunctional role in neurodegeneration. In mice, the heterozygous omission of Beclin1 increases nerve cell proliferation and reduces autophagy of nerve cells *in vivo* (Qu et al., 2003), as well as disruption of lysosomes and subsequent neurodegeneration (Pickford et al., 2008). In the brain of adult mice overexpression of Beclin1 defends the nerve cells from *Sindbis* virus infection-induced apoptosisand prevents accumulation of misfolded proteins within neurons (Liang et al., 1998; Jaeger et al., 2010).

In the course of the development of neurodegenerative disorders, aging is the greatest risk factor because of the impairment of autophagic activity upon aging. It has also been well reported and documented that upon aging or in certain neurodegenerative disorders, levels of Beclin1 decrease (Nascimento-Ferreira et al., 2011; Shibata et al., 2006). Overexpression of Beclin1 confers neuroprotection by clearing the misfolded alpha-synuclein, (Spencer et al., 2009) amyloid-βand ataxin 3 proteins (Jaeger et al., 2010; Nascimento-Ferreira et al., 2011). Thus, sufficient evidence suggests that disruption of Beclin 1 level within the body and impairment of autophagy are the underlying principles for the progression of some neurodegenerative disorders.

Alzheimer's Disease

Alzheimer's disease (AD) is specified as impairment or loss of memory and cognitive behavior. The most common pathological characteristic of AD is the accumulation of neurotoxic extracellular beta-amyloid (Aβ) plaques and hyper-phosphorylated tau in intracellular neurofibrillary tangles (NFTs) in the brain (Wang et al., 2017). Sequential cleavage of amyloid precursor protein (APP) by β-secretase and γ-secretase generates Aβ. Tau is hyper-phosphorylated and self-aggregates to form NFTs that are unable to attach with tubulin leading to microtubule network disruption (Li, et al., 2007). Several studies have provided evidence of pivotal roles played by various clearance mechanisms in the initiation and further development of AD (Nixon et al., 2005).

The accumulated Aβ and APP are normally degraded by autophagy, however, in pathological state accumulation of autophagic components occur due to altered endocytic pathway, reduced fusion, and impeded turnover of increased autophagic vacuoles (Nixon, 2007). The disruptive changes in the autolysosome based mechanism of degradation for cytosolic components are evidenced from the brains of AD patients and buildup of autophagosomes expression in dystrophic neurites (Yu et al., 2005). Further, the presenilin1 enzyme (part of the γ-secretase complex) cleaves APP and plays a pivotal role in the maturation of V-ATPase, an enzyme essential for lysosomes acidification. Presenilin1 enzyme dysfunction in AD results in impaired lysosomes acidification. This leads to disruption of autolysosome formation and inhibition of proteolysis which results in the accumulation of autophagosomes as well as accumulation of Aβ peptides within autophagosomes (Lee et al., 2010).

The core protein complex (Beclin1/VPS34 complex) for autophagosome formation is very important for initiating the autophagy process. The deficiency of Beclin1, either due to non-functional sequestration or underexpression, causes impairment of APP processing and autophagy in Alzheimer's disease. Conversely, Beclin1 overexpression enhances APP processing and reduces amyloid deposition in transgenictransgenic AD mice both in and outside the cell. Several neuronal inhibitory proteins like Bcl-2, Bcl-XL, and inflammasomal receptors can inhibit the functions of Beclin1 by interacting with it. Many neurovirulent proteins (herpexsimplexICP34.5 protein) also block Beclin1 complex formation (Salminen et al., 2013). Furthermore, mTOR signaling pathway inhibition by drugs like rapamycin causes a reduction in Aβ peptides level and improvement of cognitive behavior in AD mouse models (Spilman et al., 2010).

Accumulation of Aβ peptides in the neurons is still considered as the most detectable pathological incident in case of AD. This whole accumulation depends directly on the balance genesis and clearance rate of Aβ peptides or turnover rate. In normal physiological conditions, genesis rate equals clearance rate but in case of AD this balance gets disturbed as the rate of clearance gets slowed down to a significant extent. Moreover, as AD is more prevalent in elder age people this makes the situation more difficult to deal with as the biological half-life of Aβ peptides gets increased up to 2.5 times after the age of 50 years thus, promoting accumulation and reducing clearance of Aβ. In AD the turnover rate gets altered due to an increase in the biological half-life of Aβ which is about 9 hours in a normal situation. This causes accumulation of Aβ aggregates and these accumulated protein aggregates are susceptible to misfolding or changes in conformations (Patterson et al., 2015)

Parkinson's Disease

Parkinson's disease (PD) affects the dopaminergic neurons in the substantia nigra of the brain. In terms of prevalence, it comes after AD and ranks at the number in two in the list of most prevalent neurodegenerative disease. The most observable characteristics of PD are rigidity of muscles, bradykinesia, neurocognitive deficit and the resting tremor (Poewe et al., 2017). The most common, detectable and identifiable neurological characteristic in PD is the existence of Lewy bodies or Lewy neurites (intraneuronal proteinaceous inclusions). Lewy bodies enclose synaptic protein α-synuclein in filamentous form and regulatory protein ubiquitin within themselves. Ubiquitin monomer or its polymeric chains are very important for the proper intracellular transportation and disposal of various cellular proteins. Thus, axonal transport gets impaired due to ubiquitin accumulation within the Lewy bodies (Abeliovich et al., 2016). In PD altered functioning of lysosomes and chaperone-mediated autophagy is related to the accumulation of α-synuclein (Cerri et al., 2018). Moreover, mutations in specific genes like LRRK2, PARK2, PARK7, PINK1 and SNCA have been reported in cases of inherited Parkinson's disease (Lin et al., 2014).

Mitophagy, a quality-control mechanism or pathway for selective elimination of damaged mitochondria, also gets disrupted in PD. PARKIN (ligase) and PINK1(kinase) have significant roles mitophagy process and their mutations cause the development of the autosomal recessive form of PD. In normal conditions processing of PINK 1 protein by presenilin-associated rhomboid-like (PARL) protease causes degradation of PINK1 by the ubiquitin-proteasome system. In pathological conditions, processing of PINK1 gets inhibited and PINK 1 gets deposited on the external membrane of the damaged mitochondria. Then autophosphorylation of PINK1 occurs and the recruitment of PARKIN on damaged mitochondria is done. PINK1 further phosphorylates ubiquitin, which forms the phagophore around mitochondria by ubiquitinating mitochondrial surface proteins and hence leading to lysosomal degradation. PINK1 and PARKIN genes mutations hamper the lysosomal degradation of damaged mitochondria leading to mitophagy impairment, an important characteristic of PD (Deas et al., 2011).

It is also found that Beclin1 interacts only with full-length PINK1. Basal and starvation-induced autophagy was significantly enhanced by PINK1. This enhanced autophagy was then reduced either by overexpression of Beclin1 or by inhibition of VPS34. All these findings proved a consolidated association between autophagy and proteins implicated in neurodegenerative disorders (Michiorri et al., 2010).

Amyotrophic Lateral Sclerosis

Amyotrophic lateral sclerosis (ALS) is marked by the continuing loss of motor neurons of both the brain and spinal cord. This loss of nerve ultimately causes deterioration in functioning and activity of voluntary muscles controlled by lost neurons which makes the disorder a fatal one. Mutations of over 20 genes are reported to play a significant part in the pathology of ALS. Important genes whose mutations have been established to have a role in ALS are superoxide dismutase 1 (SOD1), hexanucleotide repeat expansions in the intronic region of chromosome 9 open reading frame (*C9orf72)*, fused in sarcoma (FUS) and AR-DBP/TDP43 (TAR DNA binding protein) (Chen et al., 2013). Various studies reported the involvement of autophagy in the pathology of ALS. In normal disease-free mice, the motor defects were produced due to the loss of neuro-specific Atg5 and Atg7 proteins. This suggests that in ALS the generation of toxic protein aggregates may be mainly due to impaired autophagosome formation (Nguyen et al., 2019).

In mutant SOD1 transgenic mice, delay in the induction of the experimental ALS is observed because of the activation of mTOR-independent autophagy (Castillo et al., 2013). Beclin1 acts as an integrator of stimuli engaging autophagy as it regulates Atg5-Atg7 independent autophagy. Autophagy, activation by abnormally interacting with mutant SOD1 (upstream alteration) and promotion of misfolded protein degradation (downstream effect) are the reported actions of Beclin1 that are implicated as evidence of the protective role of Beclin1 in neurodegenerative disorders. Therefore, pointing out the contribution of Beclin1 in the pathological pathway of ALS (Nassif et al., 2014).

Polyglutamine Diseases

Polyglutamine diseases (polyQ diseases) in total are a cluster of nine genetically inherited neurodegenerative disorders. polyQ diseases include Huntington's disease (HD), dentatorubral-pallidoluysian atrophy (DRPLA), Kennedy disease, spinocerebellar ataxia (SCA) subtypes (SCA1, SCA2, SCA3, SCA6, SCA7, and SCA17) and HD-like 2 diseases. In all these diseases, except SCA6, the liberation of toxic polyglutamine containing fragments due to proteolytic cleavage is observed. Repeated expansion of cytosine-adenine-guanine (CAG) triplet in translated regions of otherwise unrelated genes results in the formation of proteins with expanded polyglutamine domains. These expanded polyglutamine proteins are prone to aggregation and within a target protein, the expanded polyglutamine tracts facilitate the transition to a novel, toxic conformation. They can initiate neurodegeneration by interacting with specific transcription factors which might upset the gene expression (Shao et al., 2007).

Atypical extension of CAG triplet repeat length for a polyQ tail in the huntingtin gene is the underlying principle for Huntington disease. The mutant huntington protein (HTT) in HD and ataxins in SCA are cleared by autophagy and thereby reducing the neurotoxicity is also reduced. Beclin1 is very important for the regulation of intracellular levels of mutant HTT. Degradation of damaged, abnormal, long-lived proteins by Beclin1 is reduced due to impaired autophagy and enhances the accumulation of mutant HTT. As age is a major risk factor for neurodegeneration, with growing age expression of Beclin1 within the neurons decreases due to which the regulation of autophagic activity gets decreased. This enhances the buildup of mutant HTT within the neurons and thus causes the further progression of the disease. Thus, Beclin1-mediated autophagy pathway play is very crucial in both initiation and further development of HD (Shibata et al., 2006). Furthermore, HTT also competes with mTOR to bind with ULK1 and induces autophagy by releasing UKL1 from mTORC1 (inhibitor of ULK1/2 complex activity). Binding of HTT with p62, promotes the formation of autophagosome by interlinking ubiquitinated complexes and LC3 complex. Autophagy gets enhanced in neurons after polyQ tails are deleted in HTT. Thus, conformational changes in HTT due to polyQ tail can cause neurodegeneration because of the impairment of autophagy in neurons (Croce et al., 2019).

Static Encephalopathy of Childhood with Neurodegeneration in Adulthood

Static encephalopathy of childhood with neurodegeneration in adulthood (SENDA) a neurodegenerative syndrome caused by excessive deposition of iron molecules within the globus pallidus and substantia nigraof the brain (Haack et al., 2012). Major characteristics are delayed intellectual development with serious intellectual disability, the fast progression of dementia and dystonia-parkinsonism in early adulthood. Mutations of WDR45 gene were identified in SNEDA patients by exome sequencing (Haack et al., 2012). WIPI proteins, a novel autophagy factor homologous to ATG18, attaches with PI3P to generate the early autophagic vacuoles. Mutations in the WIPI4 gene causes the production of unstable WIPI4 proteins due to which extreme reduction in autophagy was observed in lymphoblastoid cell lines that are derived from SENDA patients (Saitsu et al., 2013). Thus, WIPI4 mutations lead to the accumulation of early autophagosomal membranes as well as disruption of the autophagy process in SNEDA patients. The identification of these mutations furnishes direct evidence of defective autophagy in causing the neurodegenerative disorders in humans (Proikas-Cezanne et al., 2015).

Tauopathies

Tau, a neuronal microtubule-associated protein that is expressed mostly in axons. It stabilizes the microtubule assembly and is involved in anterograde axonal transport. Tauopathies are the cluster of heterogenous neurodegenerative disorders, which include frontotemporal lobar degeneration (FTLD), Steele-Richardson-Olszewski syndrome, corticobasal degeneration, argyrophilic grain disease (AgD), and Pick's disease. Alteration in concentrations anomalous aggregation, mutations, phosphorylation and changed expression of some isoforms of tau proteins are the factors responsible for pathological changes in the neuronal environment in tauopathies (Hernández et al., 2007).

Delay in tau clearance due to the perturbation of autophagy and heat shock proteins results in aggregation of tau. Aggregates of tau proteins in glial cells are ubiquitinated indicating impairment of the ubiquitin/proteasome pathway. The tau pathology affects both astrocytes as well as oligodendrocytes in the brain (Leyk et al., 2015). Furthermore, the presence of phosphorylated tau has also been confirmed in brain specimens examined after the death of the patients suffering AD, progressive supranuclear palsy and corticobasal degeneration, in association with LC3 and p62 proteins. Phosphorylated tau proteins are detected within the brain specimens because of engulfment of these proteins into autophagosomes occurs but degradation by autolysosome does not take place (Piras et al., 2016). Thus, dysfunctional autophagy has a critical and significant contribution to the advancement of tauopathies whereas normally functioning autophagic mechanism degrades phosphorylated tau proteins and maintains their low cellular levels.

The role of tauopathy was not very well established and researched in case of PD. Initially, the contribution of tau in the pathology of PD is undermined but in recent studies showed the existence of tau aggregates in nearly half of the PD brains taken under consideration in the study and it is proposed that these deposits are being transported from neuron to neuron (Zhang et al., 2018). Assembling or aggregation of tau proteins may be due to mutations of *MAPT* gene (microtubule-associated protein tau). A gene wide association study revealed a close connection between *MAPT* and sporadic PD (Nalls et al., 2015). Low soluble tau concentration in the substantia nigraof the brain of the PD patients indicates a link between healthy tau protein and neurodegeneration in PD (Lei et al., 2012). The ratio of phosphorylated tau concentration concerning total tau protein concentration can be used as a biomarker for ALS. A low ratio suggests the presence of the pathology of ALS.

RECENT DEVELOPMENTS AND FUTURE RESEARCH DIRECTIONS

Autophagy has been correlated with numerous cellular processes and pathological conditions like cancer and neurodegeneration. The quest for the search for the application of autophagy within the cellular environment began nearly half a century ago since the discovery of autophagy. Various research studies have investigated the functions of autophagy and confirmed its role in cancer, tumor suppression, neurodegeneration, antigen presentation, immune responses, inflammation and aging process. These research works provide a deep insight into the mechanism of autophagy at the cellular, molecular and genetic levels. This helps to discover the basic proteins or macromolecules involved in the process of autophagy and the genes behind these proteins. These studies also show how these proteins regulate the autophagy and the consequences if these macromolecules undergo alterations and mutations. The connection between the various pathological conditions with the deviations from the normal autophagy process was also revealed by these studies conducted in the recent past. Beclin1 is one such protein involved in the autophagy process and also reported to have an involvement in various neurodegenerative conditions.

In recent studies the Beclin1 has been reported to have a role in cancer immunity, the myeloid deficient Beclin1 mice have shown neutrophilia and are at the high risk for developing spontaneous precursor β cell lymphoma and can be related to autophagic dysfunction in absence of Beclin1. The same finding can be correlated with the low expression of Beclin1 in neutrophils of patients with pre-B acute lymphoblastic lymphoma (ALL) and its relation with PDl1 levels in neutrophils (Ten et al., 2019). Thus, myeloid Beclin1 can act as a regulator of cancer immunity and may help to provide a therapeutic target for efficient therapy. Application of knowledge of Beclin1 interaction with antiandrogen enzalutamide in castration-resistant prostate cancer cell lines was studied. The study provided evidence for the regulation of growth factor signaling by Beclin1 independent of autophagy and also opens a new dimension for the treatment of castration-resistant prostate cancer (Zhang et al., 2019). Some studies have reported that the activation of Beclin1 can attenuate HIV infection. Therefore, based on this finding, the delivery systems like intranasal systems have been developed to target the Beclin1 through SiRNA for activating Beclin1 at sites that are not easily accessible (like the brain) and thus help in treating HIV (Rodriguez et al., 2017). The relation between Beclin1 and autophagy is well established and recent findings have revealed that Beclin1 mediated autophagy protects the heart during sepsis (Sun et al. 2018). Enhanced Beclin1 mediated autophagy has been reported in case of type II diabetes mellitus. Exaggerated neurodegeneration because of proteolytic cleavage has also been reported (Munasinghe et al., 2016). Spermidine, a natural polyamine, inhibits caspase 3 mediated cleavage of Beclin1 to induce autophagy (Yang et al., 2017).

These studies how the research field of autophagy gets enlarged, initially autophagy was detected in yeast cells only and various Atg proteins were discovered subsequently. Then hunt began for discovering these proteins in cells of other living organisms including humans which resulted in the discovery of Beclin1 in mammalian cells. Beclin1 was found to have a very close association with the regulation of autophagy. The next level of research was then focused on the role that Beclin1 plays in various pathological conditions like neurodegeneration, cancer, inflammation, etc. Now the future course of action for Beclin1 is to understand the Beclin1 autophagy relationship with high precision and discover more of its autophagy-independent roles in mammalian cells. Research in autophagy is also now focused on discovering or defining a pathway by which autophagy can be induced without utilizing the mechanism involved in the formation of autophagosome i.e. independent of Beclin1 or associated cellular substrates (Mizushima et al., 2017). These are both the future as well as the challenges in this field of research.

CONCLUSION

Neuronal autophagy is the major cellular mechanism for neuroprotection. Impaired autophagy as well as excessive autophagy both contribute to neurodegeneration and neuronal cell death. Beclin1, an important structural unit of Beclin1-PI3KC3 complex, acts as the cornerstone of autophagosome nucleation. The early expression of Beclin1 can protect healthy neuronal cells by destroying the neurons with dysfunctional protein aggregates and therefore, hinders the development of neurodegenerative conditions. Beclin1 is positively as well as negatively regulated by various functional interactions. The presence of proteins like Bcl-2 and Bcl-XL suppress Beclin1 protein's autophagic activity whereas binding of Beclin1 with PI3KC3 and few other nuclear proteins activate the autophagic pathway of functional proteins. Therefore, Beclin1 is very important for the formation and maturation of autophagosome/endosome which contribute to autophagy. Moreover, altered autophagy is implicated in nearly all major neurodegenerative disorders. Thus, Beclin1 has also a role in pathology, progression, and control of neurodegeneration in various neurodegenerative disorders.

REFERENCES

Abeliovich, A., & Gitler, A. D. (2016). Defects in trafficking bridge Parkinson's disease pathology and genetics. *Nature, 539*(7628), 207–216. doi:10.1038/nature20414 PMID:27830778

Aita, V. M., Liang, X. H., Murty, V. V. V. S., Pincus, D. L., Yu, W., Cayanis, E., ... Levine, B. (1999). Cloning and Genomic Organization of Beclin 1, a Candidate Tumor Suppressor Gene on Chromosome 17q21. *Genomics, 59*(1), 59–65. doi:10.1006/geno.1999.5851 PMID:10395800

Axe, E. L., Walker, S. A., Manifava, M., Chandra, P., Roderick, H. L., Habermann, A., ... Ktistakis, N. T. (2008). Autophagosome formation from membrane compartments enriched in phosphatidylinositol 3-phosphate and dynamically connected to the endoplasmic reticulum. *The Journal of Cell Biology, 182*(4), 685–701. doi:10.1083/jcb.200803137 PMID:18725538

Bento, C. F., Renna, M., Ghislat, G., Puri, C., Ashkenazi, A., Vicinanza, M., ... Rubinsztein, D. C. (2016). Mammalian Autophagy: How Does It Work? *Annual Review of Biochemistry, 85*(1), 685–713. doi:10.1146/annurev-biochem-060815-014556 PMID:26865532

Castillo, K., Nassif, M., Valenzuela, V., Rojas, F., Matus, S., Mercado, G., ... Hetz, C. (2013). Trehalose delays the progression of amyotrophic lateral sclerosis by enhancing autophagy in motoneurons. *Autophagy, 9*(9), 1308–1320. doi:10.4161/auto.25188 PMID:23851366

Cerri, S., & Blandini, F. (2018). Role of Autophagy in Parkinson's Disease. *Current Medicinal Chemistry, 25*. doi:10.2174/0929867325666180226094351 PMID:29484979

Chen, S., Sayana, P., Zhang, X., & Le, W. (2013). Genetics of amyotrophic lateral sclerosis: An update. *Molecular Neurodegeneration, 8*(1), 28. doi:10.1186/1750-1326-8-28 PMID:23941283

Cianfanelli, V., D'Orazio, M., & Cecconi, F. (2015). AMBRA1 and BECLIN 1 interplay in the crosstalk between autophagy and cell proliferation. *Cell Cycle (Georgetown, Tex.), 14*(7), 959–963. doi:10.1080/15384101.2015.1021526 PMID:25803737

Croce, K. R., & Yamamoto, A. (2019). A role for autophagy in Huntington's disease. *Neurobiology of Disease, 122*, 16–22. doi:10.1016/j.nbd.2018.08.010 PMID:30149183

Deas, E., Wood, N. W., & Plun-Favreau, H. (2011). Mitophagy and Parkinson's disease: The PINK1–parkin link. Biochimica et Biophysica Acta (BBA) -. *Molecular Cell Research, 1813*(4), 623–633. doi:10.1016/j.bbamcr.2010.08.007 PMID:20736035

Deretic, V., & Klionsky, D. J. (2018). Autophagy and inflammation: A special review issue. *Autophagy, 14*(2), 179–180. doi:10.1080/15548627.2017.1412229 PMID:29304718

Dikic, I. (2017). Proteasomal and Autophagic Degradation Systems. *Annual Review of Biochemistry, 86*(1), 193–224. doi:10.1146/annurev-biochem-061516-044908 PMID:28460188

Dikic, I., & Elazar, Z. (2018). Mechanism and medical implications of mammalian autophagy. *Nature Reviews. Molecular Cell Biology, 19*(6), 349–364. doi:10.103841580-018-0003-4 PMID:29618831

Fujita, N., Itoh, T., Omori, H., Fukuda, M., Noda, T., & Yoshimori, T. (2008). The Atg16L Complex Specifies the Site of LC3 Lipidation for Membrane Biogenesis in Autophagy. *Molecular Biology of the Cell, 19*(5), 2092–2100. doi:10.1091/mbc.e07-12-1257 PMID:18321988

Gelmetti, V., De Rosa, P., Torosantucci, L., Marini, E. S., Romagnoli, A., Di Rienzo, M., ... Valente, E. M. (2017). PINK1 and BECN1 relocalize at mitochondria-associated membranes during mitophagy and promote ER-mitochondria tethering and autophagosome formation. *Autophagy, 13*(4), 654–669. doi:10.1080/15548627.2016.1277309 PMID:28368777

Haack, T. B., Hogarth, P., Kruer, M. C., Gregory, A., Wieland, T., Schwarzmayr, T., ... Hayflick, S. J. (2012). Exome Sequencing Reveals De Novo WDR45 Mutations Causing a Phenotypically Distinct, X-Linked Dominant Form of NBIA. *American Journal of Human Genetics, 91*(6), 1144–1149. doi:10.1016/j.ajhg.2012.10.019 PMID:23176820

Han, J., Hou, W., Lu, C., Goldstein, L. A., Stolz, D. B., Watkins, S. C., & Rabinowich, H. (2013). Interaction between Her2 and Beclin-1 Proteins Underlies a New Mechanism of Reciprocal Regulation. *The Journal of Biological Chemistry, 288*(28), 20315–20325. doi:10.1074/jbc.M113.461350 PMID:23703612

Hara, T., Nakamura, K., Matsui, M., Yamamoto, A., Nakahara, Y., Suzuki-Migishima, R., ... Mizushima, N. (2006). Suppression of basal autophagy in neural cells causes neurodegenerative disease in mice. *Nature, 441*(7095), 885–889. doi:10.1038/nature04724 PMID:16625204

Hernández, F., & Avila, J. (2007). Tauopathies. *Cellular and Molecular Life Sciences, 64*(17), 2219–2233. doi:10.100700018-007-7220-x PMID:17604998

Huang, J., Yang, J., Shen, Y., Jiang, H., Han, C., Zhang, G., ... Wang, T. (2017). HMGB1 Mediates Autophagy Dysfunction via Perturbing Beclin1-VPS34 Complex in Dopaminergic Cell Model. *Frontiers in Molecular Neuroscience, 10*. doi:10.3389/fnmol.2017.00013 PMID:28197072

Itakura, E., Kishi, C., Inoue, K., & Mizushima, N. (2008). Beclin 1 Forms Two Distinct Phosphatidylinositol 3-Kinase Complexes with Mammalian Atg14 and UVRAG. *Molecular Biology of the Cell, 19*(12), 5360–5372. doi:10.1091/mbc.e08-01-0080 PMID:18843052

Jaeger, P. A., Pickford, F., Sun, C.-H., Lucin, K. M., Masliah, E., & Wyss-Coray, T. (2010). Regulation of Amyloid Precursor Protein Processing by the Beclin 1 Complex. *PLoS One*, *5*(6), e11102. doi:10.1371/journal.pone.0011102 PMID:20559548

Jung, C. H., Jun, C. B., Ro, S.-H., Kim, Y.-M., Otto, N. M., Cao, J., ... Kim, D.-H. (2009). ULK-Atg13-FIP200 Complexes Mediate mTOR Signaling to the Autophagy Machinery. *Molecular Biology of the Cell*, *20*(7), 1992–2003. doi:10.1091/mbc.e08-12-1249 PMID:19225151

Kabeya, Y. (2000). LC3, a mammalian homologue of yeast Apg8p, is localized in autophagosome membranes after processing. *The EMBO Journal*, *19*(21), 5720–5728. doi:10.1093/emboj/19.21.5720 PMID:11060023

Kania, E., Roest, G., Vervliet, T., Parys, J. B., & Bultynck, G. (2017). IP3 Receptor-Mediated Calcium Signaling and Its Role in Autophagy in Cancer. *Frontiers in Oncology*, *7*, 140. doi:10.3389/fonc.2017.00140 PMID:28725634

Kihara, A., Kabeya, Y., Ohsumi, Y., & Yoshimori, T. (2001). Beclin-phosphatidylinositol 3-kinase complex functions at the trans -Golgi network. *EMBO Reports*, *2*(4), 330–335. doi:10.1093/embo-reports/kve061 PMID:11306555

Kim, J., Kim, Y. C., Fang, C., Russell, R. C., Kim, J. H., Fan, W., ... Guan, K.-L. (2013). Differential Regulation of Distinct VPS34 Complexes by AMPK in Nutrient Stress and Autophagy. *Cell*, *152*(1–2), 290–303. doi:10.1016/j.cell.2012.12.016 PMID:23332761

Komatsu, M., Waguri, S., Chiba, T., Murata, S., Iwata, J., Tanida, I., ... Tanaka, K. (2006). Loss of autophagy in the central nervous system causes neurodegeneration in mice. *Nature*, *441*(7095), 880–884. doi:10.1038/nature04723 PMID:16625205

Lee, J.-H., Yu, W. H., Kumar, A., Lee, S., Mohan, P. S., Peterhoff, C. M., ... Nixon, R. A. (2010). Lysosomal Proteolysis and Autophagy Require Presenilin 1 and Are Disrupted by Alzheimer-Related PS1 Mutations. *Cell*, *141*(7), 1146–1158. doi:10.1016/j.cell.2010.05.008 PMID:20541250

Lei, P., Ayton, S., Finkelstein, D. I., Spoerri, L., Ciccotosto, G. D., Wright, D. K., ... Roberts, B. R. (2012). Tau deficiency induces parkinsonism with dementia by impairing APP-mediated iron export. *Nature Medicine*, *18*(2), 291–296. doi:10.1038/nm.2613 PMID:22286308

Levine, B., Sinha, S. C., & Kroemer, G. (2008). Bcl-2 family members: Dual regulators of apoptosis and autophagy. *Autophagy*, *4*(5), 600–606. doi:10.4161/auto.6260 PMID:28186856

Leyk, J., Goldbaum, O., Noack, M., & Richter-Landsberg, C. (2015). Inhibition of HDAC6 Modifies Tau Inclusion Body Formation and Impairs Autophagic Clearance. *Journal of Molecular Neuroscience*, *55*(4), 1031–1046. doi:10.100712031-014-0460-y PMID:25434725

Li, B., Chohan, M. O., Grundke-Iqbal, I., & Iqbal, K. (2007). Disruption of microtubule network by Alzheimer abnormally hyperphosphorylated tau. *Acta Neuropathologica*, *113*(5), 501–511. doi:10.100700401-007-0207-8 PMID:17372746

Li, X., Prescott, M., Adler, B., Boyce, J. D., & Devenish, R. J. (2013). Beclin 1 Is Required for Starvation-Enhanced, but Not Rapamycin-Enhanced, LC3-Associated Phagocytosis of Burkholderiapseudomallei in RAW 264.7 Cells. *Infection and Immunity*, *81*(1), 271–277. doi:10.1128/IAI.00834-12 PMID:23115045

Liang, X. H., Kleeman, L. K., Jiang, H. H., Gordon, G., Goldman, J. E., & Berry, G. … Levine, B. (1998). Protection against fatal Sindbis virus encephalitis by beclin, a novel Bcl-2-interacting protein. *Journal of Virology*, *72*(11), 8586–8596. Retrieved from https://www.ncbi.nlm.nih.gov/pubmed/9765397

Lin, M. K., & Farrer, M. J. (2014). Genetics and genomics of Parkinson's disease. *Genome Medicine*, *6*(6), 48. doi:10.1186/gm566 PMID:25061481

Lipinski, M. M., Zheng, B., Lu, T., Yan, Z., Py, B. F., Ng, A., … Yuan, J. (2010). Genome-wide analysis reveals mechanisms modulating autophagy in normal brain aging and in Alzheimer's disease. *Proceedings of the National Academy of Sciences*, *107*(32), 14164–14169. 10.1073/pnas.1009485107

Maiuri, M. C., Le Toumelin, G., Criollo, A., Rain, J.-C., Gautier, F., Juin, P., ... Kroemer, G. (2007). Functional and physical interaction between Bcl-XL and a BH3-like domain in Beclin-1. *The EMBO Journal*, *26*(10), 2527–2539. doi:10.1038j.emboj.7601689 PMID:17446862

Maria Fimia, G., Stoykova, A., Romagnoli, A., Giunta, L., Di Bartolomeo, S., Nardacci, R., ... Cecconi, F. (2007). Ambra1 regulates autophagy and development of the nervous system. *Nature*, *447*(7148), 1121–1125. doi:10.1038/nature05925 PMID:17589504

Matsunaga, K., Morita, E., Saitoh, T., Akira, S., Ktistakis, N. T., Izumi, T., ... Yoshimori, T. (2010). Autophagy requires endoplasmic reticulum targeting of the PI3-kinase complex via Atg14L. *The Journal of Cell Biology*, *190*(4), 511–521. doi:10.1083/jcb.200911141 PMID:20713597

McKnight, N. C., Zhong, Y., Wold, M. S., Gong, S., Phillips, G. R., Dou, Z., ... Yue, Z. (2014). Beclin 1 Is Required for Neuron Viability and Regulates Endosome Pathways via the UVRAG-VPS34 Complex. *PLOS Genetics*, *10*(10), e1004626. doi:10.1371/journal.pgen.1004626 PMID:25275521

Mercer, C. A., Kaliappan, A., & Dennis, P. B. (2009). A novel, human Atg13 binding protein, Atg101, interacts with ULK1 and is essential for macroautophagy. *Autophagy*, *5*(5), 649–662. doi:10.4161/auto.5.5.8249 PMID:19287211

Michiorri, S., Gelmetti, V., Giarda, E., Lombardi, F., Romano, F., Marongiu, R., ... Casari, G. (2010). The Parkinson-associated protein PINK1 interacts with Beclin1 and promotes autophagy. *Cell Death and Differentiation*, *17*(6), 962–974. doi:10.1038/cdd.2009.200 PMID:20057503

Mizushima, N. (2017). The exponential growth of autophagy-related research: From the humble yeast to the Nobel Prize. *FEBS Letters*, *591*(5), 681–689. doi:10.1002/1873-3468.12594 PMID:28186333

Mizushima, N., & Komatsu, M. (2011). Autophagy: Renovation of Cells and Tissues. *Cell*, *147*(4), 728–741. doi:10.1016/j.cell.2011.10.026 PMID:22078875

Mizushima, N., Noda, T., & Ohsumi, Y. (1999). Apg16p is required for the function of the Apg12p–Apg5p conjugate in the yeast autophagy pathway. *The EMBO Journal*, *18*(14), 3888–3896. doi:10.1093/emboj/18.14.3888 PMID:10406794

Molejon, M. I., Ropolo, A., Lo Re, A., Boggio, V., & Vaccaro, M. I. (2013). The VMP1-Beclin 1 interaction regulates autophagy induction. *Scientific Reports*, *3*(1), 1055. doi:10.1038rep01055 PMID:23316280

Munasinghe, P. E., Riu, F., Dixit, P., Edamatsu, M., Saxena, P., Hamer, N. S., ... Lamberts, R. R. (2016). Type-2 diabetes increases autophagy in the human heart through promotion of Beclin-1 mediated pathway. *International Journal of Cardiology*, *202*, 13–20. doi:10.1016/j.ijcard.2015.08.111 PMID:26386349

Münz, C. (2011). Beclin-1 Targeting for Viral Immune Escape. *Viruses*, *3*(7), 1166–1178. doi:10.3390/v3071166 PMID:21994775

Nakatogawa, H., Suzuki, K., Kamada, Y., & Ohsumi, Y. (2009). Dynamics and diversity in autophagy mechanisms: Lessons from yeast. *Nature Reviews. Molecular Cell Biology*, *10*(7), 458–467. doi:10.1038/nrm2708 PMID:19491929

Nalls, M. A., Pankratz, N., Lill, C. M., Do, C. B., Hernandez, D. G., Saad, M., ... Schulte, C. (2014). Large-scale meta-analysis of genome-wide association data identifies six new risk loci for Parkinson's disease. *Nature Genetics*, *46*(9), 989–993. doi:10.1038/ng.3043 PMID:25064009

Nascimento-Ferreira, I., Santos-Ferreira, T., Sousa-Ferreira, L., Auregan, G., Onofre, I., Alves, S., ... Pereira de Almeida, L. (2011). Overexpression of the autophagic beclin-1 protein clears mutant ataxin-3 and alleviates Machado–Joseph disease. *Brain*, *134*(5), 1400–1415. doi:10.1093/brain/awr047 PMID:21478185

Nassif, M., Valenzuela, V., Rojas-Rivera, D., Vidal, R., Matus, S., Castillo, K., ... Hetz, C. (2014). Pathogenic role of BECN1/Beclin 1 in the development of amyotrophic lateral sclerosis. *Autophagy*, *10*(7), 1256–1271. doi:10.4161/auto.28784 PMID:24905722

Nguyen, D. K. H., Thombre, R., & Wang, J. (2019). Autophagy as a common pathway in amyotrophic lateral sclerosis. *Neuroscience Letters*, *697*, 34–48. doi:10.1016/j.neulet.2018.04.006 PMID:29626651

Nigar, S., Pottoo, F. H., Tabassum, N., Verma, S. K., & Javed, M. N. (2016). Molecular Insights into the Role of Inflammation and Oxidative Stress in Epilepsy. *Journal of Advances in Medical and Pharmaceutical Sciences*, *10*(1), 1–9. doi:10.9734/JAMPS/2016/24441

Niu, T.-K., Cheng, Y., Ren, X., & Yang, J.-M. (2010). Interaction of Beclin 1 with survivin regulates sensitivity of human glioma cells to TRAIL-induced apoptosis. *FEBS Letters*, *584*(16), 3519–3524. doi:10.1016/j.febslet.2010.07.018 PMID:20638385

Nixon, R. A. (2007). Autophagy, amyloidogenesis and Alzheimer disease. *Journal of Cell Science*, *120*(23), 4081–4091. doi:10.1242/jcs.019265 PMID:18032783

Nixon, R. A., Wegiel, J., Kumar, A., Yu, W. H., Peterhoff, C., Cataldo, A., & Cuervo, A. M. (2005). Extensive Involvement of Autophagy in Alzheimer Disease: An Immuno-Electron Microscopy Study. *Journal of Neuropathology and Experimental Neurology*, *64*(2), 113–122. doi:10.1093/jnen/64.2.113 PMID:15751225

Ohsumi, Y. (2001). Molecular dissection of autophagy: Two ubiquitin-like systems. *Nature Reviews. Molecular Cell Biology*, *2*(3), 211–216. doi:10.1038/35056522 PMID:11265251

Patterson, B. W., Elbert, D. L., Mawuenyega, K. G., Kasten, T., Ovod, V., Ma, S., ... Zhang, L. (2015). Age and amyloid effects on human central nervous system amyloid-beta kinetics. *Annals of Neurology, 78*(3), 439–453. doi:10.1002/ana.24454 PMID:26040676

Pattingre, S., Tassa, A., Qu, X., Garuti, R., Liang, X. H., Mizushima, N., ... Levine, B. (2005). Bcl-2 Antiapoptotic Proteins Inhibit Beclin 1-Dependent Autophagy. *Cell, 122*(6), 927–939. doi:10.1016/j.cell.2005.07.002 PMID:16179260

Pickford, F., Masliah, E., Britschgi, M., Lucin, K., Narasimhan, R., & Jaeger, P. A. … Wyss-Coray, T. (2008). The autophagy-related protein beclin 1 shows reduced expression in early Alzheimer disease and regulates amyloid β accumulation in mice. *The Journal of Clinical Investigation*. doi:10.1172/JCI33585 PMID:18497889

Piras, A., Collin, L., Grüninger, F., Graff, C., & Rönnbäck, A. (2016). Autophagic and lysosomal defects in human tauopathies: Analysis of post-mortem brain from patients with familial Alzheimer disease, corticobasal degeneration and progressive supranuclear palsy. *Acta Neuropathologica Communications, 4*(1), 22. doi:10.118640478-016-0292-9 PMID:26936765

Poewe, W., Seppi, K., Tanner, C. M., Halliday, G. M., Brundin, P., Volkmann, J., ... Lang, A. E. (2017). Parkinson disease. *Nature Reviews. Disease Primers, 3*(1), 17013. doi:10.1038/nrdp.2017.13 PMID:28332488

Pottoo, F. H., Bhowmik, M., & Vohora, D. (2014). Raloxifene protects against seizures and neurodegeneration in a mouse model mimicking epilepsy in postmenopausal woman. *European Journal of Pharmaceutical Sciences, 65*, 167–173. doi:10.1016/j.ejps.2014.09.002 PMID:25218046

Pottoo, F. H., Tabassum, N., Javed, M. N., Nigar, S., Rasheed, R., Khan, A., ... Ashraf, G. M. (2019). The Synergistic Effect of Raloxifene, Fluoxetine, and Bromocriptine Protects Against Pilocarpine-Induced Status Epilepticus and Temporal Lobe Epilepsy. *Molecular Neurobiology, 56*(2), 1233–1247. doi:10.100712035-018-1121-x PMID:29881945

Proikas-Cezanne, T., Takacs, Z., Donnes, P., & Kohlbacher, O. (2015). WIPI proteins: Essential PtdIns3P effectors at the nascent autophagosome. *Journal of Cell Science, 128*(2), 207–217. doi:10.1242/jcs.146258 PMID:25568150

Qu, X., Yu, J., Bhagat, G., Furuya, N., Hibshoosh, H., Troxel, A., ... Levine, B. (2003). Promotion of tumorigenesis by heterozygous disruption of the beclin 1 autophagy gene. *The Journal of Clinical Investigation, 112*(12), 1809–1820. doi:10.1172/JCI20039 PMID:14638851

Rodriguez, M., Lapierre, J., Ojha, C. R., Kaushik, A., Batrakova, E., Kashanchi, F., ... El-Hage, N. (2017). Intranasal drug delivery of small interfering RNA targeting Beclin1 encapsulated with polyethylenimine (PEI) in mouse brain to achieve HIV attenuation. *Scientific Reports, 7*(1), 1862. doi:10.103841598-017-01819-9 PMID:28500326

Ross, C. A., & Poirier, M. A. (2004). Protein aggregation and neurodegenerative disease. *Nature Medicine, 10*(S7), S10–S17. doi:10.1038/nm1066 PMID:15272267

Russell, R. C., Tian, Y., Yuan, H., Park, H. W., Chang, Y.-Y., Kim, J., ... Guan, K.-L. (2013). ULK1 induces autophagy by phosphorylating Beclin-1 and activating VPS34 lipid kinase. *Nature Cell Biology*, *15*(7), 741–750. doi:10.1038/ncb2757 PMID:23685627

Saitsu, H., Nishimura, T., Muramatsu, K., Kodera, H., Kumada, S., Sugai, K., ... Matsumoto, N. (2013). De novo mutations in the autophagy gene WDR45 cause static encephalopathy of childhood with neurodegeneration in adulthood. *Nature Genetics*, *45*(4), 445–449. doi:10.1038/ng.2562 PMID:23435086

Salminen, A., Kaarniranta, K., Kauppinen, A., Ojala, J., Haapasalo, A., Soininen, H., & Hiltunen, M. (2013). Impaired autophagy and APP processing in Alzheimer's disease: The potential role of Beclin 1 interactome. *Progress in Neurobiology*, *106–107*, 33–54. doi:10.1016/j.pneurobio.2013.06.002 PMID:23827971

Schreiber, S., Spotorno, N., Schreiber, F., Acosta-Cabronero, J., Kaufmann, J., Machts, J., ... Dengler, R. (2018). Significance of CSF NfL and tau in ALS. *Journal of Neurology*, *265*(11), 2633–2645. doi:10.100700415-018-9043-0 PMID:30187162

Shao, J., & Diamond, M. I. (2007). Polyglutamine diseases: Emerging concepts in pathogenesis and therapy. *Human Molecular Genetics*, *16*(R2), R115–R123. doi:10.1093/hmg/ddm213 PMID:17911155

Shen, H.-M., & Mizushima, N. (2014). At the end of the autophagic road: An emerging understanding of lysosomal functions in autophagy. *Trends in Biochemical Sciences*, *39*(2), 61–71. doi:10.1016/j.tibs.2013.12.001 PMID:24369758

Shen, W., & Ganetzky, B. (2009). Autophagy promotes synapse development in Drosophila. *The Journal of Cell Biology*, *187*(1), 71–79. doi:10.1083/jcb.200907109 PMID:19786572

Shibata, M., Lu, T., Furuya, T., Degterev, A., Mizushima, N., Yoshimori, T., ... Yuan, J. (2006). Regulation of Intracellular Accumulation of Mutant Huntingtin by Beclin 1. *The Journal of Biological Chemistry*, *281*(20), 14474–14485. doi:10.1074/jbc.M600364200 PMID:16522639

Son, J. H., Shim, J. H., Kim, K.-H., Ha, J.-Y., & Han, J. Y. (2012). Neuronal autophagy and neurodegenerative diseases. *Experimental & Molecular Medicine*, *44*(2), 89. doi:10.3858/emm.2012.44.2.031 PMID:22257884

Spencer, B., Potkar, R., Trejo, M., Rockenstein, E., Patrick, C., Gindi, R., ... Masliah, E. (2009). Beclin 1 Gene Transfer Activates Autophagy and Ameliorates the Neurodegenerative Pathology in -Synuclein Models of Parkinson's and Lewy Body Diseases. *The Journal of Neuroscience: The Official Journal of the Society for Neuroscience*, *29*(43), 13578–13588. doi:10.1523/JNEUROSCI.4390-09.2009 PMID:19864570

Spilman, P., Podlutskaya, N., Hart, M. J., Debnath, J., Gorostiza, O., Bredesen, D., ... Galvan, V. (2010). Inhibition of mTOR by Rapamycin Abolishes Cognitive Deficits and Reduces Amyloid-β Levels in a Mouse Model of Alzheimer's Disease. *PLoS One*, *5*(4), e9979. doi:10.1371/journal.pone.0009979 PMID:20376313

Sun, Y., Yao, X., Zhang, Q. J., Zhu, M., Liu, Z. P., Ci, B., ... Levine, B. (2018). Beclin-1-dependent autophagy protects the heart during sepsis. *Circulation*, *138*(20), 2247–2262. doi:10.1161/CIRCULATIONAHA.117.032821 PMID:29853517

Takahashi, Y., Coppola, D., Matsushita, N., Cualing, H. D., Sun, M., Sato, Y., ... Wang, H.-G. (2007). Bif-1 interacts with Beclin 1 through UVRAG and regulates autophagy and tumorigenesis. *Nature Cell Biology*, *9*(10), 1142–1151. doi:10.1038/ncb1634 PMID:17891140

Tan, P., He, L., Xing, C., Mao, J., Yu, X., Zhu, M., ... Wang, H. Y. (2019). Myeloid loss of Beclin 1 promotes PD-L1 hi precursor B Cell lymphoma development. *The Journal of Clinical Investigation*, *129*(12), 5261–5277. doi:10.1172/JCI127721 PMID:31503548

Tanida, I., Komatsu, M., Ueno, T., & Kominami, E. (2003). GATE-16 and GABARAP are authentic modifiers mediated by Apg7 and Apg3. *Biochemical and Biophysical Research Communications*, *300*(3), 637–644. doi:10.1016/S0006-291X(02)02907-8 PMID:12507496

Tini, P., Belmonte, G., Toscano, M., Miracco, C., Palumbo, S., Pastina, P., ... Pirtoli, L. (2015). … Pirtoli, L. (2015). Combined Epidermal Growth Factor Receptor and Beclin1 Autophagic Protein Expression Analysis Identifies Different Clinical Presentations, Responses to Chemo- and Radiotherapy, and Prognosis in Glioblastoma. *BioMed Research International*, *2015*, 1–13. doi:10.1155/2015/208076 PMID:25821789

Wang, D. B., Uo, T., Kinoshita, C., Sopher, B. L., Lee, R. J., Murphy, S. P., ... Morrison, R. S. (2014). Bax Interacting Factor-1 Promotes Survival and Mitochondrial Elongation in Neurons. *The Journal of Neuroscience: The Official Journal of the Society for Neuroscience*, *34*(7), 2674–2683. doi:10.1523/JNEUROSCI.4074-13.2014 PMID:24523556

Wang, J., Gu, B. J., Masters, C. L., & Wang, Y.-J. (2017). A systemic view of Alzheimer disease—Insights from amyloid-β metabolism beyond the brain. *Nature Reviews. Neurology*, *13*(10), 612–623. doi:10.1038/nrneurol.2017.111 PMID:28960209

Wirth, M., Joachim, J., & Tooze, S. A. (2013). Autophagosome formation—The role of ULK1 and Beclin1–PI3KC3 complexes in setting the stage. *Seminars in Cancer Biology*, *23*(5), 301–309. doi:10.1016/j.semcancer.2013.05.007 PMID:23727157

Xie, Y., Kang, R., Sun, X., Zhong, M., Huang, J., Klionsky, D. J., & Tang, D. (2015). Posttranslational modification of autophagy-related proteins in macroautophagy. *Autophagy*, *11*(1), 28–45. doi:10.4161/15548627.2014.984267 PMID:25484070

Yang, Y., Chen, S., Zhang, Y., Lin, X., Song, Y., Xue, Z., ... Zhang, L. (2017). Induction of autophagy by spermidine is neuroprotective via inhibition of caspase 3-mediated Beclin 1 cleavage. *Cell Death & Disease*, *8*(4), e2738. doi:10.1038/cddis.2017.161 PMID:28383560

Yu, W. H., Cuervo, A. M., Kumar, A., Peterhoff, C. M., Schmidt, S. D., Lee, J.-H., ... Nixon, R. A. (2005). Macroautophagy—A novel β-amyloid peptide-generating pathway activated in Alzheimer's disease. *The Journal of Cell Biology*, *171*(1), 87–98. doi:10.1083/jcb.200505082 PMID:16203860

Yue, Z., Horton, A., Bravin, M., DeJager, P. L., Selimi, F., & Heintz, N. (2002). A novel protein complex linking the delta 2 glutamate receptor and autophagy: Implications for neurodegeneration in lurcher mice. *Neuron*, *35*(5), 921–933. doi:10.1016/S0896-6273(02)00861-9 PMID:12372286

Zaffagnini, G., & Martens, S. (2016). Mechanisms of Selective Autophagy. *Journal of Molecular Biology*, *428*(9), 1714–1724. doi:10.1016/j.jmb.2016.02.004 PMID:26876603

Zhang, M., Sun, Y., Meng, J., Zhang, L., Liang, C., & Chang, C. (2019). Targeting AR-Beclin 1 complex-modulated growth factor signaling increases the antiandrogen Enzalutamide sensitivity to better suppress the castration-resistant prostate cancer growth. *Cancer Letters*, *442*, 483–490. doi:10.1016/j.canlet.2018.11.008 PMID:30423407

Zhang, X., Gao, F., Wang, D., Li, C., Fu, Y., He, W., & Zhang, J. (2018). Tau Pathology in Parkinson's Disease. *Frontiers in Neurology*, *9*, 809. doi:10.3389/fneur.2018.00809 PMID:30333786

Zhong, Y., Morris, D. H., Jin, L., Patel, M. S., Karunakaran, S. K., Fu, Y.-J., ... Wang, Q. J. (2014). Nrbf2 Protein Suppresses Autophagy by Modulating Atg14L Protein-containing Beclin 1-VPS34 Complex Architecture and Reducing Intracellular Phosphatidylinositol-3 Phosphate Levels. *The Journal of Biological Chemistry*, *289*(38), 26021–26037. doi:10.1074/jbc.M114.561134 PMID:25086043

Chapter 10
Multifarious Role of BAG3 in Neurodegenerative Disorders

Sujata Basu
Enago-Publication Support, Crimson Interactive Pvt. Ltd., India

Manisha Singh
Jaypee Institute of Information Technology, India

Mansi Verma
Jaypee Institute of Information Technology, India

Rachana R.
Jaypee Institute of Information Technology, India

ABSTRACT

The glial cells along with cells of hematopoietic origin and microvascular endothelia work together to maintain the normal development and/or functioning of the nervous system. Disruption in functional coordination among these cells interrupts the efficiency of the nervous system, leading to neurodegeneration. Various proteins in the nerve cells maintain the normal signaling mechanism with these cells and throughout the body. Structural/functional disorganization of these proteins causes neurodegenerative disorders. The molecular mechanisms involved in these phenomena are yet to be explored extensively from therapeutic perspectives. Through this chapter, the authors have elaborated on less known protein Bcl-2 associated athanogene 3 (BAG3) involved in neurodegeneration. They have explored BAG3 protein and its role in neurodegeneration, protein homeostasis, its mechanism of action, its uses as a drug target, and its uses as a possible diagnostic marker of neurodegeneration.

INTRODUCTION

In today's world of medical advancement, state-of-art medical technology has advanced the chances of life expectancy on an average scale in mankind. In spite of many improvements in this sphere, medical science is still in the struggling era of controlling aging. The cause behind this tussle between medical

DOI: 10.4018/978-1-7998-1317-0.ch010

science and control of the aging phenomenon lies in several concerns, including dealing with neurodegenerative disorders (NDDs). The progression of neurological ailment from its initial stage to final is linked to injuries in the brain, its associated neurons and the neuronal circuit in our body (Pekna, 2012). During neurodegeneration, a progressive deterioration of neuronal structures and functions occurs which ultimately leads to cognitive disability and dementia (Ramanan and Saykin, 2013). Various proteins play unique functional roles in different regions of healthy neurons while maintaining several functions like neuronal migration and differentiation, neuronal growth, and organized neural connectivity from axon to synapse (Tau and Peterson, 2010; de Wit et al., 2011). These include proteins involved in ion channels, proton pump, scaffolding, structural integrity of the cell and proteins responsible for postsynaptic density and anti-apoptosis (Vinothkumar and Henderson, 2010; Cherry, 2018).

The microglia (myeloid cells of the hematopoietic system) or astroglia and oligodendrocytes (neuroepithelial progenitor cells of hematopoietic origin), along with glial cells, and microvascular endothelia, work together to maintain the normal development and/or functioning of the nervous system (Eglitis and Mezey, 1997). An interruption in the coordinated functioning of these cells disrupts the functional ability of the nervous system, which thereby can lead to neurodegeneration (National Research Council, 1989). The list of neurodegenerative disorders is long, however, only a few of them have drawn the attention of researchers. These include Alzheimer's disease (AD), Huntington disease (HD), Parkinson's disease (PD) and amyotrophic lateral sclerosis (ALS), etc. The other neurodegenerative disorders are equally devastating; however, their exact cause remained essentially unexplored so far. Several scientists have attempted to examine the molecular interplays in cells for an individual neurodegenerative disease, at pathway level as well as, at the genetic level (Arneson et al., 2018; Narayanan, 2015) which are genetically controlled in neuronal cells. Though the neurodegenerative disorders differ from each other in terms of their symptoms and treatment modality yet, they are likely to be interconnected in some way through a dynamic molecular interplay inside the cells.

The modulation of neurodegenerative processes has been forecasted to occur due to strong interplay among intracellular mechanisms, local tissue environment, systemic environment, and mechanisms associated with cell development and aging as shown in Figure 1. This chapter aims to present the molecular interplay in neuronal cells under stressed conditions, with a focus on the upcoming possible role player, biomarker and drug target Bcl2-associated athanogene 3 (BAG3). The literature explored to predict the role of BAG3 in the prevention of cellular stress. Also, the probable mechanism of BAG3 to regain the protein homeostasis of neuronal cells through close interaction with various chaperones, anti-apoptotic factors, and other cellular components/proteins/organelles is discussed.

BAG FAMILY

The proteins encoded under the BAG family are highly conserved throughout the eukaryotes. Its homologs exist in reptiles, invertebrates, silkworms, yeast, and flowering plants, as well (Myers et al., 2018). It has been reported in several organisms like mouse, drosophila, silk moth and Baker's yeast, etc. (IPR003103). All mammalian cells have BAG3, in whom this protein is evolutionarily highly conserved. It is most explicitly expressed in the heart, skeletal muscle, and in several types of cancers (Myers et al., 2018). In normal conditions, BAG3 is reported to be essential in the development of both the glial cells and neuronal cells of central nervous system (CNS) and BAG3 pathway has been reported to be associated with age-related NDDs.

Figure 1. Conceptual model of probable molecular and cellular pathways that contribute to neurodegeneration.
Causal factors and factors influencing an imbalance between neuronal survival and degeneration are shown through four categories of mode of action – Intracellular mechanisms (due to formation of misfolded proteins that disrupt functional ability of cellular organelles through a probable effect of oxidant/antioxidant balance), local tissue environment, systemic environment, and mechanisms related to neurodevelopment and aging. An overlapping of interactions between various above components can collectively define the modulation behind neurodegenerative processes.
(Adapted from Ramanan and Saykin, 2013).

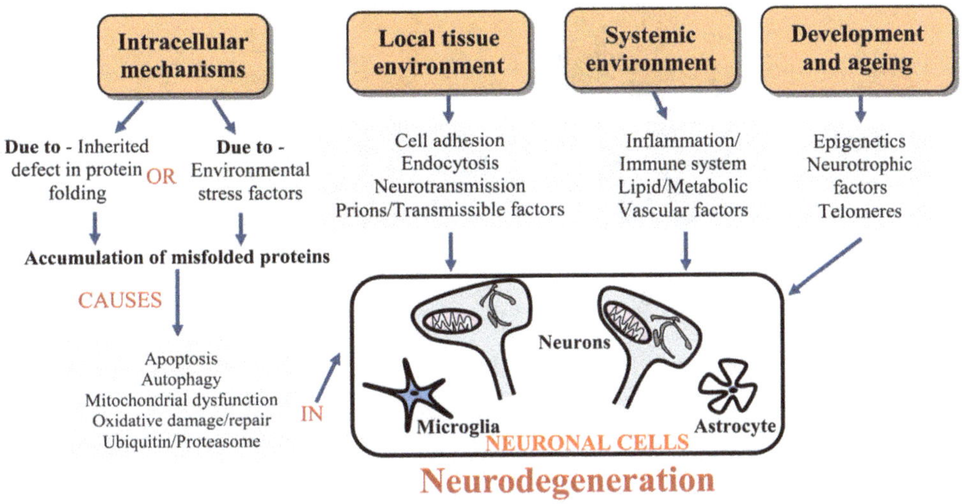

BAG3 Protein

BAG3 is also referred to as CAI stressed- (1CAIR-1) or Bcl-2 interacting death suppressor (BIS) (Sturner and Behl, 2017). It was identified in 1999 by Takayama et al., (Sturner and Behl, 2017). It is a 575 amino acids long multimodal protein (molecular weight 75 kDa) that interacts with several intracellular signaling molecules and is anti-apoptotic (Myers et al., 2018; Behl, 2016). Figure 2a depicts the solution structure of the BAG domain of Bcl2-associated athanogene 3 obtained from Protein Data Bank (PDB), deposited in PDB (10.2210/pdb1UK5/pdb; O95817; UniProtKB - O95817) and the verified structure was released in the year 2004. BAG proteins behave as regulators of several cellular pathways and this property is achieved at its conserved motif near the C-terminal end called BAG domain (BD).

The BAG3 gene is located at 10q26.11 of chromosome 10 and its molecular location is 119,651,370 to 119,677,819 base pairs on it. The BAG family of proteins are subdivided into two subfamilies based on short and long BAG domains (BD). The shorter ones are BAG3, 4, and 5 whereas, BAG1 is the only long BAG domain-containing BAG protein (Briknarova et al., 2002). In humans, the BAG family is comprised of 6 members from BAG1 to BAG6. All these human BAG proteins bind to Hsp70/Hsc70 and control the chaperone activities (Sturner and Behl, 2017). Interaction of BAG1, BAG3, and BAG4 with Hsc70-binding faces (helices α2 and α3) involves a similar charge distribution. Through multidimensional nuclear magnetic resonance (NMR) technique, it was found that BAG4 BD is a triple helix bundle with shorter residue length of three helices (380–399 [α1], 407–423 [α2], and 432–456 [α3]) in comparison to that of BAG1. The Bcl-2-associated athanogene 3 domains of BAG3 and BAG4 are about

Figure 2. A comparative pictorial representation of (a) Murine BAG domain of Bcl2-associated athanogene 3 and (b) BAG-3 from Homo sapiens (Human). The blue, green, and red color codes refer to α1, α2 and α3 domains of BAG-3 protein.

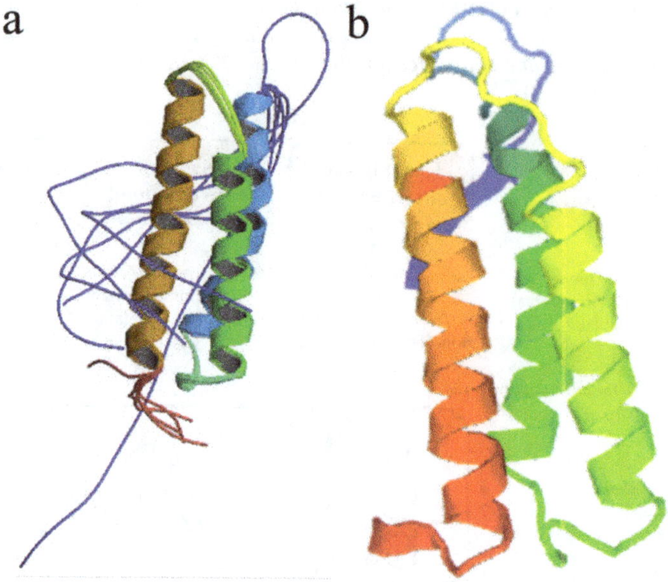

60% identical to each other, thereby exhibiting a great extent of similarity (Briknarova et al., 2002) as shown in Figure 3.

BAG domain of BAG proteins acts as regulators of cellular pathways. BAG3 binds to the ATPase domain of Hsp70/Hsc70 chaperones, interacts with Bcl2 and phospholipase C-γ proteins. The binding of BAG proteins to Hsp70/Hsc70 leads to a controlled mechanism of overall chaperone activity (Rosati et al., 2011). BAG3 is the only BAG protein that is induced by stressful stimuli, mainly through interaction with HSF1 on the bag3 gene promoter (Rosati et al., 2011). The domains of almost all the BAG proteins have about 45 amino acids near the C-terminus, however, they markedly differ from each other in amino acid composition in the N-terminal region. Apart from the BAG domain, the BAG3 protein contains a proline-rich sequence (PXXP) and WW domain that helps its binding to proteins other than Hsp70 (Rosati et al., 2011) under different stressed cellular conditions as shown in Figure 4.

Stressors like oxidants, heavy metals, high temperature, electrophile imbalance, and hemodialysis treatment, etc. induce BAG3 expression in normal cells. The WW domain of this protein interacts with other proteins at their proline-containing peptide sequence and controls several cellular functions (Otte et al., 2003). This domain is non-catalytic in nature. Besides this, the two conserved isoleucine-proline-valine (IPV) motifs placed between W- and proline-rich domains of co-chaperone mediates the binding of BAG3 to two small heat shock proteins HspB8 and HspB6 that controls macroautophagy in cells (Myers et al., 2018). In case of the deletion of the IPV motif, the HspB8 activity is suppressed and this is probably involved in defective autophagy (Rosati et al. 2011). The WW domain is made up of 40 amino acids. In some proteins, this domain may be repeated up to four times. This chain of amino acid folds into a triple-stranded beta-sheet and is stable. The domain derived its name due to the presence of two tryptophan residues that are separated from each other by 20-23 amino acids. It assists in

Figure 3. A comparative pictorial representation of (a) BAG3 and (b) BAG4 proteins.
From the two forms such as long and short length forms of BAG3– one being a product of BAG3 gene having an apparent mass of 74 kDa and a shorter BAG3 protein to be associated with neuronal synaptosomes (Rosati et al., 2011). Synaptosomes are nerve terminals that stay separated from nerve cells. They are useful to study synaptic transmission and its physiology. The lengthy form of BAG3 is located in the rough endoplasmic reticulum. The shorter form (40 kDa) was identified through immunoprecipitation from neuronal synaptosomes homogenates followed by mass spectrophotometry.
(Rosati et al., 2011).

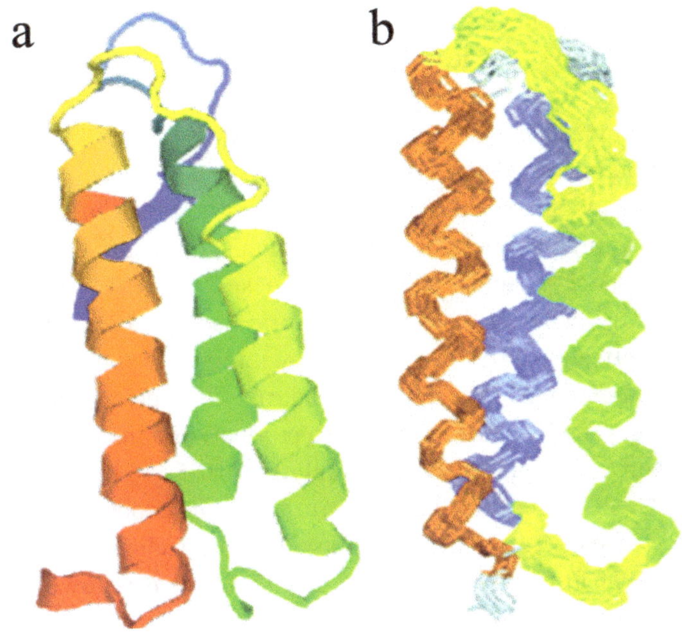

Figure 4. Multiple domains of BAG3 protein for its multifunctional and integrated interplay between various biological pathways towards the maintenance of protein homeostasis.
(Adapted from Myers et al., 2018).

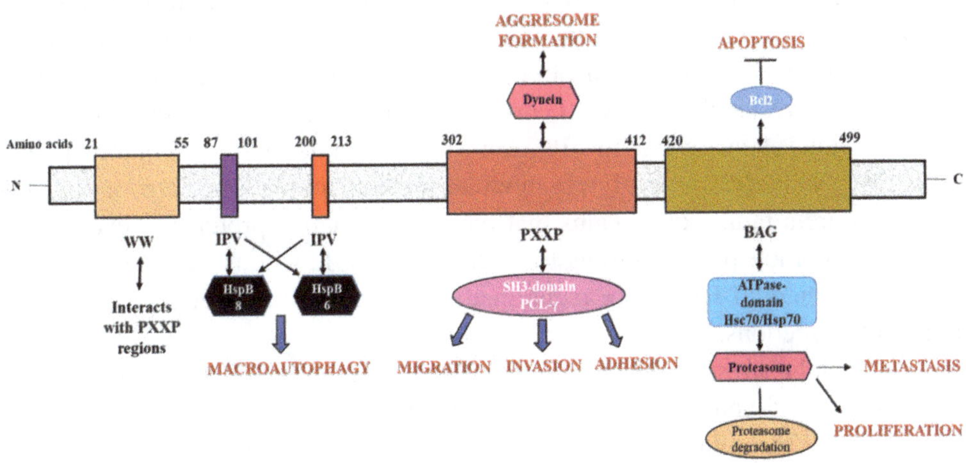

chaperone-mediated autophagy. It also combines with the proline-rich PXXP region and modifies the three-dimensional structure of proteins in nerve cells (Myers et al., 2018).

BAG3 Proteins as Molecular Chaperons

Cells can undergo protein homeostasis imbalance due to high temperature and other proteotoxic stimuli like free radicals etc. However, the cells have an inner evolutionary conserved functional ability to overcome the stress. Sometimes, cellular stress can lead to the genetic alteration of its proteins that are molecular chaperones in nature. The heat shock response in mammals is a highly conserved process regulated by heat shock transcription factors. These factors regulate the expression of several heat shock genes that prevent an imbalance in protein homeostasis in cells due to stressors. The heat shock transcription factors (HSF) are present in the cytoplasm as monomers in inactive forms in association with heat shock proteins. Exposure of cells to high temperatures, heavy metals, or accumulation of non-native proteins causes dissociation of heat shock proteins from HSF1. These dissociated HSF1 are transported to the nucleus (Kim and Yenari, 2017) wherein they are phosphorylated by protein kinase C (probably) to form activated trimeric forms. Activated HSF1 trimers thereby interact with two heat shock responsive elements within the BAG3 promoter gene. This binding leads to HSF1 interaction with the promoter region of 'heat shock protein' genes thereby causing increased production of heat shock proteins. BAG3 being a stress-inducible co-chaperone protein, the above binding of HSF1 to heat shock responsive elements also causes induction of BAG3 expression. This leads to increased production of BAG3 protein. BAG3 protein has also been reported to interact with HSF1 through the BAG domain (about 110-124 amino acids long) (Sturner and Behl, 2017). Further, overexpression of BAG3 causes down-regulation of HSF1 level in the nucleus, thereby exporting it to the cytoplasm (Jin et al, 2015). The above involvement of HSF1 in the nucleus as well as, in the cytoplasm is probably due to efficient shuttling ability with the help of shuttling protein, also called nuclear pore complex (Gama-Carvalho and Carmo-Fonseca, 2001; Jin et al., 2015). Besides heat-stressed cells, in oxidatively stressed cells, the extracellular signaling-regulated kinase (ERK) dependent phosphorylation of BAG3 helps the cells to recover from stress through overexpression of BAG3. This shuttling mechanism is important to regulate the balance between nuclear and cytoplasmic proteins in normal as well as, stressed cells. For instance, this shuttling helps to regulate the p53 activity in normal cells where it is maintained at lesser concentrations inside the nucleus. On the other hand, a rise in the level of this transcription factor inside the nucleus of stressed cells leads to the activation of cell cycle arrest and apoptosis (Gama-Carvalho and Carmo-Fonseca, 2001).

The above discussion is focussed on how BAG3 is involved in the cellular apoptotic pathway. On the other hand, this HSF1 induced BAG3 protein modulates the chaperone-mediated autophagic pathway too. The chaperones are a family of proteins that bind to non-native proteins, prevent aggregation of cellular proteins, and promote protein folding for normal cellular functions. A list of various chaperones and co-chaperones partners of BAG3 are mentioned in Table 1 (Klimek et al., 2017) and have been discussed in the below sections.

The presence of cellular stressors causes activation of two domains of BAG3 protein - PxxP (protein-protein interaction domains) and 2 IPV motifs. The PxxP domain-mediated interaction of co-chaperone with dynein causes BAG3 to induce back transport of Hsp70 to perinuclear sites. The second and third helices of overexpressed BAG3 protein interacts with the chaperone family protein Hsp70. Interaction of BAG3 with Hsp70 causes the latter to associate with the exposed hydrophobic regions of non-native proteins, thereby recognizing them for degradation. Also, BAG3 along with Hsp70 and CHIP (Hsp70-

Table 1. List of chaperones and co-chaperones associated with BAG3.

Protein	Role of Protein
HSPA8 (HSC70, HSP73)	HSP70 family members in the mammalian cytosol and nucleus cooperates with BAG3 in target molecule/protein processing and autophagic degradation
CHIP (STUB1)	HSP70-associated ubiquitin ligase conjoins with BAG3 during target molecule selection for autophagic degradation
Parkin	HSP70-associated ubiquitin ligase same as CHIP, but conjoins with BAG3 in mitophagy
HSPB2, HSPB5, HSPB6	Small heat shock proteins that can associate weakly or moderately to BAG3
HSPB8	Small heat shock protein that has a high affinity for BAG3, conjoins with BAG3 in muscle proteostasis and stress granule organization

associated ubiquitin ligase) induces the labeling of selective non-native proteins with ubiquitin. These ubiquitylated proteins come together to form aggresome at the perinuclear region thus, bringing together almost all misfolded proteins from all over the cell in one place. This activity minimizes their toxic effects on other cellular processes and initiates regaining of protein cellular homeostasis. The cell's attempt to get back to normal state simultaneously initiates clearance of misfolded plasma membrane proteins and ubiquitylated proteins. The misfolded proteins along with other degradation prone substrates accumulate and are engulfed by membrane structures to form autophagosomes. Lysosomal hydrolytic enzymes through the process called macroautophagy/autophagy degrade these engulfed materials. The ubiquitylated proteins undergo proteasomal degradation.

On the other hand, the two IPV domains of BAG3 protein mediates binding of BAG3 to HspB6 and HspB8, thereby triggering the formation of oligomeric assemblies. The latter sequester misfolded proteins inside cells and promotes their re-folding and further processing. Kim et al., (2016) have reported that over-expressed BAG3 decreases Hsp70 levels and lowers HSF1 levels in the nucleus than in the cytoplasm. This mechanism finally reduces cellular stress levels and regains normal cellular activity.

ROLE OF BAG3 IN NERVOUS SYSTEM

Several reports are explaining the role of BAG3 in the nervous system starting from their involvement in development until inhibition of apoptosis.

Development of Neurons

Immunohistochemical analysis revealed a transient expression of BAG3 in the developing central nervous system of rats, mainly in the cerebral cortex and hippocampus, wherein an abrupt rise in BAG3 expression was observed during the first postnatal week, with a decline thereafter (Rosati et al. 2011). BAG3 is also expressed in neural progenitor cells, i.e. cells that are more specific in function than stem cells. The glial progenitor cells give rise to neurons and glial cells during embryonic development in animals. BAG3 expression was reported to be altered in the cerebellum of hypothyroid juvenile mice in another study (Rosati et al. 2011).

Regulation of Actin Folding and Cytoskeletal Organization

BAG3 has been reported to interact with cytosolic chaperon in CCT (chaperonin containing TCP-1) thereby regulating actin folding and maintaining cytoskeletal organization of cells (Rosati et al. 2011). This indicates the probable role of BAG3 in regulating normal trafficking of various proteins and molecules through the cell membrane as well as, retention of organelle dynamics.

Inhibition of Mitochondrial Translocation of Pro-Apoptotic BAX

In glioblastoma cells, BAG3 prevented mitochondrial translocation of pro-apoptotic BAX protein by retaining it in the cytosol (Rosati et al. 2011). It is well known that in normal cells, BAX is always located in the cytosol. Translocation of BAX to the outer mitochondrial membrane triggers the mitochondrial apoptotic process releasing cytochrome c. This indicates a probable role of BAG3 in preventing apoptosis through the retention of BAX in the cytosol.

Macroautophagic Activity

Through the macroautophagic activity, BAG3 assists in the clearance of aggregated proteins in cells of people suffering from neurodegenerative disorders like HD (mutated huntingtin/polyQ proteins), AD (tau protein), and ALS (mutated SOD1) (Klimek et al., 2017). Macroautophagy/autophagy is a process that degrades excess or damaged cell components including proteins through lysosomal hydrolytic enzymes (Klimek et al., 2017). There are several autophagic pathways, which are very selective in choosing the cargo they intend to degrade and remove from cells. This selection often depends on the ubiquitin labeling of the cargo, which is the degradation marker. Ubiquitin labeling of the cargos are recognized by their respective adaptors that link them to autophagosome precursor membranes, also called as phagophores (Klimek et al., 2017). BAG3 encourages the formation of ubiquitin labeled chaperones along with Hsp70-associated ubiquitin ligase CHIP for removal of unwanted chaperones from cells (Klimek et al., 2017). Echaniz-Laguna et al., (2017) and Ghaoui et al., (2016) confirmed that the major function of BAG3 was to maintain proteostasis.

Neuropathic Dysfunction

Neuropathic dysfunction in the peripheral sensory nerve fibers occurs due to certain cytoskeletal and mitochondrial mechanisms in the central nervous system. It should be remembered that neuronal cells are highly susceptive to hypoxia and ischemia-induced damage (Akhtar et al., 2004). In preclinical models of common peripheral neuropathic pain conditions, it has been reported that cellular mechanisms of neurodegeneration can produce painful hyperactivity in primary afferent nociceptors (Reichling and Levine, 2011). Clinical studies on four BAG3 mutated patients with sensorimotor neuropathy have confirmed the association between giant axonal neuropathy with BAG3-associated myofibrillar myopathy (Jaffer et al., 2012). Neuropathic pain is a usual and important clinical manifestation (Reichling and Levine, 2011) that often goes unnoticed. Mutations in human BAG3, and its associated proteins like HspB8 and DNAJB6 (J-domain co-chaperone) are reported to induce motor axonal neuropathy (Ghaoui et al., 2016).

MECHANISM OF ACTION OF BAG3 IN MAJOR BIOLOGICAL PROCESSES

BAG3 protein is induced by stressful stimuli (like heat stress, mental stress, and proteotoxic stress, etc.), mainly through the activity of heat shock factor 1 (HSF-1) (Klimek et al., 2017). Major biological processes like cytoskeleton organization, apoptosis, autophagy, and mitophagy, etc. are modulated by this induction.

BAG3 as a Molecular Chaperone and Autophagic Function

To maintain cellular homeostasis under stress, BAG3 mediates selective macroautophagy as an adaptive mechanism (Klimek et al., 2017). Nerve cells adopt the process of autophagy/macroautophagy to eliminate non-functional and pathogenic proteins along with other cellular constituents to retain their functional capacity and homeostasis. In this process, the dysfunctional proteins get sequestered through microtubules inside a vesicle called as autophagosome and are transferred to lysosomes for ultimate destruction (Klimek et al., 2017).

As shown in Figure 4, IPV domain in BAG3 protein binds to a small heat shock protein family (Hsp20) member called αB-crystallin, and this causes inhibition of protein aggregation. The proline containing protein sequence PXXP brings about the binding of BAG3 to Src homology 3-containing protein phospholipase C-γ. This binding initiates the transport of misfolded proteins to aggresomes situated around the nucleus (Sturner and Behl, 2017). BAG3 has two highly protected motifs made up of a combination of isoleucine-proline-valine. Through this motif, BAG3 binds with two small heat shock proteins HspB8 and HspB6 that controls macroautophagy in cells. The misfolded and damaged proteins along with damaged organelles arising due to macroautophagy are engulfed in autophagosomes and destroyed (Myers et al., 2018). On the other hand, the toxicity caused due to the aggregation of misfolded proteins in cells is also recognized by LTN1 and VCP components of the ribosome quality control system. They interact with Hsp70-BAG3 to form a complex that regulates the signaling pathways to bring the protein homeostasis under control (Meriin et al., 2018).

BAG3 also controls the selective degradation of misfolded proteins like mutant SOD1 (superoxide dismutase 1, soluble) and polyglutamine (polyQ)-expanded HTT (huntingtin) (Rodriguez et al., 2016). BAG3 and dynein selectively get associated and transport aggregate proteins to aggresomes thus, facilitate their clearance (Figure 4). BAG3 (Sturner and Behl, 2017) gets integrated with chaperones Hsc70 and Hsp73 (members of Hsp70 group) in autophagic destruction thereby maintaining the quality of protein in neurons as shown in Figure 4. It is to be noted that the autophagy triggered by BAG3 differs in its mechanistic pathway in comparison to the chaperone-mediated pathway. Hence, two different terms are being used to distinguish them as chaperone-assisted selective autophagy (CASA) for BAG3 mediated pathway and the other being chaperone-mediated autophagy (CMA). The latter is not dependent on autophagosome formation. BAG3 mediated autophagy plays a critical role in maintaining the protein homeostasis in post-mitotic cells of differentiated neurons (Klimek et al., 2017).

In addition to this mechanism, SYNPO2 (synaptopodin 2, an actin-binding protein) associates itself with BAG3 and facilitates the fusion of phagophore membranes to promote autophagic engulfment of BAG3-protein complexes (Klimek et al., 2017). This process is however accomplished through the involvement of vacuolar protein sorting factors like VPS18 and VPS16 in addition to syntaxin-7. Thus, the BAG3:BAG1 ratio is an important determinant of the fate of aggregate proteins inside cells as depicted in Figure 5. In addition to this, mitochondrial Superoxide dismutase, a cellular antioxidant,

Figure 5. The cellular protein quality control mechanism, molecular chaperone systems in coordination with transport and degradation systems of proteins inside cells to ensure cellular proteostasis (Adapted from Myers et al., 2018).

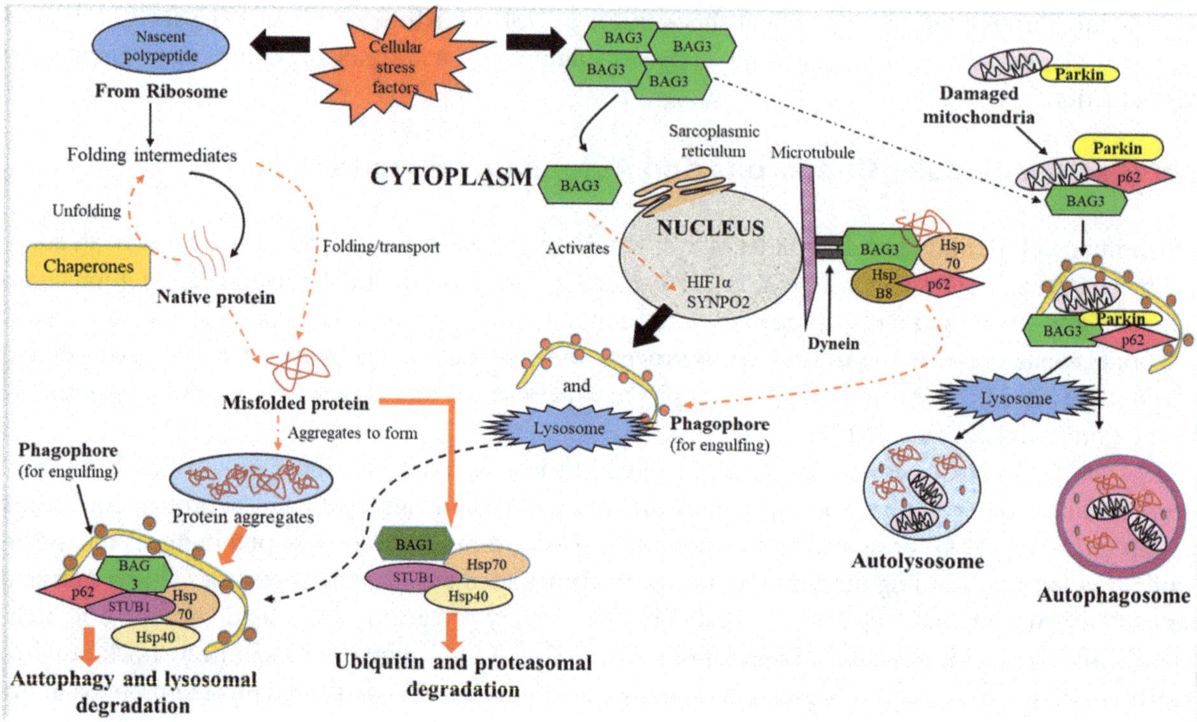

is found to be related to the aggresome pathway through functional cooperation between BAG3 and Hsp70 (Gamerdinger et al., 2011). Studies have shown that misfolded proteins are directed by BAG3 to the aggresome pathway by charging Hsp70 substrates on to the dynein motor complex. Dynein is a cytoskeletal protein that transports through microtubules in cells.

Mitophagic Activity

Among other chaperone and co-chaperone partners of BAG3, Parkin (PRKN) is an Hsp70-associated ubiquitin ligase, which plays a major role in mitophagy (Klimek et al., 2017). Besides being an efficient modulator of apoptosis and autophagy in stressed cells, BAG3 is also reported to play a critical role in maintaining the quality of mitochondrial proteins (Tahrir et al. 2017). Together, BAG3 and Parkin enters depolarized mitochondria and plays a role in autophagic removal of non-functional/degraded mitochondria.

Cytoskeleton Organization

BAG3 is also reported to regulate cell adhesion, which is dependent on its multiple interactions with other cellular proteins through various structural domains (Rosati et al. 2011). Recent studies have shown that BAG3 interacts with cytosolic chaperonin CCT (chaperonin containing TCP-1) which further regulates cytoskeleton organization and controls cellular trafficking and organelle structural integrity (Rosati et al. 2011).

Inhibition of Apoptosis

Coupling of the BAG domain with anti-apoptotic protein Bcl2 results in inhibition of apoptosis. Bcl-2 family regulates cell death and survival in neuronal cells of the mature nervous system (Reichling and Levine, 2011). This Bcl-2 protein is at a comparatively higher level in sympathetic neurons and sensory in the peripheral nervous system (PNS) of adults. Alteration of this programmed cellular activity may lead to a wide variety of neurodevelopmental anomalies. Neuronal overexpression of either Bcl-2 or B-cell lymphoma-extra large (Bcl-xL) or both are responsible for neuroprotective function because Bcl-xL is anti-apoptotic in nature (Akhtar et al., 2004). In addition to these, BAG3 has been reported to protect IKK-γ from proteasome delivery thereby causing sustained NF-kB activation and cell survival (Rosati et al., 2011).

In vitro studies have shown that the heat shock response, the NFκB signaling pathway, and the unfolded protein response, all participate to ensure BAG3 is upregulated upon proteasome inhibition (Minoia et al., 2014). An interesting shift from proteasome-to-autophagy has thus been observed in various *in vitro* studies. This is mediated through BAG3's high sensitivity towards the amount of protein necessary to be degraded inside cells (Rodriguez et al., 2016). This ultimately restores protein homeostasis.

The anti-apoptotic activity of BAG3 is observed due to its competitive nature with BAG1 to positively cooperate with Hsp70 and CHIP (C-terminus of the Hsc70-interacting protein) (Rosati et al., 2011). Indeed, BAG3 competes with BaG1 for the HspA1A-bound (poly) ubiquitinated substrates and directs the aggregate proteins to cytoplasmic puncta and autophagosomes.

Regulation of mTOR

In addition to various pathways and processes mediated by BAG3, its intriguing role in association with mammalian target of rapamycin (mTOR) inhibitor TSC1, brings more insight into the mechanism of regulating anabolic and catabolic processes in mammals (Klimek et al., 2017). One of the forms of mTOR, namely mTORC1 is the major regulator of both anabolic and catabolic processes in mammals. In starved or stressed cells, the inactivation of mTORC1 leads to the cessation of the protein translation process and activation of autophagy. Besides, phosphorylation of the α-subunit of translation initiator factor eIF2 shuts the translational process and stimulates autophagy, as well (Rosati et al., 2011).

BAG3 AND NEURODEGENERATIVE DISORDERS

As discussed above, BAG3 participates in the rerouting of (poly)-ubiquitinated proteins to autophagy for degradation upon proteasome inhibition (Minoia et al., 2014). The major cause of NDDs is aggregated proteins, which should undergo autophagic destruction initiated by BAG3. An example of such proteins includes extended poly-glutamine stretches of Huntington protein that cause HD. Interaction between BECN1 (Beclin1, an autophagy-related protein) and Bcl2 is stabilized by BAG3 expression (Rodriguez et al., 2016).

BAG proteins regulate Hsp70 by binding it via ATPase domain through BAG domain (110–124 amino acids) (Rosati et al., 2011). BAG3 domain has a crystal appearance and consists of 3 anti-parallel α helices (Kumar et al., 2011) as revealed through X-ray crystallography and nuclear magnetic resonance (NMR) studies (Doong et al., 2002). The first and second α helix associated with serine/threonine kinase Raf-1 while the second and third interact with the ATP-binding pocket of Hsp70. This indicates a competitive binding nature between Raf-1 and Hsp70 with BAG3, which might play a crucial role in the cellular control of NDDs. Cummings et al., (2001) reported the protective role of Hsp70 against neurodegeneration in crossbred SCA1 mice overexpressing inducible Hsp70 (Cummings et al., 2001). Besides these, binding of BAG3 to guanine nucleotide exchange factor 2 (short form) regulates cell adhesion through activation of Rap1 by the latter. Rap1 is responsible for the regulation of cell-cell junction formation and modification thus, increasing integrin-mediated cell adhesion (Rosati et al., 2011). The PPDY motif of above factor 2 was found to bind to WW domain of BAG3 also.

In a recent study reported by Sam Eaton and Monte S. Willis (2017), it was hypothesized that the Bag3-Pro209 Leu pre-amyloid oligomers (PAOs) bring about the same mechanistic changes that are observed in Alzheimer's and other neurodegenerative disorders, wherein PAOs directly induces neuronal dysfunction and death. It is assumed that PAOs activate p38 MAPK signaling, to initiate cell damage and death. p38 inhibition reverses the neuronal proteotoxicity, and this aspect is currently in clinical trials to treat AD (Eaton and Willis, 2017).

The BAG3-pathway in some of the age-related neurodegenerative disorders is detailed below.

Various age-related neurodegenerative disorders, such as AD, Parkinson's disease, HD or ALS have been reported to have disturbed neuronal protein homeostasis.

Alzheimer's Disease

AD is also called as 'senile dementia' as it is a progressive disease that ultimately leads to the destruction of memory and related mental functions. The brain cells gradually degenerate and die, thereby causing a decline in thinking, behavioral, and social abilities of an affected person. The genetic heritability for AD is estimated to be 60 – 80% (Arneson et al., 2018). It is characterized by extracellular β-amyloid (Aβ) plaque formation, intracellular neurofibrillary tangles (NFTs), and reduced synaptic density in the brain (Sturner and Behl, 2017). The filamentous core of NFTs is made of microtubule-associated protein tau (MAPT) in a highly phosphorylated form (Sturner and Behl, 2017). In stressed neuronal cells, BAG3 has been reported to degrade tau proteins through selective macroautophagy, thereby revealing a possible therapeutic pathway. Several key AD-related proteins are localized to mitochondria or are present in the interface between mitochondria and endoplasmic reticulum (Ramanan and Saykin, 2013). This inter-association of proteins concerning mitochondria gives a possible indication of molecular interplay in and around it, which can be modulated to treat neurodegenerative disorders. Induction of cellular stress can lead to impairment of mitochondrial function, which can lead to hyperphosphorylation of MAPT. This causes accumulation of dysfunctional mitochondria and induction of apoptosis due to poor cellular energetics. In addition to these, some reports indicate a possible involvement of high levels of mutated mitochondrial DNA in AD (Ramanan and Saykin, 2013). However, this aspect needs further scientific validation.

The above facts depict the robust participation of BAG3 in the regulation of various pathways significantly extending its ability to maintain a fine balance between various cellular activities like protein folding, network of integrated pathways that involve gene expression, cell signaling, and protein degradation system.

Huntington's Disease

The molecular changes in the protein polyQ-huntingtin are involved in the progression of HD. This disease is characterized by the deposition of these protein aggregates in the cytoplasm and nucleus of brain cells. This polyQ protein aggregates also consist of sequestered transcription factors, thereby enhancing cellular toxicity. BAG3 has been reported to selectively remove the pathogenic polyQ43-huntingtin (a protein with 43 glutamines) through macroautophagy (Sturner and Behl, 2017). Also, BAG3 mediates HspB8's ability to prevent polyQ43-huntingtin aggregate formation as well as the removal of already existing polyQ43-huntingtin protein aggregates from the cells. However, it has been observed that the binding of BAG3 to Hsp70 is not a prerequisite for the clearance of polyQ43-huntingtin protein aggregates. Although the WW domain of BAG3 combines with its proline-rich PXXP region and modifies the three-dimensional structure of proteins in nerve cells (Myers et al., 2018), the PXXP region is required for autophagic degradation of polyQ43-huntingtin proteins.

Amyotrophic Lateral Sclerosis

ALS belongs to a group of rare neurological disorders known as motor neuron diseases, which are caused due to progressive deterioration and death of motor neurons. Its most common symptom is muscle weakness that occurs due to the slow death of upper and lower motor neurons. This is eventually followed by a loss of the brain's control over voluntary movements. Currently, there is no cure for ALS. Hence, there is a need for biomarkers to identify the presence or rate of progression of this disorder. Scientists have reported the accumulation of misfolded and mutant forms of SOD1 (Cu/Zn Superoxide dismutase 1) in unhealthy motoneurons of the spinal cord of ALS patients (Sturner and Behl, 2017). The functional role of BAG3 through macroautophagic activity has been discussed above. In acutely stressed cells, misfolded proteins bind to a multi-chaperone complex consisting of HspB8, HSP70, HSP40, BAG3, 14-3-3g, and STUB1 as shown in Figure 5. This binding initiates further binding of SOD1 to the cytoplasmic dynein complex and directs SOD1 to the aggresome. Besides this, in transgenic mice, removal of SOD1 aggregates through the autophagosome-lysosome system was observed in spinal cord motoneurons of transgenic mice (Sturner and Behl, 2017).

MOLECULAR INTERLINK BETWEEN GENE EXPRESSION AND BAG3 IN NEURODEGENERATION

The genetic makeup of a person regulates the stable protein environment of his/her body. Because of the interplay between genes and the local environment or environmental stress (emotional, pathological or injury, etc.), the gene expression and cellular signaling mechanisms get affected (Meo et al., 2016). It results in an altered level of various proteins due to various cytoplasmic and/or nuclear events. This may affect the homeostasis of the cellular proteins via their increased expression, sooner degradation, and faulty autophagic functions, etc. and may lead to pathogenic conditions in neurons/brain (Klimek et al., 2017). The vesicle-mediated transport was identified to be a common molecular pathway causing NDDs like AD, PD, and ALS (Klimek et al., 2017). Moving ahead, the molecular dynamic interlinking between genes and proteins (specifically BAG3) in neurodegenerative disorders is discussed in detail.

Multidimensional Interlinking Between Genomics and Proteomics in Neurodegenerative Disorders

Genome-wide association studies (GWAS) were undertaken to find the genetic makeup of cells under various such diseases (Narayanan, 2015). About a handful (86 studies) of GWAS studies that were conducted for AD, ALS, and PD, revealed the involvement of several single-nucleotide polymorphisms (SNPs) playing a role in these diseases (Narayanan, 2015). In an attempt to model the interactions among genes responsible for AD and PD, transcription factor binding networks were scientifically applied in the MetaCore software. Scientists identified highly relevant connections between GWAS genes of AD and PD. Nine out of thirteen AD genes were tightly networked with ten out of fifteen PD GWAS genes, in coordination with transcription factors SP1 and AP-1 (Arneson et al., 2018).

Techniques like microarray analysis and next-generation RNA sequencing have revealed several target gene expression profiling and interlinked theories between AD, PD, HD, and ALS. Through a meta-analytical approach, scientists have analyzed tissues from different regions of the brains of 13 patients. The brain tissues were collected from the hippocampus, substantia nigra, dorsolateral prefrontal cortex, motor cortex, cervical spinal cord, etc. based on the neurodegenerative disorder under consideration. This study identified a strong correlation between gene expression signatures of 243 genes among the above disorders. These sets of 73 up-regulated set of genes which include BAG3 (BAG3, BCL6, NUPR1, MT1H, and TNFRSF1A, etc.) and 170 down-regulated (TUBG2, CLASP2, and DYNC1LI1, etc.) genes are together termed as a common neurodegeneration module (CNM) (Li et al., 2014). These studies revealed that several metabolic pathways involved in chronic neuroinflammation, oxidative stress, altered synaptic transmission, and altered protein degradation are the common pathways observed in NDDs.

Nerve cells are a pool of proteins with varied functional role, including maintenance of signaling flow throughout the body. Some of these proteins are anti-apoptotic too. Structural disorganization of any of these proteins is linked with neurodegenerative disorders, but the molecular mechanism behind such cellular disorganization is under continuous investigation. Among such proteins, Bcl-2 associated athanogene (BAG family) protein is one of them, which is reported to play an important role in NDDs (Guo et al., 2015).

RECENT DEVELOPMENTS AND FUTURE RESEARCH DIRECTIONS

BAG3 Drug Targeting Potential

The identification of a novel drug target for the majority of the above discussed neurodegenerative disorders will involve a deeper dive into the structure and interaction properties of protein aggregates, mitochondrial proteins, voltage-gated calcium channels, oxidative stress proteins, and enzymes involved in the repair of damaged DNA. The GWAS gene database has a high chance of target identification for therapeutic drug discovery against neurodegeneration.

Comparative studies (Gamerdinger et al., 2009) conducted between young and old mice to understand the role of BAG3 in neuronal aging revealed a high concentration of this protein in different regions of old mice's brains. High levels of BAG3 and HspB8 in astrocytes may degrade the debris from dead neurons and extracellular aggregated proteins (Sturner and Behl, 2017). One of the studies conducted in patients suffering from AD, PD, HD, and spinocerebellar ataxia type 3, showed that upregulation of heat shock protein B8 and BAG3 helps astrocytes to remove abnormally aggregated proteins thus, maintaining cellular homeostasis and regaining their normal cytoskeletal structure (Seidel et al., 2012). In rats fed with curcumin, an upregulation of BAG2 was observed in primary cortical neurons (Seneci, 2015). In another report, it has been shown in rat primary neurons that the overexpression of BAG3 can lead to a drastic reduction in levels of tau and phosphor-tau proteins (Sturner and Behl, 2017).

Overexpression of BAG3 has been reported to activate the anti-apoptotic pathway in neoplastic cells (Sturner and Behl, 2017) and studies have proven that BAG3 to be a promising drug target in the treatment of cancer (Klimek et al., 2017). Drugs that behave as TNF-related apoptosis-inducing ligand, proteasome inhibitors like cytosine arabinoside, fludarabine, and etoposide, all of them increase the levels of BAG3 protein causing cellular resistance to therapy, thus making BAG3 silencing, a mechanism to improve neoplastic cell apoptotic response to drugs. For example, one study that is more clinical conducted in patients affected with brain tumor revealed high expression of BAG3 in glioblastomas and astrocytic tumors, through immunohistochemistry (Festa et al., 2011) which indicates inhibition of apoptosis in cancer cells in the result of the high level of expression of BAG3. This fact may be utilized in protecting the normal neural cells as well, from neurodegeneration. Therefore, it is suggested that pharmacological manipulation of the Bcl-2 family could improve the human neurological conditions such as stroke and NDDs.

Other than BAG3, the monoamine oxidase (MAO)-B inhibitor, rasagiline, was reported to have mild to moderate effect in relieving PD symptoms along with its other neuroprotective effects, in a study conducted by Ji-Feng Guo et al., (2015) in MPTP-lesioned mice. In this study, BAG2 and BAG5 protein levels were upregulated whereas no changes were observed in the proteins BAG1 and BAG3 (Guo et al., 2015).

Selective macroautophagy properties of BAG3 are also gaining its importance in the field of designing novel therapeutic drugs to cure NDDs. However, limitations in knowledge about the cause and mechanism through which neurons die in neurodegenerative diseases, have affected the development of protective/preventive therapies in this field. The WW domain of BAG3 assists in chaperone-mediated autophagy. It also combines with the proline-rich PXXP region and modifies the three-dimensional structure of proteins in nerve cells (Myers et al., 2018). The level of expression of this protein might also be manipulated by using the drug molecule to regulate the cellular homeostasis of the proteins to get the desired result. *In vivo* studies conducted in rodents (Klimek et al., 2017) proved that a simultaneous BAG3 expression and CASA activity due to several cellular stimuli switches the proteasomal mode of protein degradation in aging brains to autophagic destruction. This led to improved removal of degraded proteins from neuronal cells (Jaffer et al., 2012).

Scientists have attempted to artificially induce this process pharmacologically or through an altered level of oxygen and nutrients in cell culture models (Jennewein et al., 2016). In these models, a study about four key autophagosomal molecules analyzed in 350 gliomas/astrocytoma was carried out to understand the glioma environment better and use it as a therapeutic target. Besides this, the drug rapamycin has been established as an autophagy inducer. *In vivo* studies on mice revealed the potential of rilmenidine drug to induce autophagy, thereby promoting clearance of huntingtin from HD models (Hochfeld et al.,

2013). In spite of a continuous effort of scientists to discover the therapeutic modulation of BAG3 associated proteins, the mechanism of action of BAG3 is still a mystery under research and investigation in various cellular models and clinical studies. Its dynamic ability to bind/interact with multiple proteins in various cell types has placed a challenge to the scientists as to what exactly is its mechanism of action in neurodegenerative diseases.

Other than apoptosis and autophagy BAG1 has been reported to have many other functions as it interacts with several other proteins and cellular components like chaperones, Bcl-2, Raf-1, DNA, proteasome, nuclear hormone receptors, as well as proliferation, metastasis, transcription, and cell survival activities (Kumar et al., 2011). Among the 3 α helices of the BAG domain, research was initiated in tumor cells by designing to the targeted drug to the first α-helix, thereby preventing the binding of Raf-1 and promoting Hsp70 binding to the domain. This inhibited Hsp70 and promoted attenuation of DNA synthesis and cell proliferation. This therapeutic interventional study through the application of docking protocols helped to identify a potential drug to arrest tumor progression (Kumar et al., 2011). Besides this, it is known that Hsp70 is involved in protein homeostasis and cell survival pathways. Therefore, modification of Hsp70-BAG3 protein-protein interaction can be looked at as one of the drug targets

As it has been shown that autophagy is inactivated by TSC1 (Rapamycin [mTOR] inhibitor tuberous sclerosis 1 [TSC1]) through inhibition of mTORC1, the serine/threonine kinase RPS6KB1 that responds to mTOR signaling BAG3 can also be a highly potential drug target in the treatment initiative of neurodegenerative disorders.

Apart from the above targets, about 30 proteins associated with multiple sclerosis, frontotemporal dementia, PD, etc. have been under research for their possible therapeutic targets (Narayanan, 2015). These proteins include myosin/tubulin-binding proteins, cytoplasmic and nuclear transporters, lysosomal transmembrane enzyme, etc.

BAG3 as a Possible Diagnostic Marker of Neurodegeneration

As we have seen that BAG3 is involved in many pathways in neurons and other associated cells, earlier identification of their levels and associated functions and proteins might help us to determine the initiation or progression of NDDs. In patients suffering from neurodegeneration, the nerve cells get affected earlier than the time when neurodegenerative symptoms become identifiable/prominent. This period is dependent on the speed of neurodegeneration, which can vary from one month to several years. Thus, the number of residual neurons in that particular site falls below the number required to maintain normal neurological function (Przedborski et al., 2003). Intracellular proteinaceous bodies and degraded proteins have been under continuous research for the identification of a potential diagnostic marker for any type of neurodegenerative disorder. For instance, the Lewy bodies have been well established as markers of PD. Only a specific subpopulation of neurons is indeed affected at the specific structural site(s) in neurodegenerative disorders (Przedborski et al., 2003). In olivopontocerebellar atrophy, several brain structures inside the nervous system get affected at different sites, however, the neurons nearby remain unaffected. This brings in a challenge to detect a single diagnostic marker in such type of neurodegeneration. A lack of pre-symptomatic markers makes it almost impossible to determine the disease onset. Neuropathic pain is a typical clinical manifestation (Reichling and Levine, 2011) that often goes unnoticed, and if identified, can help in early diagnosis of neuropathy.

CONCLUSION

The metabolic and cellular pathways involved in neurodegenerative disorders have been continuously researched to identify the molecular mechanisms involved in each of them. Scientists have attempted to find a correlation between these pathways and have succeeded to an extent in finding drug targets to treat these disorders. Though an overlap exists in the molecular activity between them, yet the core genes, proteins, and molecules are diverse, thereby challenging researchers in drug discovery. Several factors are responsible for influencing the transcriptional module of BAG3 and its gene expression under various cellular conditions. With structure-activity relationship studies, GWAS tools, and several NeuroGenes database being identified, it is forecasted that stronger association between BAG3 and cellular proteins will enhance the drug discovery research for treatment of neurodegenerative disorders in future.

REFERENCES

O95817. (n.d.). *3D model from MODBASE based on 1uk5A – Protein Model Portal*. Retrieved from https://www.proteinmodelportal.org/?pid=modelDetail&provider=MODBASE&template=1uk5A&pmpuid=1001228397804&range_from=1&range_to=575&ref_ac=O95817&mapped_ac=O95817&zid=async

Akhtar, R. S., Ness, J. M., & Roth, K. A. (2004). Bcl-2 family regulation of neuronal development and neurodegeneration. *Biochimica et Biophysica Acta (BBA) - Molecular Cell Research*, *1644*(2–3), 189–203. PMID:14996503

Arneson, D., Zhang, Y., Yang, X., & Narayanan, M. (2018). Shared mechanisms among neurodegenerative diseases: From genetic factors to gene networks. *Journal of Genetics*, *97*(3), 795–806. doi:10.100712041-018-0963-3 PMID:30027910

BAG3 gene. BCL2 associated athanogene 3. Genetics Home Reference. (n.d.). US National Library of Medicine. Retrieved from https://ghr.nlm.nih.gov/gene/BAG3#location

BAG domain (IPR003103). (n.d.). *InterPro*. Retrieved from http://www.ebi.ac.uk/interpro/entry/IPR003103

Behl, C. (2016). Breaking BAG: The Co-Chaperone BAG3 in health and disease. *Trends in Pharmacological Sciences*, *37*(8), 672–688. doi:10.1016/j.tips.2016.04.007 PMID:27162137

Briknarova, K., Takayama, S., Homma, S., Baker, K., Cabezas, E., Hoyt, D. W., ... Ely, K. R. (2002). BAG4/SODD protein contains a short BAG domain. *The Journal of Biological Chemistry*, *277*(34), 31172–31178. doi:10.1074/jbc.M202792200 PMID:12058034

Cellular response to heat stress. (n.d.). Retrieved from https://reactome.org/PathwayBrowser/#/R-HSA-3371556&PATH=R-HSA-8953897,R-HSA-2262752&FLG=O95817

Cherry, K. (2018). An overview of the different parts of a neuron. *Theories. Biological Psychology*. Retrieved from https://www.verywellmind.com/structure-of-a-neuron-2794896

Cummings, C. J., Sun, Y., Opal, P., Antalffy, B., Mestril, R., Orr, H. T., ... Zoghbi, H. Y. (2001). Overexpression of inducible HSP70 chaperone suppresses neuropathology and improves motor function in SCA1 mice. *Human Molecular Genetics*, *10*(14), 1511–1518. doi:10.1093/hmg/10.14.1511 PMID:11448943

de Wit, J., Hong, W., Luo, L., & Ghosh, A. (2011). Role of leucine-rich repeat proteins in the development and function of neural circuits. *Annual Review of Cell and Developmental Biology*, *27*(1), 697–729. doi:10.1146/annurev-cellbio-092910-154111 PMID:21740233

Doong, H., Vrailas, A., & Kohn, E. C. (2002). What's in the 'BAG'? – A functional domain analysis of the BAG-family proteins. *Cancer Letters*, *188*(1–2), 25–32. doi:10.1016/S0304-3835(02)00456-1 PMID:12406544

Eaton, S., & Willis, M. S. (2017, Feb. 28). Discovery of Bag3-P209L signaling mechanism brings hope for therapeutic options currently in development. Atlas of Science.

Echaniz-Laguna, A., Geuens, T., Petiot, P., Pereon, Y., Adriaenssens, E., Haidar, M., ... Stojkovic, T. (2017). Axonal neuropathies due to mutations in small heat shock proteins: Clinical, genetic, and functional insights into novel mutations. *Human Mutation*, *38*(5), 556–568. doi:10.1002/humu.23189 PMID:28144995

Eglitis, M. A., & Mezey, E. (1997). Hematopoietic cells differentiate into both microglia and macroglia in the brains of adult mice. *Proceedings of the National Academy of Sciences of the United States of America*, *94*(8), 4080–4085. doi:10.1073/pnas.94.8.4080 PMID:9108108

Festa, M., Valle, L. D., Khalili, K., Franco, R., Scognamiglio, G., Graziano, V., ... Rosati, A. (2011). BAG3 protein is overexpressed in human glioblastoma and is a potential target for therapy. *American Journal of Pathology*, *178*(6), 2504–2512. doi:10.1016/j.ajpath.2011.02.002 PMID:21561597

Gama-Carvalho, M., & Carmo-Fonseca, M. (2001). The rules and roles of nucleocytoplasmic shuttling proteins. *FEBS Letters*, *498*(2–3), 157–163. doi:10.1016/S0014-5793(01)02487-5 PMID:11412848

Gamerdinger, M., Kaya, A. M., Wolfrum, U., Clement, A. M., & Behl, C. (2011). BAG3 mediates chaperone-based aggresome-targeting and selective autophagy of misfolded proteins. *EMBO Reports*, *12*(2), 149–156. doi:10.1038/embor.2010.203 PMID:21252941

Ghaoui, R., Palmio, J., Brewer, J., Lek, M., Needham, M., Evila, A., ... Udd, B. (2016). Mutations in HSPB8 causing a new phenotype of distal myopathy and motor neuropathy. *Neurology*, *86*(4), 391–398. doi:10.1212/WNL.0000000000002324 PMID:26718575

Guo, J., He, S., Kang, J., Xu, Q., Hu, Y., Zhang, H., ... Tang, B. (2015). Involvement of Bcl-2-associated athanogene (BAG)-family proteins in the neuroprotection by Rasagiline. *International Journal of Clinical and Experimental Medicine*, *8*(10), 18158–18164. PMID:26770414

Hochfeld, W. E., Lee, S., & Rubinsztein, D. C. (2013). Therapeutic induction of autophagy to modulate neurodegenerative disease progression. *Acta Pharmacologica Sinica*, *34*(5), 600–604. doi:10.1038/aps.2012.189 PMID:23377551

Jaffer, F., Murphy, S. M., Scoto, M., Healy, E., Rossor, A. M., Brandner, S., ... Reilly, M. M. (2012). BAG3 mutations: Another cause of giant axonal neuropathy. *Journal of the Peripheral Nervous System, 17*(2), 210–216. doi:10.1111/j.1529-8027.2012.00409.x PMID:22734908

Jennewein, L., Ronellenfitsch, M. W., Antonietti, P., Ilina, E. I., Jung, J., Stadel, D., ... Mittelbronn, M. (2016). Diagnostic and clinical relevance of the autophago-lysosomal network in human gliomas. *Oncotarget, 7*(15), 20016–20032. doi:10.18632/oncotarget.7910 PMID:26956048

Jin, Y., Ahn, S., & Kim, S. (2015). BAG3 affects the nucleocytoplasmic shuttling of HSF1 upon heat stress. *Biochemical and Biophysical Research Communications, 464*(2), 561–567. doi:10.1016/j.bbrc.2015.07.006 PMID:26159920

Kim, H. Y., Kim, Y., Yun, H. H., Im, C., Ko, J., & Lee, J. (2016). ERK-mediated phosphorylation of BIS regulates nuclear translocation of HSF1 under oxidative stress. *Experimental & Molecular Medicine, 48*(9), e260. doi:10.1038/emm.2016.84 PMID:27659916

Kim, J. Y., & Yenari, M. (2017). Heat shock proteins and the stress response in primer on cerebrovascular diseases. In *Primer on cerebrovascular diseases* (pp. 273–275). Cambridge, UK: Elsevier. doi:10.1016/B978-0-12-803058-5.00056-4

Klimek, C., Kathage, B., Wordehoff, J., & Hohfeld, J. (2017). BAG3-mediated proteostasis at a glance. *Journal of Cell Science, 130*(17), 2781–2788. doi:10.1242/jcs.203679 PMID:28808089

Kumar, A., Verma, K., & Sinha, A. (2011). Designing of anti-cancer drug targeted to Bcl-2 associated athanogene (BAG1) protein. In M. A. Mahdavi (Ed.), *Bioinformatics - Trends and Methodologies* (pp. 707–722). London, UK: IntechOpen Limited. doi:10.5772/18409

Li, M. D., Burns, T. C., Morgan, A. A., & Khatri, P. (2014). Integrated multi-cohort transcriptional meta-analysis of neurodegenerative diseases. *Acta Neuropathologica Communications, 2*(1), 93. doi:10.118640478-014-0093-y PMID:25187168

Meo, S. D., Reed, T. T., Venditti, P., & Victor, V. M. (2016). Role of ROS and RNS sources in physiological and pathological conditions. *Oxidative Medicine and Cellular Longevity, 2016*, 1245049. PMID:27478531

Meriin, A. B., Narayanan, A., Meng, L., Alexandrov, I., Varelas, X., Cisse, I. I., & Sherman, M. Y. (2018). Hsp70–Bag3 complex is a hub for proteotoxicity-induced signaling that controls protein aggregation. *Proceedings of the National Academy of Sciences of the United States of America, 115*(30), E7043–E7052. doi:10.1073/pnas.1803130115 PMID:29987014

Minoia, M., Boncoraglio, A., Vinet, J., Morelli, F. F., Brunsting, J. F., Poletti, A., ... Carra, S. (2014). BAG3 induces the sequestration of proteasomal clients into cytoplasmic puncta: Implications for a proteasome-to-autophagy switch. *Autophagy, 9*(9), 1603–1621. doi:10.4161/auto.29409 PMID:25046115

Myers, V. D., McClung, J. M., Wang, J., Tahrir, F. G., Gupta, M. K., Gordon, J., ... Feldman, A. M. (2018). The multifunctional protein BAG3: A novel therapeutic target in cardiovascular disease. *JACC. Basic to Translational Science, 3*(1), 122–131. doi:10.1016/j.jacbts.2017.09.009 PMID:29938246

Narayanan, R. (2015). Neurodegenerative diseases: Phenome to genome analysis. *MOJ Proteomics & Bioinformatics*, *2*(1), 1–8.

National Research Council, Division on Earth and Life Studies, Board on Biology, Commission on Life Sciences, Committee on Research Opportunities in Biology. (1989). The nervous system and behavior. In Opportunities in Biology (pp. 175-223). Washington, DC: National Academies Press.

Otte, L., Wiedemann, U., Schlegel, B., Pires, J. R., Beyermann, M., Schmieder, P., ... Oschkinat, H. (2003). WW domain sequence activity relationships identified using ligand recognition propensities of 42 WW domains. *Protein Science*, *12*(3), 491–500. doi:10.1110/ps.0233203 PMID:12592019

Pekna, M., & Pekny, M. (2012). The neurobiology of brain injury. *Cerebrum*, *2012*, 9. PMID:23447795

Przedborski, S., Vila, M., & Jackson-Lewis, V. (2003). Series Introduction: Neurodegeneration: What is it and where are we? *The Journal of Clinical Investigation*, *111*(1), 3–10. doi:10.1172/JCI200317522 PMID:12511579

Ramanan, V. K., & Saykin, A. J. (2013). Pathways to neurodegeneration: Mechanistic insights from GWAS in Alzheimer's disease, Parkinson's disease, and related disorders. *American Journal of Neurodegenerative Disease*, *2*(3), 145–175. PMID:24093081

Reichling, D. B., & Levine, J. D. (2011). Pain and death: Neurodegenerative disease mechanisms in the nociceptor. *Annals of Neurology*, *69*(1), 13–21. doi:10.1002/ana.22351 PMID:21280072

Rodriguez, A. E., Lopez-Crisosto, C., Pena-Oyarzun, D., Salas, D., Parra, V., Quiroga, C., ... Lavandero, S. (2016). BAG3 regulates total MAP1LC3B protein levels through a translational but not transcriptional mechanism. *Autophagy*, *12*(2), 287–296. doi:10.1080/15548627.2015.1124225 PMID:26654586

Rosati, A., Graziano, V., Laurenzi, V., & De. (2011). BAG3: A multifaceted protein that regulates major cell pathways. *Cell Death & Disease*, *2*(4), e141. doi:10.1038/cddis.2011.24 PMID:21472004

Seidel, K., Vinet, J., den Dunnen, W. F. A., Brunt, E. R., Meister, M., Boncoraglio, A., ... Carra, S. (2012). The HSPB8-BAG3 chaperone complex is upregulated in astrocytes in the human brain affected by protein aggregation diseases. *Neuropathology and Applied Neurobiology*, *38*(1), 39–53. doi:10.1111/j.1365-2990.2011.01198.x PMID:21696420

Seneci, P. (Ed.). (2015). *Chemical modulators of protein misfolding and neurodegenerative disease*. Bethesda, MD: Academic Press.

Sturner, E., & Behl, C. (2017). The role of the multifunctional BAG3 protein in cellular Protein Quality Control and in disease. *Frontiers in Molecular Neuroscience*, *10*, 177. doi:10.3389/fnmol.2017.00177 PMID:28680391

Tahrir, F. G., Knezevic, T., Gupta, M. K., Gordon, J., Cheung, J. Y., Feldman, A. M., & Khalili, K. (2017). Evidence for the role of BAG3 in mitochondrial quality control in cardiomyocytes. *Journal of Cellular Physiology*, *232*(4), 797–805. doi:10.1002/jcp.25476 PMID:27381181

Tau, G. Z., & Peterson, B. S. (2010). Normal development of brain circuits. *Neuropsychopharmacology*, *35*(1), 147–168. doi:10.1038/npp.2009.115 PMID:19794405

1. UK5. (n.d). *Solution structure of the Murine BAG domain of Bcl2-associated athanogene 3*. Retrieved from https://www.rcsb.org/structure/1UK5

UniProtKB - O95817 (BAG3_HUMAN). (n.d.). Retrieved fromhttps://www.uniprot.org/uniprot/O95817

Vinothkumar, K. R., & Henderson, R. (2010). Structures of membrane proteins. *Quarterly Reviews of Biophysics*, *43*(1), 65–158. doi:10.1017/S0033583510000041 PMID:20667175

WW domain (IPR001202). (n.d.). *InterPro, Protein sequence analysis & classification*. Retrieved from http://www.ebi.ac.uk/interpro/entry/IPR001202

Chapter 11
PINK1/Parkin in Neurodegenerative Disorders:
Crosstalk Between Mitochondrial Stress and Neurodegeneration

Mukesh Pandey
https://orcid.org/0000-0003-4474-0417
Department of Pharmaceutics, Delhi Institute of Pharmaceutical Sciences and Research, India

Shakir Saleem
https://orcid.org/0000-0003-2340-7885
Department of Public Health, College of Health Science, Saudi Electronic University, Riyadh, Saudi Arabia

Himani Nautiyal
Department of Pharmacology, Siddhartha Institute of Pharmacy, India

Faheem Hyder Pottoo
Department of Pharmacology, College of Clinical Pharmacy, Imam Abdulrahman Bin Faisal University, Saudi Arabia

Md. Noushad Javed
https://orcid.org/0000-0002-5775-8256
School of Pharmaceutical Sciences, Apeejay Stya University, India

ABSTRACT

PTEN-induced kinase 1 (PINK1), a mitochondrial serine/threonine-protein kinase encoded by the PINK1 gene, is thought to protect cells from stress-induced mitochondrial dysfunction. The activity of PINK1 facilitates the binding of Parkin protein with depolarized mitochondria to induce autophagy. Mutations of PINK1causes a type of autosomal recessive early-onset Parkinson's disease. Cell depends on the surveillance systems or mechanisms like protein quality control to handle the alterations in the proteins that are induced because of these mutations. These mutant proteins are found to be pathogenic and are reported to be related to various neurodegenerative disorders. This chapter focuses on the role of PINK1/Parkin in mitochondria quality control and its subsequent effect in neurodegeneration.

DOI: 10.4018/978-1-7998-1317-0.ch011

INTRODUCTION

The pathology of neurodegenerative disorders comprises the neuronal loss progressively in the central nervous system (Pottoo et al, 2014; Pottoo et al., 2019; Herrero et al., 2015). The reason behind this loss is the generation of pathogenic misfolded proteins that are required to be removed from the neurons. These misfolded proteins can be transported from one neuron to other thus affecting many other neurons and once the system for controlling the population of such proteins gets compromised then ultimately the destruction of neurons remains the last option. The occurrence of this disease at older ages indicates that the main risk factor for neurodegenerative disorders is aging. Sometimes genetic and environmental components are also involved.

Mitophagy, a dynamic process through which damaged mitochondria undergo deterioration by lysosomes, and in this process, *PINK1* and Parkin work together. Several lines of evidence are available in this favor of fact that mitochondria function loss has a role in the development of Parkinson's disease (PD) and other neurodegenerative disorders. Exposure to the chemical substance called MPTP (1-methyl-4-phenyl-1,2,3,6-tetrahydropyridine) developed the symptoms of PD in patients taking certain intravenous (IV) drugs (Greenamyre et al., 2001). A chemical reaction takes place in the dopaminergic neuron responsible for disease generation i.e. PD. MPTP inside the cellular environment gets transformed into MPP^+ (1-methyl-4-phenylpyridinium) and then transported to dopaminergic neurons. Inside the neurons, MPP^+ inhibits mitochondrial oxidative phosphorylation complex 1. Greenamyre et al., (2001) reported a poison, called rotenone can induce and cause further progression of PD pathology when studied in mice. Thus, exposure to this pesticide might induce PD on exposure. Both rotenone and MPTP acts as mitochondrial poisons and destroys dopaminergic neurons of the midbrain in particular (Jin et al., 2012). The alterations in *PINK1* or Parkin regulated mitophagy either due to mutations and mitochondrial defects induce due to stress can cause PD by the accumulation of damaged or dysfunctional mitochondria (Valente et al., 2004; Kitada et al., 1998).

The familial, autosomal recessive inherited form of PD is a result of *PINK1* gene mutations. The mutations in *PINK1*/Parkin generally lose their functions and hampers the removal of damaged mitochondria. Inflammation was considered as a crucial factor for PD pathology, but the involvement of mito-inflammation in PD pathology is still under investigation (Herrero et al., 2015). The current chapter focuses on the involvement of mitochondria and mitophagy in neurodegeneration with emphasis on the contribution of *PINK1*/Parkin in the pathogenesis of neurodegenerative disorders.

PROTEIN QUALITY CONTROL BY AUTOPHAGY

Inside the cellular environment processes essential for the survival of the cells as well as the organism as a whole goes on continuously. These processes include energy generation mechanism, biological macromolecules utilization, synthesis of proteins, intercellular or intracellular transportation of essential molecules and new cells as well as new cell organelles generation. Sometimes these processes undergo alterations or deviations from the defined procedures or mechanism leading to the formation of defective end products of the process. It is necessary to remove such products for maintaining the hemostatic condition within the cell. The cell organelles also undergo wear and tear resulting in damage and these damaged ones are again needed to remove on an urgent basis to be replaced by the new or healthy ones. The process which ensures removal of such useless biological material or molecules is called autophagy.

Table 1. Different proteins involved in various neurodegenerative disorders

Disease	Proteins Involved	Function
Alzheimer's disease	*PINK1*/Parkin	Regulate mitophagy Modulate expression of PSEN1 am jor component of APP enzyme γ-secretase
	Tau	Translocation of parkin
	Sirtuins	Protect against Aβ aggregation-induced toxicity in neurons
Parkinson's disease	*PINK1*/Parkin	Ubiquintate outer membrane proteins of mitochondria Induction of mitophagy
	A-Synuclein	Impair mitochondrial function Upregulate oxidative stress in mitochondria by activating MPTP
	DJ1	Removal of aggregates of p62 in PD Elimination of reactive oxidative species α Synuclein dimerization process
	LRRK2	Interaction with mitochondria: lysosome fusion regulators Translocation of CMA complex Regulate mitochondrial depolarization Regulate mitochondrial uptake of calcium
Huntingtin disease	*PINK1*/Parkin	Maintain morphology of mitochondria and ATP level
	GAPDH	Induce mitochondrial dysfunction because of amplified polyglutamine
	TG2	Potential loss of inner membrane of mitochondria by cross-linking with Huntingtin protein Abnormal protein accumulation in the brain
	VCP	Facilitates mitophagy in combination with LC3

(Wang et al. 2019)

Autophagy mechanism is accountable for removing dysfunctional proteins, unutilized biological macromolecules, damages or disfigured cell organelles, foreign biological material, and pathogens that invaded the cells. Different pathological proteins involved in neurodegeneration as shown in Table 1.

Three types of autophagy were reported in the literature; first is chaperone-mediated autophagy (CMA), second is macroautophagy (Figure 1) and the third one is microautophagy. Macroautophagy (simply autophagy) was reported as the most common process for removing the dysfunctional proteins and damaged cellular organelles from the cells (Deretic et al., 2018). Autophagy starts with the seizing of damaged organelles and dysfunctional proteins within an isolated fragment of the membrane (phagophore). This phagophore further engulfs a small cytoplasmic portion and then forms autophagosome which is a double layer membrane structure. Then autophagosome and lysosome fuses with each other and the hydrolases within the lysosomes degrade the material engulfed inside autophagosome. Thus, autophagy acts as a survival mechanism in various stress conditions like starvation, in which autophagy degrades unwanted cellular material to generate energy. It also serves as a defense mechanism against pathogens (Mizushima et al., 2011).

In yeast, more than 30 different autophagy-related genes Atgs, were reported by various studies. These genes are also conserved within the higher eukaryotes (Suzuki et al., 2007; Xie et al., 2007; Longatti et al., 2009). In the mammalian autophagy process, 36 different Atgs were reported to be involved along with various protein complexes which act as regulators of the whole process (Xie et al., 2015). Autophagy within the mammalian cells is a six-step process starting with the initiation step and then followed by nucleation, elongation, closure, maturation, and degradation (final step). Five necessary

Figure 1. Steps of macroautophagy in mammalian cells.

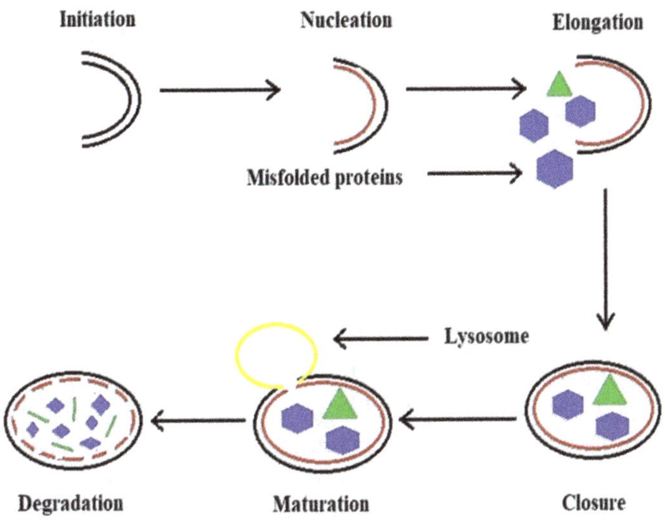

protein complexes for initiating the autophagy process are (a) Unc-51-like kinase 1 complex (ULK1); (b) transmembrane core ATG (ATG9); (c) the class III PI3K complex (PI3KC3); (d) WD repeat domain phosphoinositide- interacting proteins (WIPI) with ATG2; and (e) two ubiquitin (Ub)-like proteins such as ATG12 and ATG8 family proteins (ATG8s) and their covalent conjugation targets. ATG12 conjugates with ATG5 and ATG8 family proteins (ATG8s) (Wirth et al., 2013).Studies also suggested that autophagy is crucial for maintaining a pool for the supply of essential peptides during starvation, development of embryo before preimplantation and tumor suppression. Studies reported that various reverse genetic approaches are present in different organisms, including mice to accomplish autophagy (Rubinsztein et al., 2006; Cecconi et al., 2008; Mizushima et al. 2008; Mizushima et al., 2010; Levine et al. 2011).

The nutrient starvation process is an adaptive reaction or response against the stress conditions and it is a conserved role of autophagy. Cells derived cultures of mammalian cells, yeast cells, as well as mice can up-regulate autophagy after starvation or during the starving condition (Tsukada et al., 1993; Kuma et al. 2004). The nervous system and liver possess an important quality control system called basal or constitutive autophagy.

Role of Atg7, an important protein involved autophagosome formation during the autophagy process was discovered for the first time in liver-specific Atg7 deficient mice (Komatsu et al. 2005). Hepatomegaly, deformed cellular organelles and the presence of ubiquitin-positive aggregates within the cells are observed in such mice. An autophagy is described as the intracellular clearance system and its necessity within the biological system was again confirmed when the development of different motor defects like ataxic gait, impairment of motoric activity, difficulty in grasping by limbs, systemic tremor and even sporadic deaths were reported in research studies by analysis of mice (three weeks old) in which either *Atg7* (Atg7flox/flox; Nestin-Cre mice) or *Atg5*(Atg5flox/flox; Nestin-Cre; mice) genes specific to neurons were deleted (Komatsu et al. 2006). *Atg5* is another important gene with a role in the formation of autophagosome (Hara et al. 2006). In some test animals, neurodegenerative changes like the loss of cerebral pyramidal cells and Purkinje cells to a limited extent were observed. Swelling at the axons and generation of ubiquitin-positive protein aggregates (inclusion bodies) was detected in nerve cells of *Atg5*flox/flox; Nestin-Cre as well as*Atg7*flox/flox; Nestin-Cre mice. The same observations were recorded in mice in which the liver-specific *Atg7* gene was deleted. Hara et al. revealed due to systemic Atg5 deletion some lethal changes were developed in neonates. Deletion of Atg5 causes accumulation of ubiquitin proteins aggregates in hepatocytes, pituitary gland (anterior lobe), adrenal gland and the subset of neurons (Hara et al. 2006). Therefore, from these studies, as reported it is very clear about autophagy that it is a very crucial process for cellular homeostasis.

THE *PINK1*-PARKIN PATHWAY

PINK1 and Parkin are linked to autophagy through their contribution in the mitophagy (degradation and elimination of non-functioning mitochondria). Mitochondria is an important organelle of the eukaryotic cell which performs numerous critical functions like energy production, cytoplasmic calcium flux buffering, promotion of metabolism of lipids, and cell death. Autophagy removes damaged mitochondria by activating *PINK1* (kinase) and Parkin (E3 ubiquitin ligase) (Narendra et al 2010). This *PINK1* and Parkin facilitated the autophagy process for eliminating damaged and dysfunctional mitochondria are referred to as mitophagy. The functional failure of regulatory mitochondria quality control (mitoQC) mechanisms was considered to be a major driving force for normal aging. This fact has been known for more than a decade now (Harman et al., 1972). Furthermore, recent studies have added more information to it. Numerous conclusive studies reported that elevated oxidative stress due to failed mitochondrial mechanisms quality control mechanisms (Figure 2) was strongly related to the older age conditions like neurodegeneration (Schon et al., 2011; Wallace et al., 2005).

Mitochondria perform the function of an energy generator and supplier within the cells. The energy is generated as adenosine triphosphate (ATP) molecules that are necessary to keep the cell alive. This energy-providing molecule ATP is generated through oxidative phosphorylation which relies on the electron transport chain (ETC) mechanism for propagating electron flow. The main components of the ETC are found at the inner membrane of mitochondria. This ETC mechanism generates potential across the inner membrane and reduces the oxygen level within the matrix of mitochondria (Whitworth et al., 2017). This process also generates highly reactive oxygen species (ROS). These generated ROS cause serious damages to the mitochondrial components and disturbs it's functioning. However, against these reactive oxygen species, several protective mechanisms are present with the mitochondria (Whitworth et al., 2017). Primary defendants of mitochondria against these reactive species are the antioxidative enzymes like superoxide dismutase and glutathione which neutralizes the hazardous effects of these species. Secondary defendants include mitochondrial proteases, chaperones, ubiquitin protease system, and DNA repair enzymes (Whitworth et al., 2017). These defendants remove the dysfunctional and damaged mitochondrial components or proteins. But when the damage to the mitochondria due to the degradative species is too expensive or irreversible ultimately the mitochondria as a whole organelle are removed by means of lysosomes through the mitophagy process. The understanding with regards to the mitophagy was very less until the relation of *PINK1* and Parkin in PD was investigated by the Richard Youle in 2008. After the findings of these two proteins having an association with PD then various models have been developed to understand their relation with PD and the mitophagy. The current models reveal that under normal conditions there is a high import rate and degradation rate of *PINK1* within the mitochondria which causes the suppression of degradation signals for mitochondria (Matsuda et al., 2010; Narendra et al., 2010). In the damaged condition the generation of membrane potential within the mitochondria gets

Figure 2. Steps involved in mitochondria quality control (i.e. mitophagy).

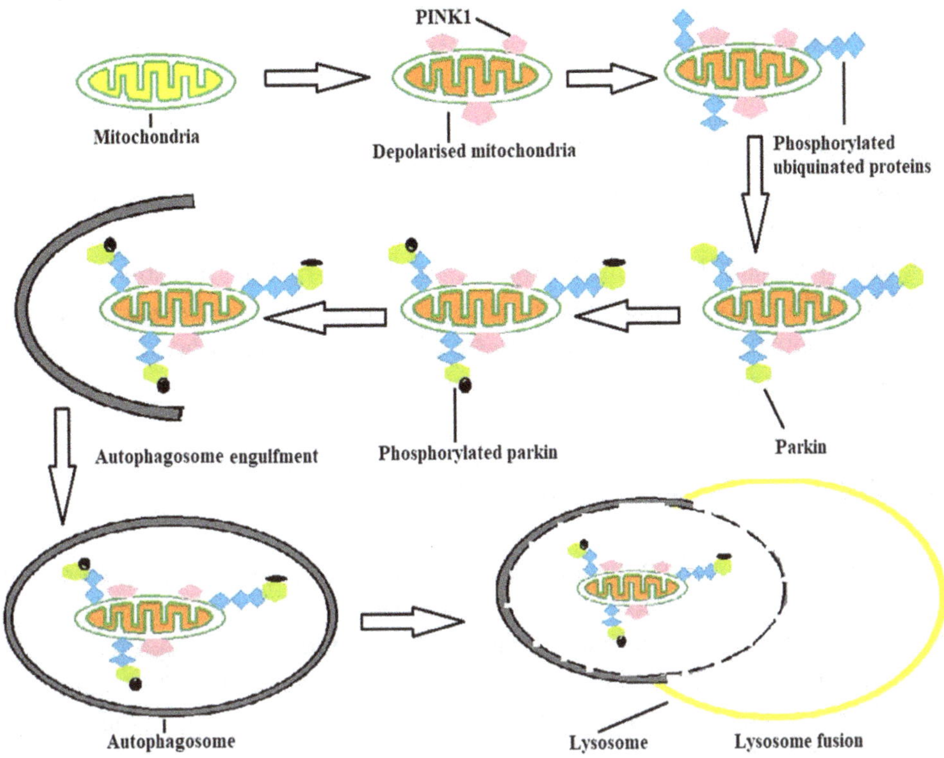

disrupted either because of mitochondrial poisons, uncoupling agents and other factors. The import of *PINK1* within the damaged mitochondria gets blocked due to which *PINK1* gets deposited on its outer membrane (Kazlauskaite et al., 2014). Then *PINK1* guides the phosphorylation of ubiquitin at ser65 residue and this stimulates the recruitment of Parkin on the damaged mitochondria surface. *PINK1* then phosphorylate the recruited Parkin at ser65 residue in the ubiquitin-like domain of Parkin (Kane et al., 2014; Koyano et al., 2014; Narendra et al., 2008; Kondapalli et al., 2012).

The recruitment and phosphorylation of Parkin stimulate its ubiquitin ligase activity. This stimulation of activity causes Parkin to ubiquitinate the various target proteins present on the mitochondria's outer membrane (Chan et al., 2011). The local supply of ubiquitin further acts as a substrate for *PINK1* which further helps to deploy more Parkin on the outer membrane which results in the high concentration of ubiquitinated proteins on the outer membrane of the mitochondria (Ordureau et al., 2014). This presence of an excessive concentration of ubiquitinated proteins leads to the deployment of ubiquitin adapter proteins which promotes engulfment of damaged, depolarized and dysfunctional by autophagosomes (Yamano et al., 2016; Pickrell et al., 2015). The process through which *PINK1*/Parkin induces mitophagy is still under investigation, but the above-reported mechanism provides explanations and answers regarding various questions linked with mitochondrial quality control and contribution of *PINK1*/Parkin in neurodegeneration. First of all, the proposed mechanism of mitophagy by Richard Youle et al., provides evidence for the role of mitochondrial defects in the pathology of neurodegeneration (Dias et al., 2013; Bender et al., 2006; Schapira et al., 1990). The study also provides an insight into how *PINK1* and Parkin protect the mitochondria from the deteriorating effects of mitotoxins (Rosen et al., 2006; Haque et al.,

2008; Paterna et al., 2007). This model also puts some light on the physiology of neurons of substantia nigra that requires high energy supply continuously and why these neurons are highly susceptible to the loss of mitoQC (Sulzer et al., 2013). In the initial phase of research on mitophagy, *PINK1*, and Parkin the failure to provide the evidence for Parkin translocation on depolarised mitochondria, damage of mitochondria due to depolarization and dysfunction of mitochondria due to mutations of Parkin (Van Laar et al., 2010; Sterky et al., 2011). But later studies on primary neurons revealed Parkin translocation on depolarised mitochondria (Cai et al., 2012). Parkin recruitment and autophagosome formation in the distal axons of the cultured neurons thus providing evidence in support of role on *PINK1*/Parkin in mitophagy (Ashrafi et al., 2014). Some recent studies also carried out investigations to find out the agents that trigger the *PINK1*-Parkin mediated autophagy without changing the mitochondrial membrane polarization state. It is found some of the unfolded matrix proteins of mitochondria can trigger *PINK1*-Parkin mediated autophagy (Jin et al., 2013; De Castro et al., 2012) .

MITOPHAGY AND ASSOCIATED FACTORS

Autophagy degrades the entire dysfunctional cell organelles as well as dysfunctional cytoplasmic proteins. Mitochondrial quality control is studied extensively and investigated to explore its role specifically in the pathogenicity of PD. Parkin is an E3 ubiquitin ligase, and its mutations were reported to be accountable for the onset and pathology of autosomal recessive juvenile PD (Kitada et al., 1998). Parkin also reported playing a part in the mitophagy (mechanism for removing damaged and dysfunctional mitochondria by autophagic degradation process) in various recent studies (Youleet al., 2011).

The initiation of mitophagy occurs with the Parkin's translocation from cytoplasm to the outer membrane of the depolarized mitochondria. This phenomenon was reported for the first time in 2008 (Narendra et al., 2008). *PINK1* another genetic factor that helps in translocation and targeting of Parkin to mitochondria (Gegg et al., 2010; Geisler et al., 2010; Kawajiri et al., 2010; Matsuda et al., 2010; Narendra et al., 2010; Rakovic et al. 2010; Vives-Bauza et al. 2010; Ziviani et al. 2010). However, *PINK1* protein has high instability and undergoes regular degradation. Many studies showed upon depolarization of mitochondria Pink 1 gets stabilized and then only recruitment of Parkin occurs (Jin et al., 2010; Shi et al., 2011). Still, the way how Parkin initiates the induction of mitophagy is not clearly understood. Now, several proteins of mitochondria are ubiquitinated by Parkin and examples of which are like voltage-dependent anion channel 1 (VDAC1), mitofusin (a mitochondrial pro-fusion factor), Bcl-2, and Drp1 (Geisler et al., 2010; Gegg et al., 2010; Poole et al., 2010; Tanaka et al., 2010; Ziviani et al., 2010; Chen et al., 2010; Wang et al., 2011). The autophagy adaptor p62 may recruit these ubiquitinated proteins (Ding et al., 2010; Geisler et al., 2010; Lee et al., 2010). Although, the role of p62 remains controversial as it was reported in the studies that p62 is important for grouping mitochondria in the perinuclear region, itself not directly involved in mitophagy (Narendra et al., 2010a; Okatsu et al., 2010). The findings of some scientific experiments reported that interaction between LC3-p62 was not the first indicator of disrupted mitochondria, and Atg proteins can also be deployed to the damaged mitochondria in LC3-independent manner (Itakura et al., 2012).

Mitochondria of yeast contain Atg32 (cargo receptor) that interacts with both Cvt11 and Atg8 (Kanki et al., 2009; Okamoto et al., 2009). However, in mammals, its counterpart was not reported yet. Several reports revealed that for proteasomal degradation of various types of proteins associated with the external membrane of mitochondria Parkin is very important (Tanaka et al., 2010; Chan et al., 2011; Yoshii et al., 2011). Proteins present at the outer membrane and in intermembrane space of mitochondria undergo proteasome-dependent degradation induced by mitochondrial depolarization in Parkin-overexpressing cells. While, the proteins present on the inner membrane and in the matrix mostly undergo mitophagy (Chan et al., 2011; Yoshii et al., 2011). Initially, on the damaged mitochondria p97/VCP and proteasome were recruited (Tanaka et al., 2010; Chan et al., 2011; Yoshii et al., 2011). Electron microscopy of these damaged mitochondria showed a ruptured outer membrane (Yoshii et al., 2011). Still question about the essentiality of protease based degradation of proteins present on the outer membrane of mitochondria for initiating mitophagy is still under investigation. It is however observed that mitophagy is not altered due to the inhibition of the proteasome, indicating that Parkin perform two different independent functions; (a) at molecular level it regulates turnover of outer membrane proteins and (b) at organelle level regulate the turnover of entire mitochondria (Yoshii et al., 2011). However, studies by various researchers showed that the degradation of outer membrane proteins of mitochondria facilitates mitophagy (Tanaka et al. 2010; Chan et al., 2011). Mitofusin protein ubiquitination and degradation facilitate mitophagy by affecting mitochondrial fusion (Tanaka et al., 2010).

PINK1/PARKIN ASSOCIATION WITH MITOCHONDRIAL STRESS AND NEURODEGENERATION

Mitochondria-induced inflammation (mitoflammation as given in Figure 3) is reported to be involved in common neuronal degeneration diseases like PD and Alzheimer's disease (AD). Mitochondrial damage and dysfunction are normally associated with age or this may due to some type of intracellular stress generated within the cell because of various stress creating agents and phenomena (Nigar et al., 2016). The strong connection or link between mitophagy and PD was not very well-investigated or established because of the absence of suitable animal models. *PINK1* and Parkin gene knockouts in *Drosophila melanogaster* were reported to produce disbalances in mitochondrial steady-state, oxidative stress, functions of dopaminergic neurons, locomotive activity as well as in immune responses (Whitworth et al., 2017). Recent findings suggest that age or environmental stressors lead to damage and functional defects in the mitochondria. A striking question comes into the mind about the association of chronic inflammation and age-related diseases like AD and PD. These problems induced by age have a cause behind called mitochondria-induced inflammation mito-flammation. In PD loss of dopaminergic neurons is reflected in the symptoms like changes in postures, resting tremors, rigidity in muscles, impaired motor coordination, and weakness. Mitochondria were considered to play a part in PD through various polypeptides like Parkin, *PINK1*, DJ-1, and OMI, (Abou-Sleiman et al., 2006; Zhu et al., 2010). Sliter et al., (2018), performed crossed breeding of Parkin$^{-/-}$ mice with mtDNA polymerase mutator mice which have mtDNA mutations but show no expression neurodegenerative phenotypes (Sliter et al., 2018). The results of the crossbreeding showed the degeneration of dopaminergic neurons and motor function defects in the new mice. This suggests that PD like the pathological state may be the result of the inability to eliminate mutated mtDNA through mitophagy. To illustrate the connection between mito-flammation and PD pathology, Sliter et al., (2018) reported an exciting phenomenon for the first time that the *PINK1* or Parkin absence along with the presence of mtDNA mutational stress and under exhaustive exercise the DNA-sensing cGAS–STING pathway becomes activated (Sliter et al., 2018). Sliter et al., (2018), subjected wild-type, *PINK1*$^{-/-}$, and *Parkin*$^{-/-}$ mice to exhaustive exercise, which increases mitophagy in cardiac cells. Levels of various cytokines were elevated in the serum of both the experimental mice

Figure 3. The phenomenon of mitoflammation that lead to inflammation and associated pathology.

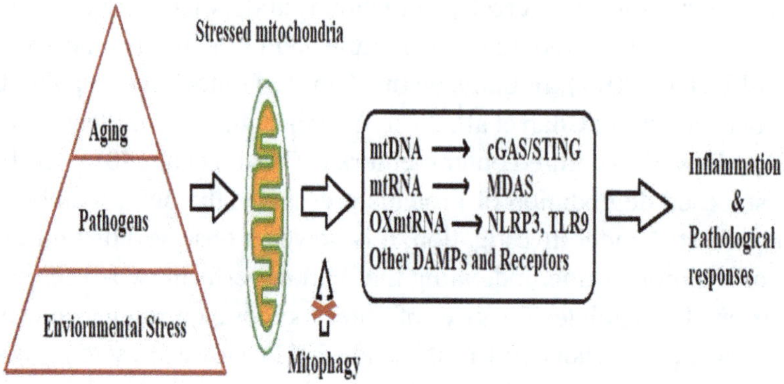

models, which continued for several days after exercise. A wider variety of cytokines were found in Mutator/*Parkin*$^{-/-}$ mice which shows neurodegeneration as observed in the case of PD (Sliter et al., 2018).

Studies showed extensive exercise-related and age-related mitophagy and production of cytokine can be blocked in the *Parkin*$^{-/-}$ and Mutator/*Parkin*$^{-/-}$ mice through genetically engineered inactivation of STING. Lack of STING in the Mutator/*Parkin*$^{-/-}$ mice rescued against the neurodegenerative and locomotor defects. They also found the elevated level of mtDNA in the circulation activates cGAS–STING and this may be an important factor responsible for the pathology of inflammation in PD. Thus, they concluded that cGAS–STING pathway inhibition or any other mechanism able to reduce mtDNA release in cytoplasm or circulation might be useful in the treatment of PD (Sliter et al., 2018).

Various studies showed that mtDNA can enter the cytoplasm of cells and can activate DNA-sensing cGAS–STING pathway-based innate immune responses. This results in pro-inflammatory type I interferons (IFNs) and NF-κB (nuclear factor kappa-light-chain-enhancer of activated B cells) signaling, and cytokine production (West et al., 2015; West et al., 2017). Research also revealed that during the anti-viral innate immune signaling process, the endoplasmic reticulum and mitochondria work as a platform for RNA and DNA sensing. Mitochondrial DNA because of their bacterial origin or pathogen like composition can stimulate immune receptors for promoting inflammation (West et al. 2017).

Impaired memory is the most recognizable character/symptom of Alzheimer's disease, which is a result of pathological conditions like amyloid-β protein accumulation and damaged mitochondria. Still, the Link between *PINK1* and Alzheimer's disease is not very clear but a research study showed decreased *PINK1* expression is related to the pathology of AD. Restoration of *PINK1* functions and proper expression significantly lowered the elevated amyloid-β levels and pathologies related to it. Restoration of the *PINK1* function also normalizes mitochondrial dysfunction and also lowered the oxidative stress. The animal model (*PINK1*-deficient mAPP mice) is diagnosed with the presence of amyloid-β protein aggregates in the cerebrum, abnormal mitochondrial functions, impaired in cognitive behavior and memory along with synaptic plasticity when compared normal mAPP mice (Du et al., 2017). However, treatment with gene therapy for restoring *PINK1* overexpression increases autophagy-mediated disposing of damaged mitochondria and the autophagic receptors involved are optineurin (OPTN) and nuclear dot protein 52 kDa (NDP52). This helps in recovering from neuronal loss and cognitive impairment. The study also showed *PINK1* activity loss or blockade of *PINK1*-mediated signaling onsets the neuronal destruction and the destruction cannot be reversed. Thus, a novel *PINK1* based signaling mechanism provides defense against Alzheimer's disease and give a new therapeutic strategy to treat AD (Du et al., 2017).

Huntington's disease (HD) is the result of a mutation in the huntingtin gene which includes the expansion of CAG repeat in the resultant protein. Alteration of various cellular processes, neuronal dysfunction and death are the consequences of the mutant huntingtin protein (mHtt). Impairment of mitochondria metabolism is one of the major pathological consequences of mHtt mutation and was observed profoundly in HD. *Drosophila* model has been used as a system to pointing out the effect of mitochondrial damages in HD. In photoreceptor neurons' effect of mHtt on mitochondrial morphology revealed that it causes the formation of mitochondrial spheroids (abnormal ring-shaped mitochondria). These spheroids were also reported to be present in cells with impaired mitophagy. Upon analyzing the result it was found that *PINK1* overexpression causes a decrease in mitochondrial spheroid formation. *PINK1* also improved the ATP levels, integrity of neurons and increases the survival rate of the HD flies, Thus, confirming the protective role of *PINK1* against the neurotoxicity induced because of mHtt (Khalil et al., 2015).

This protection provided by *PINK1* was Parkin-dependent and proteins like mitofusin and voltage-dependent anion channel were also required to accomplish this protective act of *PINK1*. Improved removal of defective and damaged mitochondria in HD striatal cells derived from HdhQ111 knock-in mice was observed due to *PINK1* overexpression that causes partial restoration of mitophagy. Mutant huntingtin (mHtt) can decrease the transfer of defective mitochondria to autophagosomes. Therefore, findings of the study proved the involvement of defected mitochondria and mitophagy in the neurodegeneration as well as the protective role of *PINK1*/Parkin in disorders like HD. The study also gives an insight into the *PINK1* based therapeutic dimensions to treat neurodegenerative disorders (Khalil et al., 2015).

PINK1/PARKIN RELATED AUTOPHAGY AND TUMORIGENESIS IN NEURODEGENERATION

Recently, mitochondria mesmerized the scientist with its contribution in triggering immune responses and immune system regulation. The findings have paved a way for the new exciting domain of research in mitochondrial and cellular biology. Mitochondria working in close collaboration with endoplasmic reticulum can promote inflammation through the stimulation of many intrinsic immune receptors (West et al., 2017). This action of mitochondria was supported by the fact that mitochondria are the organelles involved in DNA and RNA sensing for the immune response against viruses at the time of viral infection. Mitochondria-induced inflammation (mito-flammation) was considered to be playing a pivotal role in the pathogenesis of neurodegenerative disorders. The reasons behind this are damage and dysfunction of mitochondria either due to age or by environmental stress-causing agents.

The various scientific studies found and concluded that autophagy might have involvement in tumor suppression (Levine et al., 2008; Chen et al., 2010; White et al., 2010). Based on these studies several tumor-suppressive mechanisms have been proposed using cell cultures and allografted tumor models. The basic underlying principles of such mechanisms include:

- Suppression of tumor-related inflammation (Degenhardt et al., 2006),
- Metabolic stress reduction and mitigation of genome damage (Karantza-Wadsworth et al., 2007; Mathew et al., 2007),
- Degradation of p62 (SQSTM1) (Mathew et al., 2009).

Although, the results from studies based on such *in vivo* models generated a very limited set of data. However, higher frequency of development of spontaneous lung, liver and lymphoid tissue cancers has been reported in Beclin1 heterozygous mutant mice (Qu et al., 2003; Yue et al., 2003). AMBRA1 (Fimia et al., 2007), Bif-1 (Takahashi et al., 2007), and UVRAG (Liang et al., 2006) are several Beclin1-interacting proteins, which also have tumor suppressive or antiproliferation effects. However, various studies reported that these factors are not highly autophagy specific (Thoresen et al., 2010; He et al.,2010; Funderburk et al., 2010).

A mouse model with the systemic mosaic deletion of Atg5 was developed (Takamura et al., 2011). The growth of several benign tumors in the liver was reported in such mice models. An earlier study reported neonatal lethality in animal models after the deletion of Atg5 (Kuma et al., 2004). The hepatic tumor cells showed the presence of swollen mitochondria, reactive oxidative species, and damages to genomic makeup. The various studies regarding the deletion of liver-specific Atg7 in mice showed that the mice models developed liver tumors. Although a reduction in tumor size was observed after concomitant deletion of p62 this does not lead to complete suppression of tumor formation. Various studies also reported that p62 accumulation may have a contribution to tumor progression but not in tumorigenesis (Takamura et al. 2011). Thus, from the findings of these studies, it can be inferred that proteins, as well as organelles, are crucial for preventing spontaneous tumorigenesis and also in the progression of the tumor.

RECENT DEVELOPMENTS AND FUTURE RESEARCH DIRECTIONS

Many studies revealed that *PINK1*/Parkin has a role in mitoQC and few models also have been designed to investigate the exact contribution of these two have in mitophagy. However, different models lead to different conclusions therefore, still their exact role in the quality control system for mitochondria is not very consolidated and well documented. Still, the researchers are developing models solve this scientific puzzle. Mitochondrial derived vesicles (MDV) and their formation and role in mitochondria are now under investigation. Some studies pointed out their role in the transportation of oxidized mitochondrial components to the lysosomes. These MDVs are cargo selective and involved in the degradation of damaged components and not of the whole organelle. This MDV based selection process is called as piecemeal degradation of mitochondria. *PINK1* and Parkin along with Vsp35 are reported to be linked with the formation and transportation of these MDV (McLelland et al., 2010; Wang et al., 2016; Braschi et al., 2010). However, this MDV based model has challenges as it is difficult to carry out *in vivo* experimentations and validation for such model as well the mechanism that helps to select which degradation pathway is to be chosen for degradation is not yet defined.

CONCLUSION

Numerous literatures reported a crucial link between mitochondrial defects and neurodegenerative disorders. The findings of this literature paved the way for further studies for deep exploration of *PINK1*/Parkin and mitophagy role in neurodegenerative disorders. The literature in this regard revealed how the *PINK1*/Parkin expression protects the mitochondria of neurons from destructive factors like mitopoisons or genetic mutations. Thus, protecting the neurons from degeneration. However, still much more research is needed in this area to completely understand the exact pathway by which *PINK1*/Parkin induces this neuroprotective effect. This protective role of *PINK1*/Parkin at the cellular level can be exploited to develop effective gene-based therapy or therapeutics to counter the challenges of neurodegenerative disorders.

REFERENCES

Abou-Sleiman, P. M., Muqit, M. M., & Wood, N. W. (2006). Expanding insights of mitochondrial dysfunction in Parkinson's disease. *Nature Reviews. Neuroscience*, *7*(3), 207–219. doi:10.1038/nrn1868 PMID:16495942

Ashrafi, G., Schlehe, J. S., LaVoie, M. J., & Schwarz, T. L. (2014). Mitophagy of damaged mitochondria occurs locally in distal neuronal axons and requires *PINK1* and Parkin. *The Journal of Cell Biology*, *206*(5), 655–670. doi:10.1083/jcb.201401070 PMID:25154397

Bender, A., Krishnan, K. J., Morris, C. M., Taylor, G. A., Reeve, A. K., Perry, R. H., ... Taylor, R. W. (2006). High levels of mitochondrial DNA deletions in substantia nigra neurons in aging and Parkinson disease. *Nature Genetics*, *38*(5), 515–517. doi:10.1038/ng1769 PMID:16604074

Braschi, E., Goyon, V., Zunino, R., Mohanty, A., Xu, L., & McBride, H. M. (2010, July 27). Vps35 mediates vesicle transport between the mitochondria and peroxisomes. *Current Biology*, *20*(14), 1310–1315. doi:10.1016/j.cub.2010.05.066 PMID:20619655

Cai, Q., Zakaria, H. M., Simone, A., & Sheng, Z. H. (2012). Spatial parkin translocation and degradation of damaged mitochondria via mitophagy in live cortical neurons. *Current Biology*, *22*(6), 545–552. doi:10.1016/j.cub.2012.02.005 PMID:22342752

Cecconi, F., & Levine, B. (2008). The role of autophagy in mammalian development: Cell makeover rather than cell death. *Developmental Cell*, *15*(3), 344–357. doi:10.1016/j.devcel.2008.08.012 PMID:18804433

Chan, N. C., Salazar, A. M., Pham, A. H., Sweredoski, M. J., Kolawa, N. J., Graham, R. L., ... Chan, D. C. (2011). Broad activation of the ubiquitin–proteasome system by Parkin is critical for mitophagy. *Human Molecular Genetics*, *20*(9), 1726–1737. doi:10.1093/hmg/ddr048 PMID:21296869

Chen, D., Gao, F., Li, B., Wang, H., Xu, Y., Zhu, C., & Wang, G. (2010). Parkin mono-ubiquitinates Bcl-2 and regulates autophagy. *The Journal of Biological Chemistry*, *285*(49), 38214–38223. doi:10.1074/jbc.M110.101469 PMID:20889974

Chen, N., & Debnath, J. (2010). Autophagy and tumorigenesis. *FEBS Letters*, *584*(7), 1427–1435. doi:10.1016/j.febslet.2009.12.034 PMID:20035753

De Castro, I. P., Costa, A. C., Lam, D., Tufi, R., Fedele, V., Moisoi, N., ... Martins, L. M. (2012). Genetic analysis of mitochondrial protein misfolding in Drosophila melanogaster. *Cell Death and Differentiation*, *19*(8), 1308–1316. doi:10.1038/cdd.2012.5 PMID:22301916

Degenhardt, K., Mathew, R., Beaudoin, B., Bray, K., Anderson, D., Chen, G., ... Nelson, D. A. (2006). Autophagy promotes tumor cell survival and restricts necrosis, inflammation, and tumorigenesis. *Cancer Cell*, *10*(1), 51–64. doi:10.1016/j.ccr.2006.06.001 PMID:16843265

Deretic, V., & Daniel, J. (2018). Klionsky. Autophagy and inflammation: A special review issue. *Autophagy, 14*(2), 179–180. doi:10.1080/15548627.2017.1412229 PMID:29304718

Dias, V., Junn, E., & Mouradian, M. M. (2013). The role of oxidative stress in Parkinson's disease. *Journal of Parkinson's Disease, 3*(4), 461–491. doi:10.3233/JPD-130230 PMID:24252804

Ding, W. X., Ni, H. M., Li, M., Liao, Y., Chen, X., Stolz, D. B., ... Yin, X. M. (2010). Nix is critical to two distinct phases of mitophagy, reactive oxygen species-mediated autophagy induction and Parkin-ubiquitin-p62-mediated mitochondrial priming. *The Journal of Biological Chemistry, 285*(36), 27879–27890. doi:10.1074/jbc.M110.119537 PMID:20573959

Du, F., Yu, Q., Yan, S., Hu, G., Lue, L. F., Walker, D. G., ... Yan, S. S. (2017). *PINK1* signalling rescues amyloid pathology and mitochondrial dysfunction in Alzheimer's disease. *Brain, 140*(12), 3233–3251. doi:10.1093/brain/awx258 PMID:29077793

Fimia, G. M., Stoykova, A., Romagnoli, A., Giunta, L., Di Bartolomeo, S., Nardacci, R., ... Gruss, P. (2007). Ambra1 regulates autophagy and development of the nervous system. *Nature, 447*(7148), 1121–1125. doi:10.1038/nature05925 PMID:17589504

Funderburk, S. F., Wang, Q. J., & Yue, Z. (2010). The Beclin 1–VPS34 complex–at the crossroads of autophagy and beyond. *Trends in Cell Biology, 20*(6), 355–362. doi:10.1016/j.tcb.2010.03.002 PMID:20356743

Gegg, M. E., Cooper, J. M., Chau, K. Y., Rojo, M., Schapira, A. H., & Taanman, J. W. (2010). Mitofusin 1 and mitofusin 2 are ubiquitinated in a *PINK1*/parkin-dependent manner upon induction of mitophagy. *Human Molecular Genetics, 19*(24), 4861–4870. doi:10.1093/hmg/ddq419 PMID:20871098

Geisler, S., Holmström, K. M., Skujat, D., Fiesel, F. C., Rothfuss, O. C., Kahle, P. J., & Springer, W. (2010). *PINK1*/Parkin-mediated mitophagy is dependent on VDAC1 and p62/SQSTM1. *Nature Cell Biology, 12*(2), 119–131. doi:10.1038/ncb2012 PMID:20098416

Greenamyre, J. T., Sherer, T. B., Betarbet, R., & Panov, A. V. (2001). Complex I and Parkinson's disease. *IUBMB Life, 52*(3-5), 135–141. doi:10.1080/15216540152845939 PMID:11798025

Haque, M. E., Thomas, K. J., D'Souza, C., Callaghan, S., Kitada, T., Slack, R. S., ... Park, D. S. (2008). Cytoplasmic *PINK1* activity protects neurons from dopaminergic neurotoxin MPTP. *Proceedings of the National Academy of Sciences of the United States of America, 105*(5), 1716–1721. doi:10.1073/pnas.0705363105 PMID:18218782

Hara, T., Nakamura, K., Matsui, M., Yamamoto, A., Nakahara, Y., Suzuki-Migishima, R., ... Mizushima, N. (2006). Suppression of basal autophagy in neural cells causes neurodegenerative disease in mice. *Nature, 441*(7095), 885–889. doi:10.1038/nature04724 PMID:16625204

Harman, D. (1972). The biologic clock: The mitochondria? *Journal of the American Geriatrics Society, 20*(4), 145–147. doi:10.1111/j.1532-5415.1972.tb00787.x PMID:5016631

He, C., & Levine, B. (2010). The beclin 1 interactome. *Current Opinion in Cell Biology, 22*(2), 140–149. doi:10.1016/j.ceb.2010.01.001 PMID:20097051

Herrero, M. T., Estrada, C., Maatouk, L., & Vyas, S. (2015). Inflammation in Parkinson's disease: Role of glucocorticoids. *Frontiers in Neuroanatomy, 9*, 32. doi:10.3389/fnana.2015.00032 PMID:25883554

Itakura, E., Kishi-Itakura, C., Koyama-Honda, I., & Mizushima, N. (2012). Structures containing Atg9A and the ULK1 complex independently target depolarized mitochondria at initial stages of Parkin-mediated mitophagy. *Journal of Cell Science, 125*(6), 1488–1499. doi:10.1242/jcs.094110 PMID:22275429

Jin, S. M., Lazarou, M., Wang, C., Kane, L. A., Narendra, D. P., & Youle, R. J. (2010). Mitochondrial membrane potential regulates *PINK1* import and proteolytic destabilization by PARL. *The Journal of Cell Biology, 191*(5), 933–942. doi:10.1083/jcb.201008084 PMID:21115803

Jin, S. M., & Youle, R. J. (2012). *PINK1*-and Parkin-mediated mitophagy at a glance. *Journal of Cell Science, 125*(4), 795–799. doi:10.1242/jcs.093849 PMID:22448035

Jin, S. M., & Youle, R. J. (2013). The accumulation of misfolded proteins in the mitochondrial matrix is sensed by *PINK1* to induce PARK2/Parkin-mediated mitophagy of polarized mitochondria. *Autophagy, 9*(11), 1750–1757. doi:10.4161/auto.26122 PMID:24149988

Kane, L. A., Lazarou, M., Fogel, A. I., Li, Y., Yamano, K., Sarraf, S. A., ... Youle, R. J. (2014). *PINK1* phosphorylates ubiquitin to activate Parkin E3 ubiquitin ligase activity. *The Journal of Cell Biology, 205*(2), 143–153. doi:10.1083/jcb.201402104 PMID:24751536

Kanki, T., Wang, K., Cao, Y., Baba, M., & Klionsky, D. J. (2009). Atg32 is a mitochondrial protein that confers selectivity during mitophagy. *Developmental Cell, 17*(1), 98–109. doi:10.1016/j.devcel.2009.06.014 PMID:19619495

Karantza-Wadsworth, V., Patel, S., Kravchuk, O., Chen, G., Mathew, R., Jin, S., & White, E. (2007). Autophagy mitigates metabolic stress and genome damage in mammary tumorigenesis. *Genes & Development, 21*(13), 1621–1635. doi:10.1101/gad.1565707 PMID:17606641

Kawajiri, S., Saiki, S., Sato, S., Sato, F., Hatano, T., Eguchi, H., & Hattori, N. (2010). *PINK1* is recruited to mitochondria with parkin and associates with LC3 in mitophagy. *FEBS Letters, 584*(6), 1073–1079. doi:10.1016/j.febslet.2010.02.016 PMID:20153330

Kazlauskaite, A., Kondapalli, C., Gourlay, R., Campbell, D. G., Ritorto, M. S., Hofmann, K., ... Muqit, M. M. (2014). Parkin is activated by *PINK1*-dependent phosphorylation of ubiquitin at Ser65. *The Biochemical Journal, 460*(1), 127–141. doi:10.1042/BJ20140334 PMID:24660806

Khalil, B., El Fissi, N., Aouane, A., Cabirol-Pol, M. J., Rival, T., & Liévens, J. C. (2015). *PINK1*-induced mitophagy promotes neuroprotection in Huntington's disease. *Cell Death & Disease, 6*(1), e1617. doi:10.1038/cddis.2014.581 PMID:25611391

Kitada, T., Asakawa, S., Hattori, N., Matsumine, H., Yamamura, Y., Minoshima, S., ... Shimizu, N. (1998). Mutations in the parkin gene cause autosomal recessive juvenile parkinsonism. *Nature, 392*(6676), 605–608. doi:10.1038/33416 PMID:9560156

Komatsu, M., Waguri, S., Chiba, T., Murata, S., Iwata, J. I., Tanida, I., ... Tanaka, K. (2006). Loss of autophagy in the central nervous system causes neurodegeneration in mice. *Nature, 441*(7095), 880–884. doi:10.1038/nature04723 PMID:16625205

Komatsu, M., Waguri, S., Ueno, T., Iwata, J., Murata, S., Tanida, I., ... Kominami, E. (2005). Impairment of starvation-induced and constitutive autophagy in Atg7-deficient mice. *The Journal of Cell Biology, 169*(3), 425–434. doi:10.1083/jcb.200412022 PMID:15866887

Kondapalli, C., Kazlauskaite, A., Zhang, N., Woodroof, H. I., Campbell, D. G., Gourlay, R., ... Knebel, A. (2012). *PINK1* is activated by mitochondrial membrane potential depolarization and stimulates Parkin E3 ligase activity by phosphorylating Serine 65. *Open Biology, 2*(5), 120080. doi:10.1098/rsob.120080 PMID:22724072

Koyano, F., Okatsu, K., Kosako, H., Tamura, Y., Go, E., Kimura, M., ... Endo, T. (2014). Ubiquitin is phosphorylated by *PINK1* to activate parkin. *Nature, 510*(7503), 162–166. doi:10.1038/nature13392 PMID:24784582

Kuma, A., Hatano, M., Matsui, M., Yamamoto, A., Nakaya, H., Yoshimori, T., ... Mizushima, N. (2004). The role of autophagy during the early neonatal starvation period. *Nature, 432*(7020), 1032–1036. doi:10.1038/nature03029 PMID:15525940

Lee, J. Y., Nagano, Y., Taylor, J. P., Lim, K. L., & Yao, T. P. (2010). Disease-causing mutations in parkin impair mitochondrial ubiquitination, aggregation, and HDAC6-dependent mitophagy. *The Journal of Cell Biology, 189*(4), 671–679. doi:10.1083/jcb.201001039 PMID:20457763

Levine, B., & Kroemer, G. (2008). Autophagy in the pathogenesis of disease. *Cell, 132*(1), 27–42. doi:10.1016/j.cell.2007.12.018 PMID:18191218

Levine, B., Mizushima, N., & Virgin, H. W. (2011). Autophagy in immunity and inflammation. *Nature, 469*(7330), 323–335. doi:10.1038/nature09782 PMID:21248839

Liang, C., Feng, P., Ku, B., Dotan, I., Canaani, D., Oh, B. H., & Jung, J. U. (2006, July). Autophagic and tumour suppressor activity of a novel Beclin1-binding protein UVRAG. *Nature Cell Biology, 8*(7), 688–699. doi:10.1038/ncb1426 PMID:16799551

Longatti, A., & Tooze, S. A. (2009). Vesicular trafficking and autophagosome formation. *Cell Death and Differentiation, 16*(7), 956–965. doi:10.1038/cdd.2009.39 PMID:19373247

Mathew, R., Karp, C. M., Beaudoin, B., Vuong, N., Chen, G., Chen, H. Y., ... DiPaola, R. S. (2009). Autophagy suppresses tumorigenesis through elimination of p62. *Cell, 137*(6), 1062–1075. doi:10.1016/j.cell.2009.03.048 PMID:19524509

Mathew, R., Kongara, S., Beaudoin, B., Karp, C. M., Bray, K., Degenhardt, K., ... White, E. (2007). Autophagy suppresses tumor progression by limiting chromosomal instability. *Genes & Development, 21*(11), 1367–1381. doi:10.1101/gad.1545107 PMID:17510285

Matsuda, N., Sato, S., Shiba, K., Okatsu, K., Saisho, K., Gautier, C. A., ... Kimura, M. (2010). *PINK1* stabilized by mitochondrial depolarization recruits Parkin to damaged mitochondria and activates latent Parkin for mitophagy. *The Journal of Cell Biology, 189*(2), 211–221. doi:10.1083/jcb.200910140 PMID:20404107

McLelland, G. L., Soubannier, V., Chen, C. X., McBride, H. M., & Fon, E. A. (2014, February 18). Parkin and *PINK1* function in a vesicular trafficking pathway regulating mitochondrial quality control. *The EMBO Journal, 33*(4), 282–295. doi:10.1002/embj.201385902 PMID:24446486

Mizushima, N., & Komatsu, M. (2011). Autophagy: Renovation of cells and tissues. *Cell, 147*(4), 728–741. doi:10.1016/j.cell.2011.10.026 PMID:22078875

Mizushima, N., Levine, B., Cuervo, A. M., & Klionsky, D. J. (2008). Autophagy fights disease through cellular self-digestion. *Nature, 451*(7182), 1069–1075. doi:10.1038/nature06639 PMID:18305538

Mizushima, N., Yoshimori, T., & Levine, B. (2010). Methods in mammalian autophagy research. *Cell, 140*(3), 313–326. doi:10.1016/j.cell.2010.01.028 PMID:20144757

Narendra, D., Kane, L. A., Hauser, D. N., Fearnley, I. M., & Youle, R. J. (2010). p62/SQSTM1 is required for Parkin-induced mitochondrial clustering but not mitophagy; VDAC1 is dispensable for both. *Autophagy, 6*(8), 1090–1106. doi:10.4161/auto.6.8.13426 PMID:20890124

Narendra, D., Tanaka, A., Suen, D. F., & Youle, R. J. (2008). Parkin is recruited selectively to impaired mitochondria and promotes their autophagy. *J C Biol., 183*(5), 795–803. doi:10.1083/jcb.200809125 PMID:19029340

Narendra, D. P., Jin, S. M., Tanaka, A., Suen, D. F., Gautier, C. A., Shen, J., ... Youle, R. J. (2010). *PINK1* is selectively stabilized on impaired mitochondria to activate Parkin. *PLoS Biology, 8*(1), e1000298. doi:10.1371/journal.pbio.1000298 PMID:20126261

Nigar, S., Pottoo, F. H., Tabassum, N., Verma, S. K., & Javed, M. N. (2016). Molecular Insights into the Role of Inflammation and Oxidative Stress in Epilepsy. *Journal of Advances in Medical and Pharmaceutical Sciences., 10*(1), 1–9. doi:10.9734/JAMPS/2016/24441

Okamoto, K., Kondo-Okamoto, N., & Ohsumi, Y. (2009). Mitochondria-anchored receptor Atg32 mediates degradation of mitochondria via selective autophagy. *Developmental Cell, 17*(1), 87–97. doi:10.1016/j.devcel.2009.06.013 PMID:19619494

Okatsu, K., Saisho, K., Shimanuki, M., Nakada, K., Shitara, H., Sou, Y. S., ... Tanaka, K. (2010). p62/SQSTM1 cooperates with Parkin for perinuclear clustering of depolarized mitochondria. *Genes to Cells, 15*(8), 887–900. doi:10.1111/j.1365-2443.2010.01426.x PMID:20604804

Ordureau, A., Sarraf, S. A., Duda, D. M., Heo, J. M., Jedrychowski, M. P., Sviderskiy, V. O., ... Wells, J. A. (2014). Quantitative proteomics reveal a feedforward mechanism for mitochondrial PARKIN translocation and ubiquitin chain synthesis. *Molecular Cell, 56*(3), 360–375. doi:10.1016/j.molcel.2014.09.007 PMID:25284222

Paterna, J. C., Leng, A., Weber, E., Feldon, J., & Büeler, H. (2007). DJ-1 and Parkin modulate dopamine-dependent behavior and inhibit MPTP-induced nigral dopamine neuron loss in mice. *Molecular Therapy, 15*(4), 698–704. doi:10.1038j.mt.6300067 PMID:17299411

Pickrell, A. M., & Youle, R. J. (2015). The roles of *PINK1*, parkin, and mitochondrial fidelity in Parkinson's disease. *Neuron, 85*(2), 257–273. doi:10.1016/j.neuron.2014.12.007 PMID:25611507

Poole, A. C., Thomas, R. E., Yu, S., Vincow, E. S., & Pallanck, L. (2010). The mitochondrial fusion-promoting factor mitofusin is a substrate of the *PINK1*/parkin pathway. *PLoS One*, *5*(4), e10054. doi:10.1371/journal.pone.0010054 PMID:20383334

Pottoo, F. H., Bhowmik, M., & Vohora, D. (2014). Raloxifene protects against seizures and neurodegeneration in a mouse model mimicking epilepsy in postmenopausal woman. *European Journal of Pharmaceutical Sciences*, *65*, 167–173. doi:10.1016/j.ejps.2014.09.002 PMID:25218046

Pottoo, F. H., Tabassum, N., Javed, M. N., Nigar, S., Rasheed, R., Khan, A., ... Ashraf, G. M. (2019). The Synergistic Effect of Raloxifene, Fluoxetine, and Bromocriptine Protects Against Pilocarpine-Induced Status Epilepticus and Temporal Lobe Epilepsy. *Molecular Neurobiology*, *56*(2), 1233–1247. doi:10.100712035-018-1121-x PMID:29881945

Qu, X., Yu, J., Bhagat, G., Furuya, N., Hibshoosh, H., Troxel, A., ... Cattoretti, G. (2003). Promotion of tumorigenesis by heterozygous disruption of the beclin 1 autophagy gene. *The Journal of Clinical Investigation*, *112*(12), 1809–1820. doi:10.1172/JCI20039 PMID:14638851

Rakovic, A., Grünewald, A., Seibler, P., Ramirez, A., Kock, N., Orolicki, S., ... Klein, C. (2010). Effect of endogenous mutant and wild-type *PINK1* on Parkin in fibroblasts from Parkinson disease patients. *Human Molecular Genetics*, *19*(16), 3124–3137. doi:10.1093/hmg/ddq215 PMID:20508036

Rosen, K. M., Veereshwarayya, V., Moussa, C. E., Fu, Q., Goldberg, M. S., Schlossmacher, M. G., ... Querfurth, H. W. (2006). Parkin protects against mitochondrial toxins and β-amyloid accumulation in skeletal muscle cells. *The Journal of Biological Chemistry*, *281*(18), 12809–12816. doi:10.1074/jbc.M512649200 PMID:16517603

Rubinsztein, D. C. (2006). The roles of intracellular protein-degradation pathways in neurodegeneration. *Nature*, *443*(7113), 780–786. doi:10.1038/nature05291 PMID:17051204

Schapira, A. H., Cooper, J. M., Dexter, D., Clark, J. B., Jenner, P., & Marsden, C. D. (1990). Mitochondrial complex I deficiency in Parkinson's disease. *Journal of Neurochemistry*, *54*(3), 823–827. doi:10.1111/j.1471-4159.1990.tb02325.x PMID:2154550

Schon, E. A., & Przedborski, S. (2011). Mitochondria: The next (neurode) generation. *Neuron*, *70*(6), 1033–1053. doi:10.1016/j.neuron.2011.06.003 PMID:21689593

Shi, G., Lee, J. R., Grimes, D. A., Racacho, L., Ye, D., Yang, H., ... Bulman, D. E. (2011). Functional alteration of PARL contributes to mitochondrial dysregulation in Parkinson's disease. *Human Molecular Genetics*, *20*(10), 1966–1974. doi:10.1093/hmg/ddr077 PMID:21355049

Sliter, D. A., Martinez, J., Hao, L., Chen, X., Sun, N., Fischer, T. D., ... Cai, H. (2018). Parkin and *PINK1* mitigate STING-induced inflammation. *Nature*, *561*(7722), 258–262. doi:10.103841586-018-0448-9 PMID:30135585

Sterky, F. H., Lee, S., Wibom, R., Olson, L., & Larsson, N. G. (2011). Impaired mitochondrial transport and Parkin-independent degeneration of respiratory chain-deficient dopamine neurons in vivo. *Proceedings of the National Academy of Sciences of the United States of America*, *108*(31), 12937–12942. doi:10.1073/pnas.1103295108 PMID:21768369

Sulzer, D., & Surmeier, D. J. (2013). Neuronal vulnerability, pathogenesis, and Parkinson's disease. *Movement Disorders*, *28*(6), 715–724. doi:10.1002/mds.25187 PMID:23589357

Suzuki, K., & Ohsumi, Y. (2007). Molecular machinery of autophagosome formation in yeast, Saccharomyces cerevisiae. *FEBS Letters*, *581*(11), 2156–2161. doi:10.1016/j.febslet.2007.01.096 PMID:17382324

Takahashi, Y., Coppola, D., Matsushita, N., Cualing, H. D., Sun, M., Sato, Y., ... Pledger, W. J. (2007). Bif-1 interacts with Beclin 1 through UVRAG and regulates autophagy and tumorigenesis. *Nature Cell Biology*, *9*(10), 1142–1151. doi:10.1038/ncb1634 PMID:17891140

Takamura, A., Komatsu, M., Hara, T., Sakamoto, A., Kishi, C., Waguri, S., ... Mizushima, N. (2011). Autophagy-deficient mice develop multiple liver tumors. *Genes & Development*, *25*(8), 795–800. doi:10.1101/gad.2016211 PMID:21498569

Tanaka, A., Cleland, M. M., Xu, S., Narendra, D. P., Suen, D. F., Karbowski, M., & Youle, R. J. (2010). Proteasome and p97 mediate mitophagy and degradation of mitofusins induced by Parkin. *The Journal of Cell Biology*, *191*(7), 1367–1380. doi:10.1083/jcb.201007013 PMID:21173115

Thoresen, S. B., Pedersen, N. M., Liestøl, K., & Stenmark, H. (2010). A phosphatidylinositol 3-kinase class III sub-complex containing VPS15, VPS34, Beclin 1, UVRAG and BIF-1 regulates cytokinesis and degradative endocytic traffic. *Experimental Cell Research*, *316*(20), 3368–3378. doi:10.1016/j.yexcr.2010.07.008 PMID:20643123

Tsukada, M., & Ohsumi, Y. (1993). Isolation and characterization of autophagy-defective mutants of Saccharomyces cerevisiae. *FEBS Letters*, *333*(1-2), 169–174. doi:10.1016/0014-5793(93)80398-E PMID:8224160

Valente, E. M., Abou-Sleiman, P. M., Caputo, V., Muqit, M. M., Harvey, K., Gispert, S., ... Albanese, A. (2004). Hereditary early-onset Parkinson's disease caused by mutations in *PINK1*. *Science*, *304*(5674), 1158–1160. doi:10.1126cience.1096284 PMID:15087508

Van Laar, V. S., Arnold, B., Cassady, S. J., Chu, C. T., Burton, E. A., & Berman, S. B. (2010). Bioenergetics of neurons inhibit the translocation response of Parkin following rapid mitochondrial depolarization. *Human Molecular Genetics*, *20*(5), 927–940. doi:10.1093/hmg/ddq531 PMID:21147754

Vives-Bauza, C., Zhou, C., Huang, Y., Cui, M., De Vries, R. L., Kim, J., ... Magrané, J. (2010). *PINK1*-dependent recruitment of Parkin to mitochondria in mitophagy. *Proceedings of the National Academy of Sciences of the United States of America*, *107*(1), 378–383. doi:10.1073/pnas.0911187107 PMID:19966284

Wallace, D. C. (2005). A mitochondrial paradigm of metabolic and degenerative diseases, aging, and cancer: A dawn for evolutionary medicine. *Annual Review of Genetics*, *39*(1), 359–407. doi:10.1146/annurev.genet.39.110304.095751 PMID:16285865

Wang, H., Song, P., Du, L., Tian, W., Yue, W., Liu, M., ... Zhou, J. (2011). Parkin ubiquitinates Drp1 for proteasome-dependent degradation implication of dysregulated mitochondrial dynamics in parkinson disease. *The Journal of Biological Chemistry*, *286*(13), 11649–11658. doi:10.1074/jbc.M110.144238 PMID:21292769

Wang, W., Wang, X., Fujioka, H., Hoppel, C., Whone, A. L., Caldwell, M. A., ... Zhu, X. (2016, January). Parkinson's disease–associated mutant VPS35 causes mitochondrial dysfunction by recycling DLP1 complexes. *Nature Medicine*, *22*(1), 54–63. doi:10.1038/nm.3983 PMID:26618722

Wang, Y., Liu, N., & Lu, B. (2019). Mechanisms and roles of mitophagy in neurodegenerative diseases. *CNS Neuroscience & Therapeutics*, *25*(7), 859–875. doi:10.1111/cns.13140 PMID:31050206

West, A. P., Khoury-Hanold, W., Staron, M., Tal, M. C., Pineda, C. M., Lang, S. M., ... Kaech, S. M. (2015). Mitochondrial DNA stress primes the antiviral innate immune response. *Nature*, *520*(7548), 55–57. doi:10.1038/nature14156 PMID:25642965

West, A. P., & Shadel, G. S. (2017). Mitochondrial DNA in innate immune responses and inflammatory pathology. *Nature Reviews. Immunology*, *17*(6), 363–375. doi:10.1038/nri.2017.21 PMID:28393922

White, E., Karp, C., Strohecker, A. M., Guo, Y., & Mathew, R. (2010). Role of autophagy in suppression of inflammation and cancer. *Current Opinion in Cell Biology*, *22*(2), 212–217. doi:10.1016/j.ceb.2009.12.008 PMID:20056400

Whitworth, A. J., & Pallanck, L. J. (2017). *PINK1*/Parkin mitophagy and neurodegeneration—What do we really know in vivo? *Current Opinion in Genetics & Development*, *44*, 47–53. doi:10.1016/j.gde.2017.01.016 PMID:28213158

Wirth, M., Joachim, J., & Tooze, S. A. (2013). Autophagosome formation—The role of ULK1 and Beclin1–PI3KC3 complexes in setting the stage. *Seminars in Cancer Biology*, *23*(5), 301–309. doi:10.1016/j.semcancer.2013.05.007 PMID:23727157

Xie, Y., Kang, R., Sun, X., Zhong, M., Huang, J., Klionsky, D. J., & Tang, D. (2015). Posttranslational modification of autophagy-related proteins in macroautophagy. *Autophagy*, *11*(1), 28–45. doi:10.4161/15548627.2014.984267 PMID:25484070

Xie, Z., & Klionsky, D. J. (2007). Autophagosome formation: Core machinery and adaptations. *Nature Cell Biology*, *9*(10), 1102–1109. doi:10.1038/ncb1007-1102 PMID:17909521

Yamano, K., Matsuda, N., & Tanaka, K. (2016). The ubiquitin signal and autophagy: An orchestrated dance leading to mitochondrial degradation. *EMBO Reports*, *17*(3), 300–316. doi:10.15252/embr.201541486 PMID:26882551

Yoshii, S. R., Kishi, C., Ishihara, N., & Mizushima, N. (2011). Parkin mediates proteasome-dependent protein degradation and rupture of the outer mitochondrial membrane. *The Journal of Biological Chemistry*, *286*(22), 19630–19640. doi:10.1074/jbc.M110.209338 PMID:21454557

Youle, R. J., & Narendra, D. P. (2011). Mechanisms of mitophagy. *Nature Reviews. Molecular Cell Biology*, *12*(1), 9–14. doi:10.1038/nrm3028 PMID:21179058

Yue, Z., Jin, S., Yang, C., Levine, A. J., & Heintz, N. (2003). Beclin 1, an autophagy gene essential for early embryonic development, is a haploinsufficient tumor suppressor. *Proceedings of the National Academy of Sciences of the United States of America*, *100*(25), 15077–15082. doi:10.1073/pnas.2436255100 PMID:14657337

Zhu, J., & Chu, C. T. (2010). Mitochondrial dysfunction in Parkinson's disease. *Journal of Alzheimer's Disease*, *20*(2), S325–S334. doi:10.3233/JAD-2010-100363 PMID:20442495

Ziviani, E., Tao, R. N., & Whitworth, A. J. (2010). Drosophila parkin requires *PINK1* for mitochondrial translocation and ubiquitinates mitofusin. *Proceedings of the National Academy of Sciences of the United States of America*, *107*(11), 5018–5023. doi:10.1073/pnas.0913485107 PMID:20194754

Section 4
Regulation of Neuronal Proteostasis in Neurodegeneration

Chapter 12
Cellular Cysteine Network and Neurodegeneration

Shubhangi H. Pawar
MGV Pharmacy College, Panchavati, India

Vishal S. Gulecha
School of Pharmaceutical Sciences, Sandip University, India

Manoj S. Mahajan
SNJB's SSDJ College of Pharmacy, Chandwad, India

Aman B Upaganiawar
SNJB's SSDJ College of Pharmacy, Chandwad, India

Chandrashekhar D. Upasani
SNJB's SSDJ College of Pharmacy, Chandwad, India

ABSTRACT

Oxidative stress is strongly linked to neurodegeneration and oxidative species can modify many amino acids and proteins in the brain. Cysteine amino acid is most susceptible to oxidative post-translational modifications (PTMs). Reversible or irreversible cysteine PTMs can cause dyshomeostasis, which further continued to cellular damage. Many cysteine dependent proteins and many non-proteins using cysteine as their structural components are affected by oxidative stress. Several cysteine dependent enzymes are acting as antioxidants. Cysteine is a major contributor to glutathione (GSH) and superoxide dismutase (SOD) synthesis. Cysteine precursor N-acetylcysteine (NAC) supplementation is proven as a potent free radical scavenger and increase brain antioxidants and subsequently potentiates the natural antioxidant cellular defense mechanism. Thus, in this chapter, the authors explore the linkage of cellular cysteine networks and neurodegenerative disorders.

DOI: 10.4018/978-1-7998-1317-0.ch012

INTRODUCTION

Cellular metabolic reactions are consequently producing free radicals. Amongst all, chemically reactive species of nitrogen (RNS), oxygen (ROS) as well as sulfur (RSS) are produced as a part of cellular signal transductions (Jones, 2006). These reactive radicals have a modifying effect on susceptible amino acids, proteins and other cellular components to get structural and functional changes. Specifically, cysteine residues are more sensitive to redox alterations. Each cell always has a limited amount of ROS or RNS but, when their basal level goes beyond the limit, antioxidant systems are stimulated as a defense mechanism. The body is provided with various pathways to compensate for increased free radicals and redox-active molecules (Sbodio et al., 2017).

Increased levels of ROS have adverse or damaging effects on cell leading to pathological conditions. To counteract the ROS mediated oxidative damage, the cell has provided with the chain of active antioxidant substances regulating redox homeostasis, i.e. superoxide dismutase (SOD), catalase, peroxidases, and heme oxygenase like enzymatic as well as glutathione and vitamin C like non-enzymatic antioxidants (Calabrese et al. 2010). Disturbed balance of damaging oxidative stress and protective antioxidants can result in increased oxidative stress which is not counterbalanced by the cellular antioxidant system. If this disturbed balance of oxidant-antioxidant shifted in favor of the former, the condition is called oxidative stress which also concerned with changing redox signaling and control. Redox dysregulation can affect proteins, lipids, nucleic acids and carbohydrates and many more components of the cell, which is a major contributing factor in the pathophysiology of neurodegenerative disorders viz. amyotrophic lateral sclerosis (ALS), Alzheimer's disease (AD), Huntington's disease (HD), and Parkinson's disease (PD). Increased oxidative stress along with associated cellular damage, deals with disease progression (Sbodio et al., 2017).

Normally, protein delivery to the endoplasmic reticulum (ER) is balanced by cell using unfolded protein. During proteins misfolding reaction, either abnormal protein aggregates are broken or protein refolds; or recycled by proteasome if proteins cannot be reversed by refolding. Abnormal protein aggregation, modifying conformations of proteins, is associated with many disorders or diseases. AD, PD, HD, ALS and Friedreich's ataxia (FRDA) are considered as protein conformational diseases (Tabner et al., 2001; Calabrese et al., 2010). In the cell, these abnormal protein aggregates may arise from ER dysfunction, mitochondrial dysfunction, abnormally increased in reactive oxygen species (ROS), leading to oxidative stress and cellular antioxidants and anti-apoptotic substances activates pro-survival pathway to combat this increased oxidative stress (Calabrese et al., 2010).

The production of ROS is directly proportional to metabolic rate and inversely proportional to lifespan. Many cell organelles, biochemical and physiological processes produce ROS (Adelman et al., 1988). Amongst all, the major contributors are mitochondria. Mitochondrial components like nuclear DNA, mitochondrial DNA, and associated proteins are also damaged by oxidative stress which can further damage the cell (Calabrese et al., 2000). This chapter focuses on the effect of oxidative stress on the antioxidant system of the brain specifically cysteine modifications due to redox dysregulation and its role in various neurodegenerative disorders.

BRAIN AND OXIDATIVE STRESS

In the entire body, amongst many metabolically active organs, the brain is the one. From total body weight, only 2% is comprised of the brain but it consumes 20% of inhaled oxygen (Sbodio et al., 2017). More free radicals can be generated by the brain and so it is one of the prone tissues to oxidative stress (Gadoth and Goebel, 2013). Furthermore, defense pathways of the brain are not that much effectively working as an antioxidant as those in other surrounding organs like the liver (Floyd and Carney, 1992; Sbodio et al., 2017). Redox dysregulation in the brain can affect proteins, nucleic acids, lipids and carbohydrates and other many cellular components, which is a major contributing factor in the pathophysiology of neurodegenerative disorders (Sbodio et al., 2017). Targeting these endogenous cellular defense mechanisms via modification in stress response signaling is a novel therapeutic approach in tissue damage for example neurodegeneration. Physical well-being and quality of life can be efficiently maintained by such repair processes (Guttmann, 2010).

Reactive Oxygen Species

The large proportion of physiologically reactive free radical species are O_2 containing species viz. hydroxyl radical (OH•), superoxide anion (O_2^-), and singlet oxygen ($1O_2$). Many biological enzymes generate ROS like xanthine oxidoreductases (XOR) catalyzes hypoxanthine conversion to xanthine and additionally during purine metabolism converts it to uric acid, generating hydrogen peroxide (H_2O_2) (Berry and Hare, 2004). The nicotinamide adenine dinucleotide phosphate (NADPH) oxidase utilizes cytosolic NADPH to reduce oxygen to O_2^- (Lambeth and Neish, 2014).

Although redox dysregulation is selectively attributed to ROS, many more free radicals like RNS which also contribute to altered redox homeostasis. Reactive nitrogen centered species are peroxynitrite ($ONOO^-$), nitric oxide (NO) and nitrogen dioxide (NO_2). The major RNS in cell signaling processes is NO. It is a secondary messenger to control various physiologic processes. It regulates vascular, smooth muscle functions, inflammation, and neuroplasticity. Another highly reactive RNS is $ONOO^-$ produced during the interaction of NO and O_2^- mediating many adverse effects in cells (Radi, 2013). Though $ONOO^-$ is having a short life, it causes extensive cellular damage. NO_2 is an environmental pollutant, also endogenously produced by enzymes like myeloperoxidases (Luc and Vergely, 2008). It can reduce endogenous antioxidants like ascorbic acid, alpha-tocopherol, which are radical scavengers (Hanzen et al., 2016).

Here only chemically reactive chemical oxygen and nitrogen species are focused but chemically active sulfur species also equally play a vital role in redox regulation. Sulfur is a major component of proteins and many other small biomolecules viz. glutathione (GSH). During oxidative stress, generated RSS (Figure 1) are thiyl radical (RS•), sulfenic acids (RSOH), disulfides (RSSR), thiosulfinate (RS[O]SR), thiosulfonate (RS[O]2SR) and S-nitrosothiols (SNT) (Giles et al., 2001). The oxidizing function of these RSS was firstly proposed in 2001 and the rich source is mitochondria generating RSS during sulfide oxidation reaction (Sbodio et al., 2017).

Figure 1. Reactive sulphur species
(Giles et al., 2001).

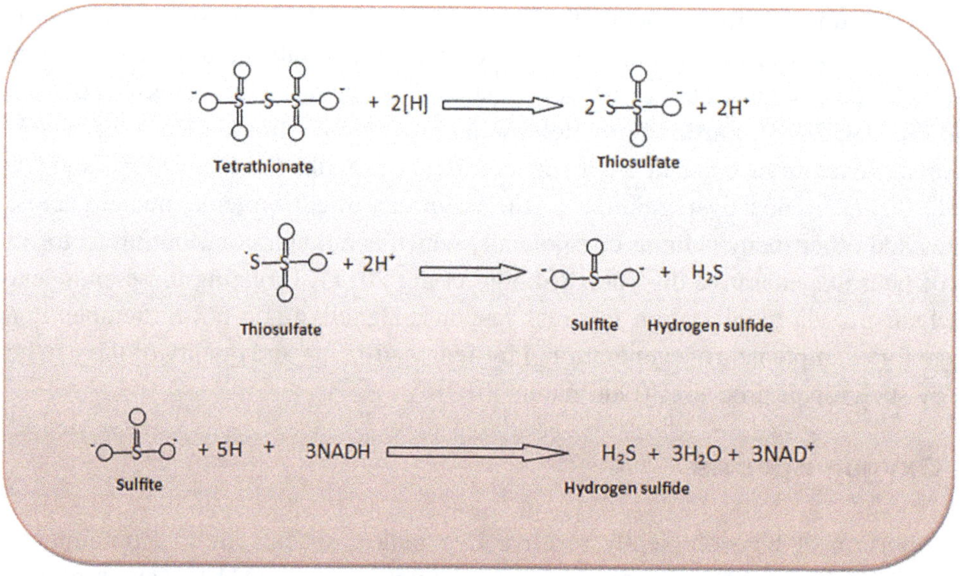

Effect of Free Radicals on Cysteines

The three major oxidizing reactive species that may modify cysteines are ROS, RNS, and RSS. Thiol oxidation of cysteine molecules forms disulfides, sulfenic acids, sulfinic acids, and sulfonic acids. Each of these oxidants has different reaction rates with thiols (Tabner et al., 2001). A non-radical, hydrogen peroxide, is an important physiological redox regulator peroxide oxidation also contributes to the bridging of the disulfide bond and formation of sulfonic acids, sulfenic and sulfinic acids. These oxidizing molecules lead to cysteine oxidation. Due to oxidation, reversible modifications occur in cysteine like S-nitrosylation, S-glutathionylation, the formation of thiyl radicals, sulfonic acids, sulfenic and sulfinic acids (Pacher et al., 2007). These modifications in cysteine-containing proteins or peptides mediated by ROS, RNS, RSS alter redox microenvironment. In the body, several mechanisms generate thiol-reactive species. The thiol group within cysteine is highly sensitive and thiol oxidation is a significant reaction. Various enzymes involved in energy production, transcription, apoptosis, and cell signaling, in their active sites, have cysteines and oxidation of which may give cellular dysfunction (Jacob et al., 2006).

Cysteines exist in many different oxidation states and thus can catalyze various reactions. At physiological pH, cysteine can generate anionic sulfur and this is the main contribution for cysteine sensitivity to oxidation. Amongst many cysteine protease enzymes, participating histidine is catalyzing the generation of the anion of sulfur within the catalytic dyad or triad site (Witt et al., 2008). New prominent techniques, to prevent adverse effects of oxidative damage, can be developed by studying oxidation of major cysteine dependent enzymes. Many cysteine-dependent enzymatic and non-enzymatic constituents have proved cysteine as one of the important redox regulator (Guttmann, 2010).

ANTIOXIDANT SYSTEM

Antioxidants or reducing agents get oxidized to reduce the protein in a protective reaction. Thus to remove oxidants, the cell will generate water or oxygen by sharing the electron(s) from the species, which are oxidized.

The cell is provided with a battery of defense mechanisms to control ROS formation. Major antioxidant systems include various enzymes like catalase and superoxide dismutase (SOD) and non-enzymatic components viz. cysteine, GSH, and vitamins (e.g., vitamins C and E):

- Superoxide dismutases (SOD) are found in mammalian cells in two forms. SOD1 is intracellularly located and it is CuZnSOD. SOD_2 is MnSOD localized to mitochondria. SOD converts superoxide to H_2O_2, slowly and spontaneously. Peroxynitrite generation during the chemical interaction of superoxide and NO essentially can be prevented by increased SOD activity. H_2O_2 can be removed by enzymes like Catalase and peroxidases. The reaction of peroxide with ferrous iron (i.e. Fenton reaction) converts peroxide into short-lived oxidant i.e. hydroxyl radical so removal of H_2O_2 is essential (Pacher et al., 2007; Tabner et al., 2001).
- Thioredoxin (TRx) is a 105 amino acids containing protein, of size about 12 kDa. The active site of its C-terminal is containing selenocysteine residue. TRx enzyme is participating in hydroxyl radicals scavenging (Ren et al., 1993).
- Glutaredoxins (GRxs) are 10 to 16 kDa sized small proteins containing cysteine. This antioxidant GRx system is constituted by GSH, enzyme GSH reductase, NADPH, and GRx (Giles et al., 2003).
- Glutathione peroxidase (GPx) is the general name of an enzyme family with peroxidase activity. They are proteins family comprised of selenocysteine, mainly participating in the lipid peroxides and reduction of hydrogen peroxides. Its Isoforms, GPx5, playing a role to protect oxidative damage of DNA in sperm and GPx6, present in the olfactory system (Guttmann, 2010).
- Peroxiredoxins (Prx) contains cysteine at its active site; mediates H_2O_2 removal, lipid hydroperoxide removal and also peroxynitrite removal. In Prx system, cysteine residue is oxidized to form sulfenic acid, at its active-site and later through reaction, it forms disulphide (Park et al., 2009).
- Glutathione (GSH) is γ-glutamylcysteinyl-glycine, a tripeptide, serves as a resource for the formation of cysteine. Along with glutathione disulfide (GSSG), it has a major contribution to antioxidant defense. The redox ratio (GSH to GSSG) is a major indicator of the intracellular redox state (Jones, 2008).
- Cysteine is the major extracellular antioxidant. On oxidation, it forms its disulfide CySS, or other oxidized thiols (Go and Jones, 2008; Guttmann, 2010).

CYSTEINES AS MAJOR ANTIOXIDANT

Cysteine-dependent protein is a unique group of proteins involved in neurodegeneration. These include antioxidant enzymes, phosphatases, proteases, kinases, and many other enzymes as well as other non-enzymatic proteins that utilize cysteine in their structure itself, besides using it as a catalytic site (Guttmann and Powell, 2012). Cysteine is a non-essential amino acid. Our body may get it from daily dietary components, also it can be synthesized endogenously by an enzyme cystathionine γ-lyase (CSE)

Figure 2. Biosynthesis and metabolism of cysteine (Mani et al., 2011).

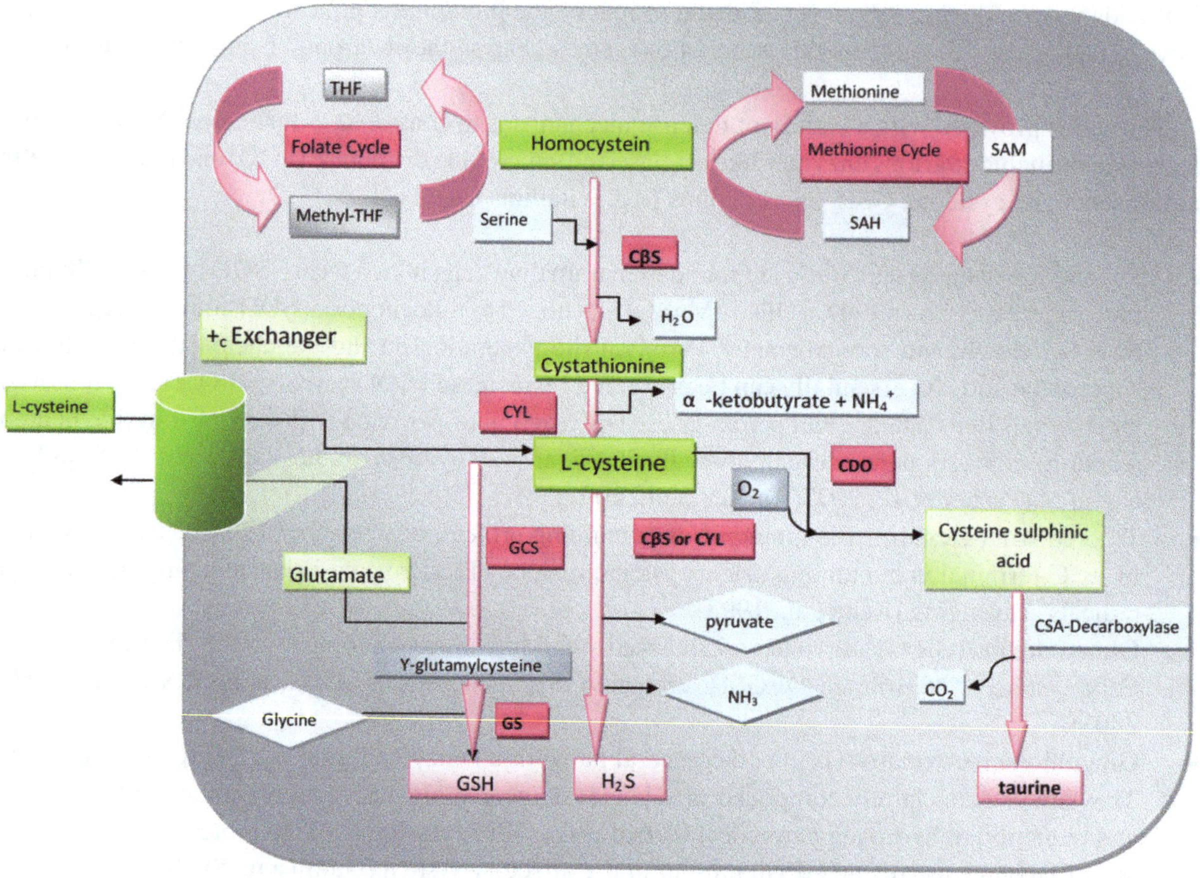

(Figure 2) or from amino acid methionine through reverse trans-sulfuration routeway. Some proteins can generate cysteine by process of autophagy and also a breakdown of glutathione generates cysteine in the cell. Thus the cell is getting a continuous supply of cysteine by various mechanisms. Endogenously cysteine is produced from cystathionine by the CSE, which is formed by the process of condensation of serine and homocysteine by another enzyme i.e. cystathionine β-synthase (CBS). CSE is the main element in the biosynthesis of cysteine and its lesser availability may create a need for exogenous cysteine (Mani et al., 2011).

Deficiency of cystathionine γ-lyase (CSE) causes aberrant stress responses vascular deficits and an increase in oxidative stress (Yang et al., 2008). Cysteine is readily oxidized to cysteine. At physiological pH cysteine exists as thiolate anions that are undergoing multiple modification reactions viz. glutathionylation, cysteinylation, sulfhydration, palmitoylation, guanylation, sumoylation, farnesylation, and nitrosylation. Cysteine, acting as a substrate, generates the gasotransmitter H_2S, which impacts many physiological processes (Paul et al., 2018).

Several Cysteine-Dependent Enzymes as Antioxidants

The oxidants and antioxidants work as a complete system, which has implications for the physiological regulation and pathological impact of numerous proteins. Although several of the key antioxidant systems are cysteine-dependent enzymes, the following section highlights a few of them.

- PTK (protein tyrosine kinases) are kinases containing a conserved active-site cysteine (His-Cys-X-X-Gly-X-X-Arg-Ser/Thr). Due to low pKa, at its active-site cysteine becomes susceptible to redox regulation and thiolate anion is formed at physiological pH, which is highly sensitive. Reduced cysteine causes phosphorylation of PTK, acting as the nucleophile in its active site. Such PTK redox regulations are classically represented by an example of enzyme protein tyrosine phosphatase 1B. This protein tyrosine phosphatase 1B is oxidized by the process of glutathionylation to form sulfenic acid by several oxidizing agents e.g. H_2O_2. In this case, an increase in tyrosine phosphorylation occurs as a net result of oxidative stress under physiological conditions (Guttmann, 2010).
- Calpains, involved in several physiological pathways, are cysteine-dependent proteases. Calpains are found to be participating in various pathological conditions (Goll et al., 2003). They have a cysteine, histidine, and asparagine containing catalytic triad. There are numerous isoforms of mainly found in various muscles. Here also like other thiol-dependent enzymes, oxidative stress catalyzes the formation of the anionic sulfur group at the active site of cysteine. Many oxidant molecules viz. peroxide and NO, can inactivate various members of this enzyme's family (Guttmann et al., 1997).

Importantly it should be considered that oxidative stress modifies protein function by targeting only 1 or 2 cysteine molecules from a quarternary structure of the protein. All oxidized cysteines within a given protein will not contribute to change in any activity (Guttmann, 2010). The therapy which can reverse this oxidative stress-mediated damage in DNA, lipids or proteins will be useful. Antioxidant systems, using cysteine as a major component act as redox sensors and indicating the importance of cysteine as an active center of protein (Guttmann and Powell, 2012).

Cysteine-Dependent Enzymes Affected by Oxidative Stress

One particular group of proteins that appear to be intimately involved in the neurodegenerative processes is the cysteine-dependent proteins. This group includes various proteases, antioxidant enzymes, kinases, phosphatases, and other types of enzymes as well as other non-enzymatic proteins such as those that use cysteine as a structural component rather than as part of a catalytic site.

- Janus kinase 2 (JAK2) activates the transcription pathway and is a component of the JAK2/signal transducers that has participation in neurotransmission, cell proliferation, synaptic plasticity, migration, and apoptosis. JAK2 contains Cys866 and Cys917. These two cysteine residues act as a redox-sensitive switch for JAK2 activity (Guttmann and Powell, 2012; Smith et al., 2012).
- Redox state also regulates enzymes of the caspase family. Caspases enzymes regulate the process of apoptosis and many neurodegenerative conditions. S-nitrosylation is a major reaction to inhibit these enzymes by oxidation at their active-site cysteine (Guttmann and Powell, 2012; Haendeler et al., 1997).

- Phosphatase and tensin homolog (PTEN) dephosphorylates phosphatidylinositol (3,4,5)-trisphosphate (PIP3) to phosphatidylinositol (4,5)- bisphosphate (PIP2), acting as antagonist of the kinase activity of phosphatidylinositide-3kinase. PTEN activity is inhibited by the process of S-nitrosylation at its Cys-83 residue resulting in increased activity of AKT and thus assisting the survival of cells (Numajiri et al., 2011).
- Glyceraldehyde 3-phosphate dehydrogenase (GAPDH) is a glycolytic enzyme that participates in apoptosis. GAPDH is highly sensitive to the action of reactive oxidative species and it is found in affected brain regions of AD (Cumming and Schubert, 2005).
- Another cysteine and calcium-dependent endoproteases category enzyme are calpains. It is inactivated at its active sites by oxidation. Calpains have a major role in AD and besides, calpains play a role in many diseases (Guttmann et al., 1997).
- Many other related cysteine-dependent enzymes are proved to have a role in PD. These reactive species are modifying these enzymes at key cysteine residues. Increased oxidative stress also found to alter DJ-1 activity. DJ-1 is a 20kDa, having several protease and antioxidant like activities. Its mutations lead to an autosomal recessive early-onset form of PD, thus DJ-1 is strongly associated with PD. Under oxidizing conditions, the mixed disulfide bond is created between DJ-1 and apoptosis signal-regulating kinase 1 (ASK1), responsible for the neuroprotective effect of DJ-1 (Guttmann and Powell, 2012; Wilson, 2011).
- Parkin is another cysteine-containing enzyme that is modified in PD. This enzyme belongs to a ubiquitin E3 ligase family contributing to ubiquitinate, a chain of proteins. Parkin is having many cysteine residues required for full activity. Its mutation is responsible for early-onset autosomal recessive juvenile parkinsonism (Meng et al., 2011).
- Tyrosine hydroxylase (TH) is required for dopamine (DA) and norepinephrine biosynthesis (Kuhn et al., 1999). Seven cysteine molecules are present in this enzyme, some of them are essential for full TH activity.
- SOD1 undergoes oxidative modification leading to misfolding of SOD1 which is observed in ALS. Glutathionylation process causes oxidation of Cys-111and its specific modification. This mechanism destabilizes the SOD1 dimer increasing the unfolding of the monomer and subsequent aggregation. This leads to loss of SOD1 activity and cell death (Guttmann and Powell, 2012; Redler et al., 2011).

Apart from these, several other cysteine-dependent enzymes are having a crucial contribution in the different stages of chronic diseases and therapies also impact these enzymes (Guttmann and Powell, 2012).

CYSTEINES AND GLUTATHIONE

The role of mitochondrial oxidative stress in the neuronal damage is well known and various neurodegenerative diseases. Antioxidant, glutathione (GSH) is an endogenous tripeptide that has a major role to inhibit oxidative stress, to preserve mitochondrial function and avert cellular apoptosis. GSH levels have been reduced in many neurodegenerative diseases while combating increased ROS. Correlation between GSH depletion and neurodegeneration is very prominent, and that's why increased GSH levels may give a neuroprotective effect (Winter et al., 2017). Nitrosative and oxidative stress are majorly generated components after injury processes. During these stresses an important antioxidant, GSH is generated to remove hydroperoxides and other oxidative species produced during stress. GSH is working in co-ordination with GSH transferases, GSH peroxidases, and peroxiredoxins. Many research studies have shown increased brain tissue and mitochondrial GSH levels reduced damage of oxidative stress, and neuronal survival by administrating GSH precursor i.e. N-acetylcysteine (NAC) (Ignowski et al., 2018). NAC has shown neuroprotection in PD by lowering ROS generation, normalizing GSH levels, and protecting dopaminergic neurons from cell death. In ALS, NAC has shown to delay motor deficits and prolonged neuronal survival may be by elevating GSH. In AD, NAC treatment has increased cognitive performance (Winter et al., 2017). NAC is helpful in the treatment of hepatotoxicity induced by acetaminophen as it increases the production of GSH in the liver or has its direct antioxidant effect. The systemic administration of NAC crosses blood–brain barrier (BBB) and plasma membrane and can improve neuronal GSH levels (Aoyama and Nakaki, 2013). Cystine is resistant to trypsin proteolysis and circulates along with the blood circulation to its target cell. In the target cell, two cysteine molecules are readily formed by its reduction which further synthesizes GSH (Winter et al., 2017).

In the cell, GSH is formed by using three amino acids i.e. glutamate, cysteine, and glycine. Synthesis of GSH occurs by two consecutive steps which are catalyzed by an enzyme GSH synthetase (GS) and γ-glutamyl cysteine ligase (GCL, also known as γ-glutamyl cysteine synthetase) (Richman and Meister, 1975).Cysteine is stored in the form of GSH, 10–100 times more than that of cysteine in mammalian tissues. In the CSF, the cystine levels are comparatively low than cysteine and GSH, whose levels are higher in the blood. Cysteine does not have an acidic omega side chain so it is unable to cross the BBB. This side chain of acidic omega helps cysteine to the transport across the BBB. In contrast, the disulfide form with two cysteines, called cystine, is transported from blood into the endothelial cells at the BBB via a cystine transporter, called system xc−, and is subsequently transported out of the endothelial cells into the CSF via the L-type amino acid transporter LAT1 at the BBB (Aoyama and Nakaki, 2013). One formulation, prepared from non-denatured whey protein named as Immunocal® acts as a cysteine delivery system to increase GSH levels. It has preserved the ratio of GSH/GSSG, reduced lipid peroxidation in the brain, and attenuated neuronal demyelination and degeneration and shown promising improvements in motor and cognitive deficits (Ignowski et al, 2018).

CYSTEINE STRING PROTEIN IN THE NERVOUS SYSTEM

Cysteine string protein alpha (CSPα) is containing 25 amino acids and having a chain of 13–15 cysteine residues. It is distinguished from other J proteins in several aspects. It is a 34 kDa protein abundant in neural tissue and characteristically localized in synaptic vesicles. Non-neuronal secretory cells are also possessing CSPa, suggesting its role to regulate secretion. Thus, CSPa has unique anti-neurodegenerative properties. Misfolding of cysteine string proteins leading to neurodegenerative diseases e.g. observed in AD, transmissible spongiform encephalopathies, PD, HD, ALS, frontotemporal dementia and spinocerebellar ataxia type 1. Alpha-synuclein modulates the CSPa neurodegeneration pathway selectively. It is a small neural protein whose biological function is unclear. Being an abundant synaptic vesicle protein, CSPa has much potential as a new therapeutic target as it is neuroprotective (Jadah et al., 2010). Cysteine string protein (CSP) is an abundant, evolutionarily conserved synaptic vesicle protein that contains a string of C-terminal cysteine residues (Zinsmaier et al., 1990). CSP may be acting through three different ways:

- CSP is involved in Ca^{2+} channel function. CSP antisense RNA inhibited the expression of N-type Ca^{2+} channels. (Gundersen and Umbach, 1992).
- CSP directly regulates exocytosis.
- CSP activates the ATPase activity of Hsc70 and forms a trimeric complex with Hsc70 and a tetratricopeptide repeat protein called SGT (Fernandez-Chacon et al, 2004).

Cysteine string protein (CSP) was first discovered in Drosophila melanogaster. Majorly, CSP is well known for its neuroprotective chaperone at the synapse. It is observed that the absence of CSPα in animals has developed progressive sensorimotor disorders, rapid degeneration of retinal photoreceptors and difference in sensitivity of GABAnergic neurons. Mutational changes in the CSPα encoding *DNAJC5* gene lead to the onset of adult neuronal ceroid lipofuscinosis (ANCL). Apart from ANCL, CSP mutations and alterations in the levels or activity of CSP could potentially affect other neurodegenerative conditions also. CSP is expressed in all synapses and prevents neurodegeneration in humans, mice flies, and worms. Thus, drugs that either increase CSP activity or bypass the requirement of CSP could present potential therapies (Burgoyne and Morgan, 2015). Neurodegeneration can be resulted from impaired CSP function due to of misfolding of client proteins involved in neurotransmission. Mammalian CSP is phosphorylated in vivo on ser10, and this modulated its protein interactions and effect on neurotransmission (Patel et al., 2016).

CYSTEINE AND REDOX HOMEOSTASIS

Appropriate protein structure, its functions, and stability can be maintained by cysteine residues containing thiol which forms covalent disulfide bridges during the process of oxidative folding. Apart from this role, cysteine residues have a crucial role to maintain a correct redox balance in the cell (Meister, 1988).

Through GSH

Chain of complicated enzyme-catalyzed reactions is catalyzed by tripeptide glutathione cysteine residue (GSH, γ-L-Glutamyl-L-cysteinyl glycine). In mammalian cells, glutathione is the major thiol-containing antioxidant. Being an electron donor, it reduces formed disulfide bonds within cytoplasmic proteins. During this process, oxidation of glutathione occurs to form glutathione disulfide (GSSG), its oxidized form. An enzyme, glutathione reductase by using electron donor NADPH, can reduce GSSG. Several antioxidant enzymes viz. glutathione peroxidases, glutathione S-transferases, glutathione reductases uses GSH as a cofactor and all together these are involved in maintaining redox balance (Figure 3). Thus cellular GSH:GSSG ratio is commonly considered to indicate oxidative stress (Valle and Carri, 2017; Meister, 1988). Secondly, the role of redox-sensitive cysteine thiols in receptors activation, signal transduction, transcription factor binding to DNA (e.g., Nrf-2, NF-κB), and other processes is well known. Signal transduction and redox biology are important pathways to maintain redox homeostasis (Valle and Carri, 2017).

Figure 3. Redox homeostasis to maintain GSH/GSSG ratio
(Valle and Carri, 2017; Meister, 1988).

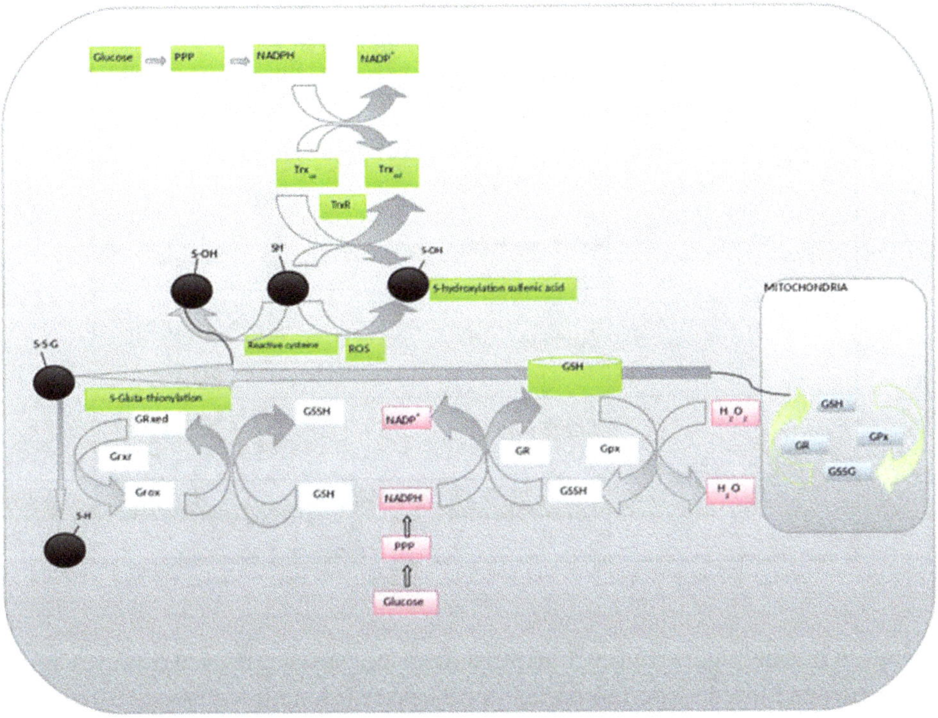

Through SOD

Four cysteine residues are present in homodimeric SOD1; out of them Cys57 and Cys146 form an intramonomer disulfide bond, and the remaining two Cys6 and Cys111 are unabridged. Protein surface exposes Cys111 near its dimer interface. Cys111 mediates covalent disulfide cross-linking that leads to oligomerization which is the mechanism of mutant SOD1 aggregation. The redox state of cysteine and SOD1 aggregation along with mitochondria has a significant role in the pathology of neurodegeneration (Valle and Carri, 2017; Cozzolino et al., 2008).

Changes in GSH/GSSG ratio, increased activity of cytosolic glutaredoxin 1 or mitochondrial glutaredoxin 2 and treatment with cisplatin that binds Cys111 are various approaches to reduce SOD1 aggregation. In short, the imbalance of the cysteine redox state seems to mediate various mechanisms to maintain appropriate protein folding and activity in neurodegeneration.Cysteine oxidation in proteins is important for metabolism and also for the survival of motor neurons e.g. cysteines oxidation in AMP-activated protein kinase (AMPK) that increases its activity. Mutation of SOD1 or TDP43 increases AMPK activity in motor neuron cells (Lim et al., 2012).

In summary, changes in the redox state of cysteines and genetic changes of proteins involved in cysteine homeostasis could be novel therapeutic approaches for neurodegenerative diseases like ALS (Valle and Carri, 2017). In metabolism, thiol-based redox regulation is a significant process. Disrupted redox homeostasis of thiols is contributing to aging and many neurodegenerative and other diseases like cardiovascular diseases, diabetes, and cancer. Detail study of the redox proteome of thiol can throw

Figure 4. Oxidative post-translational modifications of cysteine residues in proteins (Couvertier et al., 2014; Yang and Lee, 2015).

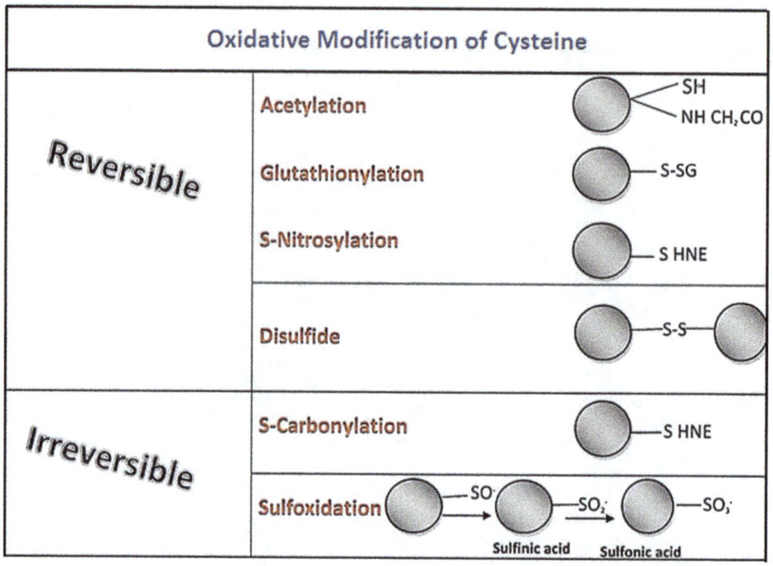

light on alterations in biochemical events occurring during the pathogenesis of diseases and may provide important diagnostics and therapeutic biomarkers of diseases (Gu and Robinson, 2016).

OXIDATIVE MODIFICATIONS IN CYSTEINES

Reactive oxygen species (ROS) and reactive nitrogen species (RNS) are significantly regulating many physiological activities of the cell. In oxidative stress, during defense response oxidative damage of cells can occur. Signal transduction mechanisms of the cell can do oxidative post-translational modifications (PTMs) of biomolecules like RNA, DNA, and proteins. Among many amino acids, cysteine is the most sensitive amino acids to undergo oxidative PTMs. Cysteine naturally occurs about 2.3% among all amino acids in the mammals. Cysteine is highly nucleophilic and redox-sensitive compared with other amino acid side chains. It has a role in enzymatic catalysis, redox homeostasis, signal conduction, metal binding, and structural stabilization. The pKa value of the cysteine thiol is ~ 8.0 but electrostatic interactions and hydrogen bonding in some proteins may lower it as 3.5 (Mossner et al., 2000). The low pKa results in endogenous reactions of cysteine with lipid-derived electrophiles and/or other oxidizing molecules (Chung et al., 2013). Cysteine will undergo reversible alterations (Figure 4) e.g. formation of S-glutathionylation (SSG), S-nitrosylation (SNO), sulfenic acid (SOH), S-palmitoylation, and disulfide bonds (Couvertier et al., 2014).

It has been proven that 10% of cysteine residues are reversibly oxidized. These reversible PTMs (Figure 5) prevent the formation of irreversible oxidative modifications and thus have significant biological roles to maintain homeostasis. These modifications are significant in the signaling mechanism of the cell. Irreversible or reversible cysteine PTMs can cause dyshomeostasis which further continued to cellular damage. During enzymatic catalysis, SOH formed as an intermediate state of cysteines e.g.

Figure 5. Oxidative modifications of cysteine residues by ROS/RNS
(Gu and Robinson, 2016).

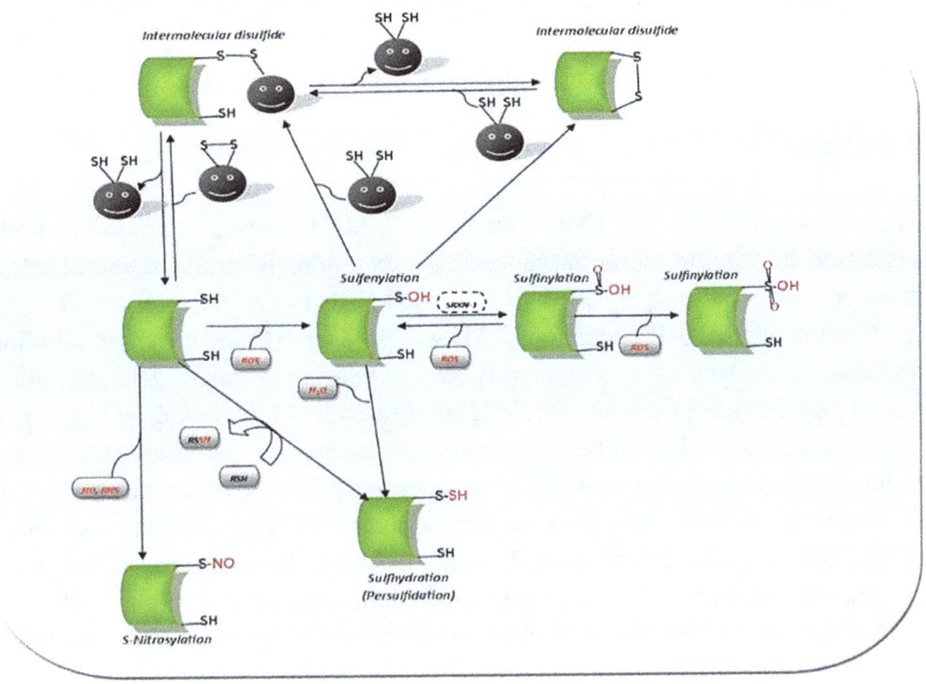

Cys96 residue from peroxiredoxin-5 converted to intermediate SOH which then catalyzes the degradation of peroxides. When free cysteines are attacked by endogenous nitric oxide (NO), SNO is generated, having a significant role in phosphorylation like cellular signal transduction pathways. SSG is the reversible formation of protein disulfides with glutathione (GSH) and can modulate protein activity. SSG can be generated by the reduction of glutaredoxin maintains thiol homeostasis by catalyzing SOH and SNO derivatives. Covalent lipid modification of cysteine with the 16-carbon fatty acid palmitate ($CH_3[CH_2]14COOH$) is S-palmitoylation, and this PTM regulates protein trafficking and subcellular localization. Disulfide bond formation is important to maintain the tertiary structure of the protein and to regulate protein function. Cysteine can be modified by electrophiles, where common PTMs are lipid peroxidation products such as 4-hydroxynonenal (HNE). This can be mediated through both enzymatic and non-enzymatic reactions (Gu and Robinson, 2016).

REVERSIBLE CYSTEINE DYSREGULATION IN AGING AND NEURODEGENERATION

Aging

With aging cysteine oxidation content is increased which furthermore consistently increases carbonylated proteins, and thus cysteine oxidation is considered as an indicator for oxidative stress. In skeletal muscle metabolic proteins (e.g., glucose 6-phosphate isomerase, glycogen phosphorylase, phosphofructokinase, phosphoglycerate mutase, and phosphoglucomutase) undergo cysteine oxidation which is age-dependent (McDonagh et al., 2014). During aging and related neurodegenerative diseases, oxidative protein modifications to cysteine are involved ubiquitously and dynamically (Fu et al., 2006). Cysteine to cystine ratio and related redox potential is shifted towards a more oxidizing one during aging. In aging, a decrease in GSH with a decline in redox-buffering capacity has been observed (Paul et al., 2018).

Alzheimer's Disease

SNO, one of the products of oxidative PTMs, is increased in AD. In some proteins, SNO modifications are required for neurotransmission and neuronal survival. For example, NO production is negatively regulated due to decreased enzymatic activity by SNO of N-methyl-D-aspartate receptor (NMDAR) (Choi et al., 2000). Hyperactivation of NMDAR occurring in AD will produce NO and causes misfolding of proteins, fragmentation of mitochondrial and consequently loss of neurons. Neurodegenerative diseases can be specifically treated by blocking site-specific SNO modification which could be a promising approach (Nakamura and Lipton, 2016). In AD three proteins i.e. superoxide dismutase, fructose-bisphosphate aldolase C, voltage-dependent anion-selective channel protein 2, were observed as SNO-modified proteins. Due to these modifications, cell adenosine triphosphate (ATP) production, glycolytic metabolism, and cell detoxification were found to be altered. As the oxidative stress increased, SNO-modification is continued and glucose metabolism is decreased in AD brain (Gu and Robinson, 2016).

In AD, SNO modifications significantly alter glutamine synthetase and citrate synthase. Hyperammonemia can enhance neurotoxicity and glutamine synthetase function to control the level of ammonia. Citrate synthase has a role in the tricarboxylic acid cycle (TCA) cycle, and SNO-modified citrate synthase may dysregulate metabolism in AD brain. Another PTM form of cysteine is SSG which also plays a significant role in neurodegenerative diseases. In AD, SSG modifies GAPDH, α-crystallin B, deoxyhemoglobin and α-enolase significantly (Gu and Robinson, 2016). Disrupted metabolism of cysteine and GSH is observed in AD. In AD patients, plasma hydrogen sulfide (H_2S) levels are found to be decreased than normal individuals and the negative correlation of H_2S levels with the severity of the disease is found. Thus, multiple aspects of cysteine metabolism are affected in AD (Paul et al., 2018).

Parkinson's Disease

GAPDH catalyzes glycolysis/gluconeogenesis, and some non-metabolic processes as well. Apoptosis and cell death may result as SNO-modified GAPDH is getting bound with E3 ubiquitin-protein ligase SIAH1 (Siah1) and gets translocated into the nucleus. Cys220 residue of an enzyme Ubiquitin carboxyl-terminal hydrolase L1 (UCH-L1) is found to be oxidized during the pathogenesis of PD. Thus in the treatment of PD, inhibition of SNO modification is developing approach (Kragten et al., 1998). In PD H_2S metabolism is altered. The formation of Lewy bodies by aggregation of α-synuclein is mainly observed in this disease, which further leads to motor deficits showing shaking and tremors in patients. The sulfhydration process activates enzyme parkin which degrades misfolded proteins and enhances neuroprotection. After the post-mortem of brains of Parkinson's patients, diminished sulfhydration of parkin and increased nitrosylation has been observed. Parkin sulfhydration and E3 ubiquitin ligase activity increases by H_2S, generated by overexpression of CBS. Stimulation of H_2S-mediated sulfhydration may be beneficial in PD (Paul et al., 2018).

In PD progressive dopaminergic (DA) loss in the substantia nigra deals with alterations in a-synuclein and parkin. Oxidative attack on parkin leads to processes like sulfination (-SO2H)/sulfonation (-SO3H) reactions. Scientists have examined putative sulfonation of several cysteine-containing proteins after oxidative stress engendered by H_2O_2 *in vitro*. Oxidative stress causes parkin sulfonation producing Lewy bodies (LBs)-like aggregates. Mutations of many cysteine thiol sulfinated/sulfonated peptides of parkin alter parkin solubility. Cysteine PTMs may alter tertiary structure as rare hereditary mutations, thus providing a mechanistic link between genetic and sporadic forms of PD (Meng et al., 2011). Around, 1509 endogenous SNO-proteins are identified, so proteomic studies of cysteine oxidation are becoming most important (Gu and Robinson, 2016).

Huntington's Disease

In oxidative stress, cysteine is observed to be depleted and dysregulated cysteine and cysteine transporters are also observed in HD. The CSE is also found to be decreased in HD. Oxidative stress-mediated CSE loss was contributing to the failure of corrective responses to cysteine deprivation, which enhances neurotoxicity and progression of the disease (Paul et al., 2018). Oxidative stress, neurotoxicity, and motor and behavioral changes are elicited by a mutation in the gene encoding huntingtin (Htt). Though the particular molecular mechanism by which mHtt elicits neurodegeneration is unclear, majorly depletion in cysteine biosynthetic enzyme CSE has been found. The defect occurs at the transcriptional level and seems to reflect the influences of mHtt on specificity protein, a transcriptional activator for CSE. Cysteine supplementation is found to reverse abnormalities in tissue cultures of Huntington's disease, bringing the attention of researchers towards this therapeutic approach (Paul et al., 2014).

Amyotrophic Lateral Sclerosis

ALS is a motor neuron disorder that affects the upper and lower neurons and leads to paralysis. In ALS, SOD1 aggregation and motor neuron degeneration are observed. Aggregation of SOD1 is observed to involve Cysteine residues on SOD1. Cys6 and Cys111, Cys57 and Cys146 are four cysteine residues in human SOD1. Cys57 and Cys146 are two residues involved in the formation of an intra-molecular disulfide bond and the other two Cys6 and Cys111, not forming disulfide bonds. When the enzyme is in its metal-free form, Cys6 and Cys111 are oxidized and aggregation of SOD1 is increased (Paul et al., 2018).

The cysteine amino acid is integrated with structure and signaling features of the cell and a wide variety of molecules like gas H_2S to coenzyme A can be formed from cysteine. PTMs occurring in cysteine has a major contribution in controlling protein function. Disturbed cellular redox balance in neurodegenerative disorders is also regulated by it. Thus, cysteine stresses can disrupt cysteine metabolism to have a widespread impact on neurodegeneration (Paul et al., 2018).

BENEFITS OF N-ACETYLCYSTEINE SUPPLEMENTATION

Apoptosis is an important physiological process in the cell. This form of predetermined cell death is occurring through many morphological and biochemical changes. The intrinsic and extrinsic pathways are major signaling pathways mediating apoptosis. The extrinsic pathway is dependent on cell-surface death receptors such as the first apoptosis signal (FAS) and the intrinsic pathway is initiated within mitochondria. NAC supplements have shown to reduce cellular oxidative stress and thereby prevent apoptosis and cell necrosis. Various reactive oxygen species along with many external stimuli are contributing to apoptosis. Specifically, oxygen-free radicals change DNA sequencing and induce rearrangements that may trigger apoptotic cell death of neuronal cells. In vitro studies have evidence that free radical scavengers such as NAC can significantly suppress cell death (Okamoto et al, 2016). The antioxidant capacity of glutathione is scavenging the oxygen free radicals via the thiol group of cysteine. NAC crosses the blood-brain barrier (BBB) and elevates the content of glutathione in the brain. NAC treatment has shown improvement in patients with neurodegenerative diseases, such as Alzheimer's disease and Parkinson's disease (Lee et al., 2018). NAC, as a precursor source of the rate-limiting step in GSH synthesis, cysteine could restore depleted levels of GSH and additionally restored GSH depletion after amyloid infusion and in the model of pesticide-induced neurotoxicity (Christopher-Choy et al., 2010).

In AD, amyloid β-peptide (Aβ), the central constituent of senile plaques, is a source of reactive free radicals and oxidative stress that gives neurodegeneration. NAC, having small thiol molecule and is freely filterable with ready access to BBB, has antioxidant properties. NAC is known to serve as cysteine donors. NAC is a direct and potent free radical scavenger that has proved itself benefits in AD. Firstly, intracellular cysteine level is increased by NAC, it improves GSH biosynthesis and then subsequently potentiates the natural anti-oxidative cellular defense mechanisms. Secondly, its reducing thiol group can act by direct reaction with reactive oxygen species. Thirdly, NAC is proven to inhibit apoptosis in cultured neuronal cells. Another advantage of NAC is its less neurotoxicity compared with cysteine itself. NAC improves learning and memory deficits by regulating the cholinergic system. It scavenges free radicals formed by Aβ aggregation and inhibits cell apoptosis. Because of the ease of administration and low cost and low toxicity of NAC, it is promising supportive therapy of AD (Fu et al., 2006). Memory deficits can be prevented by the treatment with NAC due to a decrease in acetylcholinesterase (AChE) activity and oxidative stress. Some scientists have suggested that NAC protection against neurotoxicity is due to its antioxidant properties. But in other experiments, it is observed that treatment with other antioxidants e.g. vitamin E and trolox did not protect neuronal death. It means along with antioxidant property there are some unclear mechanisms of NAC which make it neuroprotective (Isaev et al., 2017).

The transgenic animal model lacking murine apolipoprotein E ($ApoE^{-/-}$ mice) was studied to understand the pathogenesis of AD. It is observed in this study, deficiency of murine apolipoprotein E ($ApoE^{-/-}$ mice) increases oxidative stress leading to AD. NAC treatment has shown to scavenge ROS and increases GSH and reverses memory impairment and oxidative damage in mice. NAC supplementation also alleviated oxidative damage and prevented cognitive impairment in $ApoE^{-/-}$ mice (Tchantchoua et.al, 2005). NAC is a well-known prescription product as a dietary supplement in the treatment of cystic fibrosis and acetaminophen overdose. In a mouse model of PD, oral administration of NAC reduced oxidative stress, and motor abnormalities and neuronal loss NAC's primary mechanism of action is hypothesized to be through its deacetylated product, cystine, which is the rate-limiting substrate for GSH synthesis. NAC regimen resulted in significant increases in cystine and the blood antioxidant measures: GSH/GSSG and catalase (Coles et al., 2017).

Thus, a possible way to boost CNS GSH level is to use NAC, a membrane-permeable cysteine precursor. NAC increases intracellular GSH in human erythrocytes. After intravenous (IV) administration of NAC brain GSH level and blood GSH redox ratios increase (Holmay et al., 2013). Oral supplement NAC is available as over-the-counter and is available in an injectable formulation. NAC can be used in neurodegenerative proteinopathies. Long-term treatment with NAC increases cytoplasmic retention of NF-kappa B in the brain. This prevents the action of NF-kappa B as a transcription factor in the nucleus.

This effect of NAC has shown that it can be used in the treatment of PD in which NF-kappa B activation is one of the pathological mechanism (Monti et al., 2016). Thus, the beneficial effects of NAC on the brain are due to its direct and potent free radical scavenger activity, increase in intracellular cysteine, GSH, SOD and catalase and which subsequently potentiates the natural antioxidant cellular defense mechanism (Garg et al., 2018).

RECENT DEVELOPMENTS AND FUTURE RESEARCH DIRECTIONS

Modulation of endogenous cellular defense mechanisms via the stress response signaling represents an innovative approach to the therapeutic intervention of diseases causing tissue damage, such as neurodegeneration. Efficient maintenance and repair processes seem to be crucial for survival, physical well-being, and quality of life (Calabrese et al., 2010). Dysregulated cysteine and hydrogen sulfide metabolism are frequently encountered in several neurodegenerative disorders (Paul et al., 2018). The redox regulation of cysteine-dependent enzymes is an important area of study. This is particularly evident in neurodegenerative conditions due to their strong association with increases in oxidative stress (Guttmann and Powell, 2012).

Besides, the identification of sensors of various forms of stress may reveal additional hubs and targets of redox regulation. Transcription factors that regulate major hubs of redox control may serve as targets of drug design and therapeutics (Sbodio et al., 2017).

CONCLUSION

Cysteine is a natural antioxidant that works in the cell as a redox sensor and redox regulator. Cysteine oxidation or cysteine PTMs disturb redox balance and may lead to oxidative damage and several pathological conditions including neurodegenerative diseases. An increase in biosynthesis of cysteine or administration of its dietary supplements is neuroprotective. Cysteine supplementation NAC is shown to increase GSH, SOD, catalase availability. Cysteine is a precursor for GSH and SOD synthesis. Many cysteine-containing proteins and non-proteins are affected during oxidative stress. Cysteine dependant enzymes are proven as an antioxidant. Cysteine metabolic pathway, its PTMs and antioxidant mechanisms are providing novel approaches for scientists to treat neurodegenerative diseases.

REFERENCES

Adelman, R., Saul, R. L., & Ames, B. N. (1988). Oxidative damage to DNA: Relation to species metabolic rate and life span. *Proceedings of the National Academy of Sciences of the United States of America*, *85*(8), 2706–2708. doi:10.1073/pnas.85.8.2706 PMID:3128794

Aoyama, K., & Nakaki, T. (2013). Impaired Glutathione Synthesis in Neurodegeneration. *International Journal of Molecular Sciences*, *14*(10), 21021–21044. doi:10.3390/ijms141021021 PMID:24145751

Berry, C. E., & Hare, J. M. (2004). Xanthine oxidoreductase and cardiovascular disease: Molecular mechanisms and pathophysiological implications. *The Journal of Physiology, 555*(3), 589–606. doi:10.1113/jphysiol.2003.055913 PMID:14694147

Burgoyne, R. D., & Morgan, A. (2015). Cysteine String Protein (CSP) and its role in preventing neurodegeneration. *Seminars in Cell & Developmental Biology, 40*, 153–159. doi:10.1016/j.semcdb.2015.03.008 PMID:25800794

Calabrese, V., Cornelius, C., Mancuso, C., Lentile, R., Giuffrida Stella, A. M., & Butterfield, D. A. (2010). Redox Homeostasis and Cellular Stress Response in Aging and Neurodegeneration. Free Radicals and Antioxidant Protocols, Methods in Molecular Biology, 610, 285-308. doi:10.1007/978-1-60327-029-8_17

Calabrese, V., Scapagnini, G., Stella, A. M., Bates, T. E., & Clark, J. B. (2000). Mitochondrial involvement in brain function and dysfunction: Relevance to aging, neurodegenerative disorders and longevity. *Neurochemical Research, 26*(6), 739–764. doi:10.1023/A:1010955807739 PMID:11519733

Choi, Y. B., Tenneti, L., Le, D. A., Ortiz, J., Bai, G., Chen, H. S., & Lipton, S. A. (2000). Molecular basis of NMDA receptor-coupled ion channel modulation by S-nitrosylation. *Nature Neuroscience, 3*(1), 15–21. doi:10.1038/71090 PMID:10607390

Christopher-Choy, K. H., Dean, O., Berk, M., Bush, A. I., & Buuse, M. (2010). Effect of N-acetyl-cysteine treatment on glutathione depletion and a short-term spatial memory deficit in 2-cyclohexene 1-one-treated rats. *European Journal of Pharmacology, 649*(1-3), 224–228. doi:10.1016/j.ejphar.2010.09.035 PMID:20868666

Chung H. S., Wang S. B., Venkatraman V., Murray C. I., & Van Eyk J. E. (2013). Cysteine oxidative posttranslational modifications: emerging regulation in the cardiovascular system. *Circu resear, 112*, 382-392.

Coles, L. D., Tuite, P. J., Gulin, O., Mishra, U. R., Kartha, R. V., Sullivan, K. M., & .... (2017). Repeated Dose Oral N-Acetyl cysteine in Parkinson's Disease: Pharmacokinetics and Effect on Brain Glutathione and Oxidative Stress. *Journal of Clinical Pharmacology, 58*(2), 158–167. doi:10.1002/jcph.1008 PMID:28940353

Couvertier, S. M., Zhou, Y., & Weerapana, E. (2014). Chemical-proteomic strategies to investigate cysteine posttranslational modifications. *Biochimica et Biophysica Acta, 1844*(12), 2315–2330. doi:10.1016/j.bbapap.2014.09.024 PMID:25291386

Cozzolino, M., Ferri, A., & Carri, M. T. (2008). Amyotrophic lateral sclerosis: From current developments in the laboratory to clinical implications. *Antioxidants & Redox Signalling, 10*(3), 405–444. doi:10.1089/ars.2007.1760 PMID:18370853

Cumming, R. C., & Schubert, D. (2005). Amyloid-β induces disulfide bonding and aggregation of GAPDH in Alzheimer's disease. *The FASEB Journal, 19*(14), 2060–2062. doi:10.1096/fj.05-4195fje PMID:16186172

Fernandez-Chacon, R., Wolfel, M., Nishimune, H., Tabares, L., Schmitz, F., Castellano, M., & .... (2004). The Synaptic Vesicle Protein CSP Prevents Presynaptic Degeneration. *Neur, 42*, 237–251. PMID:15091340

Floyd R.A., & Carney J.M. (1992). Free radical damage to protein and DNA: mechanisms involved and relevant observations on brain undergoing oxidative stress. *Ann Neurol, 32*, S22-27.

Fu, A. L., Dong, Z. H., & Sun, M. J. (2006). Protective effect of N-acetyl-cysteine on amyloid ß peptide- induced learning and memory deficits in mice. *Brain Research, 1109*(1), 201–206. doi:10.1016/j.brainres.2006.06.042 PMID:16872586

Gadoth, N., & Goebel, H. H. (2013). Oxidative stress and free radical damage in neurology. New York: Humana Press.

Garg, G., Singh, S., Singh, A., & Rizvi, S. I. (2018). N-acetyl L-cysteine attenuates oxidative damage and neurodegeneration in rat brain during aging. *Canadian Journal of Physiology and Pharmacology, 96*(12), 1189–1196. doi:10.1139/cjpp-2018-0209 PMID:30107137

Giles, G. I., Tasker, K. M., & Jacob, C. (2001). Hypothesis: The role of reactive sulfur species in oxidative stress. *Free Radical Biology & Medicine, 31*(10), 1279–1283. doi:10.1016/S0891-5849(01)00710-9 PMID:11705707

Giles, N. M., Watts, A. B., Giles, G. I., Fry, F. H., Littlechild, J. A., & Jacob, C. (2003). Metal and redox modulation of cysteine protein function. *Chemistry & Biology, 10*(8), 677–693. doi:10.1016/S1074-5521(03)00174-1 PMID:12954327

Go, Y. M., & Jones, D. P. (2008). Redox compartmentalization in eukaryotic cells. *Biochimica et Biophysica Acta, 1780*(11), 1273–1290. doi:10.1016/j.bbagen.2008.01.011 PMID:18267127

Goll, D. E., Thompson, V. F., Li, H., Wei, W., & Cong, J. (2003). The calpain system. *Physiological Reviews, 83*(3), 731–801. doi:10.1152/physrev.00029.2002 PMID:12843408

Gu, L., & Robinson, A. S. (2016). Proteomic Approaches to Quantify Cysteine Reversible Modifications in Aging and Neurodegenerative Diseases. *Proteomics. Clinical Applications, 10*(12), 1159–1177. doi:10.1002/prca.201600015 PMID:27666938

Gundersen, C. B., & Umbach, J. A. (1992). Suppression cloning of the cDNA for a candidate subunit of a presynaptic calcium channel. *Neur, 9*(3), 527–537. doi:10.1016/0896-6273(92)90190-O PMID:1326297

Guttmann, R. P. (2010). Redox regulation of cysteine-dependent enzymes. *Journal of Animal Science, 88*(4), 1297–1306. doi:10.2527/jas.2009-2381 PMID:19820057

Guttmann, R. P., Elce, J. S., Bell, P. D., Isbell, J. C., & Johnson, G. V. (1997). Oxidation inhibits substrate proteolysis by calpain I but not autolysis. *The Journal of Biological Chemistry, 272*(3), 2005–2012. doi:10.1074/jbc.272.3.2005 PMID:8999893

Guttmann, R. P., & Powell, T. J. (2012). Redox Regulation of Cysteine-Dependent Enzymes in Neurodegeneration. *International Journal of Cell Biology, 2012*, 1–8. doi:10.1155/2012/703164 PMID:22829832

Haendeler, J., Weiland, U., Zeiher, A. M., & Dimmeler, S. (1997). Effects of redox-related congeners of NO on apoptosis and caspase-3 activity. *Nit Ox, 1*(4), 282–293. doi:10.1006/niox.1997.0134 PMID:9441900

Hanzen, S., Vielfort, K., Yang, J., Roger, F., Andersson, V., Zamarbide-Fores, S., ... Nyström, T. (2016). Lifespan Control by Redox-Dependent Recruitment of Chaperones to Misfolded Proteins. *Cell, 166*(1), 140–151. doi:10.1016/j.cell.2016.05.006 PMID:27264606

Holmay, M. J., Terpstra, M., Coles, L. D., Mishra, U., Ahlskog, M., Gulin, O., & … . (2013). N-Acetyl cysteine Boosts Brain and Blood Glutathione in Gaucher and Parkinsons Disease. *Clinical Neuropharmacology*, *36*(4), 103–106. doi:10.1097/WNF.0b013e31829ae713 PMID:23860343

Ignowski, E., Winter, A. N., Dual, N., Fleming, H., Wallace, T., Manning, E., & … . (2018). The cysteine-rich whey protein supplement, Immunocal®, preserves brain glutathione and improves cognitive, motor, and histopathological indices of traumatic brain injury in a mouse model of controlled cortical impact. *Free Radical Biology & Medicine*, *124*, 328–341. doi:10.1016/j.freeradbiomed.2018.06.026 PMID:29940352

Imlay, J. A., Chin, S. M., & Linn, S. (1988). Toxic DNA damage by hydrogen peroxide through the Fenton reaction in vivo and in vitro. *Sci*, *240*(4852), 640–642. doi:10.1126cience.2834821 PMID:2834821

Isaev, N. K., Ailkina, S., Golyshev, S. A., Genrikh, E. E., Alexandrova, O. P., Kapkaeva, M. R., & … . (2017). N-acetyl-L-cysteine and Mn2+ attenuate Cd2+-induced disturbance of the intracellular free calcium homeostasis in cultured cerebellar granule neurons. *Toxicol*, *393*, 1–8. doi:10.1016/j.tox.2017.10.017 PMID:29100878

Jacob, C., Knight, I., & Winyard, P. G. (2006). Aspects of the biological redox chemistry of cysteine: From simple redox responses to sophisticated signalling pathways. *Biological Chemistry*, *387*(10/11), 1385–1397. doi:10.1515/BC.2006.174 PMID:17081111

Jadah, N. J., Eva, A., & Janice, E. A. (2010). CSPa: The neuroprotective J protein. *Biochemistry and Cell Biology*, *88*, 157–165. PMID:20453918

Jones, D. P. (2006). Redefining oxidative stress. *Antioxidants & Redox Signalling*, *8*(9-10), 1865–1879. doi:10.1089/ars.2006.8.1865 PMID:16987039

Jones, D. P. (2008). Radical-free biology of oxidative stress. *American Journal of Physiology. Cell Physiology*, *295*(4), 849–868. doi:10.1152/ajpcell.00283.2008 PMID:18684987

Kragten, E., Lalande, I., Zimmermann, K., Roggo, S., Schindler, P., Muller, D., ... Fürst, P. (1998). Glyceraldehyde-3-phosphate dehydrogenase, the putative target of the antiapoptotic compounds CGP 3466 and R-(-)-deprenyl. *The Journal of Biological Chemistry*, *273*(10), 5821–5828. doi:10.1074/jbc.273.10.5821 PMID:9488718

Kuhn, D. M., Arthur, R. E. Jr, Thomas, D. M., & Elferink, L. A. (1999). Tyrosine hydroxylase is inactivated by catechol-quinones and converted to a redox-cycling quinoprotein: Possible relevance to Parkinson's disease. *Journal of Neurochemistry*, *73*(3), 1309–1317. doi:10.1046/j.1471-4159.1999.0731309.x PMID:10461926

Lambeth, J. D., & Neish, A. S. (2014). Nox enzymes and new thinking on reactive oxygen: A doubleedged sword revisited. *Annual Review of Pathology*, *9*(1), 119–145. doi:10.1146/annurev-pathol-012513-104651 PMID:24050626

Lee, C. Y., Su, C. H., Tsai, M. L., Ho, Y. C., Lee, S. S., Chen, C. H., & ... . (2018). Cadmium nitrate-induced neuronal apoptosis is protected by N-acetyl-Lcysteine via reducing reactive oxygen species generation and mitochondria dysfunction. *Biomedicine and Pharmacotherapy*, *108*, 448–456. doi:10.1016/j.biopha.2018.09.054 PMID:30241048

Lim, M. A., Selak, M. A., Xiang, Z., Krainc, D., Neve, R. L., Kraemer, B. C., ... Kalb, R. G. (2012). Reduced activity of AMP-activated protein kinase protects against genetic models of motor neuron disease. *The Journal of Neuroscience: The Official Journal of the Society for Neuroscience*, *32*(3), 1123–1141. doi:10.1523/JNEUROSCI.6554-10.2012 PMID:22262909

Luc, R., & Vergely, C. (2008). Forgotten radicals in biology. *International Journal of Biomedical Science; IJBS*, *4*, 255–259. PMID:23675099

Mani, S., Yang, G., & Wang, R. (2011). A critical life-supporting role for cystathionine-γ-lyase in the absence of dietary cysteine supply. *Free Radical Biology & Medicine*, *50*(10), 1280–1287. doi:10.1016/j.freeradbiomed.2011.01.038 PMID:21310231

McDonagh, B., Sakellariou, G. K., Smith, N. T., Brownridge, P., & Jackson, M. J. (2014). Differential cysteine labeling and global label-free proteomics reveals an altered metabolic state in skeletal muscle aging. *Journal of Proteome Research*, *13*(11), 5008–5021. doi:10.1021/pr5006394 PMID:25181601

Meister, A. (1988). Glutathione metabolism and its selective modification. *The Journal of Biological Chemistry*, *263*, 17205–17208. PMID:3053703

Meng, F., Yao, D., Shi, Y., Kabakoff, J., Wu, W., Reicher, J., ... Gu, Z. (2011). Oxidation of the cysteine-rich regions of parkin perturbs its E3 ligase activity and contributes to protein aggregation. *Molecular Neurodegeneration*, *6*(34), 1–15. doi:10.1186/1750-1326-6-34 PMID:21595948

Monti, D. A., Zabrecky, G., Kremens, D., Liang, T. W., Wintering, N. A., Cal, J., & ... . (2016). N-Acetyl Cysteine May Support Dopamine Neuronsin Parkinson's Disease: Preliminary Clinical and Cell Line Data. *PLoS One*, *11*(6), 1–15. doi:10.1371/journal.pone.0157602 PMID:27309537

Mossner, E., Iwai, H., & Glockshuber, R. (2000). Influence of the pK(a) value of the buried, active-site cysteine on the redox properties of thioredoxin-like oxidoreductases. *FEBS Letters*, *477*(1-2), 21–26. doi:10.1016/S0014-5793(00)01738-5 PMID:10899304

Nakamura, T., & Lipton, S. A. (2016). Protein S-Nitrosylation as a Therapeutic Target for Neurodegenerative Diseases. *Trends in Pharmacological Sciences*, *37*(1), 73–84. doi:10.1016/j.tips.2015.10.002 PMID:26707925

Numajiri, N., Takasawa, K., Nishiya, T., Tanaka, H., Ohno, K., Hayakawa, W., ... Uehara, T. (2011). On-off system for PI3-kinase-Akt signaling through S-nitrosylation of phosphatase with sequence homology to tensin (PTEN). *Proceedings of the National Academy of Sciences of the United States of America*, *108*(25), 10349–10354. doi:10.1073/pnas.1103503108 PMID:21646525

Okamoto, A., Tanaka, M., Sumi, C., Oku, K., Kusunoki, M., Nishi, K., ... Hirota, K. (2016). The antioxidant N-acetyl cysteine suppresses lidocaine-induced intracellular reactive oxygen species production and cell death in neuronal SH-SY5Y cells. *BMC Anesthesiology*, *16*(104), 1–12. doi:10.118612871-016-0273-3 PMID:27776485

Pacher, P. J., Beckman, S., & Liaudet, L. (2007). Nitric oxide and peroxynitrite in health and disease. *Physiological Reviews*, *87*(1), 315–424. doi:10.1152/physrev.00029.2006 PMID:17237348

Park, J. W., Mieya, J. J., Rhee, S. G., & Chock, P. B. (2009). Deglutathionylation of 2-Cys peroxiredoxin is specifically catalyzed by sulfiredoxin. *The Journal of Biological Chemistry*, *284*(35), 23364–23374. doi:10.1074/jbc.M109.021394 PMID:19561357

Patel, P., Prescott, G. R., Burgoyne, R. D., Lian, L. Y., & Morgan, A. (2016). Phosphorylation of Cysteine String Protein Triggers a Major Conformational Switch. *Struc*, *24*(8), 1380–1386. doi:10.1016/j.str.2016.06.009 PMID:27452402

Paul, B. D., Sbodio, J. I., & Snyder, S. H. (2018). Cysteine Metabolism in Neuronal Redox Homeostasis. *Trends in Pharmacological Sciences*, *2*(7), 1–12. PMID:29530337

Paul, B. D., Sbodio, J. I., Vandier, M. S., Cha, J. Y., Snowman, A. M., & Snyder, S. H. (2014). Cystathionine c-lyase deficiency mediates neurodegeneration in Huntington's disease. *Nat*, *509*(7498), 96–103. doi:10.1038/nature13136 PMID:24670645

Radi, R. (2013). Peroxynitrite, a stealthy biological oxidant. *The Journal of Biological Chemistry*, *288*(37), 26464–26472. doi:10.1074/jbc.R113.472936 PMID:23861390

Redler, R. L., Wilcox, K. C., Proctor, E. A., Fee, L., Caplow, M., & Dokholyan, N. V. (2011). Glutathionylation at Cys-111 induces dissociation of wild type and FALS mutant SOD1 dimers. *Biochem*, *50*(32), 7057–7066. doi:10.1021/bi200614y PMID:21739997

Ren, X., Bjornstedt, M., Shen, B., Ericson, M. L., & Holmgren, A. (1993). Mutagenesis of structural half-cystine residues in human thioredoxin and effects on the regulation of activity by selenodiglutathione. *Biochem*, *32*(37), 9701–9708. doi:10.1021/bi00088a023 PMID:8373774

Richman, P. G., & Meister, A. (1975). Regulation of γ-glutamyl-cysteine synthetase by nonallosteric feedback inhibition by glutathione. *The Journal of Biological Chemistry*, *250*, 1422–1426. PMID:1112810

Sbodio, J. I., Snyder, S. H., & Paul, B. D. (2017). Redox Mechanisms In Neurodegeneration: From Disease Outcomes To Therapeutic Opportunities. *Antioxidants & Redox Signalling*, *7321*, 1–136.

Smith, J. K., Patil, C. N., Patlolla, S., Gunter, B. W., Booz, G. W., & Duhe, R. J. (2012). Identification of a redox-sensitive switch within the JAK2 catalytic domain. *Free Radical Biology & Medicine*, *52*(6), 1101–1110. doi:10.1016/j.freeradbiomed.2011.12.025 PMID:22281400

Tabner, B. J., Turnbull, S., El-Agnaf, O., & Allsop, D. (2001). Production of reactive oxygen species from aggregating proteins implicated in Alzheimer's disease, Parkinson's disease and other neurodegenerative diseases. *Current Topics in Medicinal Chemistry*, *1*, 507–517. doi:10.2174/1568026013394822 PMID:11895127

Tchantchoua, F., Graves, M., Rogers, E., Ortiz, D., & Shea, T. B. (2005). N-acteyl cysteine alleviates oxidative damage to central nervous system of ApoE-deficient mice following folate and vitamin E-deficiency. *J Alzh Dis*, *7*(2), 135–138. doi:10.3233/JAD-2005-7206 PMID:15851851

Valle, C., & Carri, M. T. (2017). Cysteine Modification in the Pathogenesis of ALS. *Frontiers in Molecular Neuroscience*, *10*(5), 1–8. doi:10.3389/fnmol.2017.00005 PMID:28167899

Wilson, M. A. (2011). The role of cysteine oxidation in DJ-1function and dysfunction. *Antioxidants & Redox Signalling*, *15*(1), 111–122. doi:10.1089/ars.2010.3481 PMID:20812780

Winter, A. N., Ross, E. K., Daliparthi, V., Sumner, W. A., Kirchhof, D. M., Manning, E., ... Linseman, D. A. (2017). A Cystine-Rich Whey Supplement (Immunocal®) Provides Neuroprotection from Diverse Oxidative Stress-Inducing Agents In Vitro by Preserving Cellular Glutathione. *Oxidative Medicine and Cellular Longevity*, *2017*, 1–15. doi:10.1155/2017/3103272 PMID:28894506

Witt, A. C., Lakshminarasimhan, M. B., Remington, S., Hasim, E., Pozharski, E., & Wilson, M. A. (2008). Cysteine pKa depression by a protonated glutamic acid in human DJ-1. *Biochem*, *47*(28), 7430–7440. doi:10.1021/bi800282d PMID:18570440

Yang, G., Wu, L., Jiang, B., Yang, W., Qi, J., Cao, K., ... Wang, R. (2008). H2S as a physiologic vasorelaxant; hypertension in mice with deletion of cystathionine-γ-lyase. *Sci*, *322*(5901), 587–590. doi:10.1126cience.1162667 PMID:18948540

Yang, H. Y., & Lee, T. H. (2015). Antioxidant enzymes as redox-based biomarkers: A brief review. *BMB Reports*, *48*(4), 200–208. doi:10.5483/BMBRep.2015.48.4.274 PMID:25560698

Zinsmaier, K. E., Hofbauer, A., Heimbeck, G., Pflugfelder, G. O., Buchner, S., & Buchner, E. (1990). A cysteine-string protein is expressed in retina and brain of Drosophila. *Journal of Neurogenetics*, *7*(1), 15–29. doi:10.3109/01677069009084150 PMID:2129171

Chapter 13
Cysteine in Alzheimer's Disease:
Redox Regulation of Protein Functions

Suvarna P. Ingale
Indira College of Pharmacy, India

Rupali Patil
Sir Dr. M. S. Gosavi College of Pharmaceutical Education and Research, India

Aman B. Upaganlawar
SNJB's SSDJ College of Pharmacy, Chandwad, Nashik, India

ABSTRACT

Alzheimer's disease (AD) is characterized by selective loss of neurons in the hippocampus and neocortex due to abnormalities in proteins, mainly Aβ peptide and tau protein, in the form of abnormal protein aggregations or depositions in neurons. Recently oxidative/nitrosative stress has been identified as an important facilitator of neurodegeneration in AD. Cysteine-dependent proteins are known to be associated with the neurodegenerative process. Such cysteine-dependent enzyme proteins are proteases, antioxidant enzymes, kinases, phosphatases, and also non-enzymatic proteins such that utilize cysteine as a structural part of the catalytic site. This chapter deals with the role of cysteine in handling reactive oxygen/nitrogen species during oxidative/nitrosative stress and posttranslational modification of proteins causing protein misfolding or protein aggregation during neurodegeneration associated with AD.

INTRODUCTION

Alzheimer's disease (AD) is a continuously advancing, devastating, prolonged neurodegenerative disorder chiefly manifested as amnesia limiting the patient's ability to perform daily activities, affecting mainly the elderly population above 60 years of age (Kumar et al., 2015; Masters et al., 2018). According to Alzheimer's Disease International (ADI), there will be one fresh incident of AD per 3 seconds glob-

DOI: 10.4018/978-1-7998-1317-0.ch013

ally. It was estimated that in 2018, globally 50 million people are suffering from AD and the cost of its management is US$ 1 trillion. By 2050, this incidence will be above threefold to 152 million with cost upswing to US$ 2 trillion (Alzheimer's Disease International, 2018). According to the age of onset, AD is of two types, early-onset AD (EOAD) and late-onset AD (LOAD). EOAD affects around 1 to 6 present of all cases in the age range of thirty to sixty years, while the onset of LOAD is after sixty or sixty-five years (David et al., 2005). AD share a strong genetic basis that distinguishes AD into two genetically diverse forms such as familial AD (FAD) and sporadic AD (SAD). FAD exhibit autosomal dominant inheritance (Ray et al., 1998). While, SAD exhibit environmental and genetic differences as major risk factors without the involvement of autosomal-dominant inheritance (Braak and Tredici, 2012).

Aging is the utmost imperative risk factor of AD. Succeeding aging, the occurrence of the apolipoprotein E ε4 (ApoE ε4) allele is one more major risk factor (Dong et al, 2012). The lifetime risk of AD in an individual without an ApoE ε4 is 9% which increases to 29% in individuals with at least 1 ε4 allele (Seshadri et al., 1995). Other suspected risk factors of AD include cardiovascular disease, traumatic brain injury, depression, lower educational and/or occupational strata, parental age at time of birth, smoking, first-degree relative with Down syndrome, low levels of folate and vitamin B12, and elevated plasma and total homocysteine levels (David et al., 2005).

AD is manifested in the form of clinical dementia syndrome which is mixed and diverse. The signs and symptoms differ with the advancement and chronicity of the illness. The hallmark manifestation of the clinical dementia syndrome is progressive intellectual deterioration with invariable memory loss leading to the inability to perform and function in the normal milieu (Burns and lliffe, 2009). Also, other cognitive functions are affected which comprise the language, visuospatial, judgment, appreciation, planning, organizing and decision-making problems. Along with cognitive impairment, AD is universally presented with non-cognitive neuropsychiatric symptoms (Förstl and Kurz, 1998) viz. behavioral disturbance (lethargy, violence, nervousness, disinhibition and abnormal motor behavior), altered mood (despair, apprehension, and irritability), psychomotor disturbance (disturbed sleep and appetite), perceptual disturbance (misconceptions, hallucinations, misapprehensions) (Salami and Lyketsos, 2011). Many researchers clinically have characterized AD by deposition of the Aβ peptide plaque mainly within and around arterioles and twisted bundles of aggregated tau within the neurons. Literature evidence supports the concept of injurious participation of oxidative and nitrosative stress in the initiation and progression of AD in the form of two most important pathological markers of AD i.e. abnormal deposition of beta amyloid (Aβ) plaques and neurofibrillary tangles (NFTs) of tau proteins.

Chemical alterations in various cellular protein conformations appear to be mainly responsible for neuronal toxicity associated with AD. Cysteine-thiol containing protein always remained the target for many post-translational modifications like S-nitrosylation (SNO), S-glutathionylation, S-palmitoylation which have been linked to AD pathology. Thus, the main objective of this chapter is to discuss the major pathological events involved in AD and the role of oxidative/nitrosative stress in AD and to link various cysteine-thiols mediated post-translational modifications in the pathogenesis of AD.

NEUROPATHOPHYSIOLOGY OF ALZHEIMER'S DISEASE

In 1906, a German psychiatrist, Dr. Alois Alzheimer, first time noticed and studied alterations in brain cells of a woman who died due to a rare psychological ailment described by memory impairment, language problems and erratic imperative behavior (Hippius, 2003). In the postmortem study, he found two abnormal structures namely extracellular Aβ plaques and NFTs. Hence, the disease was named AD (Alzheimer's Association, 2018; Perl, 2010). Subsequently, many researchers clinically characterized AD by deposition of Aβ peptide plaque mainly within and around arterioles and twisted bundles of aggregated tau within the neurons, supporting the observation of Dr. Alois Alzheimer (Masters et al., 2018; Serrano-Pozo et al., 2011). An interplay of many relating mechanisms such as aging, environmental impacts, genetic changes, cerebral infarcts, and microscopic structural changes forms the basis of pathogenesis of AD (David at al., 2005).

A healthy human brain is made up of around 100 billion long neurons with branching extensions at which one neuron connects with other neurons (called synapses) forming a neuronal network of communication via about 100 trillion synapses. This communication between the neurons occurs through tiny surges of chemicals released by one neuron and sensed by another neuron which permits the information to travel fast through a neuronal network of communication generating the cellular basis of cognition (Solomon et al., 2015). The deposition of Aβ plaques outside neurons and abnormal NFTs inside neurons are two of several pathological changes in the brain of AD patients as discussed above (Sajjad et al., 2018).

The toxic Aβ plaques were assumed to be responsible for neuronal cell death by disrupting neuronal communication at synapses and activating immune cells called microglia. The microglia try hard to get rid of toxic proteins accumulated along with the debris of dead cells. As microglia fails to do so or is not enough, chronic inflammation sets in resulting in brain atrophy due to cell death (Mokhtar et al., 2013). Moreover, tau tangles inhibit the transport and metabolism of nutrients i.e. glucose and additional vital substances inside neurons compromising normal brain function (Alzheimer's Association, 2018; National Institutes of Health, 2017; Ray et al., 1998).

Amyloid Plaques

The toxic Aβ peptides, mainly Aβ42 and Aβ40 peptides, existing extracellularly in the brain of AD patients are derivatives obtained from a precursor known as an amyloid precursor protein (APP), a cell surface protein. APP facilitates signal transduction, axonal elongation, and cell migration during normal physiological circumstances, by sequential cleavage through various proteases. APP expression, processing, and intracellular trafficking and Aβ peptides in the trans-Golgi network, endosomes and plasma membrane and toxic effects of the Aβ peptides are believed as most important players in the pathogenesis of AD (amyloid cascade hypothesis) (Dorszewska et al., 2016). Further, Ganguly et al., (2017) have suggested that Aβ peptides oligomerization can initiate mitochondrial dysfunction, Ca^{2+} dysregulation, inflammatory reactions, endoplasmic reticulum (ER) stress, and oxidative destruction to elicit the progressive neurodegeneration in AD (Ganguly et al., 2017).

APP is acted upon by various proteases in two different pathways, one being the amyloidogenic pathway (Devi and Ohno, 2012) is responsible for amyloid plaque formation and other being non-amyloidogenic (Lichtenthaler, 2012), as depicted in Figure 1. Normally almost 90 percent of APP is processed by the non-amyloidogenic pathway, and only 10 percent enters the amyloidogenic pathway. However, these proportions are altered due to various risk factors of AD viz. aging, environmental toxins, genetic changes, cerebral infarcts, etc. The ultimate products obtained from these pathways may be a major participant in neuronal growth and function (O'Brien and Wong, 2011; Zhang et al., 2011).

Figure 1. The processing of amyloid precursor protein by α- β- and γ-secretases and cleavage products (Snyder et al., 1994).

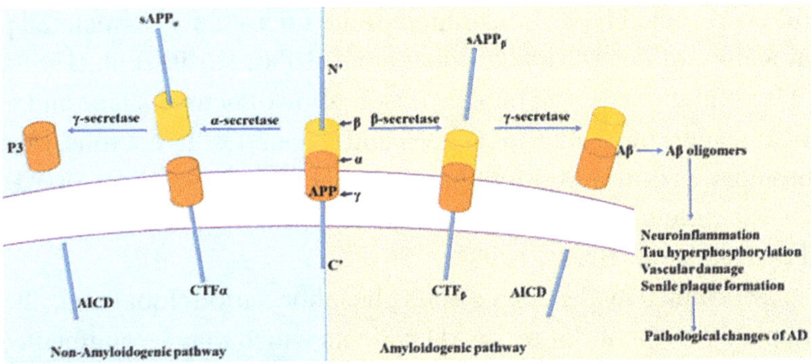

During the non-amyloidogenic pathway, APP is first sliced by the action of α-secretase to produce two fragments, a soluble N-terminal fragment (sAPP$_\alpha$) and a membrane-bound C-terminal fragment (CTFα) (Thinakaran and Koo, 2008). sAPP$_\alpha$ may facilitate the formation of synapses (synaptogenesis), neurite extension and neuronal survival during early development. Thus, sAPP$_\alpha$ plays a neuroprotective role in the brain. CTF$_\alpha$ is further cleaved by presenilin-containing γ-secretase to produce a soluble N-terminal fragment (p3) and a membrane-bound C-terminal fragment (AICD, or APP intracellular domain) (Mokhtar et al., 2013). AICD is known to be linked with nuclear signaling via regulation of transcription of various genes like APP and BACE 1 as well as axonal transport and increasing the interaction of APP with cellular nuclear factors (Bhadbhade and Cheng, 2012).

During the amyloidogenic pathway, which is the plaque-forming event in the pathogenesis of AD, APP is acted upon by another proteolytic enzyme, β-secretase, to produce two fragments: a soluble N-terminal fragment (sAPP$_\beta$) and a membrane-bound C-terminal fragment (CTF$_\beta$) (Citron et al. 1995). β-secretase cuts near to the N-terminal end of APP as compared to α-secretase producing longer CTF$_\beta$ than CTF$_\alpha$. sAPP$_\beta$ is well known to cause neuronal death and excess generation of CTF$_\beta$ to disrupt signal transduction and thereby leads to neurodegeneration (Zhang et al., 2011). CTF$_\beta$ is further cleaved by γ-secretase similar to the non-amyloidogenic pathway to produce a membrane-bound C-terminal fragment (AICD) and a soluble N-terminal fragment (β-amyloid, or Aβ), longer than p3 (Mokhtar et al., 2013; Ray et al., 1998). Many pieces of evidence reported Aβ playing role in neuronal function. However, excessive production and thereby accumulation of Aβ trigger neurodegenerative cascade responsible for disturbed synaptic function, deposition of intraneuronal fibrillary tangles, ultimately involved in neuron death (Shankar and Walsh, 2009; Selkoe,1998). γ-secretase enzyme can cut APP at diverse positions to yield various species of Aβ: 38, 40 and 42 amino acids in length, Aβ 40 being utmost abundant species generated (Seubert et al., 1992). But, Aβ 42 is the most commonly found species in plaques in AD patients as demonstrated by its extreme tendency to aggregate *in vitro* (Roher et al., 1993), which further get transformed into fibrils. Consequently, an abnormal Aβ42 production may be a major promoter of AD (Snyder et al., 1994).

Neurofibrillary Tangles

Many research coworkers revealed NFTs constituted of the microtubule-associated protein (MAP), tau, as a key pathological feature of AD (Grundke-Iqbal et al., 1986; Kosik et al., 1986; Wood et al., 1986; Binder et al., 2005). Microtubules are vital for the conservation of neuron shape and take up an important role in the axoplasmic transport of various molecules and organelles. In neurons, microtubules are made up of the globular proteins, viz. α- and β-tubulin, and of a cluster of MAP1, MAP2, and tau proteins (Schweers, 1995). Of which, tau exist copolymerized with microtubules and facilitate their assembly supporting neuronal cytoskeleton (Brion, 1998).

The tau function is determined by the degree of its phosphorylation (Iqbal et al., 2010). It was revealed that tau undergoes atypical phosphorylation in AD patients which may be augmented in the presence of toxic Aβ species through protein kinases and phosphatases normalizing tau phosphorylation (Bloom, 2014). Altered hyperphosphorylated tau not only disconnects from the microtubules, weakening and depolymerizing them but also facilitates its polymerization to initiate accumulation to form pathological intracellular NFT that may lastly block the affected neurons and facilitate cell death (Barghorn and Mandelkow, 2002). Subsequently, the neurons become susceptible to oxidative stress and other proapoptotic factors (Dorszewska et al., 2016; Gong and Iqbal, 2008).

Oxidative Stress and Alzheimer's Disease

Literature evidence supports the concept of injurious participation of oxidative stress (OS) and nitrosative stress in the initiation and progression of AD (Markesbery, 1997). However, OS induced mitochondria dysfunction was also evident in many studies as a prominent contributor in AD (Selfridge et al., 2013; Tramutola et al., 2017). OS is an obligatory product of aerobic metabolism producing reactive oxygen species (ROS) (Paget, 2003). Generation of ROS starts with the reduction of molecular oxygen in the water giving first superoxide anion radical ($O_2^{\bullet-}$), followed by the addition of an electron to produce hydrogen peroxide (H_2O_2) which further gets reduced to highly reactive hydroxyl radicals ($OH^{\bullet}$). These reactions are catalyzed by iron or copper ions (Sheldon, 2012). The ROS can deteriorate the functions of various cellular components like proteins, nucleic acids, and lipids (Paget, 2003). With aging, OS in the brain continues to increase due to the overproduction of ROS and deficient antioxidant mechanisms to neutralize ROS (Huang et al., 2016; Leeuwenburgh and Heinecke, 2001).

The brain is an organ with high oxygen demand and is made up of easily oxidizable and peroxidation-susceptible lipid cells (Halliwell, 1992). Moreover, cerebrospinal fluid (CSF) cannot combine with released iron ions. Thus, OS may extremely disturb the brain functions through various interconnected mechanisms like rising in intracellular free Ca^{2+}, the release of excitatory amino acids, and neuronal toxicity (Huang, et al., 2016). The two most important pathological markers of AD i.e. abnormal deposition of Aβ plaques and NFTs of tau proteins are results of OS (Christen, 2000).

There is increasing evidence revealing the involvement of biometals like iron (Fe), zinc (Zn), and copper (Cu) in Aβ deposition and neurodegeneration. According to this evidence, there are high-affinity binding sites on Aβ and its precursor APP for Zn and Cu. Of this, Cu plays an essential part in the generation of highly reactive hydroxyl radical ($OH^{\bullet}$) which induce OS, a major representative of AD. This is further supported by the elevated level of Cu in Aβ plaques in AD, more specifically Aβ42 fragment being most toxic to produce H_2O_2 and other ROS (Butterfield, 2002). While, high amount of Zn was found in the neocortex, amygdala, and hippocampus disturbing memory and cognitive functions, which typically occur during AD (Huang, 2004). The binding of Zn on Aβ form a predominant Aβ40 fragment, involved in the formation of neurofibrillary Aβ aggregates, which activate immunological/inflammatory response to disturb Zn homeostasis with further agitated zinc release in the brain, precipitating OS and cytotoxicity (Chen and Zhong, 2014).

The brain phospholipids membrane made up of polyunsaturated fatty acids provides susceptible double binding sites for free radical attacks causing increased lipid peroxidation (Dröge, 2002). Along with this lipid peroxidation, protein oxidation by ROS may be noteworthy in AD as enzymes harmful to neuron and glial functions are altered through protein oxidation. The two most oxidation prone enzymes glutamine synthetase and creatine kinase are distinctly decreased in AD patients (Varadaraja, 2000). Decreased glutamine synthetase alters glutamate concentrations and augments excitotoxicity. However oxidative damage of creatine kinase may decrease energy metabolism in the AD brain. Thus protein oxidation triggers their pathologic fibrillary aggregation and insolubility (Chauhan and Chauhan, 2006). NFTs are expressed as the accumulation of hyperphosphorylation of the tau protein in the form of paired helical filaments. Hyperphosphorylation is critical oxidation mediated activation of the MAP kinase pathway through stimulation of the transcription factor (Gong and Iqbal, 2008). Also, non- enzymatic interaction of proteins with monosaccharides, known as advanced glycation is also activated through protein oxidation (Finkel and Holbrook, 2000). Furthermore, oxidation can trigger a critical event like DNA filament disruptions, sister chromatid interchange, DNA-protein crosslinking, and base modification, affecting brain Thus, ROS produced during OS may damage the cellular components, establishing the most prevalent pathological events in AD (Cooke et al., 2003; Huang, 2004). In summary, the neuropathology of AD includes proteinopathies (Aβ and tau protein) which involves various enzymes like β-secretase, γ-secretase, Glycogen synthase kinase 3β, acetylcholinesterase (AChE) and butyrylcholinesterase (BuChE), Rho kinase, prolyl endopeptidase, monoglycerol lipase, and catechol-O-methyl transferase and OS leading to neurodegeneration (Choudhury et al., 2018).

Proteins, the ultimate yields of gene transcription and translation, are essential to assume appropriate tertiary and quaternary structures accountable for various cellular functions as discussed above. For some of the biologically important proteins, cysteines are involved in redox activity (Netto et al., 2006).

Antioxidant mechanisms against OS induced ROS, comprising of enzymes namely catalase, peroxiredoxin, and superoxide dismutase, as well as other antioxidant mechanisms, are largely dependent on sulfur-containing amino acids like cysteine and methionine in proteins and nonprotein cofactors (Perkins, 2015). More important is cysteine-containing tripeptide, glutathione (GSH) rich within cells (McBean, 2017). It is a well-known fact that the redox potential of sulfur regulates many cell functions viz. signaling, development, survival, and cell death and cysteine and methionine modifications (Lushchak, 2012). It disturbs intracellular sulfhydryl homeostasis affecting protein function or stability which may be responsible for anomalous cell physiology, aging, and many diseases, like arthritis, cancer, cardiovascular disorders, diabetes, and neurodegenerative diseases (Gelain, 2012; Sabens Liedhegner et al., 2012). Thus, the literature analysis reveals that reactive cysteine residues play an important role in pathogenetic events of AD.

CYSTEINE AND ITS REACTIVITY

Cysteine is sulfur (S) containing amino acid amalgamated in protein. The S atom in cysteine contributes to sensitive sulfhydryl (–SH) formation, which determines its polarity (Lu, 2009). The sulfhydryl moiety being highly reactive reducing agents have great potential of reducing many proteins, accountable for various cellular functions. Cysteine is a semi-essential amino acid, produced from methionine. Connective tissue, cell membranes, and the myelin sheaths composed of cysteines as an important building block are the protective structural components of neurons, which protects them from OS and severe environmental situations (Sameem et al., 2019). Under normal physiological circumstances, the cysteine side chain act as the most reactive nucleophile amongst all amino acids, which is associated with –SH functional group. The pKa value of cysteine-thiol is approximately 4 to 9 based on the protein structure and the surrounding. Thus, the thiol group in cysteine is mildly acidic. The pKa value of cysteine-thiol is related to its reactivity (Tajc et al., 2004). However, the reactivity of cysteine in the deprotonated thiolate anion (RS^-) form is much higher than thiol form. Thus, the thiol side chain of cysteine in proteins may undergo spontaneous oxidation, known as post-translational modifications to yield sulfenic acids, disulfides, sulphinic acids and sulphonic acids (McBean et al., 2015).

The fact that the cellular respiratory process involves sulfur compounds, explains the importance of oxidation of cysteine in it. Cysteine-thiol being highly reactive undergo oxidation to form disulfide bonds (Trivedi, 2009). This disulfide bond formation through cysteines in proteins makes the native protein conformation stable. Many such oxidative modifications through cysteines in proteins regulate human physiological processes (Ahmad et al., 2017).

Cysteine residues present in proteins can be documented in four functional forms such as structural, metal binding, catalytic, and regulatory (Fomenko, 2007). Cysteine residues in proteins via oxidative disulfide bond formation plays a leading structural role throughout protein folding. S-palmitoylation, an enzymatically regulated oxidative modification of cysteine residues controls protein site in membranes, function, and stability (Linder and Deschenes, 2007). Protein functionality and structural stability are monitored through cysteine residues coordinated metal binding (Marino and Gladyshev, 2012). Cysteine residues act as a catalyst in various protein families, namely oxidoreductases, proteases, and acyltransferases. Finally, reversible oxidation products of cysteine residues, i.e. sulfenic acids, as well covalent addition, like S-nitrosylation, S-glutathionylation, and S-sulfhydrylation enable oxidation-reduction based cell signaling regulating protein function (Poole, 2008; Paulsen, 2010; Winterbourn, 2008). Thus, the highest chemical reactivity of cysteine-thiols allows distinctive modifications that are expressively involved in the homeostatic regulation of redox sensing, signaling, protein function, stability, and trafficking in cells (Gould et al., 2013).

Disulfide bond (inter and intra-chain) formation through cysteines in the protein plays a critical part in protein (Banerjee, 2012). Usually, proteins involved in distribution or integration in the cell membrane and important in cell metabolism, protein folding, assembly, and stability contain a disulfide bond. Furthermore, it is evident that thioredoxin (Lillig and Holmgren, 2007) and glutathione (Vitvitsky et al., 2006), the cysteine-derived proteins, are a major player in the prevention of OS (Sperandio et al., 2005).

In addition to the formation of sulfenic acids, disulfides, sulphinic acids, and sulphonic acids through oxidative modification of cysteines, other post-translational modifications occurring frequently during aging are deamidation, racemization, phosphorylation, methylation, glycoxidation, and oxidation, as well as conformational changes without chemical deviations (Goto and Radak, 2013). Amongst all these

Figure 2. Chemical modifications to cysteine residues
(Paget and Buttner, 2003).

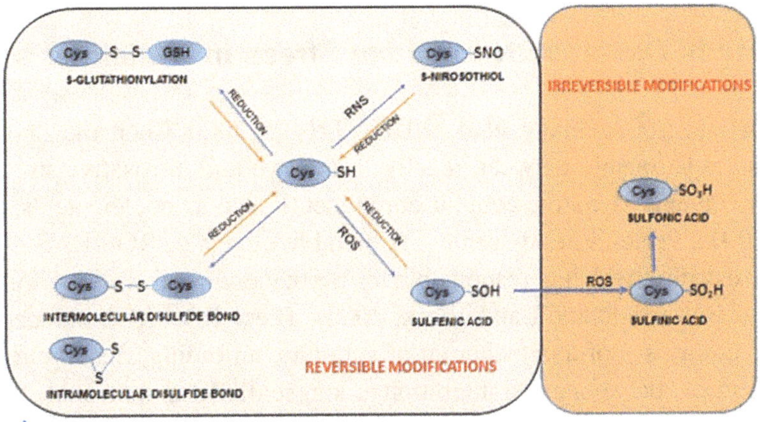

modifications, oxidation is the utmost important modification occurring in a cell in response to ROS produced in mitochondria during oxygen metabolism (Ahmad et al., 2017).

ROS/RNS (reactive nitrogen species) in cells mainly targets high nucleophilic cysteine side chains. Due to this the S atom of cysteine may adopt different oxidation states and can take part in diverse biochemical reactions. In the thiolate form, sulfur is oxidized to produce a sulfenic acid (RSOH) (Winther and Thorpe, 2014). Being the first oxidation product, RSOH is a major indicator of ROS/RNS-susceptible cysteine residues and also plays an important part in cell signaling as evident. This sulfenic acid reacts spontaneously with proximal inter-and intra-molecular thiols groups to produce disulfide bonds (R-S-S-R′). However, in the absence of such thiol groups, sulfenic acid slowly undergoes an irreversible oxidative reaction to form sulfinic (R-SO_2H) and sulfonic acid (R-SO_3H) (Lowther and Haynes, 2011). In some cases, sulfenic acids may react with its one more molecule to produce thiosulfate or may react with amine or amides to produce sulfenamides. With the formation of S-nitrosothiol (R-SNO), and persulfide through the interaction of thiolate anions with highly reactive nitrogen and sulfur species (RNS and RSS), the cysteine reactivity appears further complicated. Additionally, it must be understood that this cysteine mediated alteration depicted in Figure 2 is reactive and are interconvertible with one another (Poole et al., 2004; Poole, 2015).

The functional capacity and stability of many proteins, especially secretory proteins are regulated through the formation of disulfide bonds. RSOH, can also react with tripeptide glutathione (GSH) in a process known as S-glutathionylation (Gallogly and Mieyal, 2007). This S-glutathionylation may contribute to chronic disorders as well as aging (Moreno et al., 2014). The disulfide bonds formed with the two cysteine residues during protein maturation in the cell can be either catalytic or structural. The structural disulfide bond is responsible for the stabilization of protein structure, which remains unaffected throughout the life of the protein (Meitzler et al., 2013). While the main role of the catalytic disulfide bond is to facilitate thiol-disulfide interchange reactions in proteins. Currently, the third type of disulfide bond, involved in protein function in which they exist in i.e. allosteric disulfides, has been identified (Chiu and Hogg, 2019). Allosteric disulfide bond formation occurs during the oxidation of cysteine residues modifies protein function via reversible redox transformation. Accordingly, allosteric disulfide bonds can be a noble marker in new drug discovery of drugs for various ailments like aging, neurological disorder, and inflammation (Wouters et al., 2010).

Role of Cysteines in Oxidative/Nitrosative Stress in Alzheimer's Disease

Many studies (Butterfield, 2002; Huang, et al., 2016; Christen, 2000; Chen and Zhong, 2014) have now established the causal relationship between reactive oxidative and nitrosative stress and pathogenesis of AD, which mainly damage dynamic cellular components such as nucleic acids, lipids, and proteins (Butterfield et al., 2001). Proteins are extremely vulnerable targets of ROS/RNS disturbing secondary and tertiary protein structure, which is responsible for permanent modification of protein structure and accordingly function (Butterfield and Lauderback, 2002). These ROS/RNS induced irreversible modifications of proteins comprises of a detachment of subunits, unfolding, the revelation of hydrophobic residues, accumulation and backbone disintegration amongst all (Tramutola et al., 2017). As mentioned earlier, the cysteine residues play structural, metal binding, catalytic, and regulatory roles in proteins and are the main target of many post-translational modifications. It is well known that in response to OS, the organisms develop a response system that neutralizes ROS responsively or restore oxidative damage (Dasuri et al., 2013). For this, transducing proteins (mostly transcription factors) works like switches that can be stimulated or inhibited in response to ROS (Paget and Buttner, 2003). Proteins with redox-active amino acids, mainly and frequently cysteine, are the major component of these response system which can directly react with oxidants or oxidized cellular products (Kumsta and Jakob, 2009). Reversible oxidation of cysteine thiols in the form of sulfenic acid (RSOH), disulfide bonds (R–S–S–R), diverse disulfide bond with glutathione (R–S–SG), and over oxidation to sulfinic acids ($R\text{-}SO_2H$) are mainly involved in redox regulation of proteins which is essential for life (Paget and Buttner, 2003).

AD is known to be associated with synaptic damage and neuronal degeneration on exposure to excessive ROS/RNS. The neurodegeneration consequent to OS is due to an imbalance in oxidant production and antioxidant mechanisms (Clark, 2010). Under normal conditions, antioxidant system evolved by organisms, that reduce concentration of ROS/RNS in the neurons include cysteine- based redox proteins like glutaredoxin (Grx), peroxiredoxin (Prx), and thioredoxin (Trx) glutathione (GSH), etc., as well as transcriptional conduits exemplified by Keap1/Nrf2 and Hsp90/HSF1 (Lei et al., 2016). These low levels of reactive oxygen species stimulate explicit signaling pathways that regulate normal cell signaling and thus is neuroprotective in function. In contrast, under pathological situations as in AD and Parkinson's disease (PD), there is increased production of ROS and decreased antioxidant mechanisms responsible for cell damage in neurodegeneration (Akhtar et al., 2012).

Role of Cysteine in Post-Translational Modifications in Alzheimer's Disease

Post-translational modifications (PTMs) are chemical modifications that occur after protein biosynthesis include enzymatic covalent modification of proteins and play an important part in regulating protein function, position, and communication with other cellular proteins, nucleic acids, lipids and cofactors (Duan and Walther, 2015). Cysteine residing in thiolate anions in protein at physiological pH is the target for many PTMs including oxidation/reduction, palmitoylation, glutathionylation, guanylation, cysteinylation, sumoylation, farnesylation, prenylation, nitrosylation, and sulfhydration amongst all. Of all mentioned PTMs, prenylation, nitrosylation, palmitoylation and oxidation/reduction mainly target cysteines on proteins (Ferrer et al., 2011). In addition, Cysteine also acts as the substrate for the production of gasotransmitter hydrogen sulfide (H_2S), which controls numerous physiological mechanisms (Paul et al., 2018).

In recent times, two similar PTMs namely sulfhydration and nitrosylation have gained attention. In sulfhydration, cysteine thiol group is transformed to persulfide (SSH) through reaction with H_2S, while, in nitrosylation, thiol group of cysteine is converted to nitrosothiol group (SNO) (Mustafa et al., 2009). H_2S plays an important role in neurophysiology. Hence, controlled synthesis is required for ideal neuronal functions. It has been recognized that many neurodegenerative diseases viz. AD, PD, amyotrophic lateral sclerosis (ALS), and Huntington's disease (HD) are associated with irregular H_2S signaling (Paul et al., 2018). Oxidative post-translational modification (Ox-PTM) of cysteine residues in response to ROS/RNS is a major event that controls protein structure and function. Due to high nucleophilic nature of cysteine side chain, it exists in different oxidation states (Reddie and Carroll, 2008), which allows different Ox-PTMs of cysteine residues like S-nitrosylation, sulfhydration, S-glutathionylation, disulfide bonds, sulfenylation, sulfinic acid and sulfonic acid (Chung et al., 2013).

Cellular endogenous antioxidant mechanisms comprise enzymes such as catalase, peroxiredoxin, and superoxide dismutase that neutralize ROS (Kurutas, 2016). However, many other cellular antioxidant mechanisms mostly depend on cysteines in proteins and non-protein cofactors, mainly cysteine dependent tripeptide glutathione (GSH), which resides in cells abundantly. Accordingly, oxidation-reduction prominence of cysteine-thiols is a prime regulator of cell signaling, development, existence, and cell death. PTMs of cysteine residues modify protein function or stability and disturb thiol homeostasis inside the cell. Therefore, disturbance of the antioxidant systems and oxidation-reduction status of sulfur is interconnected to anomalous cell functioning, aging, and neurodegenerative diseases (Uttara et al., 2009). Reversible thiol mediated PTMs like intermolecular and intramolecular disulfides, sulfenic acids, and glutathione mixed disulfides (protein–SSG) safeguard the susceptible proteins from irreversible overoxidation induced PTMs, which permit the proteins to adapt normal function after a decrease in OS (Paget and Buttner, 2003). While, irreversible Ox-PTMs, including sulfinic and sulfonic acid, cause protein degradation. Some forms of the peroxiredoxins catalyzed hyper oxidation to the sulfinic acid can be reversed by sulfiredoxin. Though glutaredoxin is mainly involved in protein-SSG reduction, sulfiredoxin is known to catalyze protein deglutathionylation in exceptional circumstances (Rhee et al., 2007). Reversible Ox-PTMs of cysteine residues on proteins are restored primarily by facilitating thiol-disulfide exchange reactions via two thiol-disulfide oxidoreductase enzymes (TDORs) namely, thioredoxin (Trx) and glutaredoxin (Grx) (Bechtel and Weerapana, 2017). Trx mainly catalyzes intramolecular and intermolecular disulfide bonds reduction, while, Grx is mainly involved in protein-glutathione mixed disulfides reduction. This indicates diverse, but complementary functions of thioredoxin and glutaredoxin enzyme to act synergistically for cellular thiol status (Hanschmann et al., 2013).

It has been reported that thioredoxin reductase (and not thioredoxin) was reduced, while superoxide dismutase, catalase, and glutathione reductase, which are OS-responsive cysteine-containing enzymes, was increased in AD (Holmgren, 1989). Accordingly, it has been postulated that increased antioxidant enzyme levels due to stimulation of antioxidant genes may be opposed by Ox-PTMs which decrease thioredoxin reductase enzymes. For example, S-glutathionylation of endothelial nitric oxide synthase (eNOS) can switch its function from generation of nitric oxide (NO) to superoxide (Chen et al., 2010); also during nitrosative stress, S-nitrosylation of NADPH oxidase in neutrophils causes its inhibition to stop superoxide generation (Fujii et al., 1997). In summary, above-mentioned findings justify disturbance in intracellular redox homeostasis is a contributory factor in neurodegeneration (Sabens Liedhegner et al., 2012).

S-Nitrosylation

Many readings have offered convincing indications linking S-nitrosylation with AD pathology. Numerous proteins hazardous for synaptic function and neuronal existence and endurance have been known to undergo abnormal S-nitrosylation in AD leading to synaptic loss and ultimately neurodegeneration (Zhao et al., 2015). S-nitrosylation is one of the prime oxidative posttranslational modifications contributing to neuronal dysfunction as well as neurodegenerative changes in AD. The binding of nitric oxide (NO) and its similar form with specific protein by covalent bond through cysteine residues forming nitrosothiols (SNOs) is known as S-nitrosylation. S-nitrosylation plays a neuroprotective and /or neurodestructive role which depends on protein affected. Anomalous S-nitrosylation of proteins can cause disturbed protein conformation, mitochondrial division, synaptic injury, or apoptosis, leading to the neuropathological changes of AD (Nakamura et al., 2013; Selfridge et al., 2013).

Production of RNS or nitrosonium cation (NO^+), has been known for their ability to bring about reversible protein S-nitrosylation. Like other PTMs, S-nitrosylation also regulates various biological functions of proteins including their folding, functions, interactions with other proteins, aggregation, and localization, definitely disturbing cellular signal transduction mechanisms altering neuronal function under normal circumstances (Qu et al., 2011). Moreover, S-nitrosylation can also affect disulfide and palmitoylation reactions (Cho et al., 2009). Under pathological stress, abnormal S-nitrosylation may cause disturbed protein conformation, endoplasmic reticulum stress, and mitochondrial dysfunction, facilitating cellular obliteration and synaptic loss involved in the pathogenesis of AD (Nakamura and Lipton, 2007). Nevertheless, S-nitrosylation of N-methyl-D-aspartate receptors (NMDARs) and cysteine protease caspase may show neuroprotective characteristics in AD, as it certainly stimulates neuronal survival and inhibits apoptotic cell death. These findings are supported through SNO proteome technology coupled with bioinformatics as well as by identifying many S-nitrosylated proteins in the brains of AD patients. Thus, a comparatively low concentration of NO is neuroprotective while high concentrations of NO cause neuronal damage via abnormal S-nitrosylation of numerous cysteine-containing proteins (Zhao et al., 2015).

Fascinatingly, the S-nitrosylation of proteins could be inhibited by denitrosylation, a process which recompense S-nitrosylation with the help of S-nitrosoglutathione reductase, thioredoxins, alcohol dehydrogenase class III, and protein disulfide isomerase (PDI), which may take out NO from nitrosothiols and thus reduce nitrosative stress in pathological states (Jeon et al., 2013).

Currently, it has been recognized that S-nitrosylation enzymes take part in protein-protein transnitrosylation reactions. Transnitrosylation is the enzyme-mediated reaction to form SNO-proteins (Nakamura et al., 2013). Transnitrosylation reactions involve denitrosylation as well as S-nitrosylation, wherein NO group is taken out from one protein (donor protein) and is attached to another protein (acceptor protein) through reactive cysteine residue to yield denitrosylated and S-nitrosylated forms of protein. Ultimately, transnitrosylation reaction each time is accompanied by denitrosylation simultaneously. Thus far, proteins involved in transnitrosylation reactions during AD comprise of nitrosylated glyceraldehyde-3-phosphate dehydrogenase (GAPDH), caspase 3, and cyclin-dependent kinase 5 (Cdk5) (Zhao et al., 2015).

Cdk5 is a primarily explicit neuronal kinase, that does not take part in controlling cell cycle but then plays a vigorous role in moderating cell existence, neuronal distinction and migration, axon guidance, and synaptic plasticity (Nakamura et al., 2013). Moreover, exposure to ROS, RNS, Aβ, and neuroinflammation, has also been learned to extremely stimulate Cdk5 and activate numerous proceedings accompanying neurodegeneration. The current report has demonstrated that endogenous NO species can S-nitrosylate

Cdk5 at Cys 83 /157, which stimulates Cdk5 triggering Aβ/NMDAR-induced dendritic spine loss that is directly proportional to extent of cognitive deterioration in AD (Qu et al., 2012).

The dynamic equilibrium between mitochondrial fission and fusion is important for mitochondrial integrity, assuring sustained ATP production at critical neuronal sites. Dynamin-related protein 1 (Drp1) is a GTPase vital for typical mitochondrial fission, which regulates synaptic activities in the brain. It has been confirmed that Aβ oligomers stimulate disproportionate mitochondrial fission contributing to synaptic dysfunction and neuronal damage. S-nitrosylation of Drp1 at Cys 644 can abnormally trigger its GTPase activity facilitating dimer formation, which leads to undue mitochondrial disintegration, bioenergetics deficiency, and finally dendritic spine loss and neuronal death (Cho et al., 2009).

The endoplasmic reticulum (ER) primarily is involved in protein processing and folding. However, aggregation of undeveloped or misfit proteins precipitates ER stress during neurodegenerative progression. ER stress associated with AD makes cells to release resistance proteins such as PDI and glucose-regulated protein (GRP) to improve chaperone and isomerase activity (Roussel et al., 2013). PDI enzyme facilitates disulfide bond formation, reorganizes reactions, and preserves fundamental stability, as well as behaves as a molecular chaperone to help secretory protein development, transportation, and folding and also prevents unusual protein clumping. But, S-nitrosylation of a cysteine residue of PDI (SNO-PDI) inhibits this protective effect compromising its enzymatic activity triggering the accumulation of poly-ubiquitinated proteins (Uehara et al., 2006). Additionally, SNO-PDI can trigger ER stress enhancing neuronal cell death through misfolded proteins (Zhao et al., 2015).

The insulin-degrading enzyme (IDE) is a zinc metalloprotease that is involved in the degradation of many critical peptides like insulin and Aβ. Nevertheless, S-nitrosylation of IDE at numerous cysteine residues, Cys819, Cys789, Cys966, and Cys178 decreased its degradative activity which diminishes the degradation of both insulin and Aβ and therefore can progressively affect pathological neurodegeneration (Kruszelnicka, 2014).

ApoE an important protein involved in the regulation of lipid levels, is vital in regulating membrane integrity, synaptic function and dendritic restructuring in the brain. It has been well established that LOAD is mainly characterized by *APOE* gene polymorphisms, as a major genetic risk factor. Abrams (2011) has suggested that S-nitrosylation of ApoE may cause conformational fluctuations, which affect low-density lipoprotein (LDL) receptors binding affinity. However, more investigation is required to establish a link between S-nitrosylation of ApoE and initiation or progression of AD (Abrams et al., 2011).

MAP1B is a protein that is exceedingly expressed in developing neurons. It stabilizes the axonal cytoskeleton and regulates axonal extension. S-Nitrosylation of MAP1B at Cys2457 alters the protein folding, augmenting microtubule-binding affinity, which inhibits dynein activity to decrease the axonal extension force causing axonal retraction. Additionally, S-nitrosylated LC1 of MAP1B can enable self-degradation, inhibiting mitochondrial dysfunction and neuronal cell death (Stroissnigg et al., 2007).

Tubulins, a vital constituent of microtubules control the neuronal microtubule cytoskeleton. These tubulins are highly susceptible to various PTMs, like detyrosination, acetylation, phosphorylation, polyglutamylation, and polyglycylation. The recent findings have demonstrated that S-nitrosylation of tubulin cysteine residues via S-nitrosoglutathione may trigger interchain disulfide bond formation, which mainly inhibits microtubule polymerization, precipitating cytoskeletal abnormalities, which are major pathological findings in AD (Zahid et al., 2013).

GAPDH is a multifunctional housekeeping glycolytic enzyme which makes it a vulnerable target for ROS/RNS. S-nitrosylation of GAPDH has been detected in various brain regions, like the hippocampus, substantia nigra, and cortex in AD patients compared with the aged patients group. S-nitrosylated

GAPDH can cause apoptosis. GOSPEL, a cytosolic protein, normally binds GAPDH to preserve it in the cytoplasm, limit its access in the nucleus, and thereby prevent neuron damage. Moreover, it has been reported that S-nitrosylated of GOSPEL at Cys 47 encourage GAPDH-GOSPEL association and neuroprotective effects of GOSPEL (Sirover, 2013).

Voltage-dependent anion-selective channel (VDAC) is a channel-forming protein in the outer mitochondrial and postsynaptic membrane. It is involved in ionic movement across the mitochondrial membrane as well as mitochondria-mediated apoptosis and synaptic transmission (Manczak and Reddy, 2012). VDAC family comprises of VDAC1, VDAC2, and VDAC3. Amongst all, VADC1 and VDAC2 were expressively measured while VDAC3 was not found in AD patient's brains. Oxidative and nitrosative stress is known to alter VDAC-mediated mitochondrial membrane permeability disturbing mitochondrial calcium homeostasis and thereby ATP production, cytochrome c translocation, and activation of caspase, ultimately causing apoptotic cell death and synaptic dysfunction. The literature report has demonstrated hyper-S-nitrosylation of VDAC1 at cys 127 and 232 and VDAC2 at Cys 8 in the hippocampus, substantia nigra, and cortex of AD patients (De Pinto et al., 2016).

Superoxide dismutase (SOD2) is a mitochondrial antioxidant protein that neutralizes free radicals and thereby prevents oxidation mediated inactivation of NO. This limits peroxynitrite formation and mitochondrial dysfunction (Fukai and Ushio-Fukai, 2011). Zhao et al, (Zhao et al., 2015) has reported that S-nitrosylation of SOD2 at cysteine residues in AD disturbs mitochondrial antioxidant security enzymes.

Heat-shock protein (HSP) 90 is a molecular chaperone that is involved in protein folding, preservation, intracellular transport, and degradation along with modulating cell signaling proteins like NOS and calcineurin (Mollapour and Neckers, 2012). It is a well-known fact that aggregation of misfolded proteins can exacerbate the development of neurodegeneration due to their neurotoxicity. S-nitrosylation of HSP90 at Cys 597 may prevent its ATPase action, which makes HSP90 to work as a molecular chaperone. This S-nitrosylation of HSP90 in neurons may trigger the buildup of tau and Aβ masses, which are prominent pathologic structures of AD (Zhao et al., 2015).

Fructose bisphosphate aldolase C (ALDOC) is an enzyme regulating glucose metabolism to alter ATP generation and is highly prone to the chemical modulations. Recently S-nitrosylation of ALDOC has been reported in the hippocampus of AD patients. The significantly high concentration of ALDOC has been detected in AD, signifying a compensatory response to the above referred PTM. S-nitrosylation of ALDOC cause decreased ATP production and ultimately neuronal death. In addition, ATP reduction can precipitate hypothermia, which activates abnormal tau hyperphosphorylation further contributing to the pathogenesis of AD (Zahid et al., 2013).

Caspases are cysteine proteases that implement apoptotic cell death and facilitate excitotoxic injury during neurodegeneration associated with AD. It has been stated that S-nitrosylation of caspases can decline enzyme action, and thereby inhibit apoptotic cell death (Mannick et al., 2001). However, mitochondrial thioredoxin-2 denitrosylate caspase-3 selectively making catalytic site to function, thus exert a neuroprotective effect. In contrast, S-nitrosylation of Trx1 at Cys73 has been stated to transnitrosylate caspase 3 (Sengupta and Holmgren, 2013). Thus, S-nitrosylation of caspases prevents neuronal apoptosis and may further facilitate AD progression.

It is now recognized fact that continued stimulation of neuronal NMDA receptors can initiate intracellular signaling cascade, involving large calcium ion inflow, excess production of NO, and anomalous enzymatic action, that contribute to neurodegeneration. S-nitrosylation of the NMDARs has been reported to reduce NMDAR mediated intracellular signaling cascade, revealing pathological changes

Figure 3. Effects of S-nitrosylation of crucial proteins contributing to Alzheimer's pathology (Zhao et al., 2015).

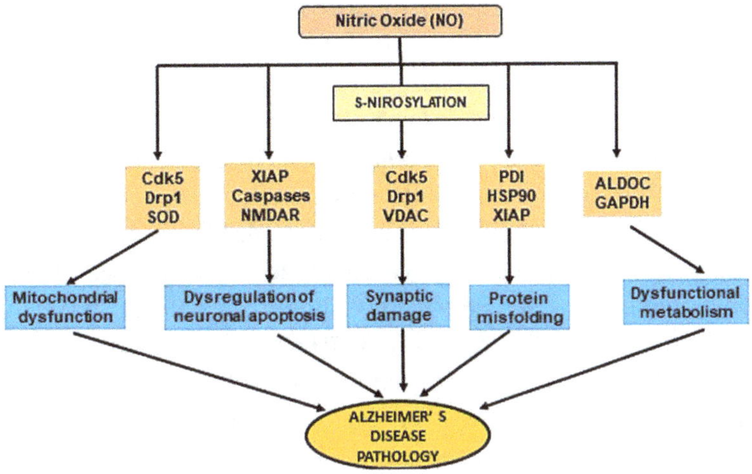

in AD. Furthermore, it has been revealed that polynitrosylation of NMDAR at Cys744, Cys798, Cys87, Cys320, and Cys399 also reduces activity (Choi et al., 2000).

Above mentioned abnormal S-nitrosylation of crucial proteins (Figure 3) contributing to AD pathology unlocks viewpoint to know AD and to discover novel therapeutic strategies. It can be summarized that several S-nitrosylated proteins including SNO-CDk5, SNO-Drp1, and SNO-NMDA may signify probable therapeutic targets to prevent the development of AD (Qu et al., 2012).

S-nitrosylation of NMDAR and caspases decrease NO production and contribute to neuroprotective effects. While S-nitrosylation of Drp1 and Cdk5 trigger undue mitochondrial division causing synaptic injury in AD. Moreover, S-nitrosylation of HSP-90 and PDI may lead to protein misfit and clump formation in aged neurons. Moreover, S-nitrosylation of ALDOC and GAPDH disturb the energy metabolism initiating apoptosis. Thus, this cysteine-thiol mediated redox reaction could alter metabolism which is involved in AD pathology (Zhao et al., 2015).

S- Glutathionylation

Glutathione (GSH) is an abundant intracellular antioxidant constituent. Maximum GSH occurs in the cell cytoplasm, and the residual amount of GSH is also present in cellular organelles. During glutathionylation, the reduced form (GSH) is transformed into the oxidized form (GSSG). It safeguards the cellular proteins from ROS, and also from ultraviolet radiation by preserving them in reduced (–SH) form, which is needed for normal protein function (Taylor et al., 1996).

Various enzymes including glutathione peroxidase (Prx), glutaredoxin (Grx), and thioredoxin (Trx), are involved in GSH mediated redox regulation reactions (Townsend et al., 2003). The three main roles of GSH are it works as the measure of the cellular redox reactions, antioxidant defense mechanisms, and the storage and transportation of cysteine (Sparaco et al., 2006). During glutathione mediated antioxidant reactions, H_2O_2 and lipid hydroperoxides are reduced in the presence of glutathione peroxidases, through oxidation of glutathione (GSH) to glutathione disulfide (GSSG). Glutathione protects cells from oxidative injury via non-enzymatic mechanisms as well. As OS has been identified as a key pathogenic factor in AD, GSH appears to be one of the important protective components during OS involved AD (Dalle-Donne et al., 2007). During OS, various cellular proteins and modulators including glutaredoxins and glutathione transferases with cysteine-thiols may undergo oxidative PTMs which helps proteins to bear OS and regulates the cellular protein functions (Cha et al., 2017).

Glutathionylation is one such protective redox modification of proteins which safeguards proteins from irreversible oxidative modifications. Glutathionylation takes place via disulfide bond formation with glutathione at the cysteine-thiol residue of cellular protein either non-enzymatically or enzymatically (Dalle-Donne et al., 2007). Glutathionylation itself is a reversible PTM wherein glutathione is freed from cysteine thiols of affected protein by glutaredoxin and thioredoxin. Till date, following cellular proteins undergoing glutathionylation have been recognized (Gallogly and Mieyal, 2007).

Glyceraldehyde-3-phosphate dehydrogenase (GAPDH) and α-ketoglutarate dehydrogenase complex (KGDHC), are involved in the glycolytic pathway and Krebs cycle, respectively which are major energy metabolic pathways in the cell. Their function is regulated by glutathionylation. S-glutathionylation of GAPDH at Cys149 in endothelial cells inactivate enzymatic activity of GADPH. Likewise, mitochondrial KGDH activity is also suppressed by glutathionylation in response to hydrogen peroxide (Cha et al., 2017).

Tau proteins play an important role in microtubule polymerization and preservation which regulate neurite extension and axonal development. Tau with its Cys 322 residue binds with microtubules. Dysregulation of tau protein has been associated with AD. It has been reported that tau Cys322 gets oxidized that inhibits tau dimer formation (Di Noto et al., 2005). Specifically, polymerization of the S-glutathionylation of three-repeat tau may alter filament assembly. Yet, further research is required to validate the role of glutathionylation to prevent aggregation of neurofibrillary tangles in AD (Liedhegner et al., 2012).

In corroboration with OS theory of AD pathology, many antioxidants like vitamin E (Cente et al., 2009), vitamin C and curcumin (Park et al., 2008) were studied against oxidative damage in AD and were reported to possess potential to halt tauopathy in AD. Joy et al., (2018) explored and evaluated the probable neuroprotective potential of N-acetyl cysteine (NAC), a glutathione precursor in colchicine-induced AD animal model and revealed that antioxidant treatment with NAC inhibits tauopathies in the neurons, thus decreasing neuronal loss and cognitive impairment in AD. Accordingly, the concept was established that increasing antioxidant cover may be helpful to protect from premature AD signs. Thus, NAC which is a glutathione precursor would be the best remedy for AD (Joy et al., 2018). NAC is a cysteine derivative, which acts as a precursor in the synthesis endogenous antioxidant substance, glutathione. NAC being membrane-permeable, does not need active transport. Inside the cell, NAC is promptly hydrolyzed to cysteine which restores total glutathione content. Also, NAC is a direct ROS scavenger (Kerksick and Willoughby, 2005). It has been well established that prolonged administration of NAC escalates ATP and GSH concentration and thereby reduce lipid and protein oxidation in presynaptic nerve terminals (Banaclocha, 2001).

S-Palmitoylation

Protein palmitoylation is PTM which forms thioester linkage between the sulfhydryl group of a cysteine residue of protein and carboxyl group of palmitic acid in the presence of palmitoyl acyltransferases with conserved Asp-His-His-Cys (DHHC) motif. Protein palmitoylation primarily makes protein hydrophobic and thus increases their interaction with cell membrane components and other hydrophobic domains of proteins (Cho and Park, 2016). Palmitoylation is the only reversible modification that influences protein function viz. cellular signaling. Fascinatingly, palmitoylation is the utmost frequently detected modification in neuronal cells regulating neuronal function (Fukata and Fukata, 2010).

Palmitoylation of synaptosomal-associated protein-25 (SNAP25), a presynaptic protein regulating vesicular exocytosis, neurite outgrowth, and long-term potentiation grasped the attention of researchers to focus on Palmitoylation (Cho and Park, 2016).

APP palmitoylation being the utmost frequent lipid modification among neuronal proteins and also recent evidence established participation of palmitoylation in the pathogenesis of AD which is known to control the amyloidogenic process. It was found that the palmitoylation of APP takes place at Cys186 and 187 in the presence of DHHC7 and DHHC21. Palmitoylated APP augmented lipid rafts act as the best substrate for β-secretase (BACE1) than α-secretase which facilitates cleavage of APP producing Aβ BACE1 has also been reported to be palmitoylated at Cys478, Cys482, and Cys485 (Bhattacharyya et al., 2013). Currently, it is unconvincing to state palmitoylation of specific proteins involved in AD. However further studies are desired to elucidate the role of palmitoylation in AD.

RECENT DEVELOPMENTS AND FUTURE RESEARCH DIRECTIONS

Based on the role of Aβ plaques and tau tangles in the pathogenesis of AD, interfering with actions of enzymes such as β-secretase, γ-secretase and kinases is a sustainable approach for the development of new drugs for AD. Recent strategies to target such enzymes have been associated with many untoward effects altering brain functions. The amyloid hypothesis is accepted as a convincing target for the development of diagnostic and therapeutic interventions, it may be proposed to design molecular clamps which will inhibit toxic effects of the enzymes like β-secretase/γ-secretase without affecting their normal physiological functions.

Many interventions like natural substances, peptidomimetics, and metal chelators were studied and used for acting at a specified target like Aβ plaques and tau tangles. But the results were not favorable. This made investigators work on the development of multifunctional intervention targeting not only Aβ plaques and tau tangles but also other pathological targets in AD-like ROS and OS, inflammation, mitochondrial dysfunction, etc. Understanding the role of cysteines in AD pathology, researchers are confident to bring new effective therapeutic strategies for AD in the future.

CONCLUSION

Cysteine-dependent proteins such as proteases, antioxidant enzymes, kinases, phosphatases, and other types of enzymes as well as other non-enzymatic proteins are known to be related to the neurodegenerative process involved in AD. Cysteine with sulfhydryl moiety is a highly reactive reducing agent, with great potential of reducing many proteins, accountable for various cellular functions in the brain. The increased production of ROS and decreased antioxidant mechanisms are responsible for cell damage and proteinopathies in AD. The fact that the cellular respiratory process involves sulfur compounds, explains the importance of oxidation of cysteine in it. ROS/RNS in cells mainly targets high nucleophilic cysteine side chains. Cysteine residing in thiolate anions in protein at physiological pH is a target for many

PTMs viz. prenylation, nitrosylation, palmitoylation, and oxidation/reduction. Thus, cysteine thiols are an important player in monitoring neurodegeneration in AD.

REFERENCES

Abrams, A. J., Farooq, A., & Wang, G. (2011). S-Nitrosylation of ApoE in Alzheimer's disease. *Biochemistry, 50*(17), 3405–3407. doi:10.1021/bi200266v PMID:21443265

Ahmad, S., Khan, H., Shahab, U., Rehman, S., Rafi, Z., Khan, M. Y., ... Uddin, M. (2017). Protein oxidation: An overview of metabolism of sulphur containing amino acid, cysteine. *Frontiers in Bioscience, 1*(1), 71–87. doi:10.2741474 PMID:27814576

Akhtar, M. W., Sunico, C. R., Nakamura, T., & Lipton, S. A. (2012). Redox Regulation of Protein Function via Cysteine S-Nitrosylation and Its Relevance to Neurodegenerative Diseases. *International Journal of Cell Biology, 9*, 1–9. doi:10.1155/2012/463756 PMID:22956959

Alzheimer's Association. (2018). Alzheimer's Disease Facts and Figures. *Alzheimer's & Dementia, 14*(3), 367–429. doi:10.1016/j.jalz.2018.02.001

Alzheimer's Disease International, World Alzheimer Report. (2018). *The state of the art of dementia research: New frontiers*. London: Author.

Banerjee, R. (2012). Redox outside the Box: Linking Extracellular Redox Remodeling with Intracellular Redox Metabolism. *The Journal of Biological Chemistry, 287*(7), 4397–4402. doi:10.1074/jbc.R111.287995 PMID:22147695

Barghorn, S., & Mandelkow, E. (2002). Toward a unified scheme for the aggregation of tau into Alzheimer paired helical filaments. *Biochemistry, 41*(50), 14885–14896. doi:10.1021/bi026469j PMID:12475237

Bechtel, T. J., & Weerapana, E. (2017). From structure to redox: The diverse functional roles of disulfides and implications in disease. *Proteomics, 17*(6). doi:10.1002/pmic.201600391

Bhadbhade, A., & Cheng, D. W. (2012). Amyloid Precursor Protein Processing in Alzheimer's Disease. *Iranian Journal of Child Neurology, 6*, 1–5.

Bhattacharyya, R., Barren, C., & Kovacs, D. M. (2013). Palmitoylation of amyloid precursor protein regulates amyloidogenic processing in lipid rafts. *The Journal of Neuroscience: The Official Journal of the Society for Neuroscience, 33*(27), 11169–11183. doi:10.1523/JNEUROSCI.4704-12.2013 PMID:23825420

Binder, L. I., Guillozet-Bongaarts, A. L., Garcia-Sierra, F., & Berry, R. W. (2005). Tau, tangles, and Alzheimer's disease. *Biochimica et Biophysica Acta 3, 1739*(2-3),216-23.

Bloom, G. S. (2014). Amyloid-β and tau: The trigger and bullet in Alzheimer disease pathogenesis. *JAMA Neurology, 71*(4), 505–508. doi:10.1001/jamaneurol.2013.5847 PMID:24493463

Braak, H., & Del, T. K. (2012). Where, when, and in what form does sporadic Alzheimer's disease begin. *Current Opinion in Neurology, 25*(6), 708–714. doi:10.1097/WCO.0b013e32835a3432 PMID:23160422

Brion, J. P. (1998). Neurofibrillary Tangles and Alzheimer's Disease. *European Neurology, 40*(3), 130–140. doi:10.1159/000007969 PMID:9748670

Burns, A., & Iliffe, S. (2009). Alzheimer's disease. *BMJ (Clinical Research Ed.), 338*(feb05 1), b158. doi:10.1136/bmj.b158 PMID:19196745

Butterfield, D. A. (2002). Amyloid β-peptide (1–42)-induced oxidative stress and neurotoxicity: Implications for neurodegeneration in Alzheimer's disease brain. A review. *Free Radical Research, 36*(12), 1307–1313. doi:10.1080/1071576021000049890 PMID:12607822

Butterfield, D. A., Drake, J., Pocernich, C., & Castegna, A. (2001). Evidence of oxidative damage in Alzheimer's disease brain: Central role for amyloid beta-peptide. *Trends in Molecular Medicine, 7*(12), 548–554. doi:10.1016/S1471-4914(01)02173-6 PMID:11733217

Butterfield, D. A., & Lauderback, C. M. (2002). Lipid peroxidation and protein oxidation in Alzheimer's disease brain: Potential causes and consequences involving amyloid betapeptide- associated free radical oxidative stress. *Free Radical Biology & Medicine, 32*, 1050–1060. doi:10.1016/S0891-5849(02)00794-3 PMID:12031889

Cha, S. J., Kim, H. C., Hyun-Jun, L., & Sanghyun, K. K. (2017). Protein Glutathionylation in the Pathogenesis of Neurodegenerative Diseases. *Oxidative Medicine and Cellular Longevity, 2017*, 1–9. doi:10.1155/2017/2818565 PMID:29456785

Chauhan, V., & Chauhan, A. (2006). Oxidative stress in Alzheimer's disease. *Pathophysiology, 13*(3), 195–208. doi:10.1016/j.pathophys.2006.05.004 PMID:16781128

Chen, C. A., Wang, T. Y., Varadharaj, S., Reyes, L. A., Hemann, C., Talukder, M. A., ... Zweier, J. L. (2010). S-glutathionylation uncouples eNOS and regulates its cellular and vascular function. *Nature, 468*(7327), 1115–1118. doi:10.1038/nature09599 PMID:21179168

Chen, Z., & Zhong, C. (2014). Oxidative stress in Alzheimer's disease. *Neuroscience Bulletin, 30*(2), 271–281. doi:10.100712264-013-1423-y PMID:24664866

Chiu, J., & Hogg, P. J. (2019). Allosteric disulfides: Sophisticated molecular structures enabling flexible protein regulation. *The Journal of Biological Chemistry, 294*(8), 2949–2960. doi:10.1074/jbc.REV118.005604 PMID:30635401

Cho, D. H., Nakamura, T., Fang, J., Cieplak, P., Godzik, A., Gu, Z., & Lipton, S. A. (2009). S-Nitrosylation of Drp1 mediates beta-amyloid-related mitochondrial fission and neuronal injury. *Science, 324*(5923), 102–105. doi:10.1126cience.1171091 PMID:19342591

Cho, D. H., Nakamura, T., Fang, J., Cieplak, P., Godzik, A., Gu, Z., & Lipton, S. A. (2009). S-Nitrosylation of Drp1 mediates beta-amyloid-related mitochondrial fission and neuronal injury. *Science, 324*(5923), 102–105. doi:10.1126cience.1171091 PMID:19342591

Cho, E., & Park, M. (2016). Palmitoylation in Alzheimer's disease and other neurodegenerative diseases. *Pharmacological Research, 111*, 133–151. doi:10.1016/j.phrs.2016.06.008 PMID:27293050

Cho, E., & Park, M. (2016). Palmitoylation in Alzheimer's disease and other neurodegenerative diseases. *Pharmacological Research, 111*, 133–151. doi:10.1016/j.phrs.2016.06.008 PMID:27293050

Choi, Y. B., Tenneti, L., Le, D. A., Ortiz, J., Bai, G., Chen, H. S., & Lipton, S. A. (2000). Molecular basis of NMDA receptor-coupled ion channel modulation by S-nitrosylation. *Nature Neuroscience*, *3*(1), 15–21. doi:10.1038/71090 PMID:10607390

Choudhury, A., Chakraborty, I., Bhattacharjee, R., Banerjee, T. S., Vana, D. R., Adapa, D., & Bhardwaj, R. (2018). An update on pathological implications of enzymatic dysregulation in Alzheimer's disease. *Biomedical Research*, *29*(10), 2215–2226. doi:10.4066/biomedicalresearch.29-18-767

Christen, Y. (2000). Oxidative stress and Alzheimer disease. *The American Journal of Clinical Nutrition*, *71*(2), 621S–629S. doi:10.1093/ajcn/71.2.621s PMID:10681270

Chung, H. S., Wang, S. B., Venkatraman, V., Murray, C. I., & Van Eyk, J. E. (2013). Cysteine oxidative posttranslational modifications: Emerging regulation in the cardiovascular system. *Circulation Research*, *112*(2), 382–392. doi:10.1161/CIRCRESAHA.112.268680 PMID:23329793

Chung, H. S., Wang, S. B., Venkatraman, V., Murray, C. I., & Van Eyk, J. E. (2013). Cysteine oxidative posttranslational modifications: Emerging regulation in the cardiovascular system. *Circulation Research*, *112*(2), 382–392. doi:10.1161/CIRCRESAHA.112.268680 PMID:23329793

Citron, M., Teplow, D. B., & Selkoe, D. J. (1995). Generation of amyloid beta protein from its precursor is sequence specific. *Neuron*, *14*(3), 661–670. doi:10.1016/0896-6273(95)90323-2 PMID:7695913

Clark, T. A., Lee, H. P., Rolston, R. K., Zhu, X., Marlatt, M. W., Castellani, R. J., ... Perry, G. (2010). Oxidative Stress and its Implications for Future Treatments and Management of Alzheimer Disease. *International Journal of Biomedical Science; IJBS*, *6*, 225–227. PMID:21765811

Cooke, M. S., Evans, M. D., Dizdaroglu, M., & Lunec, J. (2003). Oxidative DNA damage: Mechanisms, mutation, and disease. *The FASEB Journal*, *17*(10), 1195–1214. doi:10.1096/fj.02-0752rev PMID:12832285

Dalle-Donne, I., Rossi, R., Giustarini, D., Colombo, R., & Milzani, A. (2007). S-glutathionylation in protein redox regulation. *Free Radical Biology & Medicine*, *44*(6), 883–898. doi:10.1016/j.freeradbiomed.2007.06.014 PMID:17697933

Dasuri, K., Zhang, L., & Keller, J. N. (2013). Oxidative stress, neurodegeneration, and the balance of protein degradation and protein synthesis. *Free Radical Biology & Medicine*, *62*, 170–185. doi:10.1016/j.freeradbiomed.2012.09.016 PMID:23000246

David, F. T., Keith, A. J., & Ronald, C. P. (2005). Alzheimer's disease – overview. In M. F. Beal, A. E. Lang, & A. C. Ludolph (Eds.), *Neurodegenerative Diseases Neurobiology, Pathogenesis and Therapeutics* (pp. 416–432). New York: University Press.

Devi, L., & Ohno, M. (2012). Mitochondrial dysfunction and accumulation of the β-secretase-cleaved C-terminal fragment of APP in Alzheimer's disease transgenic mice. *Neurobiology of Disease*, *45*(1), 417–424. doi:10.1016/j.nbd.2011.09.001 PMID:21933711

Di, N. (2005). Structural insights into Alzheimer filament assembly pathways based on site directed mutagenesis and S-glutathionylation of three repeat neuronal Tau protein. *Microscopy Research and Technique*, *67*(3-4), 156–163. doi:10.1002/jemt.20195 PMID:16104002

Dong, S., Duan, Y., Hu, Y., & Zhao, Z. (2012). Advances in the pathogenesis of Alzheimer's disease: A re-evaluation of amyloid cascade hypothesis. *Translational Neurodegeneration*, *1*(1), 18. doi:10.1186/2047-9158-1-18 PMID:23210692

Dorszewska, J., Prendeck, M., Oczkowsk, A., Dezo, M., & Kozubski, W. (2016). Molecular Basis of Familial and Sporadic Alzheimer's Disease. *Current Alzheimer Research*, *13*(9), 1–12. doi:10.2174/1567205013666160314150501 PMID:26971934

Dröge, W. (2002). Free radicals in the physiological control of cell function. *Physiological Reviews*, *82*(1), 47–95. doi:10.1152/physrev.00018.2001 PMID:11773609

Duan, G., & Walther, D. (2015). The roles of post-translational modifications in the context of protein interaction networks. *PLoS Computational Biology*, *11*(2), e1004049. doi:10.1371/journal.pcbi.1004049 PMID:25692714

Ferrer-Sueta, G., Manta, B., Botti, H., Radi, R., Trujillo, M., & Denicola, A. (2011). Factors affecting protein thiol reactivity and specificity in peroxide reduction. *Chemical Research in Toxicology*, *24*(4), 434–450. doi:10.1021/tx100413v PMID:21391663

Finkel, T., & Holbrook, N. J. (2000). Oxidants, oxidative stress and the biology of ageing. *Nature*, *408*(6809), 239–247. doi:10.1038/35041687 PMID:11089981

Fomenko, D. E., Xing, W., Adair, B. M., Thomas, D. J., & Gladyshev, V. N. (2007). High throughput identification of catalytic redox-active cysteine residues. *Science*, *315*(5810), 387–389. doi:10.1126cience.1133114 PMID:17234949

Förstl, H., & Kurz, A. (1999). Clinical features of Alzheimer's disease. *European Archives of Psychiatry and Clinical Neuroscience*, *249*(6), 288–299. doi:10.1007004060050101 PMID:10653284

Fujii, H., Ichimori, K., Hoshiai, K., & Nakazawa, H. (1997). Nitric oxide inactivates NADPH oxidase in pig neutrophils by inhibiting its assembling process. *The Journal of Biological Chemistry*, *272*(52), 32773–32778. doi:10.1074/jbc.272.52.32773 PMID:9407051

Fukai, T., & Ushio-Fukai, M. (2011). Superoxide dismutases: Role in redox signaling, vascular function, and diseases. *Antioxidants & Redox Signalling*, *15*(6), 1583–1606. doi:10.1089/ars.2011.3999 PMID:21473702

Fukata, Y., & Fukata, M. (2010). Protein palmitoylation in neuronal development and synaptic plasticity. *Nature Reviews. Neuroscience*, *11*(3), 161–175. doi:10.1038/nrn2788 PMID:20168314

Gallogly, M. M., & Mieyal, J. J. (2007). Mechanisms of reversible protein glutathionylation in redox signaling and oxidative stress. *Current Opinion in Pharmacology*, *7*(4), 381–391. doi:10.1016/j.coph.2007.06.003 PMID:17662654

Gallogly, M. M., & Mieyal, J. J. (2007). Mechanisms of reversible protein glutathionylation in redox signaling and oxidative stress. *Current Opinion in Pharmacology*, *7*(4), 381–391. doi:10.1016/j.coph.2007.06.003 PMID:17662654

Ganguly, G., Chakrabarti, S., Chatterjee, U., & Saso, L. (2017). Proteinopathy, oxidative stress and mitochondrial dysfunction: Cross talk in Alzheimer's disease and Parkinson's disease. *Drug Design, Development and Therapy, 11*, 797–810. doi:10.2147/DDDT.S130514 PMID:28352155

Gelain, D. P., Antonio Behr, G., Birnfeld de Oliveira, R., & Trujillo, M. (2012). Antioxidant therapies for neurodegenerative diseases: Mechanisms, current trends, and perspectives. *Oxidative Medicine and Cellular Longevity, 895153*, 1–2. doi:10.1155/2012/895153 PMID:23304258

Gong, C. X., & Iqbal, K. (2008). Hyperphosphorylation of microtubule-associated protein tau: A promising therapeutic target for Alzheimer disease. *Current Medicinal Chemistry, 15*(23), 2321–2328. doi:10.2174/092986708785909111 PMID:18855662

Goto, S., & Radak, Z. (2013). Implications of oxidative damage to proteins and DNA in aging and its intervention by caloric restriction and exercise. *Journal of Sport and Health Science, 2*(2), 75–80. doi:10.1016/j.jshs.2013.03.004

Gould, N., Doulias, P. T., Tenopoulou, M., Raju, K., & Ischiropoulos, H. (2013). Regulation of Protein Function and Signaling by Reversible Cysteine S-Nitrosylation. *The Journal of Biological Chemistry, 288*(37), 26473–26479. doi:10.1074/jbc.R113.460261 PMID:23861393

Grundke-Iqbal, I., Iqbal, K., Tung, Y. C., Quinlan, M., Wisniewski, H. M., & Binder, L. I. (1986). Abnormal phosphorylation of the microtubule-associated protein tau (tau) in Alzheimer cytoskeletal pathology. *Proceedings of the National Academy of Sciences of the United States of America, 83*(13), 4913–4917. doi:10.1073/pnas.83.13.4913 PMID:3088567

Halliwell, B. (1992). Free radicals in the brain. *Journal of Neurochemistry, 59*(5), 1609–1623. doi:10.1111/j.1471-4159.1992.tb10990.x PMID:1402908

Hanschmann, E. M., Godoy, J. R., Berndt, C., Hudemann, C., & Lillig, C. H. (2013). Thioredoxins, glutaredoxins, and peroxiredoxins--molecular mechanisms and health significance: from cofactors to antioxidants to redox signaling. *Antioxidants & Redox Signaling, 19*(13), 1539–1605. doi:10.1089/ars.2012.4599

Hippius, H., & Neundörfer, G. (2003). The discovery of Alzheimer's disease. *Dialogues in Clinical Neuroscience, 5*(1), 101–108. PMID:22034141

Holmgren, A. (1989). Thioredoxin and glutaredoxin systems. *The Journal of Biological Chemistry, 264*, 13963–13966. PMID:2668278

Huang, W. J., Zhang, X., & Chen, W. W. (2016). Role of oxidative stress in Alzheimer's disease. *Biomedical Reports, 4*(5), 519–522. doi:10.3892/br.2016.630 PMID:27123241

Huang, X., Moir, R. D., Tanzi, R. E., Bush, A. I., & Rogers, J. T. (2004). Redox-active metals, oxidative stress, and Alzheimer's disease pathology. *Annals of the New York Academy of Sciences, 1012*(1), 153–163. doi:10.1196/annals.1306.012 PMID:15105262

Iqbal, K., Liu, F., Gong, C. X., & Grundke-Iqbal, I. (2010). Tau in Alzheimer Disease and Related Tauopathies. *Current Alzheimer Research, 7*(8), 656–664. doi:10.2174/156720510793611592 PMID:20678074

Jacob, C., Knight, I., & Winyard, P. G. (2006). Aspects of the biological redox chemistry of cysteine: From simple redox responses to sophisticated signalling pathways. *Biological Chemistry*, *387*(10/11), 1385–1397. doi:10.1515/BC.2006.174 PMID:17081111

Jeon, G. S., Nakamura, T., Lee, J. S., Choi, W. J., Ahn, S. W., Lee, K. W., ... Lipton, S. A. (2013). Potential effect of S-nitrosylated protein disulfide isomerase on mutant SOD1 aggregation and neuronal cell death in amyotrophic lateral sclerosis. *Molecular Neurobiology*, *49*(2), 796–807. doi:10.100712035-013-8562-z PMID:24091828

Kosik, K. S., Joachim, C. L., & Selkoe, D. J. (1986). Microtubule-associated protein tau (tau) is a major antigenic component of paired helical filaments in Alzheimer disease. *Proceedings of the National Academy of Sciences of the United States of America*, *83*(11), 4044–4048. doi:10.1073/pnas.83.11.4044 PMID:2424016

Kruszelnicka, O. (2014). Nitric oxide vs insulin secretion, action and clearance. *Diabetologia*, *57*(1), 257–258. doi:10.100700125-013-3082-y PMID:24135953

Kumar, A., Singh, A., & Ekavali. (2015). A review on Alzheimer's disease pathophysiology and its management: An update. *Pharmacological Reports*, *67*(2), 195–203. doi:10.1016/j.pharep.2014.09.004 PMID:25712639

Kumsta, C., & Jakob, U. (2009). Redox-regulated chaperones. *Biochemistry*, *48*(22), 4666–4676. doi:10.1021/bi9003556 PMID:19368357

Kurutas, E. B. (2016). The importance of antioxidants which play the role in cellular response against oxidative/nitrosative stress: Current state. *Nutrition Journal*, *15*(1), 71. doi:10.118612937-016-0186-5 PMID:27456681

Leeuwenburgh, C., & Heinecke, J. W. (2001). Oxidative stress and antioxidants in exercise. *Current Medicinal Chemistry*, *8*(7), 829–838. doi:10.2174/0929867013372896 PMID:11375753

Lei, X. G., Zhu, J. H., Cheng, W. H., Bao, Y., Ho, Y. S., Reddi, A. R., ... Arnér, E. S. (2016). Paradoxical Roles of Antioxidant Enzymes: Basic Mechanisms and Health Implications. *Physiological Reviews*, *96*(1), 307–364. doi:10.1152/physrev.00010.2014 PMID:26681794

Lichtenthaler, F. S. (2012). Alpha-secretase cleavage of the amyloid precursor protein: Proteolysis regulated by signaling pathways and protein trafficking. *Current Alzheimer Research*, *9*, 165–177. doi:10.2174/156720512799361655 PMID:21605033

Lillig, C. H., & Holmgren, A. (2007). Thioredoxin and related molecules: From biology to health and disease. *Antioxidants & Redox Signalling*, *9*(1), 25–47. doi:10.1089/ars.2007.9.25 PMID:17115886

Linder, M. E., & Deschenes, R. J. (2007). Palmitoylation: Policing protein stability and traffic. *Nature Reviews. Molecular Cell Biology*, *8*(1), 74–84. doi:10.1038/nrm2084 PMID:17183362

Lowther, W. T., & Haynes, A. C. (2011). Reduction of cysteine sulfinic acid in eukaryotic, typical 2-Cys peroxiredoxins by sulfiredoxin. *Antioxidants & Redox Signalling*, *15*(1), 99–109. doi:10.1089/ars.2010.3564 PMID:20712415

Lu, S. C. (2009). Regulation of glutathione synthesis. *Molecular Aspects of Medicine*, *30*(1-2), 42–59. doi:10.1016/j.mam.2008.05.005 PMID:18601945

Lushchak, V. I. (2012). Glutathione Homeostasis and Functions: Potential Targets for Medical Interventions. *Journal of Amino Acids*, *26*, 1–26. doi:10.1155/2012/736837 PMID:22500213

Maloney, M. T. (2015). One Hundred Years of Alzheimer's Disease: The Amyloid Cascade Hypothesis. *Nature Education*, *8*(4), 6.

Manczak, M., & Reddy, P. H. (2012). Abnormal interaction of VDAC1 with amyloid beta and phosphorylated tau causes mitochondrial dysfunction in Alzheimer's disease. *Human Molecular Genetics*, *21*(23), 5131–5146. doi:10.1093/hmg/dds360 PMID:22926141

Mannick, J. B., Schonhoff, C., Papeta, N., Ghafourifar, P., Szibor, M., Fang, K., & Gaston, B. (2001). S-Nitrosylation of mitochondrial caspases. *The Journal of Cell Biology*, *154*(6), 1111–1116. doi:10.1083/jcb.200104008 PMID:11551979

Marino, S. M., & Gladyshev, V. N. (2012). Analysis and functional prediction of reactive cysteine residues. *The Journal of Biological Chemistry*, *287*(7), 4419–4425. doi:10.1074/jbc.R111.275578 PMID:22157013

Markesbery, W. R. (1997). Oxidative stress hypothesis in Alzheimer's disease. *Free Radical Biology & Medicine*, *23*(1), 134–147. doi:10.1016/S0891-5849(96)00629-6 PMID:9165306

Masters, C. L., Gu, B., Laws, S., Lim, Y. Y., Roberts, B., Villemagne, V. L., & Beyreuther, K. (2018). Alzheimer's Disease: Toward a Quantitative Biological Approach in Describing its Natural History and Underlying Mechanisms. In M. S. Wolfe (Ed.), *The Molecular and Cellular Basis of Neurodegenerative Diseases* (pp. 57–79). London: Academic Press. doi:10.1016/B978-0-12-811304-2.00003-1

McBean, G. J. (2017). Cysteine, Glutathione, and Thiol Redox Balance in Astrocytes. *Antioxidants (Basel, Switzerland)*, *6*(3), 62. doi:10.3390/antiox6030062 PMID:28771170

McBean, G. J., Aslan, M., Griffiths, H. R., & Torrão, R. C. (2015). Thiol redox homeostasis in neurodegenerative disease. *Redox Biology*, *5*, 186–194. doi:10.1016/j.redox.2015.04.004 PMID:25974624

Meitzler, J. L., Hinde, S., Bánfi, B., Nauseef, W. M., & Ortiz de Montellano, P. R. (2013). Conserved cysteine residues provide a protein-protein interaction surface in dual oxidase (DUOX) proteins. *The Journal of Biological Chemistry*, *288*(10), 7147–7157. doi:10.1074/jbc.M112.414797 PMID:23362256

Mokhtar, S. H., Bakhuraysah, M. M., Cram, D. S., & Petratos, S. (2013). The Beta-amyloid protein of Alzheimer's disease: Communication breakdown by modifying the neuronal cytoskeleton. *International Journal of Alzheimer's Disease*, *910502*, 1–15. doi:10.1155/2013/910502 PMID:24416616

Mollapour, M., & Neckers, L. (2012). Post-translational modifications of Hsp90 and their contributions to chaperone regulation. *Biochimica et Biophysica Acta*, *1823*(3), 648–655. doi:10.1016/j.bbamcr.2011.07.018 PMID:21856339

Moreno, M. L., Escobar, J., Izquierdo-Álvarez, A., Gil, A., Pérez, S., Pereda, J., ... Sastre, J. (2014). Disulfide stress: A novel type of oxidative stress in acute pancreatitis. *Free Radical Biology & Medicine*, *70*, 265–277. doi:10.1016/j.freeradbiomed.2014.01.009 PMID:24456905

Mustafa, A. K., Gadalla, M. M., Sen, N., Kim, S., Mu, W., Gazi, S. K., ... Snyder, S. H. (2009). H2S signals through protein S-sulfhydration. *Science Signaling*, *2*(96), ra72. doi:10.1126cisignal.2000464 PMID:19903941

Nakamura, T., & Lipton, S. A. (2007). S-Nitrosylation and uncompetitive/fast off-rate (UFO) drug therapy in neurodegenerative disorders of protein misfolding. *Cell Death and Differentiation*, *14*(7), 1305–1314. doi:10.1038j.cdd.4402138 PMID:17431424

Nakamura, T., & Lipton, S. A. (2013). Emerging role of protein–protein transnitrosylation in cell signaling pathways. *Antioxidants & Redox Signalling*, *18*(3), 239–249. doi:10.1089/ars.2012.4703 PMID:22657837

Nakamura, T., Tu, S., Akhtar, M. W., Sunico, C. R., Okamoto, S., & Lipton, S. A. (2013). Aberrant protein S-nitrosylation in neurodegenerative diseases. *Neuron*, *78*(4), 596–614. doi:10.1016/j.neuron.2013.05.005 PMID:23719160

Nakamura, T., Tu, S., Akhtar, M. W., Sunico, C. R., Okamoto, S., & Lipton, S. A. (2013). Aberrant protein s-nitrosylation in neurodegenerative diseases. *Neuron*, *78*(4), 596–614. doi:10.1016/j.neuron.2013.05.005 PMID:23719160

National Institutes of Health. National Institute on Aging. (n.d.). *What happens to the Brain in the Alzheimer's Disease?* Available at https://www.nia.nih.gov/health/what-happens-brainalzheimers-disease

Netto, L. E., de Oliveira, M. A., Monteiro, G., Demasi, A. P., Cussiol, J. R., Discola, K. F., ... Horta, B. B. (2007). Reactive cysteine in proteins: Protein folding, antioxidant defense, redox signaling and more. *Comparative Biochemistry and Physiology. Toxicology & Pharmacology: CBP*, *146*(1-2), 180–193. doi:10.1016/j.cbpc.2006.07.014 PMID:17045551

O'Brien, R. J., & Wong, P. C. (2011). Amyloid precursor protein processing and Alzheimer's disease. *Annual Review of Neuroscience*, *34*(1), 185–204. doi:10.1146/annurev-neuro-061010-113613 PMID:21456963

Paget, M. S., & Buttner, M. J. (2003). Thiol-based regulatory switches. *Annual Review of Genetics*, *37*(1), 91–121. doi:10.1146/annurev.genet.37.110801.142538 PMID:14616057

Paul, B. D., Sbodio, J. I., & Snyder, S. H. (2018). Cysteine Metabolism in Neuronal Redox Homeostasis. *Trends in Pharmacological Sciences*, *39*(5), 513–524. doi:10.1016/j.tips.2018.02.007 PMID:29530337

Paulsen, C. E., & Carroll, K. S. (2010). Orchestrating redox signaling networks through regulatory cysteine switches. *ACS Chemical Biology*, *5*(1), 47–62. doi:10.1021/cb900258z PMID:19957967

Perkins, A., Nelson, K. J., Parsonage, D., Poole, L. B., & Karplus, P. A. (2015). Peroxiredoxins: Guardians against oxidative stress and modulators of peroxide signaling. *Trends in Biochemical Sciences*, *40*(8), 435–445. doi:10.1016/j.tibs.2015.05.001 PMID:26067716

Perl, D. (2010). Neuropathology of Alzheimer's Disease. *The Mount Sinai Journal of Medicine, New York*, *77*(1), 32–42. doi:10.1002/msj.20157 PMID:20101720

Poole, L. B. (2015). The basics of thiols and cysteines in redox biology and chemistry. *Free Radical Biology & Medicine*, *80*, 148–157. doi:10.1016/j.freeradbiomed.2014.11.013 PMID:25433365

Poole, L. B., Karplus, P. A., & Claiborne, A. (2004). Protein sulfenic acids in redox signalling. *Annual Review of Pharmacology and Toxicology*, *44*(1), 325–347. doi:10.1146/annurev.pharmtox.44.101802.121735 PMID:14744249

Poole, L. B., & Nelson, K. J. (2008). Discovering mechanisms of signaling-mediated cysteine oxidation. *Current Opinion in Chemical Biology*, *12*(1), 18–24. doi:10.1016/j.cbpa.2008.01.021 PMID:18282483

Qu, J., Nakamura, T., Cao, G., Holland, E. A., McKercher, S. R., & Lipton, S. A. (2011). S-Nitrosylation activates Cdk5 and contributes to synaptic spine loss induced by beta-amyloid peptide. *Proceedings of the National Academy of Sciences of the United States of America*, *108*(34), 14330–14335. doi:10.1073/pnas.1105172108 PMID:21844361

Rajasekhar, K., & Thimmaiah, G. (2018). Current progress, challenges and future prospects of diagnostic and therapeutic interventions in Alzheimer's disease. *RSC Advances*, *8*(42), 23780–23804. doi:10.1039/C8RA03620A

Ray, P. D., Huang, B. W., & Tsuji, Y. (2012). Reactive oxygen species (ROS) homeostasis and redox regulation in cellular signaling. *Cellular Signalling*, *24*(5), 981–990. doi:10.1016/j.cellsig.2012.01.008 PMID:22286106

Ray, W. J., Ashall, F., & Goate, A. M. (1998). Molecular pathogenesis of sporadic and familial forms of Alzheimer's disease. *Molecular Medicine Today*, *4*(4), 151–157. doi:10.1016/S1357-4310(98)01229-5 PMID:9572056

Reddie, K. G., & Carroll, K. S. (2008). Expanding the functional diversity of proteins through cysteine oxidation. *Current Opinion in Chemical Biology*, *12*(6), 746–754. doi:10.1016/j.cbpa.2008.07.028 PMID:18804173

Reina, S., Gupta, A., Messina, A., & Mahalakshmi, R. (2016). Role of cysteines in mammalian VDAC isoforms' function. *Biochimica et Biophysica Acta (BBA) - Bioenergetics*, 1219–1227.

Rhee, S. G., Jeong, W., Chang, T.-S., & Woo, H. A. (2007). Sulfiredoxin, the cysteine sulfinic acid reductase specific to 2-Cys peroxiredoxin: Its discovery, mechanism of action, and biological significance. *Kidney International*, *72*(106), S3–S8. doi:10.1038j.ki.5002380 PMID:17653208

Roher, A. E., Lowenson, J. D., Clarke, S., Woods, A. S., Cotter, R. J., Gowing, E., & Ball, M. J. (1993). Beta-Amyloid-(1-42) is a major component of cerebrovascular amyloid deposits: Implications for the pathology of Alzheimer disease. *Proceedings of the National Academy of Sciences of the United States of America*, *90*(22), 10836–10840. doi:10.1073/pnas.90.22.10836 PMID:8248178

Roussel, B. D., Kruppa, A. J., Miranda, E., Crowther, D. C., Lomas, D. A., & Marciniak, S. J. (2013). Endoplasmic reticulum dysfunction in neurological disease. *Lancet Neurology*, *12*(1), 105–118. doi:10.1016/S1474-4422(12)70238-7 PMID:23237905

Sabens, L. (2012). Mechanisms of Altered Redox Regulation in Neurodegenerative Diseases—Focus on S-Glutathionylation. *Antioxidants & Redox Signalling*, *16*(6), 543–566. doi:10.1089/ars.2011.4119 PMID:22066468

Sajjad, R., Arif, R., Shah, A. A., Manzoor, I., & Mustafa, G. (2018). Pathogenesis of Alzheimer's Disease: Role of Amyloid-beta and Hyperphosphorylated Tau Protein. *Indian Journal of Pharmaceutical Sciences*, *80*(4), 581–591. doi:10.4172/pharmaceutical-sciences.1000397

Salami, O., & Lyketsos, C. G. (2010). Clinical Features of Alzheimer's Disease: Cognitive and Non-Cognitive. In M. T. Abou-Saleh, C. Katona, & A. Kumar (Eds.), *Principles and Practice of Geriatric Psychiatry* (pp. 226–231). John Wiley & Sons, Inc. doi:10.1002/9780470669600.ch41

Sameem, B., Khan, F., & Niaz, K. (2019). l-Cysteine. In S. M. Nabavi & A. S. Silva (Eds.), Nonvitamin and Nonmineral Nutritional Supplements (pp. 53-58). Cambridge, MA: Academic Press.

Schweers, O., Mandelkow, E. M., Biernat, J., & Mandelkow, E. (1995). Oxidation of Cysteine-322 in the repeat domain of microtubule-associated protein Tau controls the in vitro assembly of paired helical filaments. *Proceedings of the National Academy of Sciences of the United States of America*, *92*(18), 8463–8467. doi:10.1073/pnas.92.18.8463 PMID:7667312

Selfridge, J. E., e, L., Lu, J., & Swerdlow, R. H. (2013). Role of mitochondrial homeostasis and dynamics in Alzheimer's disease. *Neurobiology of Disease*, *51*, 3–12. doi:10.1016/j.nbd.2011.12.057 PMID:22266017

Selkoe, D. J. (1998). The cell biology of beta-amyloid precursor protein and presenilin in Alzheimer's disease. *Trends in Cell Biology*, *8*(11), 447–453. doi:10.1016/S0962-8924(98)01363-4 PMID:9854312

Sengupta, R., & Holmgren, A. (2013). Thioredoxin and thioredoxin reductase in relation to reversible S-nitrosylation. *Antioxidants & Redox Signalling*, *18*(3), 259–269. doi:10.1089/ars.2012.4716 PMID:22702224

Serrano-Pozo, A., Frosch, M. P., Masliah, E., & Hyman, B. T. (2011). Neuropathological Alterations in Alzheimer Disease. *Cold Spring Harbor Perspectives in Medicine*, *1*(1), a006189. doi:10.1101/cshperspect.a006189 PMID:22229116

Seubert, P., Vigo-Pelfrey, C., Esch, F., Lee, M., Dovey, H., Davis, D., ... Schenk, D. (1992). Isolation and quantification of soluble Alzheimer's [beta]-peptide from biological fluids. *Nature*, *359*(6393), 325–327. doi:10.1038/359325a0 PMID:1406936

Shankar, G. M., & Walsh, D. M. (2009). Alzheimer's disease: Synaptic dysfunction and Abeta. *Molecular Neurodegeneration*, *4*(1), 48. doi:10.1186/1750-1326-4-48 PMID:19930651

Sheldon, R. A., & Kochi, J. K. (2012). *Metal-catalyzed oxidations of organic compounds: mechanistic principles and synthetic methodology including biochemical processes*. New York, NY: Elsevier.

Sirover, M. A. (2013). GAPDH: β-Amyloid Mediated Iron Accumulation in Alzheimer's Disease: A New Paradigm for Oxidative Stress Induction in Neurodegenerative Disorders. In D. Praticò & P. Mecocci (Eds.), *Studies on Alzheimer's Disease. Oxidative Stress in Applied Basic Research and Clinical Practice*. Totowa, NJ: Humana Press. doi:10.1007/978-1-62703-598-9_4

Snyder, S. W., Ladror, U. S., Wade, W. S., Wang, G. T., Barrett, L. W., Matayoshi, E. D., ... Holzman, T. F. (1994). Amyloid-beta aggregation: Selective inhibition of aggregation in mixtures of amyloid with different chain lengths. *Biophysical Journal*, *67*(3), 1216–1228. doi:10.1016/S0006-3495(94)80591-0 PMID:7811936

Solomon, D., Arunodhayan, S., Henri, C., Sushmitha, C., Paartha, M., & Timothy, N. (2015). Neuron the Memory Unit of the Brain. *IOSR Journal of Computer Engineering*, *17*, 48–61.

Sparaco, M., Gaeta, L. M., Tozzi, G., Bertini, E., Pastore, A., Simonati, A., ... Piemonte, F. (2006). Protein glutathionylation in human central nervous system: Potential role in redox regulation of neuronal defense against free radicals. *Journal of Neuroscience Research*, *83*(2), 256–263. doi:10.1002/jnr.20729 PMID:16385584

Sperandio, B. P., Polard, D. S., Ehrlich, P., Renault, E., & Guedon, E. (2005). Guédon: Sulfur Amino Acid Metabolism and Its Control in Lactococcus lactis IL1403. *Journal of Bacteriology*, *187*(11), 3762–3778. doi:10.1128/JB.187.11.3762-3778.2005 PMID:15901700

Stroissnigg, H., Trancikova, A., Descovich, L., Fuhrmann, J., Kutschera, W., Kostan, J., ... Propst, F. (2007). S-Nitrosylation of microtubule-associated protein 1B mediates nitric-oxide-induced axon retraction. *Nature Cell Biology*, *9*(9), 1035–1045. doi:10.1038/ncb1625 PMID:17704770

Tajc, G., Tolbert, B. S., Basavappa, R., & Miller, B. L. (2004). Direct determination of thiol pKa by isothermal titration microcalorimetry. *Journal of the American Chemical Society*, *126*(34), 10508–10509. doi:10.1021/ja047929u PMID:15327286

Taylor, C. G., Nagy, L. E., & Bray, T. M. (1996). Nutritional and hormonal regulation of glutathione homeostasis. *Current Topics in Cellular Regulation*, *34*, 189–208. doi:10.1016/S0070-2137(96)80007-0 PMID:8646848

Thinakaran, G., & Koo, E. H. (2008). Amyloid Precursor Protein Trafficking, Processing, and Function. *The Journal of Biological Chemistry*, *283*(44), 29615–29619. doi:10.1074/jbc.R800019200 PMID:18650430

Townsend, D. M., Tew, K. D., & Tapiero, H. (2003). The importance of glutathione in human disease. *Biomedicine and Pharmacotherapy*, *57*(3-4), 145–155. doi:10.1016/S0753-3322(03)00043-X PMID:12818476

Tramutola, A., Lanzillotta, C., Perluigi, M., & Butterfield, D. A. (2017). Oxidative Stress, Protein Modification and Alzheimer Disease. *Brain Research Bulletin*, *133*, 88–96. doi:10.1016/j.brainresbull.2016.06.005 PMID:27316747

Trivedi, M. V., Laurence, J. S., & Siahaan, T. J. (2009). The role of thiols and disulfides in protein chemical and physical stability. *Current Protein & Peptide Science*, *10*(6), 614–625. doi:10.2174/138920309789630534 PMID:19538140

Uehara, T., Nakamura, T., Yao, D., Shi, Z. Q., Gu, Z., Ma, Y., ... Lipton, S. A. (2006). S-Nitrosylated protein-disulphide isomerase links protein misfolding to neurodegeneration. *Nature*, *441*(7092), 513–517. doi:10.1038/nature04782 PMID:16724068

Uttara, B., Singh, A. V., Zamboni, P., & Mahajan, R. T. (2009). Oxidative stress and neurodegenerative diseases: A review of upstream and downstream antioxidant therapeutic options. *Current Neuropharmacology*, *7*(1), 65–74. doi:10.2174/157015909787602823 PMID:19721819

Varadarajan, S., Yatin, S., Aksenova, M., & Butterfield, D. A. (2000). Review: Alzheimer's amyloid β-peptide-associated free radical oxidative stress and neurotoxicity. *Journal of Structural Biology, 130*(2-3), 184–208. doi:10.1006/jsbi.2000.4274 PMID:10940225

Vitvitsky, V., Thomas, M., Ghorpade, A., Gendelman, H. E., & Banerjee, R. (2006). A functional trans-sulfuration pathway in the brain links to glutathione homeostasis. *The Journal of Biological Chemistry, 281*(47), 35785–35793. doi:10.1074/jbc.M602799200 PMID:17005561

Winterbourn, C. C., & Hampton, M. B. (2008). Thiol chemistry and specificity in redox signaling. *Free Radical Biology & Medicine, 45*(5), 549–561. doi:10.1016/j.freeradbiomed.2008.05.004 PMID:18544350

Winther, J. R., & Thorpe, C. (2014). Quantification of thiols and disulfides. *Biochimica et Biophysica Acta, 1840*(2), 838–846. doi:10.1016/j.bbagen.2013.03.031 PMID:23567800

Wood, J. G., Mirra, S. S., Pollock, N. J., & Binder, L. I. (1986). Neurofibrillary tangles of Alzheimer disease share antigenic determinants with the axonal microtubule-associated protein tau (tau). *Proceedings of the National Academy of Sciences of the United States of America, 83*(11), 4040–4043. doi:10.1073/pnas.83.11.4040 PMID:2424015

Wouters, M. A., Fan, S. W., & Haworth, N. L. (2010). Disulfides as redox switches: From molecular mechanisms to functional significance. *Antioxidants & Redox Signalling, 12*(1), 53–91. doi:10.1089/ars.2009.2510 PMID:19634988

Wouters, M. A., Fan, S. W., & Haworth, N. L. (2010). Disulfides as Redox Switches: From Molecular Mechanisms to Functional Significance. *Antioxidants & Redox Signalling, 12*(1), 53–83. doi:10.1089/ars.2009.2510 PMID:19634988

Zahid, S., Khan, R., Oellerich, M., Ahmed, N., & Asif, A. R. (2013). Differential S-nitrosylation of proteins in Alzheimer's disease. *Neuroscience, 256C*, 126–136. PMID:24157928

Zhang, Y. W., Thompson, R., Zhang, H., & Xu, H. (2011). APP processing in Alzheimer's disease. *Molecular Brain, 4*(1), 3. doi:10.1186/1756-6606-4-3 PMID:21214928

Zhao, Q. F., Yu, J. T., & Tan, L. (2015). S-Nitrosylation in Alzheimer's disease. *Molecular Neurobiology, 51*(1), 268–280. doi:10.100712035-014-8672-2 PMID:24664522

Chapter 14
Molecular Chaperones in Neurodegeneration:
Mechanisms of Regulation in Cellular Proteostasis

Mukesh Pandey
https://orcid.org/0000-0003-4474-0417
Department of Pharmaceutics, Delhi Institute of Pharmaceutical Sciences and Research, India

Jahangir Nabi
https://orcid.org/0000-0002-5212-9063
Department of Pharmaceutical Sciences (Pharmacology Division), Faculty of Applied Sciences and Technology, University of Kashmir, Srinagar, India

Nahida Tabassum
Department of Pharmaceutical Sciences (Pharmacology Division), Faculty of Applied Sciences and Technology, University of Kashmir, Srinagar, India

Faheem Hyder Pottoo
Department of Pharmacology, College of Clinical Pharmacy, Imam Abdulrahman Bin Faisal University, Saudi Arabia

Renuka Khatik
Hefei National Laboratory of Physical Sciences at the Microscale, University of Science and Technology of China, China

Niyaz Ahmad
https://orcid.org/0000-0003-1874-6006
Department of Pharmaceutics, College of Clinical Pharmacy, Imam Abdul Rahman Bin Faisal University, Saudi Arabia

ABSTRACT

Cellular chaperones are essential players to this protein quality control network that functions to prevent protein misfolding, refold misfolded proteins, or degrade them, thereby maintaining neuronal proteostasis. Moreover, overexpression of cellular chaperones is considered to inhibit protein aggregation and apoptosis in various experimental models of neurodegeneration. Alterations or downregulation of chaperone machinery by age-related decline, molecular crowding, or genetic mutations are regarded as key pathological hallmarks of neurodegenerative disorders like Alzheimer's disease (AD), Parkinson's disease (PD), amyotrophic lateral sclerosis (ALS), Huntington's disease (HD), and Prion diseases. Therefore, chaperones may serve as potential therapeutic targets in these diseases. This chapter presents a generalized view of misfolding and aggregation of proteins in neurodegeneration and then critically analyses some of the known cellular chaperones and their role in several neurodegenerative disorders.

DOI: 10.4018/978-1-7998-1317-0.ch014

INTRODUCTION

Cells are the basic unit of life and protein acts as the building block of these basic units thus, within the cell protein homeostasis (proteostasis) is very crucial. Cells have a very developed and sophisticated quality control system called the protein quality control (PQC) system to accomplish proteostasis. This PQC system promotes proteins folding in a proper manner and assists in protein degradation when required. The PQC system involves various biological pathways that are very important for post-mitotic cells, including neurons. Under normal conditions, synthesized polypeptides undergo folding for the formation of native functional conformation. Generation and accumulation of non-functional and misfolded proteins either due to mutations or proteins synthesis faults or inefficient folding of nascent proteins can cause cell toxicity leading to cell death. To prevent these various groups of specific cellular proteins called chaperones were implemented. These chaperones are the primary defense mechanism of a cell against the misfolding of proteins and their subsequent aggregation (Hartl et al., 2011).

Disruption of PQC mechanisms, presence of misfolded proteins along with cellular chaperones and degradation processes are the most common pathological hallmarks of many protein misfolding diseases, including neurodegenerative disorders (Stefani et al., 2003). The quality control machinery from chaperone has a very important role in assembling a protein molecule in three-dimensional conformation. Chaperones are present in an abundant number in the cytoplasm and within the cytoplasmic meshwork, they form complexes with various cytoskeletal components, as well as connected to a large number of other proteins (Soti et al., 2002). Driven by adenosine triphosphate (ATP) cellular chaperones function to prevent misfolding of proteins, engage in the loose folding operation, help fold and refold damaged proteins. Therefore, in protein quality control system chaperones are key components (Tiroli-Cepeda et al., 2011). Proteotoxic stress, age-related decline, molecular overcrowding, and mutations can disrupt the original folding pattern of protein molecules (Sharma et al., 2009). Chaperone complex disruption leads to instability and inaccurate subcellular signaling protein localization (Pratt et al., 2003). Thus, the failure of this essential mechanism of quality control of cellular proteins was reported to have a significant contribution to the pathology of neurodegenerative disorders (Maiti et al., 2014).

Neurodegenerative diseases are the result of progressive degeneration of neurons of the brain and spinal cord. Loss of a particular population of neurons within the central nervous system determines which type of the neurodegenerative disease will occur and describe its clinical manifestations (Pottoo et al., 2014; Nigar et al., 2016; Pottoo et al., 2019). The occurrence of diseases may be sporadic or may be due to familial history. However, in both cases, the progression of disease follows similar etiology and share common symptoms. Thus, the dysfunction and death of neurons are quite comparable with each other in sporadic and familial cases (Smith et al., 2015). Formation of aggregates of toxic proteins and their accumulation is the most common pathological feature reported in many neurodegenerative diseases which suggests the disruption of proteostasis in these diseases. Several mutations have been recognized as the reasons for misfolding of disease-related proteins and their aggregation within the cellular matrix in familial cases of these diseases. Hence, aggregation of misfolded proteins in such disorders can be associated with the failure of a chaperone based protein quality control system. In sporadic cases, aging is regarded as the most important risk factor (Smith et al., 2015). Age-related decline in PQC results in imbalanced production of misfolded proteins, that slowly overwhelm the capability of chaperones to eliminate such proteins. Thus, ultimately proteostasis collapses which results in neurodegeneration at the end (Hartl et al., 2011). Recent studies reported an aging-related decline in the ATP dependent chaperones in the human brain. This finding in combination with overall changes in the balance of the chaperone system may initiate a series of events that ultimately lead to age and neurodegeneration related disruption of proteostasis (Brehme et al., 2014). Besides this cellular senescence as observed in neurons, also disturbs proteostasis. Thus, can be one of the reasons for reduced chaperone levels upon aging and in cases of neurodegenerative disorders.

Neurons are extremely polarized cells and for the continuation of their function, they strongly rely on axonal transport between the main body and synapses. Therefore, neurons are highly susceptible to the protein aggregation which in turn places a heavy load on the cellular chaperone framework to maintain proteostasis (Smith et al., 2015). This chapter focuses and highlights the role of cellular chaperones in neurodegenerative disorders.

CLASSIFICATION AND FUNCTIONS CELLULAR CHAPERONES

In cellular biology, chaperones proteins are the one that interacts with, stabilize and provide assistance to other proteins in their covalent folding or unfolding to assemble or disassemble their three-dimensional conformation but are absent in the final structure (Ellis et al., 1987). Chaperones proteins are divided into nine major categories or families. The major criterion for this categorization is their molecular weights or functions they perform. The nine major families of chaperones are Hsp40,Hsp60, Hsp70, Hsp90, Hsp100 and small Hsps (sHsps, 15 to30 kDa); Hsp10,Hsp27) (Smith et al., 2015). Prefoldin, TPR-domain containing chaperones (Hartl et al., 2002). Organelle specific chaperones of the mitochondria (MITO) and endoplasmic reticulum (ER) (Tatsuta et al., 2005; Kleizen et al., 2004). The main focus in the chapter has only been onHsp40, Hsp60, Hsp70, Hsp90, Hsp100 and sHsps. All these types of chaperons are ATP dependent and have effects on the aging human brain as well as have involvement in neurodegenerative disorders (Table 1).

The Hsp40 System

Hsp40 system is a large protein family and also called as chaperone DnaJ. This family is the primary regulator of the activity of Hsp70 which is done by stimulating the ATP hydrolysis (Jana et al., 2000). All types of DnaJ proteins comprise a J domain and a highly conserved region consisting of 70 amino acids. This region interacts with the Hsp70 ATPase domain. Hsp40 and control its activity. It can also carry a substrate with a suitable confirmation to Hsp70. Hsp40 is mainly associated with unfolded polypeptide chains and can reduce protein aggregate formation either by translating, folding, unfolding, translocating or degrading proteins (Han et al., 2004). The human genome composed of genetic sequences that encode 49 DnaJ proteins. These DnaJ proteins depending on their domain structure were further grouped into three different classes. Class I (DNAJA) proteins composed of all the domains of *Escherichia coli* DnaJ protein along with an N-terminal J domain, a region wealthy in glycine-phenylalanine (GF), a C-terminal domain, and a zinc-binding domain. Class II (DNAJB) proteins comprise a N-terminal J domain and a region wealthy in GF, while Class III (DNAJC) proteins have only the J domain. The presence of domains other than J domain has empowered DnaJ proteins with the capability to perform a series of specific functions. For instance, domains that help the DnaJ proteins to accurately target the intracellular sites to facilitate Hsp70 interaction with explicit clients. Moreover, client binding domains of few DnaJ proteins facilitates the delivery of Hsp70's substrate-binding domain (SBD) to the clients. Specialized domains

Table 1. The functions and association of heat shock proteins (Hsps) with different neurodegenerative disorders.

Hsp	MW (kDa)	Localization	Major Function	Disease Involved	References
Hsp10	10	Cytosol, endoplasmic reticulum, mitochondria, and nucleus	Folding of proteins	AD, MS	Campanella et al., 2018 Jia et al., 2011
Hsp27	20-30	Cytosol, endoplasmic reticulum, and nucleus	Degradation of protein	AD, PD, and HD	Tóth et al., 2013 Cox et al., 2018 Stetler et al., 2009
Hsp40	40	Cytosol	Folding of proteins	HD and PD	Muchowski et al., 2000 Hasegawa et al., 2018
Hsp60	60	Mitochondria	Prevent aggregation of proteins	AD	Marino et al., 2016
Hsp70	70	Cytosol, endoplasmic reticulum, nucleus, and mitochondria	Protein folding/unfolding	AD, PD, HD, MS, and Prion,	Mayer et al., 2005
Hsp90	90	Cytosol and endoplasmic reticulum	Transcription factor and degradation of proteins	AD, PD, and HD	Lackie et al., 2017
Hsp100/104	100-110	Cytosol and endoplasmic reticulum, and nucleus	Thermotolerance	PD and Prion	Zolkiewski et al., 2012

Figure 1. Role of Hsp 40 and Hsp70 in protein folding and degradation.

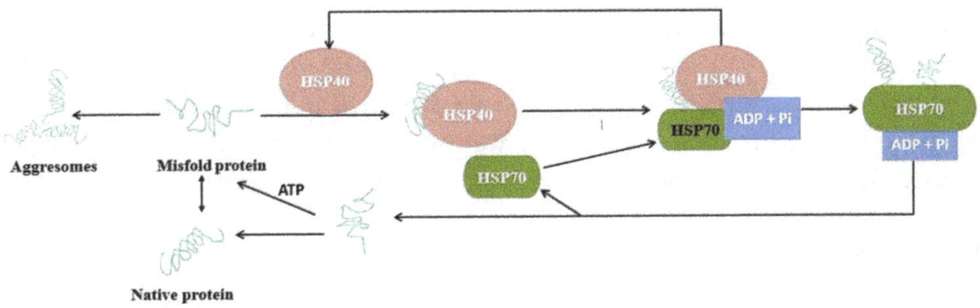

also promote client targeting for degradation, creating a significant connection between Hsp70, misfolded proteins, and their degradation (Cheetham et al., 1998; Kampinga et al., 2010) as shown in Figure 1.

The Hsp60 System

Hsp60, also called chaperonins, are a heptameric 60 kDa mitochondrial complexes with an enclosed central cavity in their core. This core cavity encapsulates the substrate proteins, typically folding intermediates. Thus offer protection to the exposed hydrophobic residues against aggregation and provide a safe environment for the substrate to fold (Hartl et al., 2011). Hsp60 chaperone is involved in folding and quality control of nearly 15-30% population of cell proteins therefore, it is a very crucial chaperone (Ranford et al., 2000). Hsp60 works in coordination with Hsp70 to accomplish protein folding. It also has an important part in the transmission, replication, distribution of mitochondrial DNA and apoptosis of proteins of mitochondria. Hsp60 helps in the protein folding and retain their structural conformation (Ranford et al., 2000). Additionally, Hsp60 carries and retains the mitochondrial proteins needed to replicate DNA (Itoh et al., 2002). Although various studies provided evidence for the interaction of Hsp60 with mutant α-synuclein in Parkinson's disease (PD), still the role of Hsp60 n the refolding of misfolded proteins (responsible for the pathology of neurodegeneration) is very poorly understood (Spillantini et al., 1998).

The Hsp70 System

Hsp70 (molecular weight: 70 kDa), is the most exhaustively and extensively studied cellular chaperone in living organisms. It is found in both cytosol and endoplasmic reticulum, and form complexes with unfolded or partly denatured proteins. Structurally Hsp70 is divided into two functional domains; one is called as the N-terminal ATPase domain also referred as nucleotide-binding domain (NBD) and the second domain is C-terminal substrate-binding domain (SBD), which in turn have α-helical structure (also called as lid of Hsp70)and a twisted β sandwich domain (Zhu et al., 1996). The operations of these domains are governed by the accessibility of the cellular ATP level. Hsp70 interacts with the unsaturated proteins through SBD at their hydrophobic sequences in an ATP dependent manner. The NBD of Hsp70 chaperon protein facilitates the transition from ATP-bound state (Hsp70 ATP) also called as the open state (low affinity for substrate) to ADP-bound state (Hsp70.ADP) which is also referred as the closed state (high affinity for substrate) (Mayer et al., 2005). Hsp40 and nucleotide exchange factors (NEFs) regulate the ATP-dependent functioning of Hsp70 chaperone (Leak et al., 2014). Hsp40 stimulates the hydrolysis of ATP present in Hsp70ATP and transforms it into the closed state (Hsp70 ADP) in which the lid (α-helical structure) closes (Bracher et al., 2015). Whereas, in the open state, the lid of Hsp70 have open conformation. Hsp40 accelerates the hydrolysis of ATP to ADP which leads to the formation of a stable peptide-binding state. Neuroprotective role of Hsp70 in protein misfolding diseases which also includes neurodegenerative disorders is very well supported by the evidence provided by various scientific investigations carried out on chaperone proteins (Fan et al., 2003; Mapa et al., 2010).

The Hsp90 System

Hsp90 (molecular weight: 90 kDa) chaperone is present in nearly all living organisms from bacteria to humans and is a highly conserved protein. It was reported to be the most abundant chaperone in cells and is mainly present in the cytosol. However, the presence in endoplasmic reticulum and mitochondria was also reported. Hsp90 exists in dimeric form i.e. there are two isomers and has highly conserved N- and C-terminal domains. Both monomers of Hsp90 are composed of three domains. The N-terminal domain is the first one and is linked to M domain, called as intermediate domain and a third domain called the C-terminal domain. The C domain is also known as the dimerization domain (Chaari et al., 2019).In human proteomics, two forms of Hsp90 were recognized; inducible Hsp90 (Hsp90α) and constitutive Hsp90 (Hsp90β). Hsp70 along with its co-chaperone were responsible for the activation of Hsp90 system (Hoter et al., 2018). They stabilize as well as activate more than 200 Hsp90-client proteins (Figure 2)

Figure 2. Role of Hsp 90 and its co-chaperones in protein folding and degradation.

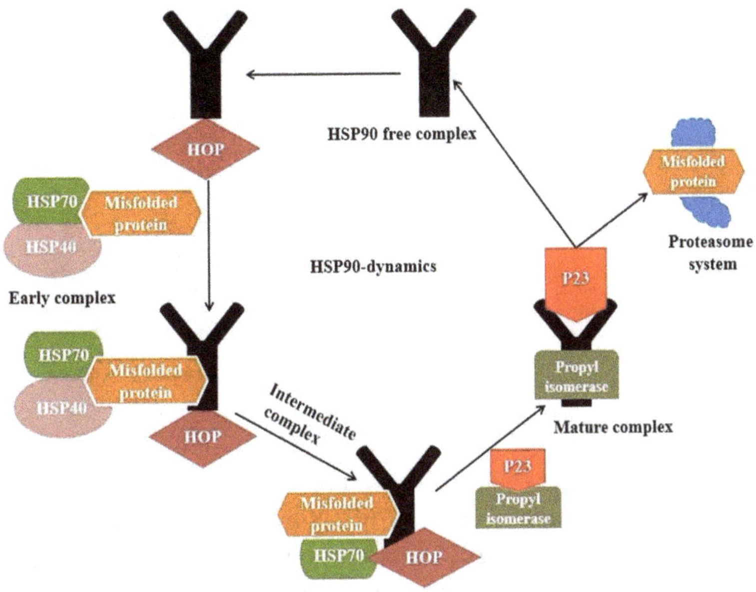

which are very necessary for cell signaling and to perform adaptive changes in response to stress conditions (Trepel et al., 2010; Zhao et al., 2005).

The Hsp100/104 System

The Hsp100 family (molecular weight: 100-110 kDa), represents heat-inducible proteins with a wide range of functions especially elevated thermotolerance (Schirmer et al., 1996). Hsp104 belongs to AAA+ Hsp100 family of motor proteins that help the organisms to adapt to extreme environmental stress conditions to ensure survival (Langklotz et al., 2012) (Figure 3). Hsp104/110 chaperones are not expressed under normal growth conditions but are induced upon exposure to extreme heat or other hostile conditions to protect the cells and to speed up the proteolytic degradation of particular cellular protein debris. Hsp104/110 has two structural domains; first is ATP binding domain, and the second one is the substrate-binding domain. Hsp104/110 system has the capability of solubilizing almost any kind of protein aggregates formed because of serious stress conditions. Furthermore, Hsp100 in coordination with Hsp70 and Hsp40 breaks large protein aggregates into smaller ones to promote rapid their proteasomal degradation (Schirmer et al., 1996). Hsp100 chaperone functions as remodeling machinery which acts to refold misfolded proteins and remove irreversible protein aggregates (Schirmer et al., 1996).

Small Heat Shock Proteins

They are lesser molecular weight (12-35kDa) chaperones. These are ATP independent chaperones and are reported to be present in the cytoplasm, nucleus and endoplasmic reticulum (Augusteyn et al., 2004). Small heat shock proteins (sHsps) contains a very conserved C terminal structure containing 90 amino acids and this terminal is also called as α- crystalline domain. Many of these sHsps units assemble to

Figure 3. Role of Hsp100 in protein folding and degradation.

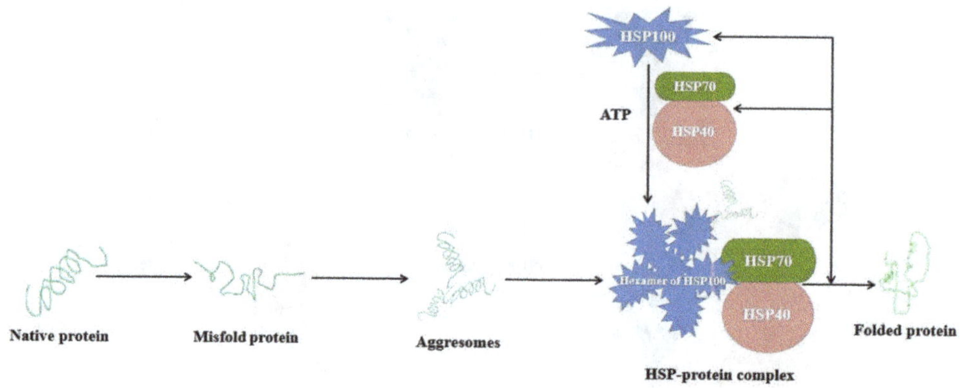

form larger multimeric units (Buchner et al., 1996; Van Montfort et al., 2001). Most of these sHsps were present in the nervous system and their expression differs from the cellular setting as well as with various neurological disorders (Kirbach et al., 2011; Pozsgai et al., 2007). The sHsps have higher chaperone activity, this causes a rapid attenuation of protein aggregation. The induction of this high expression of activity is due to cellular stress (Sarto et al., 2000; Sun et al., 2005). Apart from performing chaperone specific functions, the sHSPs have a role in thermotolerance, growth and cell differentiation, signal transduction, and inhibition of apoptosis. They act as a cochaperone for Hsp70, inhibit protein aggregation and degrades misfolded or denatured proteins by ubiquitin-proteasome (Sun et al., 2005). sHSPs protects cells from oxidative damage by increasing glutathione level (Sarto et al., 2000).

ROLE OF CELLULAR CHAPERONES IN AGING

Aging is a biological process in which with time the capacity of a living organism to perform the regular biological activities to ensure the survival diminishes. This degraded capacity is the result of alterations in the cellular organelles, processes, and mechanisms. Most of the times these alterations negatively affect the organism which is reflected by age-related problems. The aging also alters or modifies the macromolecules synthesized within the cellular environment and that have an important role in healthy survival. The most important macromolecules that are involved in carrying out various cellular processes are proteins. These proteins undergo modifications with aging and thus processes to which they are associated with undergoing modifications. Aging disrupts the cellular proteostasis and leads to various protein modifications which include main and side-chain oxidation, glycation, the formation of abnormal bonds and conformational changes which cause misfolding of proteins (Sun et al., 1999). These misfolded proteins aggregates cause toxicity and lead to cell death. However, cellular chaperones protect these misfolded aggregates but with age, this activity also diminishes. Therefore, in aged organisms to counter the misfolded proteins aggregate formation higher concentration of chaperones with elevated activity is required (Soti et al., 2002). Primarily three types of changes or defects were observed in chaperones functioning were observed in aged animals. First is the alterations in induction; several studies reported the generation of chaperones in response to stress conditions gets diminished in aged animals (Heydari et al., 2000). The second alteration in chaperones related to age is diminished functioning (Soti et al., 2003).

The third alteration is the diminished capacity of chaperones mediated mechanism to counter misfolded proteins because of aging the balance of misfolded proteins concerning chaperones gets imbalanced (Soti et al., 2003). Now because of the concentration of misfolded protein in comparison to the available chaperones the necessity for chaperones greatly exceeds this is called chaperone overload (Csermely et al., 2001). Thus, this condition can cause the problem in signal transduction, protein transport, immune function and cellular organization (Soti et al., 2003).

DEGRADATION OF MISFOLDED PROTEINS BY MOLECULAR CHAPERONES THROUGH THE UPS

The major function of chaperones is to mediate the degradation and removal of damaged and misfolded proteins. Chaperones perform this function either through ubiquitin proteasome system (UPS) or autophagy. In UPS mediated proteolysis first step is the ubiquitin activation by interaction with ubiquitin-activating enzyme E1 (Ciechanover et al., 2015). Then activated ubiquitin is transferred to another ubiquitin-conjugating enzyme E2. After that, another enzyme E3 (ubiquitin ligase) facilitates the transfer of ubiquitin from E2 to the lysine residue present on the target protein (Ciechanover et al., 2017). This ubiquitin-protein complex is then degraded by 26S proteasome (protease complex having 20S core particle associated with two 19S regulatory particles) (Ravikumar et al., 2008). However, the conjugation of ubiquitin with lysine residue is reversible and this reversal is mediated by deubiquitination enzymes (DUBs).

Many molecular chaperones work in close association with the UPS mechanism. Various ubiquitin UBR1, UBR2, San1, Hul5, E6-AP, C-terminus of Hsc70-interacting protein (CHIP) and Parkin take part in misfolded proteins degradation (Heck et al., 2010). CHIP acts as a cochaperone for Hsp70 and Hsp90 (Ballinger et al., 1999). Hsp27 binds with the Ubiquitin chain of target proteins and promotes the degradation of ubiquitinated proteins (Garrido et al., 2006). The target proteins of Hsp90 can be degraded by UPS also (Ciechanover et al., 2017). In neurodegeneration, the UPS mechanism gets downregulated at that point chaperones are the only active mechanism against misfolded proteins. The degradation of misfolded proteins by chaperones through UPS system is a multiple-step process. The first step Hsp70 identifies and binds to the hydrophobic sequences on misfolded proteins. In the second step, CHIP facilitates the ubiquitination of the chaperone protein complex. In further steps enzymes E1, E2, E3 mediates conjugation of ubiquitin with target proteins leading to the formation of ubiquitinated protein substrates which are degraded by proteasomes (Ciechanover et al. 2017).

Autophagy is another important cellular pathway employed to degrade and remove the misfolded proteins. Chaperone mediated autophagy (CMA) is the common pathway for removing terminally misfolded proteins specifically detected in neurodegenerative diseases (Ciechanover et al. 2017). Hsp70 in combination with Hsp40 initially performs refolding operation on the misfolded proteins. If refolding fails then the misfolded protein is transferred to the UPS system which degrades the soluble misfolded proteins. If misfolded proteins form insoluble aggregates then they degraded by lysosomal hydrolases through autophagy (macroautophagy or CMA). As the last step of PQC chaperones disaggregates the protein aggregates by the action of Hsp110 in combination with Hsp70 and Hsp40 (Ciechanover et al. 2017).

ROLE CELLULAR CHAPERONES IN NEURODEGENERATIVE DISEASES

Alzheimer's Disease

This disorder primarily affects older people. It ranked as the number one neurodegenerative disorder in terms of prevalence. Hippocampus and cerebral cortex are the main anatomical regions of the brain that are affected by Alzheimer's disease (AD). Neuronal loss in these regions of the brain produces problems like impaired memory, abnormalities of cognition behavior, and language problems. The main pathological characteristic observed in AD is the accumulation and deposition of two cellular proteins; Aβ and tau (Selkoe et al., 2004). The first type of protein is derived from amyloid precursor protein (APP) and is mostly present in the extracellular spaces, in the form of amyloid plaques. Tau stabilizes the microtubules and mainly accumulates in the intracellular region in neurofibrillary tangles but sometimes also present in extracellular regions (Dickey et al., 2009; Frautschy et al., 2001). Chaperone Hsp70 reduces the synthesis of both Aβ40 and Aβ42 by binding with APP and interfere with APP secretory pathway (Yang et al., 1998). Studies reported that chaperones Hsp70, as well as Hsp90 both, interact with Aβ and tau proteins. This interaction leads to the degradation of oligomers of Aβ and tau through the proteasome system (Dickey et al., 2009). Overexpression of Hsp70 chaperone reduces the load of insoluble tau, lowers the phosphorylation rate, enhances the stability of tau and binds to microtubules. Thus, finally lowers tau protein-related *in vitro* and *in vivo* toxicity (Sarkar et al., 2008). In contrast, Hsps down regulation by mediated by RNA interference (RNAi) produces a reverse effect (Dou et al., 2003).

Hsp70 or Hsp90 in combination with heat shock cognate 70 (Hsc70) can directly bind to tau irrespective of the condition whether tau is phosphorylated or not. This promotes the polymerization of microtubules and restricts the generation of aggregates of tau (Matts et al., 2011). Furthermore, Hsp90, its co-chaperones are also significant for the refolding of misfolded or denatured Aβ and tau proteins (Matts et al., 2011). The solubility of tau and its binding with microtubules were promoted by both Hsp70 and Hsp90. Both of these chaperone proteins, on the other hand, reduces the concentration of insoluble tau aggregates and also slows down the phosphorylation rate of tau (Dou et al., 2003). Many studies concluded that Hsp70protein overexpression lowered the concentration of both soluble and insoluble forms of tau in experimental mice of the age of 30 months (Dou et al., 2003; Petrucelli et al., 2004). Concentration level of Hsp90 was found to have an inverse relationship with the levels of tau oligomers and neurofibrillary tangles in AD (Sahara et al., 2007; Dou et al., 2003). In chaperone-mediated autophagy (CMA) removal of protein aggregates, Hsps avoid caspase activation, including tau, APP, and HTT substrates (Cuervo & Dice, 2000).

Parkinson's Disease

Parkinson's disease (PD) can be defined as a movement disorder induced by a pathological condition in which dopaminergic neurons of substantia nigra pars compacta (SNpc) region of the brain were progressively lost (Alexander et al., 2004). This gives rise to manifestations like motor impairment, rigidity in movement, bradykinesia, and resting tremors. PD is placed at number two in the list of most prevalent neurodegenerative disorders. Lewy bodies (intracellular protein aggregates) presence primarily of ubiquitinated α-synuclein is the most important pathological marker of PD (Spillantini et al., 1998). Careful examination of these Lewy bodies revealed that along with with α-synuclein a series of other ubiquitinated proteins and Hsps were also present in them (Dettmer et al., 2013). Hsp70 is located with α-synuclein, parkin, proteasome subunits, dopamine transporter (DAT), and ubiquitin with ubiquitin carboxy-terminal hydrolase-L1 (UCH-L1) (McLean et al., 2001). Because of its antiapoptotic operations, Hsp70 can stop dopaminergic degeneration in PD (Shulman et al., 2011). Moreover, Hsp70 can improve *in vitro* parkin binding and ubiquitination of extended polyglutamine protein. Hsp70 also facilitates the recruitment of misfolded proteins as substrates to parkin E3 ubiquitin ligase. Hsp70 can, therefore, encourage E3 ligase activity to degrade α-synuclein.

Many evidence indicates that increased Hsp70 expression decreases the aggregation and toxicity of α-synuclein protein (Auluck et al., 2002; Dedmon et al., 2005; Huang et al., 2006). Other Hsps such as Hsp40 or Hsp27 may also lower the aggregation of α-synuclein (Thakur et al., 2014). While one study showed that small-molecule inhibitors of Hsp90 decrease the formation and toxicity of α-synuclein oligomers (Preeti Putcha et al., 2010). Reports showed that Hsp90 expression is linked with alpha-synuclein filaments of brains of PD patients and in the brain of a transgenic mouse model of PD (Uryu et al., 2006).

Polyglutamine Diseases and Huntington's Disease

Huntington's disease (HD) is characterized by the progressive degeneration of the neurons present in cortex and striatum of the brain. This degeneration causes the impairment of cognitive behavior, motor activity as well as psychiatric behavior. It is an autosomal dominant genetic disorder that occurs because of the expansion of a CAG triplet repeat region in the huntingtin gene. This expansion leads to the formation of protein which comprises a long chain of polyglutamine (polyQ). This factor now triggers the formation and accumulation of aggregates of the huntingtin protein. Many other disorders involving CAG triplet expansion are also reported to have polyQ extended polypeptides that are susceptible to aggregation. These disorders show similarities in their pathogenesis.

In neurodegeneration, the role of cellular chaperones as a powerful modulator for aggregation of proteins was first reported by Cummings et al., (Cummings et al., 1998). They reported that DnaJ protein's overexpression (DnaJA1), decreased ataxin-1 aggregation. Much scientific research reports presented by various groups of researchers reported that elevation of Hsp40, Hsp70, and Hsp100 concentrations inhibit protein aggregation induced by polyglutamine and thus impede the development of disease (Leak et al., 2014). In transfected and transgenic models the presence of Hsp70 and its cochaperone is reported along with the presence of extended polyQ aggregates (Patel et al., 2005; Sharp et al., 2006). Treatment by Hsp70, Hsp40, and Hsp90 for cells expressing pathological expanded polyQ leads to a substantial reduction in polyQ aggregation (Blumen et al., 2012). *In vitro* studies reported that Hsp70 and its cochaperone interfere with the ATP dependent polyQ polypeptides aggregation. This inference diverts the process of aggregation towards the creation of non-fibrous and SDS-soluble aggregates, which are non-pathogenic (Veereshwarayya et al., 2006).

In HD, accumulation, and deposition of huntingtin's protein with pathogenic CAG expansions results in the generation of inclusion bodies in the striatum and cortex of the brain (Davies et al., 1997). Hsp70/Hsp40 complex inhibits the development of aggregates of misfolded huntingtin. The complex bind with the misfolded huntingtin protein and keep its misfolded form intact which prevents the formation of oligomeric aggregates (Jana et al., 2000). Other Hsp70 family members also influence the aggregation of mutant huntingtin (mHTT) and counter its cytotoxic effects (Bauer et al., 2010). The expression of Hsp104 in the cells of the mouse model of HD was also reported to decrease the aggregation of protein huntingtin (Shorter et al., 2008). Enhanced expression of Hsp27 in HD also protects against PolyQ-induced toxicity in neurons (Luo et al., 2006; Wilhelmus et al., 2006).

Amyotrophic Lateral Sclerosis

Amyotrophic lateral sclerosis (ALS) is a neurological disorder caused by progressive degeneration of motor neurons present in the brain, brainstem, and spinal cord. The most characteristic feature of the disorder is the muscle weakness which progresses towards paralysis and ultimately leads to cell death, generally within 2-5 years of diagnosis (Robberecht et al., 2013). Most of the ALS cases are sporadic and only about 10% cases are because of familial history related to specific gene mutations. Mutations of superoxide dismutase SOD1 are reported as a major dominant factor in 20% cases familial ALS. Majority cases of ALS are due to protein misfolding generated because of mutations in the conformation of the protein and then these mutant proteins aggregate and accumulate within the cellular matrix. In the cytoplasm of the degenerating neurons, these aggregates form inclusion bodies in ALS patients, like other neurodegenerative diseases (Kalmar et al., 2017; Okado-Matsumoto et al., 2002; Shaw et al., 2008). The formation of these protein aggregates in ALS involves various cellular chaperones. In fact, with some mutant types of SOD, Hsp40/Hsp70, Hsp25, Hsp27, and αB crystalline may form complexes.

In the various studies with ALS mouse models, it has been found that only Hsp70 overexpression was not alone enough to suppress the pathogenicity of SOD1 within the test animals (Liu et al., 2004). Studies showed that Hsp27 also protects against the pathology of ALS. Overexpression of Hsp27 along with Hsp70offers neuroprotection in ALS against neuronal death due to the expression of SOD1 G93A or G93R mutations (Patel et al., 2005). Hsp70 overexpression along with overexpression of Hsp27 has the synergistic anti-apoptotic effect (Patel et al., 2005). Inhibition of SOD1 aggregates formation by sHsps; αB-crystallin and Hsp27 were also documented in various *in vitro* studies (Yerbury et al., 2013). *In vivo* studies have reported inhibition of cell death induced due to mutant SOD1 by overexpression of Hsp27 (Patel et al., 2015). Also, HsJ1a was reported for showing a protective effect against the mutant SOD1 aggregates in the later stage of ALS and promotes its ubiquitination and proteasomal degradation (Patel et al., 2005).

Prion Diseases

It represents a cluster of progressive neurodegenerative disorders triggered by the formation of abnormal prion (a type of protein). Prion disease includes genetically inherited diseases (genetic Jakob-Creutzfeldt disease, Gerstmann-Sträussler-Scheinker syndrome, and fatal familial insomnia), sporadic diseases (Jakob-Creutzfeldt disease) and acquired diseases (kuru, variant Jakob-Creutzfeldt disease, and iatrogenic Jakob-Creutzfeldt disease) (Geschwind et al., 2015). Normal prion protein has α-helical structure and is designated as PrPC; prion related protein C where C means the cellular form of protein. Abnormal prion protein is denoted as PrPSc and Sc means scrapie, which are protein-like infectious particles that have β-pleated sheet structure (Geschwind et al., 2015). Mutations cause conversion of prion structure from α-helical structure to β-pleated sheet structure.

These abnormal prions get extracellularly deposited in the CNS and form prion plaques. These plaques then disturb the morphology of nerve cells. Hsp70 mediates degradation of abnormal prion protein aggregates through the proteasome pathway and helps in the propagation of normal prion (PrPC) (Jana et al., 2000). In yeast, Hsp100 overexpression contributes to the breakdown of large aggregates of prion into small prion seeds necessary for the propagation of fresh prion formation process (Taylor et al., 2006). Furthermore, *in vitro* studies have shown that HSP104 may also inhibit prion peptide fibrillation and disassembly (Liu et al., 2011). Hsp42 with Hsp70 prevents prion oligomers from rearranging thus, blocks these oligomeric units from self-assembling. Hsp26 stops aggregation of assembled prion oligomers by binding with them. Moreover, Hsp40, Hsp70, and Hsp104 promote disaggregation of prions by destabilizing them (Duennwald et al., 2012).

NEUROPROTECTIVE ROLE OF CELLULAR CHAPERONES

It is clearly understood from the various studies conducted previously by numerous researchers that cellular chaperones are important for neuroprotection. They act as defense machinery against these disorders by taking part in the degradation of misfolded proteins generated which are responsible for the onset and propagation of the diseases. Various recent studies further consolidated their role and provide new insights in this field of research by specifically utilizing new strategies like genetically engineered Hsp, exogenous Hsp, use of Hsp response inducers and cell-penetrating Hsp, etc. Martin et al., created a genetically engineered version of HSP70 (secHsp70) and reviewed the protective role against αβ42 aggregation in the *Drosophila* AD model. The demonstrated successful targeting of secHsp70 against Aβ42-related phenotypes in the CNS of the model can provide efficient neuroprotection and thus may give an insight into the new therapeutics for neurodegeneration (Martin et al. 2019).

The development of exogenous Hsp to counter the neurodegeneration due to misfolded proteins is a new area of research. Studies investigating the role of exogenously produced recombinant human hsp70 for the treatment of ALS were carried out by a group of researchers. They administered recombinant human Hsp in G93A SOD1 mutant mice as parenteral through intraperitoneal route three times a week and noted the effect. They observed delayed onset of symptoms, motor function preservation and delayed degeneration of motor neurons after injection. Thus, led to the conclusion that exogenously produced genetically engineered Hsp can protect against neurodegeneration (Gifondorwa et al. 2007).

Recent scientific investigations also showed biomolecules like histamine can act as an inducer of the defensive response of Hsp against neurodegeneration in SOD1 (G93A SOD1) animal model for ALS. This further confirms the importance of Hsp protection in neurodegeneration gives an insight into a new therapeutic strategy for such disorders (Apolloni et al. 2019). Kim et al., developed cell-penetrating bound Hsp70 by using trans-activating transcriptional activator (TAT, cell-penetrating peptide) bound Hsp70 (Tat-Hsp). This modified Hsp70 can easily penetrate the NSC34 cells in *the in vitro* study as well as the BBB of test animals used in the study (male New Zealand white rabbits) in *the in vivo* study. The modified Hsp70 protects neurons against the oxidative stress in NSC34 cells (Kim et al. 2019)

MUTATIONS OF CHAPERONES

The presence of mutant chaperones specifically in cases of inherited neurodegenerative disorders signifies how important the chaperones for preserving neuronal proteostasis are. The mutations in chaperones disturb this proteostasis which may cause generation and aggregation of misfolded proteins leading to neurodegeneration. However, some mutations of chaperones have been identified to have a neuroprotective effect against neurotoxicity when investigated in some animal models.

Hsp27 (HSPB1) and Hsp22 (HSPB8)

Mutations of both Hsp27 (HSPB1) and Hsp22 (HSPB8) chaperones were detected in the familial cases of distal hereditary motor neuropathies (dHMN) and Charcot-Marie-Tooth type 2 (CMT2). Sixteen different types of Hsp27 chaperone mutations were reported so far and detected mostly in α-crystallin domain as missense mutations (Evgrafov et al., 2004; Houlden et al., 2008; Ikeda et al., 2009; Kijima et al., 2005; Luigetti et al., 2010; Mandich et al., 2010). These mutations were inherited in an autosomal dominant way and cause protein instability. The dominant inheritance causes neurotoxicity and results in loss of chaperone function which contributes to the pathology of disease by means of dominant-negative impacts on chaperones. Some HSPB proteins stabilize intermediate filaments and microtubules thus can modulate the structure of the cytoskeleton. Mutant Hsp27 expression causes sequestration of intermediate filaments into Hsp27 aggregates thus, destabilizing the cytoskeleton (Evgrafov et al., 2004; Zhai et al., 2007). Also, investigation indicates that transgenic mice with Hsp27 S135F or P182L mutation showed impaired mitochondrial axonal transport and development of distal motor neuropathy (d'Ydewalle et al., 2011).

HSJ1 (DnaJB2)

Blumen et al. reported HSJ1 mutations in a consanguineous Moroccan family with dHMN (Blumen et al., 2012). A second mutation has recently been recognized with dHMN in a Turkish family (Gess et al., 2014). In both cases, mutations were homozygous and are present at donor splice sites. This causes loss of expression of HSJ1 protein. A third HSJ1 mutation was recognized in a family with CMT2 (Gess et al., 2014). All of these mutations cause loss of the function of HSJ1.

Sacsin (DnaJC29)

The progressive degeneration of Purkinje cells of the cerebellum followed by the motor neuron depletion in the spinal cord are features of autosomal recessive spastic ataxia of Charlevoix-Saguenay (ARSACS) (Bouchard et al., 1978). Mutations of DnaJ protein sacsin (DnaJC29) were reported to be responsible for ARSACS and more than 170 mutations were reported for this protein. Sacsin has the N-terminal ubiquitin-like domain that connects with the proteasome (Parfitt et al., 2009). Next to this domain, three large repeat regions of sacsin are present which are reported to have Hsp90 chaperone-like function (Anderson et al., 2010; Anderson et al., 2011). The third structural feature is a XPCB domain which interacts with protein ligases (Greer et al., 2010). Finally, a C-terminal J domain and a HEPN domain are present in and both of these acts as a mediator for the formation of sacsin dimers (Kozlov et al., 2011; Parfitt et al., 2009). These domains combined imparts a function in PQC.

Mutations within these domains cause loss of sacsin function. Sacsin is abundant in the cytoplasm but also present in mitochondria. Sacsin interacts with dynamin-related protein 1 (DRP1, a mitochondrial fission-mediating GTPase) (Girard et al., 2012). The mitochondria appeared to be over fused along with decreased mobility in fibroblasts of the patients having ARSACS and also in the sacsin knockout mouse model. This indicates sacsin have a part in mitochondrial fission. Mitochondrial dysfunction is a common hallmark observed in various types of neurodegenerative disorders and maybe the main mechanism underlying ARSAC (Girard et al., 2012).

Table 2. Vector-mediated upregulation of chaperone activity.

Viral Vector	Disease Investigated	References
AAV-H-BH (constitutively active form of HSF-1)	AAV-CDCrel-1 PD rat model	Jung et al., 2008
AAV-Hsp70	AAV-CDCrel-1 PD rat model and MPTP PD mouse model	Dong et al., 2005; Jung et al., 2008
LV-Hsp104	LV-(h)-A30P α-synuclein PD rat model	Lo Bianco et al., 2008
AAV-BAG5 (DARA)	MPTP PD mouse model	Kalia et al., 2004

THERAPEUTIC PERSPECTIVES

In neurodegenerative disorders, chaperones protect from pathogenic protein aggregates. This particular character makes them a potential target for drug development as well as an important component of a treatment strategy to combat neurodegenerative diseases. To upregulate chaperone function with adequate outcomes, an increasing body of preclinical research has explored pharmacological and gene therapy approaches. The much-researched strategy has been the activation of the thermal shock factors (HSF1), which simultaneously contributes to increased expression of various chaperones. This strategy was accomplished by inhibiting the activity of Hsp90 with agents like geldanamycin, radicicol or 17-AAG. Upon treating with one of these agents, for example, geldanamycin results in Hsp90 inhibition and HSF1 activation which results in elevation of chaperone levels. This results in subsequent inhibition of folding and stabilization of neurotoxic proteins (Boland et al., 2008; Ciechanover et al., 2017).

In PD mouse model, *Drosophila* as well as in cell culture of familial ALS geldanamycin successfully inhibited protein aggregation (Shen et al., 2005; McLean et al., 2004; Batulan et al., 2006). Reduction in α-synuclein oligomerization and its toxicity by 17-AAG was found in a cell culture model for PD. Additionally, other Hsp90 inhibitors such as SNX compounds have demonstrated their efficacy in decreasing both monomers and elevated molecular weight species of α-synuclein by encouraging α-synuclein refolding and degradation through Hsp70 induction by Hsp90 inhibition (Ebrahimi-Fakhari et al., 2011; Putcha et al., 2010).

Another approach was to modulate HSF-1 activity as it helps to promote the expression of Hsp70 and other chaperones. Arimoclomol is such an agent that has demonstrated the potential to modulate HSF-1 activity in ALS (Benn et al., 2004; Traynor et al., 2006). But still under phase II/III clinical trials for ALS (Benn et al., 2004). Finally, an alternative therapeutic strategy is to promote proteostasis under cellular stress conditions could be the vector-based gene therapy approaches (Table 2).

RECENT DEVELOPMENTS AND FUTURE RESEARCH DIRECTIONS

Chaperones are a set of proteins that play a crucial role in both the proper folding of many important proteins as well as the degradation of misfolded proteins through chaperone mediated autophagy. Thus, making them highly valuable form the research point of view to understand the pathogenesis of various misfolded proteins related disorders like neurodegenerative disorders. Recent studies on chaperones provided some new insights into some more roles of chaperones in the cellular environment. These studies along with the contribution of chaperones in autophagy and neurodegeneration also provided evidence for its role in cancer, immunotherapy, smoke-induced bronchial epithelial injury, etc. The protective role of chaperones was reported by several previous studies and recent studies also showed evidence in this regard. Molecular chaperone endoplasmic reticulum protein 29 (ERp29) has been identified as a new member of the thyroglobulin protein complex and was found to be upregulated in response to endoplasmic stress. The protein was reported to provide neuroprotection in retinal and neurodegenerative disorders (McLaughlin et al., 2018). Some studies also reported the role of chaperone-mediated autophagy in lipid metabolism, glucose metabolisms, and cell energetics irrespective of its selective nature (Tasset et al., 2016). Prefoldins a relatively newer set of cochaperones were reported to play multiple roles like cell growth cell survival, gene expression, protein transcription and also in multiple types of cancers (Djouder et al., 2018).

In addition to this chaperone GRP78 a member of Hsp70 family was reported to be involved in Chronic obstructive pulmonary disease (COPD) pathogenesis and have a role as mediator in bronchial cell inflammation and death (Wang et al., 2018). Some studies revealed the role of chaperones in cell survival at high altitudes under the hypoxic conditions and the findings showed the upregulation of chaperones under stress conditions like hypoxia protects the cells by maintaining protein hemostasis (Rathore et al., 2019). The current challenges in the field of research of chaperones are the proper screening to isolate and identify the correct chaperones. This proper screening will help to identify and allow the specific chaperone to be targeted as a potential site to counter a neuropathological condition.

CONCLUSION

Several neurodegenerative diseases are a result of disturbances or disruption of PQC. This disruption can be attributed to proteotoxic stress, molecular crowding, gene mutations and age-related decline in proteostasis. Progressive aggregation and accumulation of misfolded proteins either inside or outside the cells is the most characteristic feature of neurodegenerative diseases. Cellular chaperones such as HSPs are essential players of PQC and act as the primary defense mechanism to protect neurons from pathogenic protein aggregates by targetting misfolded proteins. These misfolded proteins either undergo refolding or gets degraded due to chaperone activity. The cellular chaperones act as a promising therapeutic target and their upregulation can potentially slow down the progression of the disease. The study of the role of chaperone mutations in neurodegenerative diseases is likely to provide significant new insights into chaperones biology and neuronal proteostasis.

REFERENCES

Alexander, G. E. (2004). Biology of Parkinson's disease: Pathogenesis and pathophysiology of a multisystem neurodegenerative disorder. *Dialogues in Clinical Neuroscience*, *6*(3), 259–280. PMID:22033559

Anderson, J. F., Siller, E., & Barral, J. M. (2010). The sacsin repeating region (SRR): A novel Hsp90-related supra-domain associated with neurodegeneration. *Journal of Molecular Biology*, *400*(4), 665–674. doi:10.1016/j.jmb.2010.05.023 PMID:20488193

Anderson, J. F., Siller, E., & Barral, J. M. (2011). The neurodegenerative-disease-related protein sacsin is a molecular chaperone. *Journal of Molecular Biology*, *411*(4), 870–880. doi:10.1016/j.jmb.2011.06.016 PMID:21726565

Augusteyn, R. (2004). α-Crystallin: A review of its structure and function. *Clinical and Experimental Optometry -. Clinical & Experimental Optometry*, *87*(6), 356–366. doi:10.1111/j.1444-0938.2004.tb03095.x PMID:15575808

Auluck, P. K., Chan, H. Y., Trojanowski, J. Q., Lee, V. M., & Bonini, N. M. (2002). Chaperone suppression of alpha-synuclein toxicity in a Drosophila model for Parkinson's disease. *Science*, *295*(5556), 865–868. doi:10.1126cience.1067389 PMID:11823645

Ballinger, C. A., Connell, P., Wu, Y., Hu, Z., Thompson, L. J., Yin, L. Y., & Patterson, C. (1999). Identification of CHIP, a novel tetratricopeptide repeat-containing protein that interacts with heat shock proteins and negatively regulates chaperone functions. *Molecular and Cellular Biology*, *19*(6), 4535–4545. doi:10.1128/MCB.19.6.4535 PMID:10330192

Batulan, Z., Taylor, D. M., Aarons, R. J., Minotti, S., Doroudchi, M. M., Nalbantoglu, J., & Durham, H. D. (2006). Induction of multiple heat shock proteins and neuroprotection in a primary culture model of familial amyotrophic lateral sclerosis. *Neurobiology of Disease*, *24*(2), 213–225. doi:10.1016/j.nbd.2006.06.017 PMID:16950627

Bauer, P. O., Goswami, A., Wong, H. K., Okuno, M., Kurosawa, M., Yamada, M., ... Nukina, N. (2010). Harnessing chaperone-mediated autophagy for the selective degradation of mutant huntingtin protein. *Nature Biotechnology*, *28*(3), 256–263. doi:10.1038/nbt.1608 PMID:20190739

Benn, S. C., & Brown, R. H. Jr. (2004). Putting the heat on ALS. *Nature Medicine*, *10*(4), 345–347. doi:10.1038/nm0404-345 PMID:15057226

Blumen, S. C., Astord, S., Robin, V., Vignaud, L., Toumi, N., Cieslik, A., ... Viollet, L. (2012). A rare recessive distal hereditary motor neuropathy with HSJ1 chaperone mutation. *Annals of Neurology*, *71*(4), 509–519. doi:10.1002/ana.22684 PMID:22522442

Boland, B., Kumar, A., Lee, S., Platt, F. M., Wegiel, J., Yu, W. H., & Nixon, R. A. (2008). Autophagy induction and autophagosome clearance in neurons: Relationship to autophagic pathology in Alzheimer's disease. *The Journal of Neuroscience: The Official Journal of the Society for Neuroscience*, *28*(27), 6926–6937. doi:10.1523/JNEUROSCI.0800-08.2008 PMID:18596167

Bouchard, J. P., Barbeau, A., Bouchard, R., & Bouchard, R. W. (1978). Autosomal recessive spastic ataxia of Charlevoix-Saguenay. *The Canadian Journal of Neurological Sciences*, *5*(1), 61–69. doi:10.1017/S0317167100024793 PMID:647499

Bracher, A., & Verghese, J. (2015). The nucleotide exchange factors of Hsp70 molecular chaperones. *Frontiers in Molecular Biosciences*, *2*, 10–10. doi:10.3389/fmolb.2015.00010 PMID:26913285

Brehme, M., Voisine, C., Rolland, T., Wachi, S., Soper, J. H., Zhu, Y., ... Morimoto, R. I. (2014). A chaperome subnetwork safeguards proteostasis in aging and neurodegenerative disease. *Cell Reports*, *9*(3), 1135–1150. doi:10.1016/j.celrep.2014.09.042 PMID:25437566

Buchner, J. (1996). Supervising the fold: Functional principles of molecular chaperones. *The FASEB Journal, 10*(1), 10–19. doi:10.1096/fasebj.10.1.8566529 PMID:8566529

Campanella, C., Pace, A., Caruso Bavisotto, C., Marzullo, P., Marino Gammazza, A., Buscemi, S., & Palumbo Piccionello, A. (2018). Heat shock proteins in Alzheimer's disease: Role and targeting. *International Journal of Molecular Sciences, 19*(9), 2603. doi:10.3390/ijms19092603 PMID:30200516

Chaari, A. (2019). Molecular chaperones biochemistry and role in neurodegenerative diseases. *International Journal of Biological Macromolecules, 131*, 396–411. doi:10.1016/j.ijbiomac.2019.02.148 PMID:30853582

Cheetham, M. E., & Caplan, A. J. (1998). Structure, function and evolution of DnaJ: Conservation and adaptation of chaperone function. *Cell Stress & Chaperones, 3*(1), 28–36. doi:10.1379/1466-1268(1998)003<0028:SFAEOD>2.3.CO;2 PMID:9585179

Ciechanover, A., & Kwon, Y. T. (2015). Degradation of misfolded proteins in neurodegenerative diseases: Therapeutic targets and strategies. *Experimental & Molecular Medicine, 47*(3), e147. doi:10.1038/emm.2014.117 PMID:25766616

Ciechanover, A., & Kwon, Y. T. (2017). Protein Quality Control by Molecular Chaperones in Neurodegeneration. *Frontiers in Neuroscience, 11*(185). doi:10.3389/fnins.2017.00185 PMID:28428740

Ciechanover, A., & Kwon, Y. T. (2017). Protein quality control by molecular chaperones in neurodegeneration. *Frontiers in Neuroscience, 11*, 185. doi:10.3389/fnins.2017.00185 PMID:28428740

Cox, D., Whiten, D. R., Brown, J. W., Horrocks, M. H., San Gil, R., Dobson, C. M., ... Ecroyd, H. (2018). The small heat shock protein Hsp27 binds α-synuclein fibrils, preventing elongation and cytotoxicity. *The Journal of Biological Chemistry, 293*(12), 4486–4497. doi:10.1074/jbc.M117.813865 PMID:29382725

Csermely, P. (2001). Chaperone overload is a possible contributor to 'civilization diseases'. *Trends in Genetics, 17*(12), 701–704. doi:10.1016/S0168-9525(01)02495-7 PMID:11718923

Cuervo, A. M., & Dice, J. F. (2000). Age-related decline in chaperone-mediated autophagy. *The Journal of Biological Chemistry, 275*(40), 31505–31513. doi:10.1074/jbc.M002102200 PMID:10806201

Cummings, C. J., Mancini, M. A., Antalffy, B., DeFranco, D. B., Orr, H. T., & Zoghbi, H. Y. (1998). Chaperone suppression of aggregation and altered subcellular proteasome localization imply protein misfolding in SCA1. *Nature Genetics, 19*(2), 148–154. doi:10.1038/502 PMID:9620770

d'Ydewalle, C., Krishnan, J., Chiheb, D. M., Van Damme, P., Irobi, J., Kozikowski, A. P., ... Van Den Bosch, L. (2011). HDAC6 inhibitors reverse axonal loss in a mouse model of mutant HSPB1-induced Charcot-Marie-Tooth disease. *Nature Medicine, 17*(8), 968–974. doi:10.1038/nm.2396 PMID:21785432

Davies, S. W., Turmaine, M., Cozens, B. A., DiFiglia, M., Sharp, A. H., Ross, C. A., ... Bates, G. P. (1997). Formation of neuronal intranuclear inclusions underlies the neurological dysfunction in mice transgenic for the HD mutation. *Cell, 90*(3), 537–548. doi:10.1016/S0092-8674(00)80513-9 PMID:9267033

Dedmon, M. M., Christodoulou, J., Wilson, M. R., & Dobson, C. M. (2005). Heat shock protein 70 inhibits alpha-synuclein fibril formation via preferential binding to prefibrillar species. *The Journal of Biological Chemistry, 280*(15), 14733–14740. doi:10.1074/jbc.M413024200 PMID:15671022

Dettmer, U., Newman, A. J., Luth, E. S., Bartels, T., & Selkoe, D. (2013). In vivo cross-linking reveals principally oligomeric forms of alpha-synuclein and beta-synuclein in neurons and non-neural cells. *The Journal of Biological Chemistry*, *288*(9), 6371–6385. doi:10.1074/jbc.M112.403311 PMID:23319586

Dickey, C., Kraft, C., Jinwal, U., Koren, J., Johnson, A., Anderson, L., ... Lewis, J. (2009). Aging Analysis Reveals Slowed Tau Turnover and Enhanced Stress Response in a Mouse Model of Tauopathy. *American Journal of Pathology*, *174*(1), 228–238. doi:10.2353/ajpath.2009.080764 PMID:19074615

Djouder, N. (Ed.). (2018). *Prefoldins: the new chaperones* (Vol. 1106). Springer. doi:10.1007/978-3-030-00737-9

Dong, Z., Wolfer, D. P., Lipp, H. P., & Bueler, H. (2005). Hsp70 gene transfer by adeno-associated virus inhibits MPTP-induced nigrostriatal degeneration in the mouse model of Parkinson disease. *Molecular Therapy*, *11*(1), 80–88. doi:10.1016/j.ymthe.2004.09.007 PMID:15585408

Dou, F., Netzer, W. J., Tanemura, K., Li, F., Hartl, F. U., Takashima, A., ... Xu, H. (2003). Chaperones increase association of tau protein with microtubules. *Proceedings of the National Academy of Sciences of the United States of America*, *100*(2), 721–726. doi:10.1073/pnas.242720499 PMID:12522269

Duennwald, M. L., Echeverria, A., & Shorter, J. (2012). Small heat shock proteins potentiate amyloid dissolution by protein disaggregases from yeast and humans. *PLoS Biology*, *10*(6), e1001346. doi:10.1371/journal.pbio.1001346 PMID:22723742

Ebrahimi-Fakhari, D., Wahlster, L., & McLean, P. J. (2011). Molecular chaperones in Parkinson's disease--present and future. *Journal of Parkinson's Disease*, *1*(4), 299–320. doi:10.3233/JPD-2011-11044 PMID:22279517

Ellis, J. (1987). Proteins as molecular chaperones. *Nature*, *328*(6129), 378–379. doi:10.1038/328378a0 PMID:3112578

Evgrafov, O. V., Mersiyanova, I., Irobi, J., Van Den Bosch, L., Dierick, I., Leung, C. L., ... Timmerman, V. (2004). Mutant small heat-shock protein 27 causes axonal Charcot-Marie-Tooth disease and distal hereditary motor neuropathy. *Nature Genetics*, *36*(6), 602–606. doi:10.1038/ng1354 PMID:15122254

Fan, C. Y., Lee, S., & Cyr, D. M. (2003). Mechanisms for regulation of Hsp70 function by Hsp40. *Cell Stress & Chaperones*, *8*(4), 309–316. doi:10.1379/1466-1268(2003)008<0309:MFROHF>2.0.CO;2 PMID:15115283

Frautschy, S. A., Hu, W., Kim, P., Miller, S. A., Chu, T., Harris-White, M. E., & Cole, G. M. (2001). Phenolic anti-inflammatory antioxidant reversal of Aβ-induced cognitive deficits and neuropathology. *Neurobiology of Aging*, *22*(6), 993–1005. doi:10.1016/S0197-4580(01)00300-1 PMID:11755008

Garrido, C., Brunet, M., Didelot, C., Zermati, Y., Schmitt, E., & Kroemer, G. (2006). Heat shock proteins 27 and 70: Anti-apoptotic proteins with tumorigenic properties. *Cell Cycle (Georgetown, Tex.)*, *5*(22), 2592–2601. doi:10.4161/cc.5.22.3448 PMID:17106261

Geschwind, M. D. (2015). Prion Diseases. *Continuum (Minneap Minn)*, *21*(6), 1612-1638. doi:10.1212/con.0000000000000251

Gess, B., Auer-Grumbach, M., Schirmacher, A., Strom, T., Zitzelsberger, M., Rudnik-Schoneborn, S., ... Senderek, J. (2014). HSJ1-related hereditary neuropathies: Novel mutations and extended clinical spectrum. *Neurology, 83*(19), 1726–1732. doi:10.1212/WNL.0000000000000966 PMID:25274842

Girard, M., Lariviere, R., Parfitt, D. A., Deane, E. C., Gaudet, R., Nossova, N., ... McPherson, P. S. (2012). Mitochondrial dysfunction and Purkinje cell loss in autosomal recessive spastic ataxia of Charlevoix-Saguenay (ARSACS). *Proceedings of the National Academy of Sciences of the United States of America, 109*(5), 1661–1666. doi:10.1073/pnas.1113166109 PMID:22307627

Greer, P. L., Hanayama, R., Bloodgood, B. L., Mardinly, A. R., Lipton, D. M., Flavell, S. W., ... Greenberg, M. E. (2010). The Angelman Syndrome protein Ube3A regulates synapse development by ubiquitinating arc. *Cell, 140*(5), 704–716. doi:10.1016/j.cell.2010.01.026 PMID:20211139

Han, W., & Christen, P. (2004). cis-Effect of DnaJ on DnaK in ternary complexes with chimeric DnaK/DnaJ-binding peptides. *FEBS Letters, 563*(1), 146–150. doi:10.1016/S0014-5793(04)00290-X PMID:15063739

Hartl, F. U., Bracher, A., & Hayer-Hartl, M. (2011). Molecular chaperones in protein folding and proteostasis. *Nature, 475*(7356), 324–332. doi:10.1038/nature10317 PMID:21776078

Hartl, F. U., & Hayer-Hartl, M. (2002). Molecular chaperones in the cytosol: From nascent chain to folded protein. *Science, 295*(5561), 1852–1858. doi:10.1126cience.1068408 PMID:11884745

Hasegawa, T., Yoshida, S., Sugeno, N., Kobayashi, J., & Aoki, M. (2018). DnaJ/Hsp40 family and Parkinson's disease. *Frontiers in Neuroscience, 11*, 743. doi:10.3389/fnins.2017.00743 PMID:29367843

Heck, J. W., Cheung, S. K., & Hampton, R. Y. (2010). Cytoplasmic protein quality control degradation mediated by parallel actions of the E3 ubiquitin ligases Ubr1 and San1. *Proceedings of the National Academy of Sciences of the United States of America, 107*(3), 1106–1111. doi:10.1073/pnas.0910591107 PMID:20080635

Heydari, A. R., You, S., Takahashi, R., Gutsmann-Conrad, A., Sarge, K. D., & Richardson, A. (2000). Age-related alterations in the activation of heat shock transcription factor 1 in rat hepatocytes. *Experimental Cell Research, 256*(1), 83–93. doi:10.1006/excr.2000.4808 PMID:10739655

Hoter, A., El-Sabban, M. E., & Naim, H. Y. (2018). The HSP90 Family: Structure, Regulation, Function, and Implications in Health and Disease. *International Journal of Molecular Sciences, 19*(9), 2560. doi:10.3390/ijms19092560 PMID:30158430

Houlden, H., Laura, M., Wavrant-De Vrieze, F., Blake, J., Wood, N., & Reilly, M. M. (2008). Mutations in the HSP27 (HSPB1) gene cause dominant, recessive, and sporadic distal HMN/CMT type 2. *Neurology, 71*(21), 1660–1668. doi:10.1212/01.wnl.0000319696.14225.67 PMID:18832141

Huang, C., Cheng, H., Hao, S., Zhou, H., Zhang, X., Gao, J., ... Wang, C. C. (2006). Heat shock protein 70 inhibits alpha-synuclein fibril formation via interactions with diverse intermediates. *Journal of Molecular Biology, 364*(3), 323–336. doi:10.1016/j.jmb.2006.08.062 PMID:17010992

Ikeda, Y., Abe, A., Ishida, C., Takahashi, K., Hayasaka, K., & Yamada, M. (2009). A clinical phenotype of distal hereditary motor neuronopathy type II with a novel HSPB1 mutation. *Journal of the Neurological Sciences, 277*(1-2), 9–12. doi:10.1016/j.jns.2008.09.031 PMID:18952241

Itoh, H., Komatsuda, A., Ohtani, H., Wakui, H., Imai, H., Sawada, K., ... Hamada, F. (2002). Mammalian HSP60 is quickly sorted into the mitochondria under conditions of dehydration. *European Journal of Biochemistry, 269*(23), 5931–5938. doi:10.1046/j.1432-1033.2002.03317.x PMID:12444982

Jana, N. R., Tanaka, M., Wang, G., & Nukina, N. (2000). Polyglutamine length-dependent interaction of Hsp40 and Hsp70 family chaperones with truncated N-terminal huntingtin: Their role in suppression of aggregation and cellular toxicity. *Human Molecular Genetics, 9*(13), 2009–2018. doi:10.1093/hmg/9.13.2009 PMID:10942430

Jia, H., Halilou, A. I., Hu, L., Cai, W., Liu, J., & Huang, B. (2011). Heat shock protein 10 (Hsp10) in immune-related diseases: One coin, two sides. *International Journal of Biochemistry and Molecular Biology, 2*(1), 47. PMID:21969171

Jung, A. E., Fitzsimons, H. L., Bland, R. J., During, M. J., & Young, D. (2008). HSP70 and constitutively active HSF1 mediate protection against CDCrel-1-mediated toxicity. *Molecular Therapy, 16*(6), 1048–1055. doi:10.1038/mt.2008.68 PMID:18398426

Kalia, S. K., Lee, S., Smith, P. D., Liu, L., Crocker, S. J., Thorarinsdottir, T. E., ... Lozano, A. M. (2004). BAG5 Inhibits Parkin and Enhances Dopaminergic Neuron Degeneration. *Neuron, 44*(6), 931–945. doi:10.1016/j.neuron.2004.11.026 PMID:15603737

Kalmar, B., & Greensmith, L. (2017). Cellular Chaperones As Therapeutic Targets in ALS to Restore Protein Homeostasis and Improve Cellular Function. *Frontiers in Molecular Neuroscience, 10*, 251. doi:10.3389/fnmol.2017.00251 PMID:28943839

Kampinga, H. H., & Craig, E. A. (2010). The HSP70 chaperone machinery: J proteins as drivers of functional specificity. *Nature Reviews. Molecular Cell Biology, 11*(8), 579–592. doi:10.1038/nrm2941 PMID:20651708

Kijima, K., Numakura, C., Goto, T., Takahashi, T., Otagiri, T., Umetsu, K., & Hayasaka, K. (2005). Small heat shock protein 27 mutation in a Japanese patient with distal hereditary motor neuropathy. *Journal of Human Genetics, 50*(9), 473–476. doi:10.100710038-005-0280-6 PMID:16155736

Kirbach, B. B., & Golenhofen, N. (2011). Differential expression and induction of small heat shock proteins in rat brain and cultured hippocampal neurons. *Journal of Neuroscience Research, 89*(2), 162–175. doi:10.1002/jnr.22536 PMID:21162124

Kleizen, B., & Braakman, I. (2004). Protein folding and quality control in the endoplasmic reticulum. *Current Opinion in Cell Biology, 16*(4), 343–349. doi:10.1016/j.ceb.2004.06.012 PMID:15261665

Kozlov, G., Denisov, A. Y., Girard, M., Dicaire, M. J., Hamlin, J., McPherson, P. S., ... Gehring, K. (2011). Structural basis of defects in the sacsin HEPN domain responsible for autosomal recessive spastic ataxia of Charlevoix-Saguenay (ARSACS). *The Journal of Biological Chemistry, 286*(23), 20407–20412. doi:10.1074/jbc.M111.232884 PMID:21507954

Lackie, R. E., Maciejewski, A., Ostapchenko, V. G., Marques-Lopes, J., Choy, W. Y., Duennwald, M. L., ... Prado, M. A. (2017). The Hsp70/Hsp90 chaperone machinery in neurodegenerative diseases. *Frontiers in Neuroscience, 11*, 254. doi:10.3389/fnins.2017.00254 PMID:28559789

Langklotz, S., Baumann, U., & Narberhaus, F. (2012). Structure and function of the bacterial AAA protease FtsH. *Biochimica et Biophysica Acta (BBA) -. Molecular Cell Research, 1823*(1), 40–48. doi:10.1016/j.bbamcr.2011.08.015 PMID:21925212

Leak, R. K. (2014). Heat shock proteins in neurodegenerative disorders and aging. *Journal of Cell Communication and Signaling, 8*(4), 293–310. doi:10.100712079-014-0243-9 PMID:25208934

Liu, J., Lillo, C., Jonsson, P. A., Velde, C. V., Ward, C. M., Miller, T. M., ... Cleveland, D. W. (2004). Toxicity of Familial ALS-Linked SOD1 Mutants from Selective Recruitment to Spinal Mitochondria. *Neuron, 43*(1), 5–17. doi:10.1016/j.neuron.2004.06.016 PMID:15233913

Liu, Y. H., Han, Y. L., Song, J., Wang, Y., Jing, Y. Y., Shi, Q., ... Dong, X. P. (2011). Heat shock protein 104 inhibited the fibrillization of prion peptide 106-126 and disassembled prion peptide 106-126 fibrils in vitro. *The International Journal of Biochemistry & Cell Biology, 43*(5), 768–774. doi:10.1016/j.biocel.2011.01.022 PMID:21296677

Lo Bianco, C., Shorter, J., Regulier, E., Lashuel, H., Iwatsubo, T., Lindquist, S., & Aebischer, P. (2008). Hsp104 antagonizes alpha-synuclein aggregation and reduces dopaminergic degeneration in a rat model of Parkinson disease. *The Journal of Clinical Investigation, 118*(9), 3087–3097. doi:10.1172/JCI35781 PMID:18704197

Luigetti, M., Fabrizi, G. M., Madia, F., Ferrarini, M., Conte, A., Del Grande, A., ... Sabatelli, M. (2010). A novel HSPB1 mutation in an Italian patient with CMT2/dHMN phenotype. *Journal of the Neurological Sciences, 298*(1-2), 114–117. doi:10.1016/j.jns.2010.09.008 PMID:20870250

Luo, G.-R., Chen, S., & Le, W.-D. (2006). Are heat shock proteins therapeutic target for Parkinson's disease? *International Journal of Biological Sciences, 3*(1), 20–26. doi:10.7150/ijbs.3.20 PMID:17200688

Maiti, P., Manna, J., Veleri, S., & Frautschy, S. (2014). Molecular chaperone dysfunction in neurodegenerative diseases and effects of curcumin. *BioMed Research International, 495091*, 1–14. doi:10.1155/2014/495091 PMID:25386560

Mandich, P., Grandis, M., Varese, A., Geroldi, A., Acquaviva, M., Ciotti, P., ... Bellone, E. (2010). Severe neuropathy after diphtheria-tetanus-pertussis vaccination in a child carrying a novel frameshift mutation in the small heat-shock protein 27 gene. *Journal of Child Neurology, 25*(1), 107–109. doi:10.1177/0883073809334387 PMID:19435728

Mapa, K., Sikor, M., Kudryavtsev, V., Waegemann, K., Kalinin, S., Seidel, C. A. M., ... Mokranjac, D. (2010). The Conformational Dynamics of the Mitochondrial Hsp70 Chaperone. *Molecular Cell, 38*(1), 89–100. doi:10.1016/j.molcel.2010.03.010 PMID:20385092

Marino Gammazza, A., Caruso Bavisotto, C., Barone, R., Macario, E. C. D., & Macario, JL, A. (. (2016). Alzheimer's disease and molecular chaperones: Current knowledge and the future of chaperonotherapy. *Current Pharmaceutical Design, 22*(26), 4040–4049. doi:10.2174/13816128226661605 18141437 PMID:27189602

Matts, R. L., Brandt, G. E. L., Lu, Y., Dixit, A., Mollapour, M., Wang, S., ... Blagg, B. S. J. (2011). A systematic protocol for the characterization of Hsp90 modulators. *Bioorganic & Medicinal Chemistry, 19*(1), 684–692. doi:10.1016/j.bmc.2010.10.029 PMID:21129982

Mayer, M. P., & Bukau, B. (2005). Hsp70 chaperones: Cellular functions and molecular mechanism. *Cellular and molecular life sciences. Cellular and Molecular Life Sciences, 62*(6), 670–684. doi:10.100700018-004-4464-6 PMID:15770419

McLaughlin, T., Falkowski, M., Wang, J. J., & Zhang, S. X. (2018). Molecular chaperone ERp29: A potential target for cellular protection in retinal and neurodegenerative diseases. In *Retinal Degenerative Diseases* (pp. 421–427). Cham: Springer; doi:10.1007/978-3-319-75402-4_52

McLean, P. J., Kawamata, H., & Hyman, B. T. (2001). Alpha-synuclein-enhanced green fluorescent protein fusion proteins form proteasome sensitive inclusions in primary neurons. *Neuroscience, 104*(3), 901–912. doi:10.1016/S0306-4522(01)00113-0 PMID:11440819

McLean, P. J., Klucken, J., Shin, Y., & Hyman, B. T. (2004). Geldanamycin induces Hsp70 and prevents alpha-synuclein aggregation and toxicity in vitro. *Biochemical and Biophysical Research Communications, 321*(3), 665–669. doi:10.1016/j.bbrc.2004.07.021 PMID:15358157

Mymrikov, E. V., Seit-Nebi, A. S., & Gusev, N. B. (2011). Large potentials of small heat shock proteins. *Physiological Reviews, 91*(4), 1123–1159. doi:10.1152/physrev.00023.2010 PMID:22013208

Nigar, S., Pottoo, F. H., Tabassum, N., Verma, S. K., & Javed, M. N. (2016). Molecular Insights into the Role of Inflammation and Oxidative Stress in Epilepsy. *Journal of Advances in Medical and Pharmaceutical Sciences, 10*(1), 1–9. doi:10.9734/JAMPS/2016/24441

Okado-Matsumoto, A., & Fridovich, I. (2002). Amyotrophic lateral sclerosis: A proposed mechanism. *Proceedings of the National Academy of Sciences of the United States of America, 99*(13), 9010–9014. doi:10.1073/pnas.132260399 PMID:12060716

Parfitt, D. A., Michael, G. J., Vermeulen, E. G., Prodromou, N. V., Webb, T. R., Gallo, J. M., ... Chapple, J. P. (2009). The ataxia protein sacsin is a functional co-chaperone that protects against polyglutamine-expanded ataxin-1. *Human Molecular Genetics, 18*(9), 1556–1565. doi:10.1093/hmg/ddp067 PMID:19208651

Patel, A., Lee, H. O., Jawerth, L., Maharana, S., Jahnel, M., Hein, M. Y., ... Alberti, S. (2015). A Liquid-to-Solid Phase Transition of the ALS Protein FUS Accelerated by Disease Mutation. *Cell, 162*(5), 1066–1077. doi:10.1016/j.cell.2015.07.047 PMID:26317470

Patel, Y. J., Payne Smith, M. D., de Belleroche, J., & Latchman, D. S. (2005). Hsp27 and Hsp70 administered in combination have a potent protective effect against FALS-associated SOD1-mutant-induced cell death in mammalian neuronal cells. *Brain Research. Molecular Brain Research, 134*(2), 256–274. doi:10.1016/j.molbrainres.2004.10.028 PMID:15836922

Patel, Y. J. K., Payne Smith, M. D., de Belleroche, J., & Latchman, D. S. (2005). Hsp27 and Hsp70 administered in combination have a potent protective effect against FALS-associated SOD1-mutant-induced cell death in mammalian neuronal cells. *Brain Research. Molecular Brain Research, 134*(2), 256–274. doi:10.1016/j.molbrainres.2004.10.028 PMID:15836922

Petrucelli, L., Dickson, D., Kehoe, K., Taylor, J., Snyder, H., Grover, A., ... Hutton, M. (2004). CHIP and Hsp70 regulate tau ubiquitination, degradation and aggregation. *Human Molecular Genetics*, *13*(7), 703–714. doi:10.1093/hmg/ddh083 PMID:14962978

Pottoo, F. H., Bhowmik, M., & Vohora, D. (2014). Raloxifene protects against seizures and neurodegeneration in a mouse model mimicking epilepsy in postmenopausal woman. *European Journal of Pharmaceutical Sciences*, *65*, 167–173. doi:10.1016/j.ejps.2014.09.002 PMID:25218046

Pottoo, F. H., Tabassum, N., Javed, M. N., Nigar, S., Rasheed, R., Khan, A., ... Ashraf, G. M. (2019). The Synergistic Effect of Raloxifene, Fluoxetine, and Bromocriptine Protects Against Pilocarpine-Induced Status Epilepticus and Temporal Lobe Epilepsy. *Molecular Neurobiology*, *56*(2), 1233–1247. doi:10.100712035-018-1121-x PMID:29881945

Pozsgai, E., Gomori, E., Szigeti, A., Boronkai, A., Gallyas, F. Jr, Sumegi, B., & Bellyei, S. (2007). Correlation between the progressive cytoplasmic expression of a novel small heat shock protein (Hsp16.2) and malignancy in brain tumors. *BMC Cancer*, *7*(1), 233. doi:10.1186/1471-2407-7-233 PMID:18154656

Pratt, W. B., & Toft, D. O. (2003). Regulation of signaling protein function and trafficking by the hsp90/hsp70-based chaperone machinery. *Experimental Biology and Medicine (Maywood, N.J.)*, *228*(2), 111–133. doi:10.1177/153537020322800201 PMID:12563018

Putcha, P., Danzer, K. M., Kranich, L. R., Scott, A., Silinski, M., Mabbett, S., ... McLean, P. J. (2010). Brain-permeable small-molecule inhibitors of Hsp90 prevent alpha-synuclein oligomer formation and rescue alpha-synuclein-induced toxicity. *The Journal of Pharmacology and Experimental Therapeutics*, *332*(3), 849–857. doi:10.1124/jpet.109.158436 PMID:19934398

Putcha, P., Danzer, K. M., Kranich, L. R., Scott, A., Silinski, M., Mabbett, S., ... McLean, P. J. (2010). Brain-permeable small-molecule inhibitors of Hsp90 prevent alpha-synuclein oligomer formation and rescue alpha-synuclein-induced toxicity. *The Journal of Pharmacology and Experimental Therapeutics*, *332*(3), 849–857. doi:10.1124/jpet.109.158436 PMID:19934398

Ranford, J. C., Coates, A. R., & Henderson, B. (2000). Chaperonins are cell-signalling proteins: The unfolding biology of molecular chaperones. *Expert Reviews in Molecular Medicine*, *2*(8), 1–17. doi:10.1017/S1462399400002015 PMID:14585136

Rathor, R., & Suryakumar, G. (2019). Upregulation of Cytoprotective Chaperones Mediate Better Tolerance to High Altitude. In *Chaperokine Activity of Heat Shock Proteins* (pp. 123–145). Cham: Springer. doi:10.1007/978-3-030-02254-9_6

Ravikumar, B., Sarkar, S., & Rubinsztein, D. C. (2008). Clearance of mutant aggregate-prone proteins by autophagy. In *Autophagosome and Phagosome* (pp. 195–211). Humana Press. doi:10.1007/978-1-59745-157-4_13

Robberecht, W., & Philips, T. (2013). The changing scene of amyotrophic lateral sclerosis. *Nature Reviews. Neuroscience*, *14*(4), 248–264. doi:10.1038/nrn3430 PMID:23463272

Sahara, N., Maeda, S., Murayama, M., Suzuki, T., Dohmae, N., Yen, S. H., & Takashima, A. (2007). Assembly of two distinct dimers and higher-order oligomers from full-length tau. *The European Journal of Neuroscience*, *25*(10), 3020–3029. doi:10.1111/j.1460-9568.2007.05555.x PMID:17561815

Sarkar, M., Kuret, J., & Lee, G. (2008). Two motifs within the tau microtubule-binding domain mediate its association with the hsc70 molecular chaperone. *Journal of Neuroscience Research, 86*(12), 2763–2773. doi:10.1002/jnr.21721 PMID:18500754

Sarto, C., Binz, P. A., & Mocarelli, P. (2000). Heat shock proteins in human cancer. *Electrophoresis, 21*(6), 1218–1226. doi:10.1002/(SICI)1522-2683(20000401)21:6<1218::AID-ELPS1218>3.0.CO;2-H PMID:10786894

Schirmer, E. C., Glover, J. R., Singer, M. A., & Lindquist, S. (1996). HSP100/Clp proteins: A common mechanism explains diverse functions. *Trends in Biochemical Sciences, 21*(8), 289–296. doi:10.1016/S0968-0004(96)10038-4 PMID:8772382

Selkoe, D. J. (2004). Cell biology of protein misfolding: The examples of Alzheimer's and Parkinson's diseases. *Nature Cell Biology, 6*(11), 1054–1061. doi:10.1038/ncb1104-1054 PMID:15516999

Sharma, S. K., Christen, P., & Goloubinoff, P. (2009). Disaggregating chaperones: An unfolding story. *Current Protein & Peptide Science, 10*(5), 432–446. doi:10.2174/138920309789351930 PMID:19538153

Sharp, P., Krishnan, M., Pullar, O., Navarrete, R., Wells, D., & Belleroche, J. (2006). Heat shock protein 27 rescues motor neurons following nerve injury and preserves muscle function. *Experimental Neurology, 198*(2), 511–518. doi:10.1016/j.expneurol.2005.12.031 PMID:16497297

Shaw, B. F., Lelie, H. L., Durazo, A., Nersissian, A. M., Xu, G., Chan, P. K., ... Whitelegge, J. P. (2008). Detergent-insoluble aggregates associated with amyotrophic lateral sclerosis in transgenic mice contain primarily full-length, unmodified superoxide dismutase-1. *The Journal of Biological Chemistry, 283*(13), 8340–8350. doi:10.1074/jbc.M707751200 PMID:18192269

Shen, H. Y., He, J. C., Wang, Y., Huang, Q. Y., & Chen, J. F. (2005). Geldanamycin induces heat shock protein 70 and protects against MPTP-induced dopaminergic neurotoxicity in mice. *The Journal of Biological Chemistry, 280*(48), 39962–39969. doi:10.1074/jbc.M505524200 PMID:16210323

Shorter, J. (2008). Hsp104: A Weapon to Combat Diverse Neurodegenerative Disorders. *Neuro-Signals, 16*(1), 63–74. doi:10.1159/000109760 PMID:18097161

Shulman, J. M., De Jager, P. L., & Feany, M. B. (2011). Parkinson's disease: Genetics and pathogenesis. *Annual Review of Pathology, 6*(1), 193–222. doi:10.1146/annurev-pathol-011110-130242 PMID:21034221

Smith, H. L., Li, W., & Cheetham, M. E. (2015). Molecular chaperones and neuronal proteostasis. *Seminars in Cell & Developmental Biology, 40*, 142–152. doi:10.1016/j.semcdb.2015.03.003 PMID:25770416

Soti, C., & Csermely, P. (2002). Chaperones and aging: Role in neurodegeneration and in other civilizational diseases. *Neurochemistry International, 41*(6), 383–389. doi:10.1016/S0197-0186(02)00043-8 PMID:12213225

Soti, C., & Csermely, P. (2003). Aging and molecular chaperones. *Experimental Gerontology, 38*(10), 1037–1040. doi:10.1016/S0531-5565(03)00185-2 PMID:14580856

Spillantini, M. G., Crowther, R. A., Jakes, R., Hasegawa, M., & Goedert, M. (1998). alpha-Synuclein in filamentous inclusions of Lewy bodies from Parkinson's disease and dementia with lewy bodies. *Proceedings of the National Academy of Sciences of the United States of America*, *95*(11), 6469–6473. doi:10.1073/pnas.95.11.6469 PMID:9600990

Stefani, M., & Dobson, C. M. (2003). Protein aggregation and aggregate toxicity: New insights into protein folding, misfolding diseases and biological evolution. *Journal of Molecular Medicine (Berlin, Germany)*, *81*(11), 678–699. doi:10.100700109-003-0464-5 PMID:12942175

Stetler, R. A., Gao, Y., Signore, A. P., Cao, G., & Chen, J. (2009). HSP27: Mechanisms of cellular protection against neuronal injury. *Current Molecular Medicine*, *9*(7), 863–872. doi:10.2174/156652409789105561 PMID:19860665

Sun, H., Gao, J., Ferrington, D. A., Biesiada, H., Williams, T. D., & Squier, T. C. (1999). Repair of oxidized calmodulin by methionine sulfoxide reductase restores ability to activate the plasma membrane Ca-ATPase. *Biochemistry*, *38*(1), 105–112. doi:10.1021/bi981295k PMID:9890888

Sun, Y., & MacRae, T. H. (2005). The small heat shock proteins and their role in human disease. *The FEBS Journal*, *272*(11), 2613–2627. doi:10.1111/j.1742-4658.2005.04708.x PMID:15943797

Tasset, I., & Cuervo, A. M. (2016). Role of chaperone-mediated autophagy in metabolism. *The FEBS Journal*, *283*(13), 2403–2413. doi:10.1111/febs.13677 PMID:26854402

Tatsuta, T., Model, K., & Langer, T. (2005). Formation of membrane-bound ring complexes by prohibitins in mitochondria. *Molecular Biology of the Cell*, *16*(1), 248–259. doi:10.1091/mbc.e04-09-0807 PMID:15525670

Taylor, D. R., & Hooper, N. M. (2006). The prion protein and lipid rafts. *Molecular Membrane Biology*, *23*(1), 89–99. doi:10.1080/09687860500449994 PMID:16611584

Thakur, P., & Nehru, B. (2014). Long-term heat shock proteins (HSPs) induction by carbenoxolone improves hallmark features of Parkinson's disease in a rotenone-based model. *Neuropharmacology*, *79*, 190–200. doi:10.1016/j.neuropharm.2013.11.016 PMID:24296154

Tiroli-Cepeda, A. O., & Ramos, C. H. (2011). An overview of the role of molecular chaperones in protein homeostasis. *Protein and Peptide Letters*, *18*(2), 101–109. doi:10.2174/092986611794475093 PMID:21121892

Tóth, M. E., Szegedi, V., Varga, E., Juhász, G., Horváth, J., Borbély, E., ... Sántha, M. (2013). Overexpression of Hsp27 ameliorates symptoms of Alzheimer's disease in APP/PS1 mice. *Cell Stress & Chaperones*, *18*(6), 759–771. doi:10.100712192-013-0428-9 PMID:23605646

Traynor, B. J., Bruijn, L., Conwit, R., Beal, F., O'Neill, G., Fagan, S. C., & Cudkowicz, M. E. (2006). Neuroprotective agents for clinical trials in ALS: A systematic assessment. *Neurology*, *67*(1), 20–27. doi:10.1212/01.wnl.0000223353.34006.54 PMID:16832072

Trepel, J., Mollapour, M., Giaccone, G., & Neckers, L. (2010). Targeting the dynamic HSP90 complex in cancer. *Nature Reviews. Cancer*, *10*(8), 537–549. doi:10.1038/nrc2887 PMID:20651736

Uryu, K., Richter-Landsberg, C., Welch, W., Sun, E., Goldbaum, O., Norris, E. H., ... Trojanowski, J. Q. (2006). Convergence of heat shock protein 90 with ubiquitin in filamentous alpha-synuclein inclusions of alpha-synucleinopathies. *American Journal of Pathology, 168*(3), 947–961. doi:10.2353/ajpath.2006.050770 PMID:16507910

Van Montfort, R., Slingsby, C., & Vierling, E. (2001). Structure and function of the small heat shock protein/alpha-crystallin family of molecular chaperones. *Advances in Protein Chemistry, 59*, 105–156. doi:10.1016/S0065-3233(01)59004-X PMID:11868270

Veereshwarayya, V., Kumar, P., Rosen, K. M., Mestril, R., & Querfurth, H. W. (2006). Differential effects of mitochondrial heat shock protein 60 and related molecular chaperones to prevent intracellular beta-amyloid-induced inhibition of complex IV and limit apoptosis. *The Journal of Biological Chemistry, 281*(40), 29468–29478. doi:10.1074/jbc.M602533200 PMID:16887805

Wang, Y., Zhou, J. S., Xu, X. C., Li, Z. Y., Chen, H. P., Ying, S. M., . . . Chen, Z. H. (2018). Endoplasmic reticulum chaperone GRP78 mediates cigarette smoke-induced necroptosis and injury in bronchial epithelium. *Int J Chron obstruct Pulmonary Dis, 13*, 571-581.

Wilhelmus, M. M., Boelens, W. C., Otte-Holler, I., Kamps, B., de Waal, R. M., & Verbeek, M. M. (2006). Small heat shock proteins inhibit amyloid-beta protein aggregation and cerebrovascular amyloid-beta protein toxicity. *Brain Research, 1089*(1), 67–78. doi:10.1016/j.brainres.2006.03.058 PMID:16635482

Yang, Y., Turner, R. S., & Gaut, J. R. (1998). The chaperone BiP/GRP78 binds to amyloid precursor protein and decreases Abeta40 and Abeta42 secretion. *The Journal of Biological Chemistry, 273*(40), 25552–25555. doi:10.1074/jbc.273.40.25552 PMID:9748217

Yerbury, J. J., Gower, D., Vanags, L., Roberts, K., Lee, J. A., & Ecroyd, H. (2013). The small heat shock proteins αB-crystallin and Hsp27 suppress SOD1 aggregation in vitro. *Cell Stress & Chaperones, 18*(2), 251–257. doi:10.100712192-012-0371-1 PMID:22993064

Zhai, J., Lin, H., Julien, J. P., & Schlaepfer, W. W. (2007). Disruption of neurofilament network with aggregation of light neurofilament protein: A common pathway leading to motor neuron degeneration due to Charcot-Marie-Tooth disease-linked mutations in NFL and HSPB1. *Human Molecular Genetics, 16*(24), 3103–3116. doi:10.1093/hmg/ddm272 PMID:17881652

Zhao, R., Davey, M., Hsu, Y. C., Kaplanek, P., Tong, A., Parsons, A. B., ... Houry, W. A. (2005). Navigating the chaperone network: An integrative map of physical and genetic interactions mediated by the hsp90 chaperone. *Cell, 120*(5), 715–727. doi:10.1016/j.cell.2004.12.024 PMID:15766533

Zhu, X., Zhao, X., Burkholder, W. F., Gragerov, A., Ogata, C. M., Gottesman, M. E., & Hendrickson, W. A. (1996). Structural analysis of substrate binding by the molecular chaperone DnaK. *Science, 272*(5268), 1606–1614. doi:10.1126cience.272.5268.1606 PMID:8658133

Zolkiewski, M., Zhang, T., & Nagy, M. (2012). Aggregate reactivation mediated by the Hsp100 chaperones. *Archives of Biochemistry and Biophysics, 520*(1), 1–6. doi:10.1016/j.abb.2012.01.012 PMID:22306514

Compilation of References

1. UK5. (n.d). *Solution structure of the Murine BAG domain of Bcl2-associated athanogene 3*. Retrieved from https://www.rcsb.org/structure/1UK5

Aamodt, E. J., & Williams, R. C. Jr. (1984). Microtubule-associated proteins connect microtubules and neurofilaments in vitro. *Biochemistry*, *23*(25), 6023–6031. doi:10.1021/bi00320a019 PMID:6543144

Abeliovich, A., & Gitler, A. D. (2016). Defects in trafficking bridge Parkinson's disease pathology and genetics. *Nature*, *539*(7628), 207–216. doi:10.1038/nature20414 PMID:27830778

Abeywardana, T., Lin, Y. H., Rott, R., Engelender, S., & Pratt, M. R. (2013). Site-specific differences in proteasome-dependent degradation of monoubiquitinated α-synuclein. *Chemistry & Biology*, *20*(10), 1207–1213. doi:10.1016/j.chembiol.2013.09.009 PMID:24210006

Abisambra, J. F., Blair, L. J., Hill, S. E., Jones, J. R., Kraft, C., Rogers, J., ... Dickey, C. A. (2010). Phosphorylation dynamics regulate Hsp27-mediated rescue of neuronal plasticity deficits in tau transgenic mice. *The Journal of Neuroscience*, *30*(46), 15374–15382. doi:10.1523/JNEUROSCI.3155-10.2010 PMID:21084594

Abisambra, J. F., Jinwal, U. K., Blair, L. J., O'Leary, J. C., Li, Q., Brady, S., ... Dickey, C. A. (2013). Tau accumulation activates the unfolded protein response by impairing endoplasmic reticulum-associated degradation. *The Journal of Neuroscience: The Official Journal of the Society for Neuroscience*, *33*(22), 9498–9507. doi:10.1523/JNEUROSCI.5397-12.2013 PMID:23719816

Abou-Sleiman, P. M., Muqit, M. M., & Wood, N. W. (2006). Expanding insights of mitochondrial dysfunction in Parkinson's disease. *Nature Reviews. Neuroscience*, *7*(3), 207–219. doi:10.1038/nrn1868 PMID:16495942

Abrams, A. J., Farooq, A., & Wang, G. (2011). S-Nitrosylation of ApoE in Alzheimer's disease. *Biochemistry*, *50*(17), 3405–3407. doi:10.1021/bi200266v PMID:21443265

Abushouk, A. I., Negida, A., Elshenawy, R. A., Zein, H., Hammad, A. M., Menshawy, A., & Mohamed, W. M. Y. (2018). C-Abl Inhibition; A Novel Therapeutic Target for Parkinson's Disease. *CNS & Neurological Disorders - Drug Targets*, *17*(1), 14–21. doi:10.2174/1871527316666170602101538 PMID:28571531

Acosta-Alvear, D., Zhou, Y., Blais, A., Tsikitis, M., Lents, N. H., Arias, C., ... Dynlacht, B. D. (2007). XBP1 controls diverse cell type- and condition- specific transcriptional regulatory networks. *Molecular Cell*, *27*(1), 53–66. doi:10.1016/j.molcel.2007.06.011 PMID:17612490

Acquarone, E., Argyrousi, E. K., van den Berg, M., Gulisano, W., Fa, M., Staniszewski, A., ... Fiorito, J. (2019). Synaptic and memory dysfunction induced by tau oligomers is rescued by up-regulation of the nitric oxide cascade. *Molecular Neurodegeneration*, *14*(1), 14. doi:10.118613024-019-0326-4 PMID:31248451

Adelman, R., Saul, R. L., & Ames, B. N. (1988). Oxidative damage to DNA: Relation to species metabolic rate and life span. *Proceedings of the National Academy of Sciences of the United States of America*, *85*(8), 2706–2708. doi:10.1073/pnas.85.8.2706 PMID:3128794

Aebi, M. (2013). N-linked protein glycosylation in the ER. *BBA-Mol Cell Res*, *1833*, 2430–2437. PMID:23583305

Agarwal, S., Tiwari, S. K., Seth, B., Yadav, A., Singh, A., Mudawal, A., ... Kumar, A. (2015). Activation of autophagic flux against xenoestrogen bisphenol-A-induced hippocampal neurodegeneration via AMP kinase (AMPK)/mammalian target of rapamycin (mTOR) pathways. *The Journal of Biological Chemistry*, *290*(34), 21163–21184. doi:10.1074/jbc.M115.648998 PMID:26139607

Ahmad, S., Khan, H., Shahab, U., Rehman, S., Rafi, Z., Khan, M. Y., ... Uddin, M. (2017). Protein oxidation: An overview of metabolism of sulphur containing amino acid, cysteine. *Frontiers in Bioscience*, *1*(1), 71–87. doi:10.2741474 PMID:27814576

Ai, S. X., Xu, Q., Hu, Y. C., Song, C. Y., Guo, J. F., Shen, L., ... Tang, B. S. (2014). Hypomethylation of SNCA in blood of patients with sporadic Parkinson's disease. *Journal of the Neurological Sciences*, *337*(1-2), 123–128. doi:10.1016/j.jns.2013.11.033 PMID:24326201

Aita, V. M., Liang, X. H., Murty, V. V. V. S., Pincus, D. L., Yu, W., Cayanis, E., ... Levine, B. (1999). Cloning and Genomic Organization of Beclin 1, a Candidate Tumor Suppressor Gene on Chromosome 17q21. *Genomics*, *59*(1), 59–65. doi:10.1006/geno.1999.5851 PMID:10395800

Ajroud-Driss, S., & Siddique, T. (2015). Sporadic and hereditary amyotrophic lateral sclerosis (ALS). *Bba-Mol Basis Dis*, *1852*(4), 679–684. doi:10.1016/j.bbadis.2014.08.010 PMID:25193032

Akhtar, M. W., Sunico, C. R., Nakamura, T., & Lipton, S. A. (2012). Redox Regulation of Protein Function via Cysteine S-Nitrosylation and Its Relevance to Neurodegenerative Diseases. *International Journal of Cell Biology*, *9*, 1–9. doi:10.1155/2012/463756 PMID:22956959

Akhtar, R. S., Ness, J. M., & Roth, K. A. (2004). Bcl-2 family regulation of neuronal development and neurodegeneration. *Biochimica et Biophysica Acta (BBA) - Molecular Cell Research*, *1644*(2–3), 189–203. PMID:14996503

Akterin, S., Cowburn, R. F., Miranda-Vizuete, A., Jiménez, A., Bogdanovic, N., Winblad, B., & Cedazo-Minguez, A. (2006). Involvement of glutaredoxin-1 and thioredoxin-1 in β-amyloid toxicity and Alzheimer's disease. *Cell Death and Differentiation*, *13*(9), 1454–1465. doi:10.1038j.cdd.4401818 PMID:16311508

Alano, C. C., Ying, W., & Swanson, R. A. (2004). Poly (ADP-ribose) polymerase-1-mediated cell death in astrocytes requires NAD+ depletion and mitochondrial permeability transition. *The Journal of Biological Chemistry*, *279*(18), 18895–18902. doi:10.1074/jbc.M313329200 PMID:14960594

Albani, D., Polito, L., Batelli, S., De Mauro, S., Fracasso, C., Martelli, G., ... Forloni, G. (2009). The SIRT1 activator resveratrol protects SK-N-BE cells from oxidative stress and against toxicity caused by alpha-synuclein or amyloid-beta (1-42) peptide. *Journal of Neurochemistry*, *110*(5), 1445–1456. doi:10.1111/j.1471-4159.2009.06228.x PMID:19558452

Alegre-Abarrategui, J., Christian, H., Lufino, M. M., Mutihac, R., Venda, L. L., Ansorge, O., & Wade-Martins, R. (2009). LRRK2 regulates autophagic activity and localizes to specific membrane microdomains in a novel human genomic reporter cellular model. *Human Molecular Genetics*, *18*(21), 4022–4034. doi:10.1093/hmg/ddp346 PMID:19640926

Alexander, G. E. (2004). Biology of Parkinson's disease: Pathogenesis and pathophysiology of a multisystem neurodegenerative disorder. *Dialogues in Clinical Neuroscience*, *6*(3), 259–280. PMID:22033559

Almeida, B., Abreu, I. A., Matos, C. A., Fraga, J. S., Fernandes, S., Macedo, M. G., . . . MacedoRibeiro, S. (2015). *SUMOylation of the brain-predominant Ataxin-3 isoform modulates its interaction with p97*. Academic Press.

Almeida, C. G., Takahashi, R. H., & Gouras, G. K. (2006). β-amyloid accumulation impairs multivesicular body sorting by inhibiting the ubiquitin-proteasome system. *The Journal of Neuroscience: The Official Journal of the Society for Neuroscience, 26*(16), 4277–4288. doi:10.1523/JNEUROSCI.5078-05.2006 PMID:16624948

Alonso, A. D., Cohen, L. S., Corbo, C., Morozova, V., ElIdrissi, A., Phillips, G., & Kleiman, F. E. (2018). Hyperphosphorylation of Tau Associates With Changes in Its Function Beyond Microtubule Stability. *Frontiers in Cellular Neuroscience, 12*, 338. doi:10.3389/fncel.2018.00338 PMID:30356756

Altuntas, S., D'Eletto, M., Rossin, F., Hidalgo, L. D., Farrace, M. G., Falasca, L., . . . Piacentini, M. (2014). *Type 2 Transglutaminase, mitochondria and Huntington's disease: menage a trois*. Academic Press.

Alvarez-Erviti, L., Rodriguez-Oroz, M. C., Cooper, J. M., Caballero, C., Ferrer, I., Obeso, J. A., & Schapira, A. H. (2010). Chaperone-mediated autophagy markers in Parkinson disease brains. *Archives of Neurology, 67*(12), 1464–1472. doi:10.1001/archneurol.2010.198 PMID:20697033

Alzheimer's Association. (2018). Alzheimer's Disease Facts and Figures. *Alzheimer's & Dementia, 14*(3), 367–429. doi:10.1016/j.jalz.2018.02.001

Alzheimer's Disease International, World Alzheimer Report. (2018). *The state of the art of dementia research: New frontiers*. London: Author.

Alzheimer, A. (1907). Uber eine eigenartige Erkrankung der Hirnrinde. *Allg Zeitschr f Psychiatrie u Psych-Gerichtl Med, 64*, 146–148. doi:10.1002/ca.980080612 PMID:8713166

Andersen, P. M., & Al-Chalabi, A. (2011). Clinical genetics of amyotrophic lateral sclerosis: What do we really know? *Nature Reviews. Neurology, 7*(11), 603–615. doi:10.1038/nrneurol.2011.150 PMID:21989245

Anderson, J. F., Siller, E., & Barral, J. M. (2010). The sacsin repeating region (SRR): A novel Hsp90-related supra-domain associated with neurodegeneration. *Journal of Molecular Biology, 400*(4), 665–674. doi:10.1016/j.jmb.2010.05.023 PMID:20488193

Anderson, J. F., Siller, E., & Barral, J. M. (2011). The neurodegenerative-disease-related protein sacsin is a molecular chaperone. *Journal of Molecular Biology, 411*(4), 870–880. doi:10.1016/j.jmb.2011.06.016 PMID:21726565

Anderson, J. P., Walker, D. E., Goldstein, J. M., de Laat, R., Banducci, K., Caccavello, R. J., ... Chilcote, T. J. (2006). Phosphorylation of Ser-129 is the dominant pathological modification of α-synuclein in familial and sporadic lewy body disease. *The Journal of Biological Chemistry, 281*(40), 29739–29752. doi:10.1074/jbc.M600933200 PMID:16847063

Anding, A. L., & Baehrecke, E. H. (2017). Cleaning House: Selective Autophagy of Organelles. *Developmental Cell, 41*(1), 10–22. doi:10.1016/j.devcel.2017.02.016 PMID:28399394

Andreadis, A. (2006). Misregulation of tau alternative splicing in neurodegeneration and dementia. In *Alternative Splicing and Disease* (pp. 89–107). Berlin: Springer. doi:10.1007/978-3-540-34449-0_5

Andres-Mateos, E., Perier, C., Zhang, L., Blanchard-Fillion, B., Greco, T. M., Thomas, B., ... Dawson, V. L. (2007). DJ-1 gene deletion reveals that DJ-1 is an atypical peroxiredoxin-like peroxidase. *Proceedings of the National Academy of Sciences of the United States of America, 104*(37), 14807–14812. doi:10.1073/pnas.0703219104 PMID:17766438

Anekonda, T. S., & Reddy, P. H. (2006). Neuronal protection by sirtuins in Alzheimer's disease. *Journal of Neurochemistry, 96*(2), 305–313. doi:10.1111/j.1471-4159.2005.03492.x PMID:16219030

Aoyama, K., & Nakaki, T. (2013). Impaired Glutathione Synthesis in Neurodegeneration. *International Journal of Molecular Sciences, 14*(10), 21021–21044. doi:10.3390/ijms141021021 PMID:24145751

Ariga, H., Takahashi-Niki, K., Kato, I., Maita, H., Niki, T., & Iguchi-Ariga, S. M. M. (2013). Neuroprotective Function of DJ-1 in Parkinson's Disease. *Oxidative Medicine and Cellular Longevity, 2013*, 683920. doi:10.1155/2013/683920 PMID:23766857

Arima, K., Mizutani, T., Alim, M. A., Tonozuka-Uehara, H., Izumiyama, Y., Hirai, S., & Uéda, K. (2000). NACP/α-synuclein and tau constitute two distinctive subsets of filaments in the same neuronal inclusions in brains from a family of parkinsonism and dementia with Lewy bodies: Double-immunolabeling fluorescence and electron microscopic studies. *Acta Neuropathologica, 100*(2), 115–121. doi:10.1007004010050002 PMID:10963357

Ariosa, A. R., & Klionsky, D. J. (2016). Autophagy core machinery: Overcoming spatial barriers in neurons. *Journal of Molecular Medicine (Berlin, Germany), 94*(11), 1217–1227. doi:10.100700109-016-1461-9 PMID:27544281

Arneson, D., Zhang, Y., Yang, X., & Narayanan, M. (2018). Shared mechanisms among neurodegenerative diseases: From genetic factors to gene networks. *Journal of Genetics, 97*(3), 795–806. doi:10.100712041-018-0963-3 PMID:30027910

Arrasate, M., & Finkbeiner, S. (2012). Protein aggregates in Huntington's disease. *Experimental Neurology, 238*(1), 1–11. doi:10.1016/j.expneurol.2011.12.013 PMID:22200539

Ashraf, G. M., Greig, N. H., Khan, T. A., Hassan, I., Tabrez, S., Shakil, S., ... Kamal, M. (2014). Protein misfolding and aggregation in Alzheimer's disease and type 2 diabetes mellitus. *CNS & Neurological Disorders - Drug Targets, 13*(7), 1280–1293. doi:10.2174/1871527313666140917095514 PMID:25230234

Ashrafi, G., Schlehe, J. S., LaVoie, M. J., & Schwarz, T. L. (2014). Mitophagy of damaged mitochondria occurs locally in distal neuronal axons and requires *PINK1* and Parkin. *The Journal of Cell Biology, 206*(5), 655–670. doi:10.1083/jcb.201401070 PMID:25154397

Atik, A., Stewart, T., & Zhang, J. (2016). Alpha-synuclein as a biomarker for Parkinson's disease. *Brain Pathology (Zurich, Switzerland), 26*(3), 410–418. doi:10.1111/bpa.12370 PMID:26940058

Atkin, G., & Paulson, H. (2014). Ubiquitin pathways in neurodegenerative disease. *Frontiers in Molecular Neuroscience*, 7. PMID:25071440

Auger, K. R., Serunian, L. A., Soltoff, S. P., Libby, P., & Cantley, L. C. (1989). PDGF-dependent tyrosine phosphorylation stimulates production of novel polyphosphoinositides in intact cells. *Cell, 57*(1), 167–175. doi:10.1016/0092-8674(89)90182-7 PMID:2467744

Augusteyn, R. (2004). α-Crystallin: A review of its structure and function. *Clinical and Experimental Optometry -. Clinical & Experimental Optometry, 87*(6), 356–366. doi:10.1111/j.1444-0938.2004.tb03095.x PMID:15575808

Auld, K. L., & Silver, P. A. (2006). Transcriptional Regulation by the Proteasome as a Mechanism for Cellular Protein. *Cell Cycle, 5*(14), 1503–1505. doi:10.4161/cc.5.14.2979 PMID:16861887

Auluck, P. K., & Bonini, N. M. (2002). Pharmacological prevention of Parkinson disease in Drosophila. *Nature Medicine, 8*(11), 1185–1186. doi:10.1038/nm1102-1185 PMID:12411925

Auluck, P. K., Chan, H. Y., Trojanowski, J. Q., Lee, V. M., & Bonini, N. M. (2002). Chaperone suppression of alpha-synuclein toxicity in a Drosophila model for Parkinson's disease. *Science, 295*(5556), 865–868. doi:10.1126cience.1067389 PMID:11823645

Axe, E. L., Walker, S. A., Manifava, M., Chandra, P., Roderick, H. L., Habermann, A., ... Ktistakis, N. T. (2008). Autophagosome formation from membrane compartments enriched in phosphatidylinositol 3-phosphate and dynamically connected to the endoplasmic reticulum. *The Journal of Cell Biology*, *182*(4), 685–701. doi:10.1083/jcb.200803137 PMID:18725538

Ayers, J. I., Diamond, J., Sari, A., Fromholt, S., Galaleldeen, A., Ostrow, L. W., ... Borchelt, D. R. (2016). Distinct conformers of transmissible misfolded SOD1 distinguish human SOD1-FALS from other forms of familial and sporadic ALS. *Acta Neuropathologica*, *132*(6), 827–840. doi:10.100700401-016-1623-4 PMID:27704280

Badadani, M. (2012). Autophagy Mechanism, Regulation, Functions, and Disorders. ISRN Cell Biology, Article ID 927064.

BAG domain (IPR003103). (n.d.). *InterPro*. Retrieved from http://www.ebi.ac.uk/interpro/entry/IPR003103

BAG3 gene. BCL2 associated athanogene 3. Genetics Home Reference. (n.d.). US National Library of Medicine. Retrieved from https://ghr.nlm.nih.gov/gene/BAG3#location

Bailey, C. D., & Johnson, G. V. (2005). *Tissue transglutaminase contributes to disease progression in the R6/2 Huntington's disease mouse model via aggregate-independent mechanisms*. Academic Press.

Bali, P., Pranpat, M., Bradner, J., Balasis, M., Fiskus, W., Guo, F., ... Bhalla, K. (2005). Inhibition of histone deacetylase 6 acetylates and disrupts the chaperone function of heat shock protein 90: A novel basis for antileukemia activity of histone deacetylase inhibitors. *The Journal of Biological Chemistry*, *280*(29), 26729–26734. doi:10.1074/jbc.C500186200 PMID:15937340

Ballinger, C. A., Connell, P., Wu, Y., Hu, Z., Thompson, L. J., Yin, L. Y., & Patterson, C. (1999). Identification of CHIP, a novel tetratricopeptide repeat-containing protein that interacts with heat shock proteins and negatively regulates chaperone functions. *Molecular and Cellular Biology*, *19*(6), 4535–4545. doi:10.1128/MCB.19.6.4535 PMID:10330192

Bandopadhyay, R., & de Belleroche, J. (2010). Pathogenesis of Parkinson's disease: Emerging role of molecular chaperones. *Trends in Molecular Medicine*, *16*(1), 27–36. doi:10.1016/j.molmed.2009.11.004 PMID:20036196

Bandyopadhyay, U., Kaushik, S., Varticovski, L., & Cuervo, A. M. (2008). The chaperone-mediated autophagy receptor organizes in dynamic protein complexes at the lysosomal membrane. *Molecular and Cellular Biology*, *28*(18), 5747–5763. doi:10.1128/MCB.02070-07 PMID:18644871

Banerjee, R. (2012). Redox outside the Box: Linking Extracellular Redox Remodeling with Intracellular Redox Metabolism. *The Journal of Biological Chemistry*, *287*(7), 4397–4402. doi:10.1074/jbc.R111.287995 PMID:22147695

Bánhegyi, G., Mandl, J., & Csala, M. (2008). Redox-based endoplasmic reticulum dysfunction in neurological diseases. *Journal of Neurochemistry*, *107*(1), 20–34. doi:10.1111/j.1471-4159.2008.05571.x PMID:18643792

Banks, C. J., & Andersen, J. L. (2019). Mechanisms of SOD1 regulation by post-translational modifications. *Redox Biology*, *26*, 26. doi:10.1016/j.redox.2019.101270 PMID:31344643

Banks, C., Rodriguez, N., Gashler, K., Bikman, B., & Andersen, J. (2016). Acylation of SOD1 Provides Tool to determine if Mitochondrial Aggregation of SOD1 is Main Driver of ALS. *The FASEB Journal*, 30.

Barbour, R., Kling, K., Anderson, J. P., Banducci, K., Cole, T., Diep, L., & Chilcote, T. J. (2008). Red blood cells are the major source of alpha-synuclein in blood. *Neurodegenerative Diseases*, *5*(2), 55–59. doi:10.1159/000112832 PMID:18182779

Barghorn, S., & Mandelkow, E. (2002). Toward a unified scheme for the aggregation of tau into Alzheimer paired helical filaments. *Biochemistry*, *41*(50), 14885–14896. doi:10.1021/bi026469j PMID:12475237

Barone, P. (2010). Neurotransmission in Parkinson's disease: Beyond dopamine. *European Journal of Neurology, 17*(3), 364–376. doi:10.1111/j.1468-1331.2009.02900.x PMID:20050885

Baron, G. S., Wehrly, K., & Dorward, D. W. (2002). Conversion of raft associated prion protein to the protease-resistant state requires insertion of PrP-res (PrPSc) into contiguous membranes. *The EMBO Journal, 21*(5), 1031–1040. doi:10.1093/emboj/21.5.1031 PMID:11867531

Bates, G. P., Dorsey, R., Gusella, J. F., Hayden, M. R., Kay, C., Leavitt, B. R., . . . Wetzel, R. (2015). *Huntington disease.* Academic Press.

Bates, G. P., Harper, P. S., & Jones, A. L. (Eds.). (2002). *Huntington's Disease*. Oxford, UK: Oxford University Press.

Batista, P., & Pereira, A. (2016). Quality of Life in Patients with Neurodegenerative Diseases. *Journal of Neurology and Neuroscience*, *7*(1), 7. doi:10.21767/2171-6625.100074

Batlevi, Y., & La Spada, A. R. (2011). Mitochondrial autophagy in neural function, neurodegenerative disease, neuron cell death, and aging. *Neurobiology of Disease*, *43*(1), 46–51. doi:10.1016/j.nbd.2010.09.009 PMID:20887789

Batulan, Z., Taylor, D. M., Aarons, R. J., Minotti, S., Doroudchi, M. M., Nalbantoglu, J., & Durham, H. D. (2006). Induction of multiple heat shock proteins and neuroprotection in a primary culture model of familial amyotrophic lateral sclerosis. *Neurobiology of Disease*, *24*(2), 213–225. doi:10.1016/j.nbd.2006.06.017 PMID:16950627

Bauer, P. O., Goswami, A., Wong, H. K., Okuno, M., Kurosawa, M., Yamada, M., ... Nukina, N. (2010). Harnessing chaperone-mediated autophagy for the selective degradation of mutant huntingtin protein. *Nature Biotechnology, 28*(3), 256–263. doi:10.1038/nbt.1608 PMID:20190739

Beach, T. G., Adler, C. H., Sue, L. I., Vedders, L., Lue, L., White, C. L. III, ... Walker, D. G. (2010). Multi-organ distribution of phosphorylated α-synuclein histopathology in subjects with Lewy body disorders. *Acta Neuropathologica, 119*(6), 689–702. doi:10.100700401-010-0664-3 PMID:20306269

Beach, T. G., Serrano, G. E., Kremer, T., Canamero, M., Dziadek, S., Sade, H., ... Chahine, L. (2018). Immunohistochemical method and histopathology judging for the systemic synuclein sampling study (S4). *Journal of Neuropathology and Experimental Neurology*, *77*(9), 793–802. doi:10.1093/jnen/nly056 PMID:30107604

Beal, M. F., Matthews, R. T., Tieleman, A., & Shults, C. W. (1998). Coenzyme Q10 attenuates the 1-Methyl-4-Phenyl-1,2,3,Tetrahydropyridine (Mptp) induced loss of striatal dopamine and dopaminergic axons in aged mice. *Brain Research, 783*(1), 109–114. doi:10.1016/S0006-8993(97)01192-X PMID:9479058

Bechtel, T. J., & Weerapana, E. (2017). From structure to redox: The diverse functional roles of disulfides and implications in disease. *Proteomics, 17*(6). doi:10.1002/pmic.201600391

Behl, C. (2016). Breaking BAG: The Co-Chaperone BAG3 in health and disease. *Trends in Pharmacological Sciences, 37*(8), 672–688. doi:10.1016/j.tips.2016.04.007 PMID:27162137

Beleza-Meireles, A., & Al-Chalabi, A. (2009). Genetic studies of amyotrophic lateral sclerosis: Controversies and perspectives. *Amyotrophic Lateral Sclerosis, 10*(1), 1–14. doi:10.1080/17482960802585469 PMID:19110986

Bellacosa, A., Kumar, C. C., Di Cristofano, A., & Testa, J. R. (2005). Activation of AKT kinases in cancer: Implications for therapeutic targeting. *Advances in Cancer Research, 94*, 29–86. doi:10.1016/S0065-230X(05)94002-5 PMID:16095999

Benatar, M. (2007). Lost in translation: Treatment trials in the SOD1 mouse and in human ALS. *Neurobiology of Disease, 26*(1), 1–13. doi:10.1016/j.nbd.2006.12.015 PMID:17300945

Bender, A., Desplats, P., Spencer, B., Rockenstein, E., Adame, A., Elstner, M., & Pham, E. (2013). TOM40 mediates mitochondrial dysfunction induced by α-synuclein accumulation in Parkinson's disease. *PLoS One*, *8*(4), e62277. doi:10.1371/journal.pone.0062277 PMID:23626796

Bender, A., Krishnan, K. J., Morris, C. M., Taylor, G. A., Reeve, A. K., Perry, R. H., ... Taylor, R. W. (2006). High levels of mitochondrial DNA deletions in substantia nigra neurons in aging and Parkinson disease. *Nature Genetics*, *38*(5), 515–517. doi:10.1038/ng1769 PMID:16604074

Bendor, J. T., Logan, T. P., & Edwards, R. H. (2013). The function of alpha-synuclein. *Neuron*, *79*(6), 1044–1066. doi:10.1016/j.neuron.2013.09.004 PMID:24050397

Bendotti, C., Atzori, C., Piva, R., Tortarolo, M., Strong, M. J., DeBiasi, S., & Migheli, A. (2004). Activated p38MAPK is a novel component of the intracellular inclusions found in human amyotrophic lateral sclerosis and mutant SOD1 transgenic mice. *Journal of Neuropathology and Experimental Neurology*, *63*(2), 113–119. doi:10.1093/jnen/63.2.113 PMID:14989597

Benhar, M., Forrester, M. T., & Stamler, J. S. (2009). Protein denitrosylation: Enzymatic mechanisms and cellular functions. *Nature Reviews. Molecular Cell Biology*, *10*(10), 721–732. doi:10.1038/nrm2764 PMID:19738628

Benn, S. C., & Brown, R. H. Jr. (2004). Putting the heat on ALS. *Nature Medicine*, *10*(4), 345–347. doi:10.1038/nm0404-345 PMID:15057226

Benskey, M. J., Perez, R. G., & Manfredsson, F. P. (2016). The contribution of alpha synuclein to neuronal survival and function - Implications for Parkinson's disease. *Journal of Neurochemistry*, *137*(3), 331–359. doi:10.1111/jnc.13570 PMID:26852372

Bento, C. F., Renna, M., Ghislat, G., Puri, C., Ashkenazi, A., Vicinanza, M., ... Rubinsztein, D. C. (2016). Mammalian Autophagy: How Does It Work? *Annual Review of Biochemistry*, *85*(1), 685–713. doi:10.1146/annurev-biochem-060815-014556 PMID:26865532

Berger, Z., Ravikumar, B., Menzies, F. M., Oroz, L. G., Underwood, B. R., Pangalos, M. N., ... Rubinsztein, D. C. (2005). Rapamycin alleviates toxicity of different aggregate-prone proteins. *Human Molecular Genetics*, *15*(3), 433–442. doi:10.1093/hmg/ddi458 PMID:16368705

Berry, C. E., & Hare, J. M. (2004). Xanthine oxidoreductase and cardiovascular disease: Molecular mechanisms and pathophysiological implications. *The Journal of Physiology*, *555*(3), 589–606. doi:10.1113/jphysiol.2003.055913 PMID:14694147

Bett, J. S. (2016). Proteostasis regulation by the ubiquitin system. *Essays in Biochemistry*, *60*(2), 143–151. doi:10.1042/EBC20160001 PMID:27744330

Beyer, K. (2007). Mechanistic aspects of Parkinson's disease: A-synuclein and the biomembrane. *Cell Biochemistry and Biophysics*, *47*(2), 285–299. doi:10.100712013-007-0014-9 PMID:17652776

Bezprozvanny, I., & Mattson, M. P. (2008). Neuronal calcium mishandling and the pathogenesis of Alzheimer's disease. *Trends in Neurosciences*, *31*(9), 454–463. doi:10.1016/j.tins.2008.06.005 PMID:18675468

Bhadbhade, A., & Cheng, D. W. (2012). Amyloid Precursor Protein Processing in Alzheimer's Disease. *Iranian Journal of Child Neurology*, *6*, 1–5.

Bhattacharyya, R., Barren, C., & Kovacs, D. M. (2013). Palmitoylation of amyloid precursor protein regulates amyloidogenic processing in lipid rafts. *The Journal of Neuroscience: The Official Journal of the Society for Neuroscience*, *33*(27), 11169–11183. doi:10.1523/JNEUROSCI.4704-12.2013 PMID:23825420

Bhujabal, Z., Birgisdottir, Å. B., Sjøttem, E., Brenne, H. B., Øvervatn, A., Habisov, S., ... Johansen, T. (2017). FKBP8 recruits LC3A to mediate Parkin-independent mitophagy. *EMBO Reports*, *18*(6), 947–961. doi:10.15252/embr.201643147 PMID:28381481

Bianco, C. L., Shorter, J., Régulier, E., Lashuel, H., Iwatsubo, T., Lindquist, S., & Aebischer, P. (2008). Hsp104 antagonizes α-synuclein aggregation and reduces dopaminergic degeneration in a rat model of Parkinson disease. *The Journal of Clinical Investigation*, *118*(9), 3087–3097. doi:10.1172/JCI35781 PMID:18704197

Bieschke, J., Russ, J., Friedrich, R. P., Ehrnhoefer, D. E., Wobst, H., Neugebauer, K., & Wanker, E. E. (2010). EGCG remodels mature α-synuclein and amyloid-β fibrils and reduces cellular toxicity. *Proceedings of the National Academy of Sciences of the United States of America*, *107*(17), 7710–7715. doi:10.1073/pnas.0910723107 PMID:20385841

Biessels, G. J., De Leeuw, F. E., Lindeboom, J., Barkhof, F., & Scheltens, P. (2006). Increased cortical atrophy in patients with Alzheimer's disease and type 2 diabetes mellitus. *Journal of Neurology, Neurosurgery, and Psychiatry*, *77*(3), 304–307. doi:10.1136/jnnp.2005.069583 PMID:16484636

Biessels, G. J., Strachan, M. W., Visseren, F. L., Kappelle, L. J., & Whitmer, R. A. (2014). Dementia and cognitive decline in type 2 diabetes and prediabetic stages: Towards targeted interventions. *The Lancet. Diabetes & Endocrinology*, *2*(3), 246–255. doi:10.1016/S2213-8587(13)70088-3 PMID:24622755

Bilen, J., & Bonini, N. M. (2007). Genome-wide screen for modifiers of ataxin-3 neurodegeneration in Drosophila. *PLOS Genetics*, *3*(10), e177. doi:10.1371/journal.pgen.0030177 PMID:17953484

Binder, L. I., Guillozet-Bongaarts, A. L., Garcia-Sierra, F., & Berry, R. W. (2005). Tau, tangles, and Alzheimer's disease. *Biochimica et Biophysica Acta 3, 1739*(2-3),216-23.

Bingol, B. (2018). Autophagy and lysosomal pathways in nervous system disorders. *Molecular and Cellular Neurosciences*, *91*, 167–208. doi:10.1016/j.mcn.2018.04.009 PMID:29729319

Bizzozero, O. A. (2009) Protein Carbonylation in Neurodegenerative and Demyelinating CNS Diseases. In Handbook of Neurochemistry and Molecular Neurobiology. Springer US. doi:10.1007/978-0-387-30375-8_23

Björkblom, B., Adilbayeva, A., Maple-Grødem, J., Piston, D., Ökvist, M., Xu, X. M., ... Møller, S. G. (2013). Parkinson Disease Protein DJ-1 Binds Metals and Protects against Metal-induced Cytotoxicity. *The Journal of Biological Chemistry*, *288*(31), 22809–22820. doi:10.1074/jbc.M113.482091 PMID:23792957

Blass, J., & Gibson, G. (1991). *The role of oxidative abnormalities in the pathophysiology of Alzheimer's disease*. Academic Press.

Blokhuis, A. M., Groen, E. J. N., Koppers, M., van den Berg, L. H., & Pasterkamp, R. J. (2013). Protein aggregation in amyotrophic lateral sclerosis. *Acta Neuropathologica*, *125*(6), 777–794. doi:10.100700401-013-1125-6 PMID:23673820

Bloom, G. S. (2014). Amyloid-β and tau: The trigger and bullet in Alzheimer disease pathogenesis. *JAMA Neurology*, *71*(4), 505–508. doi:10.1001/jamaneurol.2013.5847 PMID:24493463

Blum, D., Herrera, F., Francelle, L., Mendes, T., Basquin, M., Obriot, H., ... Buée, L. (2014). Mutant huntingtin alters Tau phosphorylation and subcellular distribution. *Human Molecular Genetics*, *24*(1), 76–85. doi:10.1093/hmg/ddu421 PMID:25143394

Blumen, S. C., Astord, S., Robin, V., Vignaud, L., Toumi, N., Cieslik, A., ... Viollet, L. (2012). A rare recessive distal hereditary motor neuropathy with HSJ1 chaperone mutation. *Annals of Neurology*, *71*(4), 509–519. doi:10.1002/ana.22684 PMID:22522442

Boellaard, J., Kao, M., Schlote, W., & Diringer, H. (1991). Neuronal autophagy in experimental scrapie. *Acta Neuropathologica*, *82*(3), 225–228. doi:10.1007/BF00294449 PMID:1927279

Boland, B., Kumar, A., Lee, S., Platt, F. M., Wegiel, J., Yu, W. H., & Nixon, R. A. (2008). Autophagy Induction and Autophagosome Clearance in Neurons:Relationship to Autophagic Pathology in Alzheimer's Disease. *The Journal of Neuroscience: The Official Journal of the Society for Neuroscience*, *28*(27), 6926–6937. doi:10.1523/JNEUROSCI.0800-08.2008 PMID:18596167

Bonda, D. J., Lee, H. G., Camins, A., Pallas, M., Casadesus, G., Smith, M. A., & Zhu, X. (2011). The sirtuin pathway in ageing and Alzheimer disease: Mechanistic and therapeutic considerations. *Lancet Neurology*, *10*(3), 275–279. doi:10.1016/S1474-4422(11)70013-8 PMID:21349442

Bonifati, V., Rizzu, P., Squitieri, F., Krieger, E., Vanacore, N., ... Van Swieten, J. C. (2003). DJ-1(PARK7), a novel gene for autosomal recessive, early onset parkinsonism. *Neurological Sciences*, *24*(3), 159–160. doi:10.100710072-003-0108-0 PMID:14598065

Boonstra, M. C., De Geus, S. W. L., & Prevoo, H. A. J. M. (2016). Selecting Targets for Tumor Imaging: An Overview of Cancer-Associated Membrane Proteins. Biomark Cancer 8. *BIC*, *S38542*. doi:10.4137/bic.s38542 PMID:27721658

Bosco, D., Fava, A., Plastino, M., Montalcini, T., & Pujia, A. (2011). Possible implications of insulin resistance and glucose metabolism in Alzheimer's disease pathogenesis. *Journal of Cellular and Molecular Medicine*, *15*(9), 1807–1821. doi:10.1111/j.1582-4934.2011.01318.x PMID:21435176

Bossy-Wetzel, E., Petrilli, A., & Knott, A. B. (2008). *Mutant huntingtin and mitochondrial dysfunction*. Academic Press.

Botella, J. A., Bayersdorfer, F., & Schneuwly, S. (2008). Superoxide dismutase overexpression protects dopaminergic neurons in a Drosophila model of Parkinson's disease. *Neurobiology of Disease*, *30*(1), 65–73. doi:10.1016/j.nbd.2007.11.013 PMID:18243716

Bouchard, J. P., Barbeau, A., Bouchard, R., & Bouchard, R. W. (1978). Autosomal recessive spastic ataxia of Charlevoix-Saguenay. *The Canadian Journal ofNeurological Sciences*, *5*(1), 61–69. doi:10.1017/S0317167100024793 PMID:647499

Boya, P., Gonzalez-Polo, R. A., Casares, N., Perfettini, J. L., Dessen, P., Larochette, N., ... Kroemer, G. (2005). Inhibition of macroautophagy triggers apoptosis. *Molecular and Cellular Biology*, *25*(3), 1025–1040. doi:10.1128/MCB.25.3.1025-1040.2005 PMID:15657430

Boyce, M., Bryant, K. F., Jousse, C., Long, K., Harding, H. P., Scheuner, D., ... Yuan, J. (2005). A selective inhibitor of elF2α dephosphorylation protects cells from ER stress. *Science*, *307*(5711), 935–939. doi:10.1126cience.1101902 PMID:15705855

Braak, H., & Braak, E. (1991). *Neuropathological stageing of Alzheimer-related changes*. Academic Press.

Braak, H., & Braak, E. (1991). Neuropathological stageing of Alzheimer-related changes. *Acta Neuropathologica*, *82*(4), 239–259. doi:10.1007/BF00308809 PMID:1759558

Braak, H., & Braak, E. V. A. (1995). Staging of Alzheimer's disease-related neurofibrillary changes. *Neurobiology of Aging*, *16*(3), 271–278. doi:10.1016/0197-4580(95)00021-6 PMID:7566337

Braak, H., & Del, T. K. (2012). Where, when, and in what form does sporadic Alzheimer's disease begin. *Current Opinion in Neurology*, *25*(6), 708–714. doi:10.1097/WCO.0b013e32835a3432 PMID:23160422

Bracher, A., & Verghese, J. (2015). The nucleotide exchange factors of Hsp70 molecular chaperones. *Frontiers in Molecular Biosciences*, *2*, 10–10. doi:10.3389/fmolb.2015.00010 PMID:26913285

Brandman, O., Stewart-Ornstein, J., Wong, D., Larson, A., Williams, C. C., Li, G.-W., ... Weissman, J. S. (2012). A ribosome-bound quality control complex triggers degradation of nascent peptides and signals translation stress. *Cell, 151*(5), 1042–1054. doi:10.1016/j.cell.2012.10.044 PMID:23178123

Braschi, E., Goyon, V., Zunino, R., Mohanty, A., Xu, L., & McBride, H. M. (2010, July 27). Vps35 mediates vesicle transport between the mitochondria and peroxisomes. *Current Biology, 20*(14), 1310–1315. doi:10.1016/j.cub.2010.05.066 PMID:20619655

Braun, R. J., Zischka, H., Madeo, F., Eisenberg, T., Wissing, S., Büttner, S., . . . Ueffing, M. (2006). *Crucial mitochondrial impairment upon CDC48 mutation in apoptotic yeast.* Academic Press.

Brehme, M., Voisine, C., Rolland, T., Wachi, S., Soper, J. H., Zhu, Y., ... Morimoto, R. I. (2014). A chaperome subnetwork safeguards proteostasis in aging and neurodegenerative disease. *Cell Reports, 9*(3), 1135–1150. doi:10.1016/j.celrep.2014.09.042 PMID:25437566

Brichta, L., & Greengard, P. (2014). Molecular determinants of selective dopaminergic vulnerability in Parkinson's disease: An update. *Frontiers in Neuroanatomy, 8*, 152–152. doi:10.3389/fnana.2014.00152 PMID:25565977

Briknarova, K., Takayama, S., Homma, S., Baker, K., Cabezas, E., Hoyt, D. W., ... Ely, K. R. (2002). BAG4/SODD protein contains a short BAG domain. *The Journal of Biological Chemistry, 277*(34), 31172–31178. doi:10.1074/jbc.M202792200 PMID:12058034

Brion, J. P. (1998). Neurofibrillary Tangles and Alzheimer's Disease. *European Neurology, 40*(3), 130–140. doi:10.1159/000007969 PMID:9748670

Browne, S. E., Bowling, A. C., Macgarvey, U., Baik, M. J., Berger, S. C., Muquit, M. M., ... Beal, M. F. (1997). Oxidative damage and metabolic dysfunction in Huntington's disease: Selective vulnerability of the basal ganglia. *Annals of Neurology: Official Journal of the American Neurological Association and the Child Neurology Society, 41*(5), 646–653. doi:10.1002/ana.410410514 PMID:9153527

Bryan, K. J., Lee, H.-g., Perry, G., Smith, M. A., & Casadesus, G. (2009). Transgenic mouse models of Alzheimer's disease: behavioral testing and considerations. In *Methods of Behavior Analysis in Neuroscience* (2nd ed.). CRC Press/Taylor and Francis.

Bryksin, A., & Laktionov, P. (2008). *Role of glyceraldehyde-3-phosphate dehydrogenase in vesicular transport from golgi apparatus to endoplasmic reticulum.* Academic Press.

Buchner, J. (1996). Supervising the fold: Functional principles of molecular chaperones. *The FASEB Journal, 10*(1), 10–19. doi:10.1096/fasebj.10.1.8566529 PMID:8566529

Buetow, L., & Huang, D. T. (2016). Structural insights into the catalysis and regulation of E3 ubiquitin ligases. *Nature Reviews. Molecular Cell Biology, 17*(10), 626–642. doi:10.1038/nrm.2016.91 PMID:27485899

Burchell, V. S., Gandhi, S., Deas, E., Wood, N. W., Abramov, A. Y., & Plun-Favreau, H. (2010). Targeting mitochondrial dysfunction in neurodegenerative disease. *Part II. Expert Opinion on Therapeutic Targets, 14*(5), 497–511. doi:10.1517/14728221003730434 PMID:20334487

Burgoyne, R. D., & Morgan, A. (2015). Cysteine String Protein (CSP) and its role in preventing neurodegeneration. *Seminars in Cell & Developmental Biology, 40*, 153–159. doi:10.1016/j.semcdb.2015.03.008 PMID:25800794

Burns, A., & Iliffe, S. (2009). Alzheimer's disease. *BMJ (Clinical Research Ed.), 338*(feb05 1), b158. doi:10.1136/bmj.b158 PMID:19196745

Bursch, W. (2001). The autophagosomal–lysosomal compartment in programmed cell death. *Cell Death and Differentiation, 8*(6), 569–581. doi:10.1038j.cdd.4400852 PMID:11536007

Butterfield, D. A. (2002). Amyloid β-peptide (1–42)-induced oxidative stress and neurotoxicity: Implications for neurodegeneration in Alzheimer's disease brain. A review. *Free Radical Research, 36*(12), 1307–1313. doi:10.1080/1071576021000049890 PMID:12607822

Butterfield, D. A., Drake, J., Pocernich, C., & Castegna, A. (2001). Evidence of oxidative damage in Alzheimer's disease brain: Central role for amyloid beta-peptide. *Trends in Molecular Medicine, 7*(12), 548–554. doi:10.1016/S1471-4914(01)02173-6 PMID:11733217

Butterfield, D. A., & Lauderback, C. M. (2002). Lipid peroxidation and protein oxidation in Alzheimer's disease brain: Potential causes and consequences involving amyloid betapeptide- associated free radical oxidative stress. *Free Radical Biology & Medicine, 32*, 1050–1060. doi:10.1016/S0891-5849(02)00794-3 PMID:12031889

Byrne, S., Walsh, C., Lynch, C., Bede, P., Elamin, M., Kenna, K., ... Hardiman, O. (2011). Rate of familial amyotrophic lateral sclerosis: A systematic review and meta-analysis. *Journal of Neurology, Neurosurgery, and Psychiatry, 82*(6), 623–627. doi:10.1136/jnnp.2010.224501 PMID:21047878

Cabin, D. E., Shimazu, K., Murphy, D., Cole, N. B., Gottschalk, W., McIlwain, K. L., ... Nussbaum, R. L. (2002). Synaptic vesicle depletion correlates with attenuated synaptic responses to prolonged repetitive stimulation in mice lacking α-synuclein. *The Journal of Neuroscience: The Official Journal of the Society for Neuroscience, 22*(20), 8797–8807. doi:10.1523/JNEUROSCI.22-20-08797.2002 PMID:12388586

Caccamo, A., Maldonado, M. A., Majumder, S., Medina, D. X., Holbein, W., Magrí, A., & Oddo, S. (2011). Naturally secreted amyloid-β increases mammalian target of rapamycin (mTOR) activity via a PRAS40-mediated mechanism. *The Journal of Biological Chemistry, 286*(11), 8924–8932. doi:10.1074/jbc.M110.180638 PMID:21266573

Caccamo, A., Oddo, S., Tran, L. X., & LaFerla, F. M. (2007). Lithium reduces tau phosphorylation but not A beta or working memory deficits in a transgenic model with both plaques and tangles. *American Journal of Pathology, 170*(5), 1669–1675. doi:10.2353/ajpath.2007.061178 PMID:17456772

Cagin, U., & Enriquez, J. A. (2015). The complex crosstalk between mitochondria and the nucleus: What goes in between? *The International Journal of Biochemistry & Cell Biology, 63*, 10–15. doi:10.1016/j.biocel.2015.01.026 PMID:25666554

Cai, Q., & Tammineni, P. (2016). Alterations in mitochondrial quality control in Alzheimer's disease. *Frontiers in Cellular Neuroscience, 10*, 24. doi:10.3389/fncel.2016.00024 PMID:26903809

Cai, Q., Zakaria, H. M., Simone, A., & Sheng, Z. H. (2012). Spatial parkin translocation and degradation of damaged mitochondria via mitophagy in live cortical neurons. *Current Biology, 22*(6), 545–552. doi:10.1016/j.cub.2012.02.005 PMID:22342752

Cai, Z., & Yan, L.-J. (2013). Protein Oxidative Modifications: Beneficial Roles in Disease and Health. *Journal of Biochemical and Pharmacological Research, 1*, 15–26. PMID:23662248

Calabrese, V., Cornelius, C., Mancuso, C., Lentile, R., Giuffrida Stella, A. M., & Butterfield, D. A. (2010). Redox Homeostasis and Cellular Stress Response in Aging and Neurodegeneration. Free Radicals and Antioxidant Protocols, Methods in Molecular Biology, 610, 285-308. doi:10.1007/978-1-60327-029-8_17

Calabrese, V., Scapagnini, G., Stella, A. M., Bates, T. E., & Clark, J. B. (2000). Mitochondrial involvement in brain function and dysfunction: Relevance to aging, neurodegenerative disorders and longevity. *Neurochemical Research, 26*(6), 739–764. doi:10.1023/A:1010955807739 PMID:11519733

Calfon, M., Zeng, H., Urano, F., Till, J. H., Hubbard, S. R., Harding, H. P., ... Ron, D. (2002). IRE1 couples endoplasmic reticulum load to secretory capacity by processing the XBP-1 mRNA. *Nature*, *415*(6867), 92–96. doi:10.1038/415092a PMID:11780124

Calkins, M. J., Manczak, M., Mao, P., Shirendeb, U., & Reddy, P. H. (2011). Impaired mitochondrial biogenesis, defective axonal transport of mitochondria, abnormal mitochondrial dynamics and synaptic degeneration in a mouse model of Alzheimer's disease. *Human Molecular Genetics*, *20*(23), 4515–4529. doi:10.1093/hmg/ddr381 PMID:21873260

Campana, V., Sarnataro, D., & Zurzolo, C. (2005). The highways and byways of prion protein trafficking. *TreNDDs Cell Biol.*, *15*(2), 102–111. doi:10.1016/j.tcb.2004.12.002 PMID:15695097

Campanella, C., Pace, A., Caruso Bavisotto, C., Marzullo, P., Marino Gammazza, A., Buscemi, S., & Palumbo Piccionello, A. (2018). Heat shock proteins in Alzheimer's disease: Role and targeting. *International Journal of Molecular Sciences*, *19*(9), 2603. doi:10.3390/ijms19092603 PMID:30200516

Campuzano, V., Montermini, L., & Moltò, M.D. (1996). Friedreich's ataxia: Autosomal recessive disease caused by an intronic GAA triplet repeat expansion. *Science*, *271*, 1423–1427. doi:10.1126cience.271.5254.1423

Canet-Avilés, R. M., Wilson, M. A., Miller, D. W., Ahmad, R., McLendon, C., ... Bandyopadhyay, S. (2004). The Parkinson's disease protein DJ-1 is neuroprotective due to cysteine-sulfnic acid-driven mitochondrial localization. *Proceedings of the National Academy of Sciences of the United States of America*, *101*(24), 9103–9108. doi:10.1073/pnas.0402959101 PMID:15181200

Carafa, V., Rotili, D., Forgione, M., Cuomo, F., Serretiello, E., Hailu, G. S., ... Altucci, L. (2016). Sirtuin functions and modulation: From chemistry to the clinic. *Clinical Epigenetics*, *8*(1), 61. doi:10.118613148-016-0224-3 PMID:27226812

Carnemolla, A., Fossale, E., Agostoni, E., Michelazzi, S., Calligaris, R., De Maso, L., ... Persichetti, F. (2009). Rrs1 Is Involved in Endoplasmic Reticulum Stress Response in Huntington Disease. *The Journal of Biological Chemistry*, *284*(27), 18167–18173. doi:10.1074/jbc.M109.018325 PMID:19433866

Carri, M. T., Valle, C., Bozzo, F., & Cozzolino, M. (2015). Oxidative stress and mitochondrial damage: Importance in non-SOD1 ALS. *Frontiers in Cellular Neuroscience*, *9*, 41. doi:10.3389/fncel.2015.00041 PMID:25741238

Carter, R. J., Lione, L. A., Humby, T., Mangiarini, L., Mahal, A., Bates, G. P., ... Morton, A. J. (1999). Characterization of progressive motor deficits in mice transgenic for the human Huntington's disease mutation. *The Journal of Neuroscience: The Official Journal of the Society for Neuroscience*, *19*(8), 3248–3257. doi:10.1523/JNEUROSCI.19-08-03248.1999 PMID:10191337

Casas-Tinto, S., Zhang, Y., Sanchez-Garcia, J., Gomez-Velazquez, M., Rincon-Limas, D. E., & Fernandez-Funez, P. (2011). The ER stress factor XBP1s prevents amyloid-β neurotoxicity. *Human Molecular Genetics*, *20*(11), 2144–2160. doi:10.1093/hmg/ddr100 PMID:21389082

Castillo, K., Nassif, M., Valenzuela, V., Rojas, F., Matus, S., Mercado, G., ... Hetz, C. (2013). Trehalose delays the progression of amyotrophic lateral sclerosis by enhancing autophagy in motoneurons. *Autophagy*, *9*(9), 1308–1320. doi:10.4161/auto.25188 PMID:23851366

Cavallarin, N., Vicario, M., & Negro, A. (2012). The Role of Phosphorylation in Synucleinopathies: Focus on Parkinsons Disease. *CNS & Neurological Disorders - Drug Targets*, *9*(4), 471–481. doi:10.2174/187152710791556140 PMID:20522010

Cawley, K., Logue, S. E., Gorman, A. M., Zeng, Q., Patterson, J., Gupta, S., & Samali, A. (2013). Disruption of microRNA biogenesis confers resistance to ER stress-induced cell death upstream of the mitochondrion. *PLoS One*, *8*(8), e73870. doi:10.1371/journal.pone.0073870 PMID:23977393

Cecconi, F., & Levine, B. (2008). The role of autophagy in mammalian development: Cell makeover rather than cell death. *Developmental Cell*, *15*(3), 344–357. doi:10.1016/j.devcel.2008.08.012 PMID:18804433

Cellular response to heat stress. (n.d.). Retrieved from https://reactome.org/PathwayBrowser/#/R-HSA-3371556&PATH=R-HSA-8953897,R-HSA-2262752&FLG=O95817

Cerri, S., & Blandini, F. (2018). Role of Autophagy in Parkinson's Disease. *Current Medicinal Chemistry*, *25*. doi:10.2174/0929867325666180226094351 PMID:29484979

Chaari, A. (2019). Molecular chaperones biochemistry and role in neurodegenerative diseases. *International Journal of Biological Macromolecules*, *131*, 396–411. doi:10.1016/j.ijbiomac.2019.02.148 PMID:30853582

Chadwick, L., Gentle, L., Strachan, J., & Layfield, R. (2012). Review: Unchained maladie - a reassessment of the role of Ubb+1-capped polyubiquitin chains in Alzheimer's disease. *Neuropathology and Applied Neurobiology*, *38*(2), 118–131. doi:10.1111/j.1365-2990.2011.01236.x PMID:22082077

Chalfant, C. E., Kishikawa, K., Mumby, M. C., Kamibayashi, C., Bielawska, A., & Hannun, Y. A. (1999). Long chain ceramides activate protein phosphatase-1 and protein phosphatase-2A Activation is stereospecific and regulated by phosphatidic acid. *The Journal of Biological Chemistry*, *274*(29), 20313–20317. doi:10.1074/jbc.274.29.20313 PMID:10400653

Chambraud, B., Sardin, E., Giustiniani, J., Dounane, O., Schumacher, M., Goedert, M., & Baulieu, E.-E. (2010). A role for FKBP52 in Tau protein function. *Proceedings of the National Academy of Sciences of the United States of America*, *107*(6), 2658–2663. doi:10.1073/pnas.0914957107 PMID:20133804

Chan, N. C., Salazar, A. M., Pham, A. H., Sweredoski, M. J., Kolawa, N. J., Graham, R. L., ... Chan, D. C. (2011). Broad activation of the ubiquitin–proteasome system by Parkin is critical for mitophagy. *Human Molecular Genetics*, *20*(9), 1726–1737. doi:10.1093/hmg/ddr048 PMID:21296869

Cha, S. J., Kim, H. C., Hyun-Jun, L., & Sanghyun, K. K. (2017). Protein Glutathionylation in the Pathogenesis of Neurodegenerative Diseases. *Oxidative Medicine and Cellular Longevity*, *2017*, 1–9. doi:10.1155/2017/2818565 PMID:29456785

Chauhan, V., & Chauhan, A. (2006). Oxidative stress in Alzheimer's disease. *Pathophysiology*, *13*(3), 195–208. doi:10.1016/j.pathophys.2006.05.004 PMID:16781128

Chava, S., Lee, C., Aydin, Y., Chandra, P. K., Dash, A., Chedid, M., ... Dash, S. (2017). Chaperone-mediated autophagy compensates for impaired macroautophagy in the cirrhotic liver to promote hepatocellular carcinoma. *Oncotarget*, *8*(25), 40019. doi:10.18632/oncotarget.16685 PMID:28402954

Cheetham, M. E., & Caplan, A. J. (1998). Structure, function and evolution of DnaJ: Conservation and adaptation of chaperone function. *Cell Stress & Chaperones*, *3*(1), 28–36. doi:10.1379/1466-1268(1998)003<0028:SFAEOD>2.3.CO;2 PMID:9585179

Chen, C. A., Wang, T. Y., Varadharaj, S., Reyes, L. A., Hemann, C., Talukder, M. A., ... Zweier, J. L. (2010). S-glutathionylation uncouples eNOS and regulates its cellular and vascular function. *Nature*, *468*(7327), 1115–1118. doi:10.1038/nature09599 PMID:21179168

Chen, D., Gao, F., Li, B., Wang, H., Xu, Y., Zhu, C., & Wang, G. (2010). Parkin mono-ubiquitinates Bcl-2 and regulates autophagy. *The Journal of Biological Chemistry*, *285*(49), 38214–38223. doi:10.1074/jbc.M110.101469 PMID:20889974

Chen, G., Han, Z., Feng, D. U., Chen, Y., Chen, L., Wu, H., ... Chen, Q. (2014). A regulatory signaling loop comprising the PGAM5 phosphatase and CK2 controls receptor-mediated mitophagy. *Molecular Cell*, *54*(3), 362–377. doi:10.1016/j.molcel.2014.02.034 PMID:24746696

Cheng, A., Wan, R., Yang, J.-L., Kamimura, N., Son, T. G., Ouyang, X., ... Mattson, M. P. (2012). Involvement of PGC-1alpha in the formation and maintenance of neuronal dendritic spines. *Nature Communications*, *3*(1), 1250. doi:10.1038/ncomms2238 PMID:23212379

Cheng, A., Yang, Y., Zhou, Y., Maharana, C., Lu, D., Peng, W., ... Bohr, V. A. (2016). Mitochondrial SIRT3 mediates adaptive responses of neurons to exercise and metabolic and excitatory challenges. *Cell Metabolism*, *23*(1), 128–142. doi:10.1016/j.cmet.2015.10.013 PMID:26698917

Cheng, G., Huang, C., Deng, H., & Wang, H. (2012). Diabetes as a risk factor for dementia and mild cognitive impairment: A meta-analysis of longitudinal studies. *Internal Medicine Journal*, *42*(5), 484–491. doi:10.1111/j.1445-5994.2012.02758.x PMID:22372522

Chen, J., Kanai, Y., Cowan, N. J., & Hirokawa, N. (1992). Projection domains of MAP2 and tau determine spacings between microtubules in dendrites and axons. *Nature*, *360*(6405), 674–677. doi:10.1038/360674a0 PMID:1465130

Chen, J., Zhou, Y., Mueller-Steiner, S., Chen, L. F., Kwon, H., Yi, S., ... Gan, L. (2005). SIRT1 protects against microglia-dependent amyloid-beta toxicity through inhibiting NF-kappaB signaling. *The Journal of Biological Chemistry*, *280*(48), 40364–40374. doi:10.1074/jbc.M509329200 PMID:16183991

Chen, M., Chen, Z., Wang, Y., Tan, Z., Zhu, C., Li, Y., ... Chen, Q. (2016). Mitophagy receptor FUNDC1 regulates mitochondrial dynamics and mitophagy. *Autophagy*, *12*(4), 689–702. doi:10.1080/15548627.2016.1151580 PMID:27050458

Chen, M., Yang, W., Li, X., Li, X., Wang, P., Yue, F., & Yu, S. (2016). Age- and brain region-dependent α-synuclein oligomerization is attributed to alterations in intrinsic enzymes regulating α-synuclein phosphorylation in aging monkey brains. *Oncotarget*, *7*(8), 8466–8480. doi:10.18632/oncotarget.6445 PMID:27032368

Chen, N., & Debnath, J. (2010). Autophagy and tumorigenesis. *FEBS Letters*, *584*(7), 1427–1435. doi:10.1016/j.febslet.2009.12.034 PMID:20035753

Chen-Plotkin, A. S. (2014). Unbiased approaches to biomarker discovery in neurodegenerative diseases. *Neuron*, *84*(3), 594–607. doi:10.1016/j.neuron.2014.10.031 PMID:25442938

Chen, Q., Zhou, Z., Zhang, L., Wang, Y., Zhang, Y. W., Zhong, M., ... Yu, Z. P. (2012). Tau protein is involved in morphological plasticity in hippocampal neurons in response to BDNF. *Neurochemistry International*, *60*(3), 233–242. doi:10.1016/j.neuint.2011.12.013 PMID:22226842

Chen, S., Owens, G. C., Makarenkova, H., & Edelman, D. B. (2010). HDAC6 Regulates Mitochondrial Transport in Hippocampal Neurons. *PLoS One*, *5*(5), e10848. doi:10.1371/journal.pone.0010848 PMID:20520769

Chen, S., Sayana, P., Zhang, X., & Le, W. (2013). Genetics of amyotrophic lateral sclerosis: An update. *Molecular Neurodegeneration*, *8*(1), 28. doi:10.1186/1750-1326-8-28 PMID:23941283

Chen, X., & Pan, W. (2014). The Strategies for neurodegenerative diseases by Integrative Medicine. *Integrative Medicine (Encinitas, Calif.)*, *1*, 223–225.

Chen, Y. G. (2018). Research Progress in the Pathogenesis of Alzheimer's Disease. *Chinese Medical Journal*, *131*(13), 1618–1624. doi:10.4103/0366-6999.235112 PMID:29941717

Chen, Y., & Brandizzi, F. (2013). IRE1: ER stress sensor and cell fate executor. *Trends in Cell Biology*, *23*(11), 547–555. doi:10.1016/j.tcb.2013.06.005 PMID:23880584

Chen, Z., Liu, L., Cheng, Q., Li, Y., Wu, H., Zhang, W., ... Wang, J. (2017). Mitochondrial E3 ligase MARCH5 regulates FUNDC1 to fine-tune hypoxic mitophagy. *EMBO Reports*, *18*(3), 495–509. doi:10.15252/embr.201643309 PMID:28104734

Chen, Z., & Zhong, C. (2014). Oxidative stress in Alzheimer's disease. *Neuroscience Bulletin, 30*(2), 271–281. doi:10.100712264-013-1423-y PMID:24664866

Cherry, K. (2018). An overview of the different parts of a neuron. *Theories. Biological Psychology*. Retrieved from https://www.verywellmind.com/structure-of-a-neuron-2794896

Chesebro, B., Trifilo, M., & Race, R. (2005). Medicine: Anchorless prion protein results in infectious amyloid disease without clinical scrapie. *Science, 308*, 1435–1439. doi:10.1126cience.1110837

Chételat, G. (2013). *Alzheimer disease: Aβ-independent processes—rethinking preclinical AD*. Academic Press.

Chételat, G., La Joie, R., Villain, N., Perrotin, A., de La Sayette, V., Eustache, F., and Vandenberghe, R. (2013). *Amyloid imaging in cognitively normal individuals, at-risk populations and preclinical Alzheimer's disease*. Academic Press.

Cheung, P., Allis, C. D., & Sassone-Corsi, P. (2000). Signaling to chromatin through historic modifications. *Cell, 103*(2), 263–271. doi:10.1016/S0092-8674(00)00118-5 PMID:11057899

Chiang, H. L., Plant, C. P., & Dice, J. F. (1989). A role for a 70-kilodalton heat shock protein in lysosomal degradation of intracellular proteins. *Science, 246*(4928), 382–385. doi:10.1126cience.2799391 PMID:2799391

Chinta, S. J., Mallajosyula, J. K., Rane, A., & Andersen, J. K. (2010). Mitochondrial alpha-synuclein accumulation impairs complex I function in dopaminergic neurons and results in increased mitophagy in vivo. *Neuroscience Letters, 486*(3), 235–239. doi:10.1016/j.neulet.2010.09.061 PMID:20887775

Chiti, F., & Dobson, C. M. (2006). Protein misfolding, functional amyloid, and human disease. *Annual Review of Biochemistry, 75*(1), 333–366. doi:10.1146/annurev.biochem.75.101304.123901 PMID:16756495

Chiu, J., & Hogg, P. J. (2019). Allosteric disulfides: Sophisticated molecular structures enabling flexible protein regulation. *The Journal of Biological Chemistry, 294*(8), 2949–2960. doi:10.1074/jbc.REV118.005604 PMID:30635401

Cho, D. H., Nakamura, T., Fang, J., Cieplak, P., Godzik, A., Gu, Z., & Lipton, S. A. (2009). S-Nitrosylation of Drp1 mediates beta-amyloid-related mitochondrial fission and neuronal injury. *Science, 324*(5923), 102–105. doi:10.1126cience.1171091 PMID:19342591

Cho, E., & Park, M. (2016). Palmitoylation in Alzheimer's disease and other neurodegenerative diseases. *Pharmacological Research, 111*, 133–151. doi:10.1016/j.phrs.2016.06.008 PMID:27293050

Choi, W., Zibaee, S., Jakes, R., Serpell, L. C., Davletov, B., Anthony Crowther, R., & Goedert, M. (2004). Mutation E46K increases phospholipid binding and assembly into filaments of human α-synuclein. *FEBS Letters, 576*(3), 363–368. doi:10.1016/j.febslet.2004.09.038 PMID:15498564

Choi, Y. B., Tenneti, L., Le, D. A., Ortiz, J., Bai, G., Chen, H. S., & Lipton, S. A. (2000). Molecular basis of NMDA receptor-coupled ion channel modulation by S-nitrosylation. *Nature Neuroscience, 3*(1), 15–21. doi:10.1038/71090 PMID:10607390

Choudhury, A., Chakraborty, I., Bhattacharjee, R., Banerjee, T. S., Vana, D. R., Adapa, D., & Bhardwaj, R. (2018). An update on pathological implications of enzymatic dysregulation in Alzheimer's disease. *Biomedical Research, 29*(10), 2215–2226. doi:10.4066/biomedicalresearch.29-18-767

Chow, V. W., Mattson, M. P., Wong, P. C., & Gleichmann, M. (2010). An Overview of APP Processing Enzymes and Products. *Neuromol Med, 12*(1), 1–12. doi:10.100712017-009-8104-z PMID:20232515

Christen, Y. (2000). Oxidative stress and Alzheimer disease. *The American Journal of Clinical Nutrition, 71*(2), 621S–629S. doi:10.1093/ajcn/71.2.621s PMID:10681270

Christopher-Choy, K. H., Dean, O., Berk, M., Bush, A. I., & Buuse, M. (2010). Effect of N-acetyl-cysteine treatment on glutathione depletion and a short-term spatial memory deficit in 2-cyclohexene 1-one-treated rats. *European Journal of Pharmacology*, *649*(1-3), 224–228. doi:10.1016/j.ejphar.2010.09.035 PMID:20868666

Chua, J. P., Reddy, S. L., Yu, Z., Giorgetti, E., Montie, H. L., Mukherjee, S., ... Lieberman, A. P. (2015). Disrupting SUMOylation enhances transcriptional function and ameliorates polyglutamine androgen receptor-mediated disease. *The Journal of Clinical Investigation*, *125*(2), 831–845. doi:10.1172/JCI73214 PMID:25607844

Chu, C. T. (2019). Mechanisms of selective autophagy and mitophagy: Implications for neurodegenerative diseases. *Neurobiology of Disease*, *122*, 23–34. doi:10.1016/j.nbd.2018.07.015 PMID:30030024

Chu, C. T., Ji, J., Dagda, R. K., Jiang, J. F., Tyurina, Y. Y., Kapralov, A. A., ... Wang, K. Z. Q. (2013). Cardiolipin externalization to the outer mitochondrial membrane acts as an elimination signal for mitophagy in neuronal cells. *Nature Cell Biology*, *15*(10), 1197–1205. doi:10.1038/ncb2837 PMID:24036476

Chung H. S., Wang S. B., Venkatraman V., Murray C. I., & Van Eyk J. E. (2013). Cysteine oxidative posttranslational modifications: emerging regulation in the cardiovascular system. *Circu resear*, *112*, 382-392.

Chung, H. S., Wang, S. B., Venkatraman, V., Murray, C. I., & Van Eyk, J. E. (2013). Cysteine oxidative posttranslational modifications: Emerging regulation in the cardiovascular system. *Circulation Research*, *112*(2), 382–392. doi:10.1161/CIRCRESAHA.112.268680 PMID:23329793

Chung, K. K. K., Dawson, T. M., & Dawson, V. L. (2005). Nitric oxide, S-nitrosylation and neurodegeneration. *Cellular and Molecular Biology*, *51*, 247–254. PMID:16191392

Ciaccioli, G., Martins, A., Rodrigues, C., Vieira, H., & Calado, P. (2013). A powerful yeast model to investigate the synergistic interaction of α-synuclein and tau in neurodegeneration. *PLoS One*, *8*(2), e55848. doi:10.1371/journal.pone.0055848 PMID:23393603

Cianfanelli, V., D'Orazio, M., & Cecconi, F. (2015). AMBRA1 and BECLIN 1 interplay in the crosstalk between autophagy and cell proliferation. *Cell Cycle (Georgetown, Tex.)*, *14*(7), 959–963. doi:10.1080/15384101.2015.1021526 PMID:25803737

Cichanover, A. (2005). Intracellular protein degradation: From a vague idea thru the lysosome and the ubiquitin-proteasome system and onto human diseases and drug targeting. *Cell Death and Differentiation*, *12*(9), 1178–1190. doi:10.1038j.cdd.4401692 PMID:16094394

Ciechanover, A., & Kwon, Y. T. (2015). Degradation of misfolded proteins in neurodegenerative diseases: Therapeutic targets and strategies. *Experimental & Molecular Medicine*, *47*(3), e147. doi:10.1038/emm.2014.117 PMID:25766616

Ciechanover, A., & Kwon, Y. T. (2017). Protein Quality Control by Molecular Chaperones in Neurodegeneration. *Frontiers in Neuroscience*, *11*, 185. doi:10.3389/fnins.2017.00185 PMID:28428740

Cissé, M., Duplan, E., Lorivel, T., Dunys, J., Bauer, C., Meckler, X., ... Checler, F. (2017). The transcription factor XBP1s restores hippocampal synaptic plasticity and memory by control of the Kalirin-7 pathway in Alzheimer model. *Molecular Psychiatry*, *22*(11), 905–917. doi:10.1038/mp.2016.152 PMID:27646263

Citron, M., Teplow, D. B., & Selkoe, D. J. (1995). Generation of amyloid beta protein from its precursor is sequence specific. *Neuron*, *14*(3), 661–670. doi:10.1016/0896-6273(95)90323-2 PMID:7695913

Clark, T. A., Lee, H. P., Rolston, R. K., Zhu, X., Marlatt, M. W., Castellani, R. J., ... Perry, G. (2010). Oxidative Stress and its Implications for Future Treatments and Management of Alzheimer Disease. *International Journal of Biomedical Science; IJBS*, *6*, 225–227. PMID:21765811

Codolo, G., Plotegher, N., Pozzobon, T., Brucale, M., Tessari, I., ... Bubacco, L. (2013). Triggering of inflammasome by aggregated alpha-synuclein, an inflammatory response in synucleinopathies. *PLoS One*, *8*(1), e55375. doi:10.1371/journal.pone.0055375 PMID:23383169

Coles, L. D., Tuite, P. J., Gulin, O., Mishra, U. R., Kartha, R. V., Sullivan, K. M., & ... . (2017). Repeated Dose Oral N-Acetyl cysteine in Parkinson's Disease: Pharmacokinetics and Effect on Brain Glutathione and Oxidative Stress. *Journal of Clinical Pharmacology*, *58*(2), 158–167. doi:10.1002/jcph.1008 PMID:28940353

Colignon, B., Delaive, E., Dieu, M., Demazy, C., Muhovski, Y., Antoine, A., ... Mauro, S. (2019). Dual coordination of the SUMOylation and phosphorylation pathways during the response to heat stress in Solanum tuberosum. *Environmental and Experimental Botany*, *162*, 192–200. doi:10.1016/j.envexpbot.2019.02.024

Comer, F. I., & Hart, G. W. (2001). Reciprocity between O-GlcNAc and O-phosphate on the carboxyl terminal domain of RNA polymerase II. *Biochemistry*, *40*(26), 7845–7852. doi:10.1021/bi0027480 PMID:11425311

Cong, J., Liu, R., Wang, X., Jiang, H., & Zhang, Y. (2016). MiR-634 decreases cell proliferation and induces apoptosis by targeting mTOR signaling pathway in cervical cancer cells. *Artificial Cells, Nanomedicine, and Biotechnology*, *44*(7), 1694–1701. doi:10.3109/21691401.2015.1080171 PMID:26367112

Conway, K. A., Harper, J. D., & Lansbury, P. T. (1998). Accelerated in vitro fibril formation by a mutant α-synuclein linked to early-onset Parkinson disease. *Nature Medicine*, *4*(11), 1318–1320. doi:10.1038/3311 PMID:9809558

Cooke, M. S., Evans, M. D., Dizdaroglu, M., & Lunec, J. (2003). Oxidative DNA damage: Mechanisms, mutation, and disease. *The FASEB Journal*, *17*(10), 1195–1214. doi:10.1096/fj.02-0752rev PMID:12832285

Cooper, A. A., Gitler, A. D., Cashikar, A., Haynes, C. M., Hill, K. J., Bhullar, B., & Cao, S. (2006). α-Synuclein blocks ER-Golgi traffic and Rab1 rescues neuron loss in Parkinson's models. *Science*, *313*(5785), 324–328. doi:10.1126cience.1129462 PMID:16794039

Cooper, G. J., Day, A. J., Willis, A. C., Roberts, A. N., Reid, K. B., & Leighton, B. (1989). Amylin and the amylin gene: Structure, function and relationship to islet amyloid and to diabetes mellitus. *Biochimica et Biophysica Acta (BBA)-. Molecular Cell Research*, *1014*(3), 247–258. PMID:2690958

Coppede, F., & Migliore, L. (2015). DNA damage in neurodegenerative diseases. *Mutation Research*, *776*, 84–97. doi:10.1016/j.mrfmmm.2014.11.010 PMID:26255941

Cornelissen, T., Haddad, D., Wauters, F., Van Humbeeck, C., Mandemakers, W., Koentjoro, B., ... Vandenberghe, W. (2014). The deubiquitinase USP15 antagonizes Parkin-mediated mitochondrial ubiquitination and mitophagy. *Human Molecular Genetics*, *23*(19), 5227–5242. doi:10.1093/hmg/ddu244 PMID:24852371

Costa, V., & Scorrano, L. (2012). *Shaping the role of mitochondria in the pathogenesis of Huntington's disease*. Academic Press.

Costa, R. O., Ferreiro, E., Cardoso, S. M., Oliveira, C. R., & Pereira, C. M. F. (2010). ER stress-mediated apoptotic pathway induced by aβ peptide requires the presence of functional mitochondria. *Journal of Alzheimer's Disease*, *20*(2), 625–636. doi:10.3233/JAD-2010-091369 PMID:20182029

Costa, R. O., Ferreiro, E., Oliveira, C. R., & Pereira, C. M. F. (2013). Inhibition of mitochondrial cytochrome c oxidase potentiates Aβ-induced ER stress and cell death in cortical neurons. *Molecular and Cellular Neurosciences*, *52*, 1–8. doi:10.1016/j.mcn.2012.09.005 PMID:23026561

Couvertier, S. M., Zhou, Y., & Weerapana, E. (2014). Chemical-proteomic strategies to investigate cysteine posttranslational modifications. *Biochimica et Biophysica Acta*, *1844*(12), 2315–2330. doi:10.1016/j.bbapap.2014.09.024 PMID:25291386

Cox, D., Whiten, D. R., Brown, J. W., Horrocks, M. H., San Gil, R., Dobson, C. M., ... Ecroyd, H. (2018). The small heat shock protein Hsp27 binds α-synuclein fibrils, preventing elongation and cytotoxicity. *The Journal of Biological Chemistry*, *293*(12), 4486–4497. doi:10.1074/jbc.M117.813865 PMID:29382725

Cozzolino, M., Ferri, A., & Carri, M. T. (2008). Amyotrophic lateral sclerosis: From current developments in the laboratory to clinical implications. *Antioxidants & Redox Signalling*, *10*(3), 405–444. doi:10.1089/ars.2007.1760 PMID:18370853

Craft, S. (2009). The role of metabolic disorders in Alzheimer disease and vascular dementia: Two roads converged. *Archives of Neurology*, *66*(3), 300–305. doi:10.1001/archneurol.2009.27 PMID:19273747

Crippa, V., Sau, D., Rusmini, P., Boncoraglio, A., Onesto, E., Bolzoni, E., ... Poletti, A. (2010). The small heat shock protein B8 (HspB8) promotes autophagic removal of misfolded proteins involved in amyotrophic lateral sclerosis (ALS). *Human Molecular Genetics*, *19*(17), 3440–3456. doi:10.1093/hmg/ddq257 PMID:20570967

Croce, K. R., & Yamamoto, A. (2019). A role for autophagy in Huntington's disease. *Neurobiology of Disease*, *122*, 16–22. doi:10.1016/j.nbd.2018.08.010 PMID:30149183

Croisier, E., Moran, L. B., Dexter, D. T., Pearce, R. K., & Graeber, M. B. (2005). Microglial inflammation in the parkinsonian substantia nigra: Relationship to alpha-synuclein deposition. *Journal of Neuroinflammation*, *2*(1), 14. doi:10.1186/1742-2094-2-14 PMID:15935098

Csermely, P. (2001). Chaperone overload is a possible contributor to 'civilization diseases'. *Trends in Genetics*, *17*(12), 701–704. doi:10.1016/S0168-9525(01)02495-7 PMID:11718923

Csibi, A., Fendt, S. M., Li, C., Poulogiannis, G., Choo, A. Y., Chapski, D. J., ... Henske, E. P. (2013). The mTORC1 pathway stimulates glutamine metabolism and cell proliferation by repressing SIRT4. *Cell*, *153*(4), 840–854. doi:10.1016/j.cell.2013.04.023 PMID:23663782

Cuervo, A. M., & Dice, J. F. (1998). *Lysosomes, a meeting point of proteins, chaperones, and proteases*. Academic Press.

Cuervo, A. M., & Dice, J. F. (2000). Age-related decline in chaperone-mediated autophagy. *The Journal of Biological Chemistry*, *275*(40), 31505–31513. doi:10.1074/jbc.M002102200 PMID:10806201

Cuervo, A. M., Stefanis, L., Fredenburg, R., Lansbury, P. T., & Sulzer, D. (2004). Impaired degradation of mutant α-synuclein by chaperone-mediated autophagy. *Science*, *305*(5688), 1292–1295. doi:10.1126cience.1101738 PMID:15333840

Cuervo, A. M., & Wong, E. (2014). Chaperone-mediated autophagy: Roles in disease and aging. *Cell Research*, *24*(1), 92–104. doi:10.1038/cr.2013.153 PMID:24281265

Cui, L., Jeong, H., Borovecki, F., Parkhurst, C. N., Tanese, N., & Krainc, D. (2006). Transcriptional repression of PGC-1α by mutant huntingtin leads to mitochondrial dysfunction and neurodegeneration. *Cell*, *127*(1), 59–69. doi:10.1016/j.cell.2006.09.015 PMID:17018277

Cui, W. G., Wang, S. T., Wang, Z. P., Wang, Z. Y., Sun, C. L., & Zhang, Y. H. (2017). Inhibition of PTEN Attenuates Endoplasmic Reticulum Stress and Apoptosis via Activation of PI3K/AKT Pathway in Alzheimer's Disease. *Neurochemical Research*, *42*(11), 3052–3060. doi:10.100711064-017-2338-1 PMID:28819903

Culjkovic, B., Tan, K., Orolicki, S., Amri, A., Meloche, S., & Borden, K. L. (2008). The eIF4E RNA regulon promotes the Akt signaling pathway. *The Journal of Cell Biology*, *181*(1), 51–63. doi:10.1083/jcb.200707018 PMID:18391071

Cullen, V., Sardi, S. P., Ng, J., Xu, Y. H., Sun, Y., Tomlinson, J. J., ... Schlossmacher, M. G. (2011). Acid β-glucosidase mutants linked to Gaucher disease, Parkinson disease, and Lewy body dementia alter α-synuclein processing. *Annals of Neurology*, *69*(6), 940–953. doi:10.1002/ana.22400 PMID:21472771

Cullinan, S. B., Zhang, D., Hannink, M., Arvisais, E., Kaufman, R. J., & Diehl, J. A. (2003). Nrf2 Is a Direct PERK Substrate and Effector of PERK-Dependent Cell Survival. *Molecular and Cellular Biology*, *23*(20), 7198–7209. doi:10.1128/MCB.23.20.7198-7209.2003 PMID:14517290

Cumming, R. C., & Schubert, D. (2005). Amyloid-β induces disulfide bonding and aggregation of GAPDH in Alzheimer's disease. *The FASEB Journal*, *19*(14), 2060–2062. doi:10.1096/fj.05-4195fje PMID:16186172

Cummings, C. J., Mancini, M. A., Antalffy, B., DeFranco, D. B., Orr, H. T., & Zoghbi, H. Y. (1998). Chaperone suppression of aggregation and altered subcellular proteasome localization imply protein misfolding in SCA1. *Nature Genetics*, *19*(2), 148–154. doi:10.1038/502 PMID:9620770

Cummings, C. J., Sun, Y., Opal, P., Antalffy, B., Mestril, R., Orr, H. T., & ... . (2001). Over-expression of inducible HSP70 chaperone suppresses neuropathology and improves motor function in SCA1 mice. *Human Molecular Genetics*, *10*(14), 1511–1518. doi:10.1093/hmg/10.14.1511 PMID:11448943

Cvetanovic, M., Rooney, R. J., Garcia, J. J., Toporovskaya, N., Zoghbi, H. Y., & Opal, P. (2007). The role of LANP and ataxin 1 in E4F-mediated transcriptional repression. *EMBO Reports*, *8*(7), 671–677. doi:10.1038j.embor.7400983 PMID:17557114

D'Amico, D., Sorrentino, V., & Auwerx, J. (2017). Cytosolic Proteostasis Networks of the Mitochondrial Stress Response. *Trends in Biochemical Sciences*, *42*(9), 712–725. doi:10.1016/j.tibs.2017.05.002 PMID:28579074

D'Ydewalle, C., Krishnan, J., Chiheb, D. M., Van Damme, P., Irobi, J., Kozikowski, A. P., ... Van Den Bosch, L. (2011). HDAC6 inhibitors reverse axonal loss in a mouse model of mutant HSPB1-induced Charcot-Marie-Tooth disease. *Nature Medicine*, *17*(8), 968–974. doi:10.1038/nm.2396 PMID:21785432

Da Cruz, S., & Cleveland, D. W. (2011). Understanding the role of TDP-43 and FUS/TLS in ALS and beyond. *Current Opinion in Neurobiology*, *21*(6), 904–919. doi:10.1016/j.conb.2011.05.029 PMID:21813273

da Silveira, S. A., Schneider, B. L., & Cifuentes-Diaz, C. (2009). Phosphorylation does not prompt, nor prevent, the formation of α-synuclein toxic species in a rat model of Parkinson's disease. *Human Molecular Genetics*, *18*, 872–887. doi:10.1093/hmg/ddn417 PMID:19074459

Dagda, R. K., & Chu, C. T. (2009). Mitochondrial quality control: Insights on how Parkinson's disease related genes PINK1, Parkin, and Omi/HtrA2 interact to maintain mitochondrial homeostasis. *Journal of Bioenergetics and Biomembranes*, *41*(6), 473–479. doi:10.100710863-009-9255-1 PMID:20012177

Dalle-Donne, I., Rossi, R., Giustarini, D., Colombo, R., & Milzani, A. (2007). S-glutathionylation in protein redox regulation. *Free Radical Biology & Medicine*, *44*(6), 883–898. doi:10.1016/j.freeradbiomed.2007.06.014 PMID:17697933

Dan, A., Takahashi, M., & Masuda-Suzukake, M. (2014). Extensive deamidation at asparagine residue 279 accounts for weak immunoreactivity of tau with RD4 antibody in Alzheimer's disease brain. *Acta Neuropathologica Communications*, *2*. doi:10.1186/2051-5960-1-54 PMID:24252707

Das, T., & Eliezer, D. (2019). Membrane interactions of intrinsically disordered proteins: The example of alpha-synuclein. *Bba-Proteins Proteom*, *1867*(10), 879–889. doi:10.1016/j.bbapap.2019.05.001 PMID:31096049

Dasuri, K., Zhang, L., & Keller, J. N. (2013). Oxidative stress, neurodegeneration, and the balance of protein degradation and protein synthesis. *Free Radical Biology & Medicine*, *62*, 170–185. doi:10.1016/j.freeradbiomed.2012.09.016 PMID:23000246

Dauer, W., & Przedborski, S. (2003). Parkinson's disease: mechanisms and models. *Neuron, 39*(6), 889-909.

David, F. T., Keith, A. J., & Ronald, C. P. (2005). Alzheimer's disease – overview. In M. F. Beal, A. E. Lang, & A. C. Ludolph (Eds.), *Neurodegenerative Diseases Neurobiology, Pathogenesis and Therapeutics* (pp. 416–432). New York: University Press.

Davie, C. A. (2008). A review of Parkinson's disease. *British Medical Bulletin, 86*(1), 109–127. doi:10.1093/bmb/ldn013 PMID:18398010

Davies, C., Mann, D., Sumpter, P., & Yates, P. (1987). *A quantitative morphometric analysis of the neuronal and synaptic content of the frontal and temporal cortex in patients with Alzheimer's disease*. Academic Press.

Davies, S. W., Turmaine, M., Cozens, B. A., DiFiglia, M., Sharp, A. H., Ross, C. A., ... Bates, G. P. (1997). Formation of neuronal intranuclear inclusions underlies the neurological dysfunction in mice transgenic for the HD mutation. *Cell, 90*(3), 537–548. doi:10.1016/S0092-8674(00)80513-9 PMID:9267033

Dawson, T. M., and Dawson, V. L. (2010). The Role of Parkin in Familial and Sporadic Parkinson's Disease. *Movement Disorders: Official Journal of the Movement Disorder Society, 25*(1), S32-S39.

Dayalu, P., & Albin, R. L. (2015). *Huntington disease: pathogenesis and treatment*. Academic Press.

De Benedetti, A., & Graff, J. R. (2004). eIF-4E expression and its role in malignancies and metastases. *Oncogene, 23*(18), 3189–3199. doi:10.1038j.onc.1207545 PMID:15094768

De Castro, I. P., Costa, A. C., Lam, D., Tufi, R., Fedele, V., Moisoi, N., ... Martins, L. M. (2012). Genetic analysis of mitochondrial protein misfolding in Drosophila melanogaster. *Cell Death and Differentiation, 19*(8), 1308–1316. doi:10.1038/cdd.2012.5 PMID:22301916

De Duve, C. (1963). The lysosome. *Scientific American, 208*(5), 64–72. doi:10.1038cientificamerican0563-64 PMID:14025755

De la Monte, S. M., & Wands, J. R. (2008). Alzheimer's disease is type 3 diabetes—Evidence reviewed. *Journal of Diabetes Science and Technology, 2*(6), 1101–1113. doi:10.1177/193229680800200619 PMID:19885299

de Oliveira, R. M., Sarkander, J., Kazantsev, A. G., & Outeiro, T. F. (2012). SIRT2 as a therapeutic target for age-related disorders. *Frontiers in Pharmacology, 3*, 3. doi:10.3389/fphar.2012.00082 PMID:22563317

de Wit, J., Hong, W., Luo, L., & Ghosh, A. (2011). Role of leucine-rich repeat proteins in the development and function of neural circuits. *Annual Review of Cell and Developmental Biology, 27*(1), 697–729. doi:10.1146/annurev-cellbio-092910-154111 PMID:21740233

De, A., Bala, N., & DasGupta, P. (2011). Alzheimer's Disease and its. *Management, 2*(4), 1439–1443.

Deas, E., Cremades, N., Angelova, P. R., Ludtmann, M. H., Yao, Z., Chen, S., ... Gissen, P. (2016). Alpha-synuclein oligomers interact with metal ions to induce oxidative stress and neuronal death in Parkinson's disease. *Antioxidants & Redox Signalling, 24*(7), 376–391. doi:10.1089/ars.2015.6343 PMID:26564470

Deas, E., Wood, N. W., & Plun-Favreau, H. (2011). Mitophagy and Parkinson's disease: The PINK1-parkin link. *Biochimica et Biophysica Acta, 1813*(4), 623–633. doi:10.1016/j.bbamcr.2010.08.007 PMID:20736035

Dedeoglu, A., Kubilus, J. K., Jeitner, T. M., Matson, S. A., Bogdanov, M., Kowall, N. W., . . . Beal, M. F. (2002). *Therapeutic effects of cystamine in a murine model of Huntington's disease*. Academic Press.

Dedmon, M. M., Christodoulou, J., Wilson, M. R., & Dobson, C. M. (2005). Heat shock protein 70 inhibits alpha-synuclein fibril formation via preferential binding to prefibrillar species. *The Journal of Biological Chemistry, 280*(15), 14733–14740. doi:10.1074/jbc.M413024200 PMID:15671022

Degenhardt, K., Mathew, R., Beaudoin, B., Bray, K., Anderson, D., Chen, G., ... Nelson, D. A. (2006). Autophagy promotes tumor cell survival and restricts necrosis, inflammation, and tumorigenesis. *Cancer Cell, 10*(1), 51–64. doi:10.1016/j.ccr.2006.06.001 PMID:16843265

DeGracia, D. J., & Montie, H. L. (2004). Cerebral ischemia and the unfolded protein response. *Journal of Neurochemistry, 91*(1), 1–8. doi:10.1111/j.1471-4159.2004.02703.x PMID:15379881

Dehay, B., Bourdenx, M., Gorry, P., Przedborski, S., Vila, M., Hunot, S., ... Meissner, W. G. (2015). Targeting α-synuclein for treatment of Parkinson's disease: Mechanistic and therapeutic considerations. *Lancet Neurology, 14*(8), 855–866. doi:10.1016/S1474-4422(15)00006-X PMID:26050140

Dehay, B., Bové, J., Rodríguez-Muela, N., Perier, C., Recasens, A., Boya, P., & Vila, M. (2010). Pathogenic lysosomal depletion in Parkinson's disease. *The Journal of Neuroscience: The Official Journal of the Society for Neuroscience, 30*(37), 12535–12544. doi:10.1523/JNEUROSCI.1920-10.2010 PMID:20844148

Del Tredici, K., & Braak, H. (2016). Sporadic Parkinson's disease: Development and distribution of α-synuclein pathology. *Neuropathology and Applied Neurobiology, 42*(1), 33–50. doi:10.1111/nan.12298 PMID:26662475

DeLaBarre, B., Brunger, A. T., & Biology, M. (2003). *Complete structure of p97/valosin-containing protein reveals communication between nucleotide domains*. Academic Press.

Delenclos, M., Burgess, J. D., Lamprokostopoulou, A., Outeiro, T. F., Vekrellis, K., & McLean, P. J. (2019). Cellular models of alpha-synuclein toxicity and aggregation. *Journal of Neurochemistry, 150*(5), 566–576. doi:10.1111/jnc.14806 PMID:31265132

D'Eletto, M., Grazia Farrace, M., Falasca, L., Reali, V., Oliverio, S., Melino, G., . . . Piacentini, M. (2009). *Transglutaminase 2 is involved in autophagosome maturation*. Academic Press.

Deng, H. X., Chen, W. J., Hong, S. T., Boycott, K. M., Gorrie, G. H., Siddique, N., ... Siddique, T. (2011). Mutations in UBQLN2 cause dominant X-linked juvenile and adult-onset ALS and ALS/dementia. *Nature, 477*(7363), 211–U113. doi:10.1038/nature10353 PMID:21857683

Deng, S., Deng, X., Yuan, L., Song, Z., Yang, Z., Xiong, W., & Deng, H. (2015). Genetic analysis of SNCA coding mutation in Chinese Han patients with Parkinson disease. *Acta Neurologica Belgica, 115*(3), 267–271. doi:10.100713760-014-0347-2 PMID:25092551

Deretic, V., & Klionsky, D. J. (2018). Autophagy and inflammation: A special review issue. *Autophagy, 14*(2), 179–180. doi:10.1080/15548627.2017.1412229 PMID:29304718

DesJarlais, R., & Tummino, P. J. (2016). Role of Histone-Modifying Enzymes and Their Complexes in Regulation of Chromatin Biology. *Biochemistry-Us, 55*(11), 1584–1599. doi:10.1021/acs.biochem.5b01210 PMID:26745824

Despa, S., Margulies, K. B., Chen, L., Knowlton, A. A., Havel, P. J., Taegtmeyer, H., ... Despa, F. (2012). Hyperamylinemia contributes to cardiac dysfunction in obesity and diabetes: A study in humans and rats. *Circulation Research, 110*(4), 598–608. doi:10.1161/CIRCRESAHA.111.258285 PMID:22275486

Dettmer, U., Newman, A. J., Luth, E. S., Bartels, T., & Selkoe, D. (2013). In vivo cross-linking reveals principally oligomeric forms of alpha-synuclein and beta-synuclein in neurons and non-neural cells. *The Journal of Biological Chemistry, 288*(9), 6371–6385. doi:10.1074/jbc.M112.403311 PMID:23319586

Devanand, D. P., Pradhaban, G., Liu, X., Khandji, A., De Santi, S., Segal, S., ... De Leon, M. J. (2007). Hippocampal and entorhinal atrophy in mild cognitive impairment: Prediction of Alzheimer disease. *Neurology, 68*(11), 828–836. doi:10.1212/01.wnl.0000256697.20968.d7 PMID:17353470

Devi, L., Alldred, M. J., Ginsberg, S. D., & Ohno, M. (2012). Mechanisms underlying insulin deficiency-induced acceleration of β-amyloidosis in a mouse model of Alzheimer's disease. *PLoS One*, *7*(3), e32792. doi:10.1371/journal.pone.0032792 PMID:22403710

Devi, L., & Ohno, M. (2012). Mitochondrial dysfunction and accumulation of the β-secretase-cleaved C-terminal fragment of APP in Alzheimer's disease transgenic mice. *Neurobiology of Disease*, *45*(1), 417–424. doi:10.1016/j.nbd.2011.09.001 PMID:21933711

Devi, L., & Ohno, M. (2014). PERK mediates eIF2α phosphorylation responsible for BACE1 elevation, CREB dysfunction and neurodegeneration in a mouse model of Alzheimer's disease. *Neurobiology of Aging*, *35*(10), 2272–2281. doi:10.1016/j.neurobiolaging.2014.04.031 PMID:24889041

Devi, L., Prabhu, B. M., Galati, D. F., Avadhani, N. G., & Anandatheerthavarada, H. K. (2006). Accumulation of amyloid precursor protein in the mitochondrial import channels of human Alzheimer's disease brain is associated with mitochondrial dysfunction. *The Journal of Neuroscience: The Official Journal of the Society for Neuroscience*, *26*(35), 9057–9068. doi:10.1523/JNEUROSCI.1469-06.2006 PMID:16943564

Di Maio, R., Barrett, P. J., Hoffman, E. K., Barrett, C. W., Zharikov, A., Borah, A., ... & Hastings, T. G. (2016). α-Synuclein binds to TOM20 and inhibits mitochondrial protein import in Parkinson's disease. *Science Translational Medicine, 8*(342), 342ra78-342ra78.

Di Napoli, P., Taccardi, A. A., Oliver, M., & de Caterina, R. (2002). Statins and stroke: Evidence for cholesterol-independent effects. *European Heart Journal*, *23*(24), 1908–1921. doi:10.1053/euhj.2002.3236 PMID:12473253

Di Paolo, G., & Kim, T. W. (2011). Linking lipids to Alzheimer's disease: Cholesterol and beyond. *Nature Reviews. Neuroscience*, *12*(5), 284–296. doi:10.1038/nrn3012 PMID:21448224

Dias, V., Junn, E., & Mouradian, M. M. (2013). The role of oxidative stress in Parkinson's disease. *Journal of Parkinson's Disease*, *3*(4), 461–491. doi:10.3233/JPD-130230 PMID:24252804

Dickey, C., Kraft, C., Jinwal, U., Koren, J., Johnson, A., Anderson, L., ... Lewis, J. (2009). Aging Analysis Reveals Slowed Tau Turnover and Enhanced Stress Response in a Mouse Model of Tauopathy. *American Journal of Pathology*, *174*(1), 228–238. doi:10.2353/ajpath.2009.080764 PMID:19074615

Dikic, I. (2017). Proteasomal and Autophagic Degradation Systems. *Annual Review of Biochemistry*, *86*(1), 193–224. doi:10.1146/annurev-biochem-061516-044908 PMID:28460188

Dikic, I., & Elazar, Z. (2018). Mechanism and medical implications of mammalian autophagy. *Nature Reviews. Molecular Cell Biology*, *19*(6), 349–364. doi:10.103841580-018-0003-4 PMID:29618831

Di, N. (2005). Structural insights into Alzheimer filament assembly pathways based on site directed mutagenesis and S-glutathionylation of three repeat neuronal Tau protein. *Microscopy Research and Technique*, *67*(3-4), 156–163. doi:10.1002/jemt.20195 PMID:16104002

Ding, W. X., Guo, F., Ni, H. M., Bockus, A., Manley, S., Stolz, D. B., ... Yin, X. M. (2012a). Parkin and mitofusins reciprocally regulate mitophagy and mitochondrial spheroid formation. *The Journal of Biological Chemistry*, *287*(50), 42379–42388. doi:10.1074/jbc.M112.413682 PMID:23095748

Ding, W. X., Li, M., Biazik, J. M., Morgan, D. G., Guo, F., Ni, H. M., ... Yin, X. M. (2012b). Electron microscopic analysis of a spherical mitochondrial structure. *The Journal of Biological Chemistry*, *287*(50), 42373–42378. doi:10.1074/jbc.M112.413674 PMID:23093403

Ding, W. X., Ni, H. M., Li, M., Liao, Y., Chen, X., Stolz, D. B., ... Yin, X. M. (2010). Nix is critical to two distinct phases of mitophagy, reactive oxygen species-mediated autophagy induction and Parkin-ubiquitin-p62-mediated mitochondrial priming. *The Journal of Biological Chemistry*, *285*(36), 27879–27890. doi:10.1074/jbc.M110.119537 PMID:20573959

Ding, W. X., & Yin, X. M. (2012). Mitophagy: Mechanisms, pathophysiological roles, and analysis. *Biological Chemistry*, *393*(7), 547–564. doi:10.1515/hsz-2012-0119 PMID:22944659

Dionísio, P. A., Amaral, J. D., Ribeiro, M. F., Lo, A. C., D'Hooge, R., & Rodrigues, C. M. P. (2015). Amyloid-β pathology is attenuated by tauroursodeoxycholic acid treatment in APP/PS1 mice after disease onset. *Neurobiology of Aging*, *36*(1), 228–240. doi:10.1016/j.neurobiolaging.2014.08.034 PMID:25443293

Diot, A., Agnew, T., Sanderson, J., Liao, C., Carver, J., Neves, R. P., ... Poulton, J. (2018). Validating the RedMIT/GFP-LC3 Mouse Model by studying mitophagy in autosomal dominant optic atrophy due to OPA1Q285STOP mutation. *Frontiers in Cell and Developmental Biology*, *6*, 103. doi:10.3389/fcell.2018.00103 PMID:30283778

Djouder, N. (Ed.). (2018). *Prefoldins: the new chaperones* (Vol. 1106). Springer. doi:10.1007/978-3-030-00737-9

Dobrowsky, R. T., Kamibayashi, C., Mumby, M. C., & Hannun, Y. A. (1993). Ceramide activates heterotrimeric protein phosphatase 2A. *The Journal of Biological Chemistry*, *268*(21), 15523–15530. PMID:8393446

Dols-Icardo, O., Garcia-Redondo, A., Rojas-Garcia, R., Borrego-Hernandez, D., Illan-Gala, I., Munoz-Blanco, J. L., ... Clarimón, J. (2018). Analysis of known amyotrophic lateral sclerosis and frontotemporal dementia genes reveals a substantial genetic burden in patients manifesting both diseases not carrying the C9orf72 expansion mutation. *J Neurol Neurosur Ps*, *89*(2), 162–168. doi:10.1136/jnnp-2017-316820 PMID:28889094

Domínguez, R. O., Pagano, M. A., Marschoff, E. R., González, S. E., Repetto, M. G., & Serra, J. A. (2014). Alzheimer disease and cognitive impairment associated with diabetes mellitus type 2: Associations and a hypothesis [English Edition]. *Neurologia (Barcelona, Spain)*, *29*(9), 567–572. PMID:24140159

Dompierre, J. P., Godin, J. D., Charrin, B. C., Cordelieres, F. P., King, S. J., Humbert, S., & Saudou, F. (2007). Histone deacetylase 6 inhibition compensates for the transport deficit in Huntington's disease by increasing tubulin acetylation. *The Journal of Neuroscience: The Official Journal of the Society for Neuroscience*, *27*(13), 3571–3583. doi:10.1523/JNEUROSCI.0037-07.2007 PMID:17392473

Dong, Z., & Cui, H.J. (2019). The Autophagy-Lysosomal Pathways and Their Emerging Roles in Modulating Proteostasis in Tumors. *Cells*, *8*.

Dong, S., Duan, Y., Hu, Y., & Zhao, Z. (2012). Advances in the pathogenesis of Alzheimer's disease: A re-evaluation of amyloid cascade hypothesis. *Translational Neurodegeneration*, *1*(1), 18. doi:10.1186/2047-9158-1-18 PMID:23210692

Dong, Z., & Cui, H. (2019). The autophagy-lysosomal pathways and their emerging roles in modulating proteostasis in tumors. *Cells*, *8*(1), 4. doi:10.3390/cells8010004 PMID:30577555

Dong, Z., Wolfer, D. P., Lipp, H. P., & Bueler, H. (2005). Hsp70 gene transfer by adeno-associated virus inhibits MPTP-induced nigrostriatal degeneration in the mouse model of Parkinson disease. *Molecular Therapy*, *11*(1), 80–88. doi:10.1016/j.ymthe.2004.09.007 PMID:15585408

Donmez, G., Arun, A., Chung, C. Y., McLean, P. J., Lindquist, S., & Guarente, L. (2012). SIRT1 protects against α-synuclein aggregation by activating molecular chaperones. *The Journal of Neuroscience: The Official Journal of the Society for Neuroscience*, *32*(1), 124–132. doi:10.1523/JNEUROSCI.3442-11.2012 PMID:22219275

Doong, H., Vrailas, A., & Kohn, E. C. (2002). What's in the 'BAG'? – A functional domain analysis of the BAG-family proteins. *Cancer Letters*, *188*(1–2), 25–32. doi:10.1016/S0304-3835(02)00456-1 PMID:12406544

Dorszewska, J., Prendeck, M., Oczkowsk, A., Dezo, M., & Kozubski, W. (2016). Molecular Basis of Familial and Sporadic Alzheimer's Disease. *Current Alzheimer Research, 13*(9), 1–12. doi:10.2174/156720501366616031415050 1 PMID:26971934

Dou, F., Netzer, W. J., Tanemura, K., Li, F., Hartl, F. U., Takashima, A., ... Xu, H. (2003). Chaperones increase association of tau protein with microtubules. *Proceedings of the National Academy of Sciences of the United States of America, 100*(2), 721–726. doi:10.1073/pnas.242720499 PMID:12522269

Doxakis, E. (2010). Post-transcriptional regulation of α-synuclein expression by mir-7 and mir-153. *The Journal of Biological Chemistry, 285*(17), 12726–12734. doi:10.1074/jbc.M109.086827 PMID:20106983

Drewes, G., Trinczek, B., Illenberger, S., Biernat, J., Schmitt-Ulms, G., Meyer, H. E., ... Mandelkow, E. (1995). Microtubule-associated protein/microtubule affinity-regulating kinase (p110mark) a novel protein kinase that regulates tau-microtubule interactions and dynamic instability by phosphorylation at the Alzheimer-specific site serine 262. *The Journal of Biological Chemistry, 270*(13), 7679–7688. doi:10.1074/jbc.270.13.7679 PMID:7706316

Dröge, W. (2002). Free radicals in the physiological control of cell function. *Physiological Reviews, 82*(1), 47–95. doi:10.1152/physrev.00018.2001 PMID:11773609

Dryden, S. C., Nahhas, F. A., Nowak, J. E., Goustin, A. S., & Tainsky, M. A. (2003). Role for human SIRT2 NAD-dependent deacetylase activity in control of mitotic exit in the cell cycle. *Molecular and Cellular Biology, 23*(9), 3173–3185. doi:10.1128/MCB.23.9.3173-3185.2003 PMID:12697818

Duan, G., & Walther, D. (2015). The roles of post-translational modifications in the context of protein interaction networks. *PLoS Computational Biology, 11*(2), e1004049. doi:10.1371/journal.pcbi.1004049 PMID:25692714

Duan, W., Ladenheim, B., Cutler, R. G., Kruman, I. I., Cadet, J. L., & Mattson, M. P. (2002). Dietary folate deficiency and elevated homocysteine levels endanger dopaminergic neurons in models of Parkinson's Disease. *Journal of Neurochemistry, 80*(1), 101–110. doi:10.1046/j.0022-3042.2001.00676.x PMID:11796748

Dubouloz, F., Deloche, O., Wanke, V., Cameroni, E., & De Virgilio, C. (2005). The TOR and EGO protein complexes orchestrate microautophagy in yeast. *Molecular Cell, 19*(1), 15–26. doi:10.1016/j.molcel.2005.05.020 PMID:15989961

Duennwald, M. L., Echeverria, A., & Shorter, J. (2012). Small heat shock proteins potentiate amyloid dissolution by protein disaggregases from yeast and humans. *PLoS Biology, 10*(6), e1001346. doi:10.1371/journal.pbio.1001346 PMID:22723742

Duennwald, M. L., & Lindquist, S. (2008). Impaired ERAD and ER stress are early and specific events in polyglutamine toxicity. *Genes & Development, 22*(23), 3308–3319. doi:10.1101/gad.1673408 PMID:19015277

Du, F., Yu, Q., Yan, S., Hu, G., Lue, L. F., Walker, D. G., ... Yan, S. S. (2017). *PINK1* signalling rescues amyloid pathology and mitochondrial dysfunction in Alzheimer's disease. *Brain, 140*(12), 3233–3251. doi:10.1093/brain/awx258 PMID:29077793

Dufey, E., Sepúlveda, D., Rojas-Rivera, D., & Hetz, C. (2014). Cellular mechanisms of endoplasmic reticulum stress signaling in health and disease. 1. An overview. *American Journal of Physiology. Cell Physiology, 307*(7), C582–C594. doi:10.1152/ajpcell.00258.2014 PMID:25143348

Dugger, B. N., & Dickson, D. W. (2017). Pathology of neurodegenerative diseases. *Cold Spring Harbor Perspectives in Biology*, 9. PMID:28062563

Du, H., Guo, L., Fang, F., Chen, D., Sosunov, A. A., McKhann, G. M., ... Yan, S. D. (2008). Cyclophilin D deficiency attenuates mitochondrial and neuronal perturbation and ameliorates learning and memory in Alzheimer's disease. *Nature Medicine, 14*(10), 1097–1105. doi:10.1038/nm.1868 PMID:18806802

Du, H., Guo, L., Yan, S., Sosunov, A. A., McKhann, G. M., & Yan, S. S. (2010). Early deficits in synaptic mitochondria in an Alzheimer's disease mouse model. *Proceedings of the National Academy of Sciences of the United States of America, 107*(43), 18670–18675. doi:10.1073/pnas.1006586107 PMID:20937894

Du, H., & Yan, S. S. (2010). Mitochondrial permeability transition pore in Alzheimer's disease: Cyclophilin D and amyloid beta. Biochimica et Biophysica Acta (BBA)-. *Molecular Basis of Disease, 1802*(1), 198–204.

Du, J., Liang, Y., Xu, F., Sun, B., & Wang, Z. (2013). Trehalose rescues Alzheimer's disease phenotypes in APP/PS1 transgenic mice. *The Journal of Pharmacy and Pharmacology, 65*(12), 1753–1756. doi:10.1111/jphp.12108 PMID:24236985

Durcan, T. M., Tang, M. Y., Pérusse, J. R., Dashti, E. A., Aguileta, M. A., McLelland, G. L., ... Fon, E. A. (2014). USP8 regulates mitophagy by removing K6-linked ubiquitin conjugates from parkin. *The EMBO Journal, 33*(21), 2473–2491.

Duyao, M. P., Auerbach, A. B., Ryan, A., Persichetti, F., Barnes, G. T., McNeil, S. M., . . . Joyner, A. (1995). *Inactivation of the mouse Huntington's disease gene homolog Hdh*. Academic Press.

Eaton, S., & Willis, M. S. (2017, Feb. 28). Discovery of Bag3-P209L signaling mechanism brings hope for therapeutic options currently in development. Atlas of Science.

Ebrahimi-Fakhari, D., Wahlster, L., & McLean, P. J. (2011). Molecular chaperones in Parkinson's disease--present and future. *Journal of Parkinson's Disease, 1*(4), 299–320. doi:10.3233/JPD-2011-11044 PMID:22279517

Ebrahimi-Fakhari, D., Wahlster, L., & McLean, P. J. (2012). Protein degradation pathways in Parkinson's disease: Curse or blessing. *Acta Neuropathologica, 124*(2), 153–172. doi:10.100700401-012-1004-6 PMID:22744791

Echaniz-Laguna, A., Geuens, T., Petiot, P., Pereon, Y., Adriaenssens, E., Haidar, M., ... Stojkovic, T. (2017). Axonal neuropathies due to mutations in small heat shock proteins: Clinical, genetic, and functional insights into novel mutations. *Human Mutation, 38*(5), 556–568. doi:10.1002/humu.23189 PMID:28144995

Edison, P., Archer, H., Hinz, R., Hammers, A., Pavese, N., Tai, Y., . . . Kennedy, A. (2007). *Amyloid, hypometabolism, and cognition in Alzheimer disease: an [11C] PIB and [18F] FDG PET study*. Academic Press.

Eglitis, M. A., & Mezey, E. (1997). Hematopoietic cells differentiate into both microglia and macroglia in the brains of adultmice. *Proceedings of the National Academy of Sciences of the United States of America, 94*(8), 4080–4085. doi:10.1073/pnas.94.8.4080 PMID:9108108

Eidenmuller, J., Thomas, F. A. T. H., Thorsten, M. A. A. S., & Madeline, P. O. O. L. (2001). Phosphorylation-mimicking glutamate clusters in the proline-rich region are sufficient to simulate the functional deficiencies of hyperphosphorylated tau protein. *The Biochemical Journal, 357*(3), 759–767. doi:10.1042/bj3570759 PMID:11463346

Eisenberg-Lerner, A., Bialik, S., Simon, H. U., & Kimchi, A. (2009). Life and death partners: Apoptosis, autophagy and the cross-talk between them. *Cell Death and Differentiation, 16*(7), 966–975. doi:10.1038/cdd.2009.33 PMID:19325568

Ellis, J. (1987). Proteins as molecular chaperones. *Nature, 328*(6129), 378–379. doi:10.1038/328378a0 PMID:3112578

Elsasser, S., & Finley, D. (2005). Delivery of ubiquitinated substrates to protein-unfolding machines. *Nature Cell Biology, 7*(8), 742–749. doi:10.1038/ncb0805-742 PMID:16056265

Engelman, J. A. (2009). Targeting PI3K signalling in cancer: Opportunities, challenges and limitations. *Nature Reviews. Cancer, 9*(8), 550–562. doi:10.1038/nrc2664 PMID:19629070

Eschbach, J., & Danzer, K. M. (2014). Alpha-Synuclein in Parkinson's disease: Pathogenic function and translation into animal models. *Neurodegenerative Diseases*, *14*(1), 1–17. doi:10.1159/000354615 PMID:24080741

Esposito, A., Dohm, C. P., Kermer, P., Bähr, M., & Wouters, F. S. (2007). α-Synuclein and its disease-related mutants interact differentially with the microtubule protein tau and associate with the actin cytoskeleton. *Neurobiology of Disease*, *26*(3), 521–531. doi:10.1016/j.nbd.2007.01.014 PMID:17408955

Esposito, E., Di Matteo, V., Benigno, A., Pierucci, M., Crescimanno, G., & Di Giovanni, G. (2007). Non-Steroidal Anti-Inflammatory Drugs in Parkinson's Disease. *Experimental Neurology*, *205*(2), 295–312. doi:10.1016/j.expneurol.2007.02.008 PMID:17433296

Evans, C. S., & Holzbaur, E. (2019). Autophagy and mitophagy in ALS. *Neurobiology of Disease*, *122*, 35–40. doi:10.1016/j.nbd.2018.07.005 PMID:29981842

Evgrafov, O. V., Mersiyanova, I., Irobi, J., Van Den Bosch, L., Dierick, I., Leung, C. L., ... Timmerman, V. (2004). Mutant small heat-shock protein 27 causes axonal Charcot-Marie-Tooth disease and distal hereditary motor neuropathy. *Nature Genetics*, *36*(6), 602–606. doi:10.1038/ng1354 PMID:15122254

Exalto, L. G., van der Flier, W. M., Scheltens, P., Vrenken, H., & Biessels, G. J. (2014). Dysglycemia, brain volume and vascular lesions on MRI in a memory clinic population. *Journal of Diabetes and Its Complications*, *28*(1), 85–90. doi:10.1016/j.jdiacomp.2012.12.003 PMID:23352495

Fagan, A. M., Mintun, M. A., Shah, A. R., Aldea, P., Roe, C. M., Mach, R. H., . . . Holtzman, D. M. (2009). *Cerebrospinal fluid tau and ptau181 increase with cortical amyloid deposition in cognitively normal individuals: Implications for future clinical trials of Alzheimer's disease*. Academic Press.

Fan, C. Y., Lee, S., & Cyr, D. M. (2003). Mechanisms for regulation of Hsp70 function by Hsp40. *Cell Stress & Chaperones*, *8*(4), 309–316. doi:10.1379/1466-1268(2003)008<0309:MFROHF>2.0.CO;2 PMID:15115283

Fang, E. F., Hou, Y., Palikaras, K., Adriaanse, B. A., Kerr, J. S., Yang, B., . . . Dan, X. (2019). *Mitophagy inhibits amyloid-β and tau pathology and reverses cognitive deficits in models of Alzheimer's disease*. Academic Press.

Fang, E. F., Kassahun, H., Croteau, D. L., Scheibye-Knudsen, M., Marosi, K., Lu, H., ... Iser, W. B. (2016). NAD+ replenishment improves lifespan and healthspan in ataxia telangiectasia models via mitophagy and DNA repair. *Cell Metabolism*, *24*(4), 566–581. doi:10.1016/j.cmet.2016.09.004 PMID:27732836

Fang, E. F., Scheibye-Knudsen, M., Chua, K. F., Mattson, M. P., Croteau, D. L., & Bohr, V. A. (2016). Nuclear DNA damage signalling to mitochondria in ageing. *Nature Reviews. Molecular Cell Biology*, *17*(5), 308–321. doi:10.1038/nrm.2016.14 PMID:26956196

Fan, W., Nassiri, A., & Zhong, Q. (2011). Autophagosome targeting and membrane curvature sensing by Barkor/Atg14(L). *Proceedings of the National Academy of Sciences of the United States of America*, *108*(19), 7769–7774. doi:10.1073/pnas.1016472108 PMID:21518905

Farazi, T. A., Waksman, G., & Gordon, J. I. (2001). The Biology and Enzymology of Protein N-Myristoylation. *The Journal of Biological Chemistry*, *276*(43), 39501–39504. doi:10.1074/jbc.R100042200 PMID:11527981

Fares, M. B., Ait-Bouziad, N., Dikiy, I., Mbefo, M. K., Jovičić, A., Kiely, A., ... Lashuel, H. A. (2014). The novel Parkinson's disease linked mutation G51D attenuates in vitro aggregation and membrane binding of α-synuclein, and enhances its secretion and nuclear localization in cells. *Human Molecular Genetics*, *23*(17), 4491–4509. doi:10.1093/hmg/ddu165 PMID:24728187

Farris, W., Mansourian, S., Chang, Y., Lindsley, L., Eckman, E. A., Frosch, M. P., ... Guénette, S. (2003). Insulin-degrading enzyme regulates the levels of insulin, amyloid β-protein, and the β-amyloid precursor protein intracellular domain in vivo. *Proceedings of the National Academy of Sciences of the United States of America, 100*(7), 4162–4167. doi:10.1073/pnas.0230450100 PMID:12634421

Fecto, F., & Siddique, T. (2012). UBQLN2/P62 cellular recycling pathways in amyotrophic lateral sclerosis and frontotemporal dementia. *Muscle & Nerve, 45*(2), 157–162. doi:10.1002/mus.23278 PMID:22246868

Fecto, F., Yan, J., Vemula, S. P., Liu, E., Yang, Y., Chen, W., & ... . (2011). SQSTM1 mutations in familial and sporadic amyotrophic lateral sclerosis. *Archives of Neurology, 68*, 1440–1446. doi:10.1001/archneurol.2011.250 PMID:22084127

Fei, E., Jia, N., Yan, M., Ying, Z., Sun, Q., Wang, H., ... Wang, G. (2006). SUMO-1 modification increases human SOD1 stability and aggregation. *Biochemical and Biophysical Research Communications, 347*(2), 406–412. doi:10.1016/j.bbrc.2006.06.092 PMID:16828461

Feige, J. N., Lagouge, M., Canto, C., Strehle, A., Houten, S. M., Milne, J. C., ... Auwerx, J. (2008). Specific SIRT1 activation mimics low energy levels and protects against diet-induced metabolic disorders by enhancing fat oxidation. *Cell Metabolism, 8*(5), 347–358. doi:10.1016/j.cmet.2008.08.017 PMID:19046567

Feligioni, M., Brambilla, E., Camassa, A., Sclip, A., Arnaboldi, A., Morelli, F., ... Borsello, T. (2011). Crosstalk between JNK and SUMO Signaling Pathways: deSUMOylation Is Protective against H2O2-Induced Cell Injury. *PLoS One, 6*(12), 6. doi:10.1371/journal.pone.0028185 PMID:22164242

Feng, H. L., Leng, Y., Ma, C. H., Zhang, J., Ren, M., & Chuang, D. M. (2008). Combined lithium and valproate treatment delays disease onset, reduces neurological deficits and prolongs survival in an amyotrophic lateral sclerosis mouse model. *Neuroscience, 155*(3), 567–572. doi:10.1016/j.neuroscience.2008.06.040 PMID:18640245

Feng, Y., He, D., Yao, Z., & Klionsky, D. J. (2014). The machinery of macroautophagy. *Cell Research, 24*(1), 24–41. doi:10.1038/cr.2013.168 PMID:24366339

Fernandez-Chacon, R., Wolfel, M., Nishimune, H., Tabares, L., Schmitz, F., Castellano, M., & ... . (2004). The Synaptic Vesicle Protein CSP Prevents Presynaptic Degeneration. *Neur, 42*, 237–251. PMID:15091340

Fernández-Nogales, M., Cabrera, J. R., Santos-Galindo, M., Hoozemans, J. J., Ferrer, I., Rozemuller, A. J., ... Lucas, J. J. (2014). Huntington's disease is a four-repeat tauopathy with tau nuclear rods. *Nature Medicine, 20*(8), 881–885. doi:10.1038/nm.3617 PMID:25038828

Fernández-Nogales, M., Santos-Galindo, M., Hernández, I. H., Cabrera, J. R., & Lucas, J. J. (2016). Faulty splicing and cytoskeleton abnormalities in H untington's disease. *Brain Pathology (Zurich, Switzerland), 26*(6), 772–778. doi:10.1111/bpa.12430 PMID:27529534

Ferrante, R. J. (2009). Mouse models of Huntington's Disease and methodological consideration for therapeutic trials. *Biochimica et Biophysica Acta, 1792*(6), 506–520. doi:10.1016/j.bbadis.2009.04.001 PMID:19362590

Ferreira, S. A., & Romero-Ramos, M. (2018). Microglia response during Parkinson's disease: Alpha-synuclein intervention. *Frontiers in Cellular Neuroscience, 12*, 247. doi:10.3389/fncel.2018.00247 PMID:30127724

Ferreira, S. T., & Klein, W. L. (2011). The Aβ oligomer hypothesis for synapse failure and memory loss in Alzheimer's disease. *Neurobiology of Learning and Memory, 96*(4), 529–543. doi:10.1016/j.nlm.2011.08.003 PMID:21914486

Ferreiro, E., Oliveira, C. R., & Pereira, C. M. F. (2008). The release of calcium from the endoplasmic reticulum induced by amyloid-beta and prion peptides activates the mitochondrial apoptotic pathway. *Neurobiology of Disease, 30*(3), 331–342. doi:10.1016/j.nbd.2008.02.003 PMID:18420416

Ferreiro, E., & Pereira, C. M. F. (2012). Endoplasmic reticulum stress: A new playER in tauopathies. *The Journal of Pathology, 226*(5), 687–692. doi:10.1002/path.3977 PMID:22190226

Ferrer-Sueta, G., Manta, B., Botti, H., Radi, R., Trujillo, M., & Denicola, A. (2011). Factors affecting protein thiol reactivity and specificity in peroxide reduction. *Chemical Research in Toxicology, 24*(4), 434–450. doi:10.1021/tx100413v PMID:21391663

Ferro, F., Servais, S., Besson, P., Roger, S., Dumas, J. F., & Brisson, L. (2019). Autophagy and mitophagy in cancer metabolic remodelling. *Seminars in Cell & Developmental Biology*. PMID:31154012

Festa, M., Valle, L. D., Khalili, K., Franco, R., Scognamiglio, G., Graziano, V., ... Rosati, A. (2011). BAG3 protein is overexpressed in human glioblastoma and is a potential target for therapy. *American Journal of Pathology, 178*(6), 2504–2512. doi:10.1016/j.ajpath.2011.02.002 PMID:21561597

Fesus, L., & Piacentini, M. (2002). *Transglutaminase 2: an enigmatic enzyme with diverse functions*. Academic Press.

Finkbeiner, S. (2019). The Autophagy Lysosomal Pathway and Neurodegeneration. *Cold Spring Harbor Perspectives in Biology*, a033993. doi:10.1101/cshperspect.a033993 PMID:30936119

Finkel, T., & Holbrook, N. J. (2000). Oxidants, oxidative stress and the biology of ageing. *Nature, 408*(6809), 239–247. doi:10.1038/35041687 PMID:11089981

Flores-Rodriguez, P., Ontiveros-Torres, M.A., Cardenas-Aguayo, M.C., Luna-Arias, J.P., Meraz-Rios, M.A., Viramontes-Pintos, A., … Floran-Garduno, B. (2015). The relationship between truncation and phosphorylation at the C-terminus of tau protein in the paired helical filaments of Alzheimer's disease. *Front Neurosci, 9*.

Floyd R.A., & Carney J.M. (1992). Free radical damage to protein and DNA: mechanisms involved and relevant observations on brain undergoing oxidative stress. *Ann Neurol, 32*, S22-27.

Fomenko, D. E., Xing, W., Adair, B. M., Thomas, D. J., & Gladyshev, V. N. (2007). High throughput identification of catalytic redox-active cysteine residues. *Science, 315*(5810), 387–389. doi:10.1126cience.1133114 PMID:17234949

Fonseca, A. C. R. G., Ferreiro, E., Oliveira, C. R., Cardoso, S. M., & Pereira, C. F. (2013). Activation of the endoplasmic reticulum stress response by the amyloid-beta 1-40 peptide in brain endothelial cells. *Biochimica et Biophysica Acta. Molecular Basis of Disease, 1832*(12), 2191–2203. doi:10.1016/j.bbadis.2013.08.007 PMID:23994613

Ford, L., Fioriti, L., & Kandel, E. R. (2019). Ubiquitination and SUMOylation of Amyloid and Amyloid-like Proteins in Health and Disease. *Current Issues in Molecular Biology, 35*, 195–230. PMID:31422940

Forloni, G., Roiter, I., & Tagliavini, F. (2019). Clinical trials of prion disease therapeutics. *Current Opinion in Pharmacology, 44*, 53–60. doi:10.1016/j.coph.2019.04.019 PMID:31108459

Förstl, H., & Kurz, A. (1999). Clinical features of Alzheimer's disease. *European Archives of Psychiatry and Clinical Neuroscience, 249*(6), 288–299. doi:10.1007004060050101 PMID:10653284

Frank, S. (2014). *Treatment of Huntington's disease*. Academic Press.

Frank, M. (2007). Intracellular amyloid-beta in Alzheimer's disease. *Nature Reviews. Neuroscience, 8*(7), 499–509. doi:10.1038/nrn2168 PMID:17551515

Frautschy, S. A., Hu, W., Kim, P., Miller, S. A., Chu, T., Harris-White, M. E., & Cole, G. M. (2001). Phenolic anti-inflammatory antioxidant reversal of Aβ-induced cognitive deficits and neuropathology. *Neurobiology of Aging, 22*(6), 993–1005. doi:10.1016/S0197-4580(01)00300-1 PMID:11755008

Frost, B., Gotz, J., & Feany, M. B. (2015). Connecting the dots between tau dysfunction and neurodegeneration. *Trends in Cell Biology*, *25*(1), 46–53. doi:10.1016/j.tcb.2014.07.005 PMID:25172552

Fruman, D. A., & Rommel, C. (2014). PI3K and cancer: Lessons, challenges and opportunities. *Nature Reviews. Drug Discovery*, *13*(2), 140–156. doi:10.1038/nrd4204 PMID:24481312

Fu, A. L., Dong, Z. H., & Sun, M. J. (2006). Protective effect of N-acetyl-cysteine on amyloid ß peptide- induced learning and memory deficits in mice. *Brain Research*, *1109*(1), 201–206. doi:10.1016/j.brainres.2006.06.042 PMID:16872586

Fu, J., Jin, J., Cichewicz, R. H., Hageman, S. A., Ellis, T. K., Xiang, L., ... Ross, C. A. (2012). trans-(−)-ϵ-Viniferin increases mitochondrial sirtuin 3 (SIRT3), activates AMP-activated protein kinase (AMPK), and protects cells in models of Huntington Disease. *The Journal of Biological Chemistry*, *287*(29), 24460–24472. doi:10.1074/jbc.M112.382226 PMID:22648412

Fujii, H., Ichimori, K., Hoshiai, K., & Nakazawa, H. (1997). Nitric oxide inactivates NADPH oxidase in pig neutrophils by inhibiting its assembling process. *The Journal of Biological Chemistry*, *272*(52), 32773–32778. doi:10.1074/jbc.272.52.32773 PMID:9407051

Fujita, K., Nakamura, Y., Oka, T., Ito, H., Tamura, T., Tagawa, K., . . . Shiwaku, H. (2013). *A functional deficiency of TERA/VCP/p97 contributes to impaired DNA repair in multiple polyglutamine diseases*. Academic Press.

Fujita, N., Hayashi-Nishino, M., Fukumoto, H., Omori, H., Yamamoto, A., Noda, T., & Yoshimori, T. (2008). An Atg4B mutant hampers the lipidation of LC3 paralogues and causes defects in autophagosome closure. *Molecular Biology of the Cell*, *19*(11), 4651–4659. doi:10.1091/mbc.e08-03-0312 PMID:18768752

Fujita, N., Itoh, T., Omori, H., Fukuda, M., Noda, T., & Yoshimori, T. (2008). The Atg16L Complex Specifies the Site of LC3 Lipidation for Membrane Biogenesis in Autophagy. *Molecular Biology of the Cell*, *19*(5), 2092–2100. doi:10.1091/mbc.e07-12-1257 PMID:18321988

Fukai, T., & Ushio-Fukai, M. (2011). Superoxide dismutases: Role in redox signaling, vascular function, and diseases. *Antioxidants & Redox Signalling*, *15*(6), 1583–1606. doi:10.1089/ars.2011.3999 PMID:21473702

Fukata, Y., & Fukata, M. (2010). Protein palmitoylation in neuronal development and synaptic plasticity. *Nature Reviews. Neuroscience*, *11*(3), 161–175. doi:10.1038/nrn2788 PMID:20168314

Funderburk, S. F., Wang, Q. J., & Yue, Z. (2010). The Beclin 1–VPS34 complex–at the crossroads of autophagy and beyond. *Trends in Cell Biology*, *20*(6), 355–362. doi:10.1016/j.tcb.2010.03.002 PMID:20356743

Fuster-Matanzo, A., de Barreda, E. G., Dawson, H. N., Vitek, M. P., Avila, J., & Hernández, F. (2009). Function of tau protein in adult newborn neurons. *FEBS Letters*, *583*(18), 3063–3068. doi:10.1016/j.febslet.2009.08.017 PMID:19695252

Fu, Y., Wu, P., Pan, Y., Sun, X., Yang, H., Difiglia, M., & Lu, B. (2017). A toxic mutant huntingtin species is resistant to selective autophagy. *Nature Chemical Biology*, *13*(11), 1152–1154. doi:10.1038/nchembio.2461 PMID:28869595

Gadoth, N., & Goebel, H. H. (2013). Oxidative stress and free radical damage in neurology. New York: Humana Press.

Gallogly, M. M., & Mieyal, J. J. (2007). Mechanisms of reversible protein glutathionylation in redox signaling and oxidative stress. *Current Opinion in Pharmacology*, *7*(4), 381–391. doi:10.1016/j.coph.2007.06.003 PMID:17662654

Galluzzi, L., & Green, D. R. (2019). Autophagy-Independent Functions of the Autophagy Machinery. *Cell*, *177*(7), 1682–1699. doi:10.1016/j.cell.2019.05.026 PMID:31199916

Gama-Carvalho, M., & Carmo-Fonseca, M. (2001). The rules and roles of nucleocytoplasmic shuttling proteins. *FEBS Letters*, *498*(2–3), 157–163. doi:10.1016/S0014-5793(01)02487-5 PMID:11412848

Gambhir, S., Czernin, J., Schwimmer, J., Silverman, D. H. S., Coleman, R. E., & Phelps, M. E. (2001). A Tabulated Summary of the FDG PET Literature. *Journal of Nuclear Medicine*, *42*(S1), 1S–93S. PMID:11483694

Gamerdinger, M., Kaya, A. M., Wolfrum, U., Clement, A. M., & Behl, C. (2011). BAG3 mediates chaperone-based aggresome-targeting and selective autophagy of misfolded proteins. *EMBO Reports*, *12*(2), 149–156. doi:10.1038/embor.2010.203 PMID:21252941

Ganguly, G., Chakrabarti, S., Chatterjee, U., & Saso, L. (2017). Proteinopathy, oxidative stress and mitochondrial dysfunction: Cross talk in Alzheimer's disease and Parkinson's disease. *Drug Design, Development and Therapy*, *11*, 797–810. doi:10.2147/DDDT.S130514 PMID:28352155

Gan, L., & Mucke, L. (2008). Paths of convergence: Sirtuins in aging and neurodegeneration. *Neuron*, *58*(1), 10–14. doi:10.1016/j.neuron.2008.03.015 PMID:18400158

Gan, L., Vargas, M. R., Johnson, D. A., & Johnson, J. A. (2012). Astrocyte-specific overexpression of Nrf2 delays motor pathology and synuclein aggregation throughout the CNS in the alpha-synuclein mutant (A53T) mouse model. *The Journal of Neuroscience: The Official Journal of the Society for Neuroscience*, *32*(49), 17775–17787. doi:10.1523/JNEUROSCI.3049-12.2012 PMID:23223297

Ganley, I. G., Lam, D. H., Wang, J., Ding, X., Chen, S., & Jiang, X. (2009). ULK1· ATG13· FIP200 complex mediates mTOR signaling and is essential for autophagy. *The Journal of Biological Chemistry*, *284*(18), 12297–12305. doi:10.1074/jbc.M900573200 PMID:19258318

Gan-Or, Z., Dion, P. A., & Rouleau, G. A. (2015). Genetic perspective on the role of the autophagy-lysosome pathway in Parkinson disease. *Autophagy*, *11*(9), 1443–1457. doi:10.1080/15548627.2015.1067364 PMID:26207393

Gao, F., Yang, J., Wang, D., Li, C., Fu, Y., Wang, H., ... Zhang, J. (2017). Mitophagy in Parkinson's Disease: Pathogenic and therapeutic implications. *Frontiers in Neurology*, *8*, 527. doi:10.3389/fneur.2017.00527 PMID:29046661

Gao, J., Shao, K., Chen, X., Li, Z., Liu, Z., Yu, Z., ... Li, P. (2019). The involvement of post-translational modifications in cardiovascular pathologies: Focus on SUMOylation, neddylation, succinylation, and prenylation. *Journal of Molecular and Cellular Cardiology*, *138*, 49–58. doi:10.1016/j.yjmcc.2019.11.146 PMID:31751566

Gardet, A., Benita, Y., Li, C., Sands, B. E., Ballester, I., Stevens, C., ... Podolsky, D. K. (2010). LRRK2 is involved in the IFN-γ response and host response to pathogens. *Journal of Immunology (Baltimore, Md.: 1950)*, *185*(9), 5577–5585. doi:10.4049/jimmunol.1000548 PMID:20921534

Garg, G., Singh, S., Singh, A., & Rizvi, S. I. (2018). N-acetyl L-cysteine attenuates oxidative damage and neurodegeneration in rat brain during aging. *Canadian Journal of Physiology and Pharmacology*, *96*(12), 1189–1196. doi:10.1139/cjpp-2018-0209 PMID:30107137

Garrido, C., Brunet, M., Didelot, C., Zermati, Y., Schmitt, E., & Kroemer, G. (2006). Heat shock proteins 27 and 70: Anti-apoptotic proteins with tumorigenic properties. *Cell Cycle (Georgetown, Tex.)*, *5*(22), 2592–2601. doi:10.4161/cc.5.22.3448 PMID:17106261

Gasperini, L., & Legname, G. (2014). Prion protein and aging. *Frontiers in Cell and Developmental Biology*, 2. PMID:25364751

Gasperini, L., Meneghetti, E., Pastore, B., Benetti, F., & Legname, G. (2015). Prion protein and copper cooperatively protect neurons by modulating NMDA receptor through S-nitrosylation. *Antioxidants & Redox Signalling*, *22*(9), 772–784. doi:10.1089/ars.2014.6032 PMID:25490055

Gazewood, D. J., Richards, R. D., & Clebak, K. (2013). Parkinson Disease: An update. *American Family Physician, 87*(4), 267–273. PMID:23418798

Gazzina, S., Premi, E., Turrone, R., Acosta-Cabronero, J., Rizzetti, M. C., Cotelli, M. S., & Borroni, B. (2016). Subcortical matter in the α-synucleinopathies spectrum: An MRI pilot study. *Journal of Neurology, 263*(8), 1575–1582. doi:10.100700415-016-8173-5 PMID:27230856

Gegg, M. E., Cooper, J. M., Chau, K. Y., Rojo, M., Schapira, A. H., & Taanman, J. W. (2010). Mitofusin 1 and mitofusin 2 are ubiquitinated in a PINK1/parkin-dependent manner upon induction of mitophagy. *Human Molecular Genetics, 19*(24), 4861–4870. doi:10.1093/hmg/ddq419 PMID:20871098

Geiger, T., & Clarke, S. (1987). Deamidation, isomerization, and racemization at asparaginyl and aspartyl residues in peptides. Succinimide-linked reactions that contribute to protein degradation. *The Journal of Biological Chemistry, 262*, 785–794. PMID:3805008

Geisler, S., Holmström, K. M., Skujat, D., Fiesel, F. C., Rothfuss, O. C., Kahle, P. J., & Springer, W. (2010). *PINK1/* Parkin-mediated mitophagy is dependent on VDAC1 and p62/SQSTM1. *Nature Cell Biology, 12*(2), 119–131. doi:10.1038/ncb2012 PMID:20098416

Gelain, D. P., Antonio Behr, G., Birnfeld de Oliveira, R., & Trujillo, M. (2012). Antioxidant therapies for neurodegenerative diseases: Mechanisms, current trends, and perspectives. *Oxidative Medicine and Cellular Longevity, 895153*, 1–2. doi:10.1155/2012/895153 PMID:23304258

Gelmetti, V., De Rosa, P., Torosantucci, L., Marini, E. S., Romagnoli, A., Di Rienzo, M., ... Valente, E. M. (2017). PINK1 and BECN1 relocalize at mitochondria-associated membranes during mitophagy and promote ER-mitochondria tethering and autophagosome formation. *Autophagy, 13*(4), 654–669. doi:10.1080/15548627.2016.1277309 PMID:28368777

George, J. M. (2001). The synucleins. *Genome Biology, 3*(1), reviews3002-1.

Gerakis, Y., & Hetz, C. (2018). Emerging roles of ER stress in the etiology and pathogenesis of Alzheimer's disease. *The FEBS Journal, 285*(6), 995–1011. doi:10.1111/febs.14332 PMID:29148236

Geschwind, M. D. (2015). Prion Diseases. *Continuum (Minneap Minn), 21*(6), 1612-1638. doi:10.1212/con.0000000000000251

Geschwind, M. D. (2015). Prion Diseases. *Continuum (Minneapolis, Minn.), 21*, 1612–1638. doi:10.1212/CON.0000000000000251 PMID:26633779

Gess, B., Auer-Grumbach, M., Schirmacher, A., Strom, T., Zitzelsberger, M., Rudnik-Schoneborn, S., ... Senderek, J. (2014). HSJ1-related hereditary neuropathies: Novel mutations and extended clinical spectrum. *Neurology, 83*(19), 1726–1732. doi:10.1212/WNL.0000000000000966 PMID:25274842

Gestwicki, J.E., & Garza, D. (2012). Protein quality control in neurodegenerative disease. In *Progress in molecular biology and translational science*. Elsevier.

Ghaemmaghami, S., May, B. C., Renslo, A. R., & Prusiner, S. B. (2010). Discovery of 2-aminothiazoles as potent antiprion compounds. *Journal of Virology, 84*(7), 3408–3412. doi:10.1128/JVI.02145-09 PMID:20032192

Ghaoui, R., Palmio, J., Brewer, J., Lek, M., Needham, M., Evila, A., ... Udd, B. (2016). Mutations in HSPB8 causing a new phenotype of distal myopathy and motor neuropathy. *Neurology, 86*(4), 391–398. doi:10.1212/WNL.0000000000002324 PMID:26718575

Ghetti, B., Oblak, A. L., Boeve, B. F., Johnson, K. A., Dickerson, B. C., & Goedert, M. (2015). Invited review: frontotemporal dementia caused by microtubule-associated protein tau gene (MAPT) mutations: a chameleon for neuropathology and neuroimaging. *Neuropathology and Applied Neurobiology*, *41*(1), 24–46. doi:10.1111/nan.12213 PMID:25556536

Ghosh, D., Mondal, M., Mohite, G. M., Singh, P. K., Ranjan, P., Anoop, A., ... Maji, S. K. (2013). The Parkinson's disease-associated H50Q mutation accelerates alpha-Synuclein aggregation *in vitro*. *Biochemistry*, *52*(40), 6925–6927. doi:10.1021/bi400999d PMID:24047453

Ghosh, S., Watanabe, R. M., Valle, T. T., Hauser, E. R., Magnuson, V. L., Langefeld, C. D., ... Chines, P. (2000). The Finland–United States investigation of non–insulin-dependent diabetes mellitus genetics (FUSION) study. I. An autosomal genome scan for genes that predispose to type 2 diabetes. *American Journal of Human Genetics*, *67*(5), 1174–1185. PMID:11032783

Giasson, B.I., Duda, J.E., & Murray, I.V.J. (2000). Oxidative damage linked to neurodegeneration by selective α-synuclein nitration in synucleinopathy lesions. *Science*, *290*, 985–989. doi:10.1126cience.290.5493.985

Giblin, W., Skinner, M. E., & Lombard, D. B. (2014). Sirtuins: Guardians of mammalian healthspan. *Trends in Genetics*, *30*(7), 271–286. doi:10.1016/j.tig.2014.04.007 PMID:24877878

Gilbert, B. J. (2013). The role of amyloid β in the pathogenesis of Alzheimer's disease. *Journal of Clinical Pathology*, *66*(5), 362–366. doi:10.1136/jclinpath-2013-201515 PMID:23526599

Giles, G. I., Tasker, K. M., & Jacob, C. (2001). Hypothesis: The role of reactive sulfur species in oxidative stress. *Free Radical Biology & Medicine*, *31*(10), 1279–1283. doi:10.1016/S0891-5849(01)00710-9 PMID:11705707

Giles, N. M., Watts, A. B., Giles, G. I., Fry, F. H., Littlechild, J. A., & Jacob, C. (2003). Metal and redox modulation of cysteine protein function. *Chemistry & Biology*, *10*(8), 677–693. doi:10.1016/S1074-5521(03)00174-1 PMID:12954327

Girard, M., Lariviere, R., Parfitt, D. A., Deane, E. C., Gaudet, R., Nossova, N., ... McPherson, P. S. (2012). Mitochondrial dysfunction and Purkinje cell loss in autosomal recessive spastic ataxia of Charlevoix-Saguenay (ARSACS). *Proceedings of the National Academy of Sciences of the United States of America*, *109*(5), 1661–1666. doi:10.1073/pnas.1113166109 PMID:22307627

Glajch, K. E., & Sadri-Vakili, G. (2015). *Epigenetic mechanisms involved in Huntington's disease pathogenesis*. Academic Press.

Glick, D., Barth, S., & Macleod, K. F. (2010). Autophagy: Cellular and molecular mechanisms. *The Journal of Pathology*, *221*(1), 3–12. doi:10.1002/path.2697 PMID:20225336

Goedert, M. (1993). *Tau protein and the neurofibrillary pathology of Alzheimer's disease*. Academic Press.

Goedert, M., Spillantini, M., Hasegawa, M., Jakes, R., Crowther, R., & Klug, A. (1996). *Molecular dissection of the neurofibrillary lesions of Alzheimer's disease*. Paper presented at the Cold Spring Harbor symposia on quantitative biology.

Goers, J., Manning-Bog, A. B., McCormack, A. L., Millett, I. S., Doniach, S., Di Monte, D. A., ... Fink, A. L. (2003). Nuclear localization of α-synuclein and its interaction with histones. *Biochemistry*, *42*(28), 8465–8471. doi:10.1021/bi0341152 PMID:12859192

Golbe, L. I., Lazzarini, A. M., Duvoisin, R. C., Iorio, G. D., Sanges, G., Bonavita, V., & la Sala, S. (1996). Clinical genetic analysis of Parkinson's disease in the Contursi kindred. *Annals of Neurology: Official Journal of the American Neurological Association and the Child Neurology Society*, *40*(5), 767–775. doi:10.1002/ana.410400513 PMID:8957018

Goldberg, M. S., Fleming, S. M., Palacino, J. J., Cepeda, C., Lam, H. A., Bhatnagar, A., ... Shen, J. (2003). Parkin-deficient mice exhibit nigrostriataldeficits but not loss of dopaminergic neurons. *The Journal of Biological Chemistry, 278*(44), 43628–43635. doi:10.1074/jbc.M308947200 PMID:12930822

Golde, T. E., Petrucelli, L., & Lewis, J. (2010). *Targeting Aβ and tau in Alzheimer's disease, an early interim report.* Academic Press.

Goldman, J. G., Andrews, H., Amara, A., Naito, A., Alcalay, R. N., Shaw, L. M., ... Kang, U. J. (2018). Cerebrospinal fluid, plasma, and saliva in the BioFIND study: Relationships among biomarkers and Parkinson's disease Features. *Movement Disorders, 33*(2), 282–288. doi:10.1002/mds.27232 PMID:29205509

Goldwurm, S., Di Fonzo, A., Simons, E. J., Rohe, C. F., Zini, M., Canesi, M., ... Meucci, N. (2005). The G6055A (G2019S) mutation in LRRK2 is frequent in both early and late onset Parkinson's disease and originates from a common ancestor. *Journal of Medical Genetics, 42*(11), e65–e65. doi:10.1136/jmg.2005.035568 PMID:16272257

Goll, D. E., Thompson, V. F., Li, H., Wei, W., & Cong, J. (2003). The calpain system. *Physiological Reviews, 83*(3), 731–801. doi:10.1152/physrev.00029.2002 PMID:12843408

Gómez-Isla, T., Hollister, R., West, H., Mui, S., Growdon, J. H., Petersen, R. C., ... Hyman, B. T. (1997). Neuronal loss correlates with but exceeds neurofibrillary tangles in Alzheimer's disease. *Annals of Neurology: Official Journal of the American Neurological Association and the Child Neurology Society, 41*(1), 17–24. doi:10.1002/ana.410410106 PMID:9005861

Goncalves, S., & Outeiro, T. F. (2013). Assessing the Subcellular Dynamics of Alpha-synuclein Using Photoactivation Microscopy. *Molecular Neurobiology, 47*(3), 1081–1092. doi:10.100712035-013-8406-x PMID:23389286

Gong, C. X., Lidsky, T., Wegiel, J., Zuck, L., Grundke-Iqbal, I., & Iqbal, K. (2000). Phosphorylation of microtubule-associated protein tau is regulated by protein phosphatase 2A in mammalian brain implications for neurofibrillary degeneration in Alzheimer's disease. *The Journal of Biological Chemistry, 275*(8), 5535–5544. doi:10.1074/jbc.275.8.5535 PMID:10681533

Gong, C.-X., & Iqbal, K. (2008). Hyperphosphorylation of Microtubule-Associated Protein Tau: A Promising Therapeutic Target for Alzheimer Disease. *Current Medicinal Chemistry, 15*(23), 2321–2328. doi:10.2174/092986708785909111 PMID:18855662

González-Polo, R. A., Boya, P., Pauleau, A. L., Jalil, A., Larochette, N., Souquère, S., ... Kroemer, G. (2005). The apoptosis/autophagy paradox: Autophagic vacuolization before apoptotic death. *Journal of Cell Science, 118*(14), 3091–3102. doi:10.1242/jcs.02447 PMID:15985464

Gorbatyuk, O. S., Li, S., Sullivan, L. F., Chen, W., Kondrikova, G., Manfredsson, F. P., ... Muzyczka, N. (2008). The phosphorylation state of Ser-129 in human α-synuclein determines neurodegeneration in a rat model of Parkinson disease. *Proceedings of the National Academy of Sciences of the United States of America, 105*(2), 763–768. doi:10.1073/pnas.0711053105 PMID:18178617

Goswami, P., Gupta, S., Biswas, J., Joshi, N., Swarnkar, S., Nath, C., & Singh, S. (2016). Endoplasmic Reticulum Stress Plays a Key Role in Rotenone-Induced Apoptotic Death of Neurons. *Molecular Neurobiology, 53*(1), 285–298. doi:10.100712035-014-9001-5 PMID:25428620

Goswami, P., Gupta, S., Biswas, J., Joshi, N., Swarnkar, S., Nath, C., & Singh, S. (2019). Correction to: Endoplasmic Reticulum Stress Plays a Key Role in Rotenone-Induced Apoptotic Death of Neurons. *Molecular Neurobiology, 56*(9), 6654–6655. doi:10.100712035-019-01674-x PMID:31230259

Goto, S., & Radak, Z. (2013). Implications of oxidative damage to proteins and DNA in aging and its intervention by caloric restriction and exercise. *Journal of Sport and Health Science*, *2*(2), 75–80. doi:10.1016/j.jshs.2013.03.004

Gould, N., Doulias, P. T., Tenopoulou, M., Raju, K., & Ischiropoulos, H. (2013). Regulation of Protein Function and Signaling by Reversible Cysteine S-Nitrosylation. *The Journal of Biological Chemistry*, *288*(37), 26473–26479. doi:10.1074/jbc.R113.460261 PMID:23861393

Go, Y. M., & Jones, D. P. (2008). Redox compartmentalization in eukaryotic cells. *Biochimica et Biophysica Acta*, *1780*(11), 1273–1290. doi:10.1016/j.bbagen.2008.01.011 PMID:18267127

Gratuze, M., Noël, A., Julien, C., Cisbani, G., Milot-Rousseau, P., Morin, F., ... Bissonnette, S. (2014). Tau hyperphosphorylation and deregulation of calcineurin in mouse models of Huntington's disease. *Human Molecular Genetics*, *24*(1), 86–99. doi:10.1093/hmg/ddu456 PMID:25205109

Greenamyre, J. T., Sherer, T. B., Betarbet, R., & Panov, A. V. (2001). Complex I and Parkinson's disease. *IUBMB Life*, *52*(3-5), 135–141. doi:10.1080/15216540152845939 PMID:11798025

Greenbaum, E. A., Graves, C. L., Mishizen-Eberz, A. J., Lupoli, M. A., Lynch, D. R., Englander, S. W., ... Giasson, B. I. (2005). The E46K mutation in α-synuclein increases amyloid fibril formation. *The Journal of Biological Chemistry*, *280*(9), 7800–7807. doi:10.1074/jbc.M411638200 PMID:15632170

Greene, A. W., Grenier, K., Aguileta, M. A., Muise, S., Farazifard, R., Haque, M. E., ... Fon, E. A. (2012). Mitochondrial processing peptidase regulates PINK1 processing, import and Parkin recruitment. *EMBO Reports*, *13*(4), 378–385. doi:10.1038/embor.2012.14 PMID:22354088

Greene, J. C., Whitworth, A. J., Kuo, I., Andrews, L. A., Feany, M. B., & Pallanck, L. J. (2003). Mitochondrial pathology and apoptotic muscle degeneration in Drosophila parkin mutants. *Proceedings of the National Academy of Sciences of the United States of America*, *100*(7), 4078–4083. doi:10.1073/pnas.0737556100 PMID:12642658

Greer, P. L., Hanayama, R., Bloodgood, B. L., Mardinly, A. R., Lipton, D. M., Flavell, S. W., ... Greenberg, M. E. (2010). The Angelman Syndrome protein Ube3A regulates synapse development by ubiquitinating arc. *Cell*, *140*(5), 704–716. doi:10.1016/j.cell.2010.01.026 PMID:20211139

Greifenhagen, U., Frolov, A., Blüher, M., & Hoffmann, R. (2016). Site-specific analysis of advanced glycation end products in plasma proteins of type 2 diabetes mellitus patients. *Analytical and Bioanalytical Chemistry*, *408*(20), 5557–5566. doi:10.100700216-016-9651-4 PMID:27236317

Grundke-Iqbal, I., Iqbal, K., Tung, Y. C., Quinlan, M., Wisniewski, H. M., & Binder, L. I. (1986). Abnormal phosphorylation of the microtubule-associated protein tau (tau) in Alzheimer cytoskeletal pathology. *Proceedings of the National Academy of Sciences of the United States of America*, *83*(13), 4913–4917. doi:10.1073/pnas.83.13.4913 PMID:3088567

Guardia-Laguarta, C., Area-Gomez, E., Rüb, C., Liu, Y., Magrané, J., Becker, D., & Przedborski, S. (2014). α-Synuclein is localized to mitochondria-associated ER membranes. *The Journal of Neuroscience: The Official Journal of the Society for Neuroscience*, *34*(1), 249–259. doi:10.1523/JNEUROSCI.2507-13.2014 PMID:24381286

Guerra de Souza, A. C., Prediger, R. D., & Cimarosti, H. (2016). SUMO-regulated mitochondrial function in Parkinson's disease. *Journal of Neurochemistry*, *137*(5), 673–686. doi:10.1111/jnc.13599 PMID:26932327

Gu, L., & Robinson, A. S. (2016). Proteomic Approaches to Quantify Cysteine Reversible Modifications in Aging and Neurodegenerative Diseases. *Proteomics. Clinical Applications*, *10*(12), 1159–1177. doi:10.1002/prca.201600015 PMID:27666938

Gundersen, C. B., & Umbach, J. A. (1992). Suppression cloning of the cDNA for a candidate subunit of a presynaptic calcium channel. *Neur, 9*(3), 527–537. doi:10.1016/0896-6273(92)90190-O PMID:1326297

Guo, X., Disatnik, M.-H., Monbureau, M., Shamloo, M., Mochly-Rosen, D., & Qi, X. (2013). *Inhibition of mitochondrial fragmentation diminishes Huntington's disease–associated neurodegeneration.* Academic Press.

Guo, J., He, S., Kang, J., Xu, Q., Hu, Y., Zhang, H., ... Tang, B. (2015). Involvement of Bcl-2-associated athanogene (BAG)-family proteins in the neuroprotection by Rasagiline. *International Journal of Clinical and Experimental Medicine, 8*(10), 18158–18164. PMID:26770414

Guo, X., Sun, X., Hu, D., Wang, Y. J., Fujioka, H., Vyas, R., ... Qi, X. (2016). VCP recruitment to mitochondria causes mitophagy impairment and neurodegeneration in models of Huntington's disease. *Nature Communications, 7*(1), 12646. doi:10.1038/ncomms12646 PMID:27561680

Gupta, S., Biswas, J., Gupta, P., Singh, A., Tiwari, S., Mishra, A., & Singh, S. (2019). Salubrinal attenuates nitric oxide mediated PERK:IRE1alpha: ATF-6 signaling and DNA damage in neuronal cells. *Neurochemistry International, 131*, 104581. doi:10.1016/j.neuint.2019.104581 PMID:31639405

Gupta, S., Giricz, Z., Natoni, A., Donnelly, N., Deegan, S., Szegezdi, E., & Samali, A. (2012). NOXA contributes to the sensitivity of PERK-deficient cells to ER stress. *FEBS Letters, 586*(22), 4023–4030. doi:10.1016/j.febslet.2012.10.002 PMID:23068609

Gustafsson, Å. B., & Gottlieb, R. A. (2008). Recycle or die: The role of autophagy in cardioprotection. *Journal of Molecular and Cellular Cardiology, 44*(4), 654–661. doi:10.1016/j.yjmcc.2008.01.010 PMID:18353358

Guttmann, R. P. (2010). Redox regulation of cysteine-dependent enzymes. *Journal of Animal Science, 88*(4), 1297–1306. doi:10.2527/jas.2009-2381 PMID:19820057

Guttmann, R. P., Elce, J. S., Bell, P. D., Isbell, J. C., & Johnson, G. V. (1997). Oxidation inhibits substrate proteolysis by calpain I but not autolysis. *The Journal of Biological Chemistry, 272*(3), 2005–2012. doi:10.1074/jbc.272.3.2005 PMID:8999893

Guttmann, R. P., & Powell, T. J. (2012). Redox Regulation of Cysteine-Dependent Enzymes in Neurodegeneration. *International Journal of Cell Biology, 2012*, 1–8. doi:10.1155/2012/703164 PMID:22829832

Haack, T. B., Hogarth, P., Kruer, M. C., Gregory, A., Wieland, T., Schwarzmayr, T., ... Hayflick, S. J. (2012). Exome Sequencing Reveals De Novo WDR45 Mutations Causing a Phenotypically Distinct, X-Linked Dominant Form of NBIA. *American Journal of Human Genetics, 91*(6), 1144–1149. doi:10.1016/j.ajhg.2012.10.019 PMID:23176820

Haas, I. G., & Wabl, M. (1983). Immunoglobulin heavy chain binding protein. *Nature, 306*(5941), 387–389. doi:10.1038/306387a0 PMID:6417546

Haendeler, J., Weiland, U., Zeiher, A. M., & Dimmeler, S. (1997). Effects of redox-related congeners of NO on apoptosis and caspase-3 activity. *Nit Ox, 1*(4), 282–293. doi:10.1006/niox.1997.0134 PMID:9441900

Hafner, A. V., Dai, J., Gomes, A. P., Xiao, C. Y., Palmeira, C. M., Rosenzweig, A., & Sinclair, D. A. (2010). Regulation of the mPTP by SIRT3-mediated deacetylation of CypD at lysine 166 suppresses age-related cardiac hypertrophy. *Aging (Albany NY), 2*(12), 914–923. doi:10.18632/aging.100252 PMID:21212461

Hageman, J., & Kampinga, H. H. (2009). Computational analysis of the human HSPH/HSPA/DNAJ family and cloning of a human HSPH/HSPA/DNAJ expression library. *Cell Stress & Chaperones, 14*(1), 1–21. doi:10.100712192-008-0060-2 PMID:18686016

Hague, S., Rogaeva, E., Hernandez, D., Gulick, C., Singleton, A., ... Hanson, M. (2003). Early-onset Parkinson's disease caused by a compound heterozygous DJ-1 mutation. *Annals of Neurology*, *54*, 271–274. doi:10.1002/ana.10663 PMID:12891685

Hahn, J. S. (2009). The Hsp90 chaperone machinery: From structure to drug development. *BMB Reports*, *42*(10), 623–630. doi:10.5483/BMBRep.2009.42.10.623 PMID:19874705

Haigis, M. C., Mostoslavsky, R., Haigis, K. M., Fahie, K., Christodoulou, D. C., Murphy, A. J., ... Wolberger, C. (2006). SIRT4 inhibits glutamate dehydrogenase and opposes the effects of calorie restriction in pancreatic β cells. *Cell*, *126*(5), 941–954. doi:10.1016/j.cell.2006.06.057 PMID:16959573

Haigis, M. C., & Sinclair, D. A. (2010). Mammalian sirtuins: Biological insights and disease relevance. *Annual Review of Pathology*, *5*(1), 253–295. doi:10.1146/annurev.pathol.4.110807.092250 PMID:20078221

Halliday, M., Radford, H., Sekine, Y., Moreno, J., Verity, N., Le Quesne, J., ... Mallucci, G. R. (2015). Partial restoration of protein synthesis rates by the small molecule ISRIB prevents neurodegeneration without pancreatic toxicity. *Cell Death & Disease*, *6*(3), e1672. doi:10.1038/cddis.2015.49 PMID:25741597

Halliday, M., Radford, H., Zents, K. A. M., Molloy, C., Moreno, J. A., Verity, N. C., ... Mallucci, G. R. (2017). Repurposed drugs targeting eIF2α-P-mediated translational repression prevent neurodegeneration in mice. *Brain*, *140*(6), 1768–1783. doi:10.1093/brain/awx074 PMID:28430857

Halliwell, B. (1992). Free radicals in the brain. *Journal of Neurochemistry*, *59*(5), 1609–1623. doi:10.1111/j.1471-4159.1992.tb10990.x PMID:1402908

Hammond, J. W., Cai, D., & Verhey, K. J. (2008). Tubulin modifications and their cellular functions. *Current Opinion in Cell Biology*, *20*(1), 71–76. doi:10.1016/j.ceb.2007.11.010 PMID:18226514

Hamos, J. E., Oblas, B., Pulaski-Salo, D., Welch, W. J., Bole, D. G., & Drachman, D. A. (1991). Expression of heat shock proteins in Alzheimer's disease. *Neurology*, *41*(3), 345–350. doi:10.1212/WNL.41.3.345 PMID:2005999

Hanada, T., Noda, N. N., Satomi, Y., Ichimura, Y., Fujioka, Y., Takao, T., ... Ohsumi, Y. (2007). The Atg12-Atg5 conjugate has a novel E3-like activity for protein lipidation in autophagy. *The Journal of Biological Chemistry*, *282*(52), 37298–37302. doi:10.1074/jbc.C700195200 PMID:17986448

Hanada, T., Satomi, Y., Takao, T., & Ohsumi, Y. (2009). The amino-terminal region of Atg3 is essential for association with phosphatidylethanolamine in Atg8 lipidation. *FEBS Letters*, *583*(7), 1078–1083. doi:10.1016/j.febslet.2009.03.009 PMID:19285500

Hanahan, D., & Weinberg, R. A. (2011). Hallmarks of cancer: the next generation. *Cell, 144*(5), 646-674.

Han, D., Lerner, A. G., Vande Walle, L., Upton, J. P., Xu, W., Hagen, A., ... Papa, F. R. (2009). IRE1α Kinase Activation Modes Control Alternate Endoribonuclease Outputs to Determine Divergent Cell Fates. *Cell*, *138*(3), 562–575. doi:10.1016/j.cell.2009.07.017 PMID:19665977

Hanger, D. P., Anderton, B. H., & Noble, W. (2009). Tau phosphorylation: The therapeutic challenge for neurodegenerative disease. *Trends in Molecular Medicine*, *15*(3), 112–119. doi:10.1016/j.molmed.2009.01.003 PMID:19246243

Hanger, D. P., Hughes, K., Woodgett, J. R., Brion, J. P., & Anderton, B. H. (1992). Glycogen synthase kinase-3 induces Alzheimer's disease-like phosphorylation of tau: Generation of paired helical filament epitopes and neuronal localisation of the kinase. *Neuroscience Letters*, *147*(1), 58–62. doi:10.1016/0304-3940(92)90774-2 PMID:1336152

Hanger, D. P., Seereeram, A., & Noble, W. (2009). Mediators of tau phosphorylation in the pathogenesis of Alzheimer's disease. *Expert Review of Neurotherapeutics*, *9*(11), 1647–1666. doi:10.1586/ern.09.104 PMID:19903024

Han, J., Hou, W., Lu, C., Goldstein, L. A., Stolz, D. B., Watkins, S. C., & Rabinowich, H. (2013). Interaction between Her2 and Beclin-1 Proteins Underlies a New Mechanism of Reciprocal Regulation. *The Journal of Biological Chemistry*, *288*(28), 20315–20325. doi:10.1074/jbc.M113.461350 PMID:23703612

Hanschmann, E. M., Godoy, J. R., Berndt, C., Hudemann, C., & Lillig, C. H. (2013). Thioredoxins, glutaredoxins, and peroxiredoxins--molecular mechanisms and health significance: from cofactors to antioxidants to redox signaling. *Antioxidants & Redox Signaling*, *19*(13), 1539–1605. doi:10.1089/ars.2012.4599

Han, W., & Christen, P. (2004). cis-Effect of DnaJ on DnaK in ternary complexes with chimeric DnaK/DnaJ-binding peptides. *FEBS Letters*, *563*(1), 146–150. doi:10.1016/S0014-5793(04)00290-X PMID:15063739

Hanzen, S., Vielfort, K., Yang, J., Roger, F., Andersson, V., Zamarbide-Fores, S., ... Nyström, T. (2016). Lifespan Control by Redox-Dependent Recruitment of Chaperones to Misfolded Proteins. *Cell*, *166*(1), 140–151. doi:10.1016/j.cell.2016.05.006 PMID:27264606

Haque, M. E., Thomas, K. J., D'Souza, C., Callaghan, S., Kitada, T., Slack, R. S., ... Park, D. S. (2008). Cytoplasmic *PINK1* activity protects neurons from dopaminergic neurotoxin MPTP. *Proceedings of the National Academy of Sciences of the United States of America*, *105*(5), 1716–1721. doi:10.1073/pnas.0705363105 PMID:18218782

Hara, M. R., & Snyder, S. H. (2006). *Nitric oxide–GAPDH–Siah: a novel cell death cascade*. Academic Press.

Hara, T., Nakamura, K., Matsui, M., Yamamoto, A., Nakahara, Y., Suzuki-Migishima, R., . . . Okano, H. (2006). *Suppression of basal autophagy in neural cells causes neurodegenerative disease in mice*. Academic Press.

Hara, T., Nakamura, K., Matsui, M., Yamamoto, A., Nakahara, Y., Suzuki-Migishima, R., ... Mizushima, N. (2006). Suppression of basal autophagy in neural cells causes neurodegenerative disease in mice. *Nature*, *441*(7095), 885–889. doi:10.1038/nature04724 PMID:16625204

Harding, R. J., & Tong, Y. F. (2018). Proteostasis in Huntington's disease: Disease mechanisms and therapeutic opportunities. *Acta Pharmacologica Sinica*, *39*(5), 754–769. doi:10.1038/aps.2018.11 PMID:29620053

Hardy, J. A., & Higgins, G. A. (1992). *Alzheimer's disease: the amyloid cascade hypothesis*. Academic Press.

Hardy, J., & Allsop, D. (1991). *Amyloid deposition as the central event in the aetiology of Alzheimer's disease*. Academic Press.

Hardy, J., & Selkoe, D. J. (2002). *The amyloid hypothesis of Alzheimer's disease: progress and problems on the road to therapeutics*. Academic Press.

Hardy, J. A., & Higgins, G. A. (1992). Alzheimer's disease: The amyloid cascade hypothesis. *Science*, *256*(5054), 184–186. doi:10.1126cience.1566067 PMID:1566067

Harhaji-Trajkovic, L., Arsikin, K., Kravic-Stevovic, T., Petricevic, S., Tovilovic, G., Pantovic, A., ... Trajkovic, V. (2012). Chloroquine-mediated lysosomal dysfunction enhances the anticancer effect of nutrient deprivation. *Pharmaceutical Research*, *29*(8), 2249–2263. doi:10.100711095-012-0753-1 PMID:22538436

Harjes, P., & Wanker, E. E. (2003). *The hunt for huntingtin function: interaction partners tell many different stories*. Academic Press.

Harman, D. (1972). The biologic clock: The mitochondria? *Journal of the American Geriatrics Society*, *20*(4), 145–147. doi:10.1111/j.1532-5415.1972.tb00787.x PMID:5016631

Harrison, I. F., & Dexter, D. T. (2013). Epigenetic targeting of histone deacetylase: Therapeutic potential in Parkinson's disease? *Pharmacology & Therapeutics*, *140*(1), 34–52. doi:10.1016/j.pharmthera.2013.05.010 PMID:23711791

Hartl, F. U., Bracher, A., & Hayer-Hartl, M. (2011). Molecular chaperones in protein folding and proteostasis. *Nature, 475*(7356), 324–332. doi:10.1038/nature10317 PMID:21776078

Hartl, F. U., & Hayer-Hartl, M. (2002). Molecular chaperones in the cytosol: From nascent chain to folded protein. *Science, 295*(5561), 1852–1858. doi:10.1126cience.1068408 PMID:11884745

Hasegawa, T., Yoshida, S., Sugeno, N., Kobayashi, J., & Aoki, M. (2018). DnaJ/Hsp40 family and Parkinson's disease. *Frontiers in Neuroscience, 11*, 743. doi:10.3389/fnins.2017.00743 PMID:29367843

Hashiguchi, M., Sobue, K., & Paudel, H. K. (2000). 14-3-3ζ is an effector of tau protein phosphorylation. *The Journal of Biological Chemistry, 275*(33), 25247–25254. doi:10.1074/jbc.M003738200 PMID:10840038

Haslbeck, M., Franzmann, T., Weinfurtner, D., & Buchner, J. (2005). Some like it hot: The structure and function of small heat-shock proteins. *Nature Structural & Molecular Biology, 12*(10), 842–846. doi:10.1038/nsmb993 PMID:16205709

Hassing, L. B., Grant, M. D., Hofer, S. M., Pedersen, N. L., Nilsson, S. E., Berg, S., ... Johansson, B. (2004). Type 2 diabetes mellitus contributes to cognitive decline in old age: A longitudinal population-based study. *Journal of the International Neuropsychological Society, 10*(4), 599–607. doi:10.1017/S1355617704104165 PMID:15327738

Haughey, N. J., Bandaru, V. V. R., Bae, M., & Mattson, M. P. (2010). Roles for dysfunctional sphingolipid metabolism in Alzheimer's disease neuropathogenesis. *Biochimica et Biophysica Acta. Molecular and Cell Biology of Lipids, 1801*(8), 878–886. doi:10.1016/j.bbalip.2010.05.003 PMID:20452460

Hayley, S., Crocker, S. J., Smith, P. D., Shree, T., Jackson-Lewis, V., Przedborski, S., ... Park, D. S. (2004). Regulation of dopaminergic loss by Fas in a 1-methyl-4-phenyl-1, 2, 3, 6-tetrahydropyridine model of Parkinson's disease. *The Journal of Neuroscience: The Official Journal of the Society for Neuroscience, 24*(8), 2045–2053. doi:10.1523/JNEUROSCI.4564-03.2004 PMID:14985447

Hay, N. (2005). The Akt-mTOR tango and its relevance to cancer. *Cancer Cell, 8*(3), 179–183. doi:10.1016/j.ccr.2005.08.008 PMID:16169463

He, C., & Klionsky, D. J. (2009). *Regulation mechanisms and signaling pathways of autophagy*. Academic Press.

He, C., & Levine, B. (2010). The beclin 1 interactome. *Current Opinion in Cell Biology, 22*(2), 140–149. doi:10.1016/j.ceb.2010.01.001 PMID:20097051

He, C., Song, H., Yorimitsu, T., Monastyrska, I., Yen, W. L., Legakis, J. E., & Klionsky, D. J. (2006). Recruitment of Atg9 to the preautophagosomal structure by Atg11 is essential for selective autophagy in budding yeast. *The Journal of Cell Biology, 175*(6), 925–935. doi:10.1083/jcb.200606084 PMID:17178909

Heck, J. W., Cheung, S. K., & Hampton, R. Y. (2010). Cytoplasmic protein quality control degradation mediated by parallel actions of the E3 ubiquitin ligases Ubr1 and San1. *Proceedings of the National Academy of Sciences of the United States of America, 107*(3), 1106–1111. doi:10.1073/pnas.0910591107 PMID:20080635

Hegarty, S. V., Sullivan, A. M., & O'Keeffe, G. W. (2016). The Epigenome as a therapeutic target for Parkinson's disease. *Neural Regeneration Research, 11*(11), 1735–1738. doi:10.4103/1673-5374.194803 PMID:28123403

Heiseke, A., Aguib, Y., Riemer, C., Baier, M., & Schatzl, H. M. (2009). Lithium induces clearance of protease resistant prion protein in prion-infected cells by induction of autophagy. *Journal of Neurochemistry, 109*(1), 25–34. doi:10.1111/j.1471-4159.2009.05906.x PMID:19183256

Hempen, B., & Brion, J. P. (1996). Reduction of Acetylated α-Tubulin Immunoreactivity in Neurofibrillary Tangle-bearing Neurons in Alzheimer's Disease. *Journal of Neuropathology and Experimental Neurology, 55*(9), 964–972. doi:10.1097/00005072-199609000-00003 PMID:8800092

Heng, M. Y., Duong, D. K., Albin, R. L., Tallaksen-Greene, S. J., Hunter, J. M., Lesort, M. J., . . . Detloff, P. J. (2010). *Early autophagic response in a novel knock-in model of Huntington disease*. Academic Press.

Hernández, F., & Avila, J. (2007). Tauopathies. *Cellular and Molecular Life Sciences*, *64*(17), 2219–2233. doi:10.100700018-007-7220-x PMID:17604998

Heron, M. (2013). Death: Leading causes for 2010. *National Vital Statistics Reports*, *62*(6). PMID:24364902

Herrero, M. T., Estrada, C., Maatouk, L., & Vyas, S. (2015). Inflammation in Parkinson's disease: Role of glucocorticoids. *Frontiers in Neuroanatomy*, *9*, 32. doi:10.3389/fnana.2015.00032 PMID:25883554

Hershko, A. (2005). The ubiquitin system for protein degradation and some of its roles in the control of the cell-division cycle (Nobel lecture). *Angewandte Chemie International Edition*, *44*(37), 5932–5943. doi:10.1002/anie.200501724 PMID:16142823

Hetz, C., Chevet, E., & Harding, H. P. (2013). Targeting the unfolded protein response in disease. *Nature Reviews. Drug Discovery*, *12*(9), 703–719. doi:10.1038/nrd3976 PMID:23989796

Hetz, C., & Mollereau, B. (2014). Disturbance of endoplasmic reticulum proteostasis in neurodegenerative diseases. *Nature Reviews. Neuroscience*, *15*(4), 233–249. doi:10.1038/nrn3689 PMID:24619348

Hetz, C., & Papa, F. R. (2018). The Unfolded Protein Response and Cell Fate Control. *Molecular Cell*, *69*(2), 169–181. doi:10.1016/j.molcel.2017.06.017 PMID:29107536

He, W., Newman, J. C., Wang, M. Z., Ho, L., & Verdin, E. (2012). Mitochondrial sirtuins: Regulators of protein acylation and metabolism. *Trends in Endocrinology and Metabolism*, *23*(9), 467–476. doi:10.1016/j.tem.2012.07.004 PMID:22902903

Heydari, A. R., You, S., Takahashi, R., Gutsmann-Conrad, A., Sarge, K. D., & Richardson, A. (2000). Age-related alterations in the activation of heat shock transcription factor 1 in rat hepatocytes. *Experimental Cell Research*, *256*(1), 83–93. doi:10.1006/excr.2000.4808 PMID:10739655

Higashiyama, H., Hirose, F., Yamaguchi, M., Inoue, Y., Fujikake, N., & Matsukage, A. (2002). *Identification of ter94, Drosophila VCP, as a modulator of polyglutamine-induced neurodegeneration*. Academic Press.

Hippius, H., & Neundörfer, G. (2003). The discovery of Alzheimer's disease. *Dialogues in Clinical Neuroscience*, *5*(1), 101–108. PMID:22034141

Hipp, M. S., Park, S. H., & Hartl, F. U. (2014). Proteostasis impairment in protein-misfolding and -aggregation diseases. *Trends in Cell Biology*, *24*(9), 506–514. doi:10.1016/j.tcb.2014.05.003 PMID:24946960

Hipp, M. S., Patel, C. N., Bersuker, K., Riley, B. E., Kaiser, S. E., Shaler, T. A., ... Kopito, R. R. (2012). Indirect inhibition of 26S proteasome activity in a cellular model of Huntington's disease. *The Journal of Cell Biology*, *196*(5), 573–587. doi:10.1083/jcb.201110093 PMID:22371559

Hirabayashi, M., Inoue, K., Tanaka, K., Nakadate, K., Ohsawa, Y., Kamei, Y., . . . Kimura, Y. (2001). *VCP/p97 in abnormal protein aggregates, cytoplasmic vacuoles, and cell death, phenotypes relevant to neurodegeneration*. Academic Press.

Hirota, Y., Yamashita, S. I., Kurihara, Y., Jin, X., Aihara, M., Saigusa, T., ... Kanki, T. (2015). Mitophagy is primarily due to alternative autophagy and requires the MAPK1 and MAPK14 signaling pathways. *Autophagy*, *11*(2), 332–343. doi:10.1080/15548627.2015.1023047 PMID:25831013

Hobson, E. V., & McDermott, C. J. (2016). Supportive and symptomatic management of amyotrophic lateral sclerosis. *Nature Reviews. Neurology*, *12*(9), 526–538. doi:10.1038/nrneurol.2016.111 PMID:27514291

Hochfeld, W. E., Lee, S., & Rubinsztein, D. C. (2013). Therapeutic induction of autophagy to modulate neurodegenerative disease progression. *Acta Pharmacologica Sinica*, *34*(5), 600–604. doi:10.1038/aps.2012.189 PMID:23377551

Hodara, R., Norris, E. H., Giasson, B. I., Mishizen-Eberz, A. J., Lynch, D. R., Lee, V. M.-Y., & Ischiropoulos, H. (2004). Functional consequences of α-synuclein tyrosine nitration: Diminished binding to lipid vesicles and increased fibril formation. *The Journal of Biological Chemistry*, *279*(46), 47746–47753. doi:10.1074/jbc.M408906200 PMID:15364911

Höhfeld, J., Cyr, D. M., & Patterson, C. (2001). From the cradle to the grave: Molecular chaperones that may choose between folding and degradation. *EMBO Reports*, *2*(10), 885–890. doi:10.1093/embo-reports/kve206 PMID:11600451

Hollien, J., Lin, J. H., Li, H., Stevens, N., Walter, P., & Weissman, J. S. (2009). Regulated Ire1-dependent decay of messenger RNAs in mammalian cells. *The Journal of Cell Biology*, *186*(3), 323–331. doi:10.1083/jcb.200903014 PMID:19651891

Holmay, M. J., Terpstra, M., Coles, L. D., Mishra, U., Ahlskog, M., Gulin, O., & … . (2013). N-Acetyl cysteine Boosts Brain and Blood Glutathione in Gaucher and Parkinsons Disease. *Clinical Neuropharmacology*, *36*(4), 103–106. doi:10.1097/WNF.0b013e31829ae713 PMID:23860343

Holmgren, A. (1989). Thioredoxin and glutaredoxin systems. *The Journal of Biological Chemistry*, *264*, 13963–13966. PMID:2668278

Hong, X. P., Peng, C. X., Wei, W., Tian, Q., Liu, Y. H., Yao, X. Q., … Wang, J. Z. (2010). Essential role of tau phosphorylation in adult hippocampal neurogenesis. *Hippocampus*, *20*(12), 1339–1349. doi:10.1002/hipo.20712 PMID:19816983

Hong, Y., Wang, X., Sun, S., Xue, G., Li, J., & Hou, Y. (2016). Progesterone exerts neuroprotective effects against Abeta-induced neuroinflammation by attenuating ER stress in astrocytes. *International Immunopharmacology*, *33*, 83–89. doi:10.1016/j.intimp.2016.02.002 PMID:26878478

Hoozemans, J. J. M., Stieler, J., van Haastert, E. S., Veerhuis, R., Rozemuller, A. J. M., Baas, F., … Scheper, W. (2006). The unfolded protein response affects neuronal cell cycle protein expression: Implications for Alzheimer's disease pathogenesis. *Experimental Gerontology*, *41*(4), 380–386. doi:10.1016/j.exger.2006.01.013 PMID:16564150

Hoozemans, J. J. M., Van Haastert, E. S., Nijholt, D. A. T., Rozemuller, A. J. M., Eikelenboom, P., & Scheper, W. (2009). The unfolded protein response is activated in pretangle neurons in alzheimer's disease hippocampus. *American Journal of Pathology*, *174*(4), 1241–1251. doi:10.2353/ajpath.2009.080814 PMID:19264902

Hoozemans, J. J., Veerhuis, R., Van Haastert, E. S., Rozemuller, J. M., Baas, F., Eikelenboom, P., & Scheper, W. (2005). The unfolded protein response is activated in Alzheimer's disease. *Acta Neuropathologica*, *110*(2), 165–172. doi:10.100700401-005-1038-0 PMID:15973543

Ho, P. W., Leung, C. T., Liu, H., Pang, S. Y., Lam, C. S., Xian, J., & … . (2019). Age-dependent accumulation of oligomeric SNCA/alpha-synuclein from impaired degradation in mutant LRRK2 knockin mouse model of Parkinson disease: Role for therapeutic activation of chaperone-mediated autophagy (CMA). *Autophagy*, 1–24. PMID:30983487

Hope, A. D., Myhre, R., Kachergus, J., Lincoln, S., Bisceglio, G., Hulihan, M., & Farrer, M. J. (2004). α-Synuclein missense and multiplication mutations in autosomal dominant Parkinson's disease. *Neuroscience Letters*, *367*(1), 97–100. doi:10.1016/j.neulet.2004.05.100 PMID:15308306

Horvath, I., Iashchishyn, I. A., Moskalenko, R. A., Wang, C., Wärmländer, S. K., Wallin, C., & Morozova-Roche, L. A. (2018). Co-aggregation of pro-inflammatory S100A9 with α-synuclein in Parkinson's disease: Ex vivo and in vitro studies. *Journal of Neuroinflammation*, *15*(1), 172. doi:10.118612974-018-1210-9 PMID:29866153

Hotamisligil, G. S. (2010). Endoplasmic Reticulum Stress and the Inflammatory Basis of Metabolic Disease. *Cell*, *140*(6), 900–917. doi:10.1016/j.cell.2010.02.034 PMID:20303879

Hoter, A., El-Sabban, M. E., & Naim, H. Y. (2018). The HSP90 Family: Structure, Regulation, Function, and Implications in Health and Disease. *International Journal of Molecular Sciences, 19*(9), 2560. doi:10.3390/ijms19092560 PMID:30158430

Houlden, H., Laura, M., Wavrant-De Vrieze, F., Blake, J., Wood, N., & Reilly, M. M. (2008). Mutations in the HSP27 (HSPB1) gene cause dominant, recessive, and sporadic distal HMN/CMT type 2. *Neurology, 71*(21), 1660–1668. doi:10.1212/01.wnl.0000319696.14225.67 PMID:18832141

Ho, Y. S., Yang, X., Lau, J. C. F., Hung, C. H. L., Wuwongse, S., Zhang, Q., ... Chang, R. C. C. (2012). Endoplasmic reticulum stress induces tau pathology and forms a vicious cycle: Implication in Alzheimer's disease pathogenesis. *Journal of Alzheimer's Disease, 28*(4), 839–854. doi:10.3233/JAD-2011-111037 PMID:22101233

Høyer-Hansen, M., & Jäättelä, M. (2007). Connecting endoplasmic reticulum stress to autophagy by unfolded protein response and calcium. *Cell Death and Differentiation, 14*(9), 1576–1582. doi:10.1038j.cdd.4402200 PMID:17612585

Hristova, V. A., Beasley, S. A., Rylett, R. J., & Shaw, G. S. (2009). Identification of a novel Zn2+-binding domain in the autosomal recessive juvenile parkinson-related E3 ligase parkin. *The Journal of Biological Chemistry, 284*(22), 14978–14986. doi:10.1074/jbc.M808700200 PMID:19339245

Hromadkova, L., & Ovsepian, S. V. (2019). Tau-Reactive Endogenous Antibodies: Origin, Functionality, and Implications for the Pathophysiology of Alzheimer's Disease. *Journal of Immunology Research, 2019*, 7406810. doi:10.1155/2019/7406810 PMID:31687413

Hsieh, C. H., Shaltouki, A., Gonzalez, A. E., da Cruz, A. B., Burbulla, L. F., Lawrence, E. S., & Wang, X. (2016). Functional impairment in miro degradation and mitophagy is a shared feature in familial and sporadic Parkinson's disease. *Cell Stem Cell, 19*(6), 709–724. doi:10.1016/j.stem.2016.08.002 PMID:27618216

Huang, C., Cheng, H., Hao, S., Zhou, H., Zhang, X., Gao, J., ... Wang, C. C. (2006). Heat shock protein 70 inhibits alpha-synuclein fibril formation via interactions with diverse intermediates. *Journal of Molecular Biology, 364*(3), 323–336. doi:10.1016/j.jmb.2006.08.062 PMID:17010992

Huang, J., Yang, J., Shen, Y., Jiang, H., Han, C., Zhang, G., ... Wang, T. (2017). HMGB1 Mediates Autophagy Dysfunction via Perturbing Beclin1-VPS34 Complex in Dopaminergic Cell Model. *Frontiers in Molecular Neuroscience, 10.* doi:10.3389/fnmol.2017.00013 PMID:28197072

Huang, W. J., Zhang, X., & Chen, W. W. (2016). Role of oxidative stress in Alzheimer's disease. *Biomedical Reports, 4*(5), 519–522. doi:10.3892/br.2016.630 PMID:27123241

Huang, X., Moir, R. D., Tanzi, R. E., Bush, A. I., & Rogers, J. T. (2004). Redox-active metals, oxidative stress, and Alzheimer's disease pathology. *Annals of the New York Academy of Sciences, 1012*(1), 153–163. doi:10.1196/annals.1306.012 PMID:15105262

Hubbard, V. M., Valdor, R., Macian, F., & Cuervo, A. M. (2012). Selective autophagy in the maintenance of cellular homeostasis in aging organisms. *Biogerontology, 13*(1), 21–35. doi:10.100710522-011-9331-x PMID:21461872

Hubbert, C., Guardiola, A., Shao, R., Kawaguchi, Y., Ito, A., Nixon, A., ... Yao, T.-P. (2002). HDAC6 is a microtubule-associated deacetylase. *Nature, 417*(6887), 455–458. doi:10.1038/417455a PMID:12024216

Huntington's Disease Foundation. (2019). *Life with Huntington's disease.*

Huotari, J., & Helenius, A. (2011). Endosome maturation. *The EMBO Journal, 30*(17), 3481–3500. doi:10.1038/emboj.2011.286 PMID:21878991

Huzil, J. T., Pannu, R., Ptak, C., Garen, G., & Ellison, M. J. (2007). Direct catalysis of lysine 48-linked polyubiquitin chains by the ubiquitin-activating enzyme. *The Journal of Biological Chemistry*, *282*(52), 37454–37460. doi:10.1074/jbc.M705242200 PMID:17951259

Hwang, S., Disatnik, M. H., & Mochly-Rosen, D. (2015). Impaired GAPDH-induced mitophagy contributes to the pathology of Huntington's disease. *EMBO Molecular Medicine*, *7*(10), 1307–1326. doi:10.15252/emmm.201505256 PMID:26268247

Ignowski, E., Winter, A. N., Dual, N., Fleming, H., Wallace, T., Manning, E., & … . (2018). The cysteine-rich whey protein supplement, Immunocal®, preserves brain glutathione and improves cognitive, motor, and histopathological indices of traumatic brain injury in a mouse model of controlled cortical impact. *Free Radical Biology & Medicine*, *124*, 328–341. doi:10.1016/j.freeradbiomed.2018.06.026 PMID:29940352

Iijima, M., Tabira, T., Poorkaj, P., Schellenberg, G. D., Trojanowski, J. Q., Lee, V. M. Y., ... Yamashita, Y. (1999). A distinct familial presenile dementia with a novel missense mutation in the tau gene. *Neuroreport*, *10*(3), 497–501. doi:10.1097/00001756-199902250-00010 PMID:10208578

Ikeda, Y., Abe, A., Ishida, C., Takahashi, K., Hayasaka, K., & Yamada, M. (2009). A clinical phenotype of distal hereditary motor neuronopathy type II with a novel HSPB1 mutation. *Journal of the Neurological Sciences*, *277*(1-2), 9–12. doi:10.1016/j.jns.2008.09.031 PMID:18952241

Imarisio, S., Carmichael, J., Korolchuk, V., Chen, C. W., Saiki, S., Rose, C., ... Rubinsztein, D. C. (2008). Huntington's disease: From pathology and genetics to potential therapies. *The Biochemical Journal*, *412*(2), 191–209. doi:10.1042/BJ20071619 PMID:18466116

Imlay, J. A., Chin, S. M., & Linn, S. (1988). Toxic DNA damage by hydrogen peroxide through the Fenton reaction in vivo and in vitro. *Sci*, *240*(4852), 640–642. doi:10.1126cience.2834821 PMID:2834821

Imran, M., & Mahmood, S. (2011). An overview of animal prion diseases. *Virology Journal*, *8*(1), 493. doi:10.1186/1743-422X-8-493 PMID:22044871

Íñigo-Marco, I., Valencia, M., Larrea, L., Bugallo, R., Martínez-Goikoetxea, M., Zuriguel, I., & Arrasate, M. (2017). E46K α-synuclein pathological mutation causes cell-autonomous toxicity without altering protein turnover or aggregation. *Proceedings of the National Academy of Sciences of the United States of America*, *114*(39), E8274–E8283. doi:10.1073/pnas.1703420114 PMID:28900007

Inoki, K., Li, Y., Zhu, T., Wu, J., & Guan, K. L. (2002). TSC2 is phosphorylated and inhibited by Akt and suppresses mTOR signalling. *Nature Cell Biology*, *4*(9), 648–657. doi:10.1038/ncb839 PMID:12172553

Intihar, T. A., Martinez, E. A., & Gomez-Pastor, R. (2019). Mitochondrial Dysfunction in Huntington's Disease; Interplay Between HSF1, p53 and PGC-1alpha Transcription Factors. *Frontiers in Cellular Neuroscience*, *13*, 103. doi:10.3389/fncel.2019.00103 PMID:30941017

Iqbal, K., Liu, F., & Gong, C.-X. (2016). *Tau and neurodegenerative disease: the story so far.* Academic Press.

Iqbal, K., Liu, F., Gong, C. X., & Grundke-Iqbal, I. (2010). Tau in Alzheimer disease and related tauopathies. *Current Alzheimer Research*, *7*(8), 656–664. doi:10.2174/156720510793611592 PMID:20678074

Irvine, G. B., El-Agnaf, O. M., Shankar, G. M., & Walsh, D. M. (2008). Protein aggregation in the brain: The molecular basis for Alzheimer's and Parkinson's diseases. *Molecular Medicine (Cambridge, Mass.)*, *14*(7-8), 451–464. doi:10.2119/2007-00100.Irvine PMID:18368143

Irwin, D. J., Lee, V. M. Y., & Trojanowski, J. Q. (2013). Parkinson's disease dementia: Convergence of α-synuclein, tau and amyloid-β pathologies. *Nature Reviews. Neuroscience*, *14*(9), 626–636. doi:10.1038/nrn3549 PMID:23900411

Isaev, N. K., Ailkina, S., Golyshev, S. A., Genrikh, E. E., Alexandrova, O. P., Kapkaeva, M. R., & … . (2017). N-acetyl-L-cysteine and Mn2+ attenuate Cd2+-induced disturbance of the intracellular free calcium homeostasis in cultured cerebellar granule neurons. *Toxicol*, *393*, 1–8. doi:10.1016/j.tox.2017.10.017 PMID:29100878

Ishizawa, T., Mattila, P., Davies, P., Wang, D., & Dickson, D. W. (2003). Colocalization of tau and alpha-synuclein epitopes in Lewy bodies. *Journal of Neuropathology and Experimental Neurology*, *62*(4), 389–397. doi:10.1093/jnen/62.4.389 PMID:12722831

Itakura, E., Kishi, C., Inoue, K., & Mizushima, N. (2008). Beclin 1 Forms Two Distinct Phosphatidylinositol 3-Kinase Complexes with Mammalian Atg14 and UVRAG. *Molecular Biology of the Cell*, *19*(12), 5360–5372. doi:10.1091/mbc.e08-01-0080 PMID:18843052

Itakura, E., Kishi-Itakura, C., Koyama-Honda, I., & Mizushima, N. (2012). Structures containing Atg9A and the ULK1 complex independently target depolarized mitochondria at initial stages of Parkin-mediated mitophagy. *Journal of Cell Science*, *125*(6), 1488–1499. doi:10.1242/jcs.094110 PMID:22275429

Itakura, E., Kishi-Itakura, C., & Mizushima, N. (2012). The hairpin-type tail-anchored SNARE syntaxin 17 targets to autophagosomes for fusion with endosomes/lysosomes. *Cell*, *151*(6), 1256–1269. doi:10.1016/j.cell.2012.11.001 PMID:23217709

Itoh, H., Komatsuda, A., Ohtani, H., Wakui, H., Imai, H., Sawada, K., … Hamada, F. (2002). Mammalian HSP60 is quickly sorted into the mitochondria under conditions of dehydration. *European Journal of Biochemistry*, *269*(23), 5931–5938. doi:10.1046/j.1432-1033.2002.03317.x PMID:12444982

Ittner, L. M., Fath, T., Ke, Y. D., Bi, M., van Eersel, J., Li, K. M., … Götz, J. (2008). Parkinsonism and impaired axonal transport in a mouse model of frontotemporal dementia. *Proceedings of the National Academy of Sciences of the United States of America*, *105*(41), 15997–16002. doi:10.1073/pnas.0808084105 PMID:18832465

Iurlaro, R., & Munoz-Pinedo, C. (2016). Cell death induced by endoplasmic reticulum stress. *The FEBS Journal*, *283*(14), 2640–2652. doi:10.1111/febs.13598 PMID:26587781

Iwawaki, T., Akai, R., Yamanaka, S., & Kohno, K. (2009). Function of IRE1 alpha in the placenta is essential for placental development and embryonic viability. *Proceedings of the National Academy of Sciences of the United States of America*, *106*(39), 16657–16662. doi:10.1073/pnas.0903775106 PMID:19805353

Jack, C. R., Petersen, R. C., Xu, Y. C., Waring, S. C., O'Brien, P. C., Tangalos, E. G., … Kokmen, E. (1997). Medial temporal atrophy on MRI in normal aging and very mild Alzheimer's disease. *Neurology*, *49*(3), 786–794. doi:10.1212/WNL.49.3.786 PMID:9305341

Jackson, K., Barisone, G. A., Diaz, E., Jin, L. W., DeCarli, C., & Despa, F. (2013). Amylin deposition in the brain: A second amyloid in Alzheimer disease? *Annals of Neurology*, *74*(4), 517–526. doi:10.1002/ana.23956 PMID:23794448

Jackson, M. P., & Hewitt, E. W. (2016). Cellular proteostasis: Degradation of misfolded proteins by lysosomes. *Proteostasis*, *60*, 173–180. PMID:27744333

Jacob, C., Knight, I., & Winyard, P. G. (2006). Aspects of the biological redox chemistry of cysteine: From simple redox responses to sophisticated signalling pathways. *Biological Chemistry*, *387*(10/11), 1385–1397. doi:10.1515/BC.2006.174 PMID:17081111

Jadah, N. J., Eva, A., & Janice, E. A. (2010). CSPa: The neuroprotective J protein. *Biochemistry and Cell Biology*, *88*, 157–165. PMID:20453918

Jaeger, P. A., Pickford, F., Sun, C.-H., Lucin, K. M., Masliah, E., & Wyss-Coray, T. (2010). Regulation of Amyloid Precursor Protein Processing by the Beclin 1 Complex. *PLoS One*, *5*(6), e11102. doi:10.1371/journal.pone.0011102 PMID:20559548

Jaffer, F., Murphy, S. M., Scoto, M., Healy, E., Rossor, A. M., Brandner, S., ... Reilly, M. M. (2012). BAG3 mutations: Another cause of giant axonal neuropathy. *Journal of the Peripheral Nervous System*, *17*(2), 210–216. doi:10.1111/j.1529-8027.2012.00409.x PMID:22734908

Jana, N. R., Tanaka, M., Wang, G., & Nukina, N. (2000). Polyglutamine length-dependent interaction of Hsp40 and Hsp70 family chaperones with truncated N-terminal huntingtin: Their role in suppression of aggregation and cellular toxicity. *Human Molecular Genetics*, *9*(13), 2009–2018. doi:10.1093/hmg/9.13.2009 PMID:10942430

Jankovic, J. (2008). Parkinson's disease: Clinical features and diagnosis. *Journal of Neurology, Neurosurgery, and Psychiatry*, *79*(4), 368–376. doi:10.1136/jnnp.2007.131045 PMID:18344392

Jauhiainen, A., Thomsen, C., Strombom, L., Grundevik, P., Andersson, C., Danielsson, A., ... Åman, P. (2012). Distinct cytoplasmic and nuclear functions of the stress induced protein DDIT3/CHOP/GADD153. *PLoS One*, *7*(4), e33208. doi:10.1371/journal.pone.0033208 PMID:22496745

Jayanthy, G., Roshana Devi, V., Ilango, K., & Subramanian, S. P. (2017). Rosmarinic acid mediates mitochondrial biogenesis in insulin resistant skeletal muscle through activation of AMPK. *Journal of Cellular Biochemistry*, *118*(7), 1839–1848. doi:10.1002/jcb.25869 PMID:28059465

Jeganathan, S., Hascher, A., Chinnathambi, S., Biernat, J., Mandelkow, E.-M., & Mandelkow, E. (2008). Proline-directed pseudo-phosphorylation at AT8 and PHF1 epitopes induces a compaction of the paperclip folding of tau and generates a pathological (MC-1) conformation. *The Journal of Biological Chemistry*, *283*(46), 32066–32076. doi:10.1074/jbc.M805300200 PMID:18725412

Jeitner, T. M., Matson, W. R., Folk, J. E., Blass, J. P., & Cooper, A. J. (2008). *Increased levels of γ-glutamylamines in Huntington disease CSF*. Academic Press.

Jennewein, L., Ronellenfitsch, M. W., Antonietti, P., Ilina, E. I., Jung, J., Stadel, D., ... Mittelbronn, M. (2016). Diagnostic and clinical relevance of the autophago-lysosomal network in human gliomas. *Oncotarget*, *7*(15), 20016–20032. doi:10.18632/oncotarget.7910 PMID:26956048

Jensen, O. N. (2004). Modification-specific proteomics: Characterization of post-translational modifications by mass spectrometry. *Current Opinion in Chemical Biology*, *8*(1), 33–41. doi:10.1016/j.cbpa.2003.12.009 PMID:15036154

Jensen, P. H., Hager, H., Nielsen, M. S., Højrup, P., Gliemann, J., & Jakes, R. (1999). α-Synuclein binds to tau and stimulates the protein kinase A-catalyzed tau phosphorylation of serine residues 262 and 356. *The Journal of Biological Chemistry*, *274*(36), 25481–25489. doi:10.1074/jbc.274.36.25481 PMID:10464279

Jensen, T. S., Baron, R., Haanpää, M., Kalso, E., Loeser, J. D., Rice, A. S., & Treede, R. D. (2011). A new definition of neuropathic pain. *Pain*, *152*(10), 2204–2205. doi:10.1016/j.pain.2011.06.017 PMID:21764514

Jeon, G. S., Nakamura, T., Lee, J. S., Choi, W. J., Ahn, S. W., Lee, K. W., ... Lipton, S. A. (2013). Potential effect of S-nitrosylated protein disulfide isomerase on mutant SOD1 aggregation and neuronal cell death in amyotrophic lateral sclerosis. *Molecular Neurobiology*, *49*(2), 796–807. doi:10.100712035-013-8562-z PMID:24091828

Jeong, H., Then, F., Melia, T. J. Jr, Mazzulli, J. R., Cui, L., Savas, J. N., ... Krainc, D. (2009). Acetylation targets mutant huntingtin to autophagosomes for degradation. *Cell, 137*(1), 60–72. doi:10.1016/j.cell.2009.03.018 PMID:19345187

Jeong, S. M., Xiao, C., Finley, L. W., Lahusen, T., Souza, A. L., Pierce, K., ... Xu, X. (2013). SIRT4 has tumor-suppressive activity and regulates the cellular metabolic response to DNA damage by inhibiting mitochondrial glutamine metabolism. *Cancer Cell, 23*(4), 450–463. doi:10.1016/j.ccr.2013.02.024 PMID:23562301

Jia, H., Halilou, A. I., Hu, L., Cai, W., Liu, J., & Huang, B. (2011). Heat shock protein 10 (Hsp10) in immune-related diseases: One coin, two sides. *International Journal of Biochemistry and Molecular Biology, 2*(1), 47. PMID:21969171

Jia, H., Kast, R. J., Steffan, J. S., & Thomas, E. A. (2012). Selective histone deacetylase (HDAC) inhibition imparts beneficial effects in Huntington's disease mice: Implications for the ubiquitin-proteasomal and autophagy systems. *Human Molecular Genetics, 21*(24), 5280–5293. doi:10.1093/hmg/dds379 PMID:22965876

Jiang, T., Yu, J. T., Zhu, X. C., Tan, M. S., Wang, H. F., Cao, L., ... Tan, L. (2014). Temsirolimus promotes autophagic clearance of amyloid-β and provides protective effects in cellular and animal models of Alzheimer's disease. *Pharmacological Research, 81*, 54–63. doi:10.1016/j.phrs.2014.02.008 PMID:24602800

Jiang, Y., Chadwick, S. R., & Lajoie, P. (2016). Endoplasmic reticulum stress: The cause and solution to Huntington's disease? *Brain Research, 1648*, 650–657. doi:10.1016/j.brainres.2016.03.034 PMID:27040914

Jimenez-Sanchez, M., Licitra, F., Underwood, B. R., & Rubinsztein, D. C. (2017). Huntington's disease: Mechanisms of pathogenesis and therapeutic strategies. *Cold Spring Harbor Perspectives in Medicine, 7*(7), a024240. doi:10.1101/cshperspect.a024240 PMID:27940602

Jin, S. M., Lazarou, M., Wang, C., Kane, L. A., Narendra, D. P., & Youle, R. J. (2010). Mitochondrial membrane potential regulates *PINK1* import and proteolytic destabilization by PARL. *The Journal of Cell Biology, 191*(5), 933–942. doi:10.1083/jcb.201008084 PMID:21115803

Jin, S. M., & Youle, R. J. (2012). *PINK1*-and Parkin-mediated mitophagy at a glance. *Journal of Cell Science, 125*(4), 795–799. doi:10.1242/jcs.093849 PMID:22448035

Jin, S. M., & Youle, R. J. (2013). The accumulation of misfolded proteins in the mitochondrial matrix is sensed by *PINK1* to induce PARK2/Parkin-mediated mitophagy of polarized mitochondria. *Autophagy, 9*(11), 1750–1757. doi:10.4161/auto.26122 PMID:24149988

Jin, Y., Ahn, S., & Kim, S. (2015). BAG3 affects the nucleocytoplasmic shuttling of HSF1 upon heat stress. *Biochemical and Biophysical Research Communications, 464*(2), 561–567. doi:10.1016/j.bbrc.2015.07.006 PMID:26159920

Joachim, C. L., Morris, J. H., Kosik, K. S., & Selkoe, D. J. (1987). Tau antisera recognize neurofibrillary tangles in a range of neurodegenerative disorders. *Annals of Neurology, 22*(4), 514–520. doi:10.1002/ana.410220411 PMID:2963585

Johansen, T., & Lamark, T. (2011). Selective autophagy mediated by autophagic adapter proteins. *Autophagy, 7*(3), 279–296. doi:10.4161/auto.7.3.14487 PMID:21189453

Johnson, J. O., Mandrioli, J., Benatar, M., Abramzon, Y., Van Deerlin, V. M., Trojanowski, J. Q., . . . Wuu, J. (2010). *Exome sequencing reveals VCP mutations as a cause of familial ALS.* Academic Press.

Johnson, E. C. B., & Kang, J. (2016). A small molecule targeting protein translation does not rescue spatial learning and memory deficits in the hAPP-J20 mouse model of Alzheimer's disease. *PeerJ, 4*, e2565. doi:10.7717/peerj.2565 PMID:27781164

Johnson, R., Noble, W., Tartaglia, G. G., & Buckley, N. J. (2012). Neurodegeneration as an RNA disorder. *Progress in Neurobiology, 99*(3), 293–315. doi:10.1016/j.pneurobio.2012.09.006 PMID:23063563

Jones, D. P. (2006). Redefining oxidative stress. *Antioxidants & Redox Signalling*, *8*(9-10), 1865–1879. doi:10.1089/ars.2006.8.1865 PMID:16987039

Jones, D. P. (2008). Radical-free biology of oxidative stress. *American Journal of Physiology. Cell Physiology*, *295*(4), 849–868. doi:10.1152/ajpcell.00283.2008 PMID:18684987

Ju, J.-S., Fuentealba, R. A., Miller, S. E., Jackson, E., Piwnica-Worms, D., Baloh, R. H., Weihl, C. C. (2009). *Valosin-containing protein (VCP) is required for autophagy and is disrupted in VCP disease.* Academic Press.

Juenemann, K., Jansen, A. H. P., van Riel, L., Merkx, R., Mulder, M. P. C., An, H., ... Reits, E. A. (2018). Dynamic recruitment of ubiquitin to mutant huntingtin inclusion bodies. *Scientific Reports*, *8*(1), 1405. doi:10.103841598-018-19538-0 PMID:29362455

Julien, C., Tremblay, C., Emond, V., Lebbadi, M., Salem, N. Jr, Bennett, D. A., & Calon, F. (2009). Sirtuin 1 reduction parallels the accumulation of tau in Alzheimer disease. *Journal of Neuropathology and Experimental Neurology*, *68*(1), 48–58. doi:10.1097/NEN.0b013e3181922348 PMID:19104446

Jung, A. E., Fitzsimons, H. L., Bland, R. J., During, M. J., & Young, D. (2008). HSP70 and constitutively active HSF1 mediate protection against CDCrel-1-mediated toxicity. *Molecular Therapy*, *16*(6), 1048–1055. doi:10.1038/mt.2008.68 PMID:18398426

Jung, C. H., Jun, C. B., Ro, S. H., Kim, Y. M., Otto, N. M., Cao, J., ... Kim, D. H. (2009). ULK-Atg13-FIP200 complexes mediate mTOR signaling to the autophagy machinery. *Molecular Biology of the Cell*, *20*(7), 1992–2003. doi:10.1091/mbc.e08-12-1249 PMID:19225151

Kabeya, Y., Mizushima, N., Ueno, T., Yamamoto, A., Kirisako, T., Noda, T., ... Yoshimori, T. (2000). LC3, a mammalian homologue of yeast Apg8p, is localized in autophagosome membranes after processing. *The EMBO Journal*, *19*(21), 5720–5728. doi:10.1093/emboj/19.21.5720 PMID:11060023

Kabuta, T., Furuta, A., Aoki, S., Furuta, K., & Wada, K. (2008). Aberrant interaction between Parkinson disease-associated mutant UCH-L1 and the lysosomal receptor for chaperone-mediated autophagy. *The Journal of Biological Chemistry*, *283*(35), 23731–23738. doi:10.1074/jbc.M801918200 PMID:18550537

Kaeberlein, M. (2010). Resveratrol and rapamycin: Are they anti-aging drugs? *BioEssays*, *32*(2), 96–99. doi:10.1002/bies.200900171 PMID:20091754

Kagan, V. E., Jiang, J., Huang, Z., Tyurina, Y. Y., Desbourdes, C., Cottet-Rousselle, C., ... Cheikhi, A. (2016). NDPK-D (NM23-H4)-mediated externalization of cardiolipin enables elimination of depolarized mitochondria by mitophagy. *Cell Death and Differentiation*, *23*(7), 1140. doi:10.1038/cdd.2015.160 PMID:26742431

Kahle, P. J. (2008). α-synucleinopathy models and human neuropathology: Similarities and differences. *Acta Neuropathologica*, *115*(1), 87–95. doi:10.100700401-007-0302-x PMID:17932682

Kakizawa, S. (2013). *Nitric oxide-induced calcium release: activation of type 1 ryanodine receptor, a calcium release channel, through non-enzymatic post-translational modification by nitric oxide.* doi:10.3389/fendo.2013.00142

Kakizuka, A. (2008). *Roles of VCP in human neurodegenerative disorders.* Portland Press Limited. doi:10.1042/BST0360105

Kalchman, M. A., Graham, R. K., Xia, G., Koide, H. B., Hodgson, J. G., Graham, K. C., ... Hayden, M. R. (1996). Huntingtin is ubiquitinated and interacts with a specific ubiquitin- conjugating enzyme. *The Journal of Biological Chemistry*, *271*(32), 19385–19394. doi:10.1074/jbc.271.32.19385 PMID:8702625

Kalia, L. V., Kalia, S. K., McLean, P. J., Lozano, A. M., & Lang, A. E. (2013). α-Synuclein oligomers and clinical implications for Parkinson disease. *Annals of Neurology*, *73*(2), 155–169. doi:10.1002/ana.23746 PMID:23225525

Kalia, S. K., Lee, S., Smith, P. D., Liu, L., Crocker, S. J., Thorarinsdottir, T. E., ... Lozano, A. M. (2004). BAG5 Inhibits Parkin and Enhances Dopaminergic Neuron Degeneration. *Neuron, 44*(6), 931–945. doi:10.1016/j.neuron.2004.11.026 PMID:15603737

Kaliszewski, M., Kennedy, A. K., Blaes, S. L., Shaffer, R. S., Knott, A. B., Song, W., ... Bossy-Wetzel, E. (2016). SOD1 Lysine 123 Acetylation in the Adult Central Nervous System. *Frontiers in Cellular Neuroscience, 10*, 287. doi:10.3389/fncel.2016.00287 PMID:28066183

Kalmar, B., & Greensmith, L. (2017). Cellular Chaperones As Therapeutic Targets in ALS to Restore Protein Homeostasis and Improve Cellular Function. *Frontiers in Molecular Neuroscience, 10*, 251. doi:10.3389/fnmol.2017.00251 PMID:28943839

Kametani, F., & Hasegawa, M. (2018). *Reconsideration of amyloid hypothesis and tau hypothesis in Alzheimer's disease*. Academic Press.

Kamp, F., Exner, N., Lutz, A. K., Wender, N., Hegermann, J., Brunner, B., ... Eimer, S. (2010). Inhibition of mitochondrial fusion by α-synuclein is rescued by PINK1, Parkin and DJ-1. *The EMBO Journal, 29*(20), 3571–3589. doi:10.1038/emboj.2010.223 PMID:20842103

Kampinga, H. H., & Craig, E. A. (2010). The HSP70 chaperone machinery: J proteins as drivers of functional specificity. *Nature Reviews. Molecular Cell Biology, 11*(8), 579–592. doi:10.1038/nrm2941 PMID:20651708

Kanagaraj, N., Beiping, H., Dheen, S. T., & Tay, S. S. W. (2014). Downregulation of miR-124 in MPTP-treated mouse model of Parkinson's disease and MPP iodide-treated MN9D cells modulates the expression of the calpain/cdk5 pathway proteins. *Neuroscience, 272*, 167–179. doi:10.1016/j.neuroscience.2014.04.039 PMID:24792712

Kanekura, K., Suzuki, H., Aiso, S., & Matsuoka, M. (2009). ER stress and unfolded protein response in amyotrophic lateral sclerosis. *Molecular Neurobiology, 39*(2), 81–89. doi:10.100712035-009-8054-3 PMID:19184563

Kane, L. A., Lazarou, M., Fogel, A. I., Li, Y., Yamano, K., Sarraf, S. A., ... Youle, R. J. (2014). PINK1 phosphorylates ubiquitin to activate Parkin E3 ubiquitin ligase activity. *The Journal of Cell Biology, 205*(2), 143–153. doi:10.1083/jcb.201402104 PMID:24751536

Kang, L., Wu, K. P., Vendruscolo, M., & Baum, J. (2011). The A53T mutation is key in defining the differences in the aggregation kinetics of human and mouse α-synuclein. *Journal of the American Chemical Society, 133*(34), 13465–13470. doi:10.1021/ja203979j PMID:21721555

Kang, S. P., & Saif, M. W. (2008). Pharmacogenomics and pancreatic cancer treatment. Optimizing current therapy and individualizing future therapy. *JOP, 9*, 251–266. PMID:18469437

Kania, E., Roest, G., Vervliet, T., Parys, J. B., & Bultynck, G. (2017). IP3 Receptor-Mediated Calcium Signaling and Its Role in Autophagy in Cancer. *Frontiers in Oncology, 7*, 140. doi:10.3389/fonc.2017.00140 PMID:28725634

Kanki, T., Kurihara, Y., Jin, X., Goda, T., Ono, Y., Aihara, M., ... Kang, D. (2013). Casein kinase 2 is essential for mitophagy. *EMBO Reports, 14*(9), 788–794. doi:10.1038/embor.2013.114 PMID:23897086

Kanki, T., Wang, K., Cao, Y., Baba, M., & Klionsky, D. J. (2009). Atg32 is a mitochondrial protein that confers selectivity during mitophagy. *Developmental Cell, 17*(1), 98–109. doi:10.1016/j.devcel.2009.06.014 PMID:19619495

Kapahi, P., Chen, D., Rogers, A. N., Katewa, S. D., Li, P. W. L., Thomas, E. L., & Kockel, L. (2010). With TOR, less is more: A key role for the conserved nutrient-sensing TOR pathway in aging. *Cell Metabolism, 11*(6), 453–465. doi:10.1016/j.cmet.2010.05.001 PMID:20519118

Kapogiannis, D., & Mattson, M. P. (2011). *Disrupted energy metabolism and neuronal circuit dysfunction in cognitive impairment and Alzheimer's disease*. Academic Press.

Karantza-Wadsworth, V., Patel, S., Kravchuk, O., Chen, G., Mathew, R., Jin, S., & White, E. (2007). Autophagy mitigates metabolic stress and genome damage in mammary tumorigenesis. *Genes & Development, 21*(13), 1621–1635. doi:10.1101/gad.1565707 PMID:17606641

Karpuj, M. V., Becher, M. W., Springer, J. E., Chabas, D., Youssef, S., Pedotti, R., . . . Steinman, L. (2002). *Prolonged survival and decreased abnormal movements in transgenic model of Huntington disease, with administration of the transglutaminase inhibitor cystamine*. Academic Press.

Karpuj, M. V., Garren, H., Slunt, H., Price, D. L., Gusella, J., Becher, M. W., & Steinman, L. (1999). *Transglutaminase aggregates huntingtin into nonamyloidogenic polymers, and its enzymatic activity increases in Huntington's disease brain nuclei*. Academic Press.

Katayama, T., Imaizumi, K., Manabe, T., Hitomi, J., Kudo, T., & Tohyama, M. (2004). Induction of neuronal death by ER stress in Alzheimer's disease. *Journal of Chemical Neuroanatomy, 28*(1–2), 67–78. doi:10.1016/j.jchemneu.2003.12.004 PMID:15363492

Katayama, T., Imaizumi, K., Sato, N., Miyoshi, K., Kudo, T., Hitomi, J., ... Tohyama, M. (1999). Presenilin-1 mutations downregulate the signalling pathway of the unfolded-protein response. *Nature Cell Biology, 1*(8), 479–485. doi:10.1038/70265 PMID:10587643

Kaur, S. J., McKeown, S. R., & Rashid, S. (2016). Mutant SOD1 mediated pathogenesis of amyotrophic lateral sclerosis. *Gene, 577*(2), 109–118. doi:10.1016/j.gene.2015.11.049 PMID:26657039

Kaushik, S., & Cuervo, A. M. (2018). The coming of age of chaperone-mediated autophagy. *Nature Reviews. Molecular Cell Biology, 19*(6), 1. doi:10.103841580-018-0001-6 PMID:29626215

Kawajiri, S., Saiki, S., Sato, S., Sato, F., Hatano, T., Eguchi, H., & Hattori, N. (2010). PINK1 is recruited to mitochondria with parkin and associates with LC3 in mitophagy. *FEBS Letters, 584*(6), 1073–1079. doi:10.1016/j.febslet.2010.02.016 PMID:20153330

Kazlauskaite, A., Kondapalli, C., Gourlay, R., Campbell, D. G., Ritorto, M. S., Hofmann, K., ... Muqit, M. M. (2014). Parkin is activated by *PINK1*-dependent phosphorylation of ubiquitin at Ser65. *The Biochemical Journal, 460*(1), 127–141. doi:10.1042/BJ20140334 PMID:24660806

Kazlauskaite, A., Martinez-Torres, R. J., Wilkie, S., Kumar, A., Peltier, J., Gonzalez, A., ... Muqit, M. M. K. (2015). Binding to serine 65-phosphorylated ubiquitin primes Parkin for optimal PINK1-dependent phosphorylation and activation. *EMBO Reports, 16*(8), 939–954. doi:10.15252/embr.201540352 PMID:26116755

Kazlauskaite, A., & Muqit, M. M. (2015). PINK1 and Parkin–mitochondrial interplay between phosphorylation and ubiquitylation in Parkinson's disease. *The FEBS Journal, 282*(2), 215–223. doi:10.1111/febs.13127 PMID:25345844

Kerr, J. S., Adriaanse, B. A., Greig, N. H., Mattson, M. P., Cader, M. Z., Bohr, V. A., & Fang, E. F. (2017). *Mitophagy and Alzheimer's disease: cellular and molecular mechanisms*. Academic Press.

Kerr, J. S., Adriaanse, B. A., Greig, N. H., Mattson, M. P., Cader, M. Z., Bohr, V. A., & Fang, E. F. (2017). Mitophagy and Alzheimer's disease: Cellular and molecular mechanisms. *Trends in Neurosciences, 40*(3), 151–166. doi:10.1016/j.tins.2017.01.002 PMID:28190529

Khalaf, O., Fauvet, B., Oueslati, A., Dikiy, I., Mahul-Mellier, A. L., Ruggeri, F. S., ... Eliezer, D. (2014). The H50Q mutation enhances α-synuclein aggregation, secretion, and toxicity. *The Journal of Biological Chemistry*, *289*(32), 21856–21876. doi:10.1074/jbc.M114.553297 PMID:24936070

Khalil, B., El Fissi, N., Aouane, A., Cabirol-Pol, M. J., Rival, T., & Liévens, J. C. (2015). PINK1-induced mitophagy promotes neuroprotection in Huntington's disease. *Cell Death & Disease*, *6*(1), e1617. doi:10.1038/cddis.2014.581 PMID:25611391

Khaminets, A., Behl, C., & Dikic, I. (2016). Ubiquitin-dependent and independent signals in selective autophagy. *Trends in Cell Biology*, *26*(1), 6–16. doi:10.1016/j.tcb.2015.08.010 PMID:26437584

Khan, S. S., LaCroix, M., Boyle, G., Sherman, M. A., Brown, J. L., Amar, F., ... Lesne, S. E. (2018). Bidirectional modulation of Alzheimer phenotype by alpha-synuclein in mice and primary neurons. *Acta Neuropathologica*, *136*(4), 589–605. doi:10.100700401-018-1886-z PMID:29995210

Khoury, G. A., Baliban, R. C., & Floudas, C. A. (2011). Proteome-wide post-translational modification statistics: Frequency analysis and curation of the swiss-prot database. *Scientific Reports*, *1*(1), 90. doi:10.1038rep00090 PMID:22034591

Kiely, A. P., Ling, H., Asi, Y. T., Kara, E., Proukakis, C., Schapira, A. H., ... Holton, J. L. (2015). Distinct clinical and neuropathological features of G51D SNCA mutation cases compared with SNCA duplication and H50Q mutation. *Molecular Neurodegeneration*, *10*(1), 41. doi:10.118613024-015-0038-3 PMID:26306801

Kieran, D., Kalmar, B., Dick, J. R., Riddoch-Contreras, J., Burnstock, G., & Greensmith, L. (2004). Treatment with arimoclomol, a coinducer of heat shock proteins, delays disease progression in ALS mice. *Nature Medicine*, *10*(4), 402–405. doi:10.1038/nm1021 PMID:15034571

Kihara, A., Kabeya, Y., Ohsumi, Y., & Yoshimori, T. (2001). Beclin-phosphatidylinositol 3-kinase complex functions at the trans -Golgi network. *EMBO Reports*, *2*(4), 330–335. doi:10.1093/embo-reports/kve061 PMID:11306555

Kijima, K., Numakura, C., Goto, T., Takahashi, T., Otagiri, T., Umetsu, K., & Hayasaka, K. (2005). Small heat shock protein 27 mutation in a Japanese patient with distal hereditary motor neuropathy. *Journal of Human Genetics*, *50*(9), 473–476. doi:10.100710038-005-0280-6 PMID:16155736

Kim, S.-Y., Marekov, L., Bubber, P., Browne, S. E., Stavrovskaya, I., Lee, J., . . . Gibson, G. E. (2005). *Mitochondrial aconitase is a transglutaminase 2 substrate: transglutamination is a probable mechanism contributing to high-molecular-weight aggregates of aconitase and loss of aconitase activity in Huntington disease brain*. Academic Press.

Kim, D. S., Li, B., Rhew, K. Y., Oh, H. W., Lim, H. D., Lee, W., ... Kim, H. R. (2012). The regulatory mechanism of 4-phenylbutyric acid against ER stress-induced autophagy in human gingival fibroblasts. *Archives of Pharmacal Research*, *35*(7), 1269–1278. doi:10.100712272-012-0718-2 PMID:22864750

Kim, E., Napierala, M., & Dent, S. Y. R. (2011). Hyperexpansion of GAA repeats affects post-initiation steps of FXN transcription in Friedreich's ataxia. *Nucleic Acids Research*, *39*(19), 8366–8377. doi:10.1093/nar/gkr542 PMID:21745819

Kim, H. Y., Kim, Y., Yun, H. H., Im, C., Ko, J., & Lee, J. (2016). ERK-mediated phosphorylation of BIS regulates nuclear translocation of HSF1 under oxidative stress. *Experimental & Molecular Medicine*, *48*(9), e260. doi:10.1038/emm.2016.84 PMID:27659916

Kim, J. Y., & Yenari, M. (2017). Heat shock proteins and the stress response in primer on cerebrovascular diseases. In *Primer on cerebrovascular diseases* (pp. 273–275). Cambridge, UK: Elsevier. doi:10.1016/B978-0-12-803058-5.00056-4

Kim, J., Inoue, K., Ishii, J., Vanti, W. B., Voronov, S. V., Murchison, E., ... Abeliovich, A. (2007). AmicroRNA feedback circuit in midbrain dopamine neurons. *Science*, *317*(5842), 1220–1224. doi:10.1126cience.1140481 PMID:17761882

Kim, J., Kim, Y. C., Fang, C., Russell, R. C., Kim, J. H., Fan, W., ... Guan, K.-L. (2013). Differential Regulation of Distinct VPS34 Complexes by AMPK in Nutrient Stress and Autophagy. *Cell, 152*(1–2), 290–303. doi:10.1016/j.cell.2012.12.016 PMID:23332761

Kim, J., Kundu, M., Viollet, B., & Guan, K. L. (2011). AMPK and mTOR regulate autophagy through direct phosphorylation of Ulk1. *Nature Cell Biology, 13*(2), 132–141. doi:10.1038/ncb2152 PMID:21258367

Kim, M., Hersh, L. B., Leissring, M. A., Ingelsson, M., Matsui, T., Farris, W., ... Tanzi, R. E. (2007). Decreased catalytic activity of the insulin-degrading enzyme in chromosome 10-linked Alzheimer disease families. *The Journal of Biological Chemistry, 282*(11), 7825–7832. doi:10.1074/jbc.M609168200 PMID:17244626

Kim, N. C., Tresse, E., Kolaitis, R. M., Molliex, A., Thomas, R. E., Alami, N. H., ... Winborn, B. J. (2013). VCP is essential for mitochondrial quality control by PINK1/Parkin and this function is impaired by VCP mutations. *Neuron, 78*(1), 65–80. doi:10.1016/j.neuron.2013.02.029 PMID:23498974

Kim, W., Lee, Y., McKenna, N. D., Yi, M., Simunovic, F., Wang, Y., ... Sonntag, K. C. (2014). miR-126 contributes to Parkinson's disease by dysregulating the insulin-like growth factor/phosphoinositide 3-kinase signaling. *Neurobiology of Aging, 35*(7), 1712–1721. doi:10.1016/j.neurobiolaging.2014.01.021 PMID:24559646

Kim, Y., Park, J., Kim, S., Song, S., Kwon, S. K., Lee, S. H., ... Chung, J. (2008). PINK1 controls mitochondrial localization of Parkin through direct phosphorylation. *Biochemical and Biophysical Research Communications, 377*(3), 975–980. doi:10.1016/j.bbrc.2008.10.104 PMID:18957282

Kins, S., Crameri, A., Evans, D. R. H., Hemmings, B. A., Nitsch, R. M., & Götz, J. (2001). Reduced Protein Phosphatase 2A Activity Induces Hyperphosphorylation and Altered Compartmentalization of Tau in Transgenic Mice. *The Journal of Biological Chemistry, 276*(41), 38193–38200. PMID:11473109

Kirbach, B. B., & Golenhofen, N. (2011). Differential expression and induction of small heat shock proteins in rat brain and cultured hippocampal neurons. *Journal of Neuroscience Research, 89*(2), 162–175. doi:10.1002/jnr.22536 PMID:21162124

Kishimoto, Y., & Kirino, Y. (2013). Presenilin 2 mutation accelerates the onset of impairment in trace eyeblink conditioning in a mouse model of Alzheimer's disease overexpressing human mutant amyloid precursor protein. *Neuroscience Letters, 538*, 15–19. doi:10.1016/j.neulet.2013.01.025 PMID:23370287

Kitada, T., Asakawa, S., Hattori, N., Matsumine, H., Yamamura, Y., Minoshima, S., ... and Shimizu, N. (1998). Mutations in the parkin gene cause autosomal recessive juvenile parkinsonism. *Nature, 392*(6676), 605.

Kitada, T., Asakawa, S., Hattori, N., Matsumine, H., Yamamura, Y., Minoshima, S., ... Shimizu, N. (1998). Mutations in the parkin gene cause autosomal recessive juvenile parkinsonism. *Nature, 392*(6676), 605–608. doi:10.1038/33416 PMID:9560156

Kitada, T., Tong, Y., Gautier, C. A., & Shen, J. (2009). Absence of nigral degeneration in aged parkin/DJ-1/PINK1 triple knockout mice. *Journal of Neurochemistry, 111*(3), 696–702. doi:10.1111/j.1471-4159.2009.06350.x PMID:19694908

Klein, C. J., & Benarroch, E. E. (2014). Epigenetic regulation: Basic concepts and relevance to neurologic disease. *Neurology, 82*(20), 1833–1840. doi:10.1212/WNL.0000000000000440 PMID:24759839

Klein, C., & Westenberger, A. (2012). Genetics of Parkinson's Disease. *Cold Spring Harbor Perspectives in Medicine, 2*(1), a008888. doi:10.1101/cshperspect.a008888 PMID:22315721

Kleizen, B., & Braakman, I. (2004). Protein folding and quality control in the endoplasmic reticulum. *Current Opinion in Cell Biology, 16*(4), 343–349. doi:10.1016/j.ceb.2004.06.012 PMID:15261665

Klemann, C. J. H. M., Visser, J. E., Van den Bosch, L., Martens, G. J. M., & Poelmans, G. (2018). Integrated molecular landscape of amyotrophic lateral sclerosis provides insights into disease etiology. *Brain Pathology (Zurich, Switzerland), 28*(2), 203–211. doi:10.1111/bpa.12485 PMID:28035716

Klimek, C., Kathage, B., Wordehoff, J., & Hohfeld, J. (2017). BAG3-mediated proteostasis at a glance. *Journal of Cell Science, 130*(17), 2781–2788. doi:10.1242/jcs.203679 PMID:28808089

Klionsky, D. J., Abdalla, F. C., Abeliovich, H., Abraham, R. T., Acevedo-Arozena, A., Adeli, K., ... Zuckerbraun, B. (2012). Guidelines for the use and interpretation of assays for monitoring autophagy. *Autophagy, 8*(8), 445–544. doi:10.4161/auto.19496 PMID:22966490

Klionsky, D. J., & Emr, S. D. (2000). Autophagy as a regulated pathway of cellular degradation. *Science, 290*(5497), 1717–1721. doi:10.1126cience.290.5497.1717 PMID:11099404

Klucken, J., Shin, Y., Masliah, E., Hyman, B. T., & McLean, P. J. (2004). Hsp70 reduces α-synuclein aggregation and toxicity. *The Journal of Biological Chemistry, 279*(24), 25497–25502. doi:10.1074/jbc.M400255200 PMID:15044495

Komatsu, M., Waguri, S., Chiba, T., Murata, S., Iwata, J.-i., Tanida, I., . . . Kominami, E. (2006). *Loss of autophagy in the central nervous system causes neurodegeneration in mice*. Academic Press.

Komatsu, M., Ueno, T., Waguri, S., Uchiyama, Y., Kominami, E., & Tanaka, K. (2007). Constitutive autophagy: Vital role in clearance of unfavorable proteins in neurons. *Cell Death and Differentiation, 14*(5), 887–894. doi:10.1038j.cdd.4402120 PMID:17332773

Komatsu, M., Waguri, S., Chiba, T., Murata, S., Iwata, J. I., Tanida, I., ... Tanaka, K. (2006). Loss of autophagy in the central nervous system causes neurodegeneration in mice. *Nature, 441*(7095), 880–884. doi:10.1038/nature04723 PMID:16625205

Komatsu, M., Waguri, S., Ueno, T., Iwata, J., Murata, S., Tanida, I., ... Kominami, E. (2005). Impairment of starvation-induced and constitutive autophagy in Atg7-deficient mice. *The Journal of Cell Biology, 169*(3), 425–434. doi:10.1083/jcb.200412022 PMID:15866887

Kondapalli, C., Kazlauskaite, A., Zhang, N., Woodroof, H. I., Campbell, D. G., Gourlay, R., ... Knebel, A. (2012). *PINK1* is activated by mitochondrial membrane potential depolarization and stimulates Parkin E3 ligase activity by phosphorylating Serine 65. *Open Biology, 2*(5), 120080. doi:10.1098/rsob.120080 PMID:22724072

Kontopoulos, E., Parvin, J. D., & Feany, M. B. (2006). α-synuclein acts in the nucleus to inhibit histone acetylation and promote neurotoxicity. *Human Molecular Genetics, 15*(20), 3012–3023. doi:10.1093/hmg/ddl243 PMID:16959795

Kopke, E., Tung, Y. C., Shaikh, S., Alonso, A. D. C., Iqbal, K., & Grundke-Iqbal, I. (1993). Microtubule-associated protein tau. Abnormal phosphorylation of a non-paired helical filament pool in Alzheimer disease. *The Journal of Biological Chemistry, 268*(32), 24374–24384. PMID:8226987

Korecka, J. A., Talbot, R., Osborn, T. M., de Leeuw, S. M., Levy, S. A., Ferrari, E. J., ... Isacson, O. (2019). Neurite Collapse and Altered ER Ca2+ Control in Human Parkinson Disease Patient iPSC-Derived Neurons with LRRK2 G2019S Mutation. *Stem Cell Reports, 12*(1), 29–41. doi:10.1016/j.stemcr.2018.11.021 PMID:30595548

Korolchuk, V. I., Menzies, F. M., & Rubinsztein, D. C. (2010). Mechanisms of cross-talk between the ubiquitin-proteasome and autophagy-lysosome systems. *FEBS Letters, 584*(7), 1393–1398. doi:10.1016/j.febslet.2009.12.047 PMID:20040365

Korolev, I. O. (2014). Alzheimer 's Disease : A Clinical and Basic Science Review. *Medical Student Research Journal, 4*, 24–33.

Kosik, K. S., Joachim, C. L., & Selkoe, D. J. (1986). Microtubule-associated protein tau (tau) is a major antigenic component of paired helical filaments in Alzheimer disease. *Proceedings of the National Academy of Sciences of the United States of America*, *83*(11), 4044–4048. doi:10.1073/pnas.83.11.4044 PMID:2424016

Koyano, F., Okatsu, K., Kosako, H., Tamura, Y., Go, E., Kimura, M., ... Endo, T. (2014). Ubiquitin is phosphorylated by *PINK1* to activate parkin. *Nature*, *510*(7503), 162–166. doi:10.1038/nature13392 PMID:24784582

Kozlov, G., Denisov, A. Y., Girard, M., Dicaire, M. J., Hamlin, J., McPherson, P. S., ... Gehring, K. (2011). Structural basis of defects in the sacsin HEPN domain responsible for autosomal recessive spastic ataxia of Charlevoix-Saguenay (ARSACS). *The Journal of Biological Chemistry*, *286*(23), 20407–20412. doi:10.1074/jbc.M111.232884 PMID:21507954

Kragh, C. L., Lund, L. B., Febbraro, F., Hansen, H. D., Gai, W.-P., El-Agnaf, O., ... Jensen, P. H. (2009). α-synuclein aggregation and Ser-129 phosphorylation-dependent cell death in oligodendroglial cells. *The Journal of Biological Chemistry*, *284*(15), 10211–10222. doi:10.1074/jbc.M809671200 PMID:19203998

Kragten, E., Lalande, I., Zimmermann, K., Roggo, S., Schindler, P., Muller, D., ... Fürst, P. (1998). Glyceraldehyde-3-phosphate dehydrogenase, the putative target of the antiapoptotic compounds CGP 3466 and R-(-)-deprenyl. *The Journal of Biological Chemistry*, *273*(10), 5821–5828. doi:10.1074/jbc.273.10.5821 PMID:9488718

Kraskiewicz, H., & Fitzgerald, U. (2012). InterfERing with endoplasmic reticulum stress. *Trends in Pharmacological Sciences*, *33*(2), 53–63. doi:10.1016/j.tips.2011.10.002 PMID:22112465

Kruger, U., Wang, Y., Kumar, S., & Mandelkow, E. M. (2012). Autophagic degradation of tau in primary neurons and its enhancement by trehalose. *Neurobiology of Aging*, *33*(10), 2291–2305. doi:10.1016/j.neurobiolaging.2011.11.009 PMID:22169203

Krumova, P., Meulmeester, E., Garrido, M., Tirard, M., Hsiao, H. H., Bossis, G., ... Weishaupt, J. H. (2011). Sumoylation inhibits alpha-synuclein aggregation and toxicity. *The Journal of Cell Biology*, *194*(1), 49–60. doi:10.1083/jcb.201010117 PMID:21746851

Kruszelnicka, O. (2014). Nitric oxide vs insulin secretion, action and clearance. *Diabetologia*, *57*(1), 257–258. doi:10.100700125-013-3082-y PMID:24135953

Ksiezak-Reding, H., Pyo, H. K., Feinstein, B., & Pasinetti, G. M. (2003). Akt/PKB kinase phosphorylates separately Thr212 and Ser214 of tau protein in vitro. *Biochimica et Biophysica Acta (BBA)-. Molecular Basis of Disease*, *1639*(3), 159–168. doi:10.1016/j.bbadis.2003.09.001 PMID:14636947

Kudo, T., Kanemoto, S., Hara, H., Morimoto, N., Morihara, T., Kimura, R., ... Takeda, M. (2008). A molecular chaperone inducer protects neurons from ER stress. *Cell Death and Differentiation*, *15*(2), 364–375. doi:10.1038j.cdd.4402276 PMID:18049481

Kuhn, D. M., Arthur, R. E. Jr, Thomas, D. M., & Elferink, L. A. (1999). Tyrosine hydroxylase is inactivated by catechol-quinones and converted to a redox-cycling quinoprotein: Possible relevance to Parkinson's disease. *Journal of Neurochemistry*, *73*(3), 1309–1317. doi:10.1046/j.1471-4159.1999.0731309.x PMID:10461926

Kuhn, M. L., Zemaitaitis, B., Hu, L. I., Sahu, A., Sorensen, D., Minasov, G., ... Wolfe, A. J. (2014). Structural, kinetic and proteomic characterization of acetyl phosphate-dependent bacterial protein acetylation. *PLoS One*, *9*(4), e94816. doi:10.1371/journal.pone.0094816 PMID:24756028

Kulczycki, J, Bertrand, E, Lojkowska, W, Dowjat, W, & Wisniewski, T, and L.-Z. M. (2001). Familial Alzheimer's disease connected with mutation in presenilin gene. *Neurologia i Neurochirurgia Polska*, *35*(2), 213–224. PMID:11599221

Kuma, A., Hatano, M., Matsui, M., Yamamoto, A., Nakaya, H., Yoshimori, T., ... Mizushima, N. (2004). The role of autophagy during the early neonatal starvation period. *Nature*, *432*(7020), 1032–1036. doi:10.1038/nature03029 PMID:15525940

Kuma, A., Mizushima, N., Ishihara, N., & Ohsumi, Y. (2002). Formation of the~ 350-kDa Apg12-Apg5· Apg16 multimeric complex, mediated by Apg16 oligomerization, is essential for autophagy in yeast. *The Journal of Biological Chemistry*, *277*(21), 18619–18625. doi:10.1074/jbc.M111889200 PMID:11897782

Kumar, A., Singh, A., & Ekavali. (2015). A review on Alzheimer's disease pathophysiology and its management: An update. *Pharmacological Reports*, *67*(2), 195–203. doi:10.1016/j.pharep.2014.09.004 PMID:25712639

Kumar, A., Verma, K., & Sinha, A. (2011). Designing of anti-cancer drug targeted to Bcl-2 associated athanogene (BAG1) protein. In M. A. Mahdavi (Ed.), *Bioinformatics - Trends and Methodologies* (pp. 707–722). London, UK: IntechOpen Limited. doi:10.5772/18409

Kumari, D., Biacsi, R. E., & Usdin, K. (2011). Repeat expansion affects both transcription initiation and elongation in Friedreich ataxia cells. *The Journal of Biological Chemistry*, *286*(6), 4209–4215. doi:10.1074/jbc.M110.194035 PMID:21127046

Kumsta, C., & Jakob, U. (2009). Redox-regulated chaperones. *Biochemistry*, *48*(22), 4666–4676. doi:10.1021/bi9003556 PMID:19368357

Kuroda, Y., Mitsui, T., Kunishige, M., Shono, M., Akaike, M., Azuma, H., & Matsumoto, T. (2006). Parkin enhances mitochondrial biogenesis in proliferating cells. *Human Molecular Genetics*, *15*(6), 883–895. doi:10.1093/hmg/ddl006 PMID:16449237

Kurtishi, A., Rosen, B., Patil, K. S., Alves, G. W., & Moller, S. G. (2019). Cellular Proteostasis in Neurodegeneration. *Molecular Neurobiology*, *56*(5), 3676–3689. doi:10.100712035-018-1334-z PMID:30182337

Kurutas, E. B. (2016). The importance of antioxidants which play the role in cellular response against oxidative/nitrosative stress: Current state. *Nutrition Journal*, *15*(1), 71. doi:10.118612937-016-0186-5 PMID:27456681

L'episcopo, F., Drouin-Ouellet, J., Tirolo, C., Pulvirenti, A., Giugno, R., Testa, N., ... Cicchetti, F. (2016). GSK-3β-induced Tau pathology drives hippocampal neuronal cell death in Huntington's disease: Involvement of astrocyte–neuron interactions. *Cell Death & Disease*, *7*(4), e2206. doi:10.1038/cddis.2016.104 PMID:27124580

Labbadia, J., & Morimoto, R. I. (2013). Huntington's disease: Underlying molecular mechanisms and emerging concepts. *Trends in Biochemical Sciences*, *38*(8), 378–385. doi:10.1016/j.tibs.2013.05.003 PMID:23768628

Labbadia, J., & Morimoto, R. I. (2015). The Biology of Proteostasis in Aging and Disease. *Annual Review of Biochemistry*, *84*(1), 435–464. doi:10.1146/annurev-biochem-060614-033955 PMID:25784053

Lackie, R. E., Maciejewski, A., Ostapchenko, V. G., Marques-Lopes, J., Choy, W. Y., Duennwald, M. L., ... Prado, M. A. (2017). The Hsp70/Hsp90 chaperone machinery in neurodegenerative diseases. *Frontiers in Neuroscience*, *11*, 254. doi:10.3389/fnins.2017.00254 PMID:28559789

Lahiri, V., Hawkins, W. D., & Klionsky, D. J. (2019). Watch what you (self-) eat: Autophagic mechanisms that modulate metabolism. *Cell Metabolism*, *29*(4), 803–826. doi:10.1016/j.cmet.2019.03.003 PMID:30943392

Lai, C. S. W., Preisler, J., Baum, L., Lee, D. H. S., Ng, H. K., Hugon, J., ... Chang, R. C. C. (2009). Low molecular weight Aβ induces collapse of endoplasmic reticulum. *Molecular and Cellular Neurosciences*, *41*(1), 31–43. doi:10.1016/j.mcn.2009.01.006 PMID:19386229

Lambeth, J. D., & Neish, A. S. (2014). Nox enzymes and new thinking on reactive oxygen: A doubleedged sword revisited. *Annual Review of Pathology*, *9*(1), 119–145. doi:10.1146/annurev-pathol-012513-104651 PMID:24050626

Lam, Y. A., Pickart, C. M., Alban, A., Landon, M., Jamieson, C., Ramage, R., ... Layfield, R. (2000). Inhibition of the ubiquitin-proteasome system in Alzheimer's disease. *Proceedings of the National Academy of Sciences of the United States of America*, *97*(18), 9902–9906. doi:10.1073/pnas.170173897 PMID:10944193

Langklotz, S., Baumann, U., & Narberhaus, F. (2012). Structure and function of the bacterial AAA protease FtsH. *Biochimica et Biophysica Acta (BBA) -. Molecular Cell Research*, *1823*(1), 40–48. doi:10.1016/j.bbamcr.2011.08.015 PMID:21925212

Larson, M. E., Sherman, M. A., Greimel, S., Kuskowski, M., Schneider, J. A., Bennett, D. A., & Lesné, S. E. (2012). Soluble α-synuclein is a novel modulator of Alzheimer's disease pathophysiology. *The Journal of Neuroscience: The Official Journal of the Society for Neuroscience*, *32*(30), 10253–10266. doi:10.1523/JNEUROSCI.0581-12.2012 PMID:22836259

Lastres-Becker, I., Ulusoy, A., Innamorato, N. G., Sahin, G., Rabano, A., ... Kirik, D. (2012). Alpha-synuclein expression and Nrf2 deficiency cooperate to aggravate protein aggregation, neuronal death and inflammation in early-stage Parkinson's disease. *Human Molecular Genetics*, *21*(14), 3173–3192. doi:10.1093/hmg/dds143 PMID:22513881

Lau, D. H., Hogseth, M., Phillips, E. C., O'Neill, M. J., Pooler, A. M., Noble, W., & Hanger, D. P. (2016). Critical residues involved in tau binding to fyn: Implications for tau phosphorylation in Alzheimer's disease. *Acta Neuropathologica Communications*, *4*(1), 49. doi:10.118640478-016-0317-4 PMID:27193083

Lawson, V. A., Collins, S. J., Masters, C. L., & Hill, A. F. (2005). Prion protein glycosylation. *Journal of Neurochemistry*, *93*(4), 793–801. doi:10.1111/j.1471-4159.2005.03104.x PMID:15857383

Lawson, V. A., Priola, S. A., Meade-White, K., Lawson, M. K., & Chesebro, B. (2004). Flexible N-terminal Region of Prion Protein Influences Conformation of Protease-resistant Prion Protein Isoforms Associated with Cross-species Scrapie Infection in Vivo and in Vitro. *The Journal of Biological Chemistry*, *279*(14), 13689–13695. doi:10.1074/jbc.M303697200 PMID:14736880

Lazarou, M., Narendra, D. P., Jin, S. M., Tekle, E., Banerjee, S., & Youle, R. J. (2013). PINK1 drives Parkin self-association and HECT-like E3 activity upstream of mitochondrial binding. *The Journal of Cell Biology*, *200*(2), 163–172. doi:10.1083/jcb.201210111 PMID:23319602

Lazarou, M., Sliter, D. A., Kane, L. A., Sarraf, S. A., Wang, C., Burman, J. L., & Youle, R. J. (2015). The ubiquitin kinase PINK1 recruits autophagy receptors to induce mitophagy. *Nature*, *524*(7565), 309–314. doi:10.1038/nature14893 PMID:26266977

Leak, R. K. (2014). Heat shock proteins in neurodegenerative disorders and aging. *Journal of Cell Communication and Signaling*, *8*(4), 293–310. doi:10.100712079-014-0243-9 PMID:25208934

Lee, A. H., Iwakoshi, N. N., & Glimcher, L. H. (2003). XBP-1 regulates a subset of endoplasmic reticulum resident chaperone genes in the unfolded protein response. *Molecular and Cellular Biology*, *23*(21), 7448–7459. doi:10.1128/MCB.23.21.7448-7459.2003 PMID:14559994

Lee, A. S. (2001). The glucose-regulated proteins: Stress induction and clinical applications. *Trends in Biochemical Sciences*, *26*(8), 504–510. doi:10.1016/S0968-0004(01)01908-9 PMID:11504627

Lee, C. Y., Su, C. H., Tsai, M. L., Ho, Y. C., Lee, S. S., Chen, C. H., & ... . (2018). Cadmium nitrate-induced neuronal apoptosis is protected by N-acetyl-Lcysteine via reducing reactive oxygen species generation and mitochondria dysfunction. *Biomedicine and Pharmacotherapy*, *108*, 448–456. doi:10.1016/j.biopha.2018.09.054 PMID:30241048

Lee, G., Newman, S. T., Gard, D. L., Band, H., & Panchamoorthy, G. (1998). Tau interacts with src-family non-receptor tyrosine kinases. *Journal of Cell Science*, *111*(21), 3167–3177. PMID:9763511

Lee, H. J., Khoshaghideh, F., Patel, S., & Lee, S. J. (2004). Clearance of α-synuclein oligomeric intermediates via the lysosomal degradation pathway. *The Journal of Neuroscience: The Official Journal of the Society for Neuroscience*, *24*(8), 1888–1896. doi:10.1523/JNEUROSCI.3809-03.2004 PMID:14985429

Lee, H. K., & Marzella, L. (1994). Regulation of intracellular protein degradation with special reference to lysosomes: Role in cell physiology and pathology. *International Review of Experimental Pathology*, *35*, 39–147. PMID:7860191

Lee, H., Noh, J. Y., Oh, Y., Kim, Y., Chang, J. W., Chung, C. W., ... Jung, Y.-K. (2012). IRE1 plays an essential role in ER stress-mediated aggregation of mutant huntingtin via the inhibition of autophagy flux. *Human Molecular Genetics*, *21*(1), 101–114. doi:10.1093/hmg/ddr445 PMID:21954231

Lee, I. H., Cao, L., Mostoslavsky, R., Lombard, D. B., Liu, J., Bruns, N. E., ... Finkel, T. (2008). A role for the NAD-dependent deacetylase Sirt1 in the regulation of autophagy. *Proceedings of the National Academy of Sciences of the United States of America*, *105*(9), 3374–3379. doi:10.1073/pnas.0712145105 PMID:18296641

Lee, I., Hüttemann, M., Kruger, A., Bollig-Fischer, A., & Malek, M. H. (2015). (–)–Epicatechin combined with 8 weeks of treadmill exercise is associated with increased angiogenic and mitochondrial signaling in mice. *Frontiers in Pharmacology*, *6*, 43. doi:10.3389/fphar.2015.00043 PMID:25821434

Lee, J. H., Won, S. M., Suh, J., Son, S. J., Moon, G. J., Park, U. J., & Gwag, B. J. (2010). Induction of the unfolded protein response and cell death pathway in alzheimer's disease, but not in aged Tg2576 mice. *Experimental & Molecular Medicine*, *42*(5), 386–394. doi:10.3858/emm.2010.42.5.040 PMID:20368688

Lee, J. H., Yu, W. H., Kumar, A., Lee, S., Mohan, P. S., Peterhoff, C. M., ... Uchiyama, Y. (2010). Lysosomal proteolysis and autophagy require presenilin 1 and are disrupted by Alzheimer-related PS1 mutations. *Cell*, *141*(7), 1146–1158. doi:10.1016/j.cell.2010.05.008 PMID:20541250

Lee, J. R., Kim, J. K., Lee, S. J., & Kim, K. P. (2009). Role of protein tyrosine nitration in neurodegenerative diseases and atherosclerosis. *Archives of Pharmacal Research*, *32*(8), 1109–1118. doi:10.100712272-009-1802-0 PMID:19727603

Lee, J. Y., Nagano, Y., Taylor, J. P., Lim, K. L., & Yao, T. P. (2010). Disease-causing mutations in parkin impair mitochondrial ubiquitination, aggregation, and HDAC6-dependent mitophagy. *The Journal of Cell Biology*, *189*(4), 671–679. doi:10.1083/jcb.201001039 PMID:20457763

Lee, J., Hyeon, S. J., Im, H., Ryu, H., Kim, Y., & Ryu, H. (2016). Astrocytes and Microglia as Non-cell Autonomous Players in the Pathogenesis of ALS. *Experimental Neurobiology*, *25*(5), 233–240. doi:10.5607/en.2016.25.5.233 PMID:27790057

Lee, K.-S., Lee, H. J., Kim, D. H., Noh, Y. H., Yu, K., Jung, H.-Y., & ... . (2010). Activation of PERK signaling attenuates Aβ-mediated ER stress. *PLoS One*, *5*(5), e10489. doi:10.1371/journal.pone.0010489 PMID:20463975

Lee, S., Sato, Y., & Nixon, R. A. (2011). Lysosomal proteolysis inhibition selectively disrupts axonal transport of degradative organelles and causes an Alzheimer's-like axonal dystrophy. *The Journal of Neuroscience*, *31*(21), 7817–7830.

Lee, S., Sato, Y., & Nixon, R. A. (2011). Lysosomal Proteolysis Inhibition Selectively Disrupts Axonal Transport of Degradative Organelles and Causes an Alzheimer's-Like Axonal Dystrophy. *The Journal of Neuroscience: The Official Journal of the Society for Neuroscience*, *31*(21), 7817–7830. doi:10.1523/JNEUROSCI.6412-10.2011 PMID:21613495

Leeuwenburgh, C., & Heinecke, J. W. (2001). Oxidative stress and antioxidants in exercise. *Current Medicinal Chemistry*, *8*(7), 829–838. doi:10.2174/0929867013372896 PMID:11375753

Lee, Y. I., Seo, M., Kim, Y., Kim, S. Y., Kang, U. G., Kim, Y. S., & Juhnn, Y. S. (2005). Membrane depolarization induces the undulating phosphorylation/dephosphorylation of glycogen synthase kinase 3β, and this dephosphorylation involves protein phosphatases 2A and 2B in SH-SY5Y human neuroblastoma cells. *The Journal of Biological Chemistry*, *280*(23), 22044–22052. doi:10.1074/jbc.M413987200 PMID:15799972

Lee, Y.-K., Jun, Y.-W., Choi, H.-E., Huh, Y. H., Kaang, B.-K., Jang, D.-J., & Lee, J.-A. (2017). Development of LC3/GABARAP sensors containing a LIR and a hydrophobic domain to monitor autophagy. *The EMBO Journal*, *36*(8), 1100–1116. doi:10.15252/embj.201696315 PMID:28320742

Lefebvre, T., Guinez, C., Dehennaut, V., Beseme-Dekeyser, O., Morelle, W., & Michalski, J.-C. (2005). Does O-GlcNAc play a role in neurodegenerative diseases? *Expert Review of Proteomics*, *2*(2), 265–275. doi:10.1586/14789450.2.2.265 PMID:15892570

Lei, P., Ayton, S., Finkelstein, D. I., Spoerri, L., Ciccotosto, G. D., Wright, D. K., ... Roberts, B. R. (2012). Tau deficiency induces parkinsonism with dementia by impairing APP-mediated iron export. *Nature Medicine*, *18*(2), 291–296. doi:10.1038/nm.2613 PMID:22286308

Leissring, M. A., Farris, W., Chang, A. Y., Walsh, D. M., Wu, X., Sun, X., ... Selkoe, D. J. (2003). Enhanced proteolysis of β-amyloid in APP transgenic mice prevents plaque formation, secondary pathology, and premature death. *Neuron*, *40*(6), 1087–1093. doi:10.1016/S0896-6273(03)00787-6 PMID:14687544

Lei, X. G., Zhu, J. H., Cheng, W. H., Bao, Y., Ho, Y. S., Reddi, A. R., ... Arnér, E. S. (2016). Paradoxical Roles of Antioxidant Enzymes: Basic Mechanisms and Health Implications. *Physiological Reviews*, *96*(1), 307–364. doi:10.1152/physrev.00010.2014 PMID:26681794

Lemasters, J. J. (2005). Selective mitochondrial autophagy, or mitophagy, as a targeted defense against oxidative stress, mitochondrial dysfunction, and aging. *Rejuvenation Research*, *8*(1), 3–5. doi:10.1089/rej.2005.8.3 PMID:15798367

Lemasters, J. J., Theruvath, T. P., Zhong, Z., & Nieminen, A. L. (2009). Mitochondrial calcium and the permeability transition in cell death. *Biochimica et Biophysica Acta (BBA)-. Bioenergetics*, *1787*(11), 1395–1401. doi:10.1016/j.bbabio.2009.06.009 PMID:19576166

Lemmon, M. A. (2007, January). Pleckstrin homology (PH) domains and phosphoinositides. [). Portland Press Limited.]. *Biochemical Society Symposium*, *74*, 81–93. doi:10.1042/BSS2007c08 PMID:17233582

Le, N. T., Chang, L., Kovlyagina, I., Georgiou, P., Safren, N., Braunstein, K. E., ... Morrison, B. M. (2016). Motor neuron disease, TDP-43 pathology, and memory deficits in mice expressing ALS–FTD-linked UBQLN2 mutations. *Proceedings of the National Academy of Sciences of the United States of America*, *113*(47), E7580–E7589. doi:10.1073/pnas.1608432113 PMID:27834214

Leroy, K., Ando, K., Héraud, C., Yilmaz, Z., Authelet, M., Boeynaems, J. M., ... Brion, J. P. (2010). Lithium treatment arrests the development of neurofibrillary tangles in mutant tau transgenic mice with advanced neurofibrillary pathology. *Journal of Alzheimer's Disease*, *19*(2), 705–719. doi:10.3233/JAD-2010-1276 PMID:20110614

Leroy, K., Boutajangout, A., Authelet, M., Woodgett, J. R., Anderton, B. H., & Brion, J. P. (2002). The active form of glycogen synthase kinase-3β is associated with granulovacuolar degeneration in neurons in Alzheimer's disease. *Acta Neuropathologica*, *103*(2), 91–99. doi:10.1007004010100435 PMID:11810173

Lesne, S., Koh, M. T., Kotilinek, L., Kayed, R., Glabe, C. G., Yang, A., ... Ashe, K. H. (2006). A specific amyloid-beta protein assembly in the brain impairs memory. *Nature*, *440*(7082), 352–357. doi:10.1038/nature04533 PMID:16541076

Lesort, M., Chun, W., Johnson, G., & Ferrante, R. (1999). *Tissue transglutaminase is increased in Huntington's disease brain*. Academic Press.

Leszek, J., Trypka, E., Tarasov, V., Md, V., Ashraf, G., & Aliev, G. (2017). Type 3 diabetes mellitus: A novel implication of Alzheimers disease. *Current Topics in Medicinal Chemistry*, *17*(12), 1331–1335. doi:10.2174/1568026617666170103163403 PMID:28049395

Leverenz, J. B., Umar, I., Wang, Q., Montine, T. J., McMillan, P. J., Tsuang, D. W., ... Zhang, J. (2007). Proteomic identification of novel proteins in cortical lewy bodies. *Brain Pathology (Zurich, Switzerland)*, *17*(2), 139–145. doi:10.1111/j.1750-3639.2007.00048.x PMID:17388944

Levine, B., & Kroemer, G. (2008). Autophagy in the pathogenesis of disease. *Cell*, *132*(1), 27–42. doi:10.1016/j.cell.2007.12.018 PMID:18191218

Levine, B., Mizushima, N., & Virgin, H. W. (2011). Autophagy in immunity and inflammation. *Nature*, *469*(7330), 323–335. doi:10.1038/nature09782 PMID:21248839

Levine, B., Sinha, S., & Kroemer, G. (2008). Bcl-2 family members: Dual regulators of apoptosis and autophagy. *Autophagy*, *4*, 600–606. doi:10.4161/auto.6260

Levy-Lahad, E., Wasco, W., Poorkaj, P., Romano, D. M., Oshima, J., Pettingell, W. H., ... Wang, K. (1995). Candidate gene for the chromosome 1 familial Alzheimer's disease locus. *Science*, *269*(5226), 973–977. doi:10.1126cience.7638622 PMID:7638622

Levy, O. A., Malagelada, C., & Greene, L. A. (2009). Cell death pathways in Parkinson's disease: Proximal triggers, distal effectors, and final steps. *Apoptosis*, *14*(4), 478–500. doi:10.100710495-008-0309-3 PMID:19165601

Leyk, J., Goldbaum, O., Noack, M., & Richter-Landsberg, C. (2015). Inhibition of HDAC6 Modifies Tau Inclusion Body Formation and Impairs Autophagic Clearance. *Journal of Molecular Neuroscience*, *55*(4), 1031–1046. doi:10.100712031-014-0460-y PMID:25434725

Li, S.H., & Li, X.-J. (2004). *Huntingtin–protein interactions and the pathogenesis of Huntington's disease*. Academic Press.

Li, Y., Rinne, J. O., Mosconi, L., Pirraglia, E., Rusinek, H., DeSanti, S., . . . Tsui, W. (2008). *Regional analysis of FDG and PIB-PET images in normal aging, mild cognitive impairment, and Alzheimer's disease*. Academic Press.

Liang, X. H., Kleeman, L. K., Jiang, H. H., Gordon, G., Goldman, J. E., & Berry, G. ... Levine, B. (1998). Protection against fatal Sindbis virus encephalitis by beclin, a novel Bcl-2-interacting protein. *Journal of Virology*, *72*(11), 8586–8596. Retrieved from https://www.ncbi.nlm.nih.gov/pubmed/9765397

Liang, C., Feng, P., Ku, B., Dotan, I., Canaani, D., Oh, B. H., & Jung, J. U. (2006, July). Autophagic and tumour suppressor activity of a novel Beclin1-binding protein UVRAG. *Nature Cell Biology*, *8*(7), 688–699. doi:10.1038/ncb1426 PMID:16799551

Liang, Q. L., Ouyang, X. S., Schneider, L., & Zhang, J. H. (2011). Reduction of mutant huntingtin accumulation and toxicity by lysosomal cathepsins D and B in neurons. *Molecular Neurodegeneration*, *6*(1), 6. doi:10.1186/1750-1326-6-37 PMID:21631942

Li, B., Chohan, M. O., Grundke-Iqbal, I., & Iqbal, K. (2007). Disruption of microtubule network by Alzheimer abnormally hyperphosphorylated tau. *Acta Neuropathologica*, *113*(5), 501–511. doi:10.100700401-007-0207-8 PMID:17372746

Liberek, K., Lewandowska, A., & Zietkiewicz, S. (2008). Chaperones in control of protein disaggregation. *The EMBO Journal*, *27*(2), 328–335. doi:10.1038j.emboj.7601970 PMID:18216875

Lichtenthaler, F. S. (2012). Alpha-secretase cleavage of the amyloid precursor protein: Proteolysis regulated by signaling pathways and protein trafficking. *Current Alzheimer Research*, *9*, 165–177. doi:10.2174/156720512799361655 PMID:21605033

Li, F., Macfarlan, T., Pittman, R. N., & Chakravarti, D. (2002). Ataxin-3 is a histone-binding protein with two independent transcriptional corepressor activities. *The Journal of Biological Chemistry*, *277*(47), 45004–45012. doi:10.1074/jbc.M205259200 PMID:12297501

Li, H., Ham, A., Ma, T. C., Kuo, S.-H., Kanter, E., Kim, D., ... Tang, G. (2019). Mitochondrial dysfunction and mitophagy defect triggered by heterozygous GBA mutations. *Autophagy*, *15*(1), 113–130. doi:10.1080/15548627.2018.1509818 PMID:30160596

Li, J.-Q., Tan, L., & Yu, J.-T. (2014). The role of the LRRK2 gene in Parkinsonism. *Molecular Neurodegeneration*, *9*(1), 47. doi:10.1186/1750-1326-9-47 PMID:25391693

Lilja, T., Heldring, N., & Hermanson, O. (2013). Like a rolling histone: Epigenetic regulation of neural stem cells and brain development by factors controlling histone acetylation and methylation. *Biochimica et Biophysica Acta. G, General Subjects*, *1830*(2), 2354–2360. doi:10.1016/j.bbagen.2012.08.011 PMID:22986149

Lillig, C. H., & Holmgren, A. (2007). Thioredoxin and related molecules: From biology to health and disease. *Antioxidants & Redox Signalling*, *9*(1), 25–47. doi:10.1089/ars.2007.9.25 PMID:17115886

Li, M. D., Burns, T. C., Morgan, A. A., & Khatri, P. (2014). Integrated multi-cohort transcriptional meta-analysis of neurodegenerative diseases. *Acta Neuropathologica Communications*, *2*(1), 93. doi:10.118640478-014-0093-y PMID:25187168

Lim, M. A., Selak, M. A., Xiang, Z., Krainc, D., Neve, R. L., Kraemer, B. C., ... Kalb, R. G. (2012). Reduced activity of AMP-activated protein kinase protects against genetic models of motor neuron disease. *The Journal of Neuroscience: The Official Journal of the Society for Neuroscience*, *32*(3), 1123–1141. doi:10.1523/JNEUROSCI.6554-10.2012 PMID:22262909

Lim, Y. A., Rhein, V., Baysang, G., Meier, F., Poljak, A., & Raftery, J., M., ... and Götz, J. (. (2010). Aβ and human amylin share a common toxicity pathway via mitochondrial dysfunction. *Proteomics*, *10*(8), 1621–1633. doi:10.1002/pmic.200900651 PMID:20186753

Linder, M. E., & Deschenes, R. J. (2007). Palmitoylation: Policing protein stability and traffic. *Nature Reviews. Molecular Cell Biology*, *8*(1), 74–84. doi:10.1038/nrm2084 PMID:17183362

Lindholm, D., Wootz, H., & Korhonen, L. (2006). ER stress and neurodegenerative diseases. *Cell Death and Differentiation*, *13*(3), 385–392. doi:10.1038j.cdd.4401778 PMID:16397584

Lindquist, S. L., & Kelly, J. W. (2011). Chemical and biological approaches for adapting proteostasis to ameliorate protein misfolding and aggregation diseases-progress and prognosis. *Cold Spring Harbor Perspectives in Biology*, *3*(12), a004507. doi:10.1101/cshperspect.a004507 PMID:21900404

Lindwall, G., & Cole, R. D. (1984). Phosphorylation affects the ability of tau protein to promote microtubule assembly. *The Journal of Biological Chemistry*, *259*, 5301–5305. PMID:6425287

Lin, M. K., & Farrer, M. J. (2014). Genetics and genomics of Parkinson's disease. *Genome Medicine*, *6*(6), 48. doi:10.1186/gm566 PMID:25061481

Lin, M. Y., Cheng, X. T., Xie, Y., Cai, Q., & Sheng, Z. H. (2017). Removing dysfunctional mitochondria from axons independent of mitophagy under pathophysiological conditions. *Autophagy*, *13*(10), 1792–1794. doi:10.1080/15548627.2017.1356552 PMID:28812939

Lin, Y. W. (2018). Structure and function of heme proteins regulated by diverse post-translational modifications. *Archives of Biochemistry and Biophysics*, *641*, 1–30. doi:10.1016/j.abb.2018.01.009 PMID:29407792

Lione, L. A., Carter, R. J., Hunt, M. J., Bates, G. P., Morton, A. J., & Dunnett, S. B. (1999). Selective discrimination learning impairments in mice expressing the human Huntington's disease mutation. *The Journal of Neuroscience: The Official Journal of the Society for Neuroscience, 19*(23), 10428–10437. doi:10.1523/JNEUROSCI.19-23-10428.1999 PMID:10575040

Lipinski, M. M., Zheng, B., Lu, T., Yan, Z., Py, B. F., Ng, A., … Yuan, J. (2010). Genome-wide analysis reveals mechanisms modulating autophagy in normal brain aging and in Alzheimer's disease. *Proceedings of the National Academy of Sciences, 107*(32), 14164–14169. 10.1073/pnas.1009485107

Liu, C.W., Corboy, M.J., DeMartino, G.N., & Thomas, P.J. (2003). Endoproteolytic activity of the proteasome. *Science, 299*, 408–411.

Liu, C., Cholerton, B., Shi, M., Ginghina, C., Cain, K. C., Auinger, P., & Zhang, J. (2015). CSF tau and tau/Aβ42 predict cognitive decline in Parkinson's disease. *Parkinsonism & Related Disorders, 21*(3), 271–276. doi:10.1016/j.parkreldis.2014.12.027 PMID:25596881

Liu, D., Pitta, M., Jiang, H., Lee, J. H., Zhang, G., Chen, X., ... Mattson, M. P. (2013). Nicotinamide forestalls pathology and cognitive decline in Alzheimer mice: Evidence for improved neuronal bioenergetics and autophagy procession. *Neurobiology of Aging, 34*(6), 1564–1580. doi:10.1016/j.neurobiolaging.2012.11.020 PMID:23273573

Liu, F., Grundke-Iqbal, I., Iqbal, K., & Gong, C. X. (2005). Contributions of protein phosphatases PP1, PP2A, PP2B and PP5 to the regulation of tau phosphorylation. *The European Journal of Neuroscience, 22*(8), 1942–1950. doi:10.1111/j.1460-9568.2005.04391.x PMID:16262633

Liu, F., Li, B., Tung, E. J., Grundke-Iqbal, I., Iqbal, K., & Gong, C. X. (2007). Site-specific effects of tau phosphorylation on its microtubule assembly activity and self-aggregation. *The European Journal of Neuroscience, 26*(12), 3429–3436. doi:10.1111/j.1460-9568.2007.05955.x PMID:18052981

Liu, J., & Li, L. (2019). Targeting Autophagy for the Treatment of Alzheimer's Disease: Challenges and Opportunities. *Frontiers in Molecular Neuroscience, 12*, 12. doi:10.3389/fnmol.2019.00203 PMID:31507373

Liu, J., Lillo, C., Jonsson, P. A., Velde, C. V., Ward, C. M., Miller, T. M., ... Cleveland, D. W. (2004). Toxicity of Familial ALS-Linked SOD1 Mutants from Selective Recruitment to Spinal Mitochondria. *Neuron, 43*(1), 5–17. doi:10.1016/j.neuron.2004.06.016 PMID:15233913

Liu, L., Feng, D., Chen, G., Chen, M., Zheng, Q., Song, P., ... Huang, L. (2012). Mitochondrial outer-membrane protein FUNDC1 mediates hypoxia-induced mitophagy in mammalian cells. *Nature Cell Biology, 14*(2), 177–185. doi:10.1038/ncb2422 PMID:22267086

Liu, L., Sakakibara, K., Chen, Q., & Okamoto, K. (2014). Receptor-mediated mitophagy in yeast and mammalian systems. *Cell Research, 24*(7), 787–795. doi:10.1038/cr.2014.75 PMID:24903109

Liu, Y. H., Han, Y. L., Song, J., Wang, Y., Jing, Y. Y., Shi, Q., ... Dong, X. P. (2011). Heat shock protein 104 inhibited the fibrillization of prion peptide 106-126 and disassembled prion peptide 106-126 fibrils in vitro. *The International Journal of Biochemistry & Cell Biology, 43*(5), 768–774. doi:10.1016/j.biocel.2011.01.022 PMID:21296677

Liu, Z., Li, H., Hong, C., Yue, T., Chen, C., Wang, Z., ... Hu, R. (2018). ALS-associated E478G mutation in human OPTN (optineurin) promotes inflammation and induces neuronal cell death. *Frontiers in Immunology, 9*, 2647. doi:10.3389/fimmu.2018.02647 PMID:30519240

Li, X. C., Hu, Y., Wang, Z. H., Luo, Y., Zhang, Y., Liu, X. P., ... Wang, J. Z. (2016). Human wild-type full-length tau accumulation disrupts mitochondrial dynamics and the functions via increasing mitofusins. *Scientific Reports, 6*(1), 24756. doi:10.1038rep24756 PMID:27099072

Li, X., Alafuzoff, I., Soininen, H., Winblad, B., & Pei, J. J. (2005). Levels of mTOR and its downstream targets 4E-BP1, eEF2, and eEF2 kinase in relationships with tau in Alzheimer's disease brain. *The FEBS Journal*, *272*(16), 4211–4220. doi:10.1111/j.1742-4658.2005.04833.x PMID:16098202

Li, X., Prescott, M., Adler, B., Boyce, J. D., & Devenish, R. J. (2013). Beclin 1 Is Required for Starvation-Enhanced, but Not Rapamycin-Enhanced, LC3-Associated Phagocytosis of Burkholderiapseudomallei in RAW 264.7 Cells. *Infection and Immunity*, *81*(1), 271–277. doi:10.1128/IAI.00834-12 PMID:23115045

Li, Y., Wang, Y., Kim, E., Beemiller, P., Wang, C.-Y., Swanson, J., ... Guan, K.-L. (2007). Bnip3 mediates the hypoxia-induced inhibition on mammalian target of rapamycin by interacting with Rheb. *The Journal of Biological Chemistry*, *282*(49), 35803–35813. doi:10.1074/jbc.M705231200 PMID:17928295

Longatti, A., & Tooze, S. A. (2009). Vesicular trafficking and autophagosome formation. *Cell Death and Differentiation*, *16*(7), 956–965. doi:10.1038/cdd.2009.39 PMID:19373247

Long, Y. C., & Zierath, J. R. (2006). AMP-activated protein kinase signaling in metabolic regulation. *The Journal of Clinical Investigation*, *116*(7), 1776–1783. doi:10.1172/JCI29044 PMID:16823475

Lopert, P., & Patel, M. (2014). Brain mitochondria from DJ-1 knockout mice show increased respiration-dependent hydrogen peroxide consumption. *Redox Biology*, *2*, 667–672. doi:10.1016/j.redox.2014.04.010 PMID:24936441

Lopes da Fonseca, T., Villar-Piqué, A., & Outeiro, T. F. (2015). The Interplay between Alpha-Synuclein Clearance and Spreading. *Biomolecules*, *5*(2), 435–471. doi:10.3390/biom5020435 PMID:25874605

Lopez Salon, M., Pasquini, L., Besio Moreno, M., Pasquini, J. M., & Soto, E. (2003). Relationship between beta-amyloid degradation and the 26S proteasome in neural cells. *Experimental Neurology*, *180*(2), 131–143. doi:10.1016/S0014-4886(02)00060-2 PMID:12684027

Lotz, G. P., Legleiter, J., Aron, R., Mitchell, E. J., Huang, S.-Y., Ng, C., ... Muchowski, P. J. (2010). Hsp70 and Hsp40 functionally interact with soluble mutant huntingtin oligomers in a classic ATP-dependent reaction cycle. *The Journal of Biological Chemistry*, *285*(49), 38183–38193. doi:10.1074/jbc.M110.160218 PMID:20864533

Lowther, W. T., & Haynes, A. C. (2011). Reduction of cysteine sulfinic acid in eukaryotic, typical 2-Cys peroxiredoxins by sulfiredoxin. *Antioxidants & Redox Signalling*, *15*(1), 99–109. doi:10.1089/ars.2010.3564 PMID:20712415

Luc, R., & Vergely, C. (2008). Forgotten radicals in biology. *International Journal of Biomedical Science; IJBS*, *4*, 255–259. PMID:23675099

Ludolph, A. C., & Brettschneider, J. (2015). TDP-43 in amyotrophic lateral sclerosis - is it a prion disease? *European Journal of Neurology*, *22*(5), 753–761. doi:10.1111/ene.12706 PMID:25846565

Luigetti, M., Fabrizi, G. M., Madia, F., Ferrarini, M., Conte, A., Del Grande, A., ... Sabatelli, M. (2010). A novel HSPB1 mutation in an Italian patient with CMT2/dHMN phenotype. *Journal of the Neurological Sciences*, *298*(1-2), 114–117. doi:10.1016/j.jns.2010.09.008 PMID:20870250

Lu, J., Xie, G., Jia, W., & Jia, W. (2013). Insulin resistance and the metabolism of branched-chain amino acids. *Frontiers of Medicine*, *7*(1), 53–59. doi:10.100711684-013-0255-5 PMID:23385611

Luk, K. C., Kehm, V., Carroll, J., Zhang, B., O'Brien, P., Trojanowski, J. Q., & Lee, V. M.-Y. (2012). Pathological alpha-synuclein transmission initiates Parkinson-like neurodegeneration in nontransgenic mice. *Science*, *338*(6109), 949–953. doi:10.1126cience.1227157 PMID:23161999

Luna-Munoz, J., Chavez-Macias, L., Garcia-Sierra, F., & Mena, R. (2007). Earliest Stages of Tau Conformational Changes are Related to the Appearance of a Sequence of Specific Phospho-Dependent Tau Epitopes in Alzheimer's Disease 1. *Journal of Alzheimer's Disease*, *12*(4), 365–375. doi:10.3233/JAD-2007-12410 PMID:18198423

Lund, J., Tedesco, P., Duke, K., Wang, J., Kim, S. K., & Johnson, T. E. (2002). Transcriptional profile of aging in C. elegans. *Current Biology*, *12*(18), 1566–1573. doi:10.1016/S0960-9822(02)01146-6 PMID:12372248

Luo, G.-R., Chen, S., & Le, W.-D. (2006). Are heat shock proteins therapeutic target for Parkinson's disease? *International Journal of Biological Sciences*, *3*(1), 20–26. doi:10.7150/ijbs.3.20 PMID:17200688

Lu, S. C. (2009). Regulation of glutathione synthesis. *Molecular Aspects of Medicine*, *30*(1-2), 42–59. doi:10.1016/j.mam.2008.05.005 PMID:18601945

Lushchak, V. I. (2012). Glutathione Homeostasis and Functions: Potential Targets for Medical Interventions. *Journal of Amino Acids*, *26*, 1–26. doi:10.1155/2012/736837 PMID:22500213

Lustbader, J. W., Cirilli, M., Lin, C., Xu, H. W., Takuma, K., Wang, N., ... Wu, H. (2004). ABAD directly links Abeta to mitochondrial toxicity in Alzheimer's disease. *Science*, *304*(5669), 448–452. doi:10.1126cience.1091230 PMID:15087549

Luzio, J. P., Pryor, P. R., & Bright, N. A. (2007). Lysosomes: Fusion and function. *Nature Reviews. Molecular Cell Biology*, *8*(8), 622–632. doi:10.1038/nrm2217 PMID:17637737

Magrane, J., Cortez, C., Gan, W. B., & Manfredi, G. (2013). Abnormal mitochondrial transport and morphology are common pathological denominators in SOD1 and TDP43 ALS mouse models. *Human Molecular Genetics*, *23*(6), 1413–1424. doi:10.1093/hmg/ddt528 PMID:24154542

Magrinelli, F., Picelli, A., Tocco, P., Federico, A., Roncari, L., Smania, N., & Tamburin, S. (2016). Pathophysiology of motor dysfunction in Parkinson's disease as the rationale for drug treatment and rehabilitation. *Parkinson's Disease*, *2016*, 9832839. doi:10.1155/2016/9832839 PMID:27366343

Maguire, C. A., Ramirez, S. H., Merkel, S. F., Sena-Esteves, M., & Breakefield, X. O. (2014). Gene Therapy for the Nervous System: Challenges and New Strategies. *Neurotherapeutics; the Journal of the American Society for Experimental NeuroTherapeutics*, *11*(4), 817–839. doi:10.100713311-014-0299-5 PMID:25159276

Mahdi, A. A., Rizvi, S. H., & Parveen, A. (2016). Role of Endoplasmic Reticulum Stress and Unfolded Protein Responses in Health and Diseases. *Indian Journal of Clinical Biochemistry*, *31*(2), 127–137. doi:10.100712291-015-0502-4 PMID:27069320

Maiti, P., Manna, J., Veleri, S., & Frautschy, S. (2014). Molecular chaperone dysfunction in neurodegenerative diseases and effects of curcumin. *BioMed Research International*, *495091*, 1–14. doi:10.1155/2014/495091 PMID:25386560

Maiuri, M. C., Le Toumelin, G., Criollo, A., Rain, J. C., Gautier, F., Juin, P., ... Hickman, J. A. (2007). Functional and physical interaction between Bcl-XL and a BH3-like domain in Beclin-1. *The EMBO Journal*, *26*(10), 2527–2539. doi:10.1038j.emboj.7601689 PMID:17446862

Majcher, V., Goode, A., James, V., & Layfield, R. (2015). Autophagy receptor defects and ALS-FTLD. *Molecular and Cellular Neurosciences*, *66*, 43–52. doi:10.1016/j.mcn.2015.01.002 PMID:25683489

Majid, T., Ali, Y. O., Venkitaramani, D. V., Jang, M. K., Lu, H. C., & Pautler, R. G. (2014). In vivo axonal transport deficits in a mouse model of fronto-temporal dementia. *NeuroImage. Clinical*, *4*, 711–717. doi:10.1016/j.nicl.2014.02.005 PMID:24936422

Mak, S. K., McCormack, A. L., Manning-Bog, A. B., Cuervo, A. M., & Di Monte, D. A. (2010). Lysosomal degradation of alpha-synuclein in vivo. *The Journal of Biological Chemistry*, 285. PMID:20200163

Malhotra, J. D., & Kaufman, R. J. (2007). The endoplasmic reticulum and the unfolded protein response. *Seminars in Cell & Developmental Biology*, *18*(8), 716–731. doi:10.1016/j.semcdb.2007.09.003 PMID:18023214

Maloney, M. T. (2015). One Hundred Years of Alzheimer's Disease: The Amyloid Cascade Hypothesis. *Nature Education*, *8*(4), 6.

Maly, D. J., & Papa, F. R. (2014). Druggable sensors of the unfolded protein response. *Nature Chemical Biology*, *10*(11), 892–901. doi:10.1038/nchembio.1664 PMID:25325700

Manabe, T., Katayama, T., Sato, N., Gomi, F., Hitomi, J., Yanagita, T., ... Tohyama, M. (2003). Induced HMGA 1a expression causes aberrant splicing a Presenilin-2 pre-mRNA in sporadic Alzhiemer's disease. *Cell Death and Differentiation*, *10*(6), 698–708. doi:10.1038j.cdd.4401221 PMID:12761578

Manabe, T., Ohe, K., Katayama, T., Matsuzaki, S., Yanagita, T., Okuda, H., ... Mayeda, A. (2007). HMGA1a: Sequence-specific RNA-binding factor causing sporadic Alzheimer's disease-linked exon skipping of presenilin-2 pre-mRNA. *Genes to Cells*, *12*(10), 1179–1191. doi:10.1111/j.1365-2443.2007.01123.x PMID:17903177

Manczak, M., Calkins, M. J., & Reddy, P. H. (2011). Impaired mitochondrial dynamics and abnormal interaction of amyloid beta with mitochondrial protein Drp1 in neurons from patients with Alzheimer's disease: Implications for neuronal damage. *Human Molecular Genetics*, *20*(13), 2495–2509. doi:10.1093/hmg/ddr139 PMID:21459773

Manczak, M., & Reddy, P. H. (2012). Abnormal interaction of VDAC1 with amyloid beta and phosphorylated tau causes mitochondrial dysfunction in Alzheimer's disease. *Human Molecular Genetics*, *21*(23), 5131–5146. doi:10.1093/hmg/dds360 PMID:22926141

Mandelkow, E. M., & Mandelkow, E. (2011). Biochemistry and cell biology of Tau protein in neurofibrillary degeneration. *Cold Spring Harbor Perspectives in Biology*, *3*, 1–25.

Mandelkow, E. M., & Mandelkow, E. (2012). Biochemistry and cell biology of tau protein in neurofibrillary degeneration. *Cold Spring Harbor Perspectives in Medicine*, *2*(7), a006247. doi:10.1101/cshperspect.a006247 PMID:22762014

Mandich, P., Grandis, M., Varese, A., Geroldi, A., Acquaviva, M., Ciotti, P., ... Bellone, E. (2010). Severe neuropathy after diphtheria-tetanus-pertussis vaccination in a child carrying a novel frame-shift mutation in the small heat-shock protein 27 gene. *Journal of Child Neurology*, *25*(1), 107–109. doi:10.1177/0883073809334387 PMID:19435728

Mani, S., Yang, G., & Wang, R. (2011). A critical life-supporting role for cystathionine-γ-lyase in the absence of dietary cysteine supply. *Free Radical Biology & Medicine*, *50*(10), 1280–1287. doi:10.1016/j.freeradbiomed.2011.01.038 PMID:21310231

Mann, D. M. A. (1996). Pyramidal nerve cell loss in Alzheimer's disease. *Neurodegeneration*, *5*(4), 423–427. doi:10.1006/neur.1996.0057 PMID:9117557

Mannick, J. B., Schonhoff, C., Papeta, N., Ghafourifar, P., Szibor, M., Fang, K., & Gaston, B. (2001). S-Nitrosylation of mitochondrial caspases. *The Journal of Cell Biology*, *154*(6), 1111–1116. doi:10.1083/jcb.200104008 PMID:11551979

Manning, B. D., & Cantley, L. C. (2007). AKT/PKB signaling: Navigating downstream. *Cell*, *129*(7), 1261–1274. doi:10.1016/j.cell.2007.06.009 PMID:17604717

Manning, B. D., Tee, A. R., Logsdon, M. N., Blenis, J., & Cantley, L. C. (2002). Identification of the tuberous sclerosis complex-2 tumor suppressor gene product tuberin as a target of the phosphoinositide 3-kinase/akt pathway. *Molecular Cell*, *10*(1), 151–162. doi:10.1016/S1097-2765(02)00568-3 PMID:12150915

Mapa, K., Sikor, M., Kudryavtsev, V., Waegemann, K., Kalinin, S., Seidel, C. A. M., ... Mokranjac, D. (2010). The Conformational Dynamics of the Mitochondrial Hsp70 Chaperone. *Molecular Cell*, *38*(1), 89–100. doi:10.1016/j.molcel.2010.03.010 PMID:20385092

Maqbool, M., Mobashir, M., & Hoda, N. (2016). Pivotal role of glycogen synthase kinase-3: A therapeutic target for Alzheimer's disease. *European Journal of Medicinal Chemistry*, *107*, 63–81. doi:10.1016/j.ejmech.2015.10.018 PMID:26562543

Marciniak, S. J., Garcia-Bonilla, L., Hu, J., Harding, H. P., & Ron, D. (2006). Activation-dependent substrate recruitment by the eukaryotic translation initiation factor 2 kinase PERK. *The Journal of Cell Biology*, *172*(2), 201–209. doi:10.1083/jcb.200508099 PMID:16418533

Maria Fimia, G., Stoykova, A., Romagnoli, A., Giunta, L., Di Bartolomeo, S., Nardacci, R., ... Cecconi, F. (2007). Ambra1 regulates autophagy and development of the nervous system. *Nature*, *447*(7148), 1121–1125. doi:10.1038/nature05925 PMID:17589504

Marino Gammazza, A., Caruso Bavisotto, C., Barone, R., Macario, E. C. D., & Macario, JL, A. (. (2016). Alzheimer's disease and molecular chaperones: Current knowledge and the future of chaperonotherapy. *Current Pharmaceutical Design*, *22*(26), 4040–4049. doi:10.2174/1381612822666160518141437 PMID:27189602

Marino, S. M., & Gladyshev, V. N. (2012). Analysis and functional prediction of reactive cysteine residues. *The Journal of Biological Chemistry*, *287*(7), 4419–4425. doi:10.1074/jbc.R111.275578 PMID:22157013

Markesbery, W. R. (1997). Oxidative stress hypothesis in Alzheimer's disease. *Free Radical Biology & Medicine*, *23*(1), 134–147. doi:10.1016/S0891-5849(96)00629-6 PMID:9165306

Marks, D. S., Hopf, T. A., & Sander, C. (2012). Protein structure prediction from sequence variation. *Nature Biotechnology*, *30*(11), 1072–1080. doi:10.1038/nbt.2419 PMID:23138306

Marosi, K., & Mattson, M. P. (2014). BDNF mediates adaptive brain and body responses to energetic challenges. *Trends in Endocrinology and Metabolism*, *25*(2), 89–98. doi:10.1016/j.tem.2013.10.006 PMID:24361004

Maroteaux, L., & Scheller, R. H. (1991). The rat brain synucleins; family of proteins transiently associated with neuronal membrane. *Brain Research. Molecular Brain Research*, *11*(3-4), 335–343. doi:10.1016/0169-328X(91)90043-W PMID:1661825

Martin, D. D. O., Ahpin, C. Y., Heit, R. J., Perinpanayagam, M. A., Yap, M. C., Veldhoen, R. A., ... Berthiaume, L. G. (2012). Tandem reporter assay for myristoylated proteins posttranslationally (TRAMPP) identifies novel substrates for post-translational myristoylation: PKCε, a case study. *The FASEB Journal*, *26*(1), 13–28. doi:10.1096/fj.11-182360 PMID:21965604

Martin, D. D. O., Heit, R. J., Yap, M. C., Davidson, M. W., Hayden, M. R., & Berthiaume, L. G. (2014). Identification of a post-translationally myristoylated autophagy-inducing domain released by caspase cleavage of huntingtin. *Human Molecular Genetics*, *23*(12), 3166–3179. doi:10.1093/hmg/ddu027 PMID:24459296

Martin, D. D. O., Ladha, S., Ehrnhoefer, D. E., & Hayden, M. R. (2015). Autophagy in Huntington disease and huntingtin in autophagy. *TreNDDs Neurosci.*, *38*(1), 26–35. doi:10.1016/j.tins.2014.09.003 PMID:25282404

Martínez, G., Duran-Aniotz, C., and Hetz, C. (2016). Gene therapy to target ER stress in brain diseases. *Brain Research*, 561–570. doi:10.1016/j.brainres.2016.04.064 PMID:27131987

Martinez, J. A., Zhang, Z., Svetlov, S. I., Hayes, R. L., Wang, K. K., & Larner, S. F. (2010). Calpain and caspase processing of caspase-12 contribute to the ER stress-induced cell death pathway in differentiated PC12 cells. *Apoptosis, 15*(12), 1480–1493. doi:10.100710495-010-0526-4 PMID:20640600

Martinez, J., Almendinger, J., Oberst, A., Ness, R., Dillon, C. P., Fitzgerald, P., ... Green, D. R. (2011). for the efficient clearance of dead cells. *Proceedings of the National Academy of Sciences of the United States of America, 108*(42), 17396–17401. doi:10.1073/pnas.1113421108 PMID:21969579

Martinez-Vicente, M., & Cuervo, A. M. (2007). *Autophagy and neurodegeneration: when the cleaning crew goes on strike*. Academic Press.

Martinez-Vicente, M., Talloczy, Z., Wong, E., Tang, G., Koga, H., Kaushik, S., . . . Sulzer, D. (2010). *Cargo recognition failure is responsible for inefficient autophagy in Huntington's disease*. Academic Press.

Martinez-Vicente, M. (2017). Neuronal mitophagy in neurodegenerative diseases. *Frontiers in Molecular Neuroscience, 10*, 64. doi:10.3389/fnmol.2017.00064 PMID:28337125

Martinez-Vicente, M., Talloczy, Z., Wong, E., Tang, G., Koga, H., Kaushik, S., ... Cuervo, A. M. (2010). Cargo recognition failure is responsible for inefficient autophagy in Huntington's disease. *Nature Neuroscience, 13*(5), 567–576. doi:10.1038/nn.2528 PMID:20383138

Martini-Stoica, H., Xu, Y., Ballabio, A., & Zheng, H. (2016). The autophagy–lysosomal pathway in neurodegeneration: A TFEB perspective. *Trends in Neurosciences, 39*(4), 221–234. doi:10.1016/j.tins.2016.02.002 PMID:26968346

Martin, L., Latypova, X., & Terro, F. (2011). Post-translational modifications of tau protein: Implications for Alzheimer's disease. *Neurochemistry International, 58*(4), 458–471. doi:10.1016/j.neuint.2010.12.023 PMID:21215781

Martin-Maestro, P., Sproul, A., Martinez, H., Paquet, D., Gerges, M., Noggle, S., & Starkov, A. A. (2019). Autophagy Induction by Bexarotene Promotes Mitophagy in Presenilin 1 Familial Alzheimer's Disease iPSC-Derived Neural Stem Cells. *Molecular Neurobiology, 56*(12), 8220–8236. doi:10.100712035-019-01665-y PMID:31203573

Martyniszyn, L., Szulc, L., Boratyńska, A., & Niemiałtowski, M. G. (2011). Beclin 1 is involved in regulation of apoptosis and autophagy during replication of ectromelia virus in permissive L929 cells. *Archivum Immunologiae et Therapiae Experimentalis, 59*(6), 463–471. doi:10.100700005-011-0149-7 PMID:21972018

Marui, W., Iseki, E., Uéda, K., & Kosaka, K. (2000). Occurrence of human α-synuclein immunoreactive neurons with neurofibrillary tangle formation in the limbic areas of patients with Alzheimer's disease. *Journal of the Neurological Sciences, 174*(2), 81–84. doi:10.1016/S0022-510X(99)00327-5 PMID:10727692

Marzella, L., Ahlberg, J., & Glaumann, H. (1981). Autophagy, heterophagy, microautophagy and crinophagy as the means for intracellular degradation. *Virchows Archiv B Cell Pathology Zell-pathologie, 36*(1), 219-234.

Masliah, E. (1995). Mechanisms of synaptic dysfunction in Alzheimer's disease. *Histology and Histopathology, 10*(2), 509–519. PMID:7599445

Masters, C. L., Gu, B., Laws, S., Lim, Y. Y., Roberts, B., Villemagne, V. L., & Beyreuther, K. (2018). Alzheimer's Disease: Toward a Quantitative Biological Approach in Describing its Natural History and Underlying Mechanisms. In M. S. Wolfe (Ed.), *The Molecular and Cellular Basis of Neurodegenerative Diseases* (pp. 57–79). London: Academic Press. doi:10.1016/B978-0-12-811304-2.00003-1

Mastroberardino, P. G., & Piacentini, M. (2010). *Type 2 transglutaminase in Huntington's disease: a double-edged sword with clinical potential*. Academic Press.

Mastroberardino, P., Iannicola, C., Nardacci, R., Bernassola, F., De Laurenzi, V., Melino, G., . . . Fesus, L. (2002). *'Tissue' transglutaminase ablation reduces neuronal death and prolongs survival in a mouse model of Huntington's disease*. Academic Press.

Mathew, R., Karp, C. M., Beaudoin, B., Vuong, N., Chen, G., Chen, H. Y., ... DiPaola, R. S. (2009). Autophagy suppresses tumorigenesis through elimination of p62. *Cell*, *137*(6), 1062–1075. doi:10.1016/j.cell.2009.03.048 PMID:19524509

Mathew, R., Kongara, S., Beaudoin, B., Karp, C. M., Bray, K., Degenhardt, K., ... White, E. (2007). Autophagy suppresses tumor progression by limiting chromosomal instability. *Genes & Development*, *21*(11), 1367–1381. doi:10.1101/gad.1545107 PMID:17510285

Matsuda, N., Sato, S., Shiba, K., Okatsu, K., Saisho, K., Gautier, C. A., ... Kimura, M. (2010). PINK1 stabilized by mitochondrial depolarization recruits Parkin to damaged mitochondria and activates latent Parkin for mitophagy. *The Journal of Cell Biology*, *189*(2), 211–221. doi:10.1083/jcb.200910140 PMID:20404107

Matsumine, H., Saito, M., Shimoda-Matsubayashi, S., Tanaka, H., Ishikawa, A., Nakagawa-Hattori, Y., ... Takano, H. (1997). Localization of a gene for an autosomal recessive form of juvenile Parkinsonism to chromosome 6q25.2-27. *American Journal of Human Genetics*, *60*, 588–596. PMID:9042918

Matsumoto, L., Takuma, H., Tamaoka, A., Kurisaki, H., Date, H., Tsuji, S., & Iwata, A. (2010). CpG demethylation enhances alpha-synuclein expression and affects the pathogenesis of Parkinson's disease. *PLoS One*, *5*(11), e15522. doi:10.1371/journal.pone.0015522 PMID:21124796

Matsunaga, K., Morita, E., Saitoh, T., Akira, S., Ktistakis, N. T., Izumi, T., ... Yoshimori, T. (2010). Autophagy requires endoplasmic reticulum targeting of the PI3-kinase complex via Atg14L. *The Journal of Cell Biology*, *190*(4), 511–521. doi:10.1083/jcb.200911141 PMID:20713597

Mattson, M. P., & Chan, S. L. (2003). Neuronal and glial calcium signaling in Alzheimer's disease. *Cell Calcium*, *34*(4-5), 385–397. doi:10.1016/S0143-4160(03)00128-3 PMID:12909083

Mattson, M. P., Gleichmann, M., & Cheng, A. (2008). Mitochondria in neuroplasticity and neurological disorders. *Neuron*, *60*(5), 748–766. doi:10.1016/j.neuron.2008.10.010 PMID:19081372

Mattson, M. P., Partin, J., & Begley, J. G. (1998). Amyloid beta-peptide induces apoptosis-related events in synapses and dendrites. *Brain Research*, *807*(1-2), 167–176. doi:10.1016/S0006-8993(98)00763-X PMID:9757026

Mattson, M. P., Pedersen, W. A., Duan, W., Culmsee, C., & Camandola, S. (1999). Cellular and molecular mechanisms underlying perturbed energy metabolism and neuronal degeneration in Alzheimer's and Parkinson's diseases. *Annals of the New York Academy of Sciences*, *893*(1 OXIDATIVE/ENE), 154–175. doi:10.1111/j.1749-6632.1999.tb07824.x PMID:10672236

Matts, R. L., Brandt, G. E. L., Lu, Y., Dixit, A., Mollapour, M., Wang, S., ... Blagg, B. S. J. (2011). A systematic protocol for the characterization of Hsp90 modulators. *Bioorganic & Medicinal Chemistry*, *19*(1), 684–692. doi:10.1016/j.bmc.2010.10.029 PMID:21129982

Matus, S., Lisbona, F., Torres, M., Leon, C., Thielen, P., & Hetz, C. (2008). The Stress Rheostat: An Interplay Between the Unfolded Protein Response (UPR) and Autophagy in Neurodegeneration. *Current Molecular Medicine*, *8*(3), 157–172. doi:10.2174/156652408784221324 PMID:18473817

Mayer, M. P., & Bukau, B. (2005). Hsp70 chaperones: Cellular functions and molecular mechanism. *Cellular and molecular life sciences. Cellular and Molecular Life Sciences*, *62*(6), 670–684. doi:10.100700018-004-4464-6 PMID:15770419

Mayor, S., & Riezman, H. (2004). Sorting GPI-anchored proteins. *Nature Reviews. Molecular Cell Biology*, *5*(2), 110–120. doi:10.1038/nrm1309 PMID:15040444

Mazzola, J. L., & Sirover, M. A. (2002). *Alteration of intracellular structure and function of glyceraldehyde-3-phosphate dehydrogenase: a common phenotype of neurodegenerative disorders*. Academic Press.

McAdam, R. L., Morton, A., Gordon, S. L., Alterman, J. F., Khvorova, A., Cousin, M. A., & Smillie, K. J. (2019). Loss of huntingtin function slows synaptic vesicle endocytosis in striatal neurons from the htt(Q140/Q140) mouse model of Huntington's disease. *Neurobiology of Disease*, *134*, 104637. doi:10.1016/j.nbd.2019.104637 PMID:31614197

McAlary, L., Plotkin, S. S., & Cashman, N. R. (2019). Emerging Developments in Targeting Proteotoxicity in Neurodegenerative Diseases. *CNS Drugs*, *33*(9), 883–904. doi:10.100740263-019-00657-9 PMID:31414322

McBean, G. J. (2017). Cysteine, Glutathione, and Thiol Redox Balance in Astrocytes. *Antioxidants (Basel, Switzerland)*, *6*(3), 62. doi:10.3390/antiox6030062 PMID:28771170

McBean, G. J., Aslan, M., Griffiths, H. R., & Torrão, R. C. (2015). Thiol redox homeostasis in neurodegenerative disease. *Redox Biology*, *5*, 186–194. doi:10.1016/j.redox.2015.04.004 PMID:25974624

McClellan, A. J., Scott, M. D., & Frydman, J. (2005). Folding and quality control of the VHL tumor suppressor proceed through distinct chaperone pathways. *Cell*, *121*(5), 739–748. doi:10.1016/j.cell.2005.03.024 PMID:15935760

McConkey, D. J. (2017). The integrated stress response and proteotoxicity in cancer therapy. *Biochemical and Biophysical Research Communications*, *482*(3), 450–453. doi:10.1016/j.bbrc.2016.11.047 PMID:28212730

McCormack, A. L., Mak, S. K., Henderson, J. M., Bumcrot, D., Farrer, M. J., & Di Monte, D. A. (2010). α-synuclein suppression by targeted small interfering RNA in the primate substantianigra. *PLoS One*, *5*(8), e12122. doi:10.1371/journal.pone.0012122 PMID:20711464

McCoy, M. K., Martinez, T. N., Ruhn, K. A., Szymkowski, D. E., Smith, C. G., Botterman, B. R., ... Tansey, M. G. (2006). Blocking soluble tumor necrosis factor signaling with dominant-negative tumor necrosis factor inhibitor attenuates loss of dopaminergic neurons in models of Parkinson's disease. *The Journal of Neuroscience: The Official Journal of the Society for Neuroscience*, *26*(37), 9365–9375. doi:10.1523/JNEUROSCI.1504-06.2006 PMID:16971520

McDonagh, B., Sakellariou, G. K., Smith, N. T., Brownridge, P., & Jackson, M. J. (2014). Differential cysteine labeling and global label-free proteomics reveals an altered metabolic state in skeletal muscle aging. *Journal of Proteome Research*, *13*(11), 5008–5021. doi:10.1021/pr5006394 PMID:25181601

McFarland, M. A., Ellis, C. E., Markey, S. P., & Nussbaum, R. L. (2008). Proteomics Analysis Identifies Phosphorylation-dependent alpha-Synuclein Protein Interactions. *Molecular & Cellular Proteomics*, *7*(11), 2123–2137. doi:10.1074/mcp.M800116-MCP200 PMID:18614564

McInnes, J. (2013). Insights on altered mitochondrial function and dynamics in the pathogenesis of neurodegeneration. *Translational Neurodegeneration*, *2*(1), 12. doi:10.1186/2047-9158-2-12 PMID:23711354

McKnight, N. C., Zhong, Y., Wold, M. S., Gong, S., Phillips, G. R., Dou, Z., ... Yue, Z. (2014). Beclin 1 Is Required for Neuron Viability and Regulates Endosome Pathways via the UVRAG-VPS34 Complex. *PLOS Genetics*, *10*(10), e1004626. doi:10.1371/journal.pgen.1004626 PMID:25275521

McLaughlin, T., Falkowski, M., Wang, J. J., & Zhang, S. X. (2018). Molecular chaperone ERp29: A potential target for cellular protection in retinal and neurodegenerative diseases. In *Retinal Degenerative Diseases* (pp. 421–427). Cham: Springer; doi:10.1007/978-3-319-75402-4_52

McLean, P. J., Kawamata, H., & Hyman, B. T. (2001). Alpha-synuclein-enhanced green fluorescent protein fusion proteins form proteasome sensitive inclusions in primary neurons. *Neuroscience*, *104*(3), 901–912. doi:10.1016/S0306-4522(01)00113-0 PMID:11440819

McLean, P. J., Klucken, J., Shin, Y., & Hyman, B. T. (2004). Geldanamycin induces Hsp70 and prevents alpha-synuclein aggregation and toxicity in vitro. *Biochemical and Biophysical Research Communications*, *321*(3), 665–669. doi:10.1016/j.bbrc.2004.07.021 PMID:15358157

McLelland, G. L., Soubannier, V., Chen, C. X., McBride, H. M., & Fon, E. A. (2014, February 18). Parkin and *PINK1* function in a vesicular trafficking pathway regulating mitochondrial quality control. *The EMBO Journal*, *33*(4), 282–295. doi:10.1002/embj.201385902 PMID:24446486

McNally, K. L., Ward, A. E., & Priola, S. A. (2009). Cells expressing anchorless prion protein are resistant to scrapie infection. *Journal of Virology*, *83*(9), 4469–4475. doi:10.1128/JVI.02412-08 PMID:19225008

McNaught, K. S. P., Belizaire, R., Isacson, O., Jenner, P., & Olanow, C. W. (2003). Altered proteasomal function in sporadic Parkinson's disease. *Experimental Neurology*, *179*(1), 38–46. doi:10.1006/exnr.2002.8050 PMID:12504866

McWilliams T.G., & Ganley, I.G. (2019). *Investigating mitophagy and Mitochondrial morphology in vivo using mito-QC: A comprehensive guide*. Academic Press.

McWilliams, T. G., Prescott, A. R., Allen, G. F., Tamjar, J., Munson, M. J., Thomson, C., ... Ganley, I. G. (2016). mito-QC illuminates mitophagy and mitochondrial architecture in vivo. *The Journal of Cell Biology*, *214*(3), 333–345. doi:10.1083/jcb.201603039 PMID:27458135

McWilliams, T. G., Prescott, A. R., Montava-Garriga, L., Ball, G., Singh, F., Barini, E., ... Ganley, I. G. (2018). Basal mitophagy occurs independently of PINK1 in mouse tissues of high metabolic demand. *Cell Metabolism*, *27*(2), 439–449. doi:10.1016/j.cmet.2017.12.008 PMID:29337137

Meckler, X., Roseman, J., Das, P., Cheng, H., Pei, S., Keat, M., ... Thinakaran, G. (2010). Reduced Alzheimer's disease β-amyloid deposition in transgenic mice expressing S-palmitoylation-deficient APH1aL and nicastrin. *The Journal of Neuroscience: The Official Journal of the Society for Neuroscience*, *30*(48), 16160–16169. doi:10.1523/JNEUROSCI.4436-10.2010 PMID:21123562

Mehrpour, M., Botti, J., & Codogno, P. (2012). Mechanisms and regulation of autophagy in mammalian cells. *Atlas of Genetics and Cytogenetics in Oncology and Haematology*, (2). doi:10.4267/2042/46951

Meister, A. (1988). Glutathione metabolism and its selective modification. *The Journal of Biological Chemistry*, *263*, 17205–17208. PMID:3053703

Meitzler, J. L., Hinde, S., Bánfi, B., Nauseef, W. M., & Ortiz de Montellano, P. R. (2013). Conserved cysteine residues provide a protein-protein interaction surface in dual oxidase (DUOX) proteins. *The Journal of Biological Chemistry*, *288*(10), 7147–7157. doi:10.1074/jbc.M112.414797 PMID:23362256

Menalled, L. B., Kudwa, A. E., Oakeshott, S., Farrar, A., Paterson, N., Filippov, I., . . . Beltran, J. (2014). *Genetic deletion of transglutaminase 2 does not rescue the phenotypic deficits observed in R6/2 and zQ175 mouse models of Huntington's disease*. Academic Press.

Meng, F., Yao, D., Shi, Y., Kabakoff, J., Wu, W., Reicher, J., ... Gu, Z. (2011). Oxidation of the cysteine-rich regions of parkin perturbs its E3 ligase activity and contributes to protein aggregation. *Molecular Neurodegeneration*, *6*(34), 1–15. doi:10.1186/1750-1326-6-34 PMID:21595948

Meo, S. D., Reed, T. T., Venditti, P., & Victor, V. M. (2016). Role of ROS and RNS sources in physiological and pathological conditions. *Oxidative Medicine and Cellular Longevity*, *2016*, 1245049. PMID:27478531

Mercado, G., & Hetz, C. (2017). Drug repurposing to target proteostasis and prevent neurodegeneration: Accelerating translational efforts. *Brain*, *140*(6), 1544–1547. doi:10.1093/brain/awx107 PMID:28549133

Mercer, C. A., Kaliappan, A., & Dennis, P. B. (2009). A novel, human Atg13 binding protein, Atg101, interacts with ULK1 and is essential for macroautophagy. *Autophagy*, *5*(5), 649–662. doi:10.4161/auto.5.5.8249 PMID:19287211

Meriin, A. B., Narayanan, A., Meng, L., Alexandrov, I., Varelas, X., Cisse, I. I., & Sherman, M. Y. (2018). Hsp70–Bag3 complex is a hub for proteotoxicity-induced signaling that controls protein aggregation. *Proceedings of the National Academy of Sciences of the United States of America*, *115*(30), E7043–E7052. doi:10.1073/pnas.1803130115 PMID:29987014

Merkley, E. D., Metz, T. O., Smith, R. D., Baynes, J. W., & Frizzell, N. (2014). The succinated proteome. *Mass Spectrometry Reviews*, *33*(2), 98–109. doi:10.1002/mas.21382 PMID:24115015

Meske, V., Albert, F., & Ohm, T. G. (2008). Coupling of mammalian target of rapamycin with phosphoinositide 3-kinase signaling pathway regulates protein phosphatase 2a-and glycogen synthase kinase-3β-dependent phosphorylation of tau. *The Journal of Biological Chemistry*, *283*(1), 100–109. doi:10.1074/jbc.M704292200 PMID:17971449

Meusser, B., Hirsch, C., Jarosch, E., & Sommer, T. (2005). ERAD: The long road to destruction. *Nature Cell Biology*, *7*(8), 766–772. doi:10.1038/ncb0805-766 PMID:16056268

Meyer, H., & Weihl, C. C. (2014). The VCP/p97 system at a glance: connecting cellular function to disease pathogenesis. The Company of Biologists Ltd.

Meyer, H., Bug, M., & Bremer, S. (2012). *Emerging functions of the VCP/p97 AAA-ATPase in the ubiquitin system.* Academic Press.

Michiorri, S., Gelmetti, V., Giarda, E., Lombardi, F., Romano, F., Marongiu, R., ... Casari, G. (2010). The Parkinson-associated protein PINK1 interacts with Beclin1 and promotes autophagy. *Cell Death and Differentiation*, *17*(6), 962–974. doi:10.1038/cdd.2009.200 PMID:20057503

Middleton, E. R., & Rhoades, E. (2010). Effects of curvature and composition on α-synuclein binding to lipid vesicles. *Biophysical Journal*, *99*(7), 2279–2288. doi:10.1016/j.bpj.2010.07.056 PMID:20923663

Miklya, I., Göltl, P., Hafenscher, F., & Pencz, N. (2014). The role of parkin in Parkinson's disease. *Neuropsychopharmacologia Hungarica: a Magyar Pszichofarmakologiai Egyesulet lapja= official journal of the Hungarian Association of Psychopharmacology*, *16*(2), 67-76.

Milhavet, O., Martindale, J. L., Camandola, S., Chan, S. L., Gary, D. S., Cheng, A., ... Mattson, M. P. (2002). Involvement of Gadd153 in the pathogenic action of presenilin-1 mutations. *Journal of Neurochemistry*, *83*(3), 673–681. doi:10.1046/j.1471-4159.2002.01165.x PMID:12390529

Miller, B. C., Eckman, E. A., Sambamurti, K., Dobbs, N., Chow, K. M., Eckman, C. B., ... Thiele, D. L. (2003). Amyloid-β peptide levels in brain are inversely correlated with insulysin activity levels in vivo. *Proceedings of the National Academy of Sciences of the United States of America*, *100*(10), 6221–6226. doi:10.1073/pnas.1031520100 PMID:12732730

Miller, J., Arrasate, M., Shaby, B. A., Mitra, S., Masliah, E., & Finkbeiner, S. (2010). Quantitative Relationships between Huntingtin Levels, Polyglutamine Length, Inclusion Body Formation, and Neuronal Death Provide Novel Insight into Huntington's Disease Molecular Pathogenesis. *The Journal of Neuroscience: The Official Journal of the Society for Neuroscience*, *30*(31), 10541–10550. doi:10.1523/JNEUROSCI.0146-10.2010 PMID:20685997

Miller, S., & Muqit, M. M. K. (2019). Therapeutic approaches to enhance PINK1/Parkin mediated mitophagy for the treatment of Parkinson's disease. *Neuroscience Letters, 705*, 7–13. doi:10.1016/j.neulet.2019.04.029 PMID:30995519

Mimura, N., Fulciniti, M., Gorgun, G., Tai, Y. T., Cirstea, D., Santo, L., ... Anderson, K. C. (2012). Blockade of XBP1 splicing by inhibition of IRE1α is a promising therapeutic option in multiple myeloma. *Blood, 119*(24), 5772–5781. doi:10.1182/blood-2011-07-366633 PMID:22538852

Minoia, M., Boncoraglio, A., Vinet, J., Morelli, F. F., Brunsting, J. F., Poletti, A., ... Carra, S. (2014). BAG3 induces the sequestration of proteasomal clients into cytoplasmic puncta: Implications for a proteasome-to-autophagy switch. *Autophagy, 9*(9), 1603–1621. doi:10.4161/auto.29409 PMID:25046115

Mishra, A., Maheshwari, M., Chhangani, D., Fujimori-Tonou, N., Endo, F., Joshi, A.P., Jana, N.R., & Yamanaka, K. (2013). E6-AP association promotes SOD1 aggresomes degradation and suppresses toxicity. *Neurobiol Aging, 34*, 1311-1323.

Mitra, S., & Finkbeiner, S. (2008). The ubiquitin-proteasome pathway in Huntington's disease. *TheScientificWorldJournal, 8*, 421–433. doi:10.1100/tsw.2008.60 PMID:18454252

Mizuguchi, Y., Hatakeyama, H., Sueoka, K., Tanaka, M., & Goto, Y. I. (2017). Low dose resveratrol ameliorates mitochondrial respiratory dysfunction and enhances cellular reprogramming. *Mitochondrion, 34*, 43–48. doi:10.1016/j.mito.2016.12.006 PMID:28093354

Mizushima, N. (2017). The exponential growth of autophagy-related research: From the humble yeast to the Nobel Prize. *FEBS Letters, 591*(5), 681–689. doi:10.1002/1873-3468.12594 PMID:28186333

Mizushima, N., & Komatsu, M. (2011). Autophagy: Renovation of cells and tissues. *Cell, 147*(4), 728–741. doi:10.1016/j.cell.2011.10.026 PMID:22078875

Mizushima, N., Levine, B., Cuervo, A. M., & Klionsky, D. J. (2008). Autophagy fights disease through cellular self-digestion. *Nature, 451*(7182), 1069–1075. doi:10.1038/nature06639 PMID:18305538

Mizushima, N., Noda, T., & Ohsumi, Y. (1999). Apg16p is required for the function of the Apg12p–Apg5p conjugate in the yeast autophagy pathway. *The EMBO Journal, 18*(14), 3888–3896. doi:10.1093/emboj/18.14.3888 PMID:10406794

Mizushima, N., & Yoshimori, T. (2007). How to interpret LC3 immunoblotting. *Autophagy, 3*(6), 542–545. doi:10.4161/auto.4600 PMID:17611390

Mizushima, N., Yoshimori, T., & Levine, B. (2010). Methods in mammalian autophagy research. *Cell, 140*(3), 313–326. doi:10.1016/j.cell.2010.01.028 PMID:20144757

Moehle, E. A., Shen, K., & Dillin, A. (2019). Mitochondrial proteostasis in the context of cellular and organismal health and aging. *The Journal of Biological Chemistry, 294*(14), 5396–5407. doi:10.1074/jbc.TM117.000893 PMID:29622680

Mokhtar, S. H., Bakhuraysah, M. M., Cram, D. S., & Petratos, S. (2013). The Beta-amyloid protein of Alzheimer's disease: Communication breakdown by modifying the neuronal cytoskeleton. *International Journal of Alzheimer's Disease, 910502*, 1–15. doi:10.1155/2013/910502 PMID:24416616

Molejon, M. I., Ropolo, A., Lo Re, A., Boggio, V., & Vaccaro, M. I. (2013). The VMP1-Beclin 1 interaction regulates autophagy induction. *Scientific Reports, 3*(1), 1055. doi:10.1038rep01055 PMID:23316280

Mollapour, M., & Neckers, L. (2012). Post-translational modifications of Hsp90 and their contributions to chaperone regulation. *Biochimica et Biophysica Acta, 1823*(3), 648–655. doi:10.1016/j.bbamcr.2011.07.018 PMID:21856339

Mollenhauer, B., Trautmann, E., Otte, B., Ng, J., Spreer, A., Lange, P., & Trenkwalder, C. (2012). α-Synuclein in human cerebrospinal fluid is principally derived from neurons of the central nervous system. *Journal of Neural Transmission (Vienna, Austria)*, *119*(7), 739–746. doi:10.100700702-012-0784-0 PMID:22426833

Moller, A., Bauer, C. S., Cohen, R. N., Webster, C. P., & De Vos, K. J. (2017). Amyotrophic lateral sclerosis-associated mutant SOD1 inhibits anterograde axonal transport of mitochondria by reducing Miro1 levels. *Human Molecular Genetics*, *26*(23), 4668–4679. doi:10.1093/hmg/ddx348 PMID:28973175

Moloney, A. M., Griffin, R. J., Timmons, S., O'Connor, R., Ravid, R., & O'Neill, C. (2010). Defects in IGF-1 receptor, insulin receptor and IRS-1/2 in Alzheimer's disease indicate possible resistance to IGF-1 and insulin signalling. *Neurobiology of Aging*, *31*(2), 224–243. doi:10.1016/j.neurobiolaging.2008.04.002 PMID:18479783

Monteiro, M. C., Coleman, M. D., Hill, E. J., Prediger, R. D., & Maia, C. S. F. (2017). Neuroprotection in neurodegenerative disease from basic science to clinical applications. *Oxidative Medicine and Cellular Longevity*, *2017*, 1–3. doi:10.1155/2017/2949102 PMID:28337247

Monti, D. A., Zabrecky, G., Kremens, D., Liang, T. W., Wintering, N. A., Cal, J., & .... (2016). N-Acetyl Cysteine May Support Dopamine Neuronsin Parkinson's Disease: Preliminary Clinical and Cell Line Data. *PLoS One*, *11*(6), 1–15. doi:10.1371/journal.pone.0157602 PMID:27309537

Moore, A. S., & Holzbaur, E. L. (2016). Spatiotemporal dynamics of autophagy receptors in selective mitophagy. *Autophagy*, *12*(10), 1956–1957. doi:10.1080/15548627.2016.1212788 PMID:27484150

Morales-Corraliza, J., Wong, H., Mazzella, M. J., Che, S., Lee, S. H., Petkova, E., ... Mathews, P. M. (2016). Brain-wide insulin resistance, tau phosphorylation changes, and hippocampal neprilysin and amyloid-β alterations in a monkey model of type 1 diabetes. *The Journal of Neuroscience: The Official Journal of the Society for Neuroscience*, *36*(15), 4248–4258. doi:10.1523/JNEUROSCI.4640-14.2016 PMID:27076423

Moreno, H., Morfini, G., Buitrago, L., Ujlaki, G., Choi, S., Yu, E., ... Sugimori, M. (2016). Tau pathology-mediated presynaptic dysfunction. *Neuroscience*, *325*, 30–38. doi:10.1016/j.neuroscience.2016.03.044 PMID:27012611

Moreno, J. A., Halliday, M., Molloy, C., Radford, H., Verity, N., Axten, J. M., ... Mallucci, G. R. (2013). Oral treatment targeting the unfolded protein response prevents neurodegeneration and clinical disease in prion-infected mice. *Science Translational Medicine*, *5*(206), 206ra138. doi:10.1126citranslmed.3006767 PMID:24107777

Moreno, M. L., Escobar, J., Izquierdo-Álvarez, A., Gil, A., Pérez, S., Pereda, J., ... Sastre, J. (2014). Disulfide stress: A novel type of oxidative stress in acute pancreatitis. *Free Radical Biology & Medicine*, *70*, 265–277. doi:10.1016/j.freeradbiomed.2014.01.009 PMID:24456905

Mori, F., Tanji, K., Toyoshima, Y., Sasaki, H., Yoshida, M., Kakita, A., . . . Wakabayashi, K. (2013). *Valosin-containing protein immunoreactivity in tauopathies, synucleinopathies, polyglutamine diseases and intranuclear inclusion body disease*. Academic Press.

Mori, H., Kondo, T., Yokochi, M., Matsumine, H., Nakagawa-Hattori, Y., Miyake, T., ... Mizuno, Y. (1998). Pathologic and biochemical studies of juvenile parkinsonism linked to chromosome 6q. *Neurology*, *51*(3), 890–892. doi:10.1212/WNL.51.3.890 PMID:9748052

Morimoto, R. I. (2019). Cell-Nonautonomous Regulation of Proteostasis in Aging and Disease. *Cold Spring Harbor Perspectives in Biology*, a034074. doi:10.1101/cshperspect.a034074 PMID:30962274

Morita, S., Villalta, S. A., Feldman, H. C., Register, A. C., Rosenthal, W., Hoffmann-Petersen, I. T., ... Papa, F. R. (2017). Targeting ABL-IRE1α Signaling Spares ER-Stressed Pancreatic β Cells to Reverse Autoimmune Diabetes. *Cell Metabolism*, *25*(4), 883–897. doi:10.1016/j.cmet.2017.03.018 PMID:28380378

Moriyasu, Y., & Inoue, Y. (2008). Chapter Thirty-Two Use of Protease Inhibitors for Detecting Autophagy in Plants. *Methods in Enzymology*, *451*, 557–580. doi:10.1016/S0076-6879(08)03232-1 PMID:19185740

Morris, M., Maeda, S., Vossel, K., & Mucke, L. (2011). The many faces of tau. *Neuron*, *70*(3), 410–426. doi:10.1016/j.neuron.2011.04.009 PMID:21555069

Morrison, B., Hensley, K., Pioro, P. E., Petri, S., & Kiaei, M. (2012). Amyotrophic Lateral Sclerosis and novel therapeutic strategies. *Neurology Research International*, *2012*, 1–3. doi:10.1155/2012/798028 PMID:23316357

Morselli, E., Marino, G., Bennetzen, M. V., Eisenberg, T., Megalou, E., Schroeder, S., ... Kroemer, G. (2011). Spermidine and resveratrol induce autophagy by distinct pathways converging on the acetylproteome. *The Journal of Cell Biology*, *192*(4), 615–629. doi:10.1083/jcb.201008167 PMID:21339330

Mosconi, L. (2005). Brain glucose metabolism in the early and specific diagnosis of Alzheimer's disease. *European Journal of Nuclear Medicine and Molecular Imaging*, *32*(4), 486–510. doi:10.100700259-005-1762-7 PMID:15747152

Mossner, E., Iwai, H., & Glockshuber, R. (2000). Influence of the pK(a) value of the buried, active-site cysteine on the redox properties of thioredoxin-like oxidoreductases. *FEBS Letters*, *477*(1-2), 21–26. doi:10.1016/S0014-5793(00)01738-5 PMID:10899304

Moussaud, S., Jones, D. R., Moussaud-Lamodière, E. L., Delenclos, M., Ross, O. A., & McLean, P. J. (2014). Alpha-synuclein and tau: Teammates in neurodegeneration? *Molecular Neurodegeneration*, *9*(1), 43. doi:10.1186/1750-1326-9-43 PMID:25352339

Mukhopadhyay, D., & Riezman, H. (2007). Proteasome-independent functions of ubiquitin in endocytosis and signaling. *Science*, *315*, 201–205.

Mukrasch, M. D., Bibow, S., Korukottu, J., Jeganathan, S., Biernat, J., Griesinger, C., ... Zweckstetter, M. (2009). Structural polymorphism of 441-residue tau at single residue resolution. *PLoS Biology*, *7*(2), e1000034. doi:10.1371/journal.pbio.1000034 PMID:19226187

Müller, R., Heinrich, M., Heck, S., Blohm, D., & Richter-Landsberg, C. (1997). Expression of microtubule-associated proteins MAP2 and tau in cultured rat brain oligodendrocytes. *Cell and Tissue Research*, *288*(2), 239–249. doi:10.1007004410050809 PMID:9082959

Munasinghe, P. E., Riu, F., Dixit, P., Edamatsu, M., Saxena, P., Hamer, N. S., ... Lamberts, R. R. (2016). Type-2 diabetes increases autophagy in the human heart through promotion of Beclin-1 mediated pathway. *International Journal of Cardiology*, *202*, 13–20. doi:10.1016/j.ijcard.2015.08.111 PMID:26386349

Munday, A. D., & López, J. A. (2007). Posttranslational protein palmitoylation: Promoting platelet purpose. *Arteriosclerosis, Thrombosis, and Vascular Biology*, *27*(7), 1496–1499. doi:10.1161/ATVBAHA.106.136226 PMID:17581830

Muntane, G., Dalfo, E., Martinez, A., & Ferrer, I. (2008). Phosphorylation of tau and α-synuclein in synaptic-enriched fractions of the frontal cortex in Alzheimer's disease, and in Parkinson's disease and related α-synucleinopathies. *Neuroscience*, *152*(4), 913–923. doi:10.1016/j.neuroscience.2008.01.030 PMID:18343584

Münz, C. (2011). Beclin-1 Targeting for Viral Immune Escape. *Viruses*, *3*(7), 1166–1178. doi:10.3390/v3071166 PMID:21994775

Murakami, T., Hino, S. I., Saito, A., & Imaizumi, K. (2007). Endoplasmic reticulum stress response in dendrites of cultured primary neurons. *Neuroscience*, *146*(1), 1–8. doi:10.1016/j.neuroscience.2007.01.069 PMID:17367944

Murphy, M. P. (2009). How mitochondria produce reactive oxygen species. *The Biochemical Journal*, *417*(1), 1–13. doi:10.1042/BJ20081386 PMID:19061483

Murphy, M. P., Holmgren, A., Larsson, N.-G., Halliwell, B., Chang, C. J., Kalyanaraman, B., ... Winterbourn, C. C. (2011). Unraveling the biological roles of reactive oxygen species. *Cell Metabolism*, *13*(4), 361–366. doi:10.1016/j.cmet.2011.03.010 PMID:21459321

Mustafa, A. K., Gadalla, M. M., Sen, N., Kim, S., Mu, W., Gazi, S. K., ... Snyder, S. H. (2009). H2S signals through protein S-sulfhydration. *Science Signaling*, *2*(96), ra72. doi:10.1126cisignal.2000464 PMID:19903941

Myeku, N., & Duff, K. E. (2018). Targeting the 26S Proteasome To Protect Against Proteotoxic Diseases. *Trends in Molecular Medicine*, *24*(1), 18–29. doi:10.1016/j.molmed.2017.11.006 PMID:29233753

Myers, V. D., McClung, J. M., Wang, J., Tahrir, F. G., Gupta, M. K., Gordon, J., ... Feldman, A. M. (2018). The multifunctional protein BAG3: A novel therapeutic target in cardiovascular disease. *JACC. Basic to Translational Science*, *3*(1), 122–131. doi:10.1016/j.jacbts.2017.09.009 PMID:29938246

Mymrikov, E. V., Seit-Nebi, A. S., & Gusev, N. B. (2011). Large potentials of small heat shock proteins. *Physiological Reviews*, *91*(4), 1123–1159. doi:10.1152/physrev.00023.2010 PMID:22013208

Nair, V. D., McNaught, K. S., Gonzalez-Maeso, J., Sealfon, S. C., & Olanow, C. W. (2006). p53 mediates nontranscriptional cell death in dopaminergic cells in response to proteasome inhibition. *The Journal of Biological Chemistry*, *281*(51), 39550–39560. doi:10.1074/jbc.M603950200 PMID:17060322

Nakagawa, T., Zhu, H., Morishima, N., Li, E., Xu, J., Yankner, B. A., & Yuan, J. (2000). Caspase-12 mediates endoplasmic-reticulum-specific apoptosis and cytotoxicity by amyloid-β. *Nature*, *403*(6765), 98–103. doi:10.1038/47513 PMID:10638761

Nakahata, Y., Kaluzova, M., Grimaldi, B., Sahar, S., Hirayama, J., Chen, D., ... Sassone-Corsi, P. (2008). The NAD+-dependent deacetylase SIRT1 modulates CLOCK-mediated chromatin remodeling and circadian control. *Cell*, *134*(2), 329–340. doi:10.1016/j.cell.2008.07.002 PMID:18662547

Nakamura, K., Nemani, V. M., Azarbal, F., Skibinski, G., Levy, J. M., Egami, K., ... Sesaki, H. (2011). Direct membrane association drives mitochondrial fission by the Parkinson disease-associated protein α-synuclein. *The Journal of Biological Chemistry*, *286*(23), 20710–20726. doi:10.1074/jbc.M110.213538 PMID:21489994

Nakamura, S., & Yoshimori, T. (2017). New insights into autophagosome-lysosome fusion. *Journal of Cell Science*, *130*(7), 1209–1216. doi:10.1242/jcs.196352 PMID:28302910

Nakamura, T., Cho, D. H., & Lipton, S. A. (2012). Redox regulation of protein misfolding, mitochondrial dysfunction, synaptic damage, and cell death in neurodegenerative diseases. *Experimental Neurology*, *238*(1), 12–21. doi:10.1016/j.expneurol.2012.06.032 PMID:22771760

Nakamura, T., & Lipton, S. A. (2007). S-Nitrosylation and uncompetitive/fast off-rate (UFO) drug therapy in neurodegenerative disorders of protein misfolding. *Cell Death and Differentiation*, *14*(7), 1305–1314. doi:10.1038j.cdd.4402138 PMID:17431424

Nakamura, T., & Lipton, S. A. (2013). Emerging role of protein–protein transnitrosylation in cell signaling pathways. *Antioxidants & Redox Signalling*, *18*(3), 239–249. doi:10.1089/ars.2012.4703 PMID:22657837

Nakamura, T., & Lipton, S. A. (2016). Protein S-Nitrosylation as a Therapeutic Target for Neurodegenerative Diseases. *Trends in Pharmacological Sciences*, *37*(1), 73–84. doi:10.1016/j.tips.2015.10.002 PMID:26707925

Nakamura, T., Tu, S., Akhtar, M. W., Sunico, C. R., Okamoto, S., & Lipton, S. A. (2013). Aberrant Protein S-nitrosylation in neurodegenerative diseases. *Neuron*, *78*(4), 596–614. doi:10.1016/j.neuron.2013.05.005 PMID:23719160

Nakatogawa, H., Suzuki, K., Kamada, Y., & Ohsumi, Y. (2009). Dynamics and diversity in autophagy mechanisms: Lessons from yeast. *Nature Reviews. Molecular Cell Biology*, *10*(7), 458–467. doi:10.1038/nrm2708 PMID:19491929

Nakayama, K. I., & Nakayama, K. (2006). Ubiquitin ligases: Cell-cycle control and cancer. *Nature Reviews. Cancer*, *6*(5), 369–381. doi:10.1038/nrc1881 PMID:16633365

Nalbandian, A., Llewellyn, K. J., Badadani, M., Yin, H. Z., Nguyen, C., Katheria, V., . . . Caiozzo, V. (2013). *A progressive translational mouse model of human valosin-containing protein disease: The VCPR155H/+ mouse*. Academic Press.

Nalbandian, A., Llewellyn, K. J., Kitazawa, M., Yin, H. Z., Badadani, M., Khanlou, N., . . . Mozaffar, T. (2012). *The homozygote VCPR155H/R155H mouse model exhibits accelerated human VCP-associated disease pathology*. Academic Press.

Nalls, M. A., Pankratz, N., Lill, C. M., Do, C. B., Hernandez, D. G., Saad, M., ... Schulte, C. (2014). Large-scale meta-analysis of genome-wide association data identifies six new risk loci for Parkinson's disease. *Nature Genetics*, *46*(9), 989–993. doi:10.1038/ng.3043 PMID:25064009

Narayanan, R. (2015). Neurodegenerative diseases: Phenome to genome analysis. *MOJ Proteomics & Bioinformatics*, *2*(1), 1–8.

Narendra, D. P., Jin, S. M., Tanaka, A., Suen, D. F., Gautier, C. A., Shen, J., ... Youle, R. J. (2010). PINK1 is selectively stabilized on impaired mitochondria to activate Parkin. *PLoS Biology*, *8*(1), e1000298. doi:10.1371/journal.pbio.1000298 PMID:20126261

Narendra, D., Kane, L. A., Hauser, D. N., Fearnley, I. M., & Youle, R. J. (2010). p62/SQSTM1 is required for Parkin-induced mitochondrial clustering but not mitophagy; VDAC1 is dispensable for both. *Autophagy*, *6*(8), 1090–1106. doi:10.4161/auto.6.8.13426 PMID:20890124

Narendra, D., Tanaka, A., Suen, D. F., & Youle, R. J. (2008). Parkin is recruited selectively to impaired mitochondria and promotes their autophagy. *The Journal of Cell Biology*, *183*(5), 795–803. doi:10.1083/jcb.200809125 PMID:19029340

Nascimento-Ferreira, I., Santos-Ferreira, T., Sousa-Ferreira, L., Auregan, G., Onofre, I., Alves, S., ... Pereira de Almeida, L. (2011). Overexpression of the autophagic beclin-1 protein clears mutant ataxin-3 and alleviates Machado–Joseph disease. *Brain*, *134*(5), 1400–1415. doi:10.1093/brain/awr047 PMID:21478185

Nasir, J., Floresco, S. B., O'Kusky, J. R., Diewert, V. M., Richman, J. M., Zeisler, J., . . . Hayden, M. R. (1995). *Targeted disruption of the Huntington's disease gene results in embryonic lethality and behavioral and morphological changes in heterozygotes*. Academic Press.

Nassif, M., Valenzuela, V., Rojas-Rivera, D., Vidal, R., Matus, S., Castillo, K., ... Hetz, C. (2014). Pathogenic role of BECN1/Beclin 1 in the development of amyotrophic lateral sclerosis. *Autophagy*, *10*(7), 1256–1271. doi:10.4161/auto.28784 PMID:24905722

National Institutes of Health. National Institute on Aging. (n.d.). *What happens to the Brain in the Alzheimer's Disease?* Available at https://www.nia.nih.gov/health/what-happens-brainalzheimers-disease

National Research Council, Division on Earth and Life Studies, Board on Biology, Commission on Life Sciences, Committee on Research Opportunities in Biology. (1989). The nervous system and behavior. In Opportunities in Biology (pp. 175-223). Washington, DC: National Academies Press.

Navarro, A., & Boveris, A. (2004). Rat brain and liver mitochondria develop oxidative stress and lose enzymatic activities on aging. *American Journal of Physiology. Regulatory, Integrative and Comparative Physiology*, *287*(5), R1244–R1249. doi:10.1152/ajpregu.00226.2004 PMID:15271654

Nave, B. T., Ouwens, D. M., Withers, D. J., Alessi, D. R., & Shepherd, P. R. (1999). Mammalian target of rapamycin is a direct target for protein kinase B: Identification of a convergence point for opposing effects of insulin and amino-acid deficiency on protein translation. *The Biochemical Journal*, *344*(2), 427–431. doi:10.1042/bj3440427 PMID:10567225

Navone, F., Genevini, P., & Borgese, N. (2015). Autophagy and Neurodegeneration: Insights from a Cultured Cell Model of ALS. *Cells-Basel*, *4*(3), 354–386. doi:10.3390/cells4030354 PMID:26287246

Nedić, O., Rattan, S., Grune, T., & Trougakos, I. (2013). Molecular effects of advanced glycation end products on cell signalling pathways, ageing and pathophysiology. *Free Radical Research*, *47*(sup1), 28–38. doi:10.3109/10715762.2013.806798 PMID:23692178

Nedić, O., Rogowska-Wrzesinska, A., & Rattan, S. I. S. (2015). Standardization and quality control in quantifying non-enzymatic oxidative protein modifications in relation to ageing and disease: Why is it important and why is it hard? *Redox Biology*, *5*, 91–100. doi:10.1016/j.redox.2015.04.001 PMID:25909343

Nestler, E.J., & Greengard, P. (1999). *Protein Phosphorylation is of Fundamental Importance in Biological Regulation.* Academic Press.

Netto, L. E., de Oliveira, M. A., Monteiro, G., Demasi, A. P., Cussiol, J. R., Discola, K. F., ... Horta, B. B. (2007). Reactive cysteine in proteins: Protein folding, antioxidant defense, redox signaling and more. *Comparative Biochemistry and Physiology. Toxicology & Pharmacology: CBP*, *146*(1-2), 180–193. doi:10.1016/j.cbpc.2006.07.014 PMID:17045551

Newcombe, E., Ramdzan, Y., Sethi, A., Lee, M., Loo, D., Graham, B., ... Hatters, D. (2015). Understanding the biology of Huntington's disease via the pathogenic huntingtin monomer. *Protein Science*, *24*, 86–87.

Newton, T. M., Duce, J. A., & Bayle, E. D. (2019). The proteostasis network provides targets for neurodegeneration. *British Journal of Pharmacology*, *176*(18), 3508–3514. doi:10.1111/bph.14643 PMID:30820936

Nguyen, D. K. H., Thombre, R., & Wang, J. (2019). Autophagy as a common pathway in amyotrophic lateral sclerosis. *Neuroscience Letters*, *697*, 34–48. doi:10.1016/j.neulet.2018.04.006 PMID:29626651

Nigar, S., Pottoo, F. H., Tabassum, N., Verma, S. K., & Javed, M. N. (2016). Molecular Insights into the Role of Inflammation and Oxidative Stress in Epilepsy. *Journal of Advances in Medical and Pharmaceutical Sciences*, *10*(1), 1–9. doi:10.9734/JAMPS/2016/24441

Ni, H. M., Williams, J. A., & Ding, W. X. (2015). Mitochondrial dynamics and mitochondrial quality control. *Redox Biology*, *4*, 6–13. doi:10.1016/j.redox.2014.11.006 PMID:25479550

Nijholt, D. A. T., De Graaf, T. R., Van Haastert, E. S., Oliveira, A. O., Berkers, C. R., Zwart, R., ... Scheper, W. (2011). Endoplasmic reticulum stress activates autophagy but not the proteasome in neuronal cells: Implications for Alzheimer's disease. *Cell Death and Differentiation*, *18*(6), 1071–1081. doi:10.1038/cdd.2010.176 PMID:21252911

Nijholt, D. A. T., van Haastert, E. S., Rozemuller, A. J. M., Scheper, W., & Hoozemans, J. J. M. (2012). The unfolded protein response is associated with early tau pathology in the hippocampus of tauopathies. *The Journal of Pathology*, *226*(5), 693–702. doi:10.1002/path.3969 PMID:22102449

Nishimura, A. L., Mitne-Neto, M., Silva, H. C., Richieri-Costa, A., Middleton, S., Cascio, D., ... Zatz, M. (2004). A mutation in the vesicle-trafficking protein VAPB causes late-onset spinal muscular atrophy and amyotrophic lateral sclerosis. *American Journal of Human Genetics*, *75*(5), 822–831. doi:10.1086/425287 PMID:15372378

Niu, J., Yu, M., Wang, C., & Xu, Z. (2012). Leucine-rich repeat kinase 2 disturbs mitochondrial dynamics via Dynamin-like protein. *Journal of Neurochemistry*, *122*(3), 650–658. doi:10.1111/j.1471-4159.2012.07809.x PMID:22639965

Niu, T.-K., Cheng, Y., Ren, X., & Yang, J.-M. (2010). Interaction of Beclin 1 with survivin regulates sensitivity of human glioma cells to TRAIL-induced apoptosis. *FEBS Letters*, *584*(16), 3519–3524. doi:10.1016/j.febslet.2010.07.018 PMID:20638385

Niu, Y. J., Nie, Z. W., Shin, K. T., Zhou, W., & Cui, X. S. (2019). PINK1 regulates mitochondrial morphology via promoting mitochondrial fission in porcine preimplantation embryos. *The FASEB Journal*, *33*(7), 7882–7895. doi:10.1096/fj.201802473R PMID:30897005

Nixon, R. A. (2013). *The role of autophagy in neurodegenerative disease*. Academic Press.

Nixon, R. A., Yang, D.-S., & Lee, J.-H. (2008). *Neurodegenerative lysosomal disorders: a continuum from development to late age*. Academic Press.

Nixon, R. A. (2007). Autophagy, amyloidogenesis and Alzheimer disease. *Journal of Cell Science*, *120*(23), 4081–4091. doi:10.1242/jcs.019265 PMID:18032783

Nixon, R. A. (2013). The role of autophagy in neurodegenerative disease. *Nature Medicine*, *19*(8), 983–997. doi:10.1038/nm.3232 PMID:23921753

Nixon, R. A., Wegiel, J., Kumar, A., Yu, W. H., Peterhoff, C., Cataldo, A., & Cuervo, A. M. (2005). Extensive involvement of autophagy in Alzheimer disease: An immuno-electron microscopy study. *Journal of Neuropathology and Experimental Neurology*, *64*(2), 113–122. doi:10.1093/jnen/64.2.113 PMID:15751225

Nixon, R. A., & Yang, D. (2011). Autophagy failure in Alzheimer ' s disease-locating the primary defect. *Neurobiology of Disease*, *43*(1), 38–45. doi:10.1016/j.nbd.2011.01.021 PMID:21296668

Nomura, J., Hosoi, T., Kaneko, M., Ozawa, K., Nishi, A., & Nomura, Y. (2016). Neuroprotection by Endoplasmic Reticulum Stress-Induced HRD1 and Chaperones: Possible Therapeutic Targets for Alzheimer's and Parkinson's Disease. *Medical Sciences: Open Access Journal*, *4*(3), 4. doi:10.3390/medsci4030014 PMID:29083378

Noormohammadi, A., Calculli, G., Gutierrez-Garcia, R., Khodakarami, A., Koyuncu, S., & Vilchez, D. (2018). Mechanisms of protein homeostasis (proteostasis) maintain stem cell identity in mammalian pluripotent stem cells. *Cellular and Molecular Life Sciences*, *75*(2), 275–290. doi:10.100700018-017-2602-1 PMID:28748323

Norfray, J. F., & Provenzale, J. M. (2004). Alzheimer's Disease: Neuropathologic Findings and Recent Advances in Imaging. *AJR. American Journal of Roentgenology*, *182*(1), 3–13. doi:10.2214/ajr.182.1.1820003 PMID:14684506

North, B. J., Marshall, B. L., Borra, M. T., Denu, J. M., & Verdin, E. (2003). The human Sir2 ortholog, SIRT2, is an NAD+-dependent tubulin deacetylase. *Molecular Cell*, *11*(2), 437–444. doi:10.1016/S1097-2765(03)00038-8 PMID:12620231

Nowacek, A., Kosloski, L. M., & Gendelman, H. E. (2009). Neurodegenerative disorders and nanoformulated drug development. *Nanomedicine (London)*, *4*(5), 541–555. doi:10.2217/nnm.09.37 PMID:19572820

Nübling, G., Bader, B., Levin, J., Hildebrandt, J., Kretzschmar, H., & Giese, A. (2012). Synergistic influence of phosphorylation and metal ions on tau oligomer formation and coaggregation with α-synuclein at the single molecule level. *Molecular Neurodegeneration*, *7*(1), 35. doi:10.1186/1750-1326-7-35 PMID:22824345

Numajiri, N., Takasawa, K., Nishiya, T., Tanaka, H., Ohno, K., Hayakawa, W., ... Uehara, T. (2011). On-off system for PI3-kinase-Akt signaling through S-nitrosylation of phosphatase with sequence homology to tensin (PTEN). *Proceedings of the National Academy of Sciences of the United States of America*, *108*(25), 10349–10354. doi:10.1073/pnas.1103503108 PMID:21646525

Nunomura, A., Moreira, P. I., Lee, H. G., Zhu, X., Castellani, R. J., Smith, M. A., & Perry, G. (2007). Neuronal death and survival under oxidative stress in Alzheimer and Parkinson diseases. *CNS & Neurological Disorders - Drug Targets, 6*(6), 411–423. doi:10.2174/187152707783399201 PMID:18220780

O'Brien, R. J., & Wong, P. C. (2011). Amyloid precursor protein processing and Alzheimer's disease. *Annual Review of Neuroscience, 34*(1), 185–204. doi:10.1146/annurev-neuro-061010-113613 PMID:21456963

O'Connor, T., Sadleir, K. R., Maus, E., Velliquette, R. A., Zhao, J., Cole, S. L., ... Vassar, R. (2008). Phosphorylation of the translation initiation factor eIF2alpha increases BACE1 levels and promotes amyloidogenesis. *Neuron, 60*(6), 988–1009. doi:10.1016/j.neuron.2008.10.047 PMID:19109907

O'Leary, J. C. III, Li, Q., Marinec, P., Blair, L. J., Congdon, E. E., Johnson, A. G., ... Dickey, C. A. (2010). Phenothiazine-mediated rescue of cognition in tau transgenic mice requires neuroprotection and reduced soluble tau burden. *Molecular Neurodegeneration, 5*(1), 45. doi:10.1186/1750-1326-5-45 PMID:21040568

O95817. (n.d.). *3D model from MODBASE based on 1uk5A – Protein Model Portal.* Retrieved from https://www.proteinmodelportal.org/?pid=modelDetail&provider=MODBASE&template=1uk5A&pmpuid=1001228397804&range_from=1&range_to=575&ref_ac=O95817&mapped_ac=O95817&zid=async

Oaks, A. W., Frankfurt, M., Finkelstein, D. I., & Sidhu, A. (2013). Age-dependent effects of A53T alpha-synuclein on behavior and dopaminergic function. *PLoS One, 8*(4), e60378. doi:10.1371/journal.pone.0060378 PMID:23560093

Obara, K., & Ohsumi, Y. (2008). Dynamics and function of PtdIns (3) P in autophagy. *Autophagy, 4*(7), 952–954. doi:10.4161/auto.6790 PMID:18769109

Obeso, J. A., Marin, C., Rodriguez-Oroz, C., Blesa, J., Benitez-Temino, B., Mena-Segovia, J., ... Olanow, C. W. (2008). The basal ganglia in Parkinson's disease: Current concepts and unexplained observations. *Annals of Neurology, 64*(S2Suppl. 2), S30–S46. doi:10.1002/ana.21481 PMID:19127584

Ochalek, A., Mihalik, B., Avci, H. X., Chandrasekaran, A., Teglasi, A., Bock, I., ... Dinnyés, A. (2017). Neurons derived from sporadic Alzheimer's disease iPSCs reveal elevated TAU hyperphosphorylation, increased amyloid levels, and GSK3B activation. *Alzheimer's Research & Therapy, 9*(1), 90. doi:10.118613195-017-0317-z PMID:29191219

Ogata, M., Hino, S.-i., Saito, A., Morikawa, K., Kondo, S., Kanemoto, S., ... Imaizumi, K. (2006). Autophagy Is Activated for Cell Survival after Endoplasmic Reticulum Stress. *Molecular and Cellular Biology, 26*(24), 9220–9231. doi:10.1128/MCB.01453-06 PMID:17030611

Ohsumi, Y. (2001). Molecular dissection of autophagy: Two ubiquitin-like systems. *Nature Reviews. Molecular Cell Biology, 2*(3), 211–216. doi:10.1038/35056522 PMID:11265251

Ohsumi, Y., & Mizushima, N. (2004). Two ubiquitin-like conjugation systems essential for autophagy. *Seminars in Cell & Developmental Biology, 2*(15), 231–236. doi:10.1016/j.semcdb.2003.12.004 PMID:15209383

Okado-Matsumoto, A., & Fridovich, I. (2002). Amyotrophic lateral sclerosis: A proposed mechanism. *Proceedings of the National Academy of Sciences of the United States of America, 99*(13), 9010–9014. doi:10.1073/pnas.132260399 PMID:12060716

Okamoto, A., Tanaka, M., Sumi, C., Oku, K., Kusunoki, M., Nishi, K., ... Hirota, K. (2016). The antioxidant N-acetyl cysteine suppresses lidocaine-induced intracellular reactive oxygen species production and cell death in neuronal SH-SY5Y cells. *BMC Anesthesiology, 16*(104), 1–12. doi:10.118612871-016-0273-3 PMID:27776485

Okamoto, K., Kondo-Okamoto, N., & Ohsumi, Y. (2009). A landmark protein essential for mitophagy: Atg32 recruits the autophagic machinery to mitochondria. *Autophagy, 5*(8), 1203–1205. doi:10.4161/auto.5.8.9830 PMID:19770589

Okamoto, K., Kondo-Okamoto, N., & Ohsumi, Y. (2009). Mitochondria-anchored receptor Atg32 mediates degradation of mitochondria via selective autophagy. *Developmental Cell, 17*(1), 87–97. doi:10.1016/j.devcel.2009.06.013 PMID:19619494

Okatsu, K., Saisho, K., Shimanuki, M., Nakada, K., Shitara, H., Sou, Y., ... Matsuda, N. (2010). p62/SQSTM1 cooperates with Parkin for perinuclear clustering of depolarized mitochondria. *Genes to Cells, 15*, 887–900. doi:10.1111/j.1365-2443.2010.01426.x PMID:20604804

Okatsu, K., Sato, Y., Yamano, K., Matsuda, N., Negishi, L., Takahashi, A., ... Fukai, S. (2018). Structural insights into ubiquitin phosphorylation by PINK1. *Scientific Reports, 8*(1), 10382. doi:10.103841598-018-28656-8 PMID:29991771

Olfati, N., Shoeibi, A., & Litvan, I. (2019). Progress in the treatment of Parkinson-Plus syndromes. *Parkinsonism & Related Disorders, 59*, 101–110. doi:10.1016/j.parkreldis.2018.10.006 PMID:30314846

Oliva, O., Réz, G., Pálfia, Z., & Fellinger, E. (1992). Dynamics of vinblastine-induced autophagocytosis in murine pancreatic acinar cells: Influence of cycloheximide post-treatments. *Experimental and Molecular Pathology, 56*(1), 76–86. doi:10.1016/0014-4800(92)90025-7 PMID:1547871

Omenn, G. S., Lane, L., Lundberg, E. K., Overall, C. M., & Deutsch, E. W. (2017). Progress on the HUPO Draft Human Proteome: 2017 Metrics of the Human Proteome Project. *Journal of Proteome Research, 16*(12), 4281–4287. doi:10.1021/acs.jproteome.7b00375 PMID:28853897

Oñate, M., Catenaccio, A., Martínez, G., Armentano, D., Parsons, G., Kerr, B., ... Court, F. A. (2016). Activation of the unfolded protein response promotes axonal regeneration after peripheral nerve injury. *Scientific Reports, 6*(1), 21709. doi:10.1038rep21709 PMID:26906090

O'Neill, C., Kiely, A. P., Coakley, M. F., Manning, S., & Long-Smith, C. M. (2012). *Insulin and IGF-1 signalling: longevity, protein homoeostasis and Alzheimer's disease*. Academic Press.

Onyango, P., Celic, I., McCaffery, J. M., Boeke, J. D., & Feinberg, A. P. (2002). SIRT3, a human SIR2 homologue, is an NAD-dependent deacetylase localized to mitochondria. *Proceedings of the National Academy of Sciences of the United States of America, 99*(21), 13653–13658. doi:10.1073/pnas.222538099 PMID:12374852

Ordureau, A., Sarraf, S. A., Duda, D. M., Heo, J. M., Jedrychowski, M. P., Sviderskiy, V. O., ... Wells, J. A. (2014). Quantitative proteomics reveal a feedforward mechanism for mitochondrial PARKIN translocation and ubiquitin chain synthesis. *Molecular Cell, 56*(3), 360–375. doi:10.1016/j.molcel.2014.09.007 PMID:25284222

Orvedahl, A., Sumpter, R. Jr, Xiao, G., Ng, A., Zou, Z., Tang, Y., ... Forst, C. V. (2011). Image-based genome-wide siRNA screen identifies selective autophagy factors. *Nature, 480*(7375), 113–117. doi:10.1038/nature10546 PMID:22020285

Osafo, N., Obiri, D. D., Yeboah, O. K., & Minkah, P. A. (2019). Amyotrophic Lateral Sclerosis: A Predominant Form of Degenerative Disease of the Motor Neuron System. In M. Uddin & M. Amran (Eds.), *Handbook of Research on Critical Examinations of Neurodegenerative Disorders* (pp. 322–346). Hershey, PA: IGI Global. doi:10.4018/978-1-5225-5282-6.ch015

Oskarsson, M. E., Paulsson, J. F., Schultz, S. W., Ingelsson, M., Westermark, P., & Westermark, G. T. (2015). In vivo seeding and cross-seeding of localized amyloidosis: A molecular link between type 2 diabetes and Alzheimer disease. *American Journal of Pathology, 185*(3), 834–846. doi:10.1016/j.ajpath.2014.11.016 PMID:25700985

Otte, L., Wiedemann, U., Schlegel, B., Pires, J. R., Beyermann, M., Schmieder, P., ... Oschkinat, H. (2003). WW domain sequence activity relationships identified using ligand recognition propensities of 42 WW domains. *Protein Science, 12*(3), 491–500. doi:10.1110/ps.0233203 PMID:12592019

Outeiro, T. F., Grammatopoulos, T. N., Altmann, S., Amore, A., Standaert, D. G., Hyman, B. T., & Kazantsev, A. G. (2007). Pharmacological inhibition of PARP-1 reduces α-synuclein-and MPP+-induced cytotoxicity in Parkinson's disease in vitro models. *Biochemical and Biophysical Research Communications, 357*(3), 596–602. doi:10.1016/j.bbrc.2007.03.163 PMID:17449015

Outeiro, T. F., Marques, O., & Kazantsev, A. (2008). Therapeutic role of sirtuins in neurodegenerative disease. *Biochimica et Biophysica Acta, 1782*(6), 363–369. doi:10.1016/j.bbadis.2008.02.010 PMID:18373985

Owen, I., & Shewmaker, F. (2019). The Role of Post-Translational Modifications in the Phase Transitions of Intrinsically Disordered Proteins. *International Journal of Molecular Sciences, 20*(21), 20. doi:10.3390/ijms20215501 PMID:31694155

Ow, Y. P., Green, D. R., Hao, Z., & Mark, T. W. (2008). Cytochrome c: Functions beyond respiration. *Nature Reviews. Molecular Cell Biology, 9*(7), 532–542. doi:10.1038/nrm2434 PMID:18568041

Ozcelik, S., Fraser, G., Castets, P., Schaeffer, V., Skachokova, Z., Breu, K., ... Tolnay, M. (2013). Rapamycin attenuates the progression of tau pathology in P301S tau transgenic mice. *PLoS One, 8*(5), e62459. doi:10.1371/journal.pone.0062459 PMID:23667480

Oz-Levi, D., Ben-Zeev, B., Ruzzo, E. K., Hitomi, Y., Gelman, A., Pelak, K., ... Alkelai, A. (2012). Mutation in TECPR2 reveals a role for autophagy in hereditary spastic paraparesis. *American Journal of Human Genetics, 91*(6), 1065–1072. doi:10.1016/j.ajhg.2012.09.015 PMID:23176824

Pacher, P. J., Beckman, S., & Liaudet, L. (2007). Nitric oxide and peroxynitrite in health and disease. *Physiological Reviews, 87*(1), 315–424. doi:10.1152/physrev.00029.2006 PMID:17237348

Pagano, G., Ferrara, N., Brooks, D. J., & Pavese, N. (2016). Age at onset and Parkinson disease phenotype. *Neurology, 86*(15), 1400–1407. doi:10.1212/WNL.0000000000002461 PMID:26865518

Paget, M. S., & Buttner, M. J. (2003). Thiol-based regulatory switches. *Annual Review of Genetics, 37*(1), 91–121. doi:10.1146/annurev.genet.37.110801.142538 PMID:14616057

Paleologou, K. E., Oueslati, A., Shakked, G., Rospigliosi, C. C., Kim, H.-Y., Lamberto, G. R., ... Lashuel, H. A. (2010). Phosphorylation at S87 is enhanced in synucleinopathies, inhibits α-synuclein oligomerization, and influences synuclein-membrane interactions. *The Journal of Neuroscience: The Official Journal of the Society for Neuroscience, 30*(9), 3184–3198. doi:10.1523/JNEUROSCI.5922-09.2010 PMID:20203178

Palikaras, K., Daskalaki, I., Markaki, M., & Tavernarakis, N. (2017). Mitophagy and age-related pathologies: Development of new therapeutics by targeting mitochondrial turnover. *Pharmacology & Therapeutics, 178*, 157–174. doi:10.1016/j.pharmthera.2017.04.005 PMID:28461251

Pallas-Bazarra, N., Jurado-Arjona, J., Navarrete, M., Esteban, J. A., Hernández, F., Ávila, J., & Llorens-Martín, M. (2016). Novel function of Tau in regulating the effects of external stimuli on adult hippocampal neurogenesis. *The EMBO Journal, 35*(13), 1417–1436. doi:10.15252/embj.201593518 PMID:27198172

Pallauf, K., & Rimbach, G. (2013). Autophagy, polyphenols and healthy ageing. *Ageing Research Reviews, 12*(1), 237–252. doi:10.1016/j.arr.2012.03.008 PMID:22504405

Panda, P. K., Fahrner, A., Vats, S., Seranova, E., Sharma, V., Chipara, M., ... Sarkar, S. (2019). Chemical screening approaches enabling drug discovery of autophagy modulators for biomedical applications in human diseases. *Frontiers in Cell and Developmental Biology, 7*, 7. doi:10.3389/fcell.2019.00038 PMID:30949479

Papasozomenos, S. C., & Binder, L. I. (1987). Phosphorylation determines two distinct species of tau in the central nervous system. *Cell Motility and the Cytoskeleton, 8*(3), 210–226. doi:10.1002/cm.970080303 PMID:2446784

Paper, W. (2011). Upgrading Analog Video Surveillance with IP Technology Panasonic Video surveillance systems. doi:10.1002/elps.201000027.Mass

Parfitt, D. A., Michael, G. J., Vermeulen, E. G., Prodromou, N. V., Webb, T. R., Gallo, J. M., ... Chapple, J. P. (2009). The ataxia protein sacsin is a functional co-chaperone that protects against polyglutamine-expanded ataxin-1. *Human Molecular Genetics*, *18*(9), 1556–1565. doi:10.1093/hmg/ddp067 PMID:19208651

Parihar, M. S., Parihar, A., Fujita, M., Hashimoto, M., & Ghafourifar, P. (2009). Alpha-synuclein overexpression and aggregation exacerbates impairment of mitochondrial functions by augmenting oxidative stress in human neuroblastoma cells. *The International Journal of Biochemistry & Cell Biology*, *41*(10), 2015–2024. doi:10.1016/j.biocel.2009.05.008 PMID:19460457

Park, J. W., Mieya, J. J., Rhee, S. G., & Chock, P. B. (2009). Deglutathionylation of 2-Cys peroxiredoxin is specifically catalyzed by sulfiredoxin. *The Journal of Biological Chemistry*, *284*(35), 23364–23374. doi:10.1074/jbc.M109.021394 PMID:19561357

Park, S. W., & Ozcan, U. (2013). Potential for therapeutic manipulation of the UPR in disease. *Seminars in Immunopathology*, *35*(3), 351–373. doi:10.100700281-013-0370-z PMID:23572207

Park, W. S., Do Heo, W., Whalen, J. H., O'Rourke, N. A., Bryan, H. M., Meyer, T., & Teruel, M. N. (2008). Comprehensive identification of PIP3-regulated PH domains from C. elegans to H. sapiens by model prediction and live imaging. *Molecular Cell*, *30*(3), 381–392. doi:10.1016/j.molcel.2008.04.008 PMID:18471983

Paschen, W., & Mengesdorf, T. (2005). Cellular abnormalities linked to endoplasmic reticulum dysfunction in cerebrovascular disease-therapeutic potential. *Pharmacology & Therapeutics*, *108*(3), 362–375. doi:10.1016/j.pharmthera.2005.05.008 PMID:16140387

Patel, A., Lee, H. O., Jawerth, L., Maharana, S., Jahnel, M., Hein, M. Y., ... Alberti, S. (2015). A Liquid-to-Solid Phase Transition of the ALS Protein FUS Accelerated by Disease Mutation. *Cell*, *162*(5), 1066–1077. doi:10.1016/j.cell.2015.07.047 PMID:26317470

Patel, P., Prescott, G. R., Burgoyne, R. D., Lian, L. Y., & Morgan, A. (2016). Phosphorylation of Cysteine String Protein Triggers a Major Conformational Switch. *Struc*, *24*(8), 1380–1386. doi:10.1016/j.str.2016.06.009 PMID:27452402

Patel, Y. J., Payne Smith, M. D., de Belleroche, J., & Latchman, D. S. (2005). Hsp27 and Hsp70 administered in combination have a potent protective effect against FALS-associated SOD1-mutant-induced cell death in mammalian neuronal cells. *Brain Research. Molecular Brain Research*, *134*(2), 256–274. doi:10.1016/j.molbrainres.2004.10.028 PMID:15836922

Paterna, J. C., Leng, A., Weber, E., Feldon, J., & Büeler, H. (2007). DJ-1 and Parkin modulate dopamine-dependent behavior and inhibit MPTP-induced nigral dopamine neuron loss in mice. *Molecular Therapy*, *15*(4), 698–704. doi:10.1038j.mt.6300067 PMID:17299411

Patil, C., & Walter, P. (2001). Intracellular signaling from the endoplasmic reticulum to the nucleus: The unfolded protein response in yeast and mammals. *Current Opinion in Cell Biology*, *13*(3), 349–355. doi:10.1016/S0955-0674(00)00219-2 PMID:11343907

Patterson, B. W., Elbert, D. L., Mawuenyega, K. G., Kasten, T., Ovod, V., Ma, S., ... Zhang, L. (2015). Age and amyloid effects on human central nervous system amyloid-beta kinetics. *Annals of Neurology*, *78*(3), 439–453. doi:10.1002/ana.24454 PMID:26040676

Pattingre, S., Tassa, A., Qu, X., Garuti, R., Liang, X. H., Mizushima, N., ... Levine, B. (2005). Bcl-2 antiapoptotic proteins inhibit Beclin 1-dependent autophagy. *Cell*, *122*(6), 927–939. doi:10.1016/j.cell.2005.07.002 PMID:16179260

Paul, B. D., Sbodio, J. I., & Snyder, S. H. (2018). Cysteine Metabolism in Neuronal Redox Homeostasis. *Trends in Pharmacological Sciences*, *2*(7), 1–12. PMID:29530337

Paul, B. D., Sbodio, J. I., Vandier, M. S., Cha, J. Y., Snowman, A. M., & Snyder, S. H. (2014). Cystathionine c-lyase deficiency mediates neurodegeneration in Huntington's disease. *Nat*, *509*(7498), 96–103. doi:10.1038/nature13136 PMID:24670645

Paulsen, C. E., & Carroll, K. S. (2010). Orchestrating redox signaling networks through regulatory cysteine switches. *ACS Chemical Biology*, *5*(1), 47–62. doi:10.1021/cb900258z PMID:19957967

Paxinou, E., Chen, Q., Weisse, M., Giasson, B. I., Norris, E. H., Rueter, S. M., ... Ischiropoulos, H. (2001). Induction of α-synuclein aggregation by intracellular nitrative insult. *The Journal of Neuroscience: The Official Journal of the Society for Neuroscience*, *21*(20), 8053–8061. doi:10.1523/JNEUROSCI.21-20-08053.2001 PMID:11588178

Pei, J. J., Tanaka, T., Tung, Y. C., Braak, E. V. A., Iqbal, K., & Grundke-Iqbal, I. (1997). Distribution, levels, and activity of glycogen synthase kinase-3 in the Alzheimer disease brain. *Journal of Neuropathology and Experimental Neurology*, *56*(1), 70–78. doi:10.1097/00005072-199701000-00007 PMID:8990130

Pekna, M., & Pekny, M. (2012). The neurobiology of brain injury. *Cerebrum*, *2012*, 9. PMID:23447795

Perez-Nievas, B. G., Stein, T. D., Tai, H. C., Dols-Icardo, O., Scotton, T. C., Barroeta-Espar, I., ... Serrano-Pozo, A. (2013). Dissecting phenotypic traits linked to human resilience to Alzheimer's pathology. *Brain*, *136*(8), 2510–2526. doi:10.1093/brain/awt171 PMID:23824488

Pérez-Revuelta, B. I., Hettich, M. M., Ciociaro, A., Rotermund, C., Kahle, P. J., Krauss, S., & Di Monte, D. A. (2014). Metformin lowers Ser-129 phosphorylated α-synuclein levels via mTOR-dependent protein phosphatase 2A activation. *Cell Death & Disease*, *5*(5), e1209. doi:10.1038/cddis.2014.175 PMID:24810045

Perez, S. E., He, B., Nadeem, M., Wuu, J., Scheff, S. W., Abrahamson, E. E., ... Mufson, E. J. (2015). Resilience of precuneus neurotrophic signaling pathways despite amyloid pathology in prodromal Alzheimer's disease. *Biological Psychiatry*, *77*(8), 693–703. doi:10.1016/j.biopsych.2013.12.016 PMID:24529280

Perkins, A., Nelson, K. J., Parsonage, D., Poole, L. B., & Karplus, P. A. (2015). Peroxiredoxins: Guardians against oxidative stress and modulators of peroxide signaling. *Trends in Biochemical Sciences*, *40*(8), 435–445. doi:10.1016/j.tibs.2015.05.001 PMID:26067716

Perl, D. (2010). Neuropathology of Alzheimer's Disease. *The Mount Sinai Journal of Medicine, New York*, *77*(1), 32–42. doi:10.1002/msj.20157 PMID:20101720

Perry, G., Friedman, R., Shaw, G., & Chau, V. (1987). Ubiquitin is detected in neurofibrillary tangles and senile plaque neurites of Alzheimer disease brains. *Proceedings of the National Academy of Sciences of the United States of America*, *84*(9), 3033–3036. doi:10.1073/pnas.84.9.3033 PMID:3033674

Peterson, R. T., Desai, B. N., Hardwick, J. S., & Schreiber, S. L. (1999). Protein phosphatase 2A interacts with the 70-kDa S6 kinase and is activated by inhibition of FKBP12–rapamycinassociated protein. *Proceedings of the National Academy of Sciences of the United States of America*, *96*(8), 4438–4442. doi:10.1073/pnas.96.8.4438 PMID:10200280

Petrucelli, L., Dickson, D., Kehoe, K., Taylor, J., Snyder, H., Grover, A., & ... . (2004). CHIP and Hsp70 regulate tau ubiquitination, degradation and aggregation. *Human Molecular Genetics*, *13*(7), 703–714. doi:10.1093/hmg/ddh083 PMID:14962978

Piacentini, M., Grazia Farrace, M., Piredda, L., Matarrese, P., Ciccosanti, F., Falasca, L., . . . Griffin, M. (2002). *Transglutaminase overexpression sensitizes neuronal cell lines to apoptosis by increasing mitochondrial membrane potential and cellular oxidative stress*. Academic Press.

Picillo, M., Lizarraga, K. J., Friesen, E. L., Chau, H., Zhang, M., Sato, C., & Kalia, S. K. (2018). Parkinsonism due to A53E α-synuclein gene mutation: Clinical, genetic, epigenetic, and biochemical features. *Movement Disorders*, *33*(12), 1950–1955. doi:10.1002/mds.27506 PMID:30423204

Pickford, F., Masliah, E., Britschgi, M., Lucin, K., Narasimhan, R., Jaeger, P. A., ... Wyss-Coray, T. (2008). The autophagy-related protein beclin 1 shows reduced expression in early Alzheimer disease and regulates amyloid β accumulation in mice. *The Journal of Clinical Investigation*, *118*(6), 2190–2199. doi:10.1172/JCI33585 PMID:18497889

Pickrell, A. M., & Youle, R. J. (2015). The roles of PINK1, parkin, and mitochondrial fidelity in Parkinson's disease. *Neuron*, *85*(2), 257–273. doi:10.1016/j.neuron.2014.12.007 PMID:25611507

Pierrot, N., Ghisdal, P., Caumont, A. S., & Octave, J. N. (2004). Intraneuronal amyloid-β1-42 production triggered by sustained increase of cytosolic calcium concentration induces neuronal death. *Journal of Neurochemistry*, *88*(5), 1140–1150. doi:10.1046/j.1471-4159.2003.02227.x PMID:15009669

Pietrobono, R., Tabolacci, E., Zalfa, F., Zito, I., Terracciano, A., Moscato, U., ... Neri, G. (2005). Molecular dissection of the events leading to inactivation of the FMR1 gene. *Human Molecular Genetics*, *14*(2), 267–277. doi:10.1093/hmg/ddi024 PMID:15563507

Piras, A., Collin, L., Grüninger, F., Graff, C., & Rönnbäck, A. (2016). Autophagic and lysosomal defects in human tauopathies: Analysis of post-mortem brain from patients with familial Alzheimer disease, corticobasal degeneration and progressive supranuclear palsy. *Acta Neuropathologica Communications*, *4*(1), 22. doi:10.118640478-016-0292-9 PMID:26936765

Plastaras, J. P., Dorsey, J. F., Carroll, K., Kim, S. H., Birnbaum, M. J., & El-Deiry, W. S. (2008). Role of PI3K/Akt signaling in TRAIL-and radiation-induced gastrointestinal apoptosis. *Cancer Biology & Therapy*, *7*(12), 2047–2053. doi:10.4161/cbt.7.12.7570 PMID:19106634

Plate, L., & Wiseman, R. L. (2017). Regulating Secretory Proteostasis through the Unfolded Protein Response: From Function to Therapy. *Trends in Cell Biology*, *27*(10), 722–737. doi:10.1016/j.tcb.2017.05.006 PMID:28647092

Plaza-Zabala, A., Sierra-Torre, V., & Sierra, A. (2017). Autophagy and Microglia: Novel Partners in Neurodegeneration and Aging. *International Journal of Molecular Sciences*, *18*(598), 1–28. doi:10.3390/ijms18030598 PMID:28282924

Plotegher, N., Gratton, E., & Bubacco, L. (2014). Number and brightness analysis of alpha-synuclein oligomerization and the associated mitochondrial morphology alterations in live cells. *Biochimica et Biophysica Acta*, *1840*(6), 2014–2024. doi:10.1016/j.bbagen.2014.02.013 PMID:24561157

Poewe, W., Seppi, K., Tanner, C. M., Halliday, G. M., Brundin, P., Volkmann, J., ... Lang, A. E. (2017). Parkinson disease. *Nature Reviews. Disease Primers*, *3*(1), 17013. doi:10.1038/nrdp.2017.13 PMID:28332488

Poksay, K. S., Madden, D. T., Peter, A. K., Niazi, K., Banwait, S., Crippen, D., . . . Rao, R. V. (2011). *Valosin-containing protein gene mutations: cellular phenotypes relevant to neurodegeneration*. Academic Press.

Polajnar, M., & Zerovnik, E. (2014). Impaired autophagy: A link between neurodegenerative and neuropsychiatric diseases. *Journal of Cellular and Molecular Medicine*, *18*(9), 1705–1711. doi:10.1111/jcmm.12349 PMID:25139375

Poole, A. C., Thomas, R. E., Andrews, L. A., McBride, H. M., Whitworth, A. J., & Pallanck, L. J. (2008). The PINK1/Parkin pathway regulates mitochondrial morphology. *Proceedings of the National Academy of Sciences of the United States of America*, *105*(5), 1638–1643. doi:10.1073/pnas.0709336105 PMID:18230723

Poole, A. C., Thomas, R. E., Yu, S., Vincow, E. S., & Pallanck, L. (2010). The mitochondrial fusion-promoting factor mitofusin is a substrate of the *PINK1*/parkin pathway. *PLoS One*, *5*(4), e10054. doi:10.1371/journal.pone.0010054 PMID:20383334

Poole, L. B. (2015). The basics of thiols and cysteines in redox biology and chemistry. *Free Radical Biology & Medicine*, *80*, 148–157. doi:10.1016/j.freeradbiomed.2014.11.013 PMID:25433365

Poole, L. B., Karplus, P. A., & Claiborne, A. (2004). Protein sulfenic acids in redox signalling. *Annual Review of Pharmacology and Toxicology*, *44*(1), 325–347. doi:10.1146/annurev.pharmtox.44.101802.121735 PMID:14744249

Poole, L. B., & Nelson, K. J. (2008). Discovering mechanisms of signaling-mediated cysteine oxidation. *Current Opinion in Chemical Biology*, *12*(1), 18–24. doi:10.1016/j.cbpa.2008.01.021 PMID:18282483

Pooler, A. M., & Hanger, D. P. (2010). *Functional implications of the association of tau with the plasma membrane*. Academic Press.

Poplawski, T., Pytel, D., Dziadek, J., & Majsterek, I. (2019). Interplay between Redox Signaling, Oxidative Stress, and Unfolded Protein Response (UPR) in Pathogenesis of Human Diseases. *Oxidative Medicine and Cellular Longevity*, *2019*, 1–2. doi:10.1155/2019/6949347 PMID:31089415

Potter, C. J., Pedraza, L. G., & Xu, T. (2002). Akt regulates growth by directly phosphorylating Tsc2. *Nature Cell Biology*, *4*(9), 658–665. doi:10.1038/ncb840 PMID:12172554

Pottoo, F. H., Alshayban, D. M., Joseph, R., Al-Musa, F., Al-Jabran, O., & Aljaafari, D. (2020). Impact of adherence to antiepileptic medications on quality of life of epileptic patients in the Eastern Province of Saudi Arabia: A cross-sectional study. *Imam. J. Appl. Sci.*, *5*, 1-8. Retrieved from http://www.e-ijas.org/text.asp?2020/5/1/1/274291

Pottoo, F. H., Bhowmik, M., & Vohora, D. (2014). Raloxifene protects against seizures and neurodegeneration in a mouse model mimicking epilepsy in postmenopausal woman. *European Journal of Pharmaceutical Sciences*, *65*, 167–173. doi:10.1016/j.ejps.2014.09.002 PMID:25218046

Pottoo, F. H., Tabassum, N., Javed, M. N., Nigar, S., Rasheed, R., Khan, A., ... Ashraf, G. M. (2019). The Synergistic Effect of Raloxifene, Fluoxetine, and Bromocriptine Protects Against Pilocarpine-Induced Status Epilepticus and Temporal Lobe Epilepsy. *Molecular Neurobiology*, *56*(2), 1233–1247. doi:10.100712035-018-1121-x PMID:29881945

Poukka, H., Karvonen, U., Jänne, O. A., & Palvimo, J. J. (2000). Covalent modification of the androgen receptor by small ubiquitin-like modifier 1 (SUMO-1). *Proceedings of the National Academy of Sciences of the United States of America*, *97*(26), 14145–14150. doi:10.1073/pnas.97.26.14145 PMID:11121022

Poulopoulos, M., Levy, O. A., & Alcalay, R. N. (2012). The neuropathology of genetic Parkinson's disease. *Movement Disorders*, *27*(7), 831–842. doi:10.1002/mds.24962 PMID:22451330

Power, J. H., Barnes, O. L., & Chegini, F. (2017). Lewy Bodies and the Mechanisms of Neuronal Cell Death in Parkinson's Disease and Dementia with Lewy Bodies. *Brain Pathology (Zurich, Switzerland)*, *27*(1), 3–12. doi:10.1111/bpa.12344 PMID:26667592

Pozsgai, E., Gomori, E., Szigeti, A., Boronkai, A., Gallyas, F. Jr, Sumegi, B., & Bellyei, S. (2007). Correlation between the progressive cytoplasmic expression of a novel small heat shock protein (Hsp16.2) and malignancy in brain tumors. *BMC Cancer*, *7*(1), 233. doi:10.1186/1471-2407-7-233 PMID:18154656

Prabhudesai, S., Sinha, S., Attar, A., Kotagiri, A., Fitzmaurice, A. G., Lakshmanan, R., & Stahl, M. (2012). A novel "molecular tweezer" inhibitor of α-synuclein neurotoxicity in vitro and in vivo. *Neurotherapeutics; the Journal of the American Society for Experimental NeuroTherapeutics*, *9*(2), 464–476. doi:10.100713311-012-0105-1 PMID:22373667

Prasad, A., Bharathi, V., Sivalingam, V., Girdhar, A., & Patel, B. K. (2019). Molecular Mechanisms of TDP-43 Misfolding and Pathology in Amyotrophic Lateral Sclerosis. *Frontiers in Molecular Neuroscience*, *12*, 25. doi:10.3389/fnmol.2019.00025 PMID:30837838

Pratt, W. B., & Toft, D. O. (2003). Regulation of signaling protein function and trafficking by the hsp90/hsp70-based chaperone machinery. *Experimental Biology and Medicine (Maywood, N.J.)*, *228*(2), 111–133. doi:10.1177/153537020322800201 PMID:12563018

Pratt, W., Morishima, Y., Murphy, M., & Harrell, M. (2006). Chaperoning of glucocorticoid receptors. In *Molecular Chaperones in Health and Disease* (pp. 111–138). Springer. doi:10.1007/3-540-29717-0_5

Pridgeon, J. W., Olzmann, J. A., Chin, L. S., & Li, L. (2007). PINK1 protects against oxidative stress by phosphorylating mitochondrial chaperone TRAP1. *PLoS Biology*, *5*(7), e172. doi:10.1371/journal.pbio.0050172 PMID:17579517

Prince, M., Bryce, R., Albanese, E., Wimo, A., Ribeiro, W., & Ferri, C. P. (2013). The global prevalence of dementia: A systematic review and metaanalysis. *Alzheimer's & Dementia*, *9*(1), 63–75.e2. doi:10.1016/j.jalz.2012.11.007 PMID:23305823

Proikas-Cezanne, T., Takacs, Z., Donnes, P., & Kohlbacher, O. (2015). WIPI proteins: Essential PtdIns3P effectors at the nascent autophagosome. *Journal of Cell Science*, *128*(2), 207–217. doi:10.1242/jcs.146258 PMID:25568150

Prusiner, S. B. (1982). Novel proteinaceous infectious particles cause scrapie. *Science*, *216*(4542), 136–144. doi:10.1126cience.6801762 PMID:6801762

Prusiner, S. B. (1998). Prions. *Proceedings of the National Academy of Sciences of the United States of America*, *95*(23), 13363–13383. doi:10.1073/pnas.95.23.13363 PMID:9811807

Prusiner, S. B. (2001). Neurodegenerative Diseases and Prions. *The New England Journal of Medicine*, *344*(20), 1516–1526. doi:10.1056/NEJM200105173442006 PMID:11357156

Przedborski, S., Vila, M., & Jackson-Lewis, V. (2003). Series Introduction: Neurodegeneration: What is it and where are we? *The Journal of Clinical Investigation*, *111*(1), 3–10. doi:10.1172/JCI200317522 PMID:12511579

Puder, M., & Soberman, R. J. (1997). *Glutathione conjugates recognize the Rossmann fold of glyceraldehyde-3-phosphate dehydrogenase*. Academic Press.

Puri, C., Renna, M., Bento, C. F., Moreau, K., & Rubinsztein, D. C. (2013). Diverse autophagosome membrane sources coalesce in recycling endosomes. *Cell*, *154*(6), 1285–1299. doi:10.1016/j.cell.2013.08.044 PMID:24034251

Putcha, P., Danzer, K. M., Kranich, L. R., Scott, A., Silinski, M., Mabbett, S., ... McLean, P. J. (2010). Brain-permeable small-molecule inhibitors of Hsp90 prevent alpha-synuclein oligomer formation and rescue alpha-synuclein-induced toxicity. *The Journal of Pharmacology and Experimental Therapeutics*, *332*(3), 849–857. doi:10.1124/jpet.109.158436 PMID:19934398

Pyo, J. O., Jang, M. H., Kwon, Y. K., Lee, H. J., Jun, J. I., Woo, H. N., ... Mizushima, N. (2005). Essential roles of Atg5 and FADD in autophagic cell death dissection of autophagic cell death into vacuole formation and cell death. *The Journal of Biological Chemistry*, *280*(21), 20722–20729. doi:10.1074/jbc.M413934200 PMID:15778222

Qi, L., Zhang, X. D., Wu, J. C., Lin, F., Wang, J., DiFiglia, M., & Qin, Z.-H. (2012). The role of chaperone-mediated autophagy in huntingtin degradation. *PLoS One*, *7*(10), e46834. doi:10.1371/journal.pone.0046834 PMID:23071649

Qin, W., Chachich, M., Lane, M., Roth, G., Bryant, M., de Cabo, R., ... Pasinetti, G. M. (2006a). Calorie restriction attenuates Alzheimer's disease type brain amyloidosis in Squirrel monkeys (Saimiri sciureus). *Journal of Alzheimer's Disease*, *10*(4), 417–422. doi:10.3233/JAD-2006-10411 PMID:17183154

Qin, W., Yang, T., Ho, L., Zhao, Z., Wang, J., Chen, L., ... Pasinetti, G. M. (2006b). Neuronal SIRT1 activation as a novel mechanism underlying the prevention of Alzheimer disease amyloid neuropathology by calorie restriction. *The Journal of Biological Chemistry*, *281*(31), 21745–21754. doi:10.1074/jbc.M602909200 PMID:16751189

Qiu, W. Q., Walsh, D. M., Ye, Z., Vekrellis, K., Zhang, J., Podlisny, M. B., ... Selkoe, D. J. (1998). Insulin-degrading enzyme regulates extracellular levels of amyloid β-protein by degradation. *The Journal of Biological Chemistry*, *273*(49), 32730–32738. doi:10.1074/jbc.273.49.32730 PMID:9830016

Qi, X., Okuma, Y., Hosoi, T., & Nomura, Y. (2004c). Edaravone protects against hypoxia/ischemia-induced endoplasmic reticulum dysfunction. *Journal de Pharmacologie*, *311*(1), 388–393. doi:10.1124/jpet.104.069088 PMID:15178695

Quiros, P. M., Mottis, A., & Auwerx, J. (2016). Mitonuclear communication in homeostasis and stress. *Nature Reviews. Molecular Cell Biology*, *17*(4), 213–226. doi:10.1038/nrm.2016.23 PMID:26956194

Qu, J., Nakamura, T., Cao, G., Holland, E. A., McKercher, S. R., & Lipton, S. A. (2011). S-Nitrosylation activates Cdk5 and contributes to synaptic spine loss induced by beta-amyloid peptide. *Proceedings of the National Academy of Sciences of the United States of America*, *108*(34), 14330–14335. doi:10.1073/pnas.1105172108 PMID:21844361

Qureshi, I. A., & Mehler, M. F. (2011). Advances in epigenetics and epigenomics for neurodegenerative diseases. *Current Neurology and Neuroscience Reports*, *11*(5), 464–473. doi:10.100711910-011-0210-2 PMID:21671162

Qu, X., Yu, J., Bhagat, G., Furuya, N., Hibshoosh, H., Troxel, A., ... Levine, B. (2003). Promotion of tumorigenesis by heterozygous disruption of the beclin 1 autophagy gene. *The Journal of Clinical Investigation*, *112*(12), 1809–1820. doi:10.1172/JCI20039 PMID:14638851

Radford, H., Moreno, J. A., Verity, N., Halliday, M., & Mallucci, G. R. (2015). PERK inhibition prevents tau-mediated neurodegeneration in a mouse model of frontotemporal dementia. *Acta Neuropathologica*, *130*(5), 633–642. doi:10.100700401-015-1487-z PMID:26450683

Radi, R. (2013). Peroxynitrite, a stealthy biological oxidant. *The Journal of Biological Chemistry*, *288*(37), 26464–26472. doi:10.1074/jbc.R113.472936 PMID:23861390

Raff, M. C., Whitmore, A. V., & Finn, J. T. (2002). Axonal self-destruction and neurodegeneration. *Science*, *296*(5569), 868–871. doi:10.1126cience.1068613 PMID:11988563

Rahman, M. A., & Rhim, H. (2017). Therapeutic implication of autophagy in neurodegenerative diseases. *BMB Reports*, *50*(7), 345–354. doi:10.5483/BMBRep.2017.50.7.069 PMID:28454606

Rahman, M. M., Rumzhum, N. N., Hansbro, P. M., Morris, J. C., Clark, A. R., Verrills, N. M., & Ammit, A. J. (2016). Activating protein phosphatase 2A (PP2A) enhances tristetraprolin (TTP) anti-inflammatory function in A549 lung epithelial cells. *Cellular Signalling*, *28*(4), 325–334. doi:10.1016/j.cellsig.2016.01.009 PMID:26820662

Rai, S. N., Singh, B. K., Rathore, A. S., Zahra, W., Keswani, C., Birla, H., ... Singh, S. P. (2019). Quality Control in Huntington's Disease: A Therapeutic Target. *Neurotoxicity Research*, *36*(3), 612–626. doi:10.100712640-019-00087-x PMID:31297710

Rajasekhar, K., & Thimmaiah, G. (2018). Current progress, challenges and future prospects of diagnostic and therapeutic interventions in Alzheimer's disease. *RSC Advances*, *8*(42), 23780–23804. doi:10.1039/C8RA03620A

Rakhit, R., Cunningham, P., Furtos-Matei, A., Dahan, S., Qi, X. F., Crow, J. P., ... Chakrabartty, A. (2002). Oxidation-induced misfolding and aggregation of superoxide dismutase and its implications for amyotrophic lateral sclerosis. *The Journal of Biological Chemistry*, *277*(49), 47551–47556. doi:10.1074/jbc.M207356200 PMID:12356748

Rakovic, A., Grünewald, A., Seibler, P., Ramirez, A., Kock, N., Orolicki, S., ... Klein, C. (2010). Effect of endogenous mutant and wild-type *PINK1* on Parkin in fibroblasts from Parkinson disease patients. *Human Molecular Genetics*, *19*(16), 3124–3137. doi:10.1093/hmg/ddq215 PMID:20508036

Ramadori, G., Lee, C. E., Bookout, A. L., Lee, S., Williams, K. W., Anderson, J., ... Coppari, R. (2008). Brain SIRT1: Anatomical distribution and regulation by energy availability. *The Journal of Neuroscience: The Official Journal of the Society for Neuroscience*, *28*(40), 9989–9996. doi:10.1523/JNEUROSCI.3257-08.2008 PMID:18829956

Ramanan, V. K., & Saykin, A. J. (2013). Pathways to neurodegeneration: Mechanistic insights from GWAS in Alzheimer's disease, Parkinson's disease, and related disorders. *American Journal of Neurodegenerative Disease*, *2*(3), 145–175. PMID:24093081

Rambold, A. S., Kostelecky, B., Elia, N., & Lippincott-Schwartz, J. (2011). Tubular network formation protects mitochondria from autophagosomal degradation during nutrient starvation. *Proceedings of the National Academy of Sciences of the United States of America*, *108*(25), 10190–10195. doi:10.1073/pnas.1107402108 PMID:21646527

Ranford, J. C., Coates, A. R., & Henderson, B. (2000). Chaperonins are cell-signalling proteins: The unfolding biology of molecular chaperones. *Expert Reviews in Molecular Medicine*, *2*(8), 1–17. doi:10.1017/S1462399400002015 PMID:14585136

Rankin, C. A., Sun, Q., & Gamblin, T. C. (2007). Tau phosphorylation by GSK-3β promotes tangle-like filament morphology. *Molecular Neurodegeneration*, *2*(1), 12. doi:10.1186/1750-1326-2-12 PMID:17598919

Rao, R. V., & Bredesen, D. E. (2004). Misfolded proteins, endoplasmic reticulum stress and neurodegeneration. *Current Opinion in Cell Biology*, *16*(6), 653–662. doi:10.1016/j.ceb.2004.09.012 PMID:15530777

Rathor, R., & Suryakumar, G. (2019). Upregulation of Cytoprotective Chaperones Mediate Better Tolerance to High Altitude. In *Chaperokine Activity of Heat Shock Proteins* (pp. 123–145). Cham: Springer. doi:10.1007/978-3-030-02254-9_6

Rattan, S. I. S. (2008). Increased molecular damage and heterogeneity as the basis of aging. *Biological Chemistry*, *389*(3), 267–272. doi:10.1515/BC.2008.030 PMID:18208348

Ravikumar, B., Acevedo-Arozena, A., Imarisio, S., Berger, Z., Vacher, C., O'Kane, C. J., ... Rubinsztein, D. C. (2005). Dynein mutations impair autophagic clearance of aggregate-prone proteins. *Nature Genetics*, *37*(7), 771–776. doi:10.1038/ng1591 PMID:15980862

Ravikumar, B., Duden, R., & Rubinsztein, D. C. (2002). Aggregate-prone proteins with polyglutamine and polyalanine expansions are degraded by autophagy. *Human Molecular Genetics*, *11*(9), 1107–1117. doi:10.1093/hmg/11.9.1107 PMID:11978769

Ravikumar, B., Sarkar, S., & Rubinsztein, D. C. (2008). Clearance of mutant aggregate-prone proteins by autophagy. In *Autophagosome and Phagosome* (pp. 195–211). Humana Press. doi:10.1007/978-1-59745-157-4_13

Ravikumar, B., Vacher, C., Berger, Z., Davies, J. E., Luo, S., Oroz, L. G., ... Rubinsztein, D. C. (2004). Inhibition of mTOR induces autophagy and reduces toxicity of polyglutamine expansions in fly and mouse models of Huntington disease. *Nature Genetics*, *36*(6), 585–595. doi:10.1038/ng1362 PMID:15146184

Ray, P. D., Huang, B. W., & Tsuji, Y. (2012). Reactive oxygen species (ROS) homeostasis and redox regulation in cellular signaling. *Cellular Signalling*, *24*(5), 981–990. doi:10.1016/j.cellsig.2012.01.008 PMID:22286106

Ray, W. J., Ashall, F., & Goate, A. M. (1998). Molecular pathogenesis of sporadic and familial forms of Alzheimer's disease. *Molecular Medicine Today*, *4*(4), 151–157. doi:10.1016/S1357-4310(98)01229-5 PMID:9572056

Reddie, K. G., & Carroll, K. S. (2008). Expanding the functional diversity of proteins through cysteine oxidation. *Current Opinion in Chemical Biology*, *12*(6), 746–754. doi:10.1016/j.cbpa.2008.07.028 PMID:18804173

Redler, R. L., Wilcox, K. C., Proctor, E. A., Fee, L., Caplow, M., & Dokholyan, N. V. (2011). Glutathionylation at Cys-111 induces dissociation of wild type and FALS mutant SOD1 dimers. *Biochem*, *50*(32), 7057–7066. doi:10.1021/bi200614y PMID:21739997

Reichling, D. B., & Levine, J. D. (2011). Pain and death: Neurodegenerative disease mechanisms in the nociceptor. *Annals of Neurology*, *69*(1), 13–21. doi:10.1002/ana.22351 PMID:21280072

Reid, M. A., Wang, W. I., Rosales, K. R., Welliver, M. X., Pan, M., & Kong, M. (2013). The B55α subunit of PP2A drives a p53-dependent metabolic adaptation to glutamine deprivation. *Molecular Cell*, *50*(2), 200–211. doi:10.1016/j.molcel.2013.02.008 PMID:23499005

Reiman, E. M., Caselli, R. J., Yun, L. S., Chen, K., Bandy, D., Minoshima, S., ... Osborne, D. (1996). Preclinical evidence of Alzheimer's disease in persons homozygous for the ε4 allele for apolipoprotein E. *The New England Journal of Medicine*, *334*(12), 752–758. doi:10.1056/NEJM199603213341202 PMID:8592548

Reina, S., Gupta, A., Messina, A., & Mahalakshmi, R. (2016). Role of cysteines in mammalian VDAC isoforms' function. *Biochimica et Biophysica Acta (BBA) - Bioenergetics*, 1219–1227.

Ren, M., Phoon, C. K., & Schlame, M. (2014). Metabolism and function of mitochondrial cardiolipin. *Progress in Lipid Research*, 55, 1-16.

Renvoisé, B., Chang, J., Singh, R., Yonekawa, S., FitzGibbon, E. J., Mankodi, A., ... Mahuran, D. J. (2014). Lysosomal abnormalities in hereditary spastic paraplegia types SPG 15 and SPG 11. *Annals of Clinical and Translational Neurology*, *1*(6), 379–389. doi:10.1002/acn3.64 PMID:24999486

Ren, X., Bjornstedt, M., Shen, B., Ericson, M. L., & Holmgren, A. (1993). Mutagenesis of structural half-cystine residues in human thioredoxin and effects on the regulation of activity by selenodiglutathione. *Biochem*, *32*(37), 9701–9708. doi:10.1021/bi00088a023 PMID:8373774

Resende, R., Ferreiro, E., Pereira, C., & Oliveira, C. R. (2008). ER stress is involved in Aβ-induced GSK-3β activation and tau phosphorylation. *Journal of Neuroscience Research*, *86*(9), 2091–2099. doi:10.1002/jnr.21648 PMID:18335524

Resh, M. D. (1999). Fatty acylation of proteins: New insights into membrane targeting of myristoylated and palmitoylated proteins. *Biochimica et Biophysica Acta. Molecular Cell Research*, *1451*(1), 1–16. doi:10.1016/S0167-4889(99)00075-0 PMID:10446384

Reynolds, M. R., Berry, R. W., & Binder, L. I. (2005). Site-specific nitration differentially influences τ assembly in vitro. *Biochemistry*, *44*(42), 13997–14009. doi:10.1021/bi051028w PMID:16229489

Reynolds, M. R., Reyes, J. F., Fu, Y., Bigio, E. H., Guillozet-Bongaarts, A. L., Berry, R. W., & Binder, L. I. (2006). Tau nitration occurs at tyrosine 29 in the fibrillar lesions of Alzheimer's disease and other tauopathies. *The Journal of Neuroscience: The Official Journal of the Society for Neuroscience*, *26*(42), 10636–10645. doi:10.1523/JNEUROSCI.2143-06.2006 PMID:17050703

Rhee, S. G., Jeong, W., Chang, T.-S., & Woo, H. A. (2007). Sulfiredoxin, the cysteine sulfinic acid reductase specific to 2-Cys peroxiredoxin: Its discovery, mechanism of action, and biological significance. *Kidney International*, *72*(106), S3–S8. doi:10.1038j.ki.5002380 PMID:17653208

Rhein, V., Song, X., Wiesner, A., Ittner, L. M., Baysang, G., Meier, F., ... Eckert, A. (2009). Amyloid-beta and tau synergistically impair the oxidative phosphorylation system in triple transgenic Alzheimer's disease mice. *Proceedings of the National Academy of Sciences of the United States of America*, *106*(47), 20057–20062. doi:10.1073/pnas.0905529106 PMID:19897719

Richman, P. G., & Meister, A. (1975). Regulation of γ-glutamyl-cysteine synthetase by nonallosteric feedback inhibition by glutathione. *The Journal of Biological Chemistry*, *250*, 1422–1426. PMID:1112810

Richter-Landberg, C., & Goldbaum, O. (2003). Stress proteins in neural cells: Functional roles in health and disease. *Cellular and Molecular Life Sciences CMLS*, *60*(2), 337–349. doi:10.1007000180300028 PMID:12678498

Ricobaraza, A., Cuadrado-Tejedor, M., Pérez-Mediavilla, A., Frechilla, D., Del Río, J., & García-Osta, A. (2009). Phenylbutyrate ameliorates cognitive deficit and reduces tau pathology in an alzheimer's disease mouse model. *Neuropsychopharmacology*, *34*(7), 1721–1732. doi:10.1038/npp.2008.229 PMID:19145227

Ricotta, M., Iannuzzi, M., De Vivo, G., & Gentile, V. (2010). *Physio-pathological roles of transglutaminase-catalyzed reactions.* Academic Press.

Rideout, H. J., Dietrich, P., Wang, Q., Dauer, W. T., & Stefanis, L. (2004). α-Synuclein is required for the fibrillar nature of ubiquitinated inclusions induced by proteasomal inhibition in primary neurons. *The Journal of Biological Chemistry*, *279*(45), 46915–46920. doi:10.1074/jbc.M405146200 PMID:15322100

Riley, B. E., Zoghbi, H. Y., & Orr, H. T. (2005). SUMOylation of the polyglutamine repeat protein, ataxin-1, is dependent on a functional nuclear localization signal. *The Journal of Biological Chemistry*, *280*(23), 21942–21948. doi:10.1074/jbc.M501677200 PMID:15824120

Robberecht, W., & Philips, T. (2013). The changing scene of amyotrophic lateral sclerosis. *Nature Reviews. Neuroscience*, *14*(4), 248–264. doi:10.1038/nrn3430 PMID:23463272

Robertson, L. A., Moya, K. L., & Breen, K. C. (2004). The potential role of tau protein O-glycosylation in Alzheimer's disease. *Journal of Alzheimer's Disease*, *6*(5), 489–495. doi:10.3233/JAD-2004-6505 PMID:15505370

Rocha, E. M., De Miranda, B., & Sanders, L. H. (2018). Alpha-synuclein: Pathology, mitochondrial dysfunction and neuroinflammation in Parkinson's disease. *Neurobiology of Disease*, *109*, 249–257. doi:10.1016/j.nbd.2017.04.004 PMID:28400134

Rodger, C.E., McWilliams, T.G., & Ganley, I.G. (2017). *Mammalian mitophagy – from in vitro molecules to in vivo models.* doi:10.1111/febs.14336

Rodolfo, C., Campello, S., & Cecconi, F. (2018). Mitophagy in neurodegenerative diseases. *Neurochemistry International*, *117*, 156–166. doi:10.1016/j.neuint.2017.08.004 PMID:28797885

Rodriguez, A. E., Lopez-Crisosto, C., Pena-Oyarzun, D., Salas, D., Parra, V., Quiroga, C., ... Lavandero, S. (2016). BAG3 regulates total MAP1LC3B protein levels through a translational but not transcriptional mechanism. *Autophagy*, *12*(2), 287–296. doi:10.1080/15548627.2015.1124225 PMID:26654586

Rodriguez, M., Lapierre, J., Ojha, C. R., Kaushik, A., Batrakova, E., Kashanchi, F., ... El-Hage, N. (2017). Intranasal drug delivery of small interfering RNA targeting Beclin1 encapsulated with polyethylenimine (PEI) in mouse brain to achieve HIV attenuation. *Scientific Reports*, *7*(1), 1862. doi:10.103841598-017-01819-9 PMID:28500326

Rodríguez-Martín, T., Cuchillo-Ibáñez, I., Noble, W., Nyenya, F., Anderton, B. H., & Hanger, D. P. (2013). Tau phosphorylation affects its axonal transport and degradation. *Neurobiology of Aging*, *34*(9), 2146–2157. doi:10.1016/j.neurobiolaging.2013.03.015 PMID:23601672

Rodríguez-Muela, N., Koga, H., García-Ledo, L., de la Villa, P., De la Rosa, E. J., Cuervo, A. M., & Boya, P. (2013). Balance between autophagic pathways preserves retinal homeostasis. *Aging Cell*, *12*(3), 478–488. doi:10.1111/acel.12072 PMID:23521856

Rodriguez-Nogales, C., Garbayo, E., Carmona-Abellan, M. M., Luquin, M. R., & Blanco-Prieto, M. J. (2016). Brain aging and Parkinson's disease: New therapeutic approaches using drug delivery systems. *Maturitas*, *84*, 25–31. doi:10.1016/j.maturitas.2015.11.009 PMID:26653838

Roher, A. E., Lowenson, J. D., Clarke, S., Woods, A. S., Cotter, R. J., Gowing, E., & Ball, M. J. (1993). Beta-Amyloid-(1-42) is a major component of cerebrovascular amyloid deposits: Implications for the pathology of Alzheimer disease. *Proceedings of the National Academy of Sciences of the United States of America*, *90*(22), 10836–10840. doi:10.1073/pnas.90.22.10836 PMID:8248178

Rohn, T. T., Wirawan, E., Brown, R. J., Harris, J. R., Masliah, E., & Vandenabeele, P. (2011). Depletion of Beclin-1 due to proteolytic cleavage by caspases in the Alzheimer's disease brain. *Neurobiology of Disease*, *43*(1), 68–78. doi:10.1016/j.nbd.2010.11.003 PMID:21081164

Rojas-Rivera, D., Delvaeye, T., Roelandt, R., Nerinckx, W., Augustyns, K., Vandenabeele, P., & Bertrand, M. J. M. (2017). When PERK inhibitors turn out to be new potent RIPK1 inhibitors: Critical issues on the specificity and use of GSK2606414 and GSK2656157. *Cell Death and Differentiation*, *24*(6), 1100–1110. doi:10.1038/cdd.2017.58 PMID:28452996

Ron, D., & Walter, P. (2007). Signal integration in the endoplasmic reticulum unfolded protein response. *Nature Reviews. Molecular Cell Biology*, *8*(7), 519–529. doi:10.1038/nrm2199 PMID:17565364

Rongvaux, A., Shea, R. J., Mulks, M. H., Gigot, D., Urbain, J., Leo, O., & Andris, F. (2002). Pre-B-cell colony-enhancing factor, whose expression is up-regulated in activated lymphocytes, is a nicotinamide phosphoribosyl transferase, a cytosolic enzyme involved in NAD biosynthesis. *European Journal of Immunology*, *32*(11), 3225–3234. doi:10.1002/1521-4141(200211)32:11<3225::AID-IMMU3225>3.0.CO;2-L PMID:12555668

Rong, Y., McPhee, C. K., Deng, S., Huang, L., Chen, L., Liu, M., ... Lenardo, M. J. (2011). Spinster is required for autophagic lysosome reformation and mTOR reactivation following starvation. *Proceedings of the National Academy of Sciences of the United States of America*, *108*(19), 7826–7831. doi:10.1073/pnas.1013800108 PMID:21518918

Roodveldt, C., Christodoulou, J., & Dobson, C. M. (2008). Immunological features of α-synuclein in Parkinson's disease. *Journal of Cellular and Molecular Medicine*, *12*(5b), 1820–1829. doi:10.1111/j.1582-4934.2008.00450.x PMID:18671754

Roriz-Filho, J. S., Sa-Roriz, T. M., Rosset, I., Camozzato, A. L., Santos, A. C., Chaves, M. L., ... Roriz-Cruz, M. (2009). (Pre) diabetes, brain aging, and cognition. *Biochimica et Biophysica Acta (BBA)-. Molecular Basis of Disease*, *1792*(5), 432–443. doi:10.1016/j.bbadis.2008.12.003

Rosas, H., Koroshetz, W., Chen, Y., Skeuse, C., Vangel, M., Cudkowicz, M., . . . Makris, N. (2003). *Evidence for more widespread cerebral pathology in early HD: an MRI-based morphometric analysis*. Academic Press.

Rosati, A., Graziano, V., Laurenzi, V., & De. (2011). BAG3: A multifaceted protein that regulates major cell pathways. *Cell Death & Disease*, *2*(4), e141. doi:10.1038/cddis.2011.24 PMID:21472004

Rosborough, K., Patel, N., & Kalia, L. V. (2017). Alpha-synuclein and Parkinsonism: Updates and Future Perspectives. *Current Neurology and Neuroscience Reports*, *17*(4), 31. doi:10.100711910-017-0737-y PMID:28324300

Rose, C., Menzies, F. M., Renna, M., Acevedo-Arozena, A., Corrochano, S., Sadiq, O., ... Rubinsztein, D. C. (2010). Rilmenidine attenuates toxicity of polyglutamine expansions in a mouse model of Huntington's disease. *Human Molecular Genetics*, *19*(11), 2144–2153. doi:10.1093/hmg/ddq093 PMID:20190273

Rosenblum, L. T., & Trotti, D. (2017). EAAT2 and the Molecular Signature of Amyotrophic Lateral Sclerosis. *Advances in Neurobiology*, *16*, 117–136. doi:10.1007/978-3-319-55769-4_6 PMID:28828608

Rosen, D. R., Siddique, T., Patterson, D., Figlewicz, D. A., Sapp, P., Hentati, A., ... Brown, R. H. Jr. (1993). Mutations in Cu/Zn superoxide dismutase gene are associated with familial amyotrophic lateral sclerosis. *Nature*, *362*(6415), 59–62. doi:10.1038/362059a0 PMID:8446170

Rosen, K. M., Veereshwarayya, V., Moussa, C. E., Fu, Q., Goldberg, M. S., Schlossmacher, M. G., ... Querfurth, H. W. (2006). Parkin protects against mitochondrial toxins and β-amyloid accumulation in skeletal muscle cells. *The Journal of Biological Chemistry*, *281*(18), 12809–12816. doi:10.1074/jbc.M512649200 PMID:16517603

Ross, C. A., & Tabrizi, S. J. (2011). *Huntington's disease: from molecular pathogenesis to clinical treatment*. Academic Press.

Ross, C. A., & Poirier, M. A. (2004). Protein aggregation and neurodegenerative disease. *Nature Medicine*, *10*(S7), S10–S17. doi:10.1038/nm1066 PMID:15272267

Rossin, F., D'Eletto, M., Macdonald, D., Farrace, M. G., & Piacentini, M. J. A. a. (2012). *TG2 transamidating activity acts as a reostat controlling the interplay between apoptosis and autophagy*. Academic Press.

Rothenberg, C., Srinivasan, D., Mah, L., Kaushik, S., Peterhoff, C. M., Ugolino, J., ... Monteiro, M. J. (2010). Ubiquilin functions in autophagy and is degraded by chaperone-mediated autophagy. *Human Molecular Genetics*, *19*(16), 3219–3232. doi:10.1093/hmg/ddq231 PMID:20529957

Rothfuss, O., Fischer, H., Hasegawa, T., Maisel, M., Leitner, P., Miesel, F., ... Patenge, N. (2009). Parkin protects mitochondrial genome integrity and supports mitochondrial DNA repair. *Human Molecular Genetics*, *18*(20), 3832–3850. doi:10.1093/hmg/ddp327 PMID:19617636

Rothgiesser, K. M., Erener, S., Waibel, S., Lüscher, B., & Hottiger, M. O. (2010). SIRT2 regulates NF-κB-dependent gene expression through deacetylation of p65 Lys310. *Journal of Cell Science*, *123*(24), 4251–4258. doi:10.1242/jcs.073783 PMID:21081649

Rousseaux, M. W., de Haro, M., Lasagna-Reeves, C. A., De Maio, A., Park, J., Jafar-Nejad, P., ... Vilanova-Velez, L. (2016). TRIM28 regulates the nuclear accumulation and toxicity of both alpha-synuclein and tau. *eLife*, *5*, e19809. doi:10.7554/eLife.19809 PMID:27779468

Roussel, B. D., Kruppa, A. J., Miranda, E., Crowther, D. C., Lomas, D. A., & Marciniak, S. J. (2013). Endoplasmic reticulum dysfunction in neurological disease. *Lancet Neurology*, *12*(1), 105–118. doi:10.1016/S1474-4422(12)70238-7 PMID:23237905

Rubinsztein, D. C. (2006). The roles of intracellular protein-degradation pathways in neurodegeneration. *Nature*, *443*(7113), 780–786. doi:10.1038/nature05291 PMID:17051204

Rubinsztein, D. C., Codogno, P., & Levine, B. (2012). Autophagy modulation as a potential therapeutic target for diverse diseases. *Nature Reviews. Drug Discovery*, *11*(9), 709–U784. doi:10.1038/nrd3802 PMID:22935804

Rubinsztein, D. C., Mariño, G., & Kroemer, G. (2011). Autophagy and aging. *Cell*, *146*(5), 682–695. doi:10.1016/j.cell.2011.07.030 PMID:21884931

Rubinsztein, D. C., Shpilka, T., & Elazar, Z. (2012). Mechanisms of autophagosome biogenesis. *Current Biology*, *22*(1), R29–R34. doi:10.1016/j.cub.2011.11.034 PMID:22240478

Rugarli, E. I., & Langer, T. (2012). Mitochondrial quality control: A matter of life and death for neurons. *The EMBO Journal*, *31*(6), 1336–1349. doi:10.1038/emboj.2012.38 PMID:22354038

Ruivo, R., Anne, C., Sagné, C., & Gasnier, B. (2009). Molecular and cellular basis of lysosomal transmembrane protein dysfunction. *Biochimica et Biophysica Acta (BBA)-. Molecular Cell Research, 1793*(4), 636–649. PMID:19146888

Russell, R. C., Tian, Y., Yuan, H., Park, H. W., Chang, Y.-Y., Kim, J., ... Guan, K.-L. (2013). ULK1 induces autophagy by phosphorylating Beclin-1 and activating VPS34 lipid kinase. *Nature Cell Biology, 15*(7), 741–750. doi:10.1038/ncb2757 PMID:23685627

Rutherford, N. J., Moore, B. D., Golde, T. E., & Giasson, B. I. (2014). Divergent effects of the H50Q and G51D SNCA mutations on the aggregation of α-synuclein. *Journal of Neurochemistry, 131*(6), 859–867. doi:10.1111/jnc.12806 PMID:24984882

Ryan, B. J., Hoek, S., Fon, E. A., & Wade-Martins, R. (2015). Mitochondrial dysfunction and mitophagy in Parkinson's: From familial to sporadic disease. *Trends in Biochemical Sciences, 40*(4), 200–210. doi:10.1016/j.tibs.2015.02.003 PMID:25757399

Ryu, D., Mouchiroud, L., Andreux, P. A., Katsyuba, E., Moullan, N., Nicolet-Dit-Felix, A. A., ... Auwerx, J. (2016). Urolithin A induces mitophagy and prolongs lifespan in C. elegans and increases muscle function in rodents. *Nature Medicine, 22*(8), 879–888. doi:10.1038/nm.4132 PMID:27400265

Ryu, S. W., Han, E. C., Yoon, J., & Choi, C. (2015). The mitochondrial fusion-related proteins Mfn2 and OPA1 are transcriptionally induced during differentiation of bone marrow progenitors to immature dendritic cells. *Molecules and Cells, 38*(1), 89. PMID:25387754

Sabens, L. (2012). Mechanisms of Altered Redox Regulation in Neurodegenerative Diseases—Focus on S-Glutathionylation. *Antioxidants & Redox Signalling, 16*(6), 543–566. doi:10.1089/ars.2011.4119 PMID:22066468

Safar, J., Wille, H., Itri, V., Groth, D., Serban, H., Torchia, M., ... Prusiner, S. B. (1998). Eight prion strains have PrP(Sc) molecules with different conformations. *Nature Medicine, 4*(10), 1157–1165. doi:10.1038/2654 PMID:9771749

Saftig, P., & Klumperman, J. (2009). Lysosome biogenesis and lysosomal membrane proteins: Trafficking meets function. *Nature Reviews. Molecular Cell Biology, 10*(9), 623–635. doi:10.1038/nrm2745 PMID:19672277

Sahara, N., Maeda, S., Murayama, M., Suzuki, T., Dohmae, N., Yen, S. H., & Takashima, A. (2007). Assembly of two distinct dimers and higher-order oligomers from full-length tau. *The European Journal of Neuroscience, 25*(10), 3020–3029. doi:10.1111/j.1460-9568.2007.05555.x PMID:17561815

Sahu, R., Kaushik, S., Clement, C. C., Cannizzo, E. S., Scharf, B., Follenzi, A., ... Santambrogio, L. (2011). Microautophagy of cytosolic proteins by late endosomes. *Developmental Cell, 20*(1), 131–139. doi:10.1016/j.devcel.2010.12.003 PMID:21238931

Saita, S., Shirane, M., & Nakayama, K. I. (2013). Selective escape of proteins from the mitochondria during mitophagy. *Nature Communications, 4*(1), 1410. doi:10.1038/ncomms2400 PMID:23361001

Saitsu, H., Nishimura, T., Muramatsu, K., Kodera, H., Kumada, S., Sugai, K., ... Matsumoto, N. (2013). De novo mutations in the autophagy gene WDR45 cause static encephalopathy of childhood with neurodegeneration in adulthood. *Nature Genetics, 45*(4), 445–449. doi:10.1038/ng.2562 PMID:23435086

Sajjad, R., Arif, R., Shah, A. A., Manzoor, I., & Mustafa, G. (2018). Pathogenesis of Alzheimer's Disease: Role of Amyloid-beta and Hyperphosphorylated Tau Protein. *Indian Journal of Pharmaceutical Sciences, 80*(4), 581–591. doi:10.4172/pharmaceutical-sciences.1000397

Salami, O., & Lyketsos, C. G. (2010). Clinical Features of Alzheimer's Disease: Cognitive and Non-Cognitive. In M. T. Abou-Saleh, C. Katona, & A. Kumar (Eds.), *Principles and Practice of Geriatric Psychiatry* (pp. 226–231). John Wiley & Sons, Inc. doi:10.1002/9780470669600.ch41

Salminen, A., & Kaarniranta, K. (2009). Regulation of the aging process by autophagy. *Trends in Molecular Medicine, 15*(5), 217–224. doi:10.1016/j.molmed.2009.03.004 PMID:19380253

Salminen, A., Kaarniranta, K., Kauppinen, A., Ojala, J., Haapasalo, A., Soininen, H., & Hiltunen, M. (2013). Impaired autophagy and APP processing in Alzheimer's disease: The potential role of Beclin 1 interactome. *Progress in Neurobiology, 106–107*, 33–54. doi:10.1016/j.pneurobio.2013.06.002 PMID:23827971

Sameem, B., Khan, F., & Niaz, K. (2019). l-Cysteine. In S. M. Nabavi & A. S. Silva (Eds.), Nonvitamin and Nonmineral Nutritional Supplements (pp. 53-58). Cambridge, MA: Academic Press.

Sanderson, T. H., Deogracias, M. P., Nangia, K. K., Wang, J., Krause, G. S., & Kumar, R. (2010). PKR-like endoplasmic reticulum kinase (PERK) activation following brain ischemia is independent of unfolded nascent proteins. *Neuroscience, 169*(3), 1307–1314. doi:10.1016/j.neuroscience.2010.05.076 PMID:20538047

Sano, R., & Reed, J. C. (2013). ER stress-induced cell death mechanisms. *Biochimica et Biophysica Acta. Molecular Cell Research, 1833*(12), 3460–3470. doi:10.1016/j.bbamcr.2013.06.028 PMID:23850759

Santos, A. L., & Lindner, A. B. (2017). Protein Posttranslational Modifications: Roles in Aging and Age-Related Disease. *Oxidative Medicine and Cellular Longevity, 2017*, 2017. doi:10.1155/2017/5716409 PMID:28894508

Santos, D., & Cardoso, S. M. (2012). Mitochondrial dynamics and neuronal fate in Parkinson's disease. *Mitochondrion, 12*(4), 428–437. doi:10.1016/j.mito.2012.05.002 PMID:22609323

Santos, L. E., & Ferreira, S. T. (2018). Crosstalk between endoplasmic reticulum stress and brain inflammation in Alzheimer's disease. *Neuropharmacology, 136*, 35–360. doi:10.1016/j.neuropharm.2017.11.016 PMID:29129774

Sarbassov, D. D., Ali, S. M., Kim, D. H., Guertin, D. A., Latek, R. R., Erdjument-Bromage, H., ... Sabatini, D. M. (2004). Rictor, a novel binding partner of mTOR, defines a rapamycin-insensitive and raptor-independent pathway that regulates the cytoskeleton. *Current Biology, 14*(14), 1296–1302. doi:10.1016/j.cub.2004.06.054 PMID:15268862

Sarkar, M., Kuret, J., & Lee, G. (2008). Two motifs within the tau microtubule-binding domain mediate its association with the hsc70 molecular chaperone. *Journal of Neuroscience Research, 86*(12), 2763–2773. doi:10.1002/jnr.21721 PMID:18500754

Sarkar, S., Davies, J. E., Huang, Z., Tunnacliffe, A., & Rubinsztein, D. C. (2007). Trehalose, a novel mTOR-independent autophagy enhancer, accelerates the clearance of mutant huntingtin and alpha-synuclein. *The Journal of Biological Chemistry, 282*(8), 5641–5652. doi:10.1074/jbc.M609532200 PMID:17182613

Sarkar, S., Floto, R. A., Berger, Z., Imarisio, S., Cordenier, A., Pasco, M., ... Rubinsztein, D. C. (2005). Lithium induces autophagy by inhibiting inositol monophosphatase. *The Journal of Cell Biology, 170*(7), 1101–1111. doi:10.1083/jcb.200504035 PMID:16186256

Sarkar, S., Ravikumar, B., Floto, R. A., & Rubinsztein, D. C. (2009). Rapamycin and mTOR-independent autophagy inducers ameliorate toxicity of polyglutamine-expanded huntingtin and related proteinopathies. *Cell Death and Differentiation, 16*(1), 46–56. doi:10.1038/cdd.2008.110 PMID:18636076

Sarkar, S., Raymick, J., & Imam, R. J. (2016). Neuroprotective and therapeutic strategies against Parkinson's Disease: Recent perspective. *International Journal of Molecular Sciences, 17*(6), 904. doi:10.3390/ijms17060904 PMID:27338353

Sarkar, S., & Rubinsztein, D. C. (2008). Huntington's disease: Degradation of mutant huntingtin by autophagy. *The FEBS Journal*, *275*(17), 4263–4270. doi:10.1111/j.1742-4658.2008.06562.x PMID:18637946

Sarto, C., Binz, P. A., & Mocarelli, P. (2000). Heat shock proteins in human cancer. *Electrophoresis*, *21*(6), 1218–1226. doi:10.1002/(SICI)1522-2683(20000401)21:6<1218::AID-ELPS1218>3.0.CO;2-H PMID:10786894

Sato, N., Imaizumi, K., Manabe, T., Taniguchi, M., Hitomi, J., Katayama, T., ... Tohyama, M. (2001). Increased production of β-amyloid and vulnerability to endoplasmic reticulum stress by an aberrant spliced form of presenilin 2*. *The Journal of Biological Chemistry*, *276*(3), 2108–2114. doi:10.1074/jbc.M006886200 PMID:11031265

Sau, D., De Biasi, S., Vitellaro-Zuccarello, L., Riso, P., Guarnieri, S., Porrini, M., ... Poletti, A. (2007). Mutation of SOD1 in ALS: A gain of a loss of function. *Human Molecular Genetics*, *16*(13), 1604–1618. doi:10.1093/hmg/ddm110 PMID:17504823

Saudou, F., & Humbert, S. (2016). The Biology of Huntingtin. *Neuron*, *89*(5), 910–926. doi:10.1016/j.neuron.2016.02.003 PMID:26938440

Saunders, J. A., Estes, K. A., Kosloski, L. M., Allen, H. E., Dempsey, K. M., ... Torres-Russotto, D. R. (2012). CD4C regulatory and effector/memory T cell subsets profile motor dysfunction in Parkinson's disease. *Journal of Neuroimmune Pharmacology*, *7*(4), 927–938. doi:10.100711481-012-9402-z PMID:23054369

Savolainen, M. H., Richie, C. T., Harvey, B. K., Männistö, P. T., Maguire-Zeiss, K. A., & Myöhänen, T. T. (2014). The beneficial effect of a prolyloligopeptidase inhibitor, KYP-2047, on alpha-synuclein clearance and autophagy in A30P transgenic mouse. *Neurobiology of Disease*, *68*, 1–15. doi:10.1016/j.nbd.2014.04.003 PMID:24746855

Sayre, L. M., Moreira, P. I., Smith, M. A., & Perry, G. (2005). Metal ions and oxidative protein modification in neurological disease. *Annali dell'Istituto Superiore di Sanita*, *41*, 143–164. PMID:16244388

Sayre, L. M., Perry, G., Atwood, C. S., & Smith, M. A. (2000). The role of metals in neurodegenerative diseases. *Cellular and Molecular Biology*, *46*, 731–741. PMID:10875436

Sayre, L., Smith, M., & Perry, G. (2012). Chemistry and Biochemistry of Oxidative Stress in Neurodegenerative Disease. *Current Medicinal Chemistry*, *8*(7), 721–738. doi:10.2174/0929867013372922 PMID:11375746

Sbodio, J. I., Snyder, S. H., & Paul, B. D. (2017). Redox Mechanisms In Neurodegeneration: From Disease Outcomes To Therapeutic Opportunities. *Antioxidants & Redox Signalling*, *7321*, 1–136.

Schapansky, J., Olson, K., Van Der Ploeg, R., & Glazner, G. (2007). NF-κB activated by ER calcium release inhibits Aβ-mediated expression of CHOP protein: Enhancement by AD-linked mutant presenilin 1. *Experimental Neurology*, *208*(2), 169–176. doi:10.1016/j.expneurol.2007.04.009 PMID:17927985

Schapira, A. H. V. (2012). Targeting mitochondria for neuroprotection in Parkinson's disease. *Antioxidants & Redox Signalling*, *16*(9), 965–973. doi:10.1089/ars.2011.4419 PMID:22229791

Schapira, A. H., Cooper, J. M., Dexter, D., Clark, J. B., Jenner, P., & Marsden, C. D. (1990). Mitochondrial complex I deficiency in Parkinson's disease. *Journal of Neurochemistry*, *54*(3), 823–827. doi:10.1111/j.1471-4159.1990.tb02325.x PMID:2154550

Schaser, A.J., Osterberg, V.R., Dent, S.E., Stackhouse, T.L., Wakeham, C.M., Boutros, S., … Luna, E. (2019). Alpha-synuclein is a DNA binding protein that modulates DNA repair with implications for Lewy body disorders. *Sci Rep*, *9*.

Scheffner, M., Nuber, U., & Huibregtse, J. M. (1995). Protein ubiquitination involving an E1–E2–E3 enzyme ubiquitin thioester cascade. *Nature*, *373*(6509), 81–83. doi:10.1038/373081a0 PMID:7800044

Scheper, W., & Hoozemans, J. (2009). Endoplasmic Reticulum Protein Quality Control in Neurodegenerative Disease: The Good, the Bad and the Therapy. *Current Medicinal Chemistry, 16*(5), 615–626. doi:10.2174/092986709787458506 PMID:19199926

Scheper, W., & Hoozemans, J. J. M. (2015). The unfolded protein response in neurodegenerative diseases: A neuropathological perspective. *Acta Neuropathologica, 130*(3), 315–331. doi:10.100700401-015-1462-8 PMID:26210990

Schiesling, C., Kieper, N., Seidel, K., & Krüger, R. (2008). Familial Parkinson's disease–genetics, clinical phenotype and neuropathology in relation to the common sporadic form of the disease. *Neuropathology and Applied Neurobiology, 34*(3), 255–271. doi:10.1111/j.1365-2990.2008.00952.x PMID:18447897

Schipper-Krom, S., Juenemann, K., & Reits, E. A. J. (2012). The ubiquitin-proteasome system in huntington's disease: Are proteasomes impaired, initiators of disease, or coming to the rescue? *Biochemistry Research International, 2012*, 1–12. doi:10.1155/2012/837015 PMID:23050151

Schirmer, E. C., Glover, J. R., Singer, M. A., & Lindquist, S. (1996). HSP100/Clp proteins: A common mechanism explains diverse functions. *Trends in Biochemical Sciences, 21*(8), 289–296. doi:10.1016/S0968-0004(96)10038-4 PMID:8772382

Schmidt, R., de Reus, M. A., Scholtens, S. H., van den Berg, L. H., & van den Heuvel, M. P. (2016). Simulating disease propagation across white matter connectome reveals anatomical substrate for neuropathology staging in amyotrophic lateral sclerosis. *NeuroImage, 124*, 762–769. doi:10.1016/j.neuroimage.2015.04.005 PMID:25869856

Schmitz, A., Schneider, A., Kummer, M. P., & Herzog, V. (2004). Endoplasmic reticulum-localized amyloid β-peptide is degraded in the cytosol by two distinct degradation pathways. *Traffic (Copenhagen, Denmark), 5*(2), 89–101. doi:10.1111/j.1600-0854.2004.00159.x PMID:14690498

Schneider, J. A., Aggarwal, N. T., Barnes, L., Boyle, P., & Bennett, D. A. (2009). The neuropathology of older persons with and without dementia from community versus clinic cohorts. *Journal of Alzheimer's Disease, 18*(3), 691–701. doi:10.3233/JAD-2009-1227 PMID:19749406

Schneider, J. A., Li, J. L., Li, Y., Wilson, R. S., Kordower, J. H., & Bennett, D. A. (2006). Substantia nigra tangles are related to gait impairment in older persons. *Annals of Neurology, 59*(1), 166–173. doi:10.1002/ana.20723 PMID:16374822

Schneider, J. L., Suh, Y., & Cuervo, A. M. (2014). Deficient chaperone-mediated autophagy in liver leads to metabolic dysregulation. *Cell Metabolism, 20*(3), 417–432. doi:10.1016/j.cmet.2014.06.009 PMID:25043815

Schnell, J. D., & Hicke, L. (2003). Non-traditional functions of ubiquitin and ubiquitin-binding proteins. *The Journal of Biological Chemistry, 278*(38), 35857–35860. doi:10.1074/jbc.R300018200 PMID:12860974

Schon, E. A., & Przedborski, S. (2011). Mitochondria: The next (neurode) generation. *Neuron, 70*(6), 1033–1053. doi:10.1016/j.neuron.2011.06.003 PMID:21689593

Schreiber, S., Spotorno, N., Schreiber, F., Acosta-Cabronero, J., Kaufmann, J., Machts, J., ... Dengler, R. (2018). Significance of CSF NfL and tau in ALS. *Journal of Neurology, 265*(11), 2633–2645. doi:10.100700415-018-9043-0 PMID:30187162

Schreurs, S., Gerard, M., Derua, R., Waelkens, E., Taymans, J.-M., Baekelandt, V., & Engelborghs, Y. (2014). In vitro phosphorylation does not influence the aggregation kinetics of WT α-synuclein in contrast to its phosphorylation mutants. *International Journal of Molecular Sciences, 15*(1), 1040–1067. doi:10.3390/ijms15011040 PMID:24434619

Schweers, O., Mandelkow, E. M., Biernat, J., & Mandelkow, E. (1995). Oxidation of Cysteine-322 in the repeat domain of microtubule-associated protein Tau controls the in vitro assembly of paired helical filaments. *Proceedings of the National Academy of Sciences of the United States of America, 92*(18), 8463–8467. doi:10.1073/pnas.92.18.8463 PMID:7667312

Schweers, R. L., Zhang, J., Randall, M. S., Loyd, M. R., Li, W., Dorsey, F. C., ... Ney, P. A. (2007). NIX is required for programmed mitochondrial clearance during reticulocyte maturation. *Proceedings of the National Academy of Sciences of the United States of America, 104*(49), 19500–19505. doi:10.1073/pnas.0708818104 PMID:18048346

Seidel, K., Schols, L., Nuber, S., Petrasch-Parwez, E., Gierga, K., Wszolek, Z., & … . (2010). First appraisal of brain pathology owing to A30P mutant alpha-synuclein. *Annals of Neurology, 67*, 684–689. doi:10.1002/ana.21966 PMID:20437567

Seidel, K., Vinet, J., den Dunnen, W. F. A., Brunt, E. R., Meister, M., Boncoraglio, A., ... Carra, S. (2012). The HSPB8-BAG3 chaperone complex is upregulated in astrocytes in the human brain affected by protein aggregation diseases. *Neuropathology and Applied Neurobiology, 38*(1), 39–53. doi:10.1111/j.1365-2990.2011.01198.x PMID:21696420

Seirafi, M., Kozlov, G., & Gehring, K. (2015). Parkin structure and function. *The FEBS Journal, 282*(11), 2076–2088. doi:10.1111/febs.13249 PMID:25712550

Sekine, S., & Youle, R. J. (2018). PINK1 import regulation; a fine system to convey mitochondrial stress to the cytosol. *BMC Biology, 16*(1), 2. doi:10.118612915-017-0470-7 PMID:29325568

Selfridge, J. E., e, L., Lu, J., & Swerdlow, R. H. (2013). Role of mitochondrial homeostasis and dynamics in Alzheimer's disease. *Neurobiology of Disease, 51*, 3–12. doi:10.1016/j.nbd.2011.12.057 PMID:22266017

Selkoe, D. J. (1991). *The molecular pathology of Alzheimer's disease*. Academic Press.

Selkoe, D. J. (1998). The cell biology of beta-amyloid precursor protein and presenilin in Alzheimer's disease. *Trends in Cell Biology, 8*(11), 447–453. doi:10.1016/S0962-8924(98)01363-4 PMID:9854312

Selkoe, D. J. (2001). Alzheimer's disease: Genes, proteins, and therapy. *Physiological Reviews, 81*(2), 741–766. doi:10.1152/physrev.2001.81.2.741 PMID:11274343

Selkoe, D. J. (2004). Cell biology of protein misfolding: The examples of Alzheimer's and Parkinson's diseases. *Nature Cell Biology, 6*(11), 1054–1061. doi:10.1038/ncb1104-1054 PMID:15516999

Sena, L. A., & Chandel, N. S. (2012). Physiological roles of mitochondrial reactive oxygen species. *Molecular Cell, 48*(2), 158–167. doi:10.1016/j.molcel.2012.09.025 PMID:23102266

Seneci, P. (Ed.). (2015). *Chemical modulators of protein misfolding and neurodegenerative disease*. Bethesda, MD: Academic Press.

Sengupta, A., Kabat, J., Novak, M., Wu, Q., Grundke-Iqbal, I., & Iqbal, K. (1998). Phosphorylation of tau at both Thr 231 and Ser 262 is required for maximal inhibition of its binding to microtubules. *Archives of Biochemistry and Biophysics, 357*(2), 299–309. doi:10.1006/abbi.1998.0813 PMID:9735171

Sengupta, R., & Holmgren, A. (2013). Thioredoxin and thioredoxin reductase in relation to reversible S-nitrosylation. *Antioxidants & Redox Signalling, 18*(3), 259–269. doi:10.1089/ars.2012.4716 PMID:22702224

Seo, H., Sonntag, K. C., Kim, W., Cattaneo, E., & Isacson, O. (2007). Proteasome activator enhances survival of Huntington's disease neuronal model cells. *PLoS One, 2*(2), e238. doi:10.1371/journal.pone.0000238 PMID:17327906

Sereno, L., Coma, M., Rodriguez, M., Sanchez-Ferrer, P., Sanchez, M. B., Gich, I., ... Clarimon, J. (2009). A novel GSK-3β inhibitor reduces Alzheimer's pathology and rescues neuronal loss in vivo. *Neurobiology of Disease, 35*(3), 359–367. doi:10.1016/j.nbd.2009.05.025 PMID:19523516

Serrano-Pozo, A., Frosch, M. P., Masliah, E., & Hyman, B. T. (2011). *Neuropathological alterations in Alzheimer disease*. Academic Press.

Serrano-Pozo, A., Frosch, M. P., Masliah, E., & Hyman, B. T. (2011). Neuropathological alterations in Alzheimer disease. *Cold Spring Harbor Perspectives in Medicine*, *1*(1), a006189. doi:10.1101/cshperspect.a006189 PMID:22229116

Serviddio, G., Di Venosa, N., Federici, A., D'Agostino, D., Rollo, T., Prigigallo, F., ... Vendemiale, G. (2005). Brief hypoxia before normoxic reperfusion (postconditioning) protects the heart against ischemia-reperfusion injury by preventing mitochondria peroxyde production and glutathione depletion. *The FASEB Journal*, *19*(3), 354–361. doi:10.1096/fj.04-2338com PMID:15746178

Seubert, P., Vigo-Pelfrey, C., Esch, F., Lee, M., Dovey, H., Davis, D., ... Schenk, D. (1992). Isolation and quantification of soluble Alzheimer's [beta]-peptide from biological fluids. *Nature*, *359*(6393), 325–327. doi:10.1038/359325a0 PMID:1406936

Shacham, T., Sharma, N., & Lederkremer, G. Z. (2019). Protein Misfolding and ER Stress in Huntington's Disease. *Frontiers in Molecular Biosciences*, *6*, 6. doi:10.3389/fmolb.2019.00020 PMID:31001537

Shankar, G. M., & Walsh, D. M. (2009). Alzheimer's disease: Synaptic dysfunction and Abeta. *Molecular Neurodegeneration*, *4*(1), 48. doi:10.1186/1750-1326-4-48 PMID:19930651

Shankar, J. (2008). Coupling Endoplasmic Reticulum Stress to the Cell Death Program in Dopaminergic Cells: Effect of Paraquat. *Neuromolecular Medicine*, *10*(4), 333–342. doi:10.100712017-008-8047-9 PMID:18773310

Shan, X., Chiang, P. M., Price, D. L., & Wong, P. C. (2010). Altered distributions of Gemini of coiled bodies and mitochondria in motor neurons of TDP-43 transgenic mice. *Proceedings of the National Academy of Sciences of the United States of America*, *107*(37), 16325–16330. doi:10.1073/pnas.1003459107 PMID:20736350

Shao, J., & Diamond, M. I. (2007). Polyglutamine diseases: Emerging concepts in pathogenesis and therapy. *Human Molecular Genetics*, *16*(R2), R115–R123. doi:10.1093/hmg/ddm213 PMID:17911155

Shao, Y., Gao, Z., Feldman, T., & Jiang, X. (2007). Stimulation of ATG12-ATG5 conjugation by ribonucleic acid. *Autophagy*, *3*(1), 10–16. doi:10.4161/auto.3270 PMID:16963840

Sharma, S. K., Christen, P., & Goloubinoff, P. (2009). Disaggregating chaperones: An unfolding story. *Current Protein & Peptide Science*, *10*(5), 432–446. doi:10.2174/138920309789351930 PMID:19538153

Sharma, S., Sarathlal, K. C., & Taliyan, R. (2019). Epigenetics in Neurodegenerative Diseases: The Role of Histone Deacetylases. *Cns Neurol Disord-Dr*, *18*(1), 11–18. doi:10.2174/1871527317666181004155136 PMID:30289079

Sharma, S., & Taliyan, R. (2015). *Targeting Histone Deacetylases: A Novel Approach in Parkinson's Disease*. Parkinsons Dis.

Sharp, P., Krishnan, M., Pullar, O., Navarrete, R., Wells, D., & Belleroche, J. (2006). Heat shock protein 27 rescues motor neurons following nerve injury and preserves muscle function. *Experimental Neurology*, *198*(2), 511–518. doi:10.1016/j.expneurol.2005.12.031 PMID:16497297

Shaw, B. F., Lelie, H. L., Durazo, A., Nersissian, A. M., Xu, G., Chan, P. K., ... Whitelegge, J. P. (2008). Detergent-insoluble aggregates associated with amyotrophic lateral sclerosis in transgenic mice contain primarily full-length, unmodified superoxide dismutase-1. *The Journal of Biological Chemistry*, *283*(13), 8340–8350. doi:10.1074/jbc.M707751200 PMID:18192269

Shaw, K., Lees, A., & Stern, G. M. (1980). The impact of treatment with levodopa on Parkinson's disease. *The Quarterly Journal of Medicine*, *49*(3), 283–293. PMID:7465763

Shefa, U., Jeong, N. Y., Song, I. O., Chung, H. J., Kim, D., Jung, J., & Huh, Y. (2019). Mitophagy links oxidative stress conditions and neurodegenerative diseases. *Neural Regeneration Research*, *14*(5), 749–756. doi:10.4103/1673-5374.249218 PMID:30688256

Sheldon, R. A., & Kochi, J. K. (2012). *Metal-catalyzed oxidations of organic compounds: mechanistic principles and synthetic methodology including biochemical processes*. New York, NY: Elsevier.

Sheng, Y., Chattopadhyay, M., Whitelegge, J., & Valentine, J. S. (2012). SOD1 aggregation and ALS: Role of metallation states and disulfide status. *Current Topics in Medicinal Chemistry*, *12*(22), 2560–2572. doi:10.2174/1568026611212220010 PMID:23339308

Shen, H. Y., He, J. C., Wang, Y., Huang, Q. Y., & Chen, J. F. (2005). Geldanamycin induces heat shock protein 70 and protects against MPTP-induced dopaminergic neurotoxicity in mice. *The Journal of Biological Chemistry*, *280*(48), 39962–39969. doi:10.1074/jbc.M505524200 PMID:16210323

Shen, H.-M., & Mizushima, N. (2014). At the end of the autophagic road: An emerging understanding of lysosomal functions in autophagy. *Trends in Biochemical Sciences*, *39*(2), 61–71. doi:10.1016/j.tibs.2013.12.001 PMID:24369758

Shen, W., & Ganetzky, B. (2009). Autophagy promotes synapse development in Drosophila. *The Journal of Cell Biology*, *187*(1), 71–79. doi:10.1083/jcb.200907109 PMID:19786572

Sherrington, R., Rogaev, E. I., Liang, Y. A., Rogaeva, E. A., Levesque, G., Ikeda, M., ... Tsuda, T. (1995). Cloning of a gene bearing missense mutations in early-onset familial Alzheimer's disease. *Nature*, *375*(6534), 754–760. doi:10.1038/375754a0 PMID:7596406

Shiba-Fukushima, K., Arano, T., Matsumoto, G., Inoshita, T., Yoshida, S., Ishihama, Y., ... Imai, Y. (2014). Phosphorylation of mitochondrial polyubiquitin by PINK1 promotes Parkin mitochondrial tethering. *PLOS Genetics*, *10*(12), e1004861. doi:10.1371/journal.pgen.1004861 PMID:25474007

Shibata, M., Lu, T., Furuya, T., Degterev, A., Mizushima, N., Yoshimori, T., ... Yuan, J. (2006). Regulation of Intracellular Accumulation of Mutant Huntingtin by Beclin 1. *The Journal of Biological Chemistry*, *281*(20), 14474–14485. doi:10.1074/jbc.M600364200 PMID:16522639

Shi, G., Lee, J. R., Grimes, D. A., Racacho, L., Ye, D., Yang, H., ... Bulman, D. E. (2011). Functional alteration of PARL contributes to mitochondrial dysregulation in Parkinson's disease. *Human Molecular Genetics*, *20*(10), 1966–1974. doi:10.1093/hmg/ddr077 PMID:21355049

Shih, J., & Donmez, G. (2013). Mitochondrial Sirtuins as Therapeutic Targets for Age-Related Disorders. *Genes & Cancer*, *4*(3-4), 91–96. doi:10.1177/1947601912474931 PMID:24019999

Shi, M., Liu, C., Cook, T. J., Bullock, K. M., Zhao, Y., Ginghina, C., ... Zhang, J. (2014). Plasma exosomal α-synuclein is likely CNS-derived and increased in Parkinson's disease. *Acta Neuropathologica*, *128*(5), 639–650. doi:10.100700401-014-1314-y PMID:24997849

Shimada, K., Motoi, Y., Ishiguro, K., Kambe, T., Matsumoto, S. E., Itaya, M., ... Mizuno, Y. (2012). Long-term oral lithium treatment attenuates motor disturbance in tauopathy model mice: Implications of autophagy promotion. *Neurobiology of Disease*, *46*(1), 101–108. doi:10.1016/j.nbd.2011.12.050 PMID:22249108

Shimizu, T., Watanabe, A., Ogawara, M., Mori, H., & Shirasawa, T. (2000). Isoaspartate formation and neurodegeneration in Alzheimer's disease. *Archives of Biochemistry and Biophysics*, *381*(2), 225–234. doi:10.1006/abbi.2000.1955 PMID:11032409

Shimura, H., Hattori, N., Kubo, S., Mizuno, Y., Asakawa, S., Minoshima, S., ... Suzuki, T. (2000). Familial Parkinson disease gene product, parkin, is a ubiquitin-protein ligase. *Nature Genetics, 25*(3), 302–305. doi:10.1038/77060 PMID:10888878

Shiozawa-West, N., Dunlop, R. A., & Rodgers, K. J. (2015). Using an in vitro model to study oxidised protein accumulation in ageing fibroblasts. *Biochimica et Biophysica Acta. G, General Subjects, 1850*(11), 2177–2184. doi:10.1016/j.bbagen.2015.07.002 PMID:26187876

Shi, R.-Y., Zhu, S.-H., Li, V., Gibson, S. B., Xu, X.-S., & Kong, J.-M. (2014). BNIP3 interacting with LC3 triggers excessive mitophagy in delayed neuronal death in stroke. *CNS Neuroscience & Therapeutics, 20*(12), 1045–1055. doi:10.1111/cns.12325 PMID:25230377

Shirasaki, D. I., Greiner, E. R., Al-Ramahi, I., Gray, M., Boontheung, P., Geschwind, D. H., . . . Loo, J. A. (2012). *Network organization of the huntingtin proteomic interactome in mammalian brain.* Academic Press.

Shirendeb, U. P., Calkins, M. J., Manczak, M., Anekonda, V., Dufour, B., McBride, J. L., . . . Reddy, P. H. (2011). *Mutant huntingtin's interaction with mitochondrial protein Drp1 impairs mitochondrial biogenesis and causes defective axonal transport and synaptic degeneration in Huntington's disease.* Academic Press.

Shirendeb, U. P., Calkins, M. J., Manczak, M., Anekonda, V., Dufour, B., McBride, J. L., ... Reddy, P. H. (2011). Mutant huntingtin's interaction with mitochondrial protein Drp1 impairs mitochondrial biogenesis and causes defective axonal transport and synaptic degeneration in Huntington's disease. *Human Molecular Genetics, 21*(2), 406–420. doi:10.1093/hmg/ddr475 PMID:21997870

Shiwaku, H., & Okazawa, H. (2015). Impaired DNA damage repair as a common feature of neurodegenerative diseases and psychiatric disorders. *Current Molecular Medicine, 15*(2), 119–128. doi:10.2174/1566524015666150303002556 PMID:25732151

Shorter, J. (2008). Hsp104: A Weapon to Combat Diverse Neurodegenerative Disorders. *Neuro-Signals, 16*(1), 63–74. doi:10.1159/000109760 PMID:18097161

Shoulders, M. D., Ryno, L. M., Genereux, J. C., Moresco, J. J., Tu, P. G., Wu, C., ... Wiseman, R. L. (2013). Stress-independent activation of XBP1s and/or ATF6 reveals three functionally diverse ER proteostasis environments. *Cell Reports, 3*(4), 1279–1292. doi:10.1016/j.celrep.2013.03.024 PMID:23583182

Shpilka, T., & Elazar, Z. (2011). Shedding light on mammalian microautophagy. *Developmental Cell, 20*(1), 1–2. doi:10.1016/j.devcel.2010.12.010 PMID:21238917

Shulman, J. M., De Jager, P. L., & Feany, M. B. (2011). Parkinson's disease: Genetics and pathogenesis. *Annual Review of Pathology, 6*(1), 193–222. doi:10.1146/annurev-pathol-011110-130242 PMID:21034221

Sidransky, E., Nalls, M. A., Aasly, J. O., Aharon-Peretz, J., Annesi, G., Barbosa, E. R., ... Chen, C. M. (2009). Multicenter analysis of glucocerebrosidase mutations in Parkinson's disease. *The New England Journal of Medicine, 361*(17), 1651–1661. doi:10.1056/NEJMoa0901281 PMID:19846850

Sierra, A., Abiega, O., Shahraz, A., & Neumann, H. (2013). Janus-faced microglia: Beneficial and detrimental consequences of microglial phagocytosis. *Frontiers in Cellular Neuroscience, 7*, 7. doi:10.3389/fncel.2013.00006 PMID:23386811

Sikorska, B., Liberski, P. P., & Brown, P. (2007). Neuronal autophagy and aggresomes constitute a consistent part of neurodegeneration in experimental scrapie. *Folia Neuropathologica, 45*, 170–178. PMID:18176890

Silberman, D. M. (2018). Metabolism, neurodegeneration and epigenetics: Emerging role of Sirtuins. *Neural Regeneration Research, 13*(3), 417–418. doi:10.4103/1673-5374.228719 PMID:29623921

Silvestri, L., Caputo, V., Bellacchio, E., Atorino, L., Dallapiccola, B., Valente, E. M., & Casari, G. (2005). Mitochondrial import and enzymatic activity of PINK1 mutants associated to recessive parkinsonism. *Human Molecular Genetics, 14*(22), 3477–3492. doi:10.1093/hmg/ddi377 PMID:16207731

Sinclair, D. A., & Guarente, L. (1997). Extrachromosomal rDNA circles--a cause of aging in yeast. *Cell, 91*(7), 1033–1042. doi:10.1016/S0092-8674(00)80493-6 PMID:9428525

Singh, K. K. (2006). Mitochondria damage checkpoint, aging, and cancer. *Annals of the New York Academy of Sciences, 1067*(1), 182–190. doi:10.1196/annals.1354.022 PMID:16803984

Singleton, A., Farrer, M., Johnson, J., Singleton, A., Hague, S., & Kachergus, J. (2003). α-Synuclein locus triplication causes Parkinson's disease. *Science, 302*, 841-841.

Singleton, A., Farrer, M., Johnson, J., Singleton, A., Hague, S., Kachergus, J., ... Gwinn-Hardy, K. (2003). α-Synuclein locus triplication causes Parkinson's disease. *Science, 302*(5646), 841–841. doi:10.1126cience.1090278 PMID:14593171

Sinha, S. C., Colbert, C. L., Becker, N., Wei, Y., & Levine, B. (2008). Molecular basis of the regulation of Beclin 1-dependent autophagy by the γ-herpesvirus 68 Bcl-2 homolog M11. *Autophagy, 4*(8), 989–997. doi:10.4161/auto.6803 PMID:18797192

Sirangelo, I., & Iannuzzi, C. (2017). The Role of Metal Binding in the Amyotrophic Lateral Sclerosis-Related Aggregation of Copper-Zinc Superoxide Dismutase. *Molecules (Basel, Switzerland), 22*(9), 22. doi:10.3390/molecules22091429 PMID:28850080

Sirover, M. A. (2013). GAPDH: β-Amyloid Mediated Iron Accumulation in Alzheimer's Disease: A New Paradigm for Oxidative Stress Induction in Neurodegenerative Disorders. In D. Praticò & P. Mecocci (Eds.), *Studies on Alzheimer's Disease. Oxidative Stress in Applied Basic Research and Clinical Practice*. Totowa, NJ: Humana Press. doi:10.1007/978-1-62703-598-9_4

Sittler, A., Lurz, R., Lueder, G., Priller, J., Lehrach, H., & Hayer-Hartl, M. K. (2001). Geldanamycin activates a heat shock response and inhibits huntingtin aggregation in a cell culture model of Huntington's disease. *Human Molecular Genetics, 10*(12), 1307–1315. doi:10.1093/hmg/10.12.1307 PMID:11406612

Sladek, R., Rocheleau, G., Rung, J., Dina, C., Shen, L., Serre, D., ... Balkau, B. (2007). A genome-wide association study identifies novel risk loci for type 2 diabetes. *Nature, 445*(7130), 881–885. doi:10.1038/nature05616 PMID:17293876

Sliter, D. A., Martinez, J., Hao, L., Chen, X., Sun, N., Fischer, T. D., ... Cai, H. (2018). Parkin and *PINK1* mitigate STING-induced inflammation. *Nature, 561*(7722), 258–262. doi:10.103841586-018-0448-9 PMID:30135585

Small, G. W., Ercoli, L. M., Silverman, D. H., Huang, S. C., Komo, S., Bookheimer, S. Y., ... Mazziotta, J. C. (2000). Cerebral metabolic and cognitive decline in persons at genetic risk for Alzheimer's disease. *Proceedings of the National Academy of Sciences of the United States of America, 97*(11), 6037–6042. doi:10.1073/pnas.090106797 PMID:10811879

Smith, Y., Wichmann, T., Factor, S. A., & DeLong, M. R. (2012). Parkinson's disease therapeutics: new developments and challenges since the introduction of levodopa. *Neuropsychopharmacology, 37*(1), 213-246. doi:10.1038/npp.2011.212

Smith, H. L., Li, W., & Cheetham, M. E. (2015). Molecular chaperones and neuronal proteostasis. *Seminars in Cell & Developmental Biology, 40*, 142–152. doi:10.1016/j.semcdb.2015.03.003 PMID:25770416

Smith, H. L., & Mallucci, G. R. (2016). The unfolded protein response: Mechanisms and therapy of neurodegeneration. *Brain, 139*(8), 2113–2121. doi:10.1093/brain/aww101 PMID:27190028

Smith, J. K., Patil, C. N., Patlolla, S., Gunter, B. W., Booz, G. W., & Duhe, R. J. (2012). Identification of a redox-sensitive switch within the JAK2 catalytic domain. *Free Radical Biology & Medicine*, *52*(6), 1101–1110. doi:10.1016/j.freeradbiomed.2011.12.025 PMID:22281400

Smith, M. H., Ploegh, H. L., & Weissman, J. S. (2011). Road to ruin: Targeting proteins for degradation in the endoplasmic reticulum. *Science*, *334*(6059), 1086–1090. doi:10.1126cience.1209235 PMID:22116878

Snyder, S. W., Ladror, U. S., Wade, W. S., Wang, G. T., Barrett, L. W., Matayoshi, E. D., ... Holzman, T. F. (1994). Amyloid-beta aggregation: Selective inhibition of aggregation in mixtures of amyloid with different chain lengths. *Biophysical Journal*, *67*(3), 1216–1228. doi:10.1016/S0006-3495(94)80591-0 PMID:7811936

So, J. S., Hur, K. Y., Tarrio, M., Ruda, V., Frank-Kamenetsky, M., Fitzgerald, K., ... Lee, A. H. (2012). Silencing of lipid metabolism genes through ire1α-mediated Mrna decay lowers plasma lipids in mice. *Cell Metabolism*, *16*(4), 487–499. doi:10.1016/j.cmet.2012.09.004 PMID:23040070

Solomon, D., Arunodhayan, S., Henri, C., Sushmitha, C., Paartha, M., & Timothy, N. (2015). Neuron the Memory Unit of the Brain. *IOSR Journal of Computer Engineering*, *17*, 48–61.

Somerville, R. A. (1999). Host and transmissible spongiform encephalopathy agent strain control glycosylation of PrP. *The Journal of General Virology*, *80*(7), 1865–1872. doi:10.1099/0022-1317-80-7-1865 PMID:10423157

Song, W., Chen, J., Petrilli, A., Liot, G., Klinglmayr, E., Zhou, Y., . . . Hayden, M. R. (2011). *Mutant huntingtin binds the mitochondrial fission GTPase dynamin-related protein-1 and increases its enzymatic activity*. Academic Press.

Song, J., Park, K. A., Lee, W. T., & Lee, J. E. (2014). Apoptosis signal regulating kinase 1 (ASK1): Potential as a therapeutic target for Alzheimer's disease. *International Journal of Molecular Sciences*, *15*(2), 2119–2129. doi:10.3390/ijms15022119 PMID:24481061

Song, P., Li, S., Wu, H., Gao, R., Rao, G., Wang, D., ... Chen, Q. (2016). Parkin promotes proteasomal degradation of p62: Implication of selective vulnerability of neuronal cells in the pathogenesis of Parkinson's disease. *Protein & Cell*, *7*(2), 114–129. doi:10.100713238-015-0230-9 PMID:26746706

Song, W., Chen, J., Petrilli, A., Liot, G., Klinglmayr, E., Zhou, Y., ... Masliah, E. (2011). Mutant huntingtin binds the mitochondrial fission GTPase dynamin-related protein-1 and increases its enzymatic activity. *Nature Medicine*, *17*(3), 377–382. doi:10.1038/nm.2313 PMID:21336284

Song, Y., Ding, H., Yang, J., Lin, Q., Xue, J., Zhang, Y., ... Cai, Y. (2014). Pyrosequencing analysis of SNCA methylation levels in leukocytes from Parkinson's disease patients. *Neuroscience Letters*, *569*, 85–88. doi:10.1016/j.neulet.2014.03.076 PMID:24721670

Soni, A., Akcakanat, A., Singh, G., Luyimbazi, D., Zheng, Y., Kim, D., ... Meric-Bernstam, F. (2008). eIF4E knockdown decreases breast cancer cell growth without activating Akt signaling. *Molecular Cancer Therapeutics*, *7*(7), 1782–1788. doi:10.1158/1535-7163.MCT-07-2357 PMID:18644990

Son, J. H., Shim, J. H., Kim, K.-H., Ha, J.-Y., & Han, J. Y. (2012). Neuronal autophagy and neurodegenerative diseases. *Experimental & Molecular Medicine*, *44*(2), 89. doi:10.3858/emm.2012.44.2.031 PMID:22257884

Soti, C., & Csermely, P. (2002). Chaperones and aging: Role in neurodegeneration and in other civilizational diseases. *Neurochemistry International*, *41*(6), 383–389. doi:10.1016/S0197-0186(02)00043-8 PMID:12213225

Soti, C., & Csermely, P. (2003). Aging and molecular chaperones. *Experimental Gerontology*, *38*(10), 1037–1040. doi:10.1016/S0531-5565(03)00185-2 PMID:14580856

Sottejeau, Y., Bretteville, A., Cantrelle, F. X., Malmanche, N., Demiaute, F., Mendes, T., ... Dourlen, P. (2015). Tau phosphorylation regulates the interaction between BIN1's SH3 domain and Tau's proline-rich domain. *Acta Neuropathologica Communications*, *3*(1), 58. doi:10.118640478-015-0237-8 PMID:26395440

Sparaco, M., Gaeta, L. M., Tozzi, G., Bertini, E., Pastore, A., Simonati, A., ... Piemonte, F. (2006). Protein glutathionylation in human central nervous system: Potential role in redox regulation of neuronal defense against free radicals. *Journal of Neuroscience Research*, *83*(2), 256–263. doi:10.1002/jnr.20729 PMID:16385584

Specht, C. G., Tigaret, C. M., Rast, G. F., Thalhammer, A., Rudhard, Y., & Schoepfer, R. (2005). Subcellular localisation of recombinant α-and γ-synuclein. *Molecular and Cellular Neurosciences*, *28*(2), 326–334. doi:10.1016/j.mcn.2004.09.017 PMID:15691713

Spencer, B., Potkar, R., Trejo, M., Rockenstein, E., Patrick, C., Gindi, R., ... Masliah, E. (2009). Beclin 1 Gene Transfer Activates Autophagy and Ameliorates the Neurodegenerative Pathology in -Synuclein Models of Parkinson's and Lewy Body Diseases. *The Journal of Neuroscience: The Official Journal of the Society for Neuroscience*, *29*(43), 13578–13588. doi:10.1523/JNEUROSCI.4390-09.2009 PMID:19864570

Sperandio, B. P., Polard, D. S., Ehrlich, P., Renault, E., & Guedon, E. (2005). Guédon: Sulfur Amino Acid Metabolism and Its Control in Lactococcus lactis IL1403. *Journal of Bacteriology*, *187*(11), 3762–3778. doi:10.1128/JB.187.11.3762-3778.2005 PMID:15901700

Sperbera, B. R., Leight, S., Goedert, M., & Lee, V. Y. (1995). Glycogen synthase kinase-3β phosphorylates tau protein at multiple sites in intact cells. *Neuroscience Letters*, *197*(2), 149–153. doi:10.1016/0304-3940(95)11902-9 PMID:8552282

Sperling, R. A., Aisen, P. S., Beckett, L. A., Bennett, D. A., Craft, S., Fagan, A. M., ... Park, D. C. (2011). Toward defining the preclinical stages of Alzheimer's disease: Recommendations from the National Institute on Aging-Alzheimer's Association workgroups on diagnostic guidelines for Alzheimer's disease. *Alzheimer's & Dementia*, *7*(3), 280–292. doi:10.1016/j.jalz.2011.03.003 PMID:21514248

Spillantini, M. G., Crowther, R. A., Jakes, R., Cairns, N. J., Lantos, P. L., & Goedert, M. (1998). Filamentous α-synuclein inclusions link multiple system atrophy with Parkinson's disease and dementia with Lewy bodies. *Neuroscience Letters*, *251*(3), 205–208. doi:10.1016/S0304-3940(98)00504-7 PMID:9726379

Spillantini, M. G., Crowther, R. A., Jakes, R., Hasegawa, M., & Goedert, M. (1998). alpha-Synuclein in filamentous inclusions of Lewy bodies from Parkinson's disease and dementia with lewy bodies. *Proceedings of the National Academy of Sciences of the United States of America*, *95*(11), 6469–6473. doi:10.1073/pnas.95.11.6469 PMID:9600990

Spillantini, M. G., Schmidt, M. L., Lee, V. M. Y., Trojanowski, J. Q., Jakes, R., & Goedert, M. (1997). α-Synuclein in Lewy bodies. *Nature*, *388*(6645), 839–840. doi:10.1038/42166 PMID:9278044

Spilman, P., Podlutskaya, N., Hart, M. J., Debnath, J., Gorostiza, O., Bredesen, D., ... Galvan, V. (2010). Inhibition of mTOR by rapamycin abolishes cognitive deficits and reduces amyloid-beta levels in a mouse model of Alzheimer's disease. *PLoS One*, *5*(4), e9979. doi:10.1371/journal.pone.0009979 PMID:20376313

Spiro, R. G. (2002). Protein glycosylation: Nature, distribution, enzymatic formation, and disease implications of glycopeptide boNDDs. *Glycobiology*, *12*(4), 12. doi:10.1093/glycob/12.4.43R PMID:12042244

Srodulski, S., Sharma, S., Bachstetter, A. B., Brelsfoard, J. M., Pascual, C., Xie, X. S., ... Despa, F. (2014). Neuroinflammation and neurologic deficits in diabetes linked to brain accumulation of amylin. *Molecular Neurodegeneration*, *9*(1), 30. doi:10.1186/1750-1326-9-30 PMID:25149184

Stadel, D., Millarte, V., Tillmann, K. D., Huber, J., Tamin-Yecheskel, B. C., Akutsu, M., ... Dötsch, V. (2015). TECPR2 cooperates with LC3C to regulate COPII-dependent ER export. *Molecular Cell, 60*(1), 89–104. doi:10.1016/j.molcel.2015.09.010 PMID:26431026

Stadtman, E.R. (1992). Protein oxidation and aging. *Science, 257*, 1220–1224.

Stadtman, E. R. (2001). Protein oxidation in aging and age-related diseases. *Annals of the New York Academy of Sciences, 928*(1), 22–38. doi:10.1111/j.1749-6632.2001.tb05632.x PMID:11795513

Stastna, M., & Van Eyk, J. E. (2015). Posttranslational modifications of lysine and evolving role in heart pathologies-Recent developments. *Proteomics, 15*(5-6), 1164–1180. doi:10.1002/pmic.201400312 PMID:25430483

Steen, E., Terry, B. M., & Rivera, E., Cannon, J. L., Neely, T. R., Tavares, R., ... de la Monte, S. M. (2005). Impaired insulin and insulin-like growth factor expression and signaling mechanisms in Alzheimer's disease–is this type 3 diabetes? *Journal of Alzheimer's Disease, 7*(1), 63–80. doi:10.3233/JAD-2005-7107 PMID:15750215

Stefani, M., & Dobson, C. M. (2003). Protein aggregation and aggregate toxicity: New insights into protein folding, misfolding diseases and biological evolution. *Journal of Molecular Medicine (Berlin, Germany), 81*(11), 678–699. doi:10.100700109-003-0464-5 PMID:12942175

Stefanis, L. (2012). α-Synuclein in Parkinson's disease. *Cold Spring Harbor Perspectives in Medicine, 2*(2), a009399–a009399. doi:10.1101/cshperspect.a009399 PMID:22355802

Steffan, J.S., Agrawal, N., & Pallos, J. (2004). SUMO Modification of Huntingtin and Huntington's Disease Pathology. *Science, 304*, 100–104. doi:10.1126cience.1092194

Steiner, J. A., Angot, E., & Brundin, P. (2011). A deadly spread: Cellular mechanisms of α-synuclein transfer. *Cell Death and Differentiation, 18*(9), 1425–1433. doi:10.1038/cdd.2011.53 PMID:21566660

Sterky, F. H., Lee, S., Wibom, R., Olson, L., & Larsson, N. G. (2011). Impaired mitochondrial transport and Parkin-independent degeneration of respiratory chain-deficient dopamine neurons in vivo. *Proceedings of the National Academy of Sciences of the United States of America, 108*(31), 12937–12942. doi:10.1073/pnas.1103295108 PMID:21768369

Stetler, R. A., Gao, Y., Signore, A. P., Cao, G., & Chen, J. (2009). HSP27: Mechanisms of cellular protection against neuronal injury. *Current Molecular Medicine, 9*(7), 863–872. doi:10.2174/156652409789105561 PMID:19860665

Stevens, C. H., Rowe, D., Morel-Kopp, M. C., Orr, C., Russell, T., ... Ranola, M. (2012). Reduced T helper and B lymphocytes in Parkinson's disease. *Journal of Neuroimmunology, 252*(1-2), 95–99. doi:10.1016/j.jneuroim.2012.07.015 PMID:22910543

Stoica, R., Paillusson, S., Gomez-Suaga, P., Mitchell, J. C., Lau, D. H., Gray, E. H., ... Hanger, D. P. (2016). ALS/FTD-associated FUS activates GSK-3β to disrupt the VAPB–PTPIP51 interaction and ER–mitochondria associations. *EMBO Reports, 17*(9), 1326–1342. doi:10.15252/embr.201541726 PMID:27418313

Stojkovska, I., Krainc, D., & Mazzulli, J. R. (2018). Molecular mechanisms of α-synuclein and GBA1 in Parkinson's disease. *Cell and Tissue Research, 373*(1), 51–60. doi:10.100700441-017-2704-y PMID:29064079

Strachan, M. W. J. (2011). RD Lawrence Lecture 2010^. The brain as a target organ in Type 2 diabetes: Exploring the links with cognitive impairment and dementia. *Diabetic Medicine, 28*(2), 141–147. doi:10.1111/j.1464-5491.2010.03199.x PMID:21219420

Strappazzon, F., Nazio, F., Corrado, M., Cianfanelli, V., Romagnoli, A., Fimia, G. M., ... Cecconi, F. (2015). AMBRA1 is able to induce mitophagy via LC3 binding, regardless of PARKIN and p62/SQSTM1. *Cell Death and Differentiation, 22*(3), 419–432. doi:10.1038/cdd.2014.139 PMID:25215947

Strassnig, M., and Ganguli, M. (2005). About a peculiar disease of the cerebral cortex: Alzheimer's original case revisited. *Psychiatry, 1*, 3–8.

Stroissnigg, H., Trancikova, A., Descovich, L., Fuhrmann, J., Kutschera, W., Kostan, J., ... Propst, F. (2007). S-Nitrosylation of microtubule-associated protein 1B mediates nitric-oxide-induced axon retraction. *Nature Cell Biology, 9*(9), 1035–1045. doi:10.1038/ncb1625 PMID:17704770

Sturner, E., & Behl, C. (2017). The role of the multifunctional BAG3 protein in cellular Protein Quality Control and in disease. *Frontiers in Molecular Neuroscience, 10*, 177. doi:10.3389/fnmol.2017.00177 PMID:28680391

Suhr, S. T., Senut, M. C., Whitelegge, J. P., Faull, K. F., Cuizon, D. B., & Gage, F. H. (2001). Identities of sequestered proteins in aggregates from cells with induced polyglutamine expression. *The Journal of Cell Biology, 153*(2), 283–294. doi:10.1083/jcb.153.2.283 PMID:11309410

Sulzer, D., & Surmeier, D. J. (2013). Neuronal vulnerability, pathogenesis, and Parkinson's disease. *Movement Disorders, 28*(6), 715–724. doi:10.1002/mds.25187 PMID:23589357

Sun, H., Gao, J., Ferrington, D. A., Biesiada, H., Williams, T. D., & Squier, T. C. (1999). Repair of oxidized calmodulin by methionine sulfoxide reductase restores ability to activate the plasma membrane Ca-ATPase. *Biochemistry, 38*(1), 105–112. doi:10.1021/bi981295k PMID:9890888

Sun, N., Yun, J., Liu, J., Malide, D., Liu, C., Rovira, I. I., ... Finkel, T. (2015). Measuring in vivo mitophagy. *Molecular Cell, 60*(4), 685–696. doi:10.1016/j.molcel.2015.10.009 PMID:26549682

Sun, Y. X., Ji, X., Mao, X., Xie, L., Jia, J., Galvan, V., ... Jin, K. (2014). Differential activation of mTOR complex 1 signaling in human brain with mild to severe Alzheimer's disease. *Journal of Alzheimer's Disease, 38*(2), 437–444. doi:10.3233/JAD-131124 PMID:23979023

Sun, Y., & MacRae, T. H. (2005). The small heat shock proteins and their role in human disease. *The FEBS Journal, 272*(11), 2613–2627. doi:10.1111/j.1742-4658.2005.04708.x PMID:15943797

Sun, Y., Yao, X., Zhang, Q. J., Zhu, M., Liu, Z. P., Ci, B., ... Levine, B. (2018). Beclin-1-dependent autophagy protects the heart during sepsis. *Circulation, 138*(20), 2247–2262. doi:10.1161/CIRCULATIONAHA.117.032821 PMID:29853517

Surridge, C. D., & Burns, R. G. (1994). The difference in the binding of phosphatidylinositol distinguishes MAP2 from MAP2C and Tau. *Biochemistry, 33*(26), 8051–8057. doi:10.1021/bi00192a009 PMID:8025110

Suzuki, K., & Ohsumi, Y. (2007). Molecular machinery of autophagosome formation in yeast, Saccharomyces cerevisiae. *FEBS Letters, 581*(11), 2156–2161. doi:10.1016/j.febslet.2007.01.096 PMID:17382324

Sweeney, P., Park, H., Baumann, M., Dunlop, J., Frydman, J., Kopito, R., ... Hodgson, R. (2017). Protein misfolding in neurodegenerative diseases: Implications and strategies. *Translational Neurodegeneration, 6*(1), 6–6. doi:10.118640035-017-0077-5 PMID:28293421

Szondy, Z., Mastroberardino, P., Váradi, J., Farrace, M., Nagy, N., Bak, I., . . . Rizzuto, R. (2006). *Tissue transglutaminase (TG2) protects cardiomyocytes against ischemia/reperfusion injury by regulating ATP synthesis*. Academic Press.

Tabner, B. J., Turnbull, S., El-Agnaf, O., & Allsop, D. (2001). Production of reactive oxygen species from aggregating proteins implicated in Alzheimer's disease, Parkinson's disease and other neurodegenerative diseases. *Current Topics in Medicinal Chemistry, 1*, 507–517. doi:10.2174/1568026013394822 PMID:11895127

Tahrir, F. G., Knezevic, T., Gupta, M. K., Gordon, J., Cheung, J. Y., Feldman, A. M., & Khalili, K. (2017). Evidence for the role of BAG3 in mitochondrial quality control in cardiomyocytes. *Journal of Cellular Physiology, 232*(4), 797–805. doi:10.1002/jcp.25476 PMID:27381181

Taipale, M. (2019). Disruption of protein function by pathogenic mutations: Common and uncommon mechanisms (1). *Biochemistry and Cell Biology*, *97*(1), 46–57. doi:10.1139/bcb-2018-0007 PMID:29693415

Tajc, G., Tolbert, B. S., Basavappa, R., & Miller, B. L. (2004). Direct determination of thiol pKa by isothermal titration microcalorimetry. *Journal of the American Chemical Society*, *126*(34), 10508–10509. doi:10.1021/ja047929u PMID:15327286

Takahashi, K., Foster, J. B., & Lin, C. L. G. (2015). Glutamate transporter EAAT2: Regulation, function, and potential as a therapeutic target for neurological and psychiatric disease. *Cellular and Molecular Life Sciences*, *72*(18), 3489–3506. doi:10.100700018-015-1937-8 PMID:26033496

Takahashi, T., Yamashita, H., Nakamura, T., Nagano, Y., & Nakamura, S. (2002). Tyrosine 125 of alpha-synuclein plays a critical role for dimerization following nitrative stress. *Brain Research*, *938*(1-2), 73–80. doi:10.1016/S0006-8993(02)02498-8 PMID:12031537

Takahashi, Y., Coppola, D., Matsushita, N., Cualing, H. D., Sun, M., Sato, Y., ... Wang, H.-G. (2007). Bif-1 interacts with Beclin 1 through UVRAG and regulates autophagy and tumorigenesis. *Nature Cell Biology*, *9*(10), 1142–1151. doi:10.1038/ncb1634 PMID:17891140

Takamura, A., Komatsu, M., Hara, T., Sakamoto, A., Kishi, C., Waguri, S., ... Mizushima, N. (2011). Autophagy-deficient mice develop multiple liver tumors. *Genes & Development*, *25*(8), 795–800. doi:10.1101/gad.2016211 PMID:21498569

Takashima, A. (2006). GSK-3 is essential in the pathogenesis of Alzheimer's disease. *Journal of Alzheimer's Disease*, *9*(3), 309–317. doi:10.3233/JAD-2006-9S335 PMID:16914869

Takuma, K., Du Yan, S. S., Stern, D. M., & Yamada, K. (2005). Mitochondrial dysfunction, endoplasmic reticulum stress, and apoptosis in Alzheimer's disease. *Journal of Pharmacological Sciences*, *97*(3), 312–316. doi:10.1254/jphs.CPJ04006X PMID:15750290

Tanaka, A., Cleland, M. M., Xu, S., Narendra, D. P., Suen, D. F., Karbowski, M., & Youle, R. J. (2010). Proteasome and p97 mediate mitophagy and degradation of mitofusins induced by Parkin. *The Journal of Cell Biology*, *191*(7), 1367–1380. doi:10.1083/jcb.201007013 PMID:21173115

Tanaka, K., Suzuki, T., Hattori, N., & Mizuno, Y. (2004). Ubiquitin, proteasome and parkin. *Biochimica et Biophysica Acta (BBA)-. Molecular Cell Research*, *1695*(1-3), 235–247.

Tanida, I., Komatsu, M., Ueno, T., & Kominami, E. (2003). GATE-16 and GABARAP are authentic modifiers mediated by Apg7 and Apg3. *Biochemical and Biophysical Research Communications*, *300*(3), 637–644. doi:10.1016/S0006-291X(02)02907-8 PMID:12507496

Tanimukai, H., Grundke-Iqbal, I., & Iqbal, K. (2005). Up-regulation of inhibitors of protein phosphatase-2A in Alzheimer's disease. *American Journal of Pathology*, *166*(6), 1761–1771. doi:10.1016/S0002-9440(10)62486-8 PMID:15920161

Tan, P., He, L., Xing, C., Mao, J., Yu, X., Zhu, M., ... Wang, H. Y. (2019). Myeloid loss of Beclin 1 promotes PD-L1 hi precursor B Cell lymphoma development. *The Journal of Clinical Investigation*, *129*(12), 5261–5277. doi:10.1172/JCI127721 PMID:31503548

Taraboulos, A., Rogers, M., Borchelt, D. R., McKinley, M. P., Scott, M., Serban, D., & Prusiner, S. B. (1990). Acquisition of protease resistance by prion proteins in scrapie-infected cells does not require asparagine-linked glycosylation. *Proceedings of the National Academy of Sciences of the United States of America*, *87*(21), 8262–8266. doi:10.1073/pnas.87.21.8262 PMID:1978322

Taraboulos, A., Scott, M., & Semenov, A. (1995). Cholesterol depletion and modification of COOH-terminal targeting sequence of the prion protein inhibit formation of the scrapie isoform. *The Journal of Cell Biology*, *129*(1), 121–132. doi:10.1083/jcb.129.1.121 PMID:7698979

Tarze, A., Deniaud, A., Le Bras, M., Maillier, E., Mollé, D., Larochette, N., . . . Brenner, C. (2007). *GAPDH, a novel regulator of the pro-apoptotic mitochondrial membrane permeabilization*. Academic Press.

Tashiro, K., Hasegawa, M., Ihara, Y., & Iwatsubo, T. (1997). Somatodendritic localization of phosphorylated tau in neonatal and adult rat cerebral cortex. *Neuroreport*, *8*(12), 2797–2801. doi:10.1097/00001756-199708180-00029 PMID:9295120

Tasset, I., & Cuervo, A. M. (2016). Role of chaperone-mediated autophagy in metabolism. *The FEBS Journal*, *283*(13), 2403–2413. doi:10.1111/febs.13677 PMID:26854402

Tatsukawa, H., Furutani, Y., Hitomi, K., & Kojima, S. (2016). *Transglutaminase 2 has opposing roles in the regulation of cellular functions as well as cell growth and death*. Academic Press.

Tatsuta, T., Model, K., & Langer, T. (2005). Formation of membrane-bound ring complexes by prohibitins in mitochondria. *Molecular Biology of the Cell*, *16*(1), 248–259. doi:10.1091/mbc.e04-09-0807 PMID:15525670

Tau, G. Z., & Peterson, B. S. (2010). Normal development of brain circuits. *Neuropsychopharmacology*, *35*(1), 147–168. doi:10.1038/npp.2009.115 PMID:19794405

Taylor, C. G., Nagy, L. E., & Bray, T. M. (1996). Nutritional and hormonal regulation of glutathione homeostasis. *Current Topics in Cellular Regulation*, *34*, 189–208. doi:10.1016/S0070-2137(96)80007-0 PMID:8646848

Taylor, D. R., & Hooper, N. M. (2006). The prion protein and lipid rafts. *Molecular Membrane Biology*, *23*(1), 89–99. doi:10.1080/09687860500449994 PMID:16611584

Tchantchoua, F., Graves, M., Rogers, E., Ortiz, D., & Shea, T. B. (2005). N-acteyl cysteine alleviates oxidative damage to central nervous system of ApoE-deficient mice following folate and vitamin E-deficiency. *J Alzh Dis*, *7*(2), 135–138. doi:10.3233/JAD-2005-7206 PMID:15851851

Temme, C., Weissbach, R., Lilie, H., Wilson, C., Meinhart, A., Meyer, S., ... Wahle, E. (2009). The Drosophila melanogaster gene cg4930 encodes a high affinity inhibitor for endonuclease G. *The Journal of Biological Chemistry*, *284*(13), 8337–8348. doi:10.1074/jbc.M808319200 PMID:19129189

Tenreiro, S., Eckermann, K., & Outeiro, T. F. (2014). Protein phosphorylation in neurodegeneration: Friend or foe? *Frontiers in Molecular Neuroscience*, *7*, 7. doi:10.3389/fnmol.2014.00042 PMID:24860424

Terashima, T., Kawai, H., Fujitani, M., Maeda, K., & Yasuda, H. (2002). SUMO-I co-localized with mutant atrophin-I with expanded polyglutamines accelerates intranuclear aggregation and cell death. *Neuroreport*, *13*(17), 2359–2364. doi:10.1097/00001756-200212030-00038 PMID:12488827

Terracciano, A., An, Y., Sutin, A. R., Thambisetty, M., & Resnick, S. M. (2017). Personality change in the preclinical phase of Alzheimer disease. *JAMA Physiatry*, *74*(12), 1259–1265. doi:10.1001/jamapsychiatry.2017.2816 PMID:28975188

Thakur, P., & Nehru, B. (2014). Long-term heat shock proteins (HSPs) induction by carbenoxolone improves hallmark features of Parkinson's disease in a rotenone-based model. *Neuropharmacology*, *79*, 190–200. doi:10.1016/j.neuropharm.2013.11.016 PMID:24296154

Thangavelu, S. R., Tripathi, P. P., Arya, U., Mishra, H. K., & Subramaniam, J. R. (2011). ALS associated mutant SOD1 impairs the motor neurons and astrocytes and wild type astrocyte secreted-factors reverse the impaired motor neurons. *Annals of Neurosciences*, *18*, 48–55. PMID:25205921

The NINDS NET-PD Investigators. (2006). A randomized, double-blind, futility clinical trial of creatine and minocycline in early Parkinson disease. *Neurology, 66*(5), 664–671. doi:10.1212/01.wnl.0000201252.57661.e1 PMID:16481597

Thinakaran, G., & Koo, E. H. (2008). Amyloid Precursor Protein Trafficking, Processing, and Function. *The Journal of Biological Chemistry, 283*(44), 29615–29619. doi:10.1074/jbc.R800019200 PMID:18650430

Thomas, B., & Beal, M.F. (2007). Parkinson's disease. *Human Molecular Genetics, 16*(2), R183-194.

Thomas, B., & Flint Beal, M. (2007). Parkinson's disease. *Human Molecular Genetics*, 16. PMID:17911161

Thoresen, S. B., Pedersen, N. M., Liestøl, K., & Stenmark, H. (2010). A phosphatidylinositol 3-kinase class III sub-complex containing VPS15, VPS34, Beclin 1, UVRAG and BIF-1 regulates cytokinesis and degradative endocytic traffic. *Experimental Cell Research, 316*(20), 3368–3378. doi:10.1016/j.yexcr.2010.07.008 PMID:20643123

Tian, Y., Bustos, V., Flajolet, M., & Greengard, P. (2011). A small-molecule enhancer of autophagy decreases levels of Aβ and APP-CTF via Atg5-dependent autophagy pathway. *The FASEB Journal, 25*(6), 1934–1942. doi:10.1096/fj.10-175158 PMID:21368103

Tini, P., Belmonte, G., Toscano, M., Miracco, C., Palumbo, S., Pastina, P., ... Pirtoli, L. (2015). … Pirtoli, L. (2015). Combined Epidermal Growth Factor Receptor and Beclin1 Autophagic Protein Expression Analysis Identifies Different Clinical Presentations, Responses to Chemo- and Radiotherapy, and Prognosis in Glioblastoma. *BioMed Research International, 2015*, 1–13. doi:10.1155/2015/208076 PMID:25821789

Tiraboschi, P., Hansen, L., Thal, L., & Corey-Bloom, J. (2004). *The importance of neuritic plaques and tangles to the development and evolution of AD*. Academic Press.

Tiroli-Cepeda, A. O., & Ramos, C. H. (2011). An overview of the role of molecular chaperones in protein homeostasis. *Protein and Peptide Letters, 18*(2), 101–109. doi:10.2174/092986611794475093 PMID:21121892

Tisdale, E. J. (2002). *Glyceraldehyde-3-phosphate dehydrogenase is phosphorylated by protein kinase Cι/λ and plays a role in microtubule dynamics in the early secretory pathway*. Academic Press.

Todd, D. J., Lee, A. H., & Glimcher, L. H. (2008). The endoplasmic reticulum stress response in immunity and autoimmunity. *Nature Reviews. Immunology, 8*(9), 663–674. doi:10.1038/nri2359 PMID:18670423

Tofaris, G. K., Kim, H. T., Hourez, R., Jung, J. W., Kim, K. P., & Goldberg, A. L. (2011). Ubiquitin ligase Nedd4 promotes alpha-synuclein degradation by the endosomal-lysosomal pathway. *Proceedings of the National Academy of Sciences of the United States of America, 108*(41), 17004–17009. doi:10.1073/pnas.1109356108 PMID:21953697

Toker, A. (2012). Phosphoinositide 3-kinases—A historical perspective. In I. Phosphoinositides (Ed.), *Enzymes of Synthesis and Degradation* (pp. 95–110). Dordrecht: Springer.

Toker, A., & Cantley, L. C. (1997). Signalling through the lipid products of phosphoinositide-3-OH kinase. *Nature, 387*(6634), 673–676. doi:10.1038/42648 PMID:9192891

Tong, Y., Yamaguchi, H., Giaime, E., Boyle, S., Kopan, R., Kelleher, R. J., & Shen, J. (2010). Loss of leucine-rich repeat kinase 2 causes impairment of protein degradation pathways, accumulation of α-synuclein, and apoptotic cell death in aged mice. *Proceedings of the National Academy of Sciences of the United States of America, 107*(21), 9879–9884. doi:10.1073/pnas.1004676107 PMID:20457918

Tonoki, A., Kuranaga, E., Tomioka, T., Hamazaki, J., Murata, S., Tanaka, K., & Miura, M. (2009). Genetic Evidence Linking Age-Dependent Attenuation of the 26S Proteasome with the Aging Process. *Molecular and Cellular Biology, 29*(4), 1095–1106. doi:10.1128/MCB.01227-08 PMID:19075009

Tooze, S. A., & Schiavo, G. (2008). Liaisons dangereuses: Autophagy, neuronal survival and neurodegeneration. *Current Opinion in Neurobiology, 18*(5), 504–515. doi:10.1016/j.conb.2008.09.015 PMID:18840524

Tooze, S. A., & Yoshimori, T. (2010). The origin of the autophagosomal membrane. *Nature Cell Biology, 12*(9), 831–835. doi:10.1038/ncb0910-831 PMID:20811355

Tóth, M. E., Szegedi, V., Varga, E., Juhász, G., Horváth, J., Borbély, E., ... Sántha, M. (2013). Overexpression of Hsp27 ameliorates symptoms of Alzheimer's disease in APP/PS1 mice. *Cell Stress & Chaperones, 18*(6), 759–771. doi:10.100712192-013-0428-9 PMID:23605646

Townsend, D. M., Tew, K. D., & Tapiero, H. (2003). The importance of glutathione in human disease. *Biomedicine and Pharmacotherapy, 57*(3-4), 145–155. doi:10.1016/S0753-3322(03)00043-X PMID:12818476

Tramutola, A., Lanzillotta, C., Perluigi, M., & Butterfield, D. A. (2017). Oxidative Stress, Protein Modification and Alzheimer Disease. *Brain Research Bulletin, 133*, 88–96. doi:10.1016/j.brainresbull.2016.06.005 PMID:27316747

Tramutola, A., Triplett, J. C., Di Domenico, F., Niedowicz, D. M., Murphy, M. P., Coccia, R., ... Butterfield, D. A. (2015). Alteration of mTOR signaling occurs early in the progression of Alzheimer disease (AD): Analysis of brain from subjects with pre-clinical AD, amnestic mild cognitive impairment and late-stage AD. *Journal of Neurochemistry, 133*(5), 739–749. doi:10.1111/jnc.13037 PMID:25645581

Traynor, B. J., Bruijn, L., Conwit, R., Beal, F., O'Neill, G., Fagan, S. C., & Cudkowicz, M. E. (2006). Neuroprotective agents for clinical trials in ALS: A systematic assessment. *Neurology, 67*(1), 20–27. doi:10.1212/01.wnl.0000223353.34006.54 PMID:16832072

Trepel, J., Mollapour, M., Giaccone, G., & Neckers, L. (2010). Targeting the dynamic HSP90 complex in cancer. *Nature Reviews. Cancer, 10*(8), 537–549. doi:10.1038/nrc2887 PMID:20651736

Tripathi, M. K., Rajput, C., Mishra, S., Rasheed, M. S. U., & Singh, M. P. (2019). Malfunctioning of Chaperone-Mediated Autophagy in Parkinson's Disease: Feats, Constraints, and Flaws of Modulators. *Neurotoxicity Research, 35*(1), 260–270. doi:10.100712640-018-9917-z PMID:29949106

Trivedi, M. V., Laurence, J. S., & Siahaan, T. J. (2009). The role of thiols and disulfides in protein chemical and physical stability. *Current Protein & Peptide Science, 10*(6), 614–625. doi:10.2174/138920309789630534 PMID:19538140

Truban, D., Hou, X., Caulfield, T. R., Fiesel, F. C., & Springer, W. (2016). PINK1, Parkin, and Mitochondrial Quality Control: What can we Learn about Parkinson's Disease Pathobiology? *Journal of Parkinson's Disease, 7*(1), 13–29. doi:10.3233/JPD-160989 PMID:27911343

Tseng, A. H., Shieh, S. S., & Wang, D. L. (2013). SIRT3 deacetylates FOXO3 to protect mitochondria against oxidative damage. *Free Radical Biology & Medicine, 63*, 222–234. doi:10.1016/j.freeradbiomed.2013.05.002 PMID:23665396

Tsigelny, I. F., Sharikov, Y., Wrasidlo, W., Gonzalez, T., Desplats, P. A., Crews, L., & Masliah, E. (2012). Role of α-synuclein penetration into the membrane in the mechanisms of oligomer pore formation. *The FEBS Journal, 279*(6), 1000–1013. doi:10.1111/j.1742-4658.2012.08489.x PMID:22251432

Tsukada, M., & Ohsumi, Y. (1993). Isolation and characterization of autophagy-defective mutants of Saccharomyces cerevisiae. *FEBS Letters, 333*(1-2), 169–174. doi:10.1016/0014-5793(93)80398-E PMID:8224160

Tsuru, A., Fujimoto, N., Takahashi, S., Saito, M., Nakamura, D., & Iwano, M., … Kohno, K. (2013). Negative feedback by IRE1β optimizes mucin production in goblet cells. *Proceedings of the National Academy of Sciences of the United States of America, 110*(8), 2864–2869. 10.1073/pnas.1212484110

Uddin, M. S., Al Mamun, A., Labu, Z. K., Hidalgo-Lanussa, O., Barreto, G. E., & Ashraf, G. M. (2019a). Autophagic dysfunction in Alzheimer's disease: Cellular and molecular mechanistic approaches to halt Alzheimer's pathogenesis. *Journal of Cellular Physiology*, *234*(6), 8094–8112. doi:10.1002/jcp.27588 PMID:30362531

Uddin, M. S., Mamun, A. A., Jakaria, M., Thangapandiyan, S., Ahmad, J., Rahman, M. A., ... Aleya, L. (2019b). Emerging promise of sulforaphane-mediated Nrf2 signaling cascade against neurological disorders. *The Science of the Total Environment*, 135624. PMID:31784171

Uddin, M., Stachowiak, A., Mamun, A. A., Tzvetkov, N. T., Takeda, S., Atanasov, A. G., ... Stankiewicz, A. M. (2018). Autophagy and Alzheimer's disease: From molecular mechanisms to therapeutic implications. *Frontiers in Aging Neuroscience*, *10*, 4. doi:10.3389/fnagi.2018.00004 PMID:29441009

Uehara, T., Nakamura, T., Yao, D., Shi, Z. Q., Gu, Z., Ma, Y., ... Lipton, S. A. (2006). S-Nitrosylated protein-disulphide isomerase links protein misfolding to neurodegeneration. *Nature*, *441*(7092), 513–517. doi:10.1038/nature04782 PMID:16724068

Ulloa, L., Montejo de Garcini, E., Gómez-Ramos, P., Morán, M. A., & Avila, J. (1994). Microtubule-associated protein MAP1B showing a fetal phosphorylation pattern is present in sites of neurofibrillary degeneration in brains of Alzheimer's disease patients. *Brain Research. Molecular Brain Research*, *26*(1-2), 113–122. doi:10.1016/0169-328X(94)90081-7 PMID:7854037

Um, J. H., & Yun, J. (2017). Emerging role of mitophagy in human diseases and physiology. *BMB Reports*, *50*(6), 299–307. doi:10.5483/BMBRep.2017.50.6.056 PMID:28366191

UniProtKB - O95817 (BAG3_HUMAN). (n.d.). Retrieved fromhttps://www.uniprot.org/uniprot/O95817

Unterberger, U., Höftberger, R., Gelpi, E., Flicker, H., Budka, H., & Voigtländer, T. (2006). Endoplasmic reticulum stress features are prominent in Alzheimer disease but not in prion diseases in vivo. *Journal of Neuropathology and Experimental Neurology*, *65*(4), 348–357. doi:10.1097/01.jnen.0000218445.30535.6f PMID:16691116

Upton, J. P., Wang, L., Han, D., Wang, E. S., Huskey, N. E., Lim, L., ... Oakes, S. A. (2012). IRE1α cleaves select microRNAs during ER stress to derepress translation of proapoptotic caspase-2. *Science*, *338*(6108), 818–822. doi:10.1126cience.1226191 PMID:23042294

Urano, F., Wang, X., Bertolotti, A., Zhang, Y., Chung, P., Harding, H. P., & Ron, D. (2000). Coupling of stress in the ER to activation of JNK protein kinases by transmembrane protein kinase IRE1. *Science*, *287*(5453), 664–666. doi:10.1126cience.287.5453.664 PMID:10650002

Urra, H., Dufey, E., Avril, T., Chevet, E., & Hetz, C. (2016). Endoplasmic Reticulum Stress and the Hallmarks of Cancer. *Trends in Cancer*, *2*(5), 252–262. doi:10.1016/j.trecan.2016.03.007 PMID:28741511

Urra, H., Dufey, E., Lisbona, F., Rojas-Rivera, D., & Hetz, C. (2013). When ER stress reaches a dead end. *Biochimica et Biophysica Acta*, *1833*(12), 3507–3517. doi:10.1016/j.bbamcr.2013.07.024 PMID:23988738

Uryu, K., Richter-Landsberg, C., Welch, W., Sun, E., Goldbaum, O., Norris, E. H., ... Trojanowski, J. Q. (2006). Convergence of heat shock protein 90 with ubiquitin in filamentous alpha-synuclein inclusions of alpha-synucleinopathies. *American Journal of Pathology*, *168*(3), 947–961. doi:10.2353/ajpath.2006.050770 PMID:16507910

Usardi, A., Pooler, A. M., Seereeram, A., Reynolds, C. H., Derkinderen, P., Anderton, B., ... Williamson, R. (2011). Tyrosine phosphorylation of tau regulates its interactions with Fyn SH2 domains, but not SH3 domains, altering the cellular localization of tau. *The FEBS Journal*, *278*(16), 2927–2937. doi:10.1111/j.1742-4658.2011.08218.x PMID:21692989

Uttara, B., Singh, A., Zamboni, P., & Mahajan, R. (2009). Oxidative Stress and Neurodegenerative Diseases: A Review of Upstream and Downstream Antioxidant Therapeutic Options. *Current Neuropharmacology*, *7*(1), 65–74. doi:10.2174/157015909787602823 PMID:19721819

Valdés, P., Mercado, G., Vidal, R. L., Molina, C., Parsons, G., & Court, F. A., … Hetz, C. (2014). Control of dopaminergic neuron survival by the unfolded protein response transcription factor XBP1. *Proceedings of the National Academy of Sciences of the United States of America*, *111*, 6804–6809. 10.1073/pnas.1321845111

Valente, E. M., Abou-Sleiman, P. M., Caputo, V., Muqit, M. M., Harvey, K., Gispert, S., ... Albanese, A. (2004). Hereditary early-onset Parkinson's disease caused by mutations in PINK1. *Science*, *304*(5674), 1158–1160. doi:10.1126cience.1096284 PMID:15087508

Valenti, D., de Bari, L., de Rasmo, D., Signorile, A., Henrion-Caude, A., Contestabile, A., & Vacca, R. A. (2016). The polyphenols resveratrol and epigallocatechin-3-gallate restore the severe impairment of mitochondria in hippocampal progenitor cells from a Down syndrome mouse model. *Biochimica et Biophysica Acta (BBA)-. Molecular Basis of Disease*, *1862*(6), 1093–1104. doi:10.1016/j.bbadis.2016.03.003 PMID:26964795

Valenzuela, V., Collyer, E., Armentano, D., Parsons, G. B., Court, F. A., & Hetz, C. (2012). Activation of the unfolded protein response enhances motor recovery after spinal cord injury. *Cell Death & Disease*, *3*(2), e272. doi:10.1038/cddis.2012.8 PMID:22337234

Vale, R. D. (2003). The molecular motor toolbox for intracellular transport. *Cell*, *112*(4), 467–480. doi:10.1016/S0092-8674(03)00111-9 PMID:12600311

Valionyte, E., Yang, Y., Roberts, S. L., Kelly, J., Lu, B., & Luo, S. (2019). Lowering mutant huntingtin levels and toxicity: Autophagy-endolysosome pathways in Huntington's disease. *Journal of Molecular Biology*. doi:10.1016/j.jmb.2019.11.012 PMID:31786267

Valle, C., & Carri, M. T. (2017). Cysteine Modification in the Pathogenesis of ALS. *Frontiers in Molecular Neuroscience*, *10*(5), 1–8. doi:10.3389/fnmol.2017.00005 PMID:28167899

Van Den Steen, P., Rudd, P. M., Dwek, R. A., & Opdenakker, G. (1998). Concepts and principles of O-linked glycosylation. *Critical Reviews in Biochemistry and Molecular Biology*, *33*(3), 151–208. doi:10.1080/10409239891204198 PMID:9673446

van der Heide, L. P., Ramakers, G. M., & Smidt, M. P. (2006). Insulin signaling in the central nervous system: Learning to survive. *Progress in Neurobiology*, *79*(4), 205–221. doi:10.1016/j.pneurobio.2006.06.003 PMID:16916571

Van Duijn, E., Kingma, E., & Van der Mast, R. (2007). *Psychopathology in verified Huntington's disease gene carriers*. Academic Press.

van Humbeeck, C., Cornelissen, T., & Vandenberghe, W. (2011). Ambra1. *Autophagy*, *7*(12), 1555–1556. doi:10.4161/auto.7.12.17893 PMID:21921694

Van Laar, V. S., Arnold, B., Cassady, S. J., Chu, C. T., Burton, E. A., & Berman, S. B. (2010). Bioenergetics of neurons inhibit the translocation response of Parkin following rapid mitochondrial depolarization. *Human Molecular Genetics*, *20*(5), 927–940. doi:10.1093/hmg/ddq531 PMID:21147754

Van Laar, V. S., & Berman, S. B. (2013). The interplay of neuronal mitochondrial dynamics and bioenergetics: Implications for Parkinson's disease. *Neurobiology of Disease*, *51*, 43–55. doi:10.1016/j.nbd.2012.05.015 PMID:22668779

Van Leeuwen, F.W., De Kleijn, D.P.V., & Van Den Hurk, H.H. (1998). Frameshift mutants of β amyloid precursor protein and ubiquitin-B in Alzheimer's and Down patients. *Science*, *279*, 242–247. doi:10.1126cience.279.5348.242

Van Montfort, R., Slingsby, C., & Vierling, E. (2001). Structure and function of the small heat shock protein/alpha-crystallin family of molecular chaperones. *Advances in Protein Chemistry*, *59*, 105–156. doi:10.1016/S0065-3233(01)59004-X PMID:11868270

Vander Heiden, M. G., Cantley, L. C., & Thompson, C. B. (2009). Understanding the Warburg effect: The metabolic requirements of cell proliferation. *Science*, *324*(5930), 1029–1033. doi:10.1126cience.1160809 PMID:19460998

Vanderweyde, T., Apicco, D. J., Youmans-Kidder, K., Ash, P. E., Cook, C., da Rocha, E. L., ... Ivanov, P. (2016). Interaction of tau with the RNA-binding protein TIA1 regulates tau pathophysiology and toxicity. *Cell Reports*, *15*(7), 1455–1466. doi:10.1016/j.celrep.2016.04.045 PMID:27160897

Vanhaesebroeck, B., & Waterfield, M. D. (1999). Signaling by distinct classes of phosphoinositide 3-kinases. *Experimental Cell Research*, *253*(1), 239–254. doi:10.1006/excr.1999.4701 PMID:10579926

Vantaggiato, C., Crimella, C., Airoldi, G., Polishchuk, R., Bonato, S., Brighina, E., ... Santorelli, F. M. (2013). Defective autophagy in spastizin mutated patients with hereditary spastic paraparesis type 15. *Brain*, *136*(10), 3119–3139. doi:10.1093/brain/awt227 PMID:24030950

Varadarajan, S., Yatin, S., Aksenova, M., & Butterfield, D. A. (2000). Review: Alzheimer's amyloid β-peptide-associated free radical oxidative stress and neurotoxicity. *Journal of Structural Biology*, *130*(2-3), 184–208. doi:10.1006/jsbi.2000.4274 PMID:10940225

Varga, R. E., Khundadze, M., Damme, M., Nietzsche, S., Hoffmann, B., Stauber, T., ... Kessels, M. M. (2015). In vivo evidence for lysosome depletion and impaired autophagic clearance in hereditary spastic paraplegia type SPG11. *PLOS Genetics*, *11*(8), e1005454. doi:10.1371/journal.pgen.1005454 PMID:26284655

Varghese, M., Ho, L., Wang, J., Zhao, W., Levine, S., Ono, K., ... Pasinetti, G. (2014). Green coffee as a novel agent for Alzheimer's disease prevention by attenuating diabetes. *Translational Neuroscience*, *5*(2), 111–116. doi:10.247813380-014-0213-y

Varshavsky, A. (2005). Regulated protein degradation. *Trends in Biochemical Sciences*, *30*(6), 283–286. doi:10.1016/j.tibs.2005.04.005 PMID:15950869

Vassar, R. (2002). *β-secretase (BACE) as a drug target for Alzheimer's disease*. Academic Press.

Vaux, D. L. (2011). Apoptogenic factors released from mitochondria. *Biochimica et Biophysica Acta. Molecular Cell Research*, *1813*(4), 546–550. doi:10.1016/j.bbamcr.2010.08.002 PMID:20713095

Vaziri, H., Dessain, S. K., Eaton, E. N., Imai, S. I., Frye, R. A., Pandita, T. K., ... Weinberg, R. A. (2001). hSIR2SIRT1 functions as an NAD-dependent p53 deacetylase. *Cell*, *107*(2), 149–159. doi:10.1016/S0092-8674(01)00527-X PMID:11672523

Veeresh, P., Kaur, H., Sarmah, D., Mounica, L., Verma, G., Kotian, V., ... Bhattacharya, P. (2019). Endoplasmic reticulum-mitochondria crosstalk: From junction to function across neurological disorders. *Annals of the New York Academy of Sciences*, *1457*(1), 41–60. doi:10.1111/nyas.14212 PMID:31460675

Veereshwarayya, V., Kumar, P., Rosen, K. M., Mestril, R., & Querfurth, H. W. (2006). Differential effects of mitochondrial heat shock protein 60 and related molecular chaperones to prevent intracellular beta-amyloid-induced inhibition of complex IV and limit apoptosis. *The Journal of Biological Chemistry*, *281*(40), 29468–29478. doi:10.1074/jbc.M602533200 PMID:16887805

Vekrellis, K., Xilouri, M., Emmanouilidou, E., Rideout, H. J., & Stefanis, L. (2011). Pathological roles of α-synuclein in neurological disorders. *Lancet Neurology*, *10*(11), 1015–1025. doi:10.1016/S1474-4422(11)70213-7 PMID:22014436

Vekrellis, K., Ye, Z., Qiu, W. Q., Walsh, D., Hartley, D., Chesneau, V., ... Selkoe, D. J. (2000). Neurons regulate extracellular levels of amyloid β-protein via proteolysis by insulin-degrading enzyme. *The Journal of Neuroscience: The Official Journal of the Society for Neuroscience*, *20*(5), 1657–1665. doi:10.1523/JNEUROSCI.20-05-01657.2000 PMID:10684867

Venkatraman, A., Hu, Y. S., Didonna, A., Cvetanovic, M., Krbanjevic, A., Bilesimo, P., & Opal, P. (2014). The histone deacetylase HDAC3 is essential for Purkinje cell function, potentially complicating the use of HDAC inhibitors in SCA1. *Human Molecular Genetics*, *23*(14), 3733–3745. doi:10.1093/hmg/ddu081 PMID:24594842

Venkatramani, A., & Panda, D. (2019). Regulation of neuronal microtubule dynamics by tau: Implications for tauopathies. *International Journal of Biological Macromolecules*, *133*, 473–483. doi:10.1016/j.ijbiomac.2019.04.120 PMID:31004638

Venkatraman, P., Wetzel, R., Tanaka, M., Nukina, N., & Goldberg, A. L. (2004). Eukaryotic proteasomes cannot digest polyglutamine sequences and release them during degradation of polyglutamine-containing proteins. *Molecular Cell*, *14*(1), 95–104. doi:10.1016/S1097-2765(04)00151-0 PMID:15068806

Verhoef, L. G. G. C. (2002). Aggregate formation inhibits proteasomal degradation of polyglutamine proteins. *Human Molecular Genetics*, *11*(22), 2689–2700. doi:10.1093/hmg/11.22.2689 PMID:12374759

Verkhratsky, A. (2005). Physiology and pathophysiology of the calcium store in the endoplasmic reticulum of neurons. *Physiological Reviews*, *85*(1), 201–279. doi:10.1152/physrev.00004.2004 PMID:15618481

Vilarino-Guell, C., Wider, C., Ross, O. A., Dachsel, J. C., Kachergus, J. M., Lincoln, S. J., ... Farrer, M. J. (2011). VPS35 mutations in Parkinson disease. *American Journal of Human Genetics*, *89*(1), 162–167. doi:10.1016/j.ajhg.2011.06.001 PMID:21763482

Villar-Pique, A., Lopes da Fonseca, T., & Outeiro, T. F. (2016). Structure, function and toxicity of alpha-synuclein: The Bermuda triangle in synucleinopathies. *Journal of Neurochemistry*, *139*(Suppl. 1), 240–255. doi:10.1111/jnc.13249 PMID:26190401

Vingtdeux, V., Chandakkar, P., Zhao, H., d'Abramo, C., Davies, P., & Marambaud, P. (2011). Novel synthetic small-molecule activators of AMPK as enhancers of autophagy and amyloid-β peptide degradation. *The FASEB Journal*, *25*(1), 219–231. doi:10.1096/fj.10-167361 PMID:20852062

Vinothkumar, K. R., & Henderson, R. (2010). Structures of membrane proteins. *Quarterly Reviews of Biophysics*, *43*(1), 65–158. doi:10.1017/S0033583510000041 PMID:20667175

Virkamäki, A., Ueki, K., & Kahn, C. R. (1999). Protein–protein interaction in insulin signaling and the molecular mechanisms of insulin resistance. *The Journal of Clinical Investigation*, *103*(7), 931–943. doi:10.1172/JCI6609 PMID:10194465

Visanji, N. P., Brooks, P. L., Hazrati, L. N., & Lang, A. E. (2013). The prion hypothesis in Parkinson's disease: Braak to the future. *Acta Neuropathologica Communications*, *1*(1), 2. doi:10.1186/2051-5960-1-2 PMID:24252164

Vitvitsky, V., Thomas, M., Ghorpade, A., Gendelman, H. E., & Banerjee, R. (2006). A functional transsulfuration pathway in the brain links to glutathione homeostasis. *The Journal of Biological Chemistry*, *281*(47), 35785–35793. doi:10.1074/jbc.M602799200 PMID:17005561

Vives-Bauza, C., Zhou, C., Huang, Y., Cui, M., De Vries, R. L., Kim, J., ... Magrané, J. (2010). *PINK1*-dependent recruitment of Parkin to mitochondria in mitophagy. *Proceedings of the National Academy of Sciences of the United States of America*, *107*(1), 378–383. doi:10.1073/pnas.0911187107 PMID:19966284

Vlassenko, A. G., Vaishnavi, S. N., Couture, L., Sacco, D., Shannon, B. J., Mach, R. H., ... Mintun, M. A. (2010). Spatial correlation between brain aerobic glycolysis and amyloid-beta (Abeta) deposition. *Proceedings of the National Academy of Sciences of the United States of America*, *107*(41), 17763–17767. doi:10.1073/pnas.1010461107 PMID:20837517

Vogiatzi, T., Xilouri, M., Vekrellis, K., & Stefanis, L. (2008). Wild type α-synuclein is degraded by chaperone-mediated autophagy and macroautophagy in neuronal cells. *The Journal of Biological Chemistry*, *283*(35), 23542–23556. doi:10.1074/jbc.M801992200 PMID:18566453

Volmer, R., & Ron, D. (2015). Lipid-dependent regulation of the unfolded protein response. *Current Opinion in Cell Biology*, *33*, 67–73. doi:10.1016/j.ceb.2014.12.002 PMID:25543896

Volmer, R., Van Der Ploeg, K., & Ron, D. (2013). Membrane lipid saturation activates endoplasmic reticulum unfolded protein response transducers through their transmembrane domains. *Proceedings of the National Academy of Sciences of the United States of America*, *110*(12), 4628–4633. doi:10.1073/pnas.1217611110 PMID:23487760

Volpicelli-Daley, L. A., Gamble, K. L., Schultheiss, C. E., Riddle, D. M., West, A. B., & Lee, V. M. Y. (2014). Formation of α-synuclein Lewy neurite–like aggregates in axons impedes the transport of distinct endosomes. *Molecular Biology of the Cell*, *25*(25), 4010–4023. doi:10.1091/mbc.e14-02-0741 PMID:25298402

Von Bergen, M., Friedhoff, P., Biernat, J., Heberle, J., Mandelkow, E. M., & Mandelkow, E. (2000). Assembly of τ protein into Alzheimer paired helical filaments depends on a local sequence motif (306VQIVYK311) forming β structure. *Proceedings of the National Academy of Sciences of the United States of America*, *97*(10), 5129–5134. doi:10.1073/pnas.97.10.5129 PMID:10805776

Vonsattel, J. P. G., & DiFiglia, M. (1998). Huntington disease. *Journal of Neuropathology and Experimental Neurology*, *57*(5), 369. doi:10.1097/00005072-199805000-00001 PMID:9596408

Vuono, R., Winder-Rhodes, S., de Silva, R., Cisbani, G., Drouin-Ouellet, J., Spillantini, M. G., ... Barker, R. A. (2015). The role of tau in the pathological process and clinical expression of Huntington's disease. *Brain*, *138*(7), 1907–1918. doi:10.1093/brain/awv107 PMID:25953777

Wade-Martins, R. (2012). Genetics: The MAPT locus—a genetic paradigm in disease susceptibility. *Nature Reviews. Neurology*, *8*(9), 477–478. doi:10.1038/nrneurol.2012.169 PMID:22940644

Waelter, S., Boeddrich, A., Lurz, R., Scherzinger, E., Lueder, G., Lehrach, H., & Wanker, E. E. (2001). Accumulation of mutant huntingtin fragments in aggresome-like inclusion bodies as a result of insufficient protein degradation. *Molecular Biology of the Cell*, *12*(5), 1393–1407. doi:10.1091/mbc.12.5.1393 PMID:11359930

Wager, K., & Russell, C. (2013). Mitophagy and neurodegeneration: The zebrafish model system. *Autophagy*, *9*(11), 1693–1709. doi:10.4161/auto.25082 PMID:23939015

Wagner, J., Ryazanov, S., Leonov, A., Levin, J., Shi, S., Schmidt, F., ... Giese, A. (2013). Anle138b: A novel oligomer modulator for disease-modifying therapy of neurodegenerative diseases such as prion and Parkinson's disease. *Acta Neuropathologica*, *125*(6), 795–813. doi:10.100700401-013-1114-9 PMID:23604588

Wallace, D. C. (2005). A mitochondrial paradigm of metabolic and degenerative diseases, aging, and cancer: A dawn for evolutionary medicine. *Annual Review of Genetics*, *39*(1), 359–407. doi:10.1146/annurev.genet.39.110304.095751 PMID:16285865

Walter, P., & Ron, D. (2011). The unfolded protein response: From stress pathway to homeostatic regulation. *Science*, *334*(6059), 1081–1086. doi:10.1126cience.1209038 PMID:22116877

Wang, Y., Zhou, J. S., Xu, X. C., Li, Z. Y., Chen, H. P., Ying, S. M., . . . Chen, Z. H. (2018). Endoplasmic reticulum chaperone GRP78 mediates cigarette smoke-induced necroptosis and injury in bronchial epithelium. *Int J Chron obstruct Pulmonary Dis*, *13*, 571-581.

Wang, C., Tan, Z., Niu, B., Tsang, K. Y., Tai, A., Chan, W. C. W., & … . (2018). Inhibiting the integrated stress response pathway prevents aberrant chondrocyte differentiation thereby alleviating chondrodysplasia. *eLife*, 7. PMID:30024379

Wang, D. B., Uo, T., Kinoshita, C., Sopher, B. L., Lee, R. J., Murphy, S. P., ... Morrison, R. S. (2014). Bax Interacting Factor-1 Promotes Survival and Mitochondrial Elongation in Neurons. *The Journal of Neuroscience: The Official Journal of the Society for Neuroscience*, *34*(7), 2674–2683. doi:10.1523/JNEUROSCI.4074-13.2014 PMID:24523556

Wang, F., Wang, X., Yuan, C. G., & Ma, J. (2010). Generating a prion with bacterially expressed recombinant prion protein. *Science*, *327*(5969), 1132–1135. doi:10.1126cience.1183748 PMID:20110469

Wang, H., Song, P., Du, L., Tian, W., Yue, W., Liu, M., ... Zhou, J. (2011). Parkin ubiquitinates Drp1 for proteasome-dependent degradation implication of dysregulated mitochondrial dynamics in parkinson disease. *The Journal of Biological Chemistry*, *286*(13), 11649–11658. doi:10.1074/jbc.M110.144238 PMID:21292769

Wang, J. Z., Grundke-Iqbal, I., & Iqbal, K. (2007). Kinases and phosphatases and tau sites involved in Alzheimer neurofibrillary degeneration. *The European Journal of Neuroscience*, *25*(1), 59–68. doi:10.1111/j.1460-9568.2006.05226.x PMID:17241267

Wang, J., Fivecoat, H., Ho, L., Pan, Y., Ling, E., & Pasinetti, G. M. (2010). The role of Sirt1: At the crossroad between promotion of longevity and protection against Alzheimer's disease neuropathology. *Biochimica et Biophysica Acta*, *1804*(8), 1690–1694. doi:10.1016/j.bbapap.2009.11.015 PMID:19945548

Wang, J., Gu, B. J., Masters, C. L., & Wang, Y.-J. (2017). A systemic view of Alzheimer disease— Insights from amyloid-β metabolism beyond the brain. *Nature Reviews. Neurology*, *13*(10), 612–623. doi:10.1038/nrneurol.2017.111 PMID:28960209

Wang, J., Tung, Y. C., Wang, Y., Li, X. T., Iqbal, K., & Grundke-Iqbal, I. (2001). Hyperphosphorylation and accumulation of neurofilament proteins in Alzheimer disease brain and in okadaic acid-treated SY5Y cells. *FEBS Letters*, *507*(1), 81–87. doi:10.1016/S0014-5793(01)02944-1 PMID:11682063

Wang, L., Harris, T. E., & Lawrence, J. C. Jr. (2008). Regulation of proline-rich Akt substrate of 40 kDa (PRAS40) function by mammalian target of rapamycin complex 1 (mTORC1)-mediated phosphorylation. *The Journal of Biological Chemistry*, *283*(23), 15619–15627. doi:10.1074/jbc.M800723200 PMID:18372248

Wang, L., Perera, B. G. K., Hari, S. B., Bhhatarai, B., Backes, B. J., Seeliger, M. A., ... Maly, D. J. (2012). Divergent allosteric control of the IRE1α endoribonuclease using kinase inhibitors. *Nature Chemical Biology*, *8*(12), 982–989. doi:10.1038/nchembio.1094 PMID:23086298

Wang, M., & Kaufman, R. J. (2016). Protein misfolding in the endoplasmic reticulum as a conduit to human disease. *Nature*, *529*(7586), 326–335. doi:10.1038/nature17041 PMID:26791723

Wang, W., Wang, X., Fujioka, H., Hoppel, C., Whone, A. L., Caldwell, M. A., ... Zhu, X. (2016, January). Parkinson's disease–associated mutant VPS35 causes mitochondrial dysfunction by recycling DLP1 complexes. *Nature Medicine*, *22*(1), 54–63. doi:10.1038/nm.3983 PMID:26618722

Wang, X., Campbell, L. E., Miller, C. M., & Proud, C. G. (1998). Amino acid availability regulates p70 S6 kinase and multiple translation factors. *The Biochemical Journal*, *335*(3), 711–711. doi:10.1042/bj3350711w PMID:9693128

Wang, X., Su, B., Siedlak, S. L., Moreira, P. I., Fujioka, H., Wang, Y., ... Zhu, X. (2008). Amyloid-beta overproduction causes abnormal mitochondrial dynamics via differential modulation of mitochondrial fission/fusion proteins. *Proceedings of the National Academy of Sciences of the United States of America*, *105*(49), 19318–19323. doi:10.1073/pnas.0804871105 PMID:19050078

Wang, X., Winter, D., Ashrafi, G., Schlehe, J., Wong, Y. L., Selkoe, D., ... Schwarz, T. L. (2011). PINK1 and Parkin target Miro for phosphorylation and degradation to arrest mitochondrial motility. *Cell*, *147*(4), 893–906. doi:10.1016/j.cell.2011.10.018 PMID:22078885

Wang, Y., Liu, N., & Lu, B. (2019). Mechanisms and roles of mitophagy in neurodegenerative diseases. *CNS Neuroscience & Therapeutics*, *25*(7), 859–875. doi:10.1111/cns.13140 PMID:31050206

Wang, Y., Martinez-Vicente, M., Krüger, U., Kaushik, S., Wong, E., Mandelkow, E.-M., ... Mandelkow, E. (2010). Synergy and antagonism of macroautophagy and chaperone-mediated autophagy in a cell model of pathological tau aggregation. *Autophagy*, *6*(1), 182–183. doi:10.4161/auto.6.1.10815 PMID:20023429

Wanker, E. E., Ast, A., Schindler, F., Trepte, P., & Schnoegl, S. (2019). The pathobiology of perturbed mutant huntingtin protein-protein interactions in Huntington's disease. *Journal of Neurochemistry*, *151*(4), 507–519. doi:10.1111/jnc.14853 PMID:31418858

Warby, S. C., Visscher, H., Collins, J. A., Doty, C. N., Carter, C., Butland, S. L., ... Hayden, M. R. (2011). HTT haplotypes contribute to differences in Huntington disease prevalence between Europe and East Asia. European journal of human genetics. *European Journal of Human Genetics*, *19*(5), 561–566. doi:10.1038/ejhg.2010.229 PMID:21248742

Watanabe, A., Takio, K., & Ihara, Y. (1999). Deamidation and isoaspartate formation in smeared tau in paired helical filaments: Unusual properties of the microtubule-binding domain of tau. *The Journal of Biological Chemistry*, *274*(11), 7368–7378. doi:10.1074/jbc.274.11.7368 PMID:10066801

Waterham, H. R., Koster, J., van Roermund, C. W., Mooyer, P. A., Wanders, R. J., & Leonard, J. V. (2007). A lethal defect of mitochondrial and peroxisomal fission. *The New England Journal of Medicine*, *356*(17), 1736–1741. doi:10.1056/NEJMoa064436 PMID:17460227

Watts, G. D., Wymer, J., Kovach, M. J., Mehta, S. G., Mumm, S., Darvish, D., . . . Kimonis, V. E. (2004). *Inclusion body myopathy associated with Paget disease of bone and frontotemporal dementia is caused by mutant valosin-containing protein*. Academic Press.

Waxman, E. A., & Giasson, B. I. (2008). Specificity and regulation of casein kinase-mediated phosphorylation of alpha-synuclein. *Journal of Neuropathology and Experimental Neurology*, *67*(5), 402–416. doi:10.1097/NEN.0b013e3186fc995 PMID:18451726

Waxman, E. A., & Giasson, B. I. (2011). Characterization of kinases involved in the phosphorylation of aggregated α-synuclein. *Journal of Neuroscience Research*, *89*(2), 231–247. doi:10.1002/jnr.22537 PMID:21162130

Webb, J. L., Ravikumar, B., Atkins, J., Skepper, J. N., & Rubinsztein, D. C. (2003). Alpha-Synuclein is degraded by both autophagy and the proteasome. *The Journal of Biological Chemistry*, *278*(27), 25009–25013. doi:10.1074/jbc.M300227200 PMID:12719433

Weggen, S., Eriksen, J. L., Das, P., Sagi, S. A., Wang, R., Pietrzik, C. U., . . . Bulter, T. (2001). *A subset of NSAIDs lower amyloidogenic Aβ42 independently of cyclooxygenase activity*. Academic Press.

Weidberg, H., Shvets, E., Shpilka, T., Shimron, F., Shinder, V., & Elazar, Z. (2010). LC3 and GATE-16/GABARAP subfamilies are both essential yet act differently in autophagosome biogenesis. *The EMBO Journal*, *29*(11), 1792–1802. doi:10.1038/emboj.2010.74 PMID:20418806

Wei, H., Liu, L., & Chen, Q. (2015). Selective removal of mitochondria via mitophagy: Distinct pathways for different mitochondrial stresses. *Biochimica et Biophysica Acta*, *1853*(10), 2784–2790. doi:10.1016/j.bbamcr.2015.03.013 PMID:25840011

Weir, H. J., Murray, T. K., Kehoe, P. G., Love, S., Verdin, E. M., O'Neill, M. J., ... Balthasar, N. (2012). CNS SIRT3 expression is altered by reactive oxygen species and in Alzheimer's disease. *PLoS One*, *7*(11), e48225. doi:10.1371/journal.pone.0048225 PMID:23139766

Welzel, A. T., & Walsh, D. M. (2011). Aberrant protein structure and diseases of the brain. *Irish Journal of Medical Science*, *180*(1), 15–22. doi:10.100711845-010-0606-z PMID:20960238

Wenk, G. L. (2003). *Neuropathologic changes in Alzheimer's disease*. Academic Press.

West, A. P., Khoury-Hanold, W., Staron, M., Tal, M. C., Pineda, C. M., Lang, S. M., ... Kaech, S. M. (2015). Mitochondrial DNA stress primes the antiviral innate immune response. *Nature*, *520*(7548), 55–57. doi:10.1038/nature14156 PMID:25642965

West, A. P., & Shadel, G. S. (2017). Mitochondrial DNA in innate immune responses and inflammatory pathology. *Nature Reviews. Immunology*, *17*(6), 363–375. doi:10.1038/nri.2017.21 PMID:28393922

Westergard, T., Jensen, B. K., Wen, X., Cai, J., Kropf, E., Iacovitti, L., ... Trotti, D. (2016). Cell-to-Cell Transmission of Dipeptide Repeat Proteins Linked to C9orf72-ALS/FTD. *Cell Reports*, *17*(3), 645–652. doi:10.1016/j.celrep.2016.09.032 PMID:27732842

White, E., & DiPaola, R. S. (2009). The double-edged sword of autophagy modulation in cancer. *Clinical Cancer Research*, *15*(17), 5308–5316. doi:10.1158/1078-0432.CCR-07-5023 PMID:19706824

White, E., Karp, C., Strohecker, A. M., Guo, Y., & Mathew, R. (2010). Role of autophagy in suppression of inflammation and cancer. *Current Opinion in Cell Biology*, *22*(2), 212–217. doi:10.1016/j.ceb.2009.12.008 PMID:20056400

Whitman, M., Downes, C. P., Keeler, M., Keller, T., & Cantley, L. (1988). Type I phosphatidylinositol kinase makes a novel inositol phospholipid, phosphatidylinositol-3-phosphate. *Nature*, *332*(6165), 644–646. doi:10.1038/332644a0 PMID:2833705

Whitworth, A. J., & Pallanck, L. J. (2017). *PINK1*/Parkin mitophagy and neurodegeneration—What do we really know in vivo? *Current Opinion in Genetics & Development*, *44*, 47–53. doi:10.1016/j.gde.2017.01.016 PMID:28213158

Who Has Parkinson's? (n.d.). Available from https://www.parkinson.org/Understanding-Parkinsons/Statistics

Wilhelmus, M. M., Boelens, W. C., Otte-Holler, I., Kamps, B., de Waal, R. M., & Verbeek, M. M. (2006). Small heat shock proteins inhibit amyloid-beta protein aggregation and cerebrovascular amyloid-beta protein toxicity. *Brain Research*, *1089*(1), 67–78. doi:10.1016/j.brainres.2006.03.058 PMID:16635482

Wilkinson, K. A., & Henley, J. M. (2010). Mechanisms, regulation and consequences of protein SUMOylation. *The Biochemical Journal*, *428*(2), 133–145. doi:10.1042/BJ20100158 PMID:20462400

Williams, A. J., & Paulson, H. L. (2008). Polyglutamine neurodegeneration: Protein misfolding revisited. *Trends in Neurosciences*, *31*(10), 521–528. doi:10.1016/j.tins.2008.07.004 PMID:18778858

Williams, B. L., & Lipkin, W. I. (2006). Endoplasmic Reticulum Stress and Neurodegeneration in Rats Neonatally Infected with Borna Disease Virus. *Journal of Virology*, *80*(17), 8613–8626. doi:10.1128/JVI.00836-06 PMID:16912310

Williams, D. R. (2006). Tauopathies: Classification and clinical update on neurodegenerative diseases associated with microtubule-associated protein tau. *Internal Medicine Journal*, *36*(10), 652–660. doi:10.1111/j.1445-5994.2006.01153.x PMID:16958643

Williams, J. A., & Ding, W.-X. (2018). Mechanisms, pathophysiological roles and methods for analyzing mitophagy – recent insights. *Biological Chemistry*, *399*(2), 147–178. doi:10.1515/hsz-2017-0228 PMID:28976892

Willingham, S., Outeiro, T. F., DeVit, M. J., Lindquist, S. L., & Muchowski, P. J. (2003). Yeast genes that enhance the toxicity of a mutant huntingtin fragment or α-synuclein. *Science*, *302*(5651), 1769–1772. doi:10.1126cience.1090389 PMID:14657499

Wills, J., Credle, J., Haggerty, T., Lee, J. H., Oaks, A. W., & Sidhu, A. (2011). Tauopathic changes in the striatum of A53T α-synuclein mutant mouse model of Parkinson's disease. *PLoS One*, *6*(3), e17953. doi:10.1371/journal.pone.0017953 PMID:21445308

Wilson, R.J., Drake, J.C., Cui, D., Zhang, M., Perry, H.M., Kashatus, J.A., … Yan, Z. (2017). *Conditional Mitotimer reporter mice for assessment of mitochondrial structure, oxidative stress, and mitophagy*. Academic Press.

Wilson, M. A. (2011). The role of cysteine oxidation in DJ-1function and dysfunction. *Antioxidants & Redox Signalling*, *15*(1), 111–122. doi:10.1089/ars.2010.3481 PMID:20812780

Winklhofer, K. F., & Haass, C. (2010). Mitochondrial dysfunction in Parkinson's disease. *Biochimicaet Biophysica Acta (BBA)- Molecular Basis of Disease*, *1802*(1), 29–44. doi:10.1016/j.bbadis.2009.08.013

Winklhofer, K. F., Reintjes, A., Hoener, M. C., Voellmy, R., & Tatzelt, J. (2001). Geldanamycin Restores a Defective Heat Shock Response in Vivo. *The Journal of Biological Chemistry*, *276*(48), 45160–45167. doi:10.1074/jbc.M104873200 PMID:11574536

Winklhofer, K. F., Tatzelt, J., & Haass, C. (2008). The two faces of protein misfolding: Gain- and loss-of-function in neurodegenerative diseases. *The EMBO Journal*, *27*(2), 336–349. doi:10.1038j.emboj.7601930 PMID:18216876

Winter, A. N., Ross, E. K., Daliparthi, V., Sumner, W. A., Kirchhof, D. M., Manning, E., ... Linseman, D. A. (2017). A Cystine-Rich Whey Supplement (Immunocal®) Provides Neuroprotection from Diverse Oxidative Stress-Inducing Agents In Vitro by Preserving Cellular Glutathione. *Oxidative Medicine and Cellular Longevity*, *2017*, 1–15. doi:10.1155/2017/3103272 PMID:28894506

Winterbourn, C. C., & Hampton, M. B. (2008). Thiol chemistry and specificity in redox signaling. *Free Radical Biology & Medicine*, *45*(5), 549–561. doi:10.1016/j.freeradbiomed.2008.05.004 PMID:18544350

Winther, J. R., & Thorpe, C. (2014). Quantification of thiols and disulfides. *Biochimica et Biophysica Acta*, *1840*(2), 838–846. doi:10.1016/j.bbagen.2013.03.031 PMID:23567800

Wirth, M., Joachim, J., & Tooze, S. A. (2013). Autophagosome formation—The role of ULK1 and Beclin1–PI3KC3 complexes in setting the stage. *Seminars in Cancer Biology*, *23*(5), 301–309. doi:10.1016/j.semcancer.2013.05.007 PMID:23727157

Witt, A. C., Lakshminarasimhan, M. B., Remington, S., Hasim, E., Pozharski, E., & Wilson, M. A. (2008). Cysteine pKa depression by a protonated glutamic acid in human DJ-1. *Biochem*, *47*(28), 7430–7440. doi:10.1021/bi800282d PMID:18570440

Witt, M., Bormann, K., Gudziol, V., Pehlke, K., Barth, K., Minovi, A., ... Hummel, T. (2009). Biopsies of olfactory epithelium in patients with Parkinson's disease. *Movement Disorders*, *24*(6), 906–914. doi:10.1002/mds.22464 PMID:19205070

Wolfe, M. S. (2008). *Inhibition and modulation of γ-secretase for Alzheimer's disease*. Academic Press.

Wong, C. M., Bansal, G., Marcocci, L., & Suzuki, Y. J. (2012). Proposed role of primary protein carbonylation in cell signaling. *Redox Report*, *17*(2), 90–94. doi:10.1179/1351000212Y.0000000007 PMID:22564352

Wong, C. M., Cheema, A. K., Zhang, L., & Suzuki, Y. J. (2008). Protein carbonylation as a novel mechanism in redox signaling. *Circulation Research*, *102*(3), 310–318. doi:10.1161/CIRCRESAHA.107.159814 PMID:18079412

Wong, Y. C., & Holzbaur, E. L. (2014). Optineurin is an autophagy receptor for damaged mitochondria in parkin-mediated mitophagy that is disrupted by an ALS-linked mutation. *Proceedings of the National Academy of Sciences of the United States of America*, *111*(42), E4439–E4448. doi:10.1073/pnas.1405752111 PMID:25294927

Wong, Y. C., & Holzbaur, E. L. (2015). Temporal dynamics of PARK2/parkin and OPTN/optineurin recruitment during the mitophagy of damaged mitochondria. *Autophagy*, *11*(2), 422–424. doi:10.1080/15548627.2015.1009792 PMID:25801386

Wood, J. G., Mirra, S. S., Pollock, N. J., & Binder, L. I. (1986). Neurofibrillary tangles of Alzheimer disease share antigenic determinants with the axonal microtubule-associated protein tau (tau). *Proceedings of the National Academy of Sciences of the United States of America*, *83*(11), 4040–4043. doi:10.1073/pnas.83.11.4040 PMID:2424015

World Alzheimer Report. (2018). Available from https://www.alz.co.uk/research/WorldAlzheimerReport2018.pdf

Wouters, M. A., Fan, S. W., & Haworth, N. L. (2010). Disulfides as redox switches: From molecular mechanisms to functional significance. *Antioxidants & Redox Signalling*, *12*(1), 53–91. doi:10.1089/ars.2009.2510 PMID:19634988

Wu, J., Lin, F., & Qin, Z. (2007). *Sequestration of glyceraldehyde-3-phosphate dehydrogenase to aggregates formed by mutant huntingtin*. Academic Press.

Wu, A., Ying, Z., & Gomez-Pinilla, F. (2006). Oxidative stress modulates Sir2α in rat hippocampus and cerebral cortex. *The European Journal of Neuroscience*, *23*(10), 2573–2580. doi:10.1111/j.1460-9568.2006.04807.x PMID:16817860

Wu, W., Tian, W., Hu, Z., Chen, G., Huang, L., Li, W., ... Feng, D. (2014). ULK1 translocates to mitochondria and phosphorylates FUNDC1 to regulate mitophagy. *EMBO Reports*, *15*(5), 566–575. doi:10.1002/embr.201438501 PMID:24671035

WW domain (IPR001202). (n.d.). *InterPro, Protein sequence analysis & classification*. Retrieved from http://www.ebi.ac.uk/interpro/entry/IPR001202

Xia, D., Tang, W. K., & Ye, Y. (2016). *Structure and function of the AAA+ ATPase p97/Cdc48p*. Academic Press.

Xie, H., Li, Y. J., & Chen, Y. Y. (2011). Mitophagy. *Chinese Journal of Biochemistry and Molecular Biology*, *27*(12), 1081–1087.

Xie, W., & Chung, K. K. (2012). Alpha-synuclein impairs normal dynamics of mitochondria in cell and animal models of Parkinson's disease. *Journal of Neurochemistry*, *122*(2), 404–414. doi:10.1111/j.1471-4159.2012.07769.x PMID:22537068

Xie, Y., Kang, R., Sun, X., Zhong, M., Huang, J., Klionsky, D. J., & Tang, D. (2015). Posttranslational modification of autophagy-related proteins in macroautophagy. *Autophagy*, *11*(1), 28–45. doi:10.4161/15548627.2014.984267 PMID:25484070

Xie, Z., & Klionsky, D. J. (2007). Autophagosome formation: Core machinery and adaptations. *Nature Cell Biology*, *9*(10), 1102–1109. doi:10.1038/ncb1007-1102 PMID:17909521

Xilouri, M., Brekk, O. R., Kirik, D., & Stefanis, L. (2013). LAMP2A as a therapeutic target in Parkinson disease. *Autophagy*, *9*(12), 2166–2168. doi:10.4161/auto.26451 PMID:24145820

Xu, K., & Zhu, X. P. (2012). Endoplasmic reticulum stress and prion diseases. *Reviews in the Neurosciences*, *23*(1), 79–84. doi:10.1515/rns.2011.062 PMID:22718614

Xu, L., & Pu, J. (2016). Alpha-synuclein in Parkinson's Disease: From pathogenetic dysfunction to potential clinical application. *Parkinson's Disease*, *2016*, 1720621. doi:10.1155/2016/1720621 PMID:27610264

Xu, M. M., Ryan, P., Rudrawar, S., Quinn, R. J., Zhang, H. Y., & Mellick, G. D. (2019). Advances in the development of imaging probes and aggregation inhibitors for alpha-synuclein. *Acta Pharmacologica Sinica*. doi:10.103841401-019-0304-y PMID:31586134

Xu, Y., Deng, Y., & Qing, H. (2015). The phosphorylation of α-synuclein: Development and implication for the mechanism and therapy of the Parkinson's disease. *Journal of Neurochemistry, 135*(1), 4–18. doi:10.1111/jnc.13234 PMID:26134497

Yaffe, K., Blackwell, T., Kanaya, A. M., Davidowitz, N., Barrett-Connor, E., & Krueger, K. (2004). Diabetes, impaired fasting glucose, and development of cognitive impairment in older women. *Neurology, 63*(4), 658–663. doi:10.1212/01.WNL.0000134666.64593.BA PMID:15326238

Yaffe, K., Blackwell, T., Whitmer, R. A., Krueger, K., & Barrett-Connor, E. (2006). Glycosylated hemoglobin level and development of mild cognitive impairment or dementia in older women. *The Journal of Nutrition, Health & Aging, 10*(4), 293. PMID:16886099

Yamaguchi, K., Cochran, E. J., Murrell, J. R., Polymeropoulos, M. H., Shannon, K. M., Crowther, R. A., ... Ghetti, B. (2005). Abundant neuritic inclusions and microvacuolar changes in a case of diffuse Lewy body disease with the A53T mutation in the α-synuclein gene. *Acta Neuropathologica, 110*(3), 298–305. doi:10.100700401-005-1042-4 PMID:15981014

Yamamoto, A., Takagi, H., Kitamura, D., Tatsuoka, H., Nakano, H., Kawano, H., ... Shirasawa, T. (1998). Deficiency in protein L-isoaspartyl methyltransferase results in a fatal progressive epilepsy. *The Journal of Neuroscience: The Official Journal of the Society for Neuroscience, 18*(6), 2063–2074. doi:10.1523/JNEUROSCI.18-06-02063.1998 PMID:9482793

Yamamoto, H., Morino, K., Mengistu, L., Ishibashi, T., Kiriyama, K., Ikami, T., & Maegawa, H. (2016). Amla enhances mitochondrial spare respiratory capacity by increasing mitochondrial biogenesis and antioxidant systems in a murine skeletal muscle cell line. *Oxidative Medicine and Cellular Longevity, 2016*, 2016. doi:10.1155/2016/1735841 PMID:27340504

Yamano, K., Matsuda, N., & Tanaka, K. (2016). The ubiquitin signal and autophagy: An orchestrated dance leading to mitochondrial degradation. *EMBO Reports, 17*(3), 300–316. doi:10.15252/embr.201541486 PMID:26882551

Yamano, K., & Youle, R. J. (2013). PINK1 is degraded through the N-end rule pathway. *Autophagy, 9*(11), 1758–1769. doi:10.4161/auto.24633 PMID:24121706

Yang, H., Li, J.-J., Liu, S., Zhao, J., Jiang, Y.-J., Song, A.-X., & Hu, H.-Y. (2014). *Aggregation of polyglutamine-expanded ataxin-3 sequesters its specific interacting partners into inclusions: implication in a loss-of-function pathology*. Academic Press.

Yang, A. J., Chandswangbhuvana, D., Margol, L., & Glabe, C. G. (1998). Loss of endosomal/lysosomal membrane impermeability is an early event in amyloid Aβ1-42 pathogenesis. *Journal of Neuroscience Research, 52*(6), 691–698. doi:10.1002/(SICI)1097-4547(19980615)52:6<691::AID-JNR8>3.0.CO;2-3 PMID:9669318

Yang, D. S., Stavrides, P., Mohan, P. S., Kaushik, S., Kumar, A., Ohno, M., ... Pawlik, M. (2010). Reversal of autophagy dysfunction in the TgCRND8 mouse model of Alzheimer's disease ameliorates amyloid pathologies and memory deficits. *Brain, 134*(1), 258–277. doi:10.1093/brain/awq341 PMID:21186265

Yang, G., Wu, L., Jiang, B., Yang, W., Qi, J., Cao, K., ... Wang, R. (2008). H2S as a physiologic vasorelaxant; hypertension in mice with deletion of cystathionine-γ-lyase. *Sci, 322*(5901), 587–590. doi:10.1126cience.1162667 PMID:18948540

Yang, H. Y., & Lee, T. H. (2015). Antioxidant enzymes as redox-based biomarkers: A brief review. *BMB Reports, 48*(4), 200–208. doi:10.5483/BMBRep.2015.48.4.274 PMID:25560698

Yang, W., Zou, Y., Zhang, M., Zhao, N., Tian, Q., Gu, M., ... Yu, W. (2015). Mitochondrial Sirt3 expression is decreased in APP/PS1 double transgenic mouse model of Alzheimer's disease. *Neurochemical Research, 40*(8), 1576–1582. doi:10.100711064-015-1630-1 PMID:26045440

Yang, Y., Chen, S., Zhang, Y., Lin, X., Song, Y., Xue, Z., ... Zhang, L. (2017). Induction of autophagy by spermidine is neuroprotective via inhibition of caspase 3-mediated Beclin 1 cleavage. *Cell Death & Disease*, *8*(4), e2738. doi:10.1038/cddis.2017.161 PMID:28383560

Yang, Y., Turner, R. S., & Gaut, J. R. (1998). The chaperone BiP/GRP78 binds to amyloid precursor protein and decreases Abeta40 and Abeta42 secretion. *The Journal of Biological Chemistry*, *273*(40), 25552–25555. doi:10.1074/jbc.273.40.25552 PMID:9748217

Yang, Z., & Klionsky, D. J. (2010). Eaten alive: A history of macroautophagy. *Nature Cell Biology*, *12*(9), 814–822. doi:10.1038/ncb0910-814 PMID:20811353

Yan, L. M., Velkova, A., & Kapurniotu, A. (2014). Molecular characterization of the hetero-assembly of β-amyloid peptide with islet amyloid polypeptide. *Current Pharmaceutical Design*, *20*(8), 1182–1191. doi:10.2174/13816128113199990064 PMID:23713771

Yano, H., Baranov, S. V., Baranova, O. V., Kim, J., Pan, Y., Yablonska, S., . . . Friedlander, R. M. J. N. n. (2014). *Inhibition of mitochondrial protein import by mutant huntingtin.* Academic Press.

Yano, H., Baranov, S. V., Baranova, O. V., Kim, J., Pan, Y., Yablonska, S., ... Friedlander, R. M. (2014). Inhibition of mitochondrial protein import by mutant huntingtin. *Nature Neuroscience*, *17*(6), 822–831. doi:10.1038/nn.3721 PMID:24836077

Yao, X. Q., Zhang, X. X., Yin, Y. Y., Liu, B., Luo, D. J., Liu, D., ... Wang, J. Z. (2011). Glycogen synthase kinase-3β regulates Tyr307 phosphorylation of protein phosphatase-2A via protein tyrosine phosphatase 1B but not Src. *The Biochemical Journal*, *437*(2), 335–344. doi:10.1042/BJ20110347 PMID:21554241

Ye, Y. (2006). *Diverse functions with a common regulator: ubiquitin takes command of an AAA ATPase.* Academic Press.

Ye, J., Rawson, R. B., Komuro, R., Chen, X., Davé, U. P., Prywes, R., ... Goldstein, J. L. (2000). ER stress induces cleavage of membrane-bound ATF6 by the same proteases that process SREBPs. *Molecular Cell*, *6*(6), 1355–1364. doi:10.1016/S1097-2765(00)00133-7 PMID:11163209

Yen, W. L., & Klionsky, D. J. (2008). How to live long and prosper: Autophagy, mitochondria, and aging. *Physiology (Bethesda, MD)*, *23*(5), 248–262. doi:10.1152/physiol.00013.2008 PMID:18927201

Yen, W. L., Shintani, T., Nair, U., Cao, Y., Richardson, B. C., Li, Z., ... Klionsky, D. J. (2010). The conserved oligomeric Golgi complex is involved in double-membrane vesicle formation during autophagy. *The Journal of Cell Biology*, *188*(1), 101–114. doi:10.1083/jcb.200904075 PMID:20065092

Yerbury, J. J., Gower, D., Vanags, L., Roberts, K., Lee, J. A., & Ecroyd, H. (2013). The small heat shock proteins αB-crystallin and Hsp27 suppress SOD1 aggregation in vitro. *Cell Stress & Chaperones*, *18*(2), 251–257. doi:10.100712192-012-0371-1 PMID:22993064

Yeung, F., Hoberg, J. E., Ramsey, C. S., Keller, M. D., Jones, D. R., Frye, R. A., & Mayo, M. W. (2004). Modulation of NF-κB-dependent transcription and cell survival by the SIRT1 deacetylase. *The EMBO Journal*, *23*(12), 2369–2380. doi:10.1038j.emboj.7600244 PMID:15152190

Ye, X., Sun, X., Starovoytov, V., & Cai, Q. (2015). Parkin-mediated mitophagy in mutant hAPP neurons and Alzheimer's disease patient brains. *Human Molecular Genetics*, *24*(10), 2938–2951. doi:10.1093/hmg/ddv056 PMID:25678552

Ye, Y., & Rape, M. (2009). Building ubiquitin chains: E2 enzymes at work. *Nature Reviews. Molecular Cell Biology*, *10*(11), 755–764. doi:10.1038/nrm2780 PMID:19851334

Yin, H., Nalbandian, A., Hsu, C., Li, S., Llewellyn, K., Mozaffar, T., . . . Weiss, J. (2012). *Slow development of ALS-like spinal cord pathology in mutant valosin-containing protein gene knock-in mice*. Academic Press.

Yin, X. M., & Ding, W. X. (2013). The reciprocal roles of PARK2 and mitofusins in mitophagy and mitochondrial spheroid formation. *Autophagy*, *9*(11), 1687–1692. doi:10.4161/auto.24871 PMID:24162069

Yogalingam, G., Hwang, S., Ferreira, J. C., & Mochly-Rosen, D. (2013). *Glyceraldehyde-3-phosphate dehydrogenase (GAPDH) phosphorylation by protein kinase Cδ (PKCδ) inhibits mitochondria elimination by lysosomal-like structures following ischemia and reoxygenation-induced injury*. Academic Press.

Yokoyama, A., Katsura, S., & Sugawara, A. (2017). Biochemical analysis of histone succinylation. *Biochemistry Research International*, *2017*, 1–7. doi:10.1155/2017/8529404 PMID:29225971

Yoon, S. O., Park, D. J., Ryu, J. C., Ozer, H. G., Tep, C., Shin, Y. J., ... Huang, K. (2012). JNK3 perpetuates metabolic stress induced by Abeta peptides. *Neuron*, *75*(5), 824–837. doi:10.1016/j.neuron.2012.06.024 PMID:22958823

Yoon, S. Y., & Kim, D. H. (2016). Alzheimer's disease genes and autophagy. *Brain Research*, *1649*, 201–209. doi:10.1016/j.brainres.2016.03.018 PMID:27016058

Yoshida, H., Yoshizawa, T., Shibasaki, F., Shoji, S., & Kanazawa, I. (2002). Chemical chaperones reduce aggregate formation and cell death caused by the truncated Machado-Joseph disease gene product with an expanded polyglutamine stretch. *Neurobiology of Disease*, *10*(2), 88–99. doi:10.1006/nbdi.2002.0502 PMID:12127147

Yoshii, S. R., Kishi, C., Ishihara, N., & Mizushima, N. (2011). Parkin mediates proteasome-dependent protein degradation and rupture of the outer mitochondrial membrane. *The Journal of Biological Chemistry*, *286*(22), 19630–19640. doi:10.1074/jbc.M110.209338 PMID:21454557

Youle, R. J., & Narendra, D. P. (2011). Mechanisms of mitophagy. *Nature Reviews. Molecular Cell Biology*, *12*(1), 9–14. doi:10.1038/nrm3028 PMID:21179058

Youle, R. J., & Van der Bliek, A. M. (2012). Mitochondrial fission, fusion, and stress. *Science*, *337*(6098), 1062–1065. doi:10.1126cience.1219855 PMID:22936770

Young, K. F., Pasternak, S. H., & Rylett, R. J. (2009). Oligomeric aggregates of amyloid β peptide 1–42 activate ERK/MAPK in SH-SY5Y cells via the α7 nicotinic receptor. *Neurochemistry International*, *55*(8), 796–801. doi:10.1016/j.neuint.2009.08.002 PMID:19666073

Yousefi, S., & Simon, H. U. (2009). Autophagy in cancer and chemotherapy. In *Death Receptors and Cognate Ligands in Cancer* (pp. 183–190). Berlin: Springer. doi:10.1007/400_2008_25

You, Y., & Ikezu, T. (2019). Emerging roles of extracellular vesicles in neurodegenerative disorders. *Neurobiology of Disease*, *130*, 104512. doi:10.1016/j.nbd.2019.104512 PMID:31229685

Yue, Z., Horton, A., Bravin, M., DeJager, P. L., Selimi, F., & Heintz, N. (2002). A novel protein complex linking the delta 2 glutamate receptor and autophagy: Implications for neurodegeneration in lurcher mice. *Neuron*, *35*(5), 921–933. doi:10.1016/S0896-6273(02)00861-9 PMID:12372286

Yue, Z., Jin, S., Yang, C., Levine, A. J., & Heintz, N. (2003). Beclin 1, an autophagy gene essential for early embryonic development, is a haploinsufficient tumor suppressor. *Proceedings of the National Academy of Sciences of the United States of America*, *100*(25), 15077–15082. doi:10.1073/pnas.2436255100 PMID:14657337

Yu, L., Chen, Y., & Tooze, S. (2018). Autophagy pathway: Cellular and molecular mechanisms. *Autophagy*, *14*(2), 207–215. doi:10.1080/15548627.2017.1378838 PMID:28933638

Yu, L., McPhee, C. K., Zheng, L., Mardones, G. A., Rong, Y., Peng, J., ... Hailey, D. W. (2010). Termination of autophagy and reformation of lysosomes regulated by mTOR. *Nature*, *465*(7300), 942–946. doi:10.1038/nature09076 PMID:20526321

Yun, J., Puri, R., Yang, H., Lizzio, M. A., Wu, C., Sheng, Z. H., & Guo, M. (2014). MUL1 acts in parallel to the PINK1/parkin pathway in regulating mitofusin and compensates for loss of PINK1/parkin. *eLife*, *3*, e01958. doi:10.7554/eLife.01958 PMID:24898855

Yura, K., Shionyu, M., Hagino, K., Hijikata, A., Hirashima, Y., Nakahara, T., ... Go, M. (2006). Alternative splicing in human transcriptome: Functional and structural influence on proteins. *Gene*, *380*(2), 63–71. doi:10.1016/j.gene.2006.05.015 PMID:16872759

Yu, W. H., Cuervo, A. M., Kumar, A., Peterhoff, C. M., Schmidt, S. D., Lee, J. H., ... Jiang, Y. (2005). Macroautophagy—A novel β-amyloid peptide-generating pathway activated in Alzheimer's disease. *The Journal of Cell Biology*, *171*(1), 87–98. doi:10.1083/jcb.200505082 PMID:16203860

Yuyama, K., Mitsutake, S., & Igarashi, Y. (2014). Pathological roles of ceramide and its metabolites in metabolic syndrome and Alzheimer's disease. *Biochimica et Biophysica Acta. Molecular and Cell Biology of Lipids*, *1841*(5), 793–798. doi:10.1016/j.bbalip.2013.08.002 PMID:23948264

Zaffagnini, G., & Martens, S. (2016). Mechanisms of Selective Autophagy. *Journal of Molecular Biology*, *428*(9), 1714–1724. doi:10.1016/j.jmb.2016.02.004 PMID:26876603

Zahid, S., Khan, R., Oellerich, M., Ahmed, N., & Asif, A. R. (2013). Differential S-nitrosylation of proteins in Alzheimer's disease. *Neuroscience*, *256C*, 126–136. PMID:24157928

Zarei, S., Carr, K., Reiley, L., Diaz, K., Guerra, O., Altamirano, P. F., ... Chinea, A. (2015). A comprehensive review of amyotrophic lateral sclerosis. *Surgical Neurology International*, *6*(1), 171–171. doi:10.4103/2152-7806.169561 PMID:26629397

Zarow, C., Lyness, S. A., Mortimer, J. A., & Chui, H. C. (2003). Neuronal loss is greater in the locus coeruleus than nucleus basalis and substantia nigra in Alzheimer and Parkinson diseases. *Archives of Neurology*, *60*(3), 337–341. doi:10.1001/archneur.60.3.337 PMID:12633144

Zeggini, E., Weedon, M. N., Lindgren, C. M., Frayling, T. M., Elliott, K. S., Lango, H., ... Barrett, J. C. (2007). Replication of genome-wide association signals in UK samples reveals risk loci for type 2 diabetes. *Science*, *316*(5829), 1336–1341. doi:10.1126cience.1142364 PMID:17463249

Zeitlin, S., Liu, J.-P., Chapman, D. L., Papaioannou, V. E., & Efstratiadis, A. J. N. g. (1995). *Increased apoptosis and early embryonic lethality in mice nullizygous for the Huntington's disease gene homologue*. Academic Press.

Zeng, L. R., Vega-Sánchez, M. E., Zhu, T., & Wang, G. L. (2006). Ubiquitination-mediated protein degradation and modification: An emerging theme in plant-microbe interactions. Cell Research, 413–426

Zhai, J., Lin, H., Julien, J. P., & Schlaepfer, W. W. (2007). Disruption of neurofilament network with aggregation of light neurofilament protein: A common pathway leading to motor neuron degeneration due to Charcot-Marie-Tooth disease-linked mutations in NFL and HSPB1. *Human Molecular Genetics*, *16*(24), 3103–3116. doi:10.1093/hmg/ddm272 PMID:17881652

Zhang, M., Sun, Y., Meng, J., Zhang, L., Liang, C., & Chang, C. (2019). Targeting AR-Beclin 1 complex-modulated growth factor signaling increases the antiandrogen Enzalutamide sensitivity to better suppress the castration-resistant prostate cancer growth. *Cancer Letters*, *442*, 483–490. doi:10.1016/j.canlet.2018.11.008 PMID:30423407

Zhang, T., Xue, L., Li, L., Tang, C., Wan, Z., Wang, R., ... Billiar, T. R. (2016). BNIP3 protein suppresses PINK1 kinase proteolytic cleavage to promote mitophagy. *The Journal of Biological Chemistry*, *291*(41), 21616–21629. doi:10.1074/jbc.M116.733410 PMID:27528605

Zhang, X., Cheng, X., Yu, L., Yang, J., Calvo, R., Patnaik, S., ... Delling, M. (2016). MCOLN1 is a ROS sensor in lysosomes that regulates autophagy. *Nature Communications*, *7*(1), 12109. doi:10.1038/ncomms12109 PMID:27357649

Zhang, X., Gao, F., Wang, D., Li, C., Fu, Y., He, W., & Zhang, J. (2018). Tau Pathology in Parkinson's Disease. *Frontiers in Neurology*, *9*, 809. doi:10.3389/fneur.2018.00809 PMID:30333786

Zhang, Y. W., Thompson, R., Zhang, H., & Xu, H. (2011). APP processing in Alzheimer's disease. *Molecular Brain*, *4*(1), 3. doi:10.1186/1756-6606-4-3 PMID:21214928

Zhang, Y., Chen, X., Zhao, Y. F., Ponnusamy, M., & Liu, Y. (2017). The role of ubiquitin proteasomal system and autophagy-lysosome pathway in Alzheimer's disease. *Reviews in the Neurosciences*, *28*(8), 861–868. doi:10.1515/revneuro-2017-0013 PMID:28704199

Zhang, Y., Gao, J., Chung, K. K., Huang, H., Dawson, V. L., & Dawson, T. M. (2000). Parkin functions as an E2-dependent ubiquitin–protein ligase and promotes the degradation of the synaptic vesicle-associated protein, CDCrel-1. *Proceedings of the National Academy of Sciences of the United States of America*, *97*(24), 13354–13359. doi:10.1073/pnas.240347797 PMID:11078524

Zhang, Y., Tian, Q., Zhang, Q., Zhou, X., Liu, S., & Wang, J. Z. (2009). Hyperphosphorylation of microtubule-associated tau protein plays dual role in neurodegeneration and neuroprotection. *Pathophysiology*, *16*(4), 311–316. doi:10.1016/j.pathophys.2009.02.003 PMID:19410438

Zhao, L., Longo-Guess, C., Harris, B. S., Lee, J.-W., & Ackerman, S. L. (2005). Protein accumulation and neurodegeneration in the woozy mutant mouse is caused by disruption of SIL1, a cochaperone of BiP. *Nature Genetics*, *37*(9), 974–979. doi:10.1038/ng1620 PMID:16116427

Zhao, Q.-F., Yu, J.-T., & Tan, L. (2015). S-Nitrosylation in Alzheimer's disease. *Molecular Neurobiology*, *51*(1), 268–280. doi:10.100712035-014-8672-2 PMID:24664522

Zhao, R., Davey, M., Hsu, Y. C., Kaplanek, P., Tong, A., Parsons, A. B., ... Houry, W. A. (2005). Navigating the chaperone network: An integrative map of physical and genetic interactions mediated by the hsp90 chaperone. *Cell*, *120*(5), 715–727. doi:10.1016/j.cell.2004.12.024 PMID:15766533

Zhao, X., & Yang, J. (2010). Amyloid-B peptide is a substrate of the human 20S proteasome. *ACS Chemical Neuroscience*, *1*(10), 655–660. doi:10.1021/cn100067e PMID:21116456

Zhao, Z., Xiang, Z., Haroutunian, V., Buxbaum, J. D., Stetka, B., & Pasinetti, G. M. (2007). Insulin degrading enzyme activity selectively decreases in the hippocampal formation of cases at high risk to develop Alzheimer's disease. *Neurobiology of Aging*, *28*(6), 824–830. doi:10.1016/j.neurobiolaging.2006.05.001 PMID:16769157

Zheng, L., Roeder, R. G., & Luo, Y. (2003). *S phase activation of the histone H2B promoter by OCA-S, a coactivator complex that contains GAPDH as a key component*. Academic Press.

Zheng, L., Terman, A., Hallbeck, M., Dehvari, N., Cowburn, R. F., Benedikz, E., ... Marcusson, J. (2011). Macroautophagy-generated increase of lysosomal amyloid β-protein mediates oxidant-induced apoptosis of cultured neuroblastoma cells. *Autophagy*, *7*(12), 1528–1545. doi:10.4161/auto.7.12.18051 PMID:22108004

Zheng, Q. Y., Huang, T., Zhang, L. S., Zhou, Y., Luo, H., Xu, H. X., & Wang, X. (2016). Dysregulation of Ubiquitin-Proteasome System in Neurodegenerative Diseases. *Frontiers in Aging Neuroscience*, *8*, 8. doi:10.3389/fnagi.2016.00303 PMID:28018215

Zheng, S., Clabough, E. B., Sarkar, S., Futter, M., Rubinsztein, D. C., & Zeitlin, S. O. (2010). Deletion of the huntingtin polyglutamine stretch enhances neuronal autophagy and longevity in mice. *PLOS Genetics*, *6*(2), e1000838. doi:10.1371/journal.pgen.1000838 PMID:20140187

Zheng, Z., Shang, Y., Tao, J., Zhang, J., & Sha, B. (2019). Endoplasmic Reticulum Stress Signaling Pathways: Activation and Diseases. *Current Protein & Peptide Science*, *20*(9), 935–943. doi:10.2174/1389203720666190621103145 PMID:31223084

Zhong, Y., Morris, D. H., Jin, L., Patel, M. S., Karunakaran, S. K., Fu, Y.-J., ... Wang, Q. J. (2014). Nrbf2 Protein Suppresses Autophagy by Modulating Atg14L Protein-containing Beclin 1-VPS34 Complex Architecture and Reducing Intracellular Phosphatidylinositol-3 Phosphate Levels. *The Journal of Biological Chemistry*, *289*(38), 26021–26037. doi:10.1074/jbc.M114.561134 PMID:25086043

Zhong, Y., Wang, Q. J., Li, X., Yan, Y., Backer, J. M., Chait, B. T., ... Yue, Z. (2009). Distinct regulation of autophagic activity by Atg14L and Rubicon associated with Beclin 1–phosphatidylinositol-3-kinase complex. *Nature Cell Biology*, *11*(4), 468–476. doi:10.1038/ncb1854 PMID:19270693

Zhu, J., & Chu, C. T. (2010). Mitochondrial dysfunction in Parkinson's disease. *Journal of Alzheimer's Disease*, *20*(2), S325–S334. doi:10.3233/JAD-2010-100363 PMID:20442495

Zhu, X., Zhao, X., Burkholder, W. F., Gragerov, A., Ogata, C. M., Gottesman, M. E., & Hendrickson, W. A. (1996). Structural analysis of substrate binding by the molecular chaperone DnaK. *Science*, *272*(5268), 1606–1614. doi:10.1126cience.272.5268.1606 PMID:8658133

Zientara-Rytter, K., & Subramani, S. (2019). The Roles of Ubiquitin-Binding Protein Shuttles in the Degradative Fate of Ubiquitinated Proteins in the Ubiquitin-Proteasome System and Autophagy. *Cells*, *8*.

Zimprich, A., Benet-Pages, A., Struhal, W., Graf, E., Eck, S. H., Offman, M. N., ... Strom, T. M. (2011). A mutation in VPS35, encoding a subunit of the retromer complex, causes late-onset Parkinson disease. *American Journal of Human Genetics*, *89*(1), 168–175. doi:10.1016/j.ajhg.2011.06.008 PMID:21763483

Zinsmaier, K. E., Hofbauer, A., Heimbeck, G., Pflugfelder, G. O., Buchner, S., & Buchner, E. (1990). A cysteine-string protein is expressed in retina and brain of Drosophila. *Journal of Neurogenetics*, *7*(1), 15–29. doi:10.3109/01677069009084150 PMID:2129171

Zinzalla, V., Stracka, D., Oppliger, W., & Hall, M. N. (2011). Activation of mTORC2 by association with the ribosome. *Cell*, *144*(5), 757–768. doi:10.1016/j.cell.2011.02.014 PMID:21376236

Ziviani, E., Tao, R. N., & Whitworth, A. J. (2010). Drosophila parkin requires *PINK1* for mitochondrial translocation and ubiquitinates mitofusin. *Proceedings of the National Academy of Sciences of the United States of America*, *107*(11), 5018–5023. doi:10.1073/pnas.0913485107 PMID:20194754

Zolkiewski, M., Zhang, T., & Nagy, M. (2012). Aggregate reactivation mediated by the Hsp100 chaperones. *Archives of Biochemistry and Biophysics*, *520*(1), 1–6. doi:10.1016/j.abb.2012.01.012 PMID:22306514

Zondler, L., Miller-Fleming, L., Repici, M., Gonçalves, S., Tenreiro, S., Rosado-Ramos, R., ... Outeiro, T. F. (2014). DJ-1 interactions with α-synuclein attenuate aggregation and cellular toxicity in models of Parkinson's disease. *Cell Death*, *5*(7), e1350. doi:10.1038/cddis.2014.307 PMID:25058424

Zugel, M., Wehrstein, F., Qiu, S., Diel, P., Steinacker, M., & Schumann, U. (2019). Moderate intensity continuous training reverses the detrimental effects of ovariectomy on RyR1 phosphorylation in rat skeletal muscle. *Molecular and Cellular Endocrinology*, *481*, 1–7. doi:10.1016/j.mce.2018.11.003 PMID:30465874

Zuleta, A., Vidal, R. L., Armentano, D., Parsons, G., & Hetz, C. (2012). AAV-mediated delivery of the transcription factor XBP1s into the striatum reduces mutant Huntingtin aggregation in a mouse model of Huntington's disease. *Biochemical and Biophysical Research Communications*, *420*(3), 558–563. doi:10.1016/j.bbrc.2012.03.033 PMID:22445760

About the Contributors

Md. Sahab Uddin is a Registered Pharmacist and a Research Scholar in the Department of Pharmacy at Southeast University, Dhaka, Bangladesh. He has published copious articles in peer-reviewed international scientific journals. He has also authored and edited seven books, including Advances in Neuropharmacology: Drugs and Therapeutics; Handbook of Research on Critical Examinations of Neurodegenerative Disorders; and Oxidative Stress and Antioxidant Defense: Biomedical Value in Health and Diseases. Md. Uddin also serves as a guest editor, editorial and reviewer board member of numerous scholarly journals. He is a member of many national and international scientific societies. He has developed five neuropsychological tests for the estimation of memory, attention, and cognition. Moreover, he is the Founder and Executive Director of the Pharmakon Neuroscience Research Network, an open innovation hub bringing together neuroscientists to advance brain health. He received his BPharm in 2014 securing a first position from the Department of Pharmacy, Southeast University, Bangladesh. Md. Uddin's research interest is how neuronal operation can be restored to abate Alzheimer's dementia.

Ghulam Md. Ashraf is currently working as an Associate Professor at King Fahd Medical Research Center, King Abdulaziz University, Jeddah, Saudi Arabia. Md Ashraf received his PhD in Biochemistry from Aligarh Muslim University, India in 2010. He has to his credit 191 publications ranging from peer-reviewed journal articles, books and book chapters published with the scholarly publishers (total citations: 2473, h-index: 22, i10-index: 68). He has been involved in 13 research projects funded by various agencies of Saudi Arabia and is a member of several renowned professional associations and organizations. Md Ashraf currently focuses on investigating the expression of proteins involved in causation, progression and potential therapy of neurological disorders. He is also interested in exploring the inherent connection of neurological disorders with other chronic diseases. Md Ashraf is involved in the editorial capacity in many reputed journals, like PLOS one, Scientific Reports, Frontiers in Neuroscience, Current Proteins and Peptides, Journal of Proteomics, to mention a few. Dr. Ghulam Md Ashraf received his PhD in Biochemistry from Aligarh Muslim University in 2010. Dr. Ashraf is currently working as Associate Professor at King Fahd Medical Research Center, King Abdulaziz University, Jeddah, Saudi Arabia. His current research interest covers proteomics, neurology, nanotechnology and cancer. Dr. Ashraf has published 160 research articles in books and journals of international repute. Dr. Ashraf is involved in editorial capacity in many internationally reputed journals, and have been involved in 6 research projects funded by various agencies of Saudi Arabia.

* * *

Maxwell Kwadwo Agyeman is currently pursuing PharmD at the Faculty of Pharmacy and Pharmaceutical Sciences, Kwame Nkrumah University of Science and Technology, Ghana. His research interests are neuroinflammation, cancer, and gene therapy.

Abhilasha Ahlawat is currently pursuing PhD in Department of Pharmaceutical Sciences, Maharshi Dayanand University, India. She has done BPharm from the College of Pharmacy, Pt. Bhagwat Dayal Sharma University of Health Sciences, India and MPharm (Pharmacology) from ISF College of Pharmacy, India. Her area of expertise includes behavioral neuroscience and neurodegenerative and metabolic disorders. Her key area of focus includes diabetes, diabetic neuropathy, and Alzheimer's disease, etc. She has about 3 years of industrial, teaching and research experience. She has published 5 publications in international journals of repute. She has published 4 book chapters in international books.

Niyaz Ahmad is working as an Assistant Professor, Department of Pharmaceutics, College of Clinical Pharmacy, Imam Abdulrahman Bin Faisal University, Saudi Arabia. He had obtained his PhD degree in 2015 from the Department of Pharmaceutics, Jamia Hamdard, India. He has published more than 100 research publications in many international journals. He has been awarded as Young Scientist Award and also received the best researcher award twice from Imam Abdulrahman Bin Faisal University. Has experience in the area of pharmaceutics, nanoformulation based targeting, bioanalysis, pharmacokinetic and pharmacodynamic studies

Yasmeena Akhter is MPharm student at the Department of Pharmaceutical Sciences (Pharmacology Division), Faculty of Applied Sc. and Tech., University of Kashmir, India. Her research interest encompasses the isolation and characterization of isolated compounds from herbal sources and establishment of their therapeutic usage in disorders of the endocrine and nervous systems.

Mohd. Amir is working as an Assistant Professor at the Department of Natural Product & Alternative Medicine, College of Clinical Pharmacy, Imam Abdul Rahman Bin Faisal University, Saudi Arabia. He has obtained his PhD degree from the Department of Pharmacognosy & Phytochemistry, School of Pharmaceutical Education & Research, India. His research area includes pharmacognosy, phytochemistry, ethnopharmacology, and analytical chemistry. He is credited with 35 publications in journals of repute.

Ghulam Bader is working as an Assistant Professor (Pharmacology) at the Department of Pharm. Sc, Faculty of Applied Sc. and Tech., University of Kashmir, India. He completed his PhD in Pharmacology from the University of Kashmir, India. His research theme focuses on the isolation and characterization of isolated compounds from herbal sources and the establishment of their therapeutic usage in disorders of endocrine, cardiovascular and nervous systems. Bader has 35 publications to his credit in journals of repute.

Md. Abul Barkat is currently working as an Assistant Professor in the Department of Pharmaceutics, College of Pharmacy, University of Hafar Al-batin, Saudi Arabia. Dr. Barkat had received his Ph.D. in Pharmaceutics from Integral University Lucknow, India, in the year 2017. He had completed his M.Pharm and B.Pharm from Jamia Hamdard, India. His major areas of research interest include herbal and silver-based development and characterization of diverse nanostructured systems, controlled release drug delivery systems, bio-enhanced drug delivery systems, nanomaterials, and nanocomposites. To date, he has authored more than 23 research publications in various peer-reviewed journals, 4 book chapters & 7 books to his credit. Barkat has also participated and presented his research work at conferences held in India, Canada, and Spain.

Sujata Basu is an Assistant Manager in the Enago-Publication Support department of Crimson Interactive Pvt. Ltd., India. She received the PhD degree in Biotechnology from Jaypee Institute of Information Technology, India. Her research interest includes diabetes, molecular profiling of plant extracts, and oxidative stress at cellular as well as mitochondrial levels. She holds several book publications in areas of life sciences. She has published many papers in national and international journals. She is currently involved in enhancing the journal file quality assessment using data-based journal behavior analysis of publication trends, thereby helping thousands of authors from ESL countries like Japan, China, Turkey, Brazil, etc. She has also been associated with the teaching profession for 5 years.

Munish Garg is currently working as Dean in the Faculty of Pharmaceutical Sciences, Maharshi Dayanand University, India. He has done his BPharm from Gulbarga University in 1995, MPharm in Pharmacognosy from Dr. MGR Medical University, India in 2002 and PhD from Punjab Technical University, India in 2009. His area of expertise includes quality control and standardization of herbal plants and formulations for safety and efficacy and development of novel drug delivery applications of medicinal plants/active plant ingredients for enhanced efficacy. He has more than 15 years of teaching and research experience. He has published more than 50 research publications in various national and international journals of repute. He has published 3 national and international books. He has published 4 book chapters in international books. He has presented in more than 150 posters in national and international conferences. He has guided more than 35 MPharm and several students of PhD.

Neha Goel has completed her BTech Biotechnology from the Jaypee Institute of Information Technology, India. She is a meritorious student and has many international articles, conference publications and chapters to her credit. She is currently preparing for higher studies.

Vishal S. Gulecha is Dean at the School of Pharmaceutical Sciences, Sandip University, India. He has perused BPharm and Master's Degree in Pharmacology from Pune University, India. He has completed his PhD from Dr. MGR Medical University, India. He has 12 years of teaching experience. He has published several research articles and reviews in the national and international journals of repute. He is a recipient of different awards for the presentation of research work at various conferences and symposia. Also, he has received grants for research from Savitribai Phule Pune University, India.

Mst Ara Gulshan received BSc and MSc degree from the University of Rajshahi, Bangladesh in 2012 and 2014, respectively. After that, she worked as a Research Assistant at the University of Malaya, Malaysia (2014). She earned PhD in 2019 in the field of RNA biochemistry. Her work interest is to investigate the effects of natural and artificial activators on the structurally unstable group I ribozymes through the kinetic analysis.

Hinna Hamid is an MPharm student at the Department of Pharmaceutical Sciences (Pharmacology Division), Faculty of Applied Sc. and Tech., University of Kashmir, India. Her research interest encompasses the isolation and characterization of isolated compounds from herbal sources and the establishment of their therapeutic usage in disorders of endocrine and nervous systems.

Md. Farhad Hossain has completed his MSc degree in Physical Therapy at the Department of Inje University, South Korea, graduating with a prestigious award from the dean of graduate school and secured top scorer (1st Position) based on academic records and research expertise. His research focuses on neuroscience, molecular science, clock-regulated genes and hormones, and neuropharmacology. He is a registered pharmacist and also completed his BPharm from Southeast University, Bangladesh. He is actively involved with several national and international research projects.

Suvarna P. Ingale is an Associate Professor in the Department of Pharmacology at SCES's Indira College of Pharmacy, India. She completed her PhD (Pharmaceutical Sciences) from Jawaharlal Nehru Technological University, India. She is actively involved in research mainly in the area of neuropharmacology and herbal drug pharmacology. She also has published many articles in leading national and international journals.

Md. Noushad Javed is working as an Assistant Professor at the School of Pharmaceutical Sciences, Apeejay Stya University, India. He had received his PhD degree in 2019 from the Department of Pharmaceutics, School of Pharmaceutical Education & Research, India. Earlier he had completed his MPharm (Pharmaceutics) and BPharm from the same institution. His area of interest includes QbD integrated optimization (DoE and Artificial Intelligence) of novel drug delivery systems for neuronal disorders, cancer, and parasitic infections.

Meenakshi Kaira is currently pursuing Ph.D from the Department of Pharmaceutical Sciences, Maharshi Dayanand University, India. She has done her B.Pharm from GVM College of Pharmacy, Sonepat affiliated by Pt. BD Sharma University of Health Sciences, India, and M.Pharma from Department of Pharmaceutical Sciences, Maharshi Dayanand University, India. Her area of expertise includes enzymatic studies, behavioral neuroscience, and metabolic disorder. She has published 4 book chapters in international books. She has received the reviewer award from the Elsevier two times. She has presented more than 5 posters in national and international conferences.

Ozaifa Kareem is PhD research scholar at the Department of Pharmaceutical Sciences (Pharmacology Division), Faculty of Applied Sc. and Tech., University of Kashmir, India. She completed her MPharm (Pharmacology) at the Department of Pharmaceutical Sciences, Faculty of Applied Sc. and Tech., University of Kashmir, India. Her research interest encompasses the isolation and characterization of isolated compounds from herbal sources and the establishment of their therapeutic usage in disorders of the endocrine and nervous systems

Simrat Kaur has completed her BTech Biotechnology from the Jaypee Institute of Information Technology, India. She is a meritorious student and has many international articles, conference publications and chapters to her credit. She is currently applying abroad for higher studies.

Renuka Khatik completed her BPharm from the Department of Pharmaceutical Sciences Dr. Hari Singh Gour, a Central University by Government of India, India, and MPharm in Pharmaceutical Chemistry from the Department of Pharmaceutical Sciences at Barkatullah Vishwavidyalaya, India. She worked for her thesis at the CSIR-Central Drug Research Institute, India. She had worked at K.R. Mangalam University as an Assistant Professor. Presently she is posted as a postdoctoral fellow at the Hefei National Laboratory of Physical Sciences at the Microscale University of Science and Technology of China.

Manoj S. Mahajan is working as an Assistant Professor, Department of Pharmacology at SNJBs SSDJ College of Pharmacy, India. He has perused BPharm and MPharm in Pharmacology from Pune University, India. He has 11 years of teaching experience. He has published several research articles and reviews in the national and international journals and presented research work at various national and international conferences. He has received grants for research from Savitribai Phule Pune University, India. He is currently perusing his PhD from Savitribai Phule Pune University.

Abdullah Al Mamun has completed his BPharm degree with an outstanding result (4th position) from the Department of Pharmacy, Southeast University, Bangladesh. He is strongly involved as a "Research Scholar" at the Natural Products and Neuropharmacology Research Group under the Pharmacology Research Laboratory, of the same university. He has published many research articles in peer-reviewed international journals. In addition, he is a reviewer board member of several peer-reviewed scholarly journals.

A. N. M. Mamun-Or-Rashid is a Post-Doctoral researcher in the Anti-Aging Medical Research Center and Glycation Stress Research Center, Graduate School of Life and Medical Sciences, Doshisha University, Japan. He received BSc and MSc degree in Biotechnology and Genetic Engineering, Islamic University, Bangladesh. He obtained PhD degree in Medical Life Systems, Doshisha University, Japan. He has research experience in stem cell differentiation, modulation, etc. He is specialized in various molecular and cell biological techniques. His research interest includes bone biology, neuroscience, etc.

Jahangir Nabi is a PhD research scholar at the Department of Pharmaceutical Sciences (Pharmacology Division), Faculty of Applied Sc. and Tech., University of Kashmir, India. He has completed MPharm from Delhi Institute of Pharmaceutical Sciences, India. His research interest encompasses the isolation and characterization of isolated compounds from herbal sources and establishment of their therapeutic usage in disorders of the endocrine and nervous systems.

Himani Nautiyal completed her Master of Pharmacy from Uttarakhand Technical University, India. She is currently employed as an Assistant Professor of Pharmacology in Siddhartha Group of Institutions, India and has more than 7 years of teaching and research experience. She has been working intensively on polycystic ovarian syndrome and has carried out several research leading to good publications.

Newman Osafo is a Lecturer and Researcher with the Department of Pharmacology, Faculty of Pharmacy and Pharmaceutical Sciences, Kwame Nkrumah University of Science and Technology, Ghana. He holds a PhD in Pharmacology from the same institution with research interests in inflammation and cancer. He is a member of various international scientific research societies in the area of natural drug research and pharmacology. He has a number of publications in reputable peer-reviewed journals. Newman serves as editorial and review board member of some scholarly journals and has been a reviewer for a number of reputable journals over the years.

Joseph Owusu-Sarfo is currently pursuing PharmD at the Faculty of Pharmacy and Pharmaceutical Sciences, Kwame Nkrumah University of Science and Technology, Ghana. He has contributed to a number of book chapters. Joseph has a profound interest in neuropharmacology.

Eyitayo Adeyemi Oyindamola is currently pursuing PharmD at the Faculty of Pharmacy and Pharmaceutical Sciences, Kwame Nkrumah University of Science and Technology, Ghana. She has contributed to a book chapter. Eyitayo has the research interest in neuroscience and ethnopharmacology.

Mukesh Pandey completed his BPharm and MPharm (Pharmaceutics) from the Department of Pharmaceutical Sciences, Hemvati Nandan Bahuguna Garhwal University, India. He worked as Production Officer in Pure and Cure Healthcare Limited, India. He worked as a Senior Research Fellow in University Grant Commission Major Research Project at Hemvati Nandan Bahuguna Garhwal University, India. Presently he is working as a full-time guest faculty and pursuing PhD at Department of Pharmaceutics, Delhi Institute of Pharmaceutical Sciences and Research, India

Myung-Geol Pang is a Distinguished Professor of the Department of Animal Science and Technology and Director of BET Research Institute, Chung-Ang University, South Korea. He is also elected fellow of Korean Academy of Science and Technology (KAST). He received his PhD from Eastern Virginia Medical School, Norfolk, USA, and, MSc and BSc from Chung-Ang University, South Korea. His research interests lie in mammalian reproduction, reproductive biology, germ cell biology, andrology, spermatogenesis, (in)fertility, drug discovery, and male contraception.

Rupali A. Patil is an Associate Professor in the Department of Pharmacology at GES's Sir Dr. M. S. Gosavi College of Pharmaceutical Education and Research, India. She published 26 research papers in national and international journals, presented more than 50 posters in national and international conferences. Her areas of research interest are neuropharmacology, psychopharmacology, and metabolic disorders. She guided 51 post-graduate research scholars for dissertation work. She was organizing committee members for many national and state-level seminars, workshops and conferences.

Shubhangi H. Pawar is working as an Assistant Professor, Department of Pharmacology at MGV's Pharmacy College, Panchavati, India. She has perused BPharm from Shivaji University and MPharm in Pharmacology from Pune University, India. She has 13 years of teaching experience. She has published several research articles and reviews in the national and international journals and presented research work at various national and international conferences. She is currently perusing her PhD from Savitribai Phule Pune University.

Faheem Hyder Pottoo is working as an Assistant Professor, Department of Pharmacology, College of Clinical Pharmacy, Imam Abdulrahman Bin Faisal University, Saudi Arabia. He got his MPharm (Pharmacology) from the Department of Pharmacology, School of Pharmaceutical Education & Research, Jamia Hamdard University, New-Delhi, India and PhD (Pharmacology) in 2016 from Department of Pharm. Sc, Faculty of Applied Sc. and Tech., University of Kashmir, Srinagar, India. His research vision is to transform the treatment of neurodegenerative diseases & neuronal cancers by modulating signaling through serotonergic, dopaminergic, cholinergic, noradrenergic & oxinergic pathways. To date, he has authored more than 35 publications in various peer-reviewed journals. He serves as a review editor in the neuropharmacology section for the journals Frontiers in Neuroscience, Frontiers in Neurology & Frontiers in Pharmacology.

Rachana R. is an Associate Professor in the Department of Biotechnology at Jaypee Institute of Information Technology, India. She post graduated in Biotechnology from IIT Roorkee in 1998 and achieved her doctorate from IIT Bombay in 2006. She has qualified NET-LS and GATE (Topper in Kanpur zone and 4th rank in India) in Life Sciences. She has been the recipient of DBT fellowship during her masters and JRF and research fellowship, from MHRD during her PhD. She has two Indian patents, two textbooks and has several international and national articles to her credit. She has been working to establish the mechanism of action of the herbal products in the field of diabetes, asthma, cancer and neurological disorders.

Md. Ataur Rahman is a Post-Doctoral researcher in Neuroscience Center, Brain Science Institute, Korea Institute of Science and Technology (KIST), South Korea. He received BSc and MSc degree in Biotechnology and Genetic Engineering, Islamic University, Bangladesh. He also obtained MSc and PhD degree in Medical Science (Pharmacology), Hallym University, South Korea. His research interest includes neurodegenerative diseases and autophagy. He is also experienced and demonstrated a history of working in neuroscience with autophagy signaling of natural products. Skilled in cellular and molecular biology techniques. Strong research professional with a PhD and Post-Doc focused on natural products and anti-cancer drug development of neuroblastoma cells in autophagy and apoptosis signaling.

Md. Motiar Rahman is a Postdoctoral Research Fellow at Shenzhen Institutes of Advanced Technology (SIAT) of the Chinese Academy of Sciences (CAS), China. Dr. Rahman received BSc and MSc degree at the Department of Biochemistry and Molecular Biology from the University of Rajshahi, Bangladesh. After that, Rahman worked as a Research Assistant at the University of Malaya, Malaysia. He earned PhD in the field of Nanosciences and Biosciences. During his Ph.D. study, he developed several long-range tertiary interacting motifs and applied these motifs to expand RNA nanostructures. Dr. Rahman also analyzed the effects of molecular crowding on the activity of group I ribozyme and found that PEG exceptionally enhanced the catalytic activity of the unstable mutation of group I ribozymes. Since group I ribozymes are involved in the splicing reaction where two exons flanked by an intron are ligated. He is now interested in engineer group I ribozyme to artificially synthesize circular mRNA and utilize this mRNA to use as a prophylactic cancer vaccine.

Md. Rezanur Rahman received the BSc and MSc degrees in Biotechnology and Genetic Engineering from Islamic University, Bangladesh. He is currently a Lecturer with the Department of Biochemistry and Biotechnology, School of Biomedical Science, Khwaja Yunus Ali University, Bangladesh. His research interests include epigenetics, systems biology and bioinformatics to study the multiomics data to decipher novel biomarkers of complex diseases including neurodegenerative disease.

Md. Saidur Rahman is a Research Professor of the Department of Animal Science and Technology, Chung-Ang University, South Korea. He is also a Korea Research Fellow at the National Research Foundation of Korea. Rahman has a degree in Veterinary Medicine and specialization in animal obstetrics and gynecology. He received his PhD and postdoctoral training from Chung-Ang University, South Korea. His research interests lie in reproductive biology and toxicology, andrology, male (in)fertility, endocrine disorders, drug discovery, and veterinary medicine.

Hyewhon Rhim is a Principal Researcher, Center for Neuroscience, Korea Institute of Science and Technology (KIST), South Korea. She received BSc and MSc degree in Chemical Education and Chemistry Department, respectively from Seoul National University, South Korea. She obtained PhD degree in the Department of Physiological & Pharmacological Sciences, University of Chicago, USA. Her research of interest is O-GlcNACylated protein modification in the brain using OGA+/- heterozygous transgenic mouse; molecular mechanisms of serotonin 6 receptors in brain; autophagy studies in reactive gliosis and neurodegenerative diseases; and other molecular mechanism using stress/depressed or Alzheimer's disease animal models.

Shakir Saleem is an Assistant Professor of Pharmacology in the College of Health Sciences, Saudi Electronic University, Saudi Arabia. He completed his MPharm from Rajiv Gandhi University of Health Sciences, India, and PhD (Pharmacology) from Glocal University, India. His area of research is cancer including drug discovery and pre-clinical pharmacological and toxicological studies of CNS drugs. The successful completion of his research projects has led to several full-length publications in reputed peer-reviewed international journals.

Aashish Sharma is currently working as a Post-Doctoral Biomedical Researcher in SFI Centre for Research in Medical Devices, National University of Ireland, Galway, Ireland, where he is working on an industrial project funded by Boston Scientific and developing vascular sealant for Vascular Closure Device. He is also co-affiliated as Assistant Professor-I, with Department of Pharmacy, School of Medical and Allied Sciences, GD Goenka University, Gurgaon, India. He had completed his PhD from the Indian Institute of Science Education and Research Bhopal India. He did his Master's and Bachelor's degree in Pharmaceutical Sciences. Aashish has 11 years of teaching and research experience. His research has been published in highly reputed, peer-reviewed international journals. He also has an Indian Patent to his credit. He has also been awarded membership of the Royal Society of Chemistry (RSC), UK and Prominent Alumni Award-2019 from Nirma University, India. His research is focused on synthetic polypeptide based materials for biomedical applications.

Looniva Shrestha is a Postdoctoral Research Fellow at Shenzhen Institutes of Advanced Technology (SIAT) of the Chinese Academy of Sciences (CAS), China. She received her Master's degree in Pathology and PhD degree in Bacteriology, both from Hiroshima University, Japan. She is specialized in molecular biology. She is a passionate researcher with long experience of working with varieties of cancer and bacterial cells. Her main research area includes bacterial biofilms and cancer therapy and has a number of research publications related to these areas. She believes that any dream can be achieved through hard work, patience, and dedication to work.

Manisha Singh is working as Assistant Professor at the Department of Biotechnology, Jaypee Institute of Information Technology, India. She has also worked as Assistant Lecturer in the Institute of Applied Medicines and Research, Lecturer in Prakash Institute of Physical Rehabilitation and Applied Medicines, India. She has completed her Masters in Physiotherapy (Neurology) from CCS University in 2007 and has been awarded the fellowship in Neurological Rehabilitation, from Apollo Hospitals, India with a certified course in neurodevelopment techniques. She is a lifetime member of the Spinal Cord Society (Indian Chapter) and Indian Association of Physiotherapists. Presently, her research interests are in the area of neuropharmaceutics, effects of neurologically used drugs and analytical rehabilitative methods for various neurological conditions.

Nahida Tabassum is working as a Professor of Pharmacology at the Department of Pharm. Sc, Faculty of Applied Sc. and Tech., University of Kashmir, India. She completed her MPharm from Delhi Institute of Pharmaceutical Sciences, India and PhD focused in Pharmacology from the University of Kashmir, India. Her research theme focuses on the isolation and characterization of isolated compounds from herbal sources and the establishment of their therapeutic usage in disorders of the endocrine and nervous systems. She has 27 years of teaching and research experience. So far her lab had evaluated 20 herbs for their antihepatotoxic and antidiabetic potential and published 74 articles related to that in journals of repute.

Md. Jamal Uddin is working as Research Professor in the Department of Pharmaceutical Sciences, Graduate School of Pharmacy, Ewha Womans University, South Korea. He received his BSc and MS degree in the discipline of Animal Science and Biology, Bangladesh Agricultural University, Bangladesh. He has job experience as Scientific Officer, Biotechnology Division, Bangladesh Livestock Research Institute, Bangladesh. In addition, he worked as Research Assistant at Gyangsang National University, South Korea on Reproductive Biology. He has completed his PhD in Biological Science from the University of Ulsan, South Korea. Currently, he is working with molecular mechanisms involved with various diseases and therapeutic strategies.

Aman Upaganlawar is presently working as an Associate Professor in the Department of Pharmacology at SNJBs SSDJ College of Pharmacy, India. He has completed his master's and PhD in Pharmacy from the MS University of Baroda, India. He has 14 years of teaching and research experience. He has more than 70 scientific publications in national and international journals with a good citation index. He has published 16 book chapters in reputed publishers. He is also active as a reviewer and editorial board member for several national and international journals. He has honored with various prestigious awards and is a life member of several professional organizations. His area of interest includes the preclinical screening of bioactive molecules for various disorders, cardiovascular pharmacology, metabolic syndrome, toxicology, and antioxidant study. To date, he has guided 26 post-graduate students for their research work.

Chandrashekhar D. Upasani is Professor and Principal at the SNJBs SSDJ College of Pharmacy, India. He had more than 22 years of teaching and administrative experience. He is former Dean and BOS Chairman of faculty of Pharmacy, Savitribai Phule Pune University, India. He is also associated with other universities of repute as BOS members and on other administrative bodies. He is an approved PhD guide of Savitribai Phule Pune University. More than 15 students awarded PhD under his guideship. He has organized several national and international conferences and seminars. He has delivered several invited talks at various conferences. He has completed major research projects and published more than 5 books.

Vaibav Walia is currently working in the Faculty of Pharmacy, D.I.T University, India. He has done his BPharm, MPharm (Pharmacology) and PhD from the Department of Pharmaceutical Sciences, Maharshi Dayanand University, India. His area of expertise includes behavioral neuroscience, neuropsychiatric and neurodegenerative disorders, animal model development for drug discovery and evaluation. He has worked on various pathologies including anxiety, depression, Parkinson's disease, Alzheimer's disease, diabetes, Stress, etc. Currently, he is establishing the relationship between neuropsychiatric and neurodegenerative disorders. He has more than 6 years of teaching and research experience. He has published more than 15 research publications in various national and international journals of repute. He has published 10 book chapters in international books. He has attended more than 15 national and international conferences. He has received the best poster award from APTI, an outstanding reviewer award from the Elsevier two times and BRIC idea exposition 2019.

Oduro Kofi Yeboah is a Pharmacist and currently pursuing MPhil Pharmacology at the Department of Pharmacology, Faculty of Pharmacy and Pharmaceutical Sciences, Kwame Nkrumah University of Science and Technology, Ghana. He has co-authored a number of publications in peer-reviewed journals as well as contributed to a number of book chapters. His research interest is neuroinflammation.

Index

A

B

C

D

E

F

G

H

I

L

M

N

O

P

R

S

T

U

www.ingramcontent.com/pod-product-compliance
Lightning Source LLC
LaVergne TN
LVHW060253011026
849161LV00011B/2711